Mantissas of the Common Logarithms

N	0	1	2	3	4	5	6	7	8	9
55	.7404	.7412	.7419	.7427	.7435	.7443	.7451	.7459	.7466	.7474
56	.7482	.7490	.7497	.7505	.7513	.7520	.7528	.7536	.7543	.7551
57	.7559	.7566	.7574	.7582	.7589	.7597	.7604	.7612	.7619	.7627
58	.7634	.7642	.7649	.7657	.7664	.7672	.7679	.7686	.7694	.7701
59	.7709	.7716	.7723	.7731	.7738	.7745	.7752	.7760	.7767	.7774
60	.7782	.7789	.7796	.7803	.7810	.7818	.7825	.7832	.7839	.7846
61	.7853	.7860	.7868	.7875	.7882	.7889	.7896	.7903	.7910	.7917
62	.7924	.7931	.7938	.7945	.7952	.7959	.7966	.7973	.7980	.7987
63	.7993	.8000	.8007	.8014	.8021	.8028	.8035	.8041	.8048	.8055
64	.8062	.8069	.8075	.8082	.8089	.8096	.8102	.8109	.8116	.8122
65	.8129	.8136	.8142	.8149	.8156	.8162	.8169	.8176	.8182	.8189
66	.8195	.8202	.8209	.8215	.8222	.8228	.8235	.8241	.8248	.8254
67	.8261	.8267	.8274	.8280	.8287	.8293	.8299	.8306	.8312	.8319
68	.8325	.8331	.8338	.8344	.8351	.8357	.8363	.8370	.8376	.8382
69	.8388	.8395	.8401	.8407	.8414	.8420	.8426	.8432	.8439	.8445
70	.8451	.8457	.8463	.8470	.8476	.8482	.8488	.8494	.8500	.8506
71	.8513	.8519	.8525	.8531	.8537	.8543	.8549	.8555	.8561	.8567
72	.8573	.8579	.8585	.8591	.8597	.8603	.8609	.8615	.8621	.8627
73	.8633	.8639	.8645	.8651	.8657	.8663	.8669	.8675	.8681	.8686
74	.8692	.8698	.8704	.8710	.8716	.8722	.8727	.8733	.8739	.8745
75	.8751	.8756	.8762	.8768	.8774	.8779	.8785	.8791	.8797	.8802
76	.8808	.8814	.8820	.8825	.8831	.8837	.8842	.8848	.8854	.8859
77	.8865	.8871	.8876	.8882	.8887	.8893	.8899	.8904	.8910	.8915
78	.8921	.8927	.8932	.8938	.8943	.8949	.8954	.8960	.8965	.8971
79	.8976	.8982	.8987	.8993	.8998	.9004	.9009	.9015	.9020	.9025
80	.9031	.9036	.9042	.9047	.9053	.9058	.9063	.9069	.9074	.9079
81	.9085	.9090	.9096	.9101	.9106	.9112	.9117	.9122	.9128	.9133
82	.9138	.9143	.9149	.9154	.9159	.9165	.9170	.9175	.9180	.9186
83	.9191	.9196	.9201	.9206	.9212	.9217	.9222	.9227	.9232	.9238
84	.9243	.9248	.9253	.9258	.9263	.9269	.9274	.9279	.9284	.9289
85	.9294	.9299	.9304	.9309	.9315	.9320	.9325	.9330	.9335	.9340
86	.9345	.9350	.9355	.9360	.9365	.9370	.9375	.9380	.9385	.9390
87	.9395	.9400	.9405	.9410	.9415	.9420	.9425	.9430	.9435	.9440
88	.9445	.9450	.9455	.9460	.9465	.9469	.9474	.9479	.9484	.9489
89	.9494	.9499	.9504	.9509	.9513	.9518	.9523	.9528	.9533	.9538
90	.9542	.9547	.9552	.9557	.9562	.9566	.9571	.9576	.9581	.9586
91	.9590	.9595	.9600	.9605	.9609	.9614	.9619	.9624	.9628	.9633
92	.9638	.9643	.9647	.9652	.9657	.9661	.9666	.9671	.9675	.9680
93	.9685	.9689	.9694	.9699	.9703	.9708	.9713	.9717	.9722	.9727
94	.9731	.9736	.9741	.9745	.9750	.9754	.9759	.9763	.9768	.9773
95	.9777	.9782	.9786	.9791	.9795	.9800	.9805	.9809	.9814	.9818
96	.9823	.9827	.9832	.9836	.9841	.9845	.9850	.9854	.9859	.9863
97	.9868	.9872	.9877	.9881	.9886	.9890	.9894	.9899	.9903	.9908
98	.9912	.9917	.9921	.9926	.9930	.9934	.9939	.9943	.9948	.9952
99	.9956	.9961	.9965	.9969	.9974	.9978	.9983	.9987	.9991	.9996
N	0	1	2	3	4	5	6	7	8	9

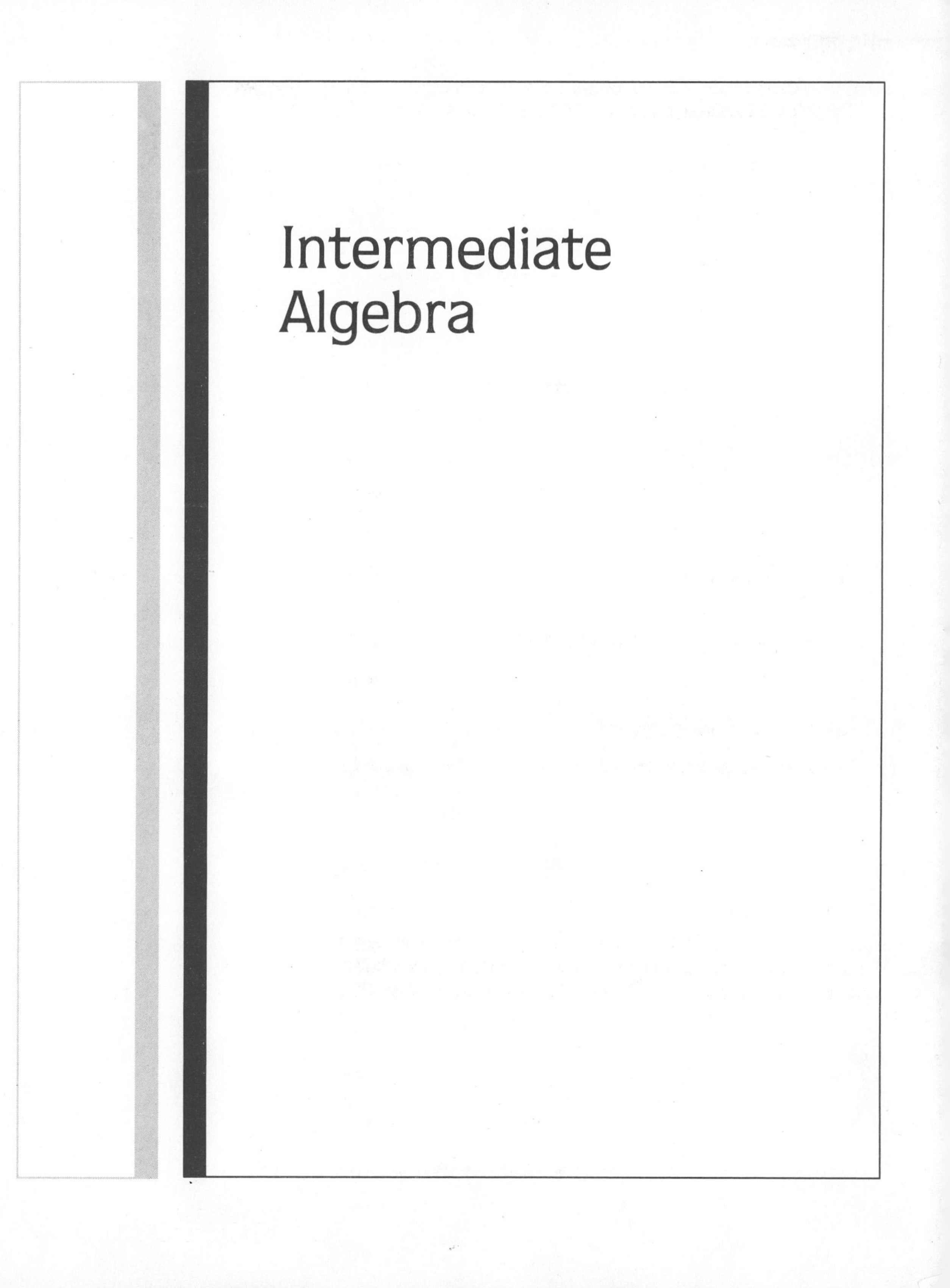
Intermediate
Algebra

To My Parents

Intermediate Algebra

Fourth Edition

Alfonse Gobran

Los Angeles Harbor College

PWS–KENT Publishing Co.

Boston

PWS-KENT
Publishing Company

20 Park Plaza
Boston, Massachusetts 02116

Copyright © 1988 by PWS-KENT Publishing Co.
Copyright © 1984 by PWS Publishers.
Copyright © 1978 by Prindle, Weber & Schmidt.

All rights reserved. No part of this book may be reproduced, stored in a retrieval system, or transcribed, in any form or by any means—electronic, mechanical, photocopying, recording, or otherwise—without the prior written permission of PWS-KENT Publishing Company.

PWS-KENT Publishing Company is a division of Wadsworth, Inc.

Library of Congress Cataloging-in-Publication Data
Gobran, Alfonse.
Intermediate algebra/Alfonse Gobran—4th ed.
P. cm.
Includes index.
ISBN 0-534-92061-6
1. Algebra. I. Title.
QA154.2.G6 1988 87-22441
512.9—dc 19 CIP

Printed in the United States of America

88 89 90 91 92 — 10 9 8 7 6 5 4 3 2 1

Production Coordinator: Ellie Connolly
Production: Sara Hunsaker ExLibris
Interior and Cover Design: Ellie Connolly
Cover Photo: Michael Furman/The Stock Market of N.Y.
Typesetting: Polyglot Pte. Ltd.
Cover Printing: New England Book Components
Printing and Binding: R.R. Donnelley & Sons Company

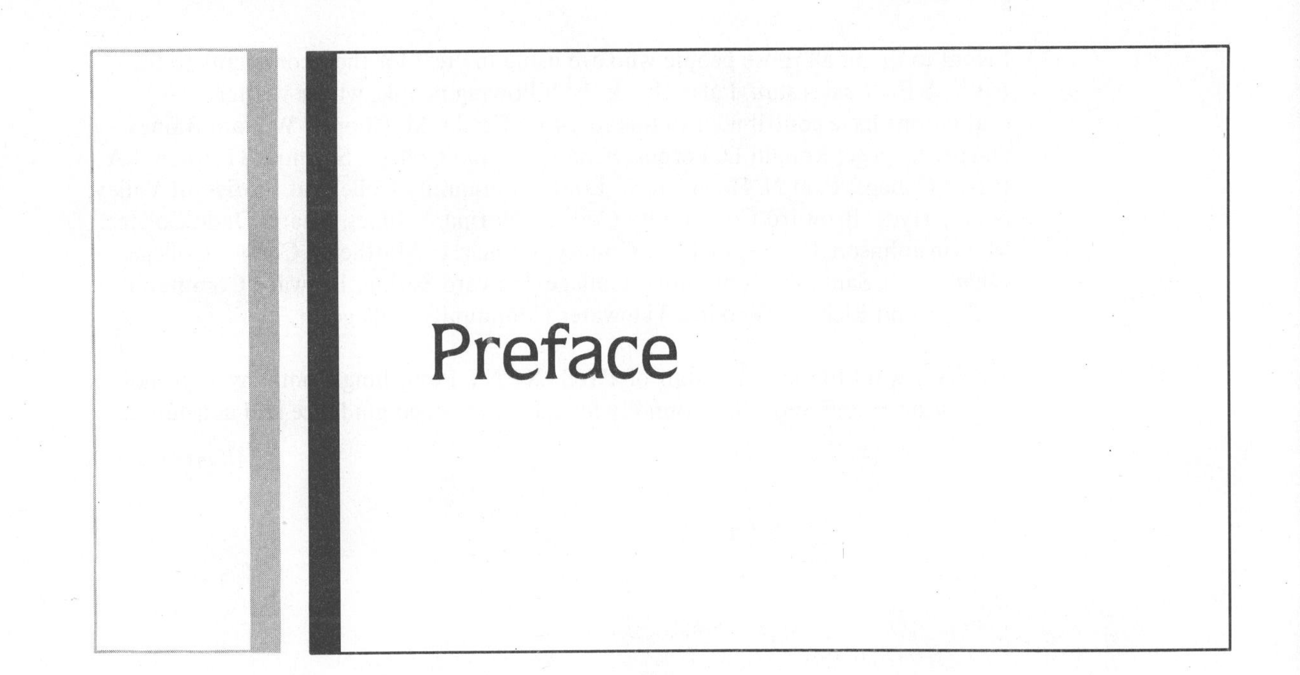

Preface

Intermediate Algebra, Fourth Edition, provides students with a sound algebraic background for further study in college mathematics. This text can be equally effective for use in a terminal course in intermediate algebra.

I believe that mathematics is best understood by applying concepts to specific examples. Consequently, the explanatory material is concise, and numerous examples and exercises are provided throughout the text.

The solving of word problems is given particular emphasis. Word problems are introduced gradually and covered repeatedly with numerous exercises. The translation of verbal expressions into algebraic equations is introduced in Chapter 3, followed by an introduction to solving equations and word problems. A wide variety of word problems are also included in Chapters 6, 9, 10, and 13.

These changes are based on responses from current users of the text as well as others teaching intermediate algebra. The revisions include the following:

- Many new examples ease the transition from elementary to more complicated concepts and methods.
- Over 1700 new exercises incorporated into the exercise sets.
- Cummulative Review sections added after Chapters 4, 8, 11, 14, and 17.
- More than 9500 exercises included in this edition.

I want to thank all those people who are using my text for their comments to the PWS–KENT sales staff. I also thank the following people, whose written evaluations have contributed to this revision: Cecilia M. Cooper, William Rainey Harper College; Ronald D. Ferguson, San Antonio College; Seymour Gottlieb, LA Pierce College; Paul N. Huchens, St. Louis Community College at Florissant Valley; Nancy Hyde, Broward Community College; Norma A. Innes, Miami Dade College; Marvin Johnson, College of Lake County; Pamela E. Matthews, Chabot College; Ellen Smith, Santa Fe Community College; Howard Sorkin, Broward Community College; and Richard Watkins, Tidewater Community College.

Finally, I want to thank the staff of PWS–KENT Publishing Company, especially the Editorial staff and Ellie Connolly for their continued guidance and assistance.

Alfonse Gobran

Contents

Chapter 1 Sets and Real Numbers 1

1.1 Introduction 2
1.2 The Number Line 2
1.3 Sets 3
Subsets 5
Operations with Sets 5
1.4 The Set of Real Numbers 8
1.5 The Set of Whole Numbers 9
Addition of Whole Numbers 9
Multiplication of Whole Numbers 10
Subtraction of Whole Numbers 12
1.6 The Set of Integers 12
Addition and Subtraction of Integers 13
Multiplication of Integers 13
Division of Integers 14
Division by Zero 15
Factoring Numbers 15
1.7 The Set of Rational Numbers 17
Reducing Fractions 18
Addition of Rational Numbers 19
Subtraction of Rational Numbers 21
Multiplication of Rational Numbers 22
Division of Rational Numbers 22

Combined Operations 23
Decimal Form of Rational Numbers 24
Mixed Numbers 26
1.8 Irrational Numbers and Real Numbers 29
Chapter 1 Review 30

Chapter 2 Basic Operations with Polynomials 32

2.1 Algebraic Notation and Terminology 33
2.2 Evaluation of Expressions 33
2.3 Addition of Polynomials 35
2.4 Subtraction of Polynomials 35
2.5 Grouping Symbols 37
2.6 Multiplication of Polynomials 39
Definition and Notation 39
Multiplication of Monomials 41
Multiplication of a Polynomial by a Monomial 45
Multiplication of Polynomials 47
2.7 Division of Polynomials 50
Division of Monomials 50
Division of a Polynomial by a Monomial 54
Division of Two Polynomials 56
Chapter 2 Review 61

Chapter 3 Linear Equations in One Variable 64

3.1 Terminology 65
3.2 Equivalent Equations 66
3.3 Solving Equations 67
3.4 Word Problems 78
Number Problems 81
Percentage Problems 85
Value Problems 90
Motion Problems 93
Temperature Problems 95
Lever Problems 96
Geometry Problems 98
Chapter 3 Review 100

Chapter 4 Linear Inequalities and Absolute Values in One Variable 104

4.1 Definitions and Notation 105
4.2 Properties of the Order Relations 106

4.3 Solution of Linear Inequalities in One Variable 108
4.4 Solution of Systems of Linear Inequalities in One Variable 113
4.5 Absolute Values 116
Properties of the Absolute Values of Real Numbers 116
4.6 Solution of Linear Equations Involving Absolute Values 117
4.7 Solution of Linear Inequalities Involving Absolute Values 124
Chapter 4 Review 126

Cumulative Review 129

Chapter 5 Factoring Polynomials 136

5.1 Factors Common to All Terms 137
5.2 Factoring a Binomial 140
Squares and Square Roots 140
Difference of Two Squares 141
Cubes and Cube Roots 143
Sum of Two Cubes 143
Difference of Two Cubes 144
5.3 Factoring a Trinomial 145
Trinomials of the Form $x^2 + bx + c, b, c \in I$ 146
Trinomials of the Form $ax^2 + bx + c, a, b, c \in I, a \neq 1$ 150
5.4 Factoring by Completing the Square 156
5.5 Factoring Four-Term Polynomials 159
Grouping as Three and One 160
Grouping in Pairs 161
Chapter 5 Review 163

Chapter 6 Algebraic Fractions 166

6.1 Simplification of Algebraic Fractions 167
6.2 Addition of Algebraic Fractions 172
Fractions with Like Denominators 172
Least Common Multiple of Polynomials 176
Fractions with Unlike Denominators 178
6.3 Multiplication of Algebraic Fractions 183
6.4 Division of Algebraic Fractions 186
6.5 Combined Operations and Complex Fractions 190
6.6 Literal Equations 196
6.7 Equations Involving Algebraic Fractions 199
6.8 Word Problems 203
Chapter 6 Review 209

Chapter 7 Exponents and Applications 214

7.1 Positive Fractional Exponents 215
7.2 Zero and Negative Exponents 224
Chapter 7 Review 234

Chapter 8 Radical Expressions 237

8.1 Definitions and Notation 238
8.2 Standard form of Radicals 239
8.3 Combination of Radical Expressions 243
8.4 Multiplication of Radical Expressions 246
8.5 Division of Radical Expressions 250
8.6 Equations Involving Radical Expressions 259
Chapter 8 Review 262

Cumulative Review 265

Chapter 9 Linear Equations and Inequalities in Two Variables 274

9.1 Ordered Pairs of Numbers 275
9.2 Rectangular or Cartesian Coordinates 275
9.3 Distance Between Two Points 280
9.4 One Linear Equation in Two Variables 285
9.5 Slope of a Line 290
9.6 Equations of Lines 295
Equation of a Line Given Two Points 295
Equation of a Line Given One Point $P_1(x_1, y_1)$ and Slope m 296
Equation of a Line Given the Intercepts 296
9.7 Two Linear Equations in Two Variables 300
Graphical Interpretations 300
9.8 Solution of Systems of Two Linear Equations in Two Variables 302
Graphical Solution 302
Algebraic Solution 304
Method of Elimination 304
Method of Substitution 308
9.9 Systems of Two Linear Equations in Two Variables Involving Grouping Symbols and Fractions 310
9.10 Fractional Equations That Can Be Made Linear 311
9.11 Word Problems 315
9.12 Graphs of Linear Inequalities in Two Variables 325
9.13 Graphical Solution of a System of Linear Inequalities in Two Variables 328
Chapter 9 Review 330

Chapter 10 Linear Equations in Three Variables 335

10.1 One Linear Equation in Three Variables 336
10.2 Systems of Two Linear Equations in Three Variables 337
Graphical Interpretations 337
Solution of a System of Two Linear Equations in Three Variables 337
10.3 Systems of Three Linear Equations in Three Variables 340
Graphical Interpretations 341
Solution of Systems of Three Linear Equations in Three Variables 342
10.4 Word Problems 346
Chapter 10 Review 350

Chapter 11 Determinants 352

11.1 Definitions and Notation 353
11.2 Properties of Determinants 353
11.3 Solution of Systems of Two Linear Equations in Two Variables by Determinants 355
11.4 Determinants of Order Higher Than Two 357
11.5 Solution of Systems of n Linear Equations in n Variables by Determinants 361
Chapter 11 Review 364

Cumulative Review 366

Chapter 12 Complex Numbers 374

12.1 Pure Imaginary Numbers 375
Addition and Subtraction of Pure Imaginary Numbers 375
Product of a Real Number and a Pure Imaginary Number 376
Products of Pure Imaginary Numbers 376
Division of Pure Imaginary Numbers 377
12.2 Complex Numbers Definition and Notation 379
12.3 Operations on Complex Numbers 381
Addition and Subtraction of Complex Numbers 381
Multiplication of Complex Numbers 383
Division of Complex Numbers 384
12.4 Graphs of Complex Numbers 386
Chapter 12 Review 388

Chapter 13 Quadratic Equations and Inequalities in One Variable 389

13.1 Introduction 390
13.2 Solution of Quadratic Equations by Factoring 390

13.3 Solution of Quadratic Equations by Completing the Square 395
13.4 Solution of Quadratic Equations by the Quadratic Formula 399
13.5 Character of the Roots 402
13.6 Properties of the Roots 406
13.7 Equations that Lead to Quadratic Equations 413
13.8 Word Problems 419
13.9 Graphs of Quadratic Equations 423
Coordinates of the Vertex and Equation of the Line of Symmetry 424
Solution of Quadratic Equations by Graphs 427
13.10 Analytic Solution of Quadratic Inequalities 429
13.11 Graphical Solution of Quadratic Inequalities 432
Chapter 13 Review 433

Chapter 14 Conic Sections and Solution of Systems of Quadratic Equations in Two Variables 438

14.1 Introduction 439
Extent 439
14.2 The Circle 440
Equation of a Circle 440
Graphing the Circle 440
14.3 The Ellipse 443
Equation of an Ellipse 443
Graphing the Ellipse 445
14.4 The Parabola 447
Equation of a Parabola 448
Another Form of the Equation of a Parabola 449
Graphing the Parabola 449
14.5 The Hyperbola 451
Equation of the Hyperbola 451
Another Form of the Equation of a Hyperbola 453
Graphing the Hyperbola 454
14.6 Solution of Systems of Quadratic Equations in Two Variables 456
Solution of a Linear and a Quadratic Equation in Two Variables 456
Solution of a System of Two Quadratic Equations of the Form $Ax^2 + Ay^2 + Dx + Ey + F = 0$ 458
Solution of a System of Two Quadratic Equations of the Form $Ax^2 + Cy^2 + F = 0$ 460
Solution of a System of Two Quadratic Equations of the Form $Ax^2 + Bxy + Cy^2 + F = 0$ 461
Chapter 14 Review 463

Cumulative Review 466

Chapter 15 Functions 473

15.1 Properties of Functions 474
15.2 Function Notation 475
15.3 Algebra of Functions 476
15.4 Functions Defined by Equations 478
15.5 Exponential Functions 481
15.6 Inverse Functions 485
Chapter 15 Review 489

Chapter 16 Logarithms 491

16.1 Logarithmic Functions 492
16.2 Graphs of Logarithmic Functions 496
16.3 Properties of Logarithms 496
16.4 Common Logarithms 502
16.5 Natural Logarithms 504
16.6 Computations with Logarithms 505
Chapter 16 Review 508

Chapter 17 Progressions 511

17.1 Sequences 512
17.2 Series 512
Summation Notation 513
17.3 Arithmetic Progressions 515
Sum of an Arithmetic Progression 516
17.4 Geometric Progressions 522
Sum of a Geometric Progression 523
17.5 Compound Interest 529
17.6 Annuities 531
Chapter 17 Review 534

Cumulative Review 536

Appendix A Interval Notation A1
Bounded Intervals A2
Infinite Intervals A2

Appendix B Binomial Expansion A3

Appendix C Synthetic Division A7

Appendix D Use of a Common Logarithms Table A12

Appendix E Theorems and Proofs A17

Answers to Odd-Numbered Exercises A26

Index A143

chapter 1

Sets and Real Numbers

1.1 Introduction

1.2 The number line

1.3 Sets

1.4 The set of real numbers

1.5 The set of whole numbers

1.6 The set of integers

1.7 The set of rational numbers

1.8 Irrational numbers and real numbers

1.1 Introduction

1/29/98

The **Hindu-Arabic system** of numeration started with nine symbols to represent the numbers 1 through 9, inclusive. The concept of zero came much later and was used to express a lack of objects. Today we use an extension and modernization of the Hindu-Arabic system. We use ten symbols: 0, 1, 2, 3, 4, 5, 6, 7, 8, and 9 to represent numbers. These symbols are combined in a **place-value system** to represent any number we need to express.

The numbers 1, 2, 3, etc. are called **counting numbers** or **natural numbers**.
The numbers 0, 1, 2, etc. are called **whole numbers**.

1.2 The Number Line

Numbers developed from the need to count. Sometimes it is convenient to use geometry to illustrate or prove some important results in algebra. Thus we would like to have a geometrical representation of the whole numbers. To accomplish this, we draw a straight line and choose a point on the line to represent the number 0. We call this point the **origin**. Take another point on the line at some distance from the origin to the right of it. Associate that point with the number 1. The segment of the line from the origin to the point representing the number 1 is our unit measurement; it is the **scale** we use on the line. Now, at a distance of 1 unit to the right of the point representing the number 1, put another point and let it represent the number 2. We continue this procedure as far as we like, thus establishing an association between the whole numbers and points on the line.

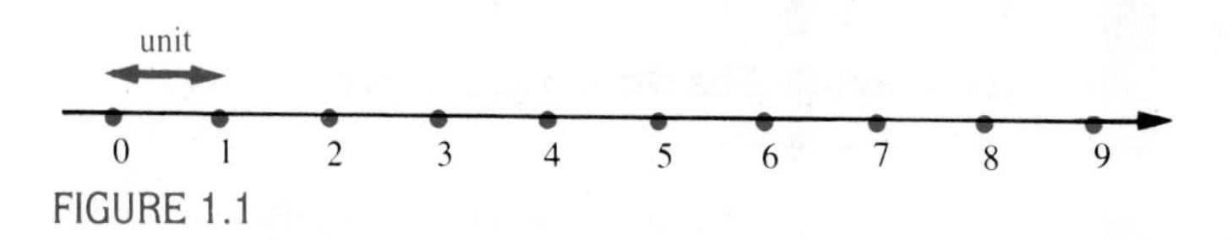

FIGURE 1.1

Figure 1.1 shows the **number line**. The arrow at the end of the line indicates that we continue in this manner and also in the direction of increasing numbers.
The segment of the line representing a unit distance, that is, the scale on the line, is taken to suit our purpose (see Figure 1.2).

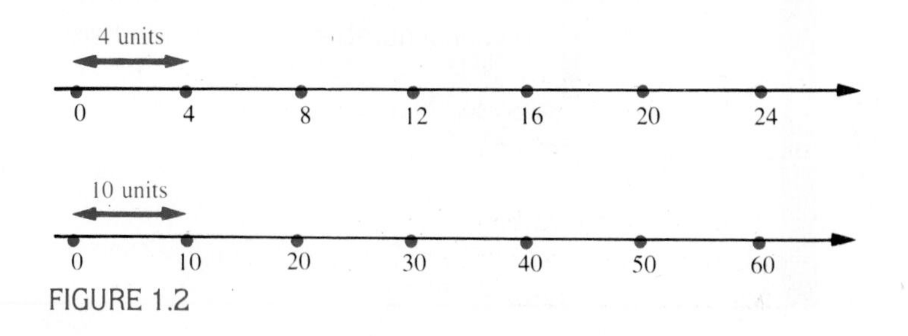

FIGURE 1.2

By taking a convenient scale and extending the line far enough, any whole number can be associated with a unique point on the line. The points on the line are called the **graph** of the numbers. The numbers are called the **coordinates of the points**.

Note If a and b are the coordinates of any two points on the line, and if the graph of b lies to the right of the graph of a, then b is **greater than** a, denoted by $b > a$, or a is **less than** b, denoted by $a < b$.

1.3 Sets

The concept of a set has been utilized so extensively throughout modern mathematics that an understanding of it is necessary for all college students. Sets are a means by which mathematicians talk of collections of things in an abstract way.

According to G. Cantor (1845–1918), the mathematician who developed set theory, "a **set** is a grouping together of single objects into a whole."

Note that no uniform property of the objects in the set is implied other than that they are grouped together to form the set.

The totality of students taking college algebra forms a set. The collection consisting of a desk, a chair, and a book constitutes a set.

The numbers 1, 2, 3, etc. constitute a set called the **set of natural numbers**, which is denoted by N. The numbers 0, 1, 2, etc. constitute the **set of whole numbers**, denoted by W.

There are two ways to define a set. The first is to list and separate by commas the objects that compose the set. We write the defining list within braces $\{\ \}$.

Thus $A = \{\text{Mars, Venus, Neptune}\}$ and $N = \{1, 2, 3, \ldots\}$.
The three dots indicate that we continue in the same manner.

Note It is usual to let capital letters signify sets and lowercase letters represent objects of sets.

If $X = \{a, b, c, d\}$, then $a, b, c,$ and d are called **members** or **elements** of the set X.
The notation $a \in X$ is read "a is an element of the set X."
To denote that e is not an element of the set X, we write $e \notin X$.

Note The order of the elements of the set is immaterial. For example, $\{1, 2, 3\}$ and $\{3, 1, 2\}$ define the same set. It is not necessary, but merely convenient, to list numbers in numerical order.

Note When we list the members of a set, each element should be listed only once, since we would otherwise be referring to the same member of the set more than once. The set of numerals in the number 83,837 is $\{3, 7, 8\}$.

The second way to define a set is to give the rule that identifies the elements of the set. We still write the defining rule within braces:

$$Y = \{\text{all natural numbers that are a multiple of } 2\}$$

DEFINITION

A **variable** is a letter that assumes various values in a problem.

When the set is defined by a rule, the rule may be written in words or, to make it short, in symbols.

To designate a generic member of a set of numbers, a variable such as $x, y, z, m, n, \ldots,$ is used.

The set X whose elements have property P is denoted by

$$X = \{x \mid x \text{ has property } P\}$$

which is read "X is the set of elements x, such that x has the property P." The vertical bar in the notation is an abbreviation for the words "such that."

EXAMPLE List the elements in the set $X = \{x \mid x = 3n + 1, n \in W\}$.

Solution

n takes the values $0, 1, 2, 3, \ldots$.

$3n$ takes the values $0, 3, 6, 9, \ldots$ ($3n$ means $3 \times n$).

$x = 3n + 1$ takes the values $1, 4, 7, 10, \ldots$.

Thus $X = \{1, 4, 7, 10, \ldots\}$

Notes

1. $\{x \mid x = 3n + 1, n \in W\}$ can be written as $\{3n + 1 \mid n \in W\}$.
2. By $5 < x < 11$, $x \in N$ is meant the natural numbers between 5 and 11. That is, x takes the values 6, 7, 8, 9, 10.

EXAMPLE List the elements in the set $A = \{2x \mid 3 < x < 10, x \in N\}$.

Solution

x takes the values 4, 5, 6, 7, 8, 9.

$2x$ takes the values 8, 10, 12, 14, 16, 18.

Thus $A = \{8, 10, 12, 14, 16, 18\}$

DEFINITION

The set that has no elements is called the **null set** and is denoted by $\varnothing$.

The set of natural numbers between 1 and 2 is the null set. The set of natural satellites of the planet Venus is the null set.

Subsets

DEFINITION

A set X is a **subset** of the set Y if every element of X is an element of Y.

If X is a subset of Y, we write $X \subset Y$.

EXAMPLE If $A = \{a, b, c\}$ and $B = \{a, b, c, d\}$, then $A \subset B$.

Note Every set is a subset of itself. Also, the null set is a subset of every set.

EXAMPLE The subsets of the set $\{1, 2, 3\}$ are

$\{1, 2, 3\}$ $\{1, 2\}$ $\{1, 3\}$ $\{2, 3\}$ $\{1\}$ $\{2\}$ $\{3\}$ $\varnothing$

The notation $A \not\subset B$ is read "A is not a subset of B." This means there exists at least one element in A that is not in B.

EXAMPLE If $A = \{1, 2, 3\}$ and $B = \{1, 3, 5, 7\}$, then $A \not\subset B$.

DEFINITION

Two sets A and B are **equal**, $A = B$, if every element of A is an element of B and every element of B is an element of A.

Note $A = B$ means that the relations $A \subset B$ and $B \subset A$ hold simultaneously.

The notation $A \neq B$, read "A is not equal to B," means that either there exists at least one element that is a member of A but not of B, or there exists at least one element that is a member of B but not of A.

EXAMPLE When $A = \{2, 4, 6\}$ and $B = \{2, 4, 6, 8\}$, then $A \neq B$.

Operations with Sets

DEFINITION

The **union** of two sets A and B, denoted by $A \cup B$, is the set of all elements that are in the set A or in the set B. It is the set of elements that belong to at least one of the two sets:

$A \cup B = \{x \mid x \in A \text{ or } x \in B\}$

EXAMPLE

Let $A = \{1, 2, 3\}$ and $B = \{1, 3, 5, 7\}$,

then $A \cup B = \{1, 2, 3, 5, 7\}$

Note For any two sets A and B,

1. $A \cup \varnothing = A$
2. $A \cup B = B \cup A$
3. $A \subset A \cup B$
4. $B \subset A \cup B$

DEFINITION

The **intersection** of two sets A and B, denoted by $A \cap B$, is the set of elements that are only in both A and B:

$A \cap B = \{x \mid x \in A \text{ and } x \in B\}$

EXAMPLES

1. If $A = \{1, 2, 3, 4\}$ and $B = \{2, 4, 6, 8\}$,
then $A \cap B = \{2, 4\}$

2. If $A = \{a, b, c\}$ and $B = \{d, e, f\}$,
then $A \cap B = \varnothing$

DEFINITION

Two sets A and B are **disjoint** if $A \cap B = \varnothing$.

Note For any two sets A and B,

1. $A \cap \varnothing = \varnothing$
2. $A \cap B = B \cap A$
3. $A \cap B \subset A$
4. $A \cap B \subset B$

When there is more than one operation with sets we use parentheses to indicate which operation should be performed first. Given the three sets A, B, and C, to calculate $(A \cap B) \cup C$ we first calculate the intersection of the sets A and B, and then find the union of the resulting set with the set C. When we have $A \cap (B \cup C)$, we first calculate the union of the two sets B and C, and then find the intersection of the set A with the resulting set.

EXAMPLE

Given $A = \{1, 2, 3\}$, $B = \{1, 3, 5\}$, and $C = \{2, 3, 4\}$, find

$A \cap (B \cup C)$ and $(A \cap B) \cup C$

Solution

To find $A \cap (B \cup C)$, we first calculate $B \cup C$:

$B \cup C = \{1, 3, 5\} \cup \{2, 3, 4\} = \{1, 2, 3, 4, 5\}$

$A \cap (B \cup C) = \{1, 2, 3\} \cap \{1, 2, 3, 4, 5\} = \{1, 2, 3\}$

To find $(A \cap B) \cup C$, we first calculate $A \cap B$:

$$A \cap B = \{1, 2, 3\} \cap \{1, 3, 5\} = \{1, 3\}$$
$$(A \cap B) \cup C = \{1, 3\} \cup \{2, 3, 4\} = \{1, 2, 3, 4\}$$

Notes

1. $A \cap (B \cup C) \neq (A \cap B) \cup C$
2. $A \cup (B \cup C) = (A \cup B) \cup C$
3. $A \cap (B \cap C) = (A \cap B) \cap C$

Exercise 1.3

List the elements in each of the following sets:

1. The names of the days in a week.
2. The names of the months of a year that have exactly 30 days.
3. The names of continents on the earth.
4. The names of the states that start with the letter A.
5. The names of the first five presidents of the United States of America.
6. The natural numbers between 18 and 32 that are divisible by 17.
7. The letters in the word Mississippi.
8. The vowels in the English alphabet.

9. $\{x \mid x = n + 1, n \in N\}$
10. $\{x \mid x = n + 3, n \in W\}$
11. $\{4n \mid n \in W\}$
12. $\{2n + 2 \mid n \in W\}$
13. $\{4n + 1 \mid n \in W\}$
14. $\{3n - 2 \mid n \in N\}$
15. $\{5n - 1 \mid n \in N\}$
16. $\{7n - 2 \mid n \in N\}$
17. $\{4x \mid 1 < x < 8, x \in W\}$
18. $\{3x \mid 0 < x < 6, x \in W\}$
19. $\{7x \mid 0 < x < 5, x \in N\}$
20. $\{5x \mid 3 < x < 9, x \in N\}$

Given the sets A, B, X and Y, check the following:

21. If every element of A, is an element of B, then is $A \subset B$?
22. If every element of A, is an element of B, then is $A = B$?
23. If $X \subset Y$ and $a \in Y$, is $a \in X$?
24. If $x \in A$, is $\{x\}$ a subset of A?
25. If $y \in B$, is y a subset of B?

26. Write all subsets of the set $\{0\}$.
27. Write all subsets of the set $\{1, 2\}$.

If $A = \{a, b\}$, use one of the symbols $\{\ \}$, $\in$, $\notin$, $\subset$, or $\not\subset$ to make the following true:

28. $a \quad A$
29. $a \subset A$
30. $c \quad A$
31. $\{a, b\} \quad A$
32. $b \subset \{a, b\}$
33. $\{c\} \quad A$

1/3/98

Let A and B be two sets:

34. If $a \in A$, then must a be an element of $A \cup B$?
35. If $a \in A$, then must a be an element of $A \cap B$?
36. If $a \in A \cup B$, then must a be an element of A?
37. If $a \in A \cup B$, then must a be an element of B?
38. If $a \in A \cup B$, then must a be an element of $A \cap B$?
39. If $a \in A \cap B$, then must a be an element of A?
40. If $a \in A \cap B$, then must a be an element of B?
41. If $a \in A \cap B$, then must a be an element of $A \cup B$?
42. If $A \not\subset B$ and $a \in A$, must a be an element of B?
43. If $A \not\subset B$ and $a \in A \cup B$, must a be an element of A?
44. If $A \not\subset B$ and $a \in A \cap B$, must a be an element of A?

List the elements in each of the following sets:

45. $\{3x \mid x < 10, x \in N\} \cap \{5x \mid 0 < x < 6, x \in N\}$
46. $\{2x \mid 1 < x < 8, x \in N\} \cap \{3x \mid x < 7, x \in N\}$
47. $\{2x \mid x \in N\} \cap \{3x \mid x \in N\}$
48. $\{2x \mid x \in N\} \cap \{5x \mid x \in N\}$
49. $\{3x \mid x \in W\} \cap \{5x \mid x \in W\}$
50. $\{2x \mid x \in W\} \cap \{7x \mid x \in W\}$

Let $A = \{2, 4, 6, 8, 10\}$, $B = \{1, 2, 3, 4, 5\}$, and $C = \{1, 3, 5, 7, 9\}$. List the elements in each of the following sets:

51. $A \cup B$
52. $B \cup C$
53. $C \cup A$
54. $A \cap B$
55. $B \cap C$
56. $C \cap A$
57. $A \cup (B \cup C)$
58. $A \cup (B \cap C)$
59. $(A \cup B) \cap C$
60. $A \cap (B \cup C)$
61. $(A \cap B) \cup C$
62. $A \cap (B \cap C)$
63. Show that $A \cup (B \cap C) = (A \cup B) \cap (A \cup C)$
64. Show that $A \cap (B \cup C) = (A \cap B) \cup (A \cap C)$

1.4 The Set of Real Numbers

The rest of this chapter deals with the development of the real number system. Properties and laws of numbers are presented so that we have the basic tools necessary to understand some algebraic concepts. To accomplish this, letters of the alphabet, referred to as **literal numbers**, are used instead of specific numbers.

The basic operations on numbers are addition, multiplication, subtraction, and division. These four operations are called **binary operations** since they are defined in terms of combining numbers only two at a time.

The symbols used to indicate these operations are

$+$	called **plus**, indicating addition
$\cdot$ or $\times$	called **times**, indicating multiplication
$-$	called **minus**, indicating subtraction
$\div$	called **divided by**, indicating division

1.5 The Set of Whole Numbers

The set of whole numbers $W = \{0, 1, 2, 3, \ldots\}$ was invented from the need to count. The discussion of the basic operations on this set will show the need for expanding this set to the set of real numbers.

Addition of Whole Numbers

For any two whole numbers a and b there exists a unique whole number called their **sum**. The sum of the two numbers a and b is denoted by $a + b$. The a and b are called **terms** of the sum.

DEFINITION

> A set X is said to be **closed under addition** if for all elements a, b of the set X, the sum $a + b$ is also an element of X.

EXAMPLES

1. The set $\{2n \mid n \in N\} = \{2, 4, 6, \ldots\}$ is closed under addition.
2. The set $\{4n + 1 \mid n \in N\} = \{5, 9, 13, 17, \ldots\}$ is not closed under addition.

For any two numbers $a, b \in W$, the sum $a + b$ is a uniquely determined number in W. Thus W is closed under addition.

The following are laws of addition of whole numbers.

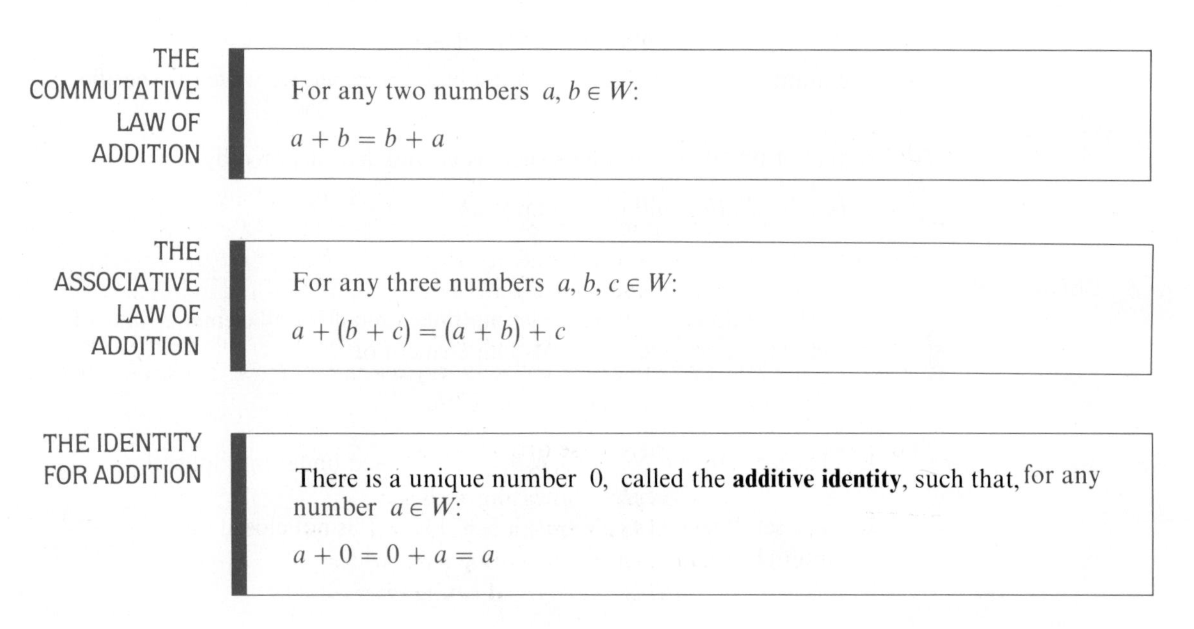

THE COMMUTATIVE LAW OF ADDITION

> For any two numbers $a, b \in W$:
>
> $a + b = b + a$

THE ASSOCIATIVE LAW OF ADDITION

> For any three numbers $a, b, c \in W$:
>
> $a + (b + c) = (a + b) + c$

THE IDENTITY FOR ADDITION

> There is a unique number 0, called the **additive identity**, such that, for any number $a \in W$:
>
> $a + 0 = 0 + a = a$

Note Although addition is a binary operation, it can be extended to find the sum of three or more numbers by adding the first two numbers and then adding each successive number to the previous sum:

$$4 + 16 + 23 = (4 + 16) + 23 = 20 + 23 = 43$$

Multiplication of Whole Numbers

DEFINITION

The **product** of two whole numbers a and b is defined to be the whole number $a \cdot b$, which is another name for the sum

$b + b + \cdots + b$ a terms of b

The a and b are called **factors** of the product.

MULTIPLICATION BY ZERO

For any number $a \in W$, $a \cdot 0 = 0$.

The product of two specific numbers such as 2 and 3 is denoted by

$2 \cdot 3$, 2×3, $2(3)$, or $(2)(3)$

The product of a specific number and a literal number such as 5 and a is denoted by

$5 \cdot a$, $5 \times a$, $5(a)$, $(5)(a)$, or simply $5a$

When we multiply a specific number and a literal number, we write the specific number as the first factor.

The product of two literal numbers such as a and b is denoted by

$a \cdot b$, $a \times b$, $a(b)$, $(a)(b)$, or simply ab

DEFINITION

A set X is said to be **closed under multiplication** if for all elements a, b of the set X, the product ab is also an element of X.

EXAMPLES

1. The set $\{3n \mid n \in N\} = \{3, 6, 9, \ldots\}$ is closed under multiplication.
2. The set $\{3n - 1 \mid n \in N\} = \{2, 5, 8, 11, \ldots\}$ is not closed under multiplication.

For any two numbers $a, b \in W$ the product ab is a uniquely determined number in W. Thus W is closed under multiplication.

The following are laws of multiplication of whole numbers:

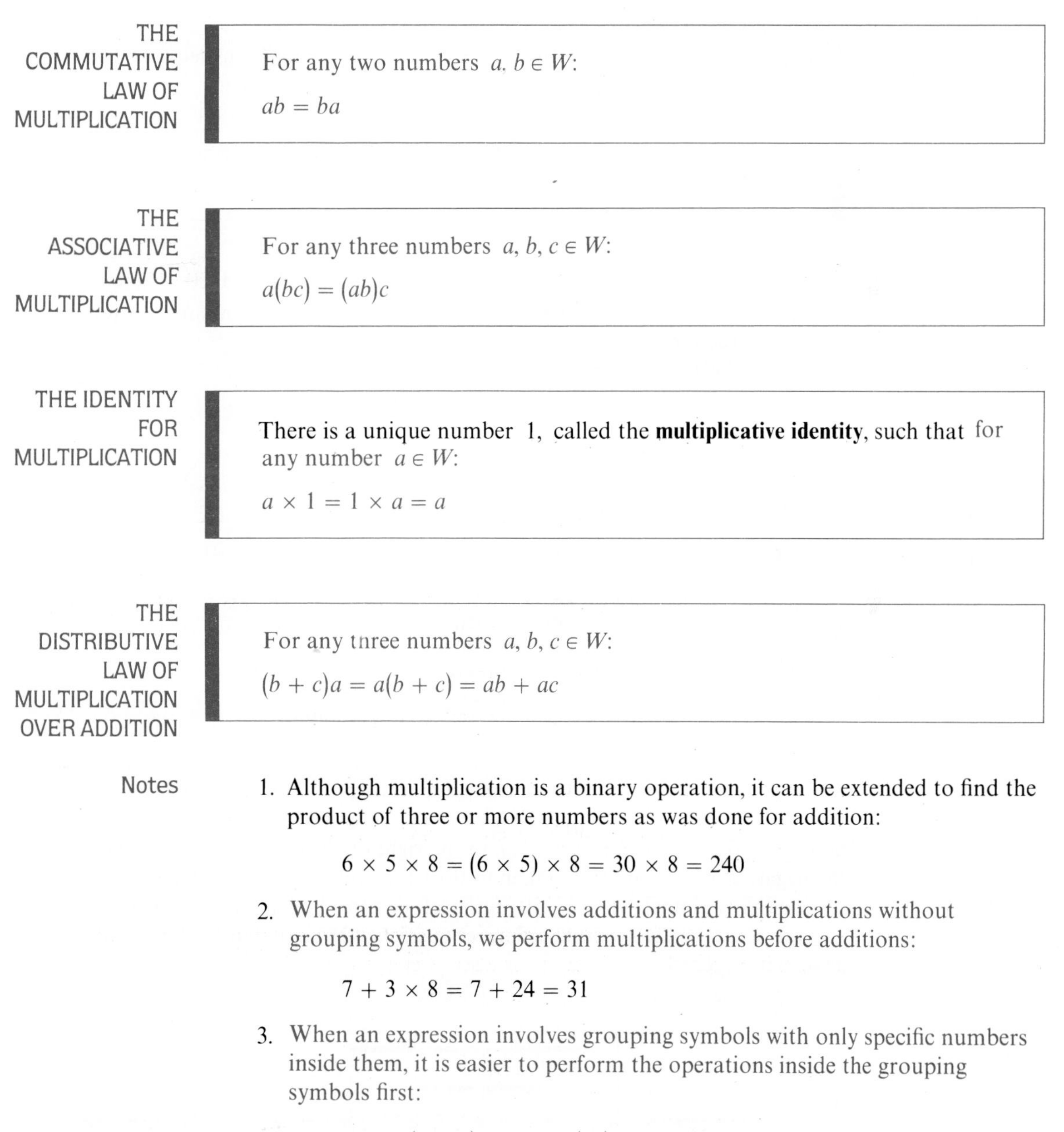

THE COMMUTATIVE LAW OF MULTIPLICATION

For any two numbers $a, b \in W$:

$ab = ba$

THE ASSOCIATIVE LAW OF MULTIPLICATION

For any three numbers $a, b, c \in W$:

$a(bc) = (ab)c$

THE IDENTITY FOR MULTIPLICATION

There is a unique number 1, called the **multiplicative identity**, such that for any number $a \in W$:

$a \times 1 = 1 \times a = a$

THE DISTRIBUTIVE LAW OF MULTIPLICATION OVER ADDITION

For any three numbers $a, b, c \in W$:

$(b + c)a = a(b + c) = ab + ac$

Notes

1. Although multiplication is a binary operation, it can be extended to find the product of three or more numbers as was done for addition:

$$6 \times 5 \times 8 = (6 \times 5) \times 8 = 30 \times 8 = 240$$

2. When an expression involves additions and multiplications without grouping symbols, we perform multiplications before additions:

$$7 + 3 \times 8 = 7 + 24 = 31$$

3. When an expression involves grouping symbols with only specific numbers inside them, it is easier to perform the operations inside the grouping symbols first:

$$20 + 8(6 + 4) = 20 + 8(10) = 20 + 80 = 100$$

Subtraction of Whole Numbers

From addition of whole numbers we have $7 + 5 = 12$. That is, 5 is the number that when added to 7 gives the result 12. The number 5 is also called the **difference** between 12 and 7.

In notation we write $12 - 7 = 5$.
The operation designated by the symbol $-$, read "minus," is called **subtraction.**

DEFINITION

> A set X is said to be **closed under subtraction** if for all elements a, b of the set X, the difference $a - b$ is also an element of X.

Consider the difference between the two whole numbers 8 and 20. There is no number $a \in W$ such that $20 + a = 8$. Thus the set of whole numbers is not closed under subtraction.
In order to have a set closed under subtraction, the set of whole numbers is enlarged by including the **negative integers** $-1, -2, -3, \ldots$.

1.6 The Set of Integers

DEFINITION

> The union of the set of negative integers and the set of whole numbers constitutes the **set of integers**, denoted by I:
>
> $I = \{\ldots, -3, -2, -1, 0, 1, 2, 3, \ldots\}$

When you are 1000 feet above sea level, this can be denoted by $+1000$ feet, while if you are 50 feet below sea level, it can be denoted by -50 feet. From this we can see that the $+$ and $-$ signs may be used to indicate two opposite directions.
Since the positive integers are taken to the right of the origin on the number line, the negative integers must be taken to the left of the origin. Thus the graph of the set of negative integers are points to the left of zero. In general, the integers a and $-a$ are coordinates of points on opposite sides of the origin and equidistant from it as seen in Figure 1.3.

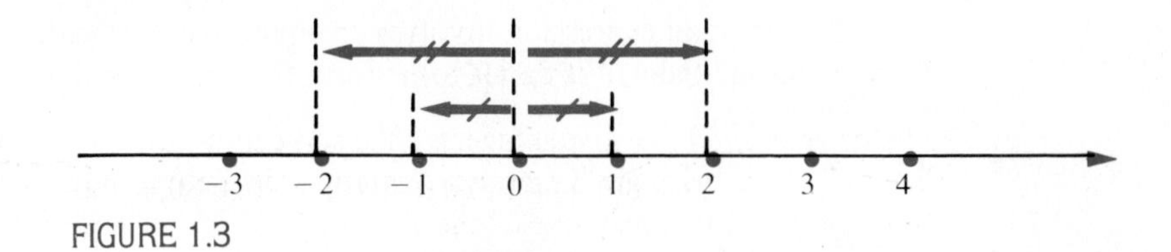

FIGURE 1.3

Note that as we move to the right along the number line, the numbers increase in value and as we move to the left, the numbers decrease in value. The **positive direction** is to the right, while the **negative direction** is to the left.

Addition and Subtraction of Integers

DEFINITION

> When the sum of two numbers is zero, the two numbers are called **additive inverses**.

For every number $a \in I$ there exists a unique number $(-a)$ in I such that
$a + (-a) = 0.$
Thus the numbers a and $(-a)$ are additive inverses.
The number $(-a)$ is sometimes called the **negative** of the number a.

THEOREM ■ If $a \in N$, then $-(-a) = a$.

DEFINITION

> If $a, b \in I$ then $a - b = a + (-b)$, that is, subtracting b from a is the same as adding the additive inverse of b to a.

Note $+(-a) = -a$

THEOREM ■ If $a, b \in I$, then $(-a) + (-b) = -a - b = -(a + b)$.

COROLLARY ■ If $a, b \in I$, then $-a + b = -(a - b)$.

Notes

1. If $a, b \in I$, $a \neq b$, then $a - b \neq b - a$.
2. The set I is closed under addition and subtraction.
3. The commutative and associative laws of addition hold for the set of integers.
4. Subtraction of integers is not commutative and is not associative.

Multiplication of Integers

Multiplication of positive integers is the same as multiplication of natural numbers. We only need to define the product of a positive integer and a negative integer, and the product of two negative integers.

THEOREM 1 If $a, b \in N$, then $a(-b) = -(ab)$.

THEOREM 2 If $a, b \in N$, then $(-a)(-b) = ab$.

THEOREM 3 If $a, b, c \in I$, then $a(b + c) = ab + ac$.

Notes

1. The set of integers is closed under multiplication.
2. The commutative and associative laws of multiplication hold for the set of integers.
3. The distributive law of multiplication holds for the set of integers.

When an expression involves additions, subtractions, and multiplications without grouping symbols, we perform multiplications before additions and subtractions.

EXAMPLE

$$\begin{aligned} \underline{3 \times 4} + \underline{7(-2)} - 6 &= 12 - 14 - 6 \\ &= 12 - 20 = -8 \end{aligned}$$

When an expression involves grouping symbols with only specific numbers inside them, it is easier to perform the operations inside the grouping symbols first.

EXAMPLE

$$4(2 - 9) - 6(3 - 8) = 4(-7) - 6(-5) = -28 + 30 = 2$$

Division of Integers

From multiplication we have $8 \times 7 = 56$; 7 is the number that, when multiplied by 8, has the result 56. The number 7 is also called the **quotient** of 56 divided by 8.

In notation we write $\quad 56 \div 8 = 7 \quad$ or $\quad \dfrac{56}{8} = 7$

The symbol $\div$, read "divided by," means division.

DEFINITION

> If $a, b, c \in I$ with $b \neq 0$ and $a = bc$, then $\dfrac{a}{b} = c$.

When $\dfrac{a}{b} = c$, the number a is called the **dividend**, b is called the **divisor**, and c or $\dfrac{a}{b}$ is called the **quotient**. The quotient $\dfrac{a}{b}$ is also called a **fraction**; a is called the **numerator** and b is called the **denominator** of the fraction. Sometimes a and b are referred to as the **terms** of the fraction.

Note The quotient of two positive or two negative numbers is a positive number. The quotient of a positive number divided by a negative number or a negative number divided by a positive number is a negative number.

Division by Zero

Division is defined from multiplication,

$$15 \div 5 = 3 \qquad \text{because} \qquad 5 \cdot 3 = 15$$

Now consider $6 \div 0$. We look for a number $a \in I$ such that $0 \cdot a = 6$. There is no such number a, since $0 \cdot a = 0$ for all $a \in I$.

Finally, consider $0 \div 0$. We look for a number $b \in I$ such that $0 \times b = 0$. This statement is true for any number $b \in I$. That is, b is not a unique number and a quotient must be unique.
Thus for any number $a \neq 0$ we have

$0 \div a = 0$

$a \div 0$ is not defined

$0 \div 0$ is not a unique number, is indeterminate

Remark Since $\frac{p}{q}$ is not defined when $q = 0$, all the denominators of the fractions will be assumed to be different from zero.

Note When an expression involves multiplications and divisions without grouping symbols, we perform the multiplications and divisions in the order in which they appear:

$$4 \times 6 \div 8 = 24 \div 8 = 3$$

$$48 \div 12 \times 2 = 4 \times 2 = 8$$

$$16 \div (-8) \times 2 = -2 \times 2 = -4$$

Note When an expression involves the four arithmetic operations without grouping symbols, we perform multiplications and divisions in the order in which they appear before additions and subtractions:

$$7 \times 6 \div 2 + 8 \div 4 \times 2 - 9 = 21 + 4 - 9 = 25 - 9 = 16$$

Factoring Numbers

DEFINITION

> The set of **prime numbers** consists of those natural numbers greater than 1 that are divisible only by themselves and 1.

DEFINITION

A natural number greater than 1 that is not prime is called **composite**.

Each composite number greater than 1 can be expressed as a product of primes in one and only one way, apart from the order of the factors. This statement is known as the **Fundamental Theorem of Arithmetic**.

To find the prime factors of a number, start with the prime numbers in order. Check to see if the number is divisible by 2; if so, divide the number by 2 and obtain the quotient. If the quotient is again divisible by 2, divide again by 2, and continue dividing by 2 until you obtain a quotient that is not divisible by 2.

Now see if the quotient is divisible by 3. When all the factors of 3 have been divided out, check to see if it is divisible by 5, and then by the successive larger primes.
Continue the division by the prime numbers until the quotient is 1.
All the divisors obtained are the **prime factors** of the number.

EXAMPLE $420 = 2 \cdot 2 \cdot 3 \cdot 5 \cdot 7$

Note We can stop testing a given number for divisibility when we reach a prime number that, if multiplied by itself, gives a product greater than the given number.

EXAMPLE 131 is prime and the only tests needed are those for 2, 3, 5, 7, and 11, but then stop, since $13 \cdot 13 = 169$.

Exercise 1.6

1. Add 12 and -16.
2. Add -15 and 6.
3. Add -7 and -4.
4. Subtract 10 from -12.
5. Subtract -5 from 7.
6. Subtract -6 from -9.

Perform the indicated operations:

7. $3 - 4 + 12$
8. $4 - 10 + 6$
9. $-8 + 3 - 11$
10. $-2 - 9 - 4$
11. $6 - (8) + (-7)$
12. $2 - (-13) + (-5)$
13. $20 - (-4) - 12$
14. $-15 - 8 - (-11)$
15. $8 + (7 - 4)$
16. $7 + (6 - 10)$
17. $3 - (8 + 2)$
18. $5 - (6 - 4)$
19. $10 - (3 - 11)$
20. $-4 + (-12 - 3)$
21. $-9 - (-4 - 13)$
22. $7(-6)$
23. $-3(9)$
24. $-8(-4)$
25. $2(-3)(4)$
26. $4(-8)(-3)$
27. $-5(4)(-2)$
28. $-2(-7)(-5)$
29. $6 \times 8 - 3$
30. $7 \times 6 - 1$

31. $12 - 7(5)$ **32.** $15 - 5(-3)$ **33.** $-9 + 4(-2)$
34. $13 + 2(7 - 10)$ **35.** $10 + 3(8 - 15)$ **36.** $11 - 8(2 - 7)$
37. $-8 + 3(-4 - 5)$ **38.** $-13 + 7(8 - 9)$ **39.** $-8 - 2(7 - 3)$
40. $10 - 4(-4) + 3(-6)$ **41.** $14 - 4(-2) + (-8)$
42. $6(5) - 4(10) + 7(-3)$ **43.** $17 - 7(-3) + 2(-13)$
44. $5(-13) - 4(-18) + 12$ **45.** $-2(19) + 6(-8) - 9(-7)$
46. $28 \div (-7)$ **47.** $-20 \div 4$ **48.** $-32 \div (-8)$
49. $10 \times 5 \div 25$ **50.** $9 \times 4 \div (-12)$ **51.** $3 \times 16 \div (-4)$
52. $12(-3) \div 6$ **53.** $4(2 - 11) \div (-6)$ **54.** $7(5 - 8) \div (-7)$
55. $-9(4 - 12) \div (-12)$ **56.** $12 \div 6 \times 2$ **57.** $81 \div 9 \times 3$
58. $8 \div (-4) \times 2$ **59.** $-34 \div 17 \times (-2)$ **60.** $72 \div 4 \div 9$
61. $12 + 20 \div (-4)$ **62.** $18 - 6 \div 3$ **63.** $20 - 15 \div (-5)$
64. $24 - 9 \div (-3)$ **65.** $10 - 24 \div (4 - 7)$ **66.** $8 - 36 \div (6 - 15)$
67. $9(3 - 5) - 16 \div (5 - 13)$ **68.** $8(4 - 9) - 32 \div (2 - 10)$
69. $24(-3) \div 9 - (15 - 3) \div 4$ **70.** $36 \div 4(-9) - (4 - 18) \div 7$

Write the following numbers in terms of their prime factors:

71. 12 **72.** 18 **73.** 26 **74.** 28
75. 36 **76.** 42 **77.** 48 **78.** 56
79. 64 **80.** 72 **81.** 84 **82.** 96
83. 108 **84.** 113 **85.** 137 **86.** 156
87. 168 **88.** 216 **89.** 252 **90.** 504

1.7 The Set of Rational Numbers

Given $a, b \in I, b \neq 0,$ the quotient $\frac{a}{b}$ does not always exist in the set of integers, for instance when $a = 2$ and $b = 3$. Thus the set of integers is not closed under division.

DEFINITION

When the set of integers is extended to include all quotients of the form $\frac{p}{q}$, where $p, q \in I,\ q \neq 0,$ we get the set of **rational numbers**, denoted by Q:

$$Q = \left\{ \frac{p}{q} \,\middle|\, p, q \in I, q \neq 0 \right\}$$

We note that $\frac{a}{1}$ in Q is the same as a in I. Similarly $\frac{2a}{2}$ in Q is the same as a in I. From this we see that the fractional representations of integers are not unique. This leads us to the following definition.

DEFINITION

If $\frac{p}{q}$ and $\frac{r}{s} \in Q$, then

$$\frac{p}{q} = \frac{r}{s} \quad \text{if and only if} \quad ps = qr$$

From the definition if $\frac{p}{q} \in Q$ and $k \in I$, $k \neq 0$, then $\frac{p}{q} = \frac{kp}{kq}$.

Note If $\frac{p}{q} \in Q$ then $\quad \frac{-p}{q} = \frac{(-1)(-p)}{(-1)(q)} = \frac{p}{-q}$.

DEFINITION

The fraction $\frac{p}{q}$ and $\frac{kp}{kq}$ are called **equivalent fractions**.

When the fraction $\frac{p}{q}$ is written in the form $\frac{kp}{kq}$, the fraction is said to be in **higher terms**.

When the fraction $\frac{kp}{kq}$ is written in the form $\frac{p}{q}$, where p and q have no common factors, it is said to be in **lowest terms**, or **reduced**.

Reducing Fractions

DEFINITION

The greatest integer that divides a set of integers is their **greatest common divisor**, denoted by GCD. Sometimes it is called their **greatest common factor**, denoted then by GCF.

The greatest common divisor of a set of numbers contains all the prime factors that are common to all members of the set and contains each prime factor the least number of times it is contained in any one of the numbers.

Note To reduce a fraction divide both numerator and denominator by their GCD.

EXAMPLE Reduce $\frac{60}{144}$ to lowest terms.

Solution We first factor the numbers into their prime factors:

$$60 = 2 \cdot 2 \cdot 3 \cdot 5$$
$$144 = 2 \cdot 2 \cdot 2 \cdot 2 \cdot 3 \cdot 3$$
$$\text{GCD} = 2 \cdot 2 \cdot 3 = 12$$

Hence $\frac{60}{144} = \frac{5(12)}{12(12)} = \frac{5}{12}$

Note We can reduce a fraction without computing the GCD. Factor both terms of the fraction and divide the numerator and the denominator by the common factors:

$$\frac{72}{96} = \frac{\overset{1}{\cancel{2}} \cdot \overset{1}{\cancel{2}} \cdot \overset{1}{\cancel{2}} \cdot \overset{1}{\cancel{3}} \cdot 3}{\underset{1}{\cancel{2}} \cdot \underset{1}{\cancel{2}} \cdot \underset{1}{\cancel{2}} \cdot 2 \cdot 2 \cdot \underset{1}{\cancel{3}}} = \frac{3}{4}$$

Note $\frac{a+b}{c}$ means $(a+b) \div c$:

$$\frac{5+2}{3} = \frac{7}{3}$$

Note $\frac{a}{b+c}$ means $a \div (b+c)$:

$$\frac{6}{4+7} = \frac{6}{11}$$

Note $\frac{a+b}{c+d}$ means $(a+b) \div (c+d)$:

$$\frac{8+7}{6+4} = \frac{15}{10} = \frac{3}{2}$$

Addition of Rational Numbers

DEFINITION

If $\frac{p}{q}, \frac{r}{q} \in Q$, then $\frac{p}{q} + \frac{r}{q} = \frac{p+r}{q}$.

The definition of addition can be extended to rational numbers with unlike denominators:

Since $\quad \dfrac{p}{q} = \dfrac{ps}{qs}$ and $\dfrac{r}{s} = \dfrac{qr}{qs}$

we have $\quad \dfrac{p}{q} + \dfrac{r}{s} = \dfrac{ps}{qs} + \dfrac{qr}{qs} = \dfrac{ps + qr}{qs}$

The number qs is a **common multiple** of q and s.

DEFINITION

> The least positive integer that is divisible by each member of a set of numbers is called their **least common multiple**, denoted by LCM.

The least common multiple of a set of numbers must contain all the prime factors each the greatest number of times it is contained in any one of the numbers.

EXAMPLE Find the LCM of 36, 48, 60.

Solution

$$36 = 2 \cdot 2 \cdot 3 \cdot 3$$
$$48 = 2 \cdot 2 \cdot 2 \cdot 2 \cdot 3$$
$$60 = 2 \cdot 2 \cdot 3 \cdot 5$$
$$\text{LCM} = 2 \cdot 2 \cdot 2 \cdot 2 \cdot 3 \cdot 3 \cdot 5 = 720$$

The least common multiple of the denominators of the fractions is called the **least common denominator**, denoted by LCD.

To add fractions with different denominators, we first find the LCD of the fractions.
We write equivalent fractions with the LCD as their denominator and then combine them using the rule $\dfrac{p}{q} + \dfrac{r}{q} = \dfrac{p + r}{q}$.

EXAMPLE Combine and simplify $\dfrac{1}{12} + \dfrac{5}{18} + \dfrac{3}{28}$.

Solution LCD = 252

$$\frac{1}{12} + \frac{5}{18} + \frac{3}{28} = \frac{21}{252} + \frac{70}{252} + \frac{27}{252}$$
$$= \frac{21 + 70 + 27}{252} = \frac{118}{252} = \frac{59}{126}$$

Note Always reduce the final fraction.

Instead of writing equivalent fractions with denominators equal to the LCD and then combining the numerators of the fractions, we write one fraction with the LCD as the denominator. Divide the LCD by the denominator of the first fraction, and then multiply the resulting quotient by the numerator of that fraction to get the first expression of the numerator. Repeat the process for each fraction, connecting them by the signs of the corresponding fractions.

EXAMPLE Combine $\frac{7}{6}+\frac{5}{8}+\frac{13}{12}$.

Solution LCD = 24:

$$\frac{7}{6}+\frac{5}{8}+\frac{13}{12}=\frac{4(7)+3(5)+2(13)}{24}$$

$$=\frac{28+15+26}{24}$$

$$=\frac{69}{24}=\frac{23}{8}$$

Subtraction of Rational Numbers

From addition we have $\frac{p}{q}+\frac{-p}{q}=\frac{p+(-p)}{q}=\frac{0}{q}=0.$

Hence $\frac{-p}{q}$ is the additive inverse of $\frac{p}{q}$.

Also $\frac{p}{q}-\frac{p}{q}=0$, that is, $-\frac{p}{q}$ is the additive inverse of $\frac{p}{q}$.

Hence $$-\frac{p}{q}=\frac{-p}{q}=\frac{p}{-q}$$

Subtraction of rational numbers is defined from addition. That is,

$$\frac{p}{q}-\frac{r}{s}=\frac{p}{q}+\frac{-r}{s}=\frac{ps+q(-r)}{qs}=\frac{ps-qr}{qs}$$

DEFINITION

If $\frac{p}{q},\frac{r}{s}\in Q$, then $\frac{p}{q}-\frac{r}{s}=\frac{ps-qr}{qs}$.

EXAMPLE Combine $\frac{7}{12} - \frac{11}{16}$.

Solution LCD = 48:

$$\frac{7}{12} - \frac{11}{16} = \frac{4(7) - 3(11)}{48} = \frac{28 - 33}{48} = -\frac{5}{48}$$

EXAMPLE Combine $\frac{5}{12} - \frac{17}{18} + \frac{13}{24}$.

Solution The LCD = 72:

$$\frac{5}{12} - \frac{17}{18} + \frac{13}{24} = \frac{6(5) - 4(17) + 3(13)}{72} = \frac{30 - 68 + 39}{72} = \frac{1}{72}$$

Multiplication of Rational Numbers

DEFINITION

If $\frac{p}{q}, \frac{r}{s} \in Q$ then $\frac{p}{q} \cdot \frac{r}{s} = \frac{pr}{qs}$

Remember that the last fraction must be in reduced form.

Division of Rational Numbers

DEFINITION

When the product of two numbers equals 1, the two numbers are called **multiplicative inverses**, sometimes called **reciprocals**.

If $\frac{p}{q} \in Q$ and $\frac{p}{q} \neq 0$, then $\frac{p}{q} \cdot \frac{q}{p} = \frac{pq}{qp} = 1$

Thus $\frac{p}{q}$ and $\frac{q}{p}$ are reciprocals.

Division of rational numbers is defined from multiplication.

If $\frac{p}{q}, \frac{r}{s} \in Q$ and $\frac{r}{s} \neq 0$, then $\frac{p}{q} \div \frac{r}{s} = \frac{\frac{p}{q}}{\frac{r}{s}} = \frac{\frac{p}{q} \cdot \frac{s}{r}}{\frac{r}{s} \cdot \frac{s}{r}} = \frac{\frac{p}{q} \cdot \frac{s}{r}}{1} = \frac{p}{q} \cdot \frac{s}{r}$

DEFINITION

If $\frac{p}{q}, \frac{r}{s} \in Q$ and $\frac{r}{s} \neq 0$, then $\frac{p}{q} \div \frac{r}{s} = \frac{p}{q} \cdot \frac{s}{r}$.

Thus dividing by a fraction is equivalent to multiplying by the reciprocal of that fraction.

EXAMPLE

$$\frac{32}{14} \div \frac{12}{21} = \frac{32}{14} \times \frac{21}{12} = 4$$

When the operations on fractions are multiplications and divisions, we change the divisions to multiplications first. Consider the numerators as one numerator and the denominators as one denominator and reduce.

EXAMPLE

Perform the indicated operations:

$$\frac{12}{11} \div \frac{21}{8} \times \frac{33}{40}$$

Solution

$$\frac{12}{11} \div \frac{21}{8} \times \frac{33}{40} = \frac{12}{11} \times \frac{8}{21} \times \frac{33}{40}$$

$$= \frac{12 \times 8 \times 33}{11 \times 21 \times 40} = \frac{12}{35}$$

Combined Operations

When we have the four arithmetic operations in one problem without grouping symbols, we perform the multiplications and divisions in the order they appear in the problem and then the additions and subtractions.

EXAMPLE

Perform the indicated operations and simplify:

$$\frac{4}{3} + \frac{9}{16} \times \frac{2}{3} - \frac{7}{6}$$

Solution

$$\frac{4}{3} + \frac{9}{16} \times \frac{2}{3} - \frac{7}{6} = \frac{4}{3} + \frac{9 \times 2}{16 \times 3} - \frac{7}{6}$$

$$= \frac{4}{3} + \frac{3}{8} - \frac{7}{6}$$

$$= \frac{32 + 9 - 28}{24} = \frac{13}{24}$$

EXAMPLE

Perform the indicated operations and simplify:

$$\frac{5}{6}-\frac{21}{40}\div\frac{3}{5}+\frac{1}{12}$$

Solution

$$\frac{5}{6}-\frac{21}{40}\div\frac{3}{5}+\frac{1}{12}=\frac{5}{6}-\frac{21}{40}\times\frac{5}{3}+\frac{1}{12}$$
$$=\frac{5}{6}-\frac{7}{8}+\frac{1}{12}$$
$$=\frac{20-21+2}{24}=\frac{1}{24}$$

EXAMPLE

Perform the indicated operations and simplify:

$$\frac{5}{4}+\frac{8}{11}\left(\frac{7}{8}-\frac{4}{3}\right)$$

Solution

$$\frac{5}{4}+\frac{8}{11}\left(\frac{7}{8}-\frac{4}{3}\right)=\frac{5}{4}+\frac{8}{11}\left(\frac{21-32}{24}\right)$$
$$=\frac{5}{4}+\frac{8}{11}\left(-\frac{11}{24}\right)$$
$$=\frac{5}{4}-\frac{1}{3}=\frac{15-4}{12}=\frac{11}{12}$$

EXAMPLE

Perform the indicated operations and simplify:

$$\frac{3}{4}-\frac{34}{27}\div\left(\frac{1}{6}+\frac{7}{9}\right)$$

Solution

$$\frac{3}{4}-\frac{34}{27}\div\left(\frac{1}{6}+\frac{7}{9}\right)=\frac{3}{4}-\frac{34}{27}\div\frac{3+14}{18}$$
$$=\frac{3}{4}-\frac{34}{27}\times\frac{18}{17}$$
$$=\frac{3}{4}-\frac{4}{3}=-\frac{7}{12}$$

Note The set of rational numbers is closed under addition, subtraction, multiplication, and division by a nonzero divisor.

Decimal Form of Rational Numbers

The Hindu-Arabic system is a place-value system. A decimal point is used to indicate the place value of a numeral. The decimal point separates the place values less than 1 from place values equal to or greater than 1. Each place value is $\frac{1}{10}$ as large as the one to its left. The first numeral to the left of the decimal point is in the units place. The second numeral to the left of the decimal point is in the tens place. The third numeral to the left of the decimal point is in the hundreds place, etc. The first numeral to the right of the decimal point is in the one-tenths place. The second numeral to the right of the decimal point is in the one-hundredths place, etc.

Thus the number 452.76 means

$$4(100) + 5(10) + 2(1) + 7 \times \frac{1}{10} + 6 \times \frac{1}{100}$$

When there are no numerals to the right of the decimal point, we usually do not write the decimal point. Thus 835 is the same as 835.0

Given a decimal number, we can find its fractional equivalent, also called a common fraction, as shown in the following examples.

EXAMPLES

1. $.125 = 1 \times \frac{1}{10} + 2 \times \frac{1}{100} + 5 \times \frac{1}{1000}$

$= \frac{1}{10} + \frac{2}{100} + \frac{5}{1000} = \frac{125}{1000} = \frac{1}{8}$

2. $.08 = \frac{8}{100} = \frac{2}{25}$

3. $2.25 = \frac{225}{100} = \frac{9}{4}$

Given a common fraction, we can find its equivalent decimal by the use of the long-division operation. By long division we find that

1. $\frac{1}{2} = .5$

2. $\frac{1}{4} = .25$

3. $\frac{3}{25} = .12$

4. $\frac{6}{125} = .048$

Again from long division $\frac{1}{3} = .33\bar{3}$; the line above the last numeral indicates that the numeral repeats without end.

Note that $\frac{1}{3} \neq .333$

When we write $\frac{1}{7} = .142857\overline{142857}$, the line on top indicates that the group of numerals repeats without end.

Mixed Numbers

DEFINITION

> A **proper fraction** is a fraction whose numerator is less than its denominator. An **improper fraction** is a fraction whose numerator is greater than or equal to its denominator.

Consider the improper fraction $\frac{23}{9}$. The numerator can be written as the sum of two numbers: One number is divisible by the denominator and the other number is less than the denominator:

$$\frac{23}{9} = \frac{18 + 5}{9}$$

Using the definition of addition of fractions $\frac{a + c}{b} = \frac{a}{b} + \frac{c}{b}$ we get

$$\frac{23}{9} = \frac{18}{9} + \frac{5}{9} = 2 + \frac{5}{9}$$

With specific numbers it has been the practice to write $2 + \frac{5}{9}$ as $2\frac{5}{9}$, which is called a **mixed number**.

Thus $\quad \frac{23}{9} = 2\frac{5}{9}$

The 23 is called the **dividend**, 9 is the **divisor**, 2 is the **quotient**, and 5 is the **remainder**. Note that the remainder is always less than the divisor.

To write an improper fraction as a mixed number, the long-division operation is used. To write $\frac{671}{39}$ as a mixed number we have

```
                 17     quotient
divisor    39 | 671     dividend
                39
                ---
                281
                273
                ---
                  8     remainder
```

Hence $\quad \frac{671}{39} = 17\frac{8}{39}$

From addition of fractions,

$$18\frac{7}{12} = \frac{18}{1} + \frac{7}{12} = \frac{18 \times 12 + 7}{12} = \frac{223}{12}$$

Thus to convert a mixed number to a fraction, multiply the quotient by the divisor, then add the remainder to the resulting product. Write the sum as the numerator of the fraction and the divisor as the denominator.

Exercise 1.7

Combine the following fractions and reduce:

1. $\frac{7}{6} + \frac{9}{8}$ **2.** $\frac{5}{12} + \frac{11}{8}$ **3.** $\frac{5}{6} + \frac{7}{10}$ **4.** $\frac{4}{9} - \frac{7}{12}$

5. $\frac{5}{6} - \frac{4}{7}$ **6.** $\frac{11}{12} - \frac{9}{16}$ **7.** $-\frac{7}{9} - \frac{3}{8}$ **8.** $-\frac{11}{24} - \frac{4}{9}$

9. $-\frac{17}{24} - \frac{13}{18}$ **10.** $\frac{3}{2} + \frac{2}{3} - \frac{5}{6}$ **11.** $\frac{1}{2} - \frac{3}{4} + \frac{5}{8}$ **12.** $\frac{1}{4} + \frac{2}{3} - \frac{3}{2}$

13. $\frac{2}{3} + \frac{5}{4} - \frac{7}{6}$ **14.** $\frac{5}{3} + \frac{1}{6} - \frac{7}{9}$ **15.** $\frac{7}{4} - \frac{5}{6} + \frac{1}{12}$

16. $\frac{7}{8} - \frac{3}{4} - \frac{1}{6}$ **17.** $\frac{5}{2} - \frac{3}{5} - \frac{7}{6}$ **18.** $\frac{4}{9} + \frac{7}{4} - \frac{5}{6}$

19. $\frac{1}{6} + \frac{7}{12} - \frac{7}{9}$ **20.** $\frac{6}{7} - \frac{5}{3} + \frac{9}{14}$ **21.** $\frac{1}{6} - \frac{9}{8} + \frac{7}{12}$

22. $\frac{7}{9} - \frac{3}{8} - \frac{5}{12}$ **23.** $\frac{7}{18} - \frac{3}{8} - \frac{8}{9}$ **24.** $\frac{11}{24} + \frac{9}{32} - \frac{7}{48}$

Perform the indicated operations and simplify:

25. $\frac{12}{20} \times \frac{45}{81}$ **26.** $\frac{33}{16} \times \frac{24}{22}$ **27.** $\frac{16}{56} \times \frac{35}{24}$ **28.** $\frac{40}{25} \times \frac{15}{28}$

29. $\frac{26}{34} \div \frac{39}{51}$ **30.** $\frac{38}{9} \div \frac{57}{27}$ **31.** $\frac{32}{27} \div \frac{24}{54}$ **32.** $\frac{68}{21} \div \frac{51}{28}$

33. $\frac{8}{10} \times \frac{15}{9} \div \frac{12}{3}$ **34.** $\frac{15}{16} \times \frac{24}{25} \div \frac{9}{10}$ **35.** $\frac{27}{34} \times \frac{38}{36} \div \frac{57}{51}$

36. $\frac{12}{18} \div \frac{16}{21} \times \frac{64}{42}$ **37.** $\frac{28}{27} \div \frac{35}{18} \times \frac{24}{32}$ **38.** $\frac{39}{16} \div \frac{26}{6} \times \frac{8}{9}$

39. $\frac{21}{44} \div \left(\frac{27}{33} \times \frac{14}{4}\right)$ **40.** $\frac{49}{36} \div \left(\frac{28}{24} \times \frac{21}{32}\right)$ **41.** $\frac{48}{32} \div \left(\frac{63}{28} \times \frac{8}{18}\right)$

42. $\frac{1}{6} - \frac{4}{3} \times \frac{5}{12} + \frac{2}{3}$ **43.** $\frac{1}{4} - \frac{3}{4} \times \frac{7}{6} + \frac{5}{6}$ **44.** $\frac{5}{6} - \frac{14}{6} \times \frac{3}{8} + \frac{1}{12}$

45. $\frac{5}{9} - \frac{21}{16} \div \frac{7}{4} + \frac{7}{12}$ **46.** $\frac{2}{3} + \frac{27}{8} \div \frac{6}{5} - \frac{1}{8}$ **47.** $\frac{5}{6} + \frac{35}{48} \div \frac{5}{4} - \frac{11}{16}$

48. $\frac{3}{8} - \frac{3}{38}\left(\frac{3}{4} + \frac{5}{6}\right)$

49. $\frac{5}{6} - \frac{3}{14}\left(\frac{4}{9} + \frac{1}{3}\right)$

50. $\frac{3}{4} - \frac{4}{17}\left(\frac{7}{6} - \frac{3}{8}\right)$

51. $\frac{11}{12} + \frac{3}{38}\left(\frac{3}{4} - \frac{7}{3}\right)$

52. $\frac{4}{9} + \frac{5}{16} \div \left(\frac{3}{4} - \frac{7}{6}\right)$

53. $\frac{7}{6} - \frac{74}{63} \div \left(\frac{7}{3} - \frac{4}{7}\right)$

54. $\frac{7}{12} - \frac{14}{9} \div \left(\frac{5}{8} + \frac{5}{6}\right)$

55. $\frac{8}{9} - \frac{11}{27} \div \left(\frac{3}{8} + \frac{1}{12}\right)$

Find the common fraction equivalent to

56. .16 **57.** .24 **58.** .064 **59.** 3.6
60. 1.8 **61.** 2.04 **62.** 4.025 **63.** 1.384

Write the following common fractions in decimal form:

64. $\frac{3}{4}$ **65.** $\frac{7}{25}$ **66.** $\frac{9}{8}$ **67.** $\frac{11}{125}$

68. $\frac{1}{6}$ **69.** $\frac{3}{7}$ **70.** $\frac{5}{9}$ **71.** $\frac{7}{11}$

Write each of the following improper fractions as mixed numbers:

72. $\frac{7}{4}$ **73.** $\frac{19}{5}$ **74.** $\frac{27}{7}$ **75.** $\frac{38}{9}$

76. $\frac{113}{4}$ **77.** $\frac{193}{6}$ **78.** $\frac{125}{8}$ **79.** $\frac{180}{7}$

80. $\frac{74}{13}$ **81.** $\frac{109}{15}$ **82.** $\frac{319}{17}$ **83.** $\frac{341}{27}$

84. $\frac{1161}{32}$ **85.** $\frac{894}{37}$ **86.** $\frac{1250}{43}$ **87.** $\frac{1553}{48}$

Write each of the following mixed numbers as common fractions:

88. $7\frac{1}{2}$ **89.** $5\frac{3}{4}$ **90.** $7\frac{1}{6}$ **91.** $4\frac{7}{8}$

92. $24\frac{2}{3}$ **93.** $26\frac{3}{5}$ **94.** $18\frac{4}{7}$ **95.** $16\frac{3}{8}$

96. $14\frac{4}{11}$ **97.** $12\frac{9}{13}$ **98.** $24\frac{11}{17}$ **99.** $34\frac{8}{19}$

100. $16\frac{7}{26}$ **101.** $27\frac{17}{37}$ **102.** $38\frac{25}{43}$ **103.** $28\frac{23}{47}$

1.8 Irrational Numbers and Real Numbers

Given any rational number we can find a point on a number line that is the graph of that number. However, given a number line there are an infinite number of points on the line whose coordinates are not rational numbers.

An example of a point on the line whose coordinate is not a rational number is shown in Figure 1.4. Draw the number line OX with O as origin. Take the point A to be the graph of the number 1. At A draw a vertical line AY. Take B on AY such that $AB = OA$. Join OB, and take C on OX such that $OC = OB$.

The coordinate of the point C is not a rational number. It is called the **square root** of 2, and is denoted by $\sqrt{2}$.

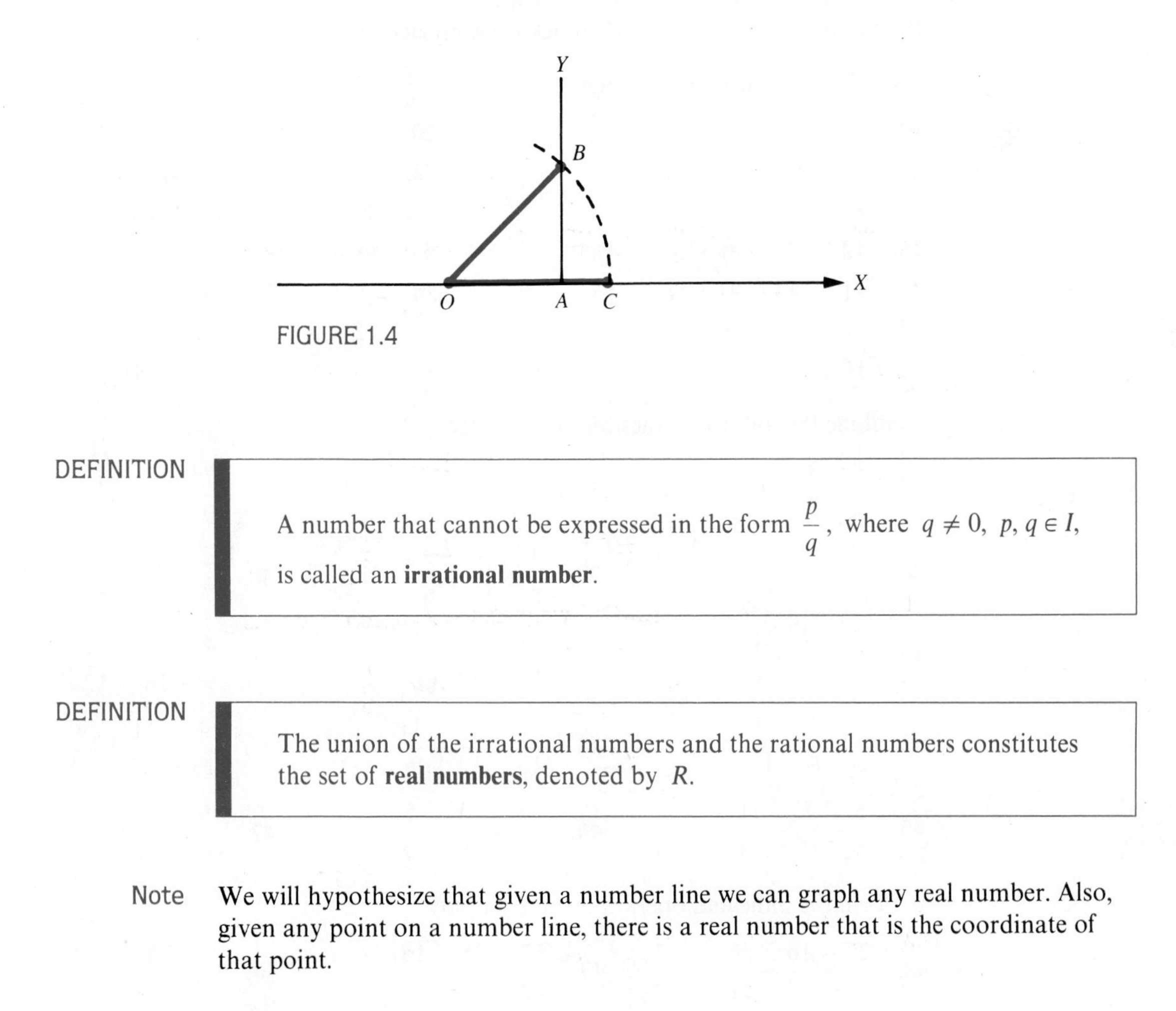

FIGURE 1.4

DEFINITION

> A number that cannot be expressed in the form $\frac{p}{q}$, where $q \neq 0$, $p, q \in I$, is called an **irrational number**.

DEFINITION

> The union of the irrational numbers and the rational numbers constitutes the set of **real numbers**, denoted by R.

Note We will hypothesize that given a number line we can graph any real number. Also, given any point on a number line, there is a real number that is the coordinate of that point.

Chapter 1 Review

Given $A = \{3, 4, 5, 6\}$, $B = \{1, 2, 3\}$, and $C = \{2, 4, 6\}$, list the elements in the following sets:

1. $A \cup B$
2. $A \cup C$
3. $B \cup C$
4. $A \cap B$
5. $A \cap C$
6. $B \cap C$
7. $(A \cup B) \cap C$
8. $(A \cap B) \cup C$
9. $(A \cap B) \cap C$
10. $B \cup (A \cap C)$
11. $C \cap (A \cup B)$
12. $B \cap (A \cup C)$

Let A and B be two sets:

13. If $A \cup B = A$, must B be a subset of A?
14. If $A \cup B = A$, must A be a subset of B?
15. If $A \cap B = B$, must B be a subset of A?
16. If $A \cap B = B$, must A be a subset of B?
17. If $a \notin A \cap B$, must a be an element of A?
18. If $a \notin A$ and $a \in A \cup B$, must a be an element of B?

Find the value of the following:

19. $6 - 3(-4) + 7(2) - 8$
20. $9 + 4(-6) - 3(-8) - 2(6)$
21. $15 + 5(-2) + 6(4) - 3$
22. $7 - 2(-3) + 5(-1) - 8$
23. $8 - 3 \times 6 + 7(-7) + 13$
24. $4 \times 3 + 6(-5) - 9 + (-4)$
25. $12 - 2(2 - 8) + 3(7 - 4)$
26. $18 - 8(7 - 11) - 3(9 + 3)$
27. $11 - 6(4 - 4) + 8(3 + 5)$
28. $8 - 7(6 + 7) - 9(3 - 8)$
29. $5 \times 10 \div 5 - 6 \times 4 - 3$
30. $12 \times 8 \div 4 - 3 \times 7 - 2$
31. $16 \div 8 \times 2 - 14 \div (-7) - 6$
32. $36 \div 9 \times 2 + 8 \div (-4) + 4$

Combine the following fractions and reduce:

33. $\frac{1}{3} - \frac{3}{2} + \frac{7}{4}$
34. $\frac{3}{4} - \frac{4}{3} - \frac{1}{6}$
35. $\frac{7}{6} - \frac{5}{4} - \frac{11}{12}$
36. $\frac{1}{4} + \frac{1}{6} - \frac{3}{8}$
37. $\frac{3}{8} + \frac{5}{12} - \frac{7}{6}$
38. $\frac{4}{7} - \frac{9}{4} + \frac{3}{14}$
39. $\frac{3}{5} - \frac{7}{15} + \frac{1}{6}$
40. $\frac{11}{12} - \frac{7}{18} - \frac{5}{6}$
41. $\frac{4}{9} + \frac{13}{18} - \frac{11}{12}$
42. $\frac{7}{5} + \frac{3}{8} - \frac{9}{10}$
43. $\frac{5}{16} - \frac{13}{12} + \frac{11}{6}$
44. $\frac{4}{15} - \frac{11}{12} + \frac{9}{10}$
45. $\frac{5}{8} - \frac{7}{18} - \frac{11}{24}$
46. $\frac{17}{24} + \frac{13}{36} - \frac{5}{18}$
47. $\frac{9}{16} + \frac{5}{12} - \frac{1}{24}$

Perform the indicated operations and simplify:

48. $\frac{27}{14} \times \frac{16}{18} \div \frac{36}{21}$
49. $\frac{52}{22} \times \frac{35}{39} \div \frac{15}{33}$
50. $\frac{14}{27} \div \frac{35}{18} \times \frac{40}{22}$

51. $\frac{46}{34} \div \frac{69}{51} \times \frac{18}{16}$

52. $\frac{65}{24} \div \frac{52}{27} \div \frac{15}{8}$

53. $\frac{81}{34} \div \frac{36}{85} \div \frac{30}{8}$

54. $\frac{3}{4} - \frac{5}{4} \times \frac{7}{15} + \frac{1}{8}$

55. $\frac{1}{6} + \frac{5}{3} \times \frac{4}{30} - \frac{5}{4}$

56. $\frac{5}{12} - \frac{21}{16} \div \frac{7}{2} + \frac{2}{3}$

57. $\frac{7}{12} - \frac{22}{9} \div \frac{4}{3} - \frac{5}{24}$

58. $\frac{5}{6} - \frac{16}{3}\left(\frac{11}{12} - \frac{7}{8}\right)$

59. $\frac{11}{12} + \frac{9}{2}\left(\frac{2}{3} - \frac{5}{8}\right)$

60. $\frac{8}{15} + \frac{3}{32} \div \left(\frac{13}{16} - \frac{11}{12}\right)$

61. $\frac{3}{4} - \frac{46}{81} \div \left(\frac{7}{18} - \frac{17}{24}\right)$

Find the common fractions equivalent to the following decimal numbers:

62. .75
63. .625
64. 1.25
65. 1.48
66. 2.28
67. .056
68. 1.024
69. .424

Write the following common fractions in decimal form:

70. $\frac{7}{4}$
71. $\frac{9}{16}$
72. $\frac{13}{25}$
73. $\frac{8}{125}$
74. $\frac{5}{6}$
75. $\frac{4}{7}$
76. $\frac{11}{9}$
77. $\frac{20}{11}$

Write the following improper fractions as mixed numbers:

78. $\frac{94}{7}$
79. $\frac{131}{9}$
80. $\frac{275}{12}$
81. $\frac{298}{15}$
82. $\frac{377}{24}$
83. $\frac{468}{29}$
84. $\frac{696}{31}$
85. $\frac{1741}{47}$

Write the following mixed numbers as common fractions:

86. $9\frac{5}{6}$
87. $12\frac{3}{8}$
88. $13\frac{6}{11}$
89. $17\frac{9}{14}$
90. $20\frac{7}{18}$
91. $19\frac{16}{21}$
92. $15\frac{17}{24}$
93. $42\frac{29}{35}$

chapter 2

Basic Operations with Polynomials

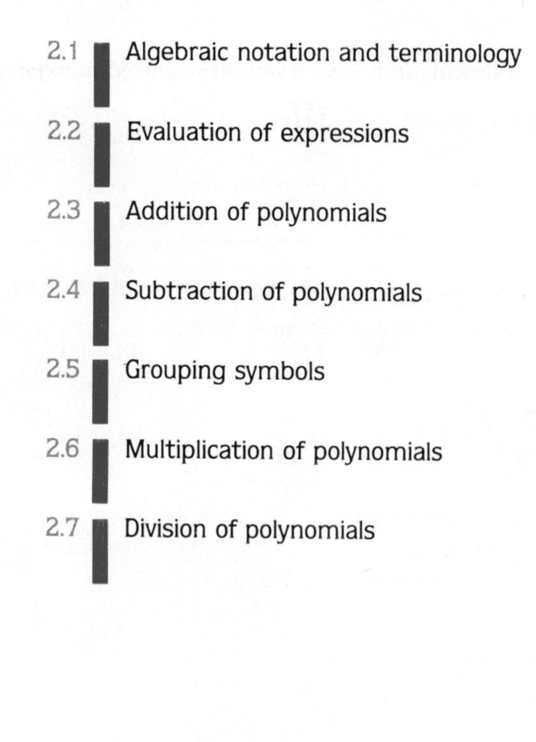

2.1 Algebraic notation and terminology

2.2 Evaluation of expressions

2.3 Addition of polynomials

2.4 Subtraction of polynomials

2.5 Grouping symbols

2.6 Multiplication of polynomials

2.7 Division of polynomials

2.1 Algebraic Notation and Terminology

Algebra deals with mathematical systems. The most basic of these is the number system. Elementary algebra is a generalization of arithmetic.
While in arithmetic we use real numbers, which are **specifics**, in algebra we use symbols, usually letters from the alphabet, referred to as **literal numbers**. Literal numbers are used so we may consider the general properties of numbers without considering their specific attributes.

Specific numbers, literal numbers, and their indicated product, quotient, or roots are called **terms**. The quantities $7, 2a, ab, -\frac{4x}{y}, \sqrt{2z}$ are examples of terms. When a term has no sign preceding it, as in $2a$, the plus sign is implied. The specific number in a term is customarily written as the first symbol in the term. It is called the **numerical coefficient** of the term. When there is no specific number in a term, as in ab, the numerical coefficient is 1.

When numbers are multiplied, each of the numbers in the product is called a **factor**. Thus some of the factors of $12xyz$ are $2, 4, 12, x, 3x, 4y, 6yz$.

The numerical coefficient of a term may be associated with any literal factor of the term, not just the first literal factor.

A term, or the sums or differences of terms, form an **expression**. The quantities $9, 5a, 6y - \frac{1}{z}, 2xyz - \frac{3a}{b} + \sqrt{xy}$ are examples of expressions.

When the literal numbers in an expression appear only in sums, differences, or products, the expression is called a **polynomial**.

A polynomial with one term, such as $7ab$, is called a **monomial**.
A polynomial with two terms, such as $xy - 4$, is called a **binomial**.
A polynomial with three terms, such as $3x + 5y - 8$, is called a **trinomial**.

2.2 Evaluation of Expressions

The numerical value of an expression can be calculated when each literal number in the expression is given a specific value. The process of calculating the numerical value of the expression is called **evaluation**.

To evaluate an expression, substitute the given specific value for each literal number. The calculations are easier, and the likelihood of an error is reduced, when the specific value for each letter is substituted using parentheses in a distinct step before the operations are performed.

Note The specific value given to a letter may vary from one problem to the next, but it remains the same for that letter throughout any one problem.

EXAMPLE Evaluate the expression $2a - b(3a - c)$ given that $a = -2, b = 3,$ and $c = 4$

Solution Substitute the specific value for each literal in the expression and then simplify:

$$\begin{aligned} 2a - b(3a - c) &= 2(-2) - (3)[3(-2) - (4)] \\ &= -4 - 3[-6 - 4] \\ &= -4 - 3[-10] \\ &= -4 + 30 = 26 \end{aligned}$$

EXAMPLE Evaluate the expression $\frac{ac}{b} + \frac{b}{ad}$ given that

$a = 2, b = -3, c = -2,$ and $d = 4$

Solution

$$\begin{aligned} \frac{ac}{b} + \frac{b}{ad} &= \frac{(2)(-2)}{(-3)} + \frac{(-3)}{(2)(4)} \\ &= \frac{4}{3} - \frac{3}{8} = \frac{32 - 9}{24} = \frac{23}{24} \end{aligned}$$

Exercise 2.2

Evaluate the following expressions, given that $a = 3, b = -2, c = 2,$ and $d = -1$.

1. $7 - a$ **2.** $4 - c$ **3.** $8 + b$ **4.** $2d - 3$
5. $a + b$ **6.** $b - c$ **7.** $a + 2d$ **8.** $2b + 3d$
9. $a + 3b - 1$ **10.** $b - 2c + 6$ **11.** $2b - 3d - 3$ **12.** $2a + 5b - 4$
13. $c - 2b + d$ **14.** $a + 2b - 4d$ **15.** $2b - c - d$
16. $2a - 5c + 3d$ **17.** $2a - (b - d)$ **18.** $b + (c - 2d)$
19. $a - (3b - c)$ **20.** $2b - (2d - 3c)$ **21.** $a + 2(3b + c)$
22. $b + 3(5c - d)$ **23.** $2b - 3(c - 2d)$ **24.** $c - 4(a - 3d)$
25. $bc + ad$ **26.** $ab + 2bc$ **27.** $ab - 3c$
28. $bd - 2ac$ **29.** $2a - a(c + d)$ **30.** $3b + d(c - a)$
31. $d - b(2c - 3a)$ **32.** $3c(2ad + 5bc)$ **33.** $2bd(7ac + 10cd)$
34. $ad(3bd + 2ad)$ **35.** $(a - 3b)(a + 3b)$
36. $(9a + 6b)(15c + 10d)$ **37.** $(5c + 2d)(2b - 4a)$
38. $(3bc - 4ad)(2bc - ad)$ **39.** $(ac - 3bd)(4ad + 5ac)$
40. $\frac{a + 3d}{a + c}$ **41.** $\frac{2c + b}{a - d}$ **42.** $\frac{ab - 2cd}{3c - bd}$ **43.** $\frac{bd - c}{5a + bc}$
44. $\frac{b}{2d} + \frac{c}{a}$ **45.** $\frac{a}{d} - \frac{2c}{b}$ **46.** $\frac{a}{4b} + \frac{bc}{2a}$ **47.** $\frac{ab}{4c} - \frac{c}{ad}$

2.3 Addition of Polynomials

The sum of the literal numbers a and b is indicated simply by $a + b$. The a and b are called **terms** of the sum.

Terms that have identical literal factors are called **like terms**: xyz, $5yxz$, $-8zyx$ are like terms.

When the terms to be added are like terms, such as $6a$ and $11a$, the sum $6a + 11a$ can be simplified by use of the **distributive law of multiplication**:

$$ac + bc = (a + b)c$$

Using the distributive law of multiplication we have

$$6a + 11a = (6 + 11)a = 17a$$

$$2x + 4x - 9x = (2 + 4 - 9)x = -3x$$

Note that $7a - 2a = (7 - 2)a = 5a$ (not just 5)

When we add polynomials, we combine only like terms in the polynomials.

EXAMPLE Add $4a - b + 2c$ and $a + 3b - 7c$.

Solution

$$(4a - b + 2c) + (a + 3b - 7c) = (4a + a) + (-b + 3b) + (2c - 7c)$$
$$= 5a + 2b - 5c$$

2.4 Subtraction of Polynomials

To subtract b from a, we write $a - b$, which is the same as $a + (-b)$. That is, to subtract b from a, we add the additive inverse (the negative) of b to a.

The additive inverse of $+4x$ is $-4x$. That is, $-(+4x) = -4x$.
The additive inverse of $-7y$ is $+7y$. That is, $-(-7y) = +7y$.
When the terms to be subtracted are like terms, the difference can be simplified by using the distributive law of multiplication.

EXAMPLE Subtract $2a$ from $-7a$.

Solution $(-7a) - (2a) = (-7a) + (-2a) = (-7 - 2)a = -9a$

EXAMPLE Subtract $-6x$ from $-4x$.

Solution $(-4x) - (-6x) = (-4x) + (+6x) = (-4 + 6)x = 2x$

The additive inverse of a polynomial is the additive inverse of every term of the polynomial.
The additive inverse of $2a - 3b + c$ is $-2a + 3b - c$. That is,

$$-(2a - 3b + c) = -2a + 3b - c$$

EXAMPLE Subtract $4a - 3b + c$ from $5a + 2b - 3c$.

Solution

$$\begin{aligned}(5a + 2b - 3c) - (4a - 3b + c) &= (5a + 2b - 3c) + (-4a + 3b - c) \\ &= (5a - 4a) + (2b + 3b) + (-3c - c) \\ &= a + 5b - 4c\end{aligned}$$

Exercises 2.3–2.4

Combine like terms in each of the following:

1. $2a + 3a - a$
2. $6b - 2b - 4$
3. $4x - 7x + 3$
4. $3xy + 2x - xy$
5. $4ab - b - 6b - 5ab$
6. $ab - bc + 2ab - 3bc$

Add the following polynomials:

7. $a + 3b,\ 2a - 5b$
8. $3a - 4b,\ 6b - 7a$
9. $2x - 4y,\ y - 2x$
10. $2x - y + 4,\ y - x - 3$
11. $6y - 3x - 2,\ x - y - 2$
12. $3a - b + 1,\ b - 3a - 6$
13. $4a - 2b,\ a + 3b,\ b - 2a$
14. $7x - 2y,\ x - y,\ 2y - 5x$
15. $3x - 2y + 1,\ 4x + 3y - 5,\ 2y + 3 - 6x$
16. $3xy - 2yz + z,\ 2yz - 3yx - z,\ xy - 2z + yz$
17. $xy + 2yz - z,\ yz - 2xy + 1,\ 5 + 3xy + 2z$
18. $3xy - 2yz,\ xy - 3y + 4yz,\ 5y - 2xy - 3yz$
19. $2(a + b) + 3(c - d),\ (a + b) - 4(c - d),\ 2(c - d) - (a + b)$
20. $3(a - b) - (c + 2d),\ (a - b) + 3(c + 2d),\ 2(c + 2d) - 4(a - b)$
21. $5(x - y) - 2(z + t),\ 3(z + t) - 2(x - y),\ (x - y) + (z + t)$
22. $4(x + 2y) + (z + 3t),\ (x + 2y) - 3(z + 3t),\ 4(z + 3t) - 5(x + 2y)$

Perform the indicated operations:

23. Subtract $6a$ from $2a$.
24. Subtract a from $7a$.
25. Subtract $3a$ from $-9a$.
26. Subtract $8a$ from $-4a$.
27. Subtract $-2a$ from $5a$.
28. Subtract $-6a$ from $3a$.
29. Subtract $-a$ from $-4a$.
30. Subtract $-10a$ from $-5a$.
31. Subtract x from xy.
32. Subtract 7 from $7a$.
33. Subtract $2a$ from $2ax$.
34. Subtract $-3a$ from $3ax$.
35. Subtract $3x - 2y$ from $2x + 5y$.
36. Subtract $x + 7y$ from $x + 7y$.
37. Subtract $3a - 2b - 3$ from $4a - 2b + 10$.

38. Subtract $a - 3b + 4$ from $7a - b + 8$.
39. Subtract $3y - 2x + 5$ from $4x - y - 6$.
40. Subtract $x - 2y + z$ from $3y - x + z$.
41. From $3x + 2y - z$ subtract $y - 2x + 1$.
42. From $9ax - 3by - 2cz$ subtract $3by - ax - 2c$.

What must be added to the first polynomial to produce the second polynomial?

43. $5, 5ab$
44. $-7a, -7ab$
45. $x - 3, x + 3$
46. $x + 1, -x - 1$
47. $x - y, 2y - x$
48. $2x + y, -2x - 3y$
49. $x - 4y + 3, 2x + y - 4$
50. $3x + y - 8, x - 2y - 1$
51. $3a + 2b - c, 4abc$
52. $2x + y - 3z, 0$
53. $4(x + 2y) + 6(a - b), 6(x + 2y) - 5(a - b)$
54. $3(a + b) - 4(x - 3y), 7(x - 3y) - 4(a + b)$

Perform the indicated operations:

55. Subtract the sum of $x + 2y - 3$ and $3y - x - 4$ from the sum of $2y - x - 6$ and $3x - y + 7$.
56. Subtract the sum of $x - 3y + 2z$ and $7y - 2z + 4x$ from the sum of $2x + y - 4z$ and $x - 3y + z$.
57. From the sum of $x + 2y - 1$ and $3x - y + 2$ subtract the sum of $x - y + 3$ and $3y - 2x + 4$.
58. From the sum of $2x - y - 4$ and $x + 3y + 1$ subtract the sum of $2y - x - 2$ and $3x - 5y + 6$.

2.5 Grouping Symbols

Grouping symbols, such as **parentheses** ( ), **braces** { }, and **brackets** [ ], are used to designate, in a simple manner, more than one operation.

When we write the binomial $3a + 5b$ as $(3a + 5b)$, we are considering the sum of $3a$ and $5b$ as one quantity. The expression $a - (b + c)$ means the sum of b and c is to be subtracted from a.

Removing the grouping symbols means performing the operations that these symbols indicate. Remove the symbols one at a time, starting with the innermost, following the proper order of operations to be performed.

EXAMPLE Remove the grouping symbols and combine like terms:

$3x - (2x - y) + (5x - 2y)$

Solution

$$
\begin{aligned}
3x - (2x - y) + (5x - 2y) &= 3x - 2x + y + 5x - 2y \\
&= (3x - 2x + 5x) + (y - 2y) \\
&= 6x - y
\end{aligned}
$$

EXAMPLE Remove the grouping symbols and combine like terms:

$2x - [3y - 2(x + 4y)]$

Solution

$$\begin{aligned} 2x - [3y - 2(x + 4y)] &= 2x - [3y - 2x - 8y] \\ &= 2x - 3y + 2x + 8y \\ &= 4x + 5y \end{aligned}$$

EXAMPLE Remove the grouping symbols and combine like terms:

$x - \{5y + [3x - 2(2x - y)]\}$

Solution

$$\begin{aligned} x - \{5y + [3x - 2(2x - y)]\} &= x - \{5y + [3x - 4x + 2y]\} \\ &= x - \{5y + 3x - 4x + 2y\} \\ &= x - 5y - 3x + 4x - 2y \\ &= 2x - 7y \end{aligned}$$

Sometimes it is necessary to group some of the terms of an expression, an operation that can be accomplished by using parentheses.

When the grouping symbol is preceded by a plus sign, we keep the signs of the terms the same; when it is preceded by a minus sign, we take the additive inverses (negatives) of the terms.

EXAMPLE Group the last three terms of the polynomial $2x - 3y + z - 1$ with a grouping symbol, the first preceded by a plus sign, the second preceded by a minus sign.

Solution

$$2x - 3y + z - 1 = 2x + (-3y + z - 1)$$
$$2x - 3y + z - 1 = 2x - (3y - z + 1)$$

Exercise 2.5

Remove the grouping symbols and combine like terms:

1. $3x - (x + 2)$
2. $x - (3x + 1)$
3. $5 - (x - 2)$
4. $2x - (3 - x)$
5. $2x + 3(x + 4)$
6. $8 + 2(2x - 1)$
7. $3x + 2(2x - y)$
8. $4x + 5(x - 2y)$
9. $8 - 2(x + 3)$
10. $x - 3(x + 1)$
11. $6 - 2(x - 3)$
12. $15 - 5(x - 1)$
13. $2(x + 1) + 3(x - 2)$
14. $5(x - 3) + 4(2x - 3)$
15. $3(x - 4) - 5(x + 2)$
16. $7(2x - 1) - 3(x + 1)$
17. $8(x - 1) - 4(x - 3)$
18. $7(2x + 1) - 6(x - 4)$
19. $2x - (x + y) + (2x - y)$
20. $x + (3x + y) - (2x - 3y)$
21. $x + 2(x - y) - 3(x - 2y)$
22. $5 - 3(x - 1) - 2(x + 2)$

23. $x + [2y - (x + 4y)]$

24. $2x + [3y - (2x - 7y)]$

25. $3x - [y + (x - 3y)]$

26. $5x - [2y + (4x - y)]$

27. $x - [6y - 2(3x - 2y)]$

28. $7x - [9y - 3(2x - 5y)]$

29. $8 - [-3x + 4(x - 2)]$

30. $9 - [-2x - 7(2x - 1)]$

31. $6 + 4[x - 3(x + 1)]$

32. $7 - 3[2x - 4(x - 1)]$

33. $3x + \{x - [3 - 4(x - 5)]\}$

34. $x + \{3x - [8 - 3(x + 2)]\}$

35. $5x - \{2x + [7 - 2(x - 4)]\}$

36. $6x - \{x + [12 - 7(x - 1)]\}$

37. $3x + \{x - [3 + 7(x - 2)]\}$

38. $2x + \{-3x + [5 - 2(x + 3)]\}$

39. $x - \{2x + [4 - 3(x - 1)]\}$

40. $4x - \{x - [3 - 2(x + 1)]\}$

41. $10 + \{x - [3 + 2(x - 3) - 3(x - 6)]\}$

42. $8 + \{2x - [4 + 8(x - 1) - 5(2x + 1)]\}$

43. $4x + \{7 - 2[3 - 2(x - 4) + 3(x - 3)]\}$

44. $x + \{6 - 3[4 - 3(x - 2) + 2(2x - 1)]\}$

45. $2x - \{8 + 2[7 + 3(2x - 1) - 4(3x - 2)]\}$

46. $x - \{-4y - [x + 3(x - y) - 4(2x - 3y)]\}$

Write equivalent polynomials wherein the last three terms are enclosed in parentheses preceded by (a) a plus sign, (b) a minus sign:

47. $3x + 5y + 6z + 7$

48. $-x + y + z - 2$

49. $6x - y + z - 4$

50. $3x - 2y + z + 5$

2.6 Multiplication of Polynomials

Definition and Notation

The product of the two natural numbers, 3 and 4, is defined by

$$3 \times 4 = 4 + 4 + 4 \qquad \text{three terms of 4}$$

Also, $\quad 4a = 4 \times a = a + a + a + a \qquad$ four terms of a

Similarly, the product of two natural numbers, a and b, is defined by

$$ab = a \times b = b + b + \cdots + b \qquad a \text{ terms of } b$$

When we have $a \cdot a \cdot a \cdot a$, that is, four factors of a, the notation a^4 is used, which reads, "a to the **power** four," or "a to the fourth power." The a is called the **base**, and the 4 is called the **exponent**.

When there is no exponent, as in x, we mean x to the power 1.

DEFINITION

> When $a \in R$, $m \in N$, then
>
> $a^m = a \cdot a \cdots a \qquad m$ factors of a

Note $(-3)^4 = (-3)(-3)(-3)(-3) = +81$

$-3^4 = -(3^4) = -(3 \cdot 3 \cdot 3 \cdot 3) = -81$

$2a^3 = 2(a \cdot a \cdot a)$

$(2a)^3 = (2a)(2a)(2a) = 8a^3$

Remark a, a^2, a^3, etc. are not like terms.

EXAMPLES

1. $3a \cdot a \cdot a \cdot a \cdot a = 3a^5$
2. $-(-2)(-2)(-2) = -(-2)^3$
3. $(x + 2)^4 = (x + 2)(x + 2)(x + 2)(x + 2)$
4. $-2^3 \cdot 3^2 = -(2 \cdot 2 \cdot 2)(3 \cdot 3) = -8 \cdot 9 = -72$
5. $3^2 + 3^3 = 3 \cdot 3 + 3 \cdot 3 \cdot 3 = 9 + 27 = 36$
6. $5^3 - 5 = 5 \cdot 5 \cdot 5 - 5 = 125 - 5 = 120$
7. $504 = 2 \cdot 2 \cdot 2 \cdot 3 \cdot 3 \cdot 7 = 2^3 \cdot 3^2 \cdot 7$

Exercise 2.6A

Write the following in exponent form:

1. $2 \cdot 2 \cdot 2 \cdot 2$ **2.** $3 \cdot 3 \cdot 3 \cdot 3 \cdot 3$ **3.** $a \cdot a \cdot a \cdot a \cdot a \cdot a$
4. $(3x)(3x)(3x)$ **5.** $(xy)(xy)(xy)(xy)$ **6.** $(-2)(-2)(-2)$
7. $(-3)(-3)(-3)(-3)$ **8.** $2a \cdot a \cdot a$ **9.** $3(-a)(-a)(-a)$
10. $(-5x)(-5x)(-5x)(-5x)$ **11.** $-(-7)(-7)(-7)(-7)(-7)$
12. $a \cdot a(-b)(-b)(-b)$ **13.** $-2(-x)(-x)(-x)y \cdot y$
14. $3 \cdot 3 \cdot 3 + 2 \cdot 2$ **15.** $5 \cdot 5 + 2 \cdot 2 \cdot 2 \cdot 2$
16. $a \cdot a - b \cdot b \cdot b$ **17.** $(-x)(-x)(-x) - y \cdot y \cdot y$
18. $(x + 1)(x + 1)(x + 1)$ **19.** $(2x - 1)(2x - 1)(2x - 1)(2x - 1)$

Write the following in expanded form:

20. 3^4 **21.** -2^5 **22.** $-(-3)^4$ **23.** -2^2a^3
24. -3^3a^2 **25.** $x(-y)^4$ **26.** $(-x)^3y^2$ **27.** $-x^2(-y)^3$
28. $(-x^2)^3$ **29.** $a(-b^3)^2$ **30.** $5(a - 2)^3$ **31.** $(x - 1)^4$
32. $(2x + 3)^3$ **33.** $x^3 + x^2$ **34.** $x^4 - x$ **35.** $x^3 - 4$.

Find the value of the following:

36. $2^2 + 2^3$ **37.** $2^3 + 3^2$ **38.** $3^4 - 2^4$ **39.** $5^2 - 3^2$
40. $(2 + 3)^3$ **41.** $(3 + 4)^2$ **42.** $(5 - 3)^2$ **43.** $(6 - 3)^3$
44. $(3 - 7)^3$ **45.** $(-5)^3$ **46.** $(-3)^4$ **47.** $-(-2)^4$

48. $2^3 \cdot 3^4$ **49.** $(-2)^2(-3)^3$ **50.** $5^2(-2)^3$

51. $(-3^2)(5)^2$ **52.** $(-2^2)(-3)^3$ **53.** $(-3)^2(-4^2)$

Factor the following numbers into their prime factors and write the answers in exponent form:

54. 28 **55.** 32 **56.** 36 **57.** 48

58. 60 **59.** 72 **60.** 84 **61.** 96

62. 108 **63.** 135 **64.** 216 **65.** 392

Multiplication of Monomials

We will discuss the multiplication of monomials, then the multiplication of a monomial and a polynomial, and finally the multiplication of two polynomials.

THEOREM 1 If $a \in R$ and $m, n \in N$, then $a^m \cdot a^n = a^{m+n}$.

Proof

$$a^m \cdot a^n = \overbrace{(a \cdot a \cdots a)}^{m \text{ factors}} \cdot \overbrace{(a \cdot a \cdots a)}^{n \text{ factors}}$$

$$= \overbrace{a \cdot a \cdot a \cdots a}^{(m+n) \text{ factors}} = a^{m+n}$$

EXAMPLES

1. $2^3 \cdot 2^4 = 2^{3+4} = 2^7$

2. $a^2 \cdot a^3 = a^{2+3} = a^5$

3. $-3^2 \cdot 3^6 = -3^{2+6} = -3^8$

4. $-2x^3 \cdot x^7 = -2x^{3+7} = -2x^{10}$

5. $x^4 \cdot x = x^{4+1} = x^5$

6. $(a-2)^2(a-2)^5 = (a-2)^{2+5} = (a-2)^7$

Remarks

1. $3^4 \cdot 3^2 \neq 9^{4+2} = 531{,}441$
   $3^4 \cdot 3^2 = 3^{4+2} = 3^6 = 729$
2. $2^3 \cdot 3^4 \neq 6^7 = 279{,}936$
   $2^3 \cdot 3^4 = (8)(81) = 648$

EXAMPLE Perform the following multiplication: $(3xy^2z)(-2x^2y^4)$.

Solution

$$(3xy^2z)(-2x^2y^4) = (3)(-2)(x \cdot x^2)(y^2 \cdot y^4)(z)$$
$$= -6x^3y^6z$$

THEOREM 2 If $a \in R$, $m, n \in N$, then $(a^m)^n = a^{mn}$.

Proof

$$(a^m)^n = \overbrace{(a^m) \cdot (a^m) \cdots (a^m)}^{n \text{ factors}}$$

$$= \overbrace{(a \cdot a \cdots a)}^{m \text{ factors}} \cdot \overbrace{(a \cdot a \cdots a)}^{m \text{ factors}} \cdots \overbrace{(a \cdot a \cdots a)}^{m \text{ factors}}$$

$$= \overbrace{a \cdot a \cdot a \cdots a}^{mn \text{ factors}} = a^{mn}$$

EXAMPLES

1. $(2^3)^2 = 2^{3 \times 2} = 2^6$
2. $(a^2)^4 = a^{2 \times 4} = a^8$

Note $3^2 \cdot 3^5 = 3^{2+5} = 3^7$ while $(3^2)^5 = 3^{2 \cdot 5} = 3^{10}$

THEOREM 3 If $a, b \in R$, $m \in N$, then $(ab)^m = a^m b^m$.

Proof

$$(ab)^m = \overbrace{(ab) \cdot (ab) \cdots (ab)}^{m \text{ factors}}$$

$$= \overbrace{(a \cdot a \cdots a)}^{m \text{ factors}} \cdot \overbrace{(b \cdot b \cdots b)}^{m \text{ factors}} = a^m b^m$$

EXAMPLES

1. $(3a)^2 = 3^2 a^2$
2. $(xy)^5 = x^5 y^5$

Remarks

1. $(ab)^4 = a^4 b^4$
2. $(a + b)^4 \neq a^4 + b^4$

   $(2 + 3)^4 = 5^4 = 625$ while $2^4 + 3^4 = 16 + 81 = 97$

   If we consider $(a + b)$ as one quantity then

   $(a + b)^4 = (a + b)(a + b)(a + b)(a + b)$

   The method of calculating the product will be explained later.

COROLLARY Applying Theorems 3 and 2, when $a, b \in R$ and $m, n, k \in N$, we get

$$(a^m b^n)^k = [(a^m)(b^n)]^k$$
$$= (a^m)^k (b^n)^k$$
$$= a^{mk} b^{nk}$$

EXAMPLES

1. $(2a^3b^2)^3 = (2)^3(a^3)^3(b^2)^3 = 8a^9b^6$

2. $(-3a^2b)^3 = (-3)^3(a^2)^3(b)^3 = -27a^6b^3$

3. $(a^{2n+1})^2 = a^{4n+2}$

EXAMPLE

Perform the following multiplication: $(2x^3y)^3(3xy^2)^2$.

Solution

$$\begin{aligned}(2x^3y)^3(3xy^2)^2 &= (2^3x^9y^3)(3^2x^2y^4)\\ &= (2^3 \cdot 3^2)(x^9 \cdot x^2)(y^3 \cdot y^4)\\ &= (8 \cdot 9)x^{11}y^7 = 72x^{11}y^7\end{aligned}$$

EXAMPLE

Perform the following multiplication: $(-3a^2b^2)^2(-2a^3b)^3(-bc^2)^4$.

Solution

$$\begin{aligned}(-3a^2b^2)^2(-2a^3b)^3(-bc^2)^4 &= (3^2a^4b^4)(-2^3a^9b^3)(b^4c^8)\\ &= (3^2)(-2^3)(a^4 \cdot a^9)(b^4 \cdot b^3 \cdot b^4)(c^8)\\ &= (9)(-8)a^{13}b^{11}c^8\\ &= -72a^{13}b^{11}c^8\end{aligned}$$

EXAMPLE

Perform the indicated operations and simplify:

$(3ab)^3(-ab^2)^2 + a^2b(-2ab^2)^3$

Solution

$$\begin{aligned}(3ab)^3(-ab^2)^2 + a^2b(-2ab^2)^3 &= (27a^3b^3)(a^2b^4) + a^2b(-8a^3b^6)\\ &= 27a^5b^7 - 8a^5b^7\\ &= 19a^5b^7\end{aligned}$$

Exercise 2.6B

Perform the indicated operations and simplify, given that $n \in N$:

1. $2 \cdot 2^3$
2. $2^3 \cdot 2^2$
3. $-2^2 \cdot 2^4$
4. $-3^4 \cdot 3^3$
5. $-2^3 \cdot 2^4$
6. $-5^3 \cdot 5^5$
7. $a \cdot a^5$
8. $a^2 \cdot a^7$
9. $-a^2 \cdot a^3$
10. $-a^4 \cdot a^6$
11. $-a^3 \cdot a^4$
12. $-a^5 \cdot a^7$
13. $3a^2 \cdot a^5$
14. $6a \cdot a^3$
15. $-3a^2 \cdot a^4$
16. $a(-b)^3$
17. $(-a^4)(-b)^2$
18. $(-a^3)(-b)^4$
19. $(-a)^4(-b)^3$
20. $(-a)^3 \cdot a^2$
21. $(-a)^2 \cdot a^4$
22. $a^2(-a^6)$
23. $a^3(-a^5)$
24. $a^5 + a$
25. $a^4 + 2a^3$
26. $a^6 - a^2$
27. $6a^3 + 3a^3$
28. $a^4 - 2a^4$
29. $9a^5 - a^5$
30. $2^n \cdot 2^3$
31. $2^n \cdot 2^4$
32. $2^{2n} \cdot 2$
33. $a^{n+2} \cdot a^2$
34. $a^{n+4} \cdot a^5$

35. $a^{n-6} \cdot a^9, n > 6$ **36.** $a^{n-3} \cdot a^7, n > 3$ **37.** $3^n \cdot 3^{4n}$

38. $a^{2n} \cdot a^n$ **39.** $a^{3n} \cdot a^n$ **40.** $a^n \cdot a^{n+2}$

41. $a^{n+1} \cdot a^{n+3}$ **42.** $(a+1)^3(a+1)$ **43.** $(a-1)^2(a-1)^4$

44. $(a+2b)^2(a+2b)^3$ **45.** $-3(2x-y)(2x-y)^4$

46. $2(x-y)^4(x-y)^3$ **47.** $5(x-2y)^3(x-2y)^5$

48. $-4(x+3y)^2(x+3y)^4$ **49.** $ab(a^3)$

50. $a^2b(b^4)$ **51.** $2ab^2(a^3)$ **52.** $3a^3(a^2b^3)$

53. $a^2(-3ab^2)$ **54.** $-a^2b^3(ab^2)$ **55.** $a^3b(-a^2b^2)$

56. $2ab(-a^3b)$ **57.** $-2^2a(3a^2b^4)$ **58.** $3a^2b(-5^2b^2c)$

59. $-2^3xy^2(3x^3y^2)$ **60.** $7x^2y^4(-x^4y)$

61. $2a^2b(-3a^3b^2)(-b^3)$ **62.** $-2^2ab^3(3a^2b)(-a^4)$

63. $6a^3b^2(-4a^2b^3)(2ab)$ **64.** $-a^2b^4(-bc)(2ac^3)$

65. $5a^2b(-3a^3b^2)(-2^2b^4c^2)$ **66.** $a^4b(-2^3b^3c)(-3ab^2c^3)$

67. $4ab^5c^2(-3^2bc^3)(-ab^6)$ **68.** $3ac^4(-a^4b^2)(-2^2ab^3c)$

69. $(2^3)^2$ **70.** $(3^2)^3$ **71.** $(a^3)^4$ **72.** $(a^2)^5$

73. $(a^3)^n$ **74.** $(a^2)^{n+1}$ **75.** $(a^3)^{n+2}$ **76.** $(a^n)^2$

77. $(a^{n+1})^3$ **78.** $(a^{n+2})^2$ **79.** $\left(a^{n^2}\right)^3$ **80.** $(a^n)^n$

81. $\left(a^{n^2}\right)^n$ **82.** $\left(a^{n^2}\right)^{2n}$ **83.** $(-2^2)^4$ **84.** $(-3^2)^3$

85. $(-2^3)^2$ **86.** $(-2^3)^3$ **87.** $(-a^2)^3$ **88.** $(-a^4)^2$

89. $(-a^3)^4$ **90.** $(-a^5)^3$ **91.** $(-a^n)^4$ **92.** $(-a^n)^3$

93. $(2a^2)^3$ **94.** $(3^2a)^2$ **95.** $(ab^2)^3$ **96.** $(a^3b)^2$

97. $(ab^2)^n$ **98.** $(a^2b^3)^n$ **99.** $(a^2b)^{n+1}$ **100.** $(a^3b^2)^{n+2}$

101. $(-a^2b^3)^2$ **102.** $(-a^2b)^3$ **103.** $(-2^2a)^3$ **104.** $(-3a^2b)^4$

105. $(-4ab^3)^3$ **106.** $(-2ab^4)^4$ **107.** $(-2^2a^3b)^2$

108. $(-a^2b^5)^4$ **109.** $-(-a^2b^3c)^4$ **110.** $-(-a^3bc^2)^3$

111. $2^2a(-a^2b)^3$ **112.** $5a^3(-3a^3b)^2$ **113.** $3a^2(-2^2a^3)^2$

114. $-a^5b^3(-ab^2)^3$ **115.** $(2ab^3)^2(a^2b)^3$ **116.** $(5a^2b)^3(b^2c)^4$

117. $(a^2b^3c)^2(a^3c)^4$ **118.** $(-ab^5c)^3(a^2c^3)^2$ **119.** $(a^2b)^3(-ab^3)^2$

120. $(-3^2a)^2(-a^2)^3$ **121.** $(-2^2a^3)^3(-a^2b^3)^2$

122. $(-a^3b)^2(-a^2b^4)^3$ **123.** $(a^2b)^3(-2ab^3c)^2(-3bc^2)$

124. $(-ab^2)^2(5a^3bc^2)^3(-bc^3)$ **125.** $(abc^2)^2(2a^2b^3)^3(-a^2c)^4$

126. $(-a^2b^4)^3(2ab^2c)^2(a^3c^2)^5$ **127.** $(-a^2b)^4(3^2ab^2c)^2(-b^4c^6)^3$

128. $(-2^2ab^3)^3(-2^3a^3c)^2(bc^2)^4$ **129.** $(-3^2a^2b)^2(-2^3bc^2)^3(ac^3)^4$

130. $[-3ab^2(3c+2d)^3]^2[2a^3b(3c+2d)^2]^3$

131. $[x^2y(x^2+y^2)^3][xy^2(x^2+y^2)^2]^3$

132. $[-x^2y^3(x^2-2y^2)^2][2^3x^4y(x^2-2y^2)^3]^2$

133. $[4a^3b^3(a-3b)^4]^2[-ab^5(a-3b)^3]^3$

134. $[-3^2ab^2(x-2y)^3]^4[-2a^3b(x-2y)^2]^3$

135. $x(x^4)+(-3^2x^2)(x^3)$

136. $(-x^2)(x^5)+(2x^3)(-x)^4$

137. $-2^2x^3(-x)^3+x(-x)^5$

138. $(-4^2xy^2)(-x^3)-(-2y^2)(x^4)$

139. $3x^4(-x^3y)-4xy(-x)^6$

140. $-2^3x^2y(-y^4)-y^3(-3x^2y^2)$

141. $2a^4(-3ab^2)^3+6a(-a^2b^2)^3$

142. $(-a^2)^3(a^4b)+a^2b(-a^4)^2$

143. $(-3^2a^3b)^2(-b^4)+(-4a^3b^3)^2$

144. $5a^4b(-2ab^3)^3-(-2^3a^2b^4)^2(-a^3b^2)$

Multiplication of a Polynomial by a Monomial

Sometimes it is necessary to use many literal numbers in one problem. In order not to use too much of the alphabet, one letter with subscripts is often used, such as a_1, read "a sub one," a_2, read "a sub two," a_3, read "a sub three," and so forth.

Remember that $a_1, a_2, a_3, \ldots,$ represent different numbers.

The **extended distributive law of multiplication**

$$a(b_1+b_2+\cdots+b_n)=ab_1+ab_2+\cdots+ab_n$$

is used to multiply a monomial and a polynomial.

EXAMPLE Multiply $ab^2-2b-3a^2$ by $2ab^2$.

Solution

$$\begin{aligned}2ab^2(ab^2-2b-3a^2)&=(2ab^2)(ab^2)+(2ab^2)(-2b)+(2ab^2)(-3a^2)\\&=2a^2b^4-4ab^3-6a^3b^2\end{aligned}$$

EXAMPLE Multiply $2x^2+x+4$ by $2x^n$ given that $n\in N$.

Solution $2x^n(2x^2+x+4)=4x^{n+2}+2x^{n+1}+8x^n$

EXAMPLE Multiply $3(x+2)^2-(x+2)+1$ by $(x+2)^2$.

Solution $(x+2)^2[3(x+2)^2-(x+2)+1]$ is of the form

$z^2(3z^2-z+1)=3z^4-z^3+z^2$

Hence $(x+2)^2[3(x+2)^2-(x+2)+1]$

$$=3(x+2)^4-(x+2)^3+(x+2)^2$$

EXAMPLE Multiply $2ab^2-5a^2b+7$ by $(-3a^2b)^2$.

Solution

$$\begin{aligned}(-3a^2b)^2(2ab^2-5a^2b+7)&=9a^4b^2(2ab^2-5a^2b+7)\\&=18a^5b^4-45a^6b^3+63a^4b^2\end{aligned}$$

EXAMPLE Multiply $\frac{2x-1}{6} - \frac{x+4}{8}$ by 24.

Solution

$$\frac{24}{1}\left[\frac{2x-1}{6} - \frac{x+4}{8}\right] = \frac{24}{1}\left[\frac{2x-1}{6}\right] - \frac{24}{1}\left[\frac{x+4}{8}\right]$$
$$= 4(2x-1) - 3(x+4)$$
$$= 8x - 4 - 3x - 12$$
$$= 5x - 16$$

Exercise 2.6C

Perform the indicated operations and simplify, given that $n \in N$.

1. $3(a+4)$ **2.** $5(2b+1)$ **3.** $8(a-3)$ **4.** $2(3a-4)$
5. $-5(3a+1)$ **6.** $-3(a+2)$ **7.** $-2(2a-3)$ **8.** $-6(3a-1)$
9. $x(y+2)$ **10.** $2x(y+1)$ **11.** $x(3y-4)$ **12.** $2x(y-6)$
13. $-3x(y+3)$ **14.** $-3x(y+2)$ **15.** $-x(2y-1)$ **16.** $-3x(y-4)$
17. $3a(a^2+1)$ **18.** $2a(a^3+3)$ **19.** $a(a^2-2)$ **20.** $5a(a^3-a)$
21. $-7a(a^2+3)$ **22.** $-5a(a^2+2)$ **23.** $-2a(a^2-3)$
24. $-a(3a^2-2a)$ **25.** $x^2(x^2-x+1)$ **26.** $2x^2(x^3-x-2)$
27. $x^3(2x^2-5x-3)$ **28.** $2x^4(x^3+2x^2-3)$ **29.** $x^2(x^2-2xy-3y^2)$
30. $2x^2y(xy^2-3x^2y+x^3)$ **31.** $-2xy(x^2-y^2+5)$
32. $-3xy^2(2x^3+y^2-2xy)$ **33.** $-5x^3y(x^2y-xy^2-1)$
34. $-xy^2(xy^2-3y-4x^2)$ **35.** $x^n(x^2+x-2)$
36. $2x^n(x^2-x+2)$ **37.** $x^{n+1}(x^2-2x-1)$ **38.** $2x^{n+3}(x^2-3x-2)$
39. $x^n(x^{2n}+x^n-1)$ **40.** $x^n(x^{n+1}+x^n+x)$ **41.** $x^{n+1}(x^{2n}-x^n-2)$
42. $x^{n+2}(x^{n+2}-x^{n+1}-x^n)$ **43.** $(x+2)^2[3(x+2)-10]$
44. $(x-1)^3[5(x-1)+2]$ **45.** $2(x+1)^2[(x+1)^2-4(x+1)]$
46. $6(x-2)^3[(x-2)^2-3(x-2)]$ **47.** $(x-y)^2[(x-y)^2-2(x-y)-3]$
48. $2(x+y)^2[3(x+y)^2+(x+y)+4]$
49. $(-2a)^2(3a-b)$ **50.** $(-3a)^3(a^2-b^2)$
51. $(3ab)^4(a^3-a^2b-b^3)$ **52.** $(a^2b)^2(a^2-2ab+b^2)$
53. $(-xy^2)^2(x^2y-7xy^2+8)$ **54.** $(-2^2a^2)^3(a^2b^3-3b^2a-1)$
55. $6\left[\frac{x+2}{3} + \frac{x-1}{2}\right]$ **56.** $12\left[\frac{x-3}{4} + \frac{2x-1}{6}\right]$
57. $12\left[\frac{2x+1}{3} - \frac{x+2}{4}\right]$ **58.** $24\left[\frac{x+4}{8} - \frac{x+3}{6}\right]$
59. $24\left[\frac{3x-1}{8} - \frac{x-3}{12}\right]$ **60.** $36\left[\frac{x-1}{4} - \frac{x-2}{9}\right]$

61. $a^3(2a^2 - a + 1) + a^4(2a + 1)$
62. $3a^3(2a^2 + a - 4) - 3a^2(a^2 - 4a - 1)$
63. $(2a)^2(a^2 - 8) - a^3(4a - 6)$
64. $(3a)^2(a^3 + 2a^2 - a + 1) - 3a^4(3a + 6)$

Multiplication of Polynomials

Multiplication of two polynomials is the same as multiplication of a polynomial by a monomial where the first polynomial is considered as one quantity.

To multiply $(x + 4)$ by $(x - 3)$, consider $(x + 4)$ as one quantity and apply the distributive law:

$$(x + 4)(x - 3) = (x + 4)(x) + (x + 4)(-3)$$

Then reapply the distributive law:

$$= x^2 + 4x - 3x - 12$$
$$= x^2 + x - 12$$

Notice that each term of the second polynomial has been multiplied by each term of the first polynomial.

The same result can be obtained by arranging the two polynomials in two rows and multiplying the upper polynomial by each term of the lower polynomial. Arrange like terms of the product in the same column so that addition of like terms is easier:

$$\begin{array}{rl} & x + 4 \\ & x - 3 \\ \hline x(x + 4) = & x^2 + 4x \\ -3(x + 4) = & \quad - 3x - 12 \\ \hline \text{add} & x^2 + x - 12 \end{array}$$

Thus $\quad (x + 4)(x - 3) = x^2 + x - 12$

Remark $\quad (x + 4)(x - 3) \neq x^2 - 12$.

EXAMPLE $\quad$ Multiply $(2x - 3)^2$.

Solution

$$(2x - 3)^2 = (2x - 3)(2x - 3)$$

$$\begin{array}{rl} & 2x - 3 \\ & 2x - 3 \\ \hline 2x(2x - 3) = & 4x^2 - 6x \\ -3(2x - 3) = & \quad - 6x + 9 \\ \hline \text{add} & 4x^2 - 12x + 9 \end{array}$$

Hence $\quad (2x - 3)^2 = 4x^2 - 12x + 9$

Remark $(2x-3)^2 \neq 4x^2+9$

Notes

1. $(a+b)^2 = a^2 + 2ab + b^2$
2. $(a-b)^2 = a^2 - 2ab + b^2$
3. $(a+b)(a-b) = a^2 - b^2$

EXAMPLE Multiply $(2x^2 - x + 3)$ by $(2x - 5)$.

Solution

$$\begin{array}{lrrrr}
 & 2x^2 & - x & + 3 & \\
 & 2x & - 5 & & \\
\hline
2x(2x^2 - x + 3) = & 4x^3 & - 2x^2 & + 6x & \\
-5(2x^2 - x + 3) = & & - 10x^2 & + 5x & - 15 \\
\hline
\text{add} & 4x^3 & - 12x^2 & + 11x & - 15
\end{array}$$

Therefore $(2x^2 - x + 3)(2x - 5) = 4x^3 - 12x^2 + 11x - 15$

EXAMPLE Multiply $3x^{n+1} + x^n - 3$ by $2x + 3$ given that $n \in N$.

Solution

$$\begin{array}{l}
3x^{n+1} + x^n - 3 \\
2x + 3 \\
\hline
6x^{n+2} + 2x^{n+1} - 6x \\
\qquad + 9x^{n+1} \qquad + 3x^n - 9 \\
\hline
6x^{n+2} + 11x^{n+1} - 6x + 3x^n - 9
\end{array}$$

Hence $(3x^{n+1} + x^n - 3)(2x + 3)$

$$= 6x^{n+2} + 11x^{n+1} + 3x^n - 6x - 9$$

EXAMPLE Perform the indicated operations and simplify:

$(3x-2)(x+2) - (2x+1)^2$

Solution

$$\begin{aligned}
(3x-2)(x+2) - (2x+1)^2 &= (3x^2 + 4x - 4) - (4x^2 + 4x + 1) \\
&= 3x^2 + 4x - 4 - 4x^2 - 4x - 1 \\
&= -x^2 - 5
\end{aligned}$$

Exercise 2.6D

Perform the indicated operations and simplify, given that $n \in N$:

1. $(x+4)(x+1)$	**2.** $(x+3)(x+5)$	**3.** $(x-3)(x+2)$
4. $(x-2)(x+6)$	**5.** $(x+4)(x-7)$	**6.** $(x+4)(x-4)$
7. $(x-4)(x-9)$	**8.** $(x-3)(x-6)$	**9.** $(2x+3)(x+4)$
10. $(4x+3)(x-3)$	**11.** $(5x+2)(x-5)$	**12.** $(4x-3)(x-2)$

13. $(2x+1)(5x+4)$ **14.** $(6x+7)(2x+3)$ **15.** $(3x-4)(3x+8)$

16. $(4x+5)(4x-5)$ **17.** $(3x-4)(5x-4)$ **18.** $(2x-7)(7x-2)$

19. $(6-x)(5+2x)$ **20.** $(4-3x)(2+9x)$ **21.** $(4+3x)(6-5x)$

22. $(2x+1)(8-3x)$ **23.** $(2x+3)(3-4x)$ **24.** $(3x+4)(3-2x)$

25. $(4x+1)(5-2x)$ **26.** $(4-3x)(5x+3)$ **27.** $(4-5x)(7x+3)$

28. $(x+6)^2$ **29.** $(4x+3)^2$ **30.** $(x-4)^2$ **31.** $(4x-7)^2$

32. $(x+2y)^2$ **33.** $(3x+y)^2$ **34.** $(2x-3y)^2$ **35.** $(5x-4y)^2$

36. $(2x+y)(x+3y)$ **37.** $(3x+y)(x-y)$ **38.** $(4x-3y)(2x-3y)$

39. $(xy-4)(3xy-4)$ **40.** $(5xy-2)(2xy+3)$ **41.** $(x-6)(x^2+3)$

42. $(x-2)(x^2-3)$ **43.** $(x^2+4)(x^2-5)$ **44.** $(3x^2-4)(x^2+6)$

45. $(2x^2+1)(3x^3+2)$ **46.** $(3x^2-1)(x^3-2)$ **47.** $(2x^3-1)(3x^2+4)$

48. $(x+2)(x^2-3x+4)$ **49.** $(x+3)(2x^2-x-4)$

50. $(x-4)(3x^2+x-2)$ **51.** $(2x-3)(x^2-4x+3)$

52. $(x-1)(x^2+x+1)$ **53.** $(3x+2)(9x^2-6x+4)$

54. $(2x+y)(4x^2-2xy+y^2)$ **55.** $(2x-3y)(4x^2+6xy+9y^2)$

56. $(x^2-2x+3)(x^2+2x-3)$ **57.** $(x^2-3x-2)(x^2+3x+2)$

58. $(x^2+3x-4)(2x^2+x+5)$ **59.** $(3x^2-2x+3)(2x^2-3x-4)$

60. $(x^2+x+1)^2$ **61.** $(2x^2-x+1)^2$ **62.** $(3x^2+2x-4)^2$

63. $3(x-4)(x+6)$ **64.** $2(x-3)(2x+1)$ **65.** $-4(x+7)(3x-2)$

66. $-2(2x-1)(4x-1)$ **67.** $-3(5-3x)(2+3x)$

68. $-2(4-x)(3+5x)$ **69.** $-4(3-2x)(6-7x)$

70. $a(3x+1)(x-4)$ **71.** $2a(x-8)(2x-3)$

72. $(x+1)(x-2)(x+3)$ **73.** $(x+3)(x-2)(x-4)$

74. $(2x-1)(x+1)(x-1)$ **75.** $(x-4)(2x-3)(2x+3)$

76. $(3x-1)(2x+1)(x-3)$ **77.** $(x+2)(3x-2)(4x-1)$

78. $(x+3)^3$ **79.** $(x-2)^3$ **80.** $(3x-1)^3$ **81.** $(2x-3)^3$

82. $(x^n+3)(x^n+1)$ **83.** $(x^n+6)(x^n-4)$ **84.** $(x^n-4)(x^n-2)$

85. $(2x^n-1)(x^n+4)$ **86.** $(2x^n-3)(3x^n-4)$ **87.** $(3x^n-2)(4x^n-5)$

88. $(x^{n+1}+3x^n+4)(x-3)$ **89.** $(x^{n+1}-2x^n+3)(x-4)$

90. $(x^{n+2}-4x^{n+1}+1)(x+4)$ **91.** $(x^{n+2}+3x^{n+1}-1)(x+3)$

92. $(x^{n+1}-3x^n-1)(x^2+3x+9)$ **93.** $(x^{n+1}+4x^n-3)(x^2-4x+16)$

94. $(x^{2n}+3x^n+4)(x^{2n}-3x^n+4)$ **95.** $(2x^{2n}+4x^n+3)(2x^{2n}-4x^n+3)$

96. $(x+3)(x+2)-x(x+6)$ **97.** $(2x-3)(3x-2)-6x(x-4)$

98. $(x+2)(x-4)+(x+1)^2$ **99.** $(x-4)(x+6)+(x-1)^2$

100. $(2x-3)(2x+5)-(2x+3)^2$ **101.** $(4x-3)(x+5)-(2x+3)^2$

102. $(x-3)(x-5)-2(x-2)^2$ **103.** $(2x-3)(x+3)-2(x+2)^2$

104. $(3x-1)(6x+1)-2(3x-2)^2$ **105.** $(x-4)^2-(x+4)^2$

106. $(2x+5)^2-(2x-5)^2$

2.7 Division of Polynomials

The following are some of the properties pertaining to fractions:

1. $\frac{a}{b} = \frac{ac}{bc}$

2. $\frac{a+b}{c} = \frac{a}{c} + \frac{b}{c}$

3. $\frac{a}{b} \cdot \frac{c}{d} = \frac{ac}{bd}$

4. $\frac{a}{b} \div \frac{c}{d} = \frac{a}{b} \cdot \frac{d}{c}$

Note Since division by zero is not defined, all the denominators of the fractions will be assumed to be different from zero.

First we will discuss division of monomials, then division of a polynomial by a monomial, and finally division of two polynomials.

Division of Monomials

The following theorems are used to divide monomials.

THEOREM 4 If $a \in R$, $a \neq 0$, and $m, n \in N$, then

$$\frac{a^m}{a^n} = \begin{cases} a^{m-n} & \text{when } m > n \\ 1 & \text{when } m = n \\ \dfrac{1}{a^{n-m}} & \text{when } m < n \end{cases}$$

Proof

$$\frac{a^m}{a^n} = \frac{a^n \cdot a^{m-n}}{a^n \cdot 1} = a^{m-n} \qquad \text{when } m > n$$

$$\frac{a^m}{a^n} = \frac{a^n}{a^n} = 1 \qquad \text{when } m = n$$

$$\frac{a^m}{a^n} = \frac{a^m \cdot 1}{a^m \cdot a^{n-m}} = \frac{1}{a^{n-m}} \qquad \text{when } m < n$$

EXAMPLES

1. $\frac{a^8}{a^4} = a^{8-4} = a^4$

2. $\frac{2^3}{2^3} = 1$

3. $\frac{a^6}{a^{11}} = \frac{1}{a^{11-6}} = \frac{1}{a^5}$

EXAMPLE

Simplify $\dfrac{20a^2b^5}{-8a^6b^2}$ by applying the laws of exponents.

Solution

$$\frac{20a^2b^5}{-8a^6b^2} = -\frac{20}{8}\cdot\frac{a^2}{a^6}\cdot\frac{b^5}{b^2}$$
$$= -\frac{5}{2}\cdot\frac{1}{a^{6-2}}\cdot\frac{b^{5-2}}{1} = -\frac{5}{2}\cdot\frac{1}{a^4}\cdot\frac{b^3}{1} = -\frac{5b^3}{2a^4}$$

Notes

1. $(a-b) = -b+a = -(b-a)$
2. $(a-b)^2 = [-(b-a)]^2 = (b-a)^2$
3. $(a-b)^3 = [-(b-a)]^3 = -(b-a)^3$

EXAMPLE

$$\frac{(x-1)^2}{(1-x)^3} = \frac{(x-1)^2}{-(x-1)^3} = \frac{1}{-(x-1)} = \frac{1}{1-x}$$

or $$\frac{(x-1)^2}{(1-x)^3} = \frac{(1-x)^2}{(1-x)^3} = \frac{1}{1-x}$$

THEOREM 5

If $a, b \in R$, $b \neq 0$, and $m \in N$, then $\left(\dfrac{a}{b}\right)^m = \dfrac{a^m}{b^m}$.

Proof

$$\left(\frac{a}{b}\right)^m = \overbrace{\frac{a}{b}\cdot\frac{a}{b}\cdots\frac{a}{b}}^{m\text{ fractions}} = \frac{\overbrace{a\cdot a\cdots a}^{m\text{ factors}}}{\underbrace{b\cdot b\cdots b}_{m\text{ factors}}} = \frac{a^m}{b^m}$$

COROLLARY

If $a, b, c, d \in R$, $c \neq 0$, $d \neq 0$, and $m, n, p, q, k \in N$, then by the use of Theorems 2, 3, and 5, we have

$$\left(\frac{a^mb^n}{c^pd^q}\right)^k = \frac{(a^mb^n)^k}{(c^pd^q)^k} = \frac{a^{mk}b^{nk}}{c^{pk}d^{qk}}$$

EXAMPLE

By applying the laws of exponents, simplify $\left[\dfrac{3x^5yz}{6x^3y^2}\right]^4$.

Solution

We can simplify the fraction first before applying the outside exponent:

$$\left[\frac{3x^5yz}{6x^3y^2}\right]^4 = \left[\frac{3}{6}\cdot\frac{x^5}{x^3}\cdot\frac{y}{y^2}\cdot\frac{z}{1}\right]^4$$
$$= \left[\frac{1}{2}\cdot\frac{x^2}{1}\cdot\frac{1}{y}\cdot\frac{z}{1}\right]^4 = \left[\frac{x^2z}{2y}\right]^4 = \frac{x^8z^4}{2^4y^4} = \frac{x^8z^4}{16y^4}$$

EXAMPLE

$$\frac{(18)^4}{(24)^3} = \frac{(2 \cdot 3^2)^4}{(2^3 \cdot 3)^3} = \frac{2^4 \cdot 3^8}{2^9 \cdot 3^3}$$
$$= \frac{2^4}{2^9} \cdot \frac{3^8}{3^3}$$
$$= \frac{1}{2^5} \cdot \frac{3^5}{1} = \frac{243}{32}$$

EXAMPLE

Simplify $\frac{(a^3b^2c)^3}{(3ab^3c)^4}$ by applying the laws of exponents.

Solution

Here we cannot simplify first, since the numerator and the denominator have different powers. Apply the outside exponents first, then simplify:

$$\frac{(a^3b^2c)^3}{(3ab^3c)^4} = \frac{a^9b^6c^3}{3^4a^4b^{12}c^4}$$
$$= \frac{a^5}{3^4b^6c} = \frac{a^5}{81b^6c}$$

EXAMPLE

Perform the indicated operations and simplify:

$24a^{14} \div (-2a^2)^3 - a^2(-a^3)^2$

Solution

$$24a^{14} \div (-2a^2)^3 - a^2(-a^3)^2 = \frac{24a^{14}}{(-2a^2)^3} - a^2(-a^3)^2$$
$$= \frac{24a^{14}}{-2^3a^6} - a^2(a^6)$$
$$= -\frac{24a^{14}}{8a^6} - a^8$$
$$= -3a^8 - a^8 = -4a^8$$

Exercise 2.7A

Perform the indicated operations and simplify, given that $n \in N$:

1. $\frac{2^6}{2^2}$ **2.** $\frac{3^9}{3^3}$ **3.** $\frac{2^4}{2^{12}}$ **4.** $\frac{5^2}{5^4}$

5. $\frac{7^{10}}{7^{10}}$ **6.** $\frac{2^9}{-2^3}$ **7.** $\frac{-3^2}{3^7}$ **8.** $\frac{(-2)^5}{2^5}$

9. $\frac{(-3)^3}{3^4}$ **10.** $\frac{(-2)^4}{2^8}$ **11.** $\frac{(-5)^6}{5^4}$ **12.** $\frac{-3^4}{(-3)^4}$

13. $\frac{a^6}{a^3}$ **14.** $\frac{a^{10}}{a^6}$ **15.** $\frac{a^3}{a^9}$ **16.** $\frac{a}{a^5}$

17. $\frac{-a^4}{a}$ **18.** $\frac{a^7}{-a^3}$ **19.** $\frac{(-a)^5}{-a^6}$ **20.** $\frac{-a^7}{(-a)^4}$

21. $\frac{-a^{11}}{(-a)^{12}}$ **22.** $\frac{(-a)^3}{-a^8}$ **23.** $\frac{-(-a)^4}{a^8}$ **24.** $\frac{-(-a)^3}{a^4}$

25. $\frac{a^{2n}}{a^n}$ **26.** $\frac{a^{3n}}{a^{6n}}$ **27.** $\frac{a^{n+2}}{a^n}$ **28.** $\frac{a^{n+5}}{a^{n+3}}$

29. $\frac{a^{n+1}}{a^{n+4}}$ **30.** $\frac{a^{n+3}}{a^{n+7}}$ **31.** $\frac{(x+y)^5}{(x+y)^3}$ **32.** $\frac{(x-2y)^7}{(x-2y)^4}$

33. $\frac{(x+2y)^2}{(x+2y)^3}$ **34.** $\frac{(2x-y)^4}{(2x-y)^6}$ **35.** $\frac{(x+y)^6}{(y+x)^9}$ **36.** $\frac{(2x+y)^2}{(y+2x)^8}$

37. $\frac{x-y}{y-x}$ **38.** $\frac{(x-y)^3}{(y-x)^2}$ **39.** $\frac{(x-y)^2}{y-x}$ **40.** $\frac{(x-y)^4}{(y-x)^5}$

41. $\frac{2^3 \cdot 3^2}{2^4 \cdot 3}$ **42.** $\frac{2^7 \cdot 3^8}{2^5 \cdot 3^{10}}$ **43.** $\frac{3^5 \cdot 5^4}{3^6 \cdot 5^2}$ **44.** $\frac{2^8 \cdot 11^4}{2^6 \cdot 11^5}$

45. $\frac{a^4b^7}{a^3b^4}$ **46.** $\frac{a^3b^4}{ab^5}$ **47.** $\frac{a^4b^6}{a^4b^3}$ **48.** $\frac{3a^2b}{6ab^3}$

49. $\frac{6a^7b}{-18ab^6}$ **50.** $\frac{24a^9b^6}{-32a^4b^2}$ **51.** $\frac{12a^5b^4c^3}{18a^4b^{10}c}$ **52.** $\frac{65a^8b^7c^4}{78a^{10}bc^4}$

53. $\frac{-4a^5b^6}{6a^3bc^2}$ **54.** $\frac{8a^3b^9c}{12a^4b^9}$ **55.** $\left(2\frac{2}{3}\right)^2$ **56.** $\left(3\frac{3}{4}\right)^2$

57. $\left(-1\frac{1}{2}\right)^3$ **58.** $\left(-3\frac{1}{4}\right)^2$ **59.** $\left(\frac{3a^4}{2a}\right)^3$ **60.** $\left(\frac{6a^5}{9a^3}\right)^4$

61. $\left(\frac{-14a^6}{21a^7}\right)^2$ **62.** $\left(\frac{12a^9}{-8a^6}\right)^3$ **63.** $\left(\frac{18a^7}{-24a^6}\right)^3$ **64.** $\left(\frac{5a^{10}b^2}{10a^8b^2}\right)^5$

65. $\left(\frac{-a^4b^6c^8}{a^3b^8c^7}\right)^4$ **66.** $\left(\frac{ab^2c^4}{-a^3b^2c}\right)^5$ **67.** $\left(\frac{22a^2b^3c}{33a^4bc^3}\right)^4$

68. $\left(\frac{16a^3bc^5}{24abc^7}\right)^3$ **69.** $\left[\frac{a^3(2b+c)^3}{a^4(2b+c)^2}\right]^4$ **70.** $\left[\frac{3a^2(b-c)}{6a(b-c)^2}\right]^2$

71. $\left[\frac{-2a^2(b+c)^3}{8a^5(b+c)}\right]^3$ **72.** $\left[\frac{a^6b(c-2d)^3}{a^4b^3(c-2d)^4}\right]^3$ **73.** $\left[\frac{a^2b(x-y)^2}{a^3(y-x)^3}\right]^3$

74. $\left[\frac{ab^3(x-2y)^3}{a^2b(2y-x)^2}\right]^2$ **75.** $\left[\frac{a^3b^2(x-y)^2}{ab^3(y-x)^4}\right]^2$ **76.** $\left[\frac{6a^3(x-y)^3}{9a^2(y-x)^5}\right]^3$

77. $\dfrac{4^4}{8^2}$ 78. $\dfrac{9^3}{6^4}$ 79. $\dfrac{6^5}{12^3}$ 80. $\dfrac{4^5}{6^5}$

81. $\dfrac{15^4}{10^5}$ 82. $\dfrac{14^5}{21^4}$ 83. $\dfrac{8^7}{12^6}$ 84. $\dfrac{18^4}{9^6}$

85. $\dfrac{10^3}{4^4}$ 86. $\dfrac{12^5}{18^4}$ 87. $\dfrac{22^3}{33^2}$ 88. $\dfrac{3 \cdot 2^3}{6^3}$

89. $\dfrac{2 \cdot 3^5}{6^5}$ 90. $\dfrac{5 \cdot 2^4}{10^3}$ 91. $\dfrac{(a^4b^2)^3}{(a^4b)^4}$ 92. $\dfrac{(2a^2b^3)^4}{(4ab^4)^3}$

93. $\dfrac{(6a^3bc^2)^3}{(3a^2b^3c)^2}$ 94. $\dfrac{(-12a^4b^3)^4}{(18a^4b^2)^3}$ 95. $\dfrac{(-15a^2b^4)^3}{(10ab^3)^5}$

96. $\dfrac{(14ab^3c^2)^4}{(-21a^2bc^4)^3}$ 97. $\dfrac{(8a^4bc^3)^4}{(-4a^6b^2c^3)^3}$ 98. $\dfrac{(-9a^2bc^5)^6}{(27abc^4)^5}$

99. $\dfrac{(-4a^2b^2c)^6}{(4a^3b^4c)^4}$ 100. $\dfrac{(-6ab^2c^4)^5}{(-9a^2b^3c)^4}$

101. $a^7 \div a^3 + a^2(-2a)^2$
102. $6a^8 \div a^3 + 4a^2(-a)^3$
103. $a^8 \div a^4 + a^2(-3a)^2$
104. $4a^9 \div a^3 + 2a^3(-a)^3$
105. $6a^{10} \div 2(-a^2)^3 - 5a^2(-2a)^2$
106. $16a^{11} \div (-4a^2)^2 - 3a(-2a^3)^2$
107. $8a^{10} \div (-2a^2)^3 - a(-2a)^3$
108. $9a^{12} \div (3a^2)^2 - 12a^2(-2a^3)^2$
109. $8a^{10}b^{12} \div (-a^2b^3)^4 + 4a^4(-b^2)^3 \div (-ab^3)^2$
110. $(2a^4b^3)^4 \div (-2a^2b)^3 - (a^3b^2)^5 \div (-a^5b)$

Division of a Polynomial by a Monomial

From the properties of fractions we have

$$\frac{a_1 + a_2 + \cdots + a_n}{a} = \frac{a_1}{a} + \frac{a_2}{a} + \cdots + \frac{a_n}{a}$$

To divide a polynomial of more than one term by a monomial, divide every term of the polynomial by the monomial.

EXAMPLE Divide $\dfrac{a^4 - 3a^3b - a^2b^2}{-ab}$ and simplify.

Solution

$$\frac{a^4 - 3a^3b - a^2b^2}{-ab} = \frac{a^4}{-ab} + \frac{-3a^3b}{-ab} + \frac{-a^2b^2}{-ab}$$

$$= -\frac{a^3}{b} + 3a^2 + ab$$

EXAMPLE

Divide $2(x-y)^3 - 4(x-y)^2$ by $2(y-x)$ and simplify.

Solution

$$\begin{aligned}\frac{2(x-y)^3 - 4(x-y)^2}{2(y-x)} &= \frac{2(x-y)^3}{2(y-x)} - \frac{4(x-y)^2}{2(y-x)} \\ &= \frac{-2(y-x)^3}{2(y-x)} - \frac{4(y-x)^2}{2(y-x)} \\ &= -(y-x)^2 - 2(y-x)\end{aligned}$$

EXAMPLE

Perform the indicated operations and simplify:

$$(6x^4 - 3x^3 + 9x^2) \div (3x^2) - (2x+3)(x-4)$$

Solution

$$\begin{aligned}&(6x^4 - 3x^3 + 9x^2) \div (3x^2) - (2x+3)(x-4) \\ &\quad= \frac{6x^4 - 3x^3 + 9x^2}{3x^2} - (2x+3)(x-4) \\ &\quad= 2x^2 - x + 3 - (2x^2 - 5x - 12) \\ &\quad= 2x^2 - x + 3 - 2x^2 + 5x + 12 \\ &\quad= 4x + 15\end{aligned}$$

Exercise 2.7B

Perform the indicated operations and simplify, given that $n \in N$:

1. $\frac{4x+2}{2}$ **2.** $\frac{3x-6}{3}$ **3.** $\frac{7-14x}{7}$ **4.** $\frac{9-27x}{9}$

5. $\frac{x^2-3x}{x}$ **6.** $\frac{x-2x^2}{x}$ **7.** $\frac{6x+4x^2}{2x}$ **8.** $\frac{5x-15x^2}{5x}$

9. $\frac{8x^3-4x}{4x}$ **10.** $\frac{9x^2+3x^3}{3x^2}$ **11.** $\frac{10x^2-5x^3}{5x^2}$ **12.** $\frac{6x^4+4ax^2}{2x^2}$

13. $\frac{6ax^2-12a^2x}{6ax}$ **14.** $\frac{2a^3b^2-4a^2b^3}{-2a^2b^2}$ **15.** $\frac{7a^2b^3-14a^3b^2}{-7a^3b^3}$

16. $\frac{5x^4-3x^3+x^2}{x^2}$ **17.** $\frac{2x^4+x^3-x^2}{-x^2}$ **18.** $\frac{6x^2-9x+3}{-3x}$

19. $\frac{4x^3-2x^2+8x}{-2x^2}$ **20.** $\frac{15x^2+10xy-5y^2}{5xy}$ **21.** $\frac{x^4-x^2y^2+y^4}{-x^2y^2}$

22. $\frac{3a^3b+6a^2b^2-9ab^3}{-3a^3b^3}$ **23.** $\frac{8a^4b^2-16a^3b^4+4a^2b^6}{4a^3b^3}$

24. $\dfrac{2(a+2b)^3+6(a+2b)^2}{2(a+2b)}$

25. $\dfrac{9(a+b)^2-3(a+b)}{3(b+a)}$

26. $\dfrac{6(x-y)^3-2(x-y)^2}{2(x-y)^2}$

27. $\dfrac{a^2b(x-5y)^3-ab^2(x-5y)^2}{-ab(x-5y)^2}$

28. $\dfrac{xy^3(a-3b)^2+x^2y^2(a-3b)}{xy^2(3b-a)}$

29. $\dfrac{(x-y)^3+(x-y)^2}{(y-x)}$

30. $\dfrac{7(2x-y)^2-14(y-2x)^3}{21(2x-y)}$

31. $\dfrac{a(x-2y)^3-b(x-2y)^4}{ab(2y-x)^2}$

32. $\dfrac{3a^6b^2+a^5b^3-6a^4b^4}{(a^2b)^2}$

33. $\dfrac{8a^8+8a^7-48a^6}{(-2a^2)^3}$

34. $\dfrac{36a^4b^4+18a^3b^3-9a^2b^2}{(-3ab)^2}$

35. $\dfrac{-64a^5b^9+64a^4b^9+128a^3b^9}{(-4ab^3)^3}$

36. $(x^{3n}-2x^{2n}-x^n)\div x^n$

37. $(x^{n+2}-3x^{n+1}+x^n)\div x^n$

38. $(x^{n+4}-x^{n+3}-x^{n+2})\div x^{n+1}$

39. $(x^{n+3}+2x^{n+4}-x^{n+5})\div x^{n+2}$

40. $(2x^4-4x^3)\div(2x^2)+x(x+3)$

41. $(8x^4-4x^3+16x^2)\div(-4x^2)+2x^2(x+1)$

42. $(12x^6+3x^5-3x^4)\div(3x^2)-(4x^2+1)(x^2-1)$

43. $(8x^5+16x^4-24x^3)\div(-8x^3)-(x-3)(x+1)$

44. $(12x^4-4x^3-16x^2)\div(4x^2)-(3x-2)(x-2)$

45. $(3x^5-2x^4+x^3)\div x^2-3(x^2+2)(x-1)$

46. $(12x^4-3x^3-9x^2)\div(3x^2)-2(x+3)(2x-1)$

47. $(x^{n+2}-3x^{n+1}-4x^n)\div x^n-(x-5)(x+2)$

48. $(2x^{n+2}+5x^{n+1}-6x^n)\div x^n-(2x+3)(x-1)$

49. $(x^{2n+1}+4x^{2n}-3x^{n+1})\div x^n+(x^n+3)(x-4)$

50. $(2x^{2n+2}-4x^{2n+1}+4x^{n+2})\div(2x^{n+1})+(x^n-2)(x+2)$

Division of Two Polynomials

DEFINITION

The **degree of a polynomial** in a literal number is the greatest exponent of that literal number in the polynomial.

For example, $x^6y-3x^4y^2+5x^2y^3+7y^4$ is a polynomial of degree 6 in x, and degree 4 in y.

The division operation is defined from the operation of multiplication. From multiplication we have $(3x - 2)(2x + 3) = 6x^2 + 5x - 6$.
Hence, when $6x^2 + 5x - 6$ is divided by $3x - 2$, the result is $2x + 3$.

The polynomial $(6x^2 + 5x - 6)$ is called the **dividend**, $(3x - 2)$ is called the **divisor**, and $(2x + 3)$ is called the **quotient**.

To divide two polynomials, we start by arranging the terms of the dividend according to the decreasing exponents of one of the literals, leaving spaces for the missing powers (or including terms with zero coefficients for the missing terms). Arrange the terms of the divisor according to the decreasing exponents of the same literal used in arranging the terms of the dividend. Divide the first term of the dividend by the first term of the divisor to get the first term of the quotient. Multiply the first term of the quotient by each term of the divisor, and write the product under the like terms in the dividend. Subtract the product from the dividend to arrive at a new dividend.

To find the next term and all subsequent terms of the quotient, treat the new dividend as if it were the original dividend. Continue this procedure until you get zero, or until the degree of the newly derived polynomial, with respect to the literal used in arranging the dividend, is at least one degree less than the degree of the divisor in that literal.
The last polynomial is called the **remainder**.

EXAMPLE Divide $8x^3 - 18x^2 + 13x - 6$ by $2x - 3$.

Solution

		$+4x^2 - 3x + 2$	quotient
	divisor $2x - 3$	$8x^3 - 18x^2 + 13x - 6$	dividend
		$\ominus \quad \oplus$	
$\dfrac{8x^2}{2x} = +4x^2$	$4x^2(2x - 3) =$	$+8x^3 - 12x^2$	subtract
		$-6x^2 + 13x - 6$	
		$\oplus \quad \ominus$	
$\dfrac{-6x^2}{2x} = -3x$	$-3x(2x - 3) =$	$-6x^2 + 9x$	subtract
		$+4x - 6$	
		$\ominus \quad \oplus$	
$\dfrac{4x}{2x} = +2$	$2(2x - 3) =$	$+4x - 6$	subtract
		0	remainder

Hence $$\frac{8x^3 - 18x^2 + 13x - 6}{2x - 3} = 4x^2 - 3x + 2$$

EXAMPLE Divide $8x^2 - 9x^3 + 2x^5 - 8x + 4$ by $2x^2 - 3 + 4x$.

Solution Write the dividend as $2x^5 + 0x^4 - 9x^3 + 8x^2 - 8x + 4$.
Write the divisor as $2x^2 + 4x - 3$.

$$
\begin{array}{r|r}
 & x^3 - 2x^2 + x - 1 \qquad\qquad\qquad\quad \\ \hline
2x^2 + 4x - 3 & 2x^5 + 0x^4 - 9x^3 + 8x^2 - 8x + 4 \\
 & \ominus \quad\;\; \ominus \quad\;\; \oplus \qquad\qquad\qquad\quad \\
 & + 2x^5 + 4x^4 - 3x^3 \qquad\qquad\qquad\quad \\ \cline{2-2}
 & - 4x^4 - 6x^3 + 8x^2 - 8x + 4 \\
 & \oplus \quad\;\; \oplus \quad\;\; \ominus \qquad\qquad\quad \\
 & - 4x^4 - 8x^3 + 6x^2 \qquad\qquad\quad \\ \cline{2-2}
 & + 2x^3 + 2x^2 - 8x + 4 \\
 & \ominus \quad\;\; \ominus \quad\;\; \oplus \qquad\quad \\
 & + 2x^3 + 4x^2 - 3x \qquad\quad \\ \cline{2-2}
 & - 2x^2 - 5x + 4 \\
 & \oplus \quad\;\; \oplus \quad\;\; \ominus \\
 & - 2x^2 - 4x + 3 \\ \cline{2-2}
 & - x + 1
\end{array}
$$

Hence

$$\frac{2x^5 - 9x^3 + 8x^2 - 8x + 4}{2x^2 + 4x - 3} = x^3 - 2x^2 + x - 1 + \frac{-x + 1}{2x^2 + 4x - 3}$$

or

$$= x^3 - 2x^2 + x - 1 - \frac{x - 1}{2x^2 + 4x - 3}$$

Note This form is similar to the form used in arithmetic when we write $\frac{17}{3} = 5 + \frac{2}{3}$.

EXAMPLE Divide $(x^2y^3 - 7x^4y + 2x^5 + y^5 - 3xy^4)$ by $(2x^2 + y^2 - xy)$.

Solution Write the dividend as $2x^5 - 7x^4y + 0x^3y^2 + x^2y^3 - 3xy^4 + y^5$. Write the divisor as $2x^2 - xy + y^2$:

$$
\begin{array}{r|r}
 & x^3 - 3x^2y - 2xy^2 + y^3 \qquad\qquad \\ \hline
2x^2 - xy + y^2 & 2x^5 - 7x^4y + 0x^3y^2 + x^2y^3 - 3xy^4 + y^5 \\
 & \ominus \quad\;\; \oplus \quad\;\; \ominus \qquad\qquad\qquad\qquad\quad \\
 & + 2x^5 - x^4y + x^3y^2 \qquad\qquad\qquad\qquad\quad \\ \cline{2-2}
 & - 6x^4y - x^3y^2 + x^2y^3 - 3xy^4 + y^5 \\
 & \oplus \quad\;\; \ominus \quad\;\; \oplus \qquad\qquad\quad \\
 & - 6x^4y + 3x^3y^2 - 3x^2y^3 \qquad\qquad\quad \\ \cline{2-2}
 & - 4x^3y^2 + 4x^2y^3 - 3xy^4 + y^5 \\
 & \oplus \quad\;\; \ominus \quad\;\; \oplus \qquad \\
 & - 4x^3y^2 + 2x^2y^3 - 2xy^4 \qquad \\ \cline{2-2}
 & + 2x^2y^3 - xy^4 + y^5 \\
 & \ominus \quad\;\; \oplus \quad\;\; \ominus \\
 & + 2x^2y^3 - xy^4 + y^5 \\ \cline{2-2}
 & 0 \qquad\quad
\end{array}
$$

Hence

$$\frac{2x^5 - 7x^4y + x^2y^3 - 3xy^4 + y^5}{2x^2 - xy + y^2} = x^3 - 3x^2y - 2xy^2 + y^3$$

Exercise 2.7C

Perform the indicated operations and simplify:

1. $\dfrac{x^2 + 10x + 16}{x + 2}$
2. $\dfrac{x^2 + 9x + 18}{x + 3}$
3. $\dfrac{x^2 + 9x + 20}{x + 5}$
4. $\dfrac{x^2 + 8x + 12}{x + 6}$
5. $\dfrac{x^2 + 4x + 4}{x + 2}$
6. $\dfrac{x^2 + 6x + 9}{x + 3}$
7. $\dfrac{x^2 + 4x - 12}{x - 2}$
8. $\dfrac{x^2 + x - 12}{x - 3}$
9. $\dfrac{x^2 - 10x + 24}{x - 6}$
10. $\dfrac{x^2 - 12x + 32}{x - 4}$
11. $\dfrac{4x^2 + 18x + 20}{2x + 5}$
12. $\dfrac{20x^2 + 39x + 18}{4x + 3}$
13. $\dfrac{12x^2 + 37x + 21}{3x + 7}$
14. $\dfrac{6x^2 - 17x + 12}{2x - 3}$
15. $\dfrac{9x^2 - 12x + 4}{3x - 2}$
16. $\dfrac{9x^2 - 18x + 8}{3x - 4}$
17. $\dfrac{12x^2 + x - 20}{4x - 5}$
18. $\dfrac{14x^2 - 53x + 14}{7x - 2}$
19. $\dfrac{2x^3 + x^2 - x + 3}{x^2 - x + 1}$
20. $\dfrac{6x^3 + 13x^2 - x - 10}{2x^2 + x - 2}$
21. $\dfrac{12x^3 + x^2 - 22x - 12}{3x^2 - 2x - 4}$
22. $\dfrac{3x^3 + 5x^2 - 37x + 21}{3x - 7}$
23. $\dfrac{18x^2 + 33x + 7}{3x + 5}$
24. $\dfrac{12x^2 + x - 1}{3x - 2}$
25. $\dfrac{6x^2 + x - 4}{3x - 4}$
26. $\dfrac{24x^2 + 73x + 21}{8x + 3}$
27. $\dfrac{9x^2 + 15x + 2}{3x + 4}$
28. $\dfrac{16x^2 - 32x + 9}{4x - 5}$
29. $\dfrac{4x^2 - 16x + 5}{2x - 5}$
30. $\dfrac{4x^2 - 12x + 1}{2x - 3}$
31. $\dfrac{6x^2 - 19x - 40}{2x - 9}$
32. $\dfrac{12x^3 - 21x^2 + 25x - 15}{4x - 3}$
33. $\dfrac{10x^4 - 7x^3 - 17x^2 + 20x - 12}{2x^2 + x - 3}$
34. $\dfrac{x^2 - 25}{x + 5}$
35. $\dfrac{9x^2 - 4}{3x + 2}$
36. $\dfrac{16x^2 - 1}{4x - 1}$
37. $\dfrac{36x^2 - 49}{6x - 7}$
38. $\dfrac{x^3 + 27}{x + 3}$
39. $\dfrac{27x^3 + 1}{3x + 1}$
40. $\dfrac{x^3 - 8}{x - 2}$
41. $\dfrac{64x^3 - 27}{4x - 3}$
42. $\dfrac{2x^4 + 3x^3 - 12x^2 + 1}{x^2 + 3x - 1}$
43. $\dfrac{x^4 - 6x^3 + 9x^2 - 4}{x^2 - 3x + 2}$
44. $\dfrac{2x^4 - 7x^3 + 38x - 48}{x^2 - 4x + 6}$
45. $\dfrac{4x^4 + 5x^2 + 5x - 4}{2x^2 + x - 1}$

46. $\dfrac{x^4 + 3x^2 + 4}{x^2 - x + 2}$

47. $\dfrac{4x^4 - 9x^2 + 4}{2x^2 + x - 2}$

48. $\dfrac{3x^4 - 10x^3 + 20x - 10}{x^2 - 3x + 1}$

49. $\dfrac{9x^4 - 4x^2 + 24x - 30}{3x^2 + 2x - 5}$

50. $\dfrac{2x^5 - x^4 - 6x^2 - 1}{x^2 - x - 1}$

51. $\dfrac{24x^4 - 20x^2 + 25x - 23}{6x^2 - 3x + 4}$

52. $\dfrac{4x^4 - 11x^2 - 40x - 20}{x^2 + 2x + 3}$

53. $\dfrac{2x^5 - 17x^3 + 7x^2 - 10x + 3}{2x^2 + 6x - 1}$

54. $\dfrac{6x^5 + 5x^4 - 12x^2 - 6x - 1}{3x^2 + x + 1}$

55. $\dfrac{6x^5 - 7x^4 - 5x - 2}{2x^2 - x - 2}$

56. $(3x^6 - 10x^4 - 5x^3 + 3x^2 + 24x - 10) \div (3x^2 - 6x + 2)$

57. $(9x^6 - 10x^4 - 5x^3 + 19x^2 + 22x + 10) \div (3x^2 + 4x + 2)$

58. $(4x^5 + 9x^4y - 5x^3y^2 - 14x^2y^3 - 11xy^4 + 7y^5) \div (x^2 + 2xy - y^2)$

59. $(7x^5 - 30x^4y + 30x^3y^2 - 11x^2y^3 + 7xy^4 - 3y^5) \div (x^2 - 4xy + 3y^2)$

60. $(10x^5 - 3x^4y + x^3y^2 + x^2y^3 - 9xy^4 + 4y^5) \div (2x^2 + xy - y^2)$

61. $(6x^5 - 19x^4y + 13x^3y^2 + 19x^2y^3 - 17xy^4 + 2y^5) \div (3x^2 + xy - 2y^2)$

62. $(x^5 - 6x^3y^2 + 13x^2y^3 - 10xy^4 + 2y^5) \div (x^2 + 3xy - y^2)$

63. $(2x^5 + x^3y^2 + 14x^2y^3 - 21xy^4 + 4y^5) \div (2x^2 + 4xy - y^2)$

64. $(3x^5 - 16x^3y^2 - 10x^2y^3 + 3xy^4 - 4y^5) \div (x^2 - xy - 4y^2)$

65. $(8x^5 - 4x^3y^2 - x^2y^3 - xy^4 + 2y^5) \div (2x^2 + xy + y^2)$

66. $\dfrac{x^3 + y^3}{x + y}$

67. $\dfrac{8x^3 + y^3}{2x + y}$

68. $\dfrac{x^3 - 8y^3}{x - 2y}$

69. $\dfrac{x^6 - 64y^6}{x^2 - 4y^2}$

70. $\dfrac{(x - y)^2 - 4(x - y) + 3}{(x - y) - 3}$

71. $\dfrac{(x + y)^2 - (x + y) - 12}{(x + y) + 3}$

72. $\dfrac{6(x + y)^2 - 17(x + y) + 12}{3(x + y) - 4}$

73. $\dfrac{6(2x + y)^2 - (2x + y) - 15}{3(2x + y) - 5}$

74. $\dfrac{8(x - y)^2 - 34(x - y) + 21}{2(x - y) - 7}$

75. $\dfrac{36(x - 3y)^2 + 7(x - 3y) - 4}{4(x - 3y) - 1}$

76. $\dfrac{x^2 + 4xy + 4y^2 - 25}{x + 2y + 5}$

77. $\dfrac{x^2 + 9y^2 - 6xy - 4}{x - 3y - 2}$

78. $\dfrac{4x^2 + 4xy + y^2 - 9}{2x + y - 3}$

79. $\dfrac{9x^2 - 12xy + 4y^2 - 16}{3x - 2y - 4}$

80. $[x^4 + 3x^3 - x^2 - 5x + 2] \div [x^2 + x - 2] + (x - 3)(x + 1)$

81. $[2x^4 + 3x^3 - 9x^2 + x + 3] \div [x^2 + 2x - 3] + (x + 2)(x - 1)$

82. $[6x^4 + 7x^3 - 2x^2 + 7x - 2] \div [3x^2 - x + 2] - (2x + 1)(x - 2)$

83. $[6x^4 - 11x^3 - x^2 - x - 2] \div [2x^2 - 3x - 2] - (3x + 2)(x - 1)$

Chapter 2 Review

Add the following polynomials:

1. $3x + 2y - 5,\ 4y - 2x + 3,\ x - y + 1$
2. $2x - 3y + z,\ 5x + y - 2z,\ 2y - 3x + 6z$
3. $xy - 2yz + 4z,\ 7yz - 2xy + z,\ 3xy - 2z + 8$
4. $x^2 + 3x - 1,\ 6x - 4x^2 + 2,\ 9x^2 - 4x - 3$
5. $2x^3 + x^2 - x,\ 5x^3 - 3x^2 + 2x,\ 4x^2 - 4x^3 - x$
6. $3x^2y - 5xy^2 - y^3,\ 4x^2y + 6xy^2 + 2y^3,\ x^2y + xy^2 + y^3$

Perform the indicated subtractions:

7. Subtract $9x$ from $8x$.
8. Subtract $2x$ from $-3x$.
9. Subtract $-4x^2$ from x^2.
10. Subtract $-5y$ from $-8y$.
11. Subtract $10x - 3y + 7$ from $6x - y - 8$.
12. Subtract $6xy + 4yz - 2$ from $8xy - yz + 1$.
13. Subtract $5x^3 - 2x^2 + 1$ from $3x^3 + x^2 - 8x$.
14. Subtract $x^2 - 2y + xy$ from $2x^2 + y^2 - 3xy$.
15. Subtract x from x^2.
16. Subtract 2 from $2x$.
17. Subtract -10 from $10x$.
18. Subtract $3x$ from x^3.
19. Subtract y from xy.
20. Subtract $-2x$ from $-2x$.

Remove the grouping symbols and combine like terms:

21. $3[x - 3(x - 1) + 4(x - 2)]$
22. $2x - [4 - 2(x + 2) + 3(2x - 1)]$
23. $7x^2 - 3[x - 4(2x + 1)] + [3 - x(5x - 2)]$
24. $8 - 2\{4x^2 + [x(x - 1) - 2x(3x + 1)]\}$

Evaluate the following expressions, given that $a = 3$, $b = -1$, $c = 2$, and $d = -2$:

25. $ab - bc$
26. $2bc - 3d$
27. $ac - d^2$
28. $a^2 - 2bd$
29. $4b^2 - a^2$
30. $ac(c^2 - bd)$
31. $3cd(d^2 - ab)$
32. $c^2 - 2d(ac - bd)$
33. $a^2 + bc(ad - b^2)$
34. $(a - bc)^2$
35. $(2b - d^2)^2$
36. $(2cd - a^2b)^5$
37. $\dfrac{2ab - c}{d - c}$
38. $\dfrac{3c + 2d}{c - d^2}$
39. $\dfrac{a^2 + b^2}{(a + b)^2}$
40. $\dfrac{a^2}{c^2} + \dfrac{b^2}{d^2}$
41. $\dfrac{2ac}{3b} - \dfrac{ad}{c^2}$
42. $\dfrac{cd}{a^2} + \dfrac{a^2}{bd}$

Perform the indicated operations and simplify, given that $n \in N$:

43. $(2x^2y)(-x^3y^2)$
44. $(3xy^2)(-4x^2y)$
45. $(-2x^4yz)(-3xz^3)$
46. $(-x^2y^2)(2x^3z^2)(-y^3z)$
47. $(-x^3y)(x^4y^3)(-2^2x^2)$
48. $(3x^2y)^3$
49. $(5x^3y^2z)^3$
50. $(-2x^3yz^2)^2$
51. $(-3^2xy^3)^2$
52. $(-x^2y^2z^4)^3$

53. $(-2^2x^4y)^3$
54. $(2xy^2)^3(-3x^2y)^2$
55. $(-xy)^4(-x^2y^2)^3$
56. $(-5x^3y)(3x^2y^3)^2(-xy^4)^3$
57. $(2x^4y)(-7y^2z)^2(-x^2z^4)^3$
58. $6(-2x^2y^3)^3 + x^2y(3xy^2)^4$
59. $4x(-xy^2)^3 + y^4(3x^2y)^2$
60. $(3x^3y^4)^3 - x^5(-2xy^3)^4$
61. $(-x^2)^3(2xy^2z)^6 - (-x^2z)^2(x^2y^3z)^4$
62. $x^3(x^{n+1} + x^n - 3)$
63. $x^2(x^{n+2} - 2x^{n+1} - x^n)$
64. $x^2(x^2 - x + 1) - x^3(x - 2)$
65. $3x^3(x^2 - 2x - 1) + x^2(6x^2 + 3x - 1)$
66. $(x + 5)(x - 6)$
67. $(3x - 2)(2x - 5)$
68. $(3x - 5)^2$
69. $(x^n - 2)(x^n + 4)$
70. $(3x^n - 1)(x^n + 2)$
71. $(2x^n - 3)(4x^n - 1)$
72. $(3x - 2)(x^2 + x - 1)$
73. $(x^2 - 4)(x^2 + x + 4)$
74. $(x + 4)(x^{n+1} - x^n + 1)$
75. $(x - 3)(x^{n+2} + 3x^{n+1} - 4)$
76. $(x^2 - 3x - 2)(x^2 - 3x + 2)$
77. $(x^2 + 2x - 1)(x^2 - 2x + 1)$
78. $(x^2 + 4x + 8)(x^2 - 4x + 8)$
79. $(x^2 + x - 5)(x^2 - x - 5)$
80. $2(3x + 1)(x - 4)$
81. $-x(3x - 2)(x + 3)$
82. $(x - 2)(x + 3)^2$
83. $(3x - 1)(x - 4)^2$
84. $x(x^2 + x - 4) - (x - 3)(x - 1)$
85. $(x + 2)(3x - 1) - (x - 1)(x + 3)$
86. $3(x - 1)(x - 2) - (3x - 1)(x + 1)$
87. $2(x - 4)(x - 2) - 2(x - 3)^2$
88. $4(2x - 1)(3x - 2) - 6(2x - 3)^2$
89. $(4x - 1)^2 - (4x + 1)^2$
90. $\dfrac{a^{n+3}}{a^{n+1}}$
91. $\dfrac{a^n}{a^{n+2}}$
92. $\dfrac{a^{9n}}{a^{3n}}$
93. $\dfrac{x^3y^2z}{x^5yz^3}$
94. $\dfrac{-10xy^3z}{15x^4yz}$
95. $\left(\dfrac{4x^2y^5z^3}{-6x^4yz^6}\right)^4$
96. $\left(\dfrac{-5x^5y^8z^2}{10xy^4z}\right)^3$
97. $\left(\dfrac{36x^7y^4z^6}{-24x^6y^8z^2}\right)^3$
98. $\left(\dfrac{x^2y^8z^6}{-x^4y^6z^8}\right)^3$
99. $\dfrac{(4x^2yz^3)^3}{(6xy^3z^3)^2}$
100. $\dfrac{(15x^4y^5z^2)^3}{(10x^3y^4z)^4}$
101. $\dfrac{(18x^6y^3z^5)^2}{(27x^2y^3z^2)^3}$
102. $\dfrac{(-6xy^3z)^5}{(36x^2y^4z^3)^3}$
103. $\dfrac{x^2 - 12x + 32}{x - 4}$
104. $\dfrac{8x^2 + 10x + 3}{2x + 1}$
105. $\dfrac{6x^2 - 5x - 6}{2x - 3}$
106. $\dfrac{12x^2 - 4x - 1}{2x - 1}$
107. $\dfrac{8x^2 - 30x + 27}{4x - 9}$
108. $\dfrac{6x^2 - 31x + 35}{2x - 7}$
109. $\dfrac{9x^2 - 24x + 16}{3x - 4}$
110. $\dfrac{x^2 - 4}{x - 2}$
111. $\dfrac{8x^3 - 1}{2x - 1}$
112. $\dfrac{x^6 - 27y^6}{x^2 - 3y^2}$
113. $\dfrac{x^4 - 4x^2 + 12x - 9}{x^2 - 2x + 3}$

114. $\dfrac{x^4 - 17x^2 + 20x - 6}{x^2 + 4x - 3}$

115. $\dfrac{6x^4 + 5x^3 + x - 12}{2x^2 + x - 3}$

116. $\dfrac{8x^4 - 6x^3 + 7x^2 - 9}{4x^2 - x - 3}$

117. $\dfrac{2x^5 - 5x^4 + 7x^2 - 3x + 6}{2x^2 - 3x + 1}$

118. $\dfrac{x^5 - 4x^3 + 10x^2 - 15x + 5}{x^2 + 2x - 3}$

119. $\dfrac{x^5 - 4x^4 + 3x^3 - 10x + 5}{x^2 - x + 2}$

120. $\dfrac{2x^5 - 5x^4 + 12x^2 - 17x + 7}{x^2 - 2x + 1}$

chapter 3

Linear Equations in One Variable

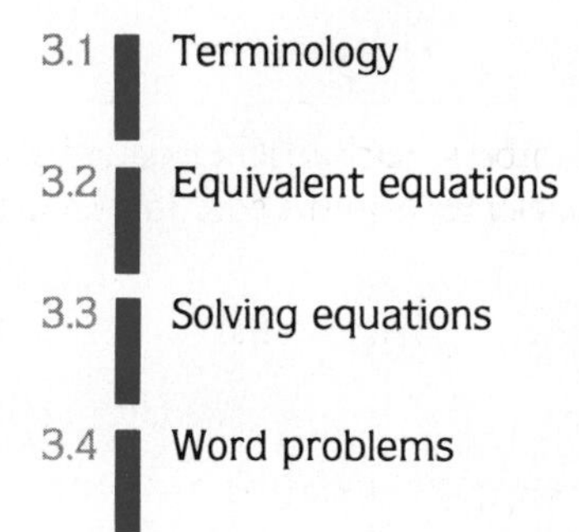

3.1 Terminology

3.2 Equivalent equations

3.3 Solving equations

3.4 Word problems

3.1 Terminology

The following are examples of statements of equality of two algebraic expressions:

1. $3(2x + 7) = 6x + 21$

2. $x^2 - \frac{2}{x} = \frac{x^3 - 2}{x}$

3. $2x + 1 = 9$

4. $3x^2 + 2x = 8$

5. $5x + 5 = 5(x - 1)$

6. $x^2 + 2x = x^2 + 2(x + 5)$

Statements 1 and 2 are true for all permissible values of x. Note that it is not permissible to assign the value 0 to x in statement 2. Such statements are called **identities**.

Statements 3 and 4 are true for some, but not all, values of x.

Statement 3 is true only if x is 4. Statement 4 is true only if x is -2 or $\frac{4}{3}$. Such statements are called **equations**.

Statements 5 and 6 are not true for any value of x and are called **false statements**.

DEFINITION

> The set of all numbers that satisfy an equation is called the equation's **solution set**. The elements in the solution set are called the **roots of the equation**.

To check whether a value of the variable is a root of the equation, substitute that value for the variable in the equation to see if the value of the right side of the equation is equal to the value of the left side of the equation.

DEFINITION

> An equation is said to be **linear** if the variables in the equation appear only with ones as exponents, and if no term of the equation has more than one variable as a factor.

The equation $2x + 3y - z = 4$ is a linear equation in x, y, and z.
The equation $3x^2 + 5x = 2$ is not a linear equation.
The equation $x - yz = 7$ is not a linear equation in x, y, and z.

This chapter deals with linear equations in one variable.

3.2 Equivalent Equations

DEFINITION

> Two equations are said to be **equivalent** if they have the same solution set.

The equations $3x + 2 = 11$ and $x = 3$ are equivalent. The two equations have the same solution set, $\{3\}$.

The solution sets of some equations are obvious by inspection. The solution set of the equation $x + 2 = 7$ is $\{5\}$, since 5 is the only number that when added to 2 equals 7.
The solution set of the equation $x - 4 = 5(x - 2)$ is not so obvious.

In order to solve an equation, that is, to find its solution set, two theorems can be applied to get an equivalent equation whose solution is obvious.

THEOREM 1

> If P, Q, and T are polynomials in the same variable and $P = Q$ is an equation, then $P = Q$ and $P + T = Q + T$ are equivalent.

Theorem 1 states that given an equation $P = Q$, we can add any polynomial T, in the same variable as P and Q, to both sides of the equation, thus obtaining an equivalent equation $P + T = Q + T$.

The two equations $2x - 5 = x + 3$ and $2x - 5 + (5 - x) = x + 3 + (5 - x)$, which simplifies to $x = 8$, are equivalent. Their solution set is $\{8\}$.

THEOREM 2

> If P and Q are polynomials in the same variable, $a \in R$, $a \neq 0$, and if $P = Q$ is an equation, then $P = Q$ and $aP = aQ$ are equivalent.

Theorem 2 states that given an equation $P = Q$, we can multiply both sides of the equation by a real number $a \neq 0$, thus obtaining an equivalent equation $aP = aQ$.

The two equations $x = 10$ and $3(x) = 3(10)$, that is, $3x = 30$, are equivalent. Their solution set is $\{10\}$.

When both sides of an equation are multiplied by a constant different from zero, the resulting equation is equivalent to the original equation.
However, when both sides of an equation are multiplied by an expression involving the variable, the resulting equation may not be equivalent to the original equation.

The two equations $3x = 5$ and $3x(x) = 5(x)$, that is, $3x^2 = 5x$, are not equivalent. The solution set of $3x = 5$ is $\left\{\frac{5}{3}\right\}$, whereas the solution set of $3x^2 = 5x$ is $\left\{0, \frac{5}{3}\right\}$.

The two equations $x = 2$ and $x(x - 1) = 2(x - 1)$ are not equivalent. The solution set of $x = 2$ is $\{2\}$, whereas the solution set of $x(x - 1) = 2(x - 1)$ is $\{1, 2\}$.

Similarly, if we raise both sides of an equation to any power different from zero or one, the resulting equation may not be equivalent to the original equation.

The two equations $x = 4$ and $(x)^2 = (4)^2$, that is, $x^2 = 16$, are not equivalent. The solution set of $x = 4$ is $\{4\}$, whereas the solution set of $x^2 = 16$ is $\{-4, 4\}$.

Note The solution set of a linear equation in one variable has exactly one element.

3.3 Solving Equations

Given a linear equation in one variable, we may use one or both of the previous two theorems to form an equivalent equation of the form $1x = a$, whose solution set is $\{a\}$.

When the coefficient of the variable in the equation is not 1, as in $\frac{b}{c}x = d$, an equivalent equation of the form $1x = a$ can be obtained by multiplying both sides of the equation by the multiplicative inverse (reciprocal) of the coefficient of x in the original equation.

The multiplicative inverse of $\frac{b}{c}$ is $\frac{c}{b}$, since $\frac{b}{c} \cdot \frac{c}{b} = 1$.

Thus when the coefficient of the variable is of the form $\frac{b}{c}$, multiply both sides of the equation by $\frac{c}{b}$.

EXAMPLE Find the solution set of the equation $6x = -9$.

Solution The coefficient of x is 6.

The multiplicative inverse of 6 is $\frac{1}{6}$.

Multiply both sides of the equation by $\frac{1}{6}$:

$$\frac{1}{6}(6x) = \frac{1}{6}(-9)$$

$$x = -\frac{9}{6} = -\frac{3}{2}$$

The solution set is $\left\{-\frac{3}{2}\right\}$.

EXAMPLE

Find the solution set of the equation $-\frac{3}{4}x = \frac{9}{2}$.

Solution

The coefficient of x is $-\frac{3}{4}$.

The multiplicative inverse of $-\frac{3}{4}$ is $-\frac{4}{3}$.

Multiply both sides of the equation by $-\frac{4}{3}$:

$$-\frac{4}{3}\left(-\frac{3}{4}x\right) = -\frac{4}{3}\left(\frac{9}{2}\right)$$

Hence $\quad x = -\frac{4 \times 9}{3 \times 2}$

$$= -6$$

The solution set is $\{-6\}$.

When the equation has more than one term containing the variable as a factor, combine the terms, utilizing the distributive law of multiplication.

EXAMPLE

Find the solution set of the equation $x - 3x + 5x = 10$.

Solution

$$x - 3x + 5x = 10$$
$$(1 - 3 + 5)x = 10$$
$$3x = 10$$

Hence $\quad x = \frac{10}{3}$

The solution set is $\left\{\frac{10}{3}\right\}$.

Sometimes both sides of the equation contain terms that have the variable as a factor and also terms that do not have the variable as a factor. In order to find the solution set of the equation, form an equivalent equation that has all terms with the variable as a factor on one side of the equation. The terms not having the variable as a factor must appear on the other side.

The equivalent equation can be formed by adding the negatives (additive inverses) of the terms to both sides of the equation.

Consider the equation:	$5x - 2 = 3x + 6$
Add $(+2)$ to both sides:	$5x - 2 + 2 = 3x + 6 + 2$
	$5x + 0 = 3x + 8$
	$5x = 3x + 8$
Add $(-3x)$ to both sides:	$5x + (-3x) = 3x + 8 + (-3x)$
	$2x = 8$
	$x = 4$

The solution set is $\{4\}$.

Remark It is important to realize the difference between the two equations

$3x = 15$ and $3 + x = 15$

In $3x = 15$, the 3 is the coefficient of x; thus to solve for x multiply both sides of the equation by $\left(\frac{1}{3}\right)$:

$$\frac{1}{3}(3x) = \frac{1}{3}(15)$$

$$x = 5$$

The solution set is $\{5\}$.

In $3 + x = 15$, the 3 is a term; thus to solve for x add (-3) to both sides of the equation:

$$3 + x + (-3) = 15 + (-3)$$

$$x = 12$$

The solution set is $\{12\}$.

EXAMPLE Solve the equation $5x + 3 - x = 7 - 2x$.

Solution Add $-3 + 2x$ to both sides of the equation:

$$5x + 3 - x - 3 + 2x = 7 - 2x - 3 + 2x$$

$$5x - x + 2x = 7 - 3$$

$$6x = 4$$

$$x = \frac{2}{3}$$

The solution set is $\left\{\frac{2}{3}\right\}$.

When some of the terms of an equation contain fractions, you can form an equivalent equation containing only integers (to facilitate combining like terms). This can be accomplished by multiplying both sides of the equation by the least common multiple of the denominators of the fractions.
Remember, when we multiply both sides of an equation by a number different from zero, an equivalent equation results.

Note The least common multiple can be obtained as follows:

1. Factor the integers into their prime factors and write the factors in the exponent form.
2. Take all the bases each to its highest exponent.

EXAMPLE Find the LCM of 12, 18, 20.

Solution

$$12 = 2 \cdot 2 \cdot 3 = 2^2 \cdot 3$$
$$18 = 2 \cdot 3 \cdot 3 = 2 \cdot 3^2$$
$$20 = 2 \cdot 2 \cdot 5 = 2^2 \cdot 5$$

The bases are 2, 3, and 5.
The highest exponent of 2 is 2, of 3 is 2, and of 5 is 1.
Hence the $\text{LCM} = 2^2 \cdot 3^2 \cdot 5 = 180$.

EXAMPLE Find the solution set of the equation $\frac{2}{9}x - \frac{5}{12}x = \frac{7}{18}$.

Solution We first find the LCM of 9, 12, and 18:

$$9 = 3^2 \qquad 12 = 2^2 \cdot 3 \qquad 18 = 2 \cdot 3^2$$
$$\text{LCM} = 2^2 \cdot 3^2 = 36$$

Multiply both sides of the equation by 36:

$$\frac{36}{1}\left(\frac{2}{9}x - \frac{5}{12}x\right) = \frac{36}{1}\left(\frac{7}{18}\right)$$
$$8x - 15x = 14$$
$$-7x = 14$$
$$x = -2$$

The solution set is $\{-2\}$.

Note When the equation contains mixed numbers, change the mixed numbers to common fractions.

EXAMPLE

Solve the equation $2\frac{1}{3}x - 1\frac{1}{5} = 2\frac{2}{5}x - \frac{2}{3}$.

Solution

First change the mixed numbers to common fractions:

$$\frac{7}{3}x - \frac{6}{5} = \frac{12}{5}x - \frac{2}{3}$$

Multiply both sides of the equation by the least common multiple of the denominators, which is 15:

$$\frac{15}{1}\left(\frac{7}{3}x - \frac{6}{5}\right) = \frac{15}{1}\left(\frac{12}{5}x - \frac{2}{3}\right)$$

$$\frac{15}{1}\left(\frac{7}{3}x\right) + \frac{15}{1}\left(-\frac{6}{5}\right) = \frac{15}{1}\left(\frac{12}{5}x\right) + \frac{15}{1}\left(-\frac{2}{3}\right)$$

$$35x - 18 = 36x - 10$$

Add $(18 - 36x)$ to both sides of the equation:

$$35x - 18 + 18 - 36x = 36x - 10 + 18 - 36x$$

$$35x - 36x = -10 + 18$$

$$-x = 8$$

$$x = -8$$

The solution set is $\{-8\}$.

EXAMPLE

List the elements in the set $\{x \mid 4x - 7x + 3x = 0, x \in R\}$.

Solution

Consider the statement

$$4x - 7x + 3x = 0$$

$$(4 - 7 + 3)x = 0$$

$$0x = 0$$

Since $0x = 0$ is true for any real value of x, we have

$$\{x \mid 4x - 7x + 3x = 0, x \in R\} = \{x \mid x \in R\}$$

EXAMPLE

List the elements in the set $\{x \mid 2x - 3x + x = 5, x \in R\}$.

Solution

Consider the statement

$$2x - 3x + x = 5$$

$$(2 - 3 + 1)x = 5$$

$$0x = 5$$

Since $0x = 5$ is not true for any real value of x, we have

$$\{x \mid 2x - 3x + x = 5, x \in R\} = \varnothing$$

Exercise 3.3A

Find the solution set of the following equations:

1. $4x = 12$
2. $2x = -8$
3. $-7x = 14$
4. $-2x = -12$
5. $6x = 0$
6. $2x = 3$
7. $4x = -7$
8. $-3x = 2$
9. $\frac{2x}{3} = -6$
10. $\frac{3x}{2} = -4$
11. $-\frac{2x}{5} = 2$
12. $-\frac{3x}{7} = 9$
13. $-\frac{5x}{4} = -5$
14. $-\frac{4}{7}x = -8$
15. $\frac{4}{3}x = \frac{2}{9}$
16. $\frac{8}{9}x = -\frac{4}{3}$
17. $-\frac{5}{4}x = \frac{15}{2}$
18. $-\frac{6}{7}x = -\frac{3}{14}$
19. $2x + 7x = 6$
20. $6x - 4x = 3$
21. $4x - 11x = -14$
22. $5x - 12x + x = 0$
23. $3x - 11x + x = 21$
24. $2x - 3x - 7x = 4$
25. $6x - 5x - 7x = -5$
26. $x + 4x - 8x = -3$
27. $4x - 7x - 12x = -20$
28. $7 - 2x = -3$
29. $6 - 5x = -9$
30. $10 - 7x = -4$
31. $18 - 11x = -4$
32. $3x + 4 = x + 6$
33. $9x - 7 = 6x + 2$
34. $4x - 5 = 2x + 3$
35. $3x + 8 = x + 2$
36. $12 - 7x = 2 - 2x$
37. $5x - 8 = 7x - 10$
38. $7x - 4x + 6 = 2x + 7$
39. $2x - 7 + 5x = 5 + 3x$
40. $9x + 3x + 4 = 8x - 8$
41. $6x + 3x + 20 = 24 - 3x$
42. $7 + x - 1 = 21x - 2$
43. $9 + 4x - 4 = 6x - 5$
44. $9x + 11 - 7x = 2 - 4x$
45. $10 + 13x - 4x = 7 - 3x$
46. $4x + 8 - 12x = 7 - 5x$
47. $3x + 4 - 6x = 6 + 5x$
48. $8x - x - 8 = 5 - 6x$
49. $\frac{x}{3} + \frac{3x}{2} = 11$
50. $\frac{2x}{3} + \frac{3x}{4} = 34$
51. $\frac{3x}{4} - \frac{x}{2} = 3$
52. $\frac{2x}{3} - \frac{4x}{9} = 1$
53. $\frac{5x}{6} - \frac{7x}{8} = 0$
54. $\frac{4x}{13} - \frac{7x}{4} = 0$
55. $\frac{5x}{3} - \frac{3x}{2} = \frac{1}{3}$
56. $\frac{7x}{12} - \frac{x}{3} = \frac{1}{4}$
57. $\frac{3x}{4} - \frac{7x}{6} = \frac{5}{3}$
58. $\frac{3x}{5} - \frac{5x}{8} = \frac{1}{20}$
59. $\frac{5x}{4} - \frac{7x}{9} = \frac{17}{18}$
60. $\frac{11x}{6} - \frac{7x}{4} = \frac{1}{3}$
61. $\frac{7x}{9} - \frac{3x}{8} = \frac{29}{12}$
62. $\frac{3x}{4} - x = -\frac{1}{8}$
63. $\frac{7x}{6} + x = -\frac{13}{3}$
64. $\frac{2x}{3} - \frac{5}{6} = \frac{1}{2}$
65. $\frac{x}{2} - \frac{1}{3} = \frac{7}{6}$
66. $\frac{8}{21} - \frac{3x}{7} = \frac{5}{3}$
67. $\frac{2x}{3} - \frac{1}{2} = \frac{x}{6} + \frac{2}{3}$
68. $\frac{7x}{6} - \frac{5}{2} = 2 - \frac{x}{3}$
69. $\frac{3x}{2} - \frac{3}{2} = \frac{9}{4} - x$

70. $\frac{x}{6} - \frac{1}{4} = \frac{1}{2} - \frac{x}{3}$

71. $\frac{x}{9} - \frac{1}{2} = \frac{1}{3} - \frac{x}{6}$

72. $\frac{x}{8} - \frac{1}{6} = \frac{1}{3} - \frac{x}{4}$

73. $\frac{x}{3} + \frac{1}{12} = \frac{1}{4} - \frac{x}{6}$

74. $\frac{x}{2} - \frac{1}{6} = \frac{3x}{8} + \frac{1}{3}$

75. $\frac{x}{7} + \frac{1}{4} = \frac{1}{2} - \frac{5x}{14}$

76. $\frac{2x}{9} - \frac{1}{2} = \frac{x}{4} + \frac{1}{6}$

77. $1\frac{2}{3}x + 1\frac{1}{2} = 1\frac{1}{6}x - 2\frac{1}{2}$

78. $1\frac{1}{3}x - \frac{5}{6} = 2x - 1\frac{1}{6}$

79. $1\frac{1}{2}x + 1\frac{2}{3} = 3x - \frac{1}{3}$

80. $\frac{7}{12}x + 2\frac{1}{2} = 1\frac{1}{3}x + 1\frac{3}{4}$

81. $1\frac{1}{4}x + 2 = \frac{2}{3} - 1\frac{3}{4}x$

82. $\frac{3x}{5} - \frac{1}{5} = 1\frac{1}{3}x - 1\frac{2}{3}$

83. $1\frac{1}{6}x - \frac{2}{9} = \frac{x}{2} - 1\frac{1}{3}$

84. $2\frac{1}{3}x - 1\frac{1}{4} = 2\frac{1}{2} - x$

List the elements in the following sets given that $x \in R$:

85. $\{x \mid 2x + 4x = 3x\}$

86. $\{x \mid 5x - x = 2x\}$

87. $\{x \mid 3x - 7x = x\}$

88. $\{x \mid x - 4x = 5x\}$

89. $\{x \mid 6x - 10x = -4x\}$

90. $\{x \mid 9x + 14x = 23x\}$

91. $\left\{x \,\middle|\, \frac{x}{4} + \frac{x}{12} = \frac{x}{3}\right\}$

92. $\left\{x \,\middle|\, \frac{x}{3} - \frac{4x}{21} = \frac{x}{7}\right\}$

93. $\{x \mid 5x - 14x = 2 - 9x\}$

94. $\{x \mid 10x - 7x = 10 + 3x\}$

95. $\left\{x \,\middle|\, \frac{x}{3} - \frac{x}{8} = \frac{5x}{24} + 1\right\}$

96. $\left\{x \,\middle|\, \frac{2x}{3} - \frac{4x}{9} = \frac{2x}{9} + 3\right\}$

When the equation involves grouping symbols, remove the grouping symbols first, utilizing the distributive law.

EXAMPLE Solve the equation $2(x - 5) - 3(2x - 3) = 3$.

Solution Apply the distributive law to remove the parentheses:

$$2x - 10 - 6x + 9 = 3$$
$$-4x = 4$$
$$x = -1$$

The solution set is $\{-1\}$.

EXAMPLE

Solve the equation $6x(x-3)-(2x-1)(3x+5)=50$.

Solution

Perform the multiplications first:

$$(6x^2-18x)-(6x^2+7x-5)=50$$

It is important to enclose the product in parentheses first, as shown above, and then apply the distributive law, in order to avoid making mistakes in the signs of some of the terms:

$$\begin{aligned} 6x^2-18x-6x^2-7x+5&=50\\ -25x&=45\\ x&=-\frac{9}{5} \end{aligned}$$

The solution set is $\left\{-\frac{9}{5}\right\}$.

The fraction $\frac{a+b}{c}$ is another way of writing $(a+b)\div c$.

When an equation contains fractions of this form, it is advisable to enclose the numerators in parentheses before multiplying by the least common denominator, as illustrated by the following examples.

EXAMPLE

Solve the equation $\frac{2x-1}{5}-\frac{x-8}{6}=\frac{2}{3}$.

Solution

Multiply both sides of the equation by 30:

$$\begin{aligned} \frac{30}{1}\left(\frac{2x-1}{5}-\frac{x-8}{6}\right)&=\frac{30}{1}\left(\frac{2}{3}\right)\\ \frac{30}{1}\cdot\frac{(2x-1)}{5}-\frac{30}{1}\cdot\frac{(x-8)}{6}&=20\\ 6(2x-1)-5(x-8)&=20\\ 12x-6-5x+40&=20\\ 7x&=-14\\ x&=-2 \end{aligned}$$

The solution set is $\{-2\}$.

EXAMPLE

Solve the equation $\frac{2}{3}(6 - x) - \frac{3}{4}(5 - 2x) = \frac{1}{6}(3 - x)$.

Solution

Multiply both sides of the equation by 12:

$$\frac{12}{1}\left[\frac{2}{3}(6 - x) - \frac{3}{4}(5 - 2x)\right] = \frac{12}{1}\left[\frac{1}{6}(3 - x)\right]$$

$$\frac{12}{1}\cdot\frac{2}{3}(6 - x) - \frac{12}{1}\cdot\frac{3}{4}(5 - 2x) = \frac{12}{1}\cdot\frac{1}{6}(3 - x)$$

$$8(6 - x) - 9(5 - 2x) = 2(3 - x)$$

$$48 - 8x - 45 + 18x = 6 - 2x$$

$$12x = 3$$

$$x = \frac{1}{4}$$

The solution set is $\left\{\frac{1}{4}\right\}$.

EXAMPLE

Solve the equation $.085x + .12(40{,}000 - x) = 4275$.

Solution

Change the decimals to common fractions:

$$\frac{85}{1000}x + \frac{12}{100}(40{,}000 - x) = 4275$$

Multiply both members of the equation by 1000:

$$85x + 120(40{,}000 - x) = 4{,}275{,}000$$

$$85x + 4{,}800{,}000 - 120x = 4{,}275{,}000$$

$$-35x = -525{,}000$$

$$x = 15{,}000$$

The solution set is $\{15{,}000\}$.

Exercise 3.3B

Solve the following equations:

1. $2(x + 1) + 5 = 9$

2. $3(x + 4) + 1 = 16$

3. $4(3x + 2) + 7 = 3$

4. $7(2x + 3) + 1 = -6$

5. $3(2x + 1) + 4 = -2$

6. $5(x + 2) + 4 = -1$

7. $6(x - 3) + 10 = 1$

8. $3(x - 4) + 2 = 5$

9. $5(2x - 1) + 6 = 11$

10. $7(x-2)-3=11$

11. $4(3x-2)-4=3$

12. $2(4x-3)-5=13$

13. $10-2(x+3)=9$

14. $4-3(x+5)=1$

15. $8-2(x+6)=-3$

16. $7-4(2x+1)=11$

17. $5-2(2x-1)=1$

18. $9-5(2x-3)=-1$

19. $13-3(3x-4)=7$

20. $11-6(2x-5)=5$

21. $3(x-2)+2(x+3)=3$

22. $4(x+1)+3(x+7)=2x$

23. $2(2x+3)+3(x+2)=5$

24. $5(3x-1)+2(x+6)=-10$

25. $3(4x-3)+2(3x+4)=23$

26. $4(3x-2)+3(2x+1)=7$

27. $2(3x-2)-3(x+4)=2$

28. $5(2x-1)-4(3x+2)=1$

29. $7(x-1)-9(x+2)=1$

30. $3(3x+4)-2(x+5)=-12$

31. $6(2x-3)-5(x-1)=8$

32. $2(4x-1)-7(x-3)=17$

33. $3(2x-5)-4(x-6)=6$

34. $3(3x-4)-2(2x-3)=4$

35. $7(2x-1)-3(4x-5)=0$

36. $6(x-4)-7(x-2)=0$

37. $8(x-3)-3(x+2)=0$

38. $4(x-5)-7(x+1)=0$

39. $3(2x-7)=4+4(3x-1)$

40. $5(2x-3)=1+6(x+2)$

41. $4(x+3)=2-3(x-1)$

42. $3(3x-1)=-4-8(x+2)$

43. $(x+3)(x-2)=x(x-3)$

44. $(x-4)(x+2)=x(x+2)$

45. $(x-3)(x-1)-x(x+2)=7$

46. $(2x+1)(x-3)-2x(x+1)=4$

47. $(2x-1)(x+3)-x(2x-1)=6$

48. $(3x+2)(x+1)-x(3x+1)=0$

49. $(3x+4)(x-2)-3x(x+3)=3$

50. $(2x-3)(2x+1)-x(4x+3)=11$

51. $(2x+1)(3x-2)-(6x-1)(x+4)=-20x$

52. $(3x-1)(x+1)-(3x+2)(x-3)=-4$

53. $(4x+3)(x-2)-(2x-1)(2x+1)=0$

54. $(3x-4)(x-3)-(3x-2)(x+3)=3$

55. $(4x-1)(x+3)-2(2x+3)(x-1)=0$

56. $(3x-2)(2x-1)-3(x-4)(2x+3)=2$

57. $(2x-5)(4x+3)-4(2x+1)(x-3)=6$

58. $(4x-3)(3x+2)-3(2x-3)(2x+3)=7$

59. $(4x-3)(x+2)-(2x+3)^2=-1$

60. $(3x-1)(6x+1)-2(3x-2)^2=19$

61. $(x+2)^2-(x-3)^2=7$

62. $(2x-1)^2-4(x-2)^2=0$

63. $\dfrac{3x+1}{2}+\dfrac{x+3}{4}=3$

64. $\dfrac{5x-2}{6}+\dfrac{2x+1}{3}=\dfrac{1}{2}$

65. $\dfrac{x+1}{2}+\dfrac{x+2}{3}=\dfrac{1}{3}$

66. $\dfrac{2x+1}{3}+\dfrac{x+1}{4}=-\dfrac{1}{3}$

67. $\frac{3x+2}{5} + \frac{x-2}{2} = \frac{1}{2}$

68. $\frac{2x-1}{5} + \frac{x+2}{3} = -1$

69. $\frac{x-3}{2} - \frac{4x-1}{6} = \frac{2}{3}$

70. $\frac{3x+2}{2} - \frac{x-5}{3} = \frac{1}{3}$

71. $\frac{3x+2}{4} - \frac{7x+2}{8} = 1$

72. $\frac{2x-3}{3} - \frac{5x+3}{9} = -1$

73. $\frac{x-2}{2} - \frac{3x+1}{9} = \frac{2}{9}$

74. $\frac{3x+1}{4} - \frac{4x-3}{5} = \frac{1}{10}$

75. $\frac{x+2}{3} - \frac{3x-1}{7} = \frac{1}{7}$

76. $\frac{x-3}{3} - \frac{4x-1}{8} = \frac{1}{8}$

77. $\frac{2x-5}{3} - \frac{3x-4}{4} = 0$

78. $\frac{x-2}{4} - \frac{2x-3}{7} = 0$

79. $\frac{2}{3}(x+1) - \frac{3}{4}(x-1) = 1$

80. $\frac{5}{3}(x-1) - \frac{4}{5}(2x-1) = \frac{2}{3}$

81. $\frac{3}{4}(x-2) - \frac{2}{5}(2x-3) = \frac{1}{5}$

82. $\frac{5}{4}(x-2) - \frac{7}{6}(x-3) = \frac{2}{3}$

83. $\frac{2}{3}(x+3) - \frac{3}{7}(x+4) = 1$

84. $\frac{3}{2}(x+1) - \frac{5}{7}(2x+3) = -\frac{1}{7}$

85. $\frac{4}{5}(x+2) - \frac{5}{6}(x+1) = \frac{1}{2}$

86. $\frac{7}{6}(x-3) - \frac{8}{9}(x-4) = \frac{1}{2}$

87. $.08x + .13(1200 - x) = 131$

88. $.11x + .09(1500 - x) = 151$

89. $.15x + .07(2000 - x) = 188$

90. $.1x + .16(2300 - x) = 314$

91. $.23x - .12(7000 - x) = 210$

92. $.18x - .07(6000 - x) = 205$

93. $.21x - .08(12{,}000 - x) = 345$

94. $.3x - .13(20{,}000 - x) = 1055$

List the elements in the following sets given that $x \in R$:

95. $\{x \mid 3(2x-8) + 4(7-x) = 2x\}$

96. $\{x \mid 7(x-1) - 2(x+3) = 5x\}$

97. $\{x \mid 9(3-x) + 5(2x+1) = x + 32\}$

98. $\{x \mid 5 - 3(x+1) = 2 - 3x\}$

99. $\{x \mid 7 - 2(5x-4) = 10 + 5(1-x)\}$

100. $\{x \mid 8(3x+1) - 3(4x-7) = 29\}$

101. $\{x \mid 2(6x-1) - 3(4x+2) = -8\}$

102. $\{x \mid (x+2)^2 - (x+1)^2 = 2x\}$

103. $\{x \mid (x-2)^2 - (x-4)^2 = 4x\}$

104. $\{x \mid (2x+1)^2 - 4(x-1)^2 = 4(3x-1)\}$

3.4 Word Problems

Word problems are statements that express relationships among numbers. Our goal is to translate the words of the problem into an algebraic equation that we may solve by known means.

Determine the unknown quantity and represent it by a variable. All other unknown quantities must be expressed in terms of the same variable. Then translate from the problem the statements relating to the variable into an algebraic equation. Solve the equation for the unknown and then find the other required quantities. Check your answer in the word problem, not in the equation.

The following are illustrations of certain verbal phrases and questions and their algebraic equivalents.

1. A number increased by 6:

$x + 6$

2. A number decreased by 3:

$x - 3$

3. One number is 8 more than a second number:

First Number	*Second Number*
$x + 8$	x

4. One number is 3 less than a second number:

First Number	*Second Number*
$x - 3$	x

5. The sum of two numbers is 20:

First Number	*Second Number*
x	$20 - x$

6. Three consecutive integers:

First Integer	*Second Integer*	*Third Integer*
x	$x + 1$	$x + 2$

7. Three consecutive odd integers:

First Integer	*Second Integer*	*Third Integer*
x	$x + 2$	$x + 4$

8. Three consecutive even integers:

First Integer	*Second Integer*	*Third Integer*
x	$x + 2$	$x + 4$

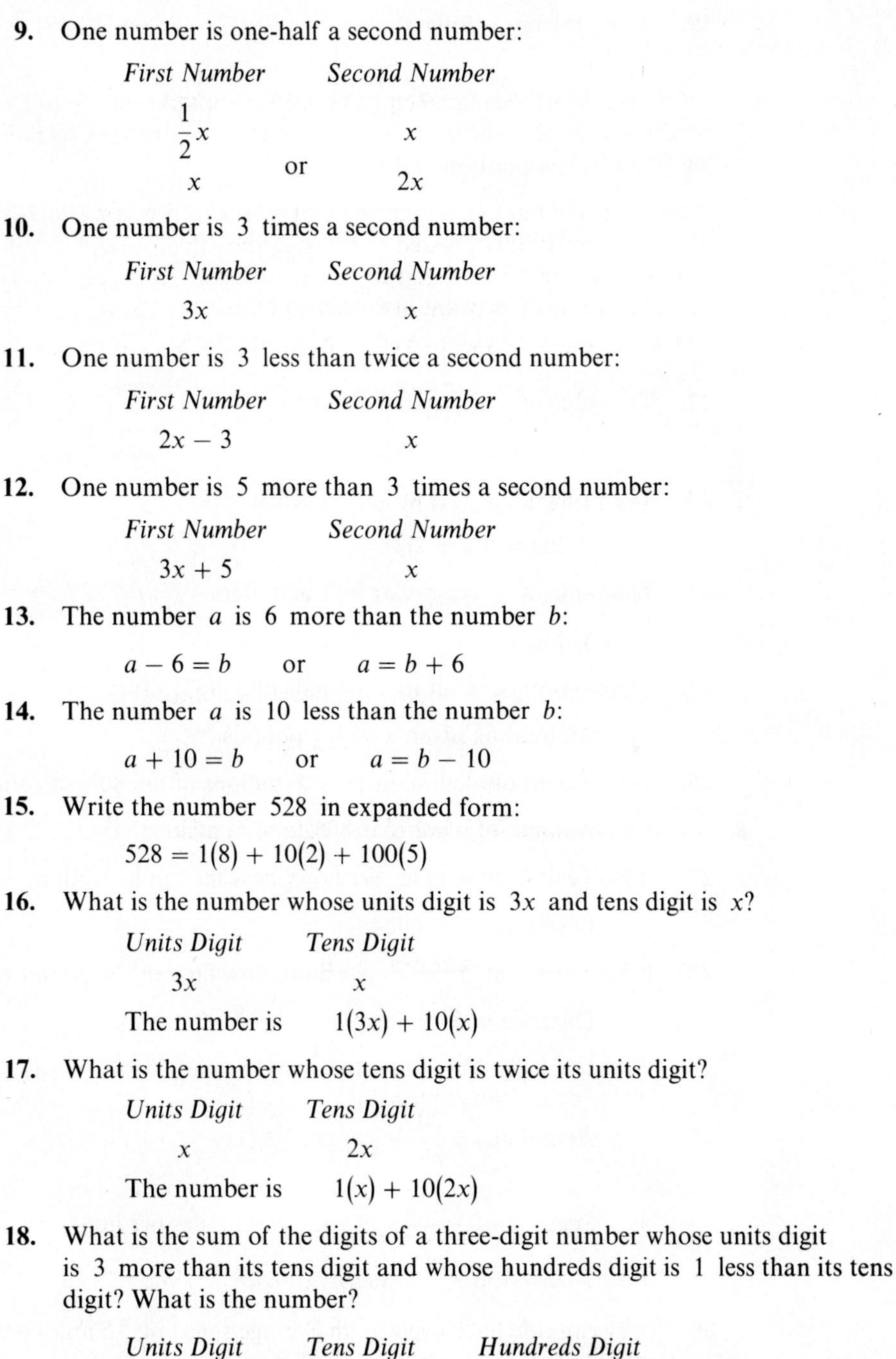

9. One number is one-half a second number:

First Number		*Second Number*
$\frac{1}{2}x$	or	x
x		$2x$

10. One number is 3 times a second number:

First Number	*Second Number*
$3x$	x

11. One number is 3 less than twice a second number:

First Number	*Second Number*
$2x - 3$	x

12. One number is 5 more than 3 times a second number:

First Number	*Second Number*
$3x + 5$	x

13. The number a is 6 more than the number b:

$a - 6 = b$ or $a = b + 6$

14. The number a is 10 less than the number b:

$a + 10 = b$ or $a = b - 10$

15. Write the number 528 in expanded form:

$528 = 1(8) + 10(2) + 100(5)$

16. What is the number whose units digit is $3x$ and tens digit is x?

Units Digit	*Tens Digit*
$3x$	x

The number is $1(3x) + 10(x)$

17. What is the number whose tens digit is twice its units digit?

Units Digit	*Tens Digit*
x	$2x$

The number is $1(x) + 10(2x)$

18. What is the sum of the digits of a three-digit number whose units digit is 3 more than its tens digit and whose hundreds digit is 1 less than its tens digit? What is the number?

Units Digit	*Tens Digit*	*Hundreds Digit*
$x + 3$	x	$x - 1$

The sum of the digits is $(x + 3) + x + (x - 1)$
The number is $1(x + 3) + 10(x) + 100(x - 1)$

19. A 6% tax on x dollars:

$$\text{Tax} = 6\%x = 6 \cdot \frac{1}{100}x \quad \text{or} \quad \frac{6}{100}x$$

20. A 15% discount on x dollars:

$$\text{Discount} = 15\%x = 15 \cdot \frac{1}{100}x \quad \text{or} \quad \frac{15}{100}x$$

21. The value of x twenty-two cent stamps:

Value $= 22(x) = 22x$¢

22. The value of x quarters in cents:

Value $= 25x$¢

23. The value of $(x + 2)$ nickels in cents:

Value $= 5(x + 2)$¢

24. The value of x five-dollar bills in dollars:

Value $= \$5x$

25. The amount of silver in x pounds of a 98% silver alloy:

Amount of silver $= 98\%x$ pounds

26. The amount of alcohol in $(x + 5)$ gallons of an 80% alcohol solution:

Amount of alcohol $= 80\%(x + 5)$ gallons

27. If Bob can walk x miles per hour, how far can he walk in 3 hours?

Distance $= 3x$ miles

28. If Kay drives at 55 miles per hour, how far can she go in t hours?

Distance $= 55t$ miles

29. If it took Jack 20 minutes to drive 15 miles, how fast was he driving?

$$20 \text{ minutes} = \frac{20}{60} = \frac{1}{3} \text{ hour}$$

$$\text{Speed} = \frac{15 \text{ miles}}{\frac{1}{3} \text{ hour}} = 15 \times \frac{3}{1} = 45 \text{ miles per hour}$$

30. Grey can ride his bicycle at an average speed of 15 miles per hour. How long will it take him to travel x miles?

$$\text{Time} = \frac{x \text{ miles}}{15 \text{ miles per hour}} = \frac{x}{15} \text{ hours}$$

31. The width of a rectangle is x feet. What is its perimeter if its length is three times its width?

Width	*Length*
x	$3x$

Perimeter $= 2(x + 3x)$ feet or $2(x) + 2(3x)$ feet

32. The width of a rectangle is x feet. What is the area of the rectangle if its length is 4 feet longer than its width?

Width	*Length*
x feet	$(x + 4)$ feet

Area $= (x)(x + 4)$ square feet

Number Problems

EXAMPLE

The sum of two numbers is 65. Eight times the smaller number is 12 less than 6 times the larger number. Find the two numbers.

Solution

Smaller Number	*Larger Number*
x	$65 - x$

$$8(x) + 12 = 6(65 - x)$$
$$8x + 12 = 390 - 6x$$
$$14x = 378$$
$$x = 27$$

First number $= 27$

Second number $= 65 - 27 = 38$

EXAMPLE

One number is 7 less than another number. Find the two numbers if the difference of their squares is 273.

Solution

First Number	*Second Number*
x	$x - 7$

$$x^2 - (x - 7)^2 = 273$$
$$x^2 - (x^2 - 14x + 49) = 273$$
$$x^2 - x^2 + 14x - 49 = 273$$
$$14x = 322$$
$$x = 23$$

First number $= 23$

Second number $= 23 - 7 = 16$

EXAMPLE Find three consecutive odd integers such that the product of the second and third is 76 more than the product of the first and second.

Solution

First Integer	*Second Integer*	*Third Integer*
x	$x + 2$	$x + 4$

$$(x + 2)(x + 4) - 76 = x(x + 2)$$
$$x^2 + 6x + 8 - 76 = x^2 + 2x$$
$$4x = 68$$
$$x = 17$$

First integer is 17

Second integer is $17 + 2 = 19$

Third integer is $17 + 4 = 21$

EXAMPLE The tens digit of a two-digit number is 5 less than the units digit. If the number is 6 less than 4 times the sum of the digits, find the number.

Solution

Units Digit	*Tens Digit*
x	$x - 5$

The number is $10(x - 5) + 1(x)$

The sum of the digits is $x + (x - 5)$

$$10(x - 5) + 1(x) + 6 = 4[x + (x - 5)]$$
$$10x - 50 + x + 6 = 4(2x - 5)$$
$$11x - 44 = 8x - 20$$
$$3x = 24$$
$$x = 8$$

The units digit is 8

The tens digit is $8 - 5 = 3$

The number is 38

EXAMPLE The sum of the digits of a three-digit number is 16. The units digit is 2 more than the tens digit. If the number is 2 more than 65 times the units digit, find the number.

Solution

Units Digit	*Tens Digit*	*Hundreds Digit*
$x + 2$	x	$16 - [(x + 2) + x]$
$x + 2$	x	$14 - 2x$

$$1(x + 2) + 10(x) + 100(14 - 2x) - 2 = 65(x + 2)$$
$$x + 2 + 10x + 1400 - 200x - 2 = 65x + 130$$
$$-254x = -1270$$
$$x = 5$$

Units digit $= 5 + 2 = 7$

Tens digit $= 5$

Hundreds digit $= 14 - 2(5) = 4$

The number is 457

Exercise 3.4A

1. Six times a number is 24 more than four times the number. Find the number.
2. Three times a number is 72 less than seven times the number. Find the number.
3. Eight times a number decreased by 50 equals the number increased by 41. Find the number.
4. Four times a number decreased by 60 equals the number increased by 21. Find the number.
5. One number is seven times a second number. The sum of the two numbers is 256. Find the two numbers.
6. One number is five times a second number. The sum of the two numbers is 288. Find the two numbers.
7. The sum of two numbers is 30. Three times the smaller number is twice the larger number. Find the two numbers.
8. The sum of two numbers is 84. Four times the smaller number is three times the larger number. Find the two numbers.
9. The sum of two numbers is 74. Four times the smaller number is 26 less than three times the larger number. Find the two numbers.
10. The sum of two numbers is 88. Five times the larger number is 28 less than eight times the smaller number. Find the two numbers.
11. The sum of two numbers is 96. Four times the larger number is 14 more than six times the smaller number. Find the two numbers.
12. The sum of two numbers is 67. Six times the larger number is 25 more than seven times the smaller number.
13. One number is 3 less than another number. Find the two numbers if 4 times the smaller number is 7 more than 3 times the larger number.
14. One number is 8 less than another number. Find the two numbers if 5 times the larger number is 22 less than 7 times the smaller number.
15. One number is 7 more than another number. Find the two numbers if 6 times the larger number is 4 less than 8 times the smaller number.
16. One number is 9 more than another number. Find the two numbers if 11 times the smaller number is 13 less than 9 times the larger number.
17. The sum of three numbers is 79. The second number is 5 more than the first number and the third number is 2 less than twice the first number. Find the numbers.
18. The sum of three numbers is 107. The second number is 7 more than the first number and the third number is 6 less than twice the second number. Find the numbers.
19. Find three consecutive integers such that the sum of the second and third integers is 5 less than 3 times the first integer.

20. Find three consecutive integers such that the sum of the first and second integers is 31 less than 3 times the third integer.
21. Find three consecutive odd integers such that 3 times the sum of the first and second integers is 7 more than 5 times the third integer.
22. Find three consecutive even integers such that 4 times the sum of the second and third integers is 18 less than 9 times the first integer.
23. Find two consecutive integers such that the difference of their squares is 63.
24. The difference of the squares of two consecutive even integers is 60. What are the two integers?
25. The product of two consecutive even integers is 112 more than the square of the first integer. Find the two integers.
26. The product of two consecutive odd integers is 38 less than the square of the second integer. Find the two integers.
27. Find two numbers whose difference is 6, and the difference of their squares is 288.
28. Find two numbers whose sum is 56, and the difference of their squares is 224.
29. Find two numbers whose difference is 5, and whose product is 195 less than the square of the larger number.
30. Find two numbers whose difference is 3, and whose product is 33 more than the square of the smaller number.
31. Find three consecutive odd integers such that the product of the first and third integers is 26 less than the product of the second and third integers.
32. Find three consecutive even integers such that the product of the second and third integers is 44 more than the product of the first and third integers.
33. The units digit of a two-digit number is 2 less than the tens digit. If the number is 2 more than 6 times the sum of the digits, find the number.
34. The units digit of a two-digit number is 3 more than the tens digit. If the number is 7 less than 5 times the sum of the digits, find the number.
35. The units digit of a two-digit number is 4 more than the tens digit. If the number is 2 more than 5 times the units digit, find the number.
36. The units digit of a two-digit number is 3 less than the tens digit. If the number is 11 more than 9 times the tens digit, find the number.
37. The sum of the digits of a three-digit number is 17. The units digit is 2 more than the hundreds digit. If the number is 54 more than 95 times the hundreds digit, find the number.
38. The sum of the digits of a three-digit number is 16. The units digit is 4 more than the tens digit. If the number is 2 less than 29 times the units digit, find the number.
39. The sum of the digits of a three-digit number is 14. The units digit is 1 less than the tens digit. If the number is 5 more than 60 times the tens digit, find the number.
40. The sum of the digits of a three-digit number is 19. The hundreds digit is 2 less than the tens digit. If the number is 1 more than 88 times the tens digit, find the number.
41. In a certain three-digit number the units digit is twice the tens digit, and the sum of the digits is 13. If the units and hundreds digits are interchanged, the number is decreased by 297. Find the original number.

42. In a certain three-digit number the units digit is 3 times the tens digit, and the sum of the digits is 13. If the units and hundreds digits are interchanged, the number is decreased by 594. Find the original number.

Percentage Problems

Sometimes the relation between two numbers is expressed as a percentage. **Per cent** means "per hundred" and is represented by the symbol %. Thus

$$30\% = 30 \div 100 = 30 \times \frac{1}{100} = \frac{30}{100}$$

$$3\frac{1}{2}\% = 3\frac{1}{2} \div 100 = \frac{7}{2} \times \frac{1}{100} = \frac{7}{200}$$

$$500\% = 500 \div 100 = 500 \times \frac{1}{100} = \frac{500}{100}$$

1. To find what per cent one number is of another, divide the first number by the second number, multiply the quotient by 100% and simplify. Note that $100\% = 1$.

EXAMPLE 167 is what per cent of 200?

Solution

$$\frac{167}{200} \times 100\% = \frac{16{,}700}{200}\% = \frac{167}{2}\% = 83\frac{1}{2}\%$$

2. To express a number as a per cent, multiply the number by 100% and simplify.

EXAMPLE Express $\frac{53}{90}$ as a per cent.

Solution

$$\frac{53}{90} = \frac{53}{90}(100\%) = \frac{5300}{90}\% = 58\frac{8}{9}\%$$

3. To find a percentage of any number, change the per cent symbol to $\frac{1}{100}$, then multiply by the number and simplify.

EXAMPLE What is $5\frac{3}{4}\%$ of 200?

Solution

$$5\frac{3}{4}\%(200) = 5\frac{3}{4} \times \frac{1}{100} \times 200 = \frac{23}{4} \times \frac{1}{100} \times 200 = 11.5$$

Most business problems and mixture problems involve percentage.

When you deposit money in a bank, the amount you deposit is called the **principal** and is denoted by P.
The **rate of interest** per year is denoted by $r\%$.
The **interest** you receive is denoted by I.
The interest you receive at the end of a year is the product of the principal and the rate of interest:

$$I = Pr$$

The above formula is helpful in solving problems involving per cent.

EXAMPLE

How much will a vacuum cleaner sell for if the tag price is \$180 and the store gives a 15% discount?

Solution

Discount = $15\%(180) = \$27$
Sales price = $180 - 27 = \$153$

EXAMPLE

A washer cost \$336. What is the selling price if the markup is 20% of the selling price?

Solution

When the markup is calculated on the cost, let the cost be x, but when the markup is calculated on the selling price, let the selling price be x.

Let the selling price = $\$x$

Markup = $20\%x$

Selling price minus the markup gives the cost.

$$x - 20\%x = 336$$

$$x - \frac{20}{100}x = 336 \qquad \text{Multiply by 100}$$

$$100x - 20x = 33{,}600$$

$$x = 420$$

Selling price = \$420

EXAMPLE

The selling price of a self-cleaning range is \$672. What is the cost if markup is 40% on the cost?

Solution

Let the cost be $\$x$. Markup is $40\%x$.
Cost plus markup on cost is equal to the selling price:

$$x + 40\%x = 672$$

$$x + \frac{40}{100}x = 672 \qquad \text{Multiply by 100}$$

$$100x + 40x = 67{,}200$$
$$140x = 67{,}200$$
$$x = 480$$

The cost is \$480

EXAMPLE

A man made two investments totaling \$21,000. On one investment he made an 8% profit, but on the other he took a 3% loss. If his net profit was \$800, how much was in each investment?

Solution

First Investment	*Second Investment*
$\$x$	$\$(21{,}000 - x)$
gain of 8%	loss of 3%

Amount gained minus amount lost is equal to the net profit:

$$8\%x - 3\%(21{,}000 - x) = 800$$
$$\frac{8}{100}x - \frac{3}{100}(21{,}000 - x) = 800 \qquad \text{Multiply by 100}$$
$$8x - 3(21{,}000 - x) = 80{,}000$$
$$8x - 63{,}000 + 3x = 80{,}000$$
$$11x = 143{,}000$$
$$x = 13{,}000$$

First investment = \$13,000

Second investment = \$8000

EXAMPLE

The amount of annual interest earned by \$12,000 is \$270 less than that earned by \$18,000 at .5% less interest per year. What is the rate of interest on each amount of money?

Solution

	First Amount	*Second Amount*
Principal	\$12,000	\$18,000
Rate	$x\%$	$(x - .5)\%$

$$18{,}000(x - .5)\% - 12{,}000(x\%) = 270$$
$$18{,}000(x - .5) \times \frac{1}{100} - 12{,}000(x) \cdot \frac{1}{100} = 270 \qquad \text{Multiply by 100}$$
$$18{,}000(x - .5) - 12{,}000x = 27{,}000$$
$$18{,}000x - 9000 - 12{,}000x = 27{,}000$$
$$6000x = 36{,}000$$
$$x = 6$$

Rates are 6% and 5.5%

EXAMPLE

How many ounces of a 3% iodine solution must be added to 20 ounces of a 12% iodine solution to make an 8% iodine solution?

Solution

x Ounces	*20 Ounces*	*(x + 20) Ounces*
3%	12%	8%

The amount of iodine in the first solution plus the amount of iodine in the second solution must equal the amount of iodine in the final solution:

$$3\%x + 12\%(20) = 8\%(x + 20)$$

$$\frac{3}{100}x + \frac{12}{100}(20) = \frac{8}{100}(x + 20) \qquad \text{Multiply by 100}$$

$$3x + 12(20) = 8(x + 20)$$

$$3x + 240 = 8x + 160$$

$$5x = 80$$

$$x = 16$$

The amount of 3% iodine solution to be added is 16 ounces.

Exercise 3.4B

1. A certain car sold for $6900 four years ago. The same model sells this year for $11,040. What is the percentage increase in the purchase price of the car?
2. The retail price of a refrigerator is $780. If the sale price is $639.60 what is the percentage reduction in the price of the refrigerator?
3. The discount on a washer was $154 based on a 22% rate of discount. What was the regular price of the washer?
4. The discount on a radio was $12 based on a 15% rate of discount. What was the regular price of the radio?
5. How much will a shirt sell for if the tag price is $18 and the store gives a 12% discount?
6. How much will a suit sell for if the tag price is $268 and the store gives an 18% discount?
7. After an 8% increase in salary Petra makes $1777.68 a month. How much money did she earn per month before the increase?
8. Gene bought a car with 6% tax included for $10,464.32 What was the selling price of the car without the tax?
9. A sofa was sold for $1628 after a 12% reduction sale. What was the regular price of the sofa?
10. A desk was sold for $520 after a 35% reduction sale. What was the regular price of the desk?
11. The cost of a color television set is $630. What is the selling price if the markup is 40% of the selling price of the TV?
12. The cost of a dining room set is $2450. What is the selling price if the markup is 30% of the selling price of the set?

13. The selling price of a suit is \$454.40. What is the cost of the suit if the markup is 42% of the cost?
14. The selling price of a rug is \$2035. What is the cost if the markup is 48% of the cost?
15. Gina has \$8000 invested at 8.25%. How much money must she invest at 10.4% so that the interest from both investments makes her income \$1336?
16. Alison has \$5400 invested at 7.6%. How much money must she invest at 11.3% so that the interest from both investments makes her income \$2060.20?
17. Two sums of money totaling \$20,000 earn respectively 8% and 11.5% interest per year. Find the two amounts of money if together they earn \$2037.50.
18. Two sums of money totaling \$35,000 earn respectively 10.8% and 9.7% interest per year. Find the two amounts of money if together they earn \$3670.
19. A man made two investments whose difference is \$14,000. The smaller investment is at 9.6% and the larger at 12.4%. Find the two investments if his total annual income is \$4376.
20. Ann made two investments whose difference is \$8000. The smaller investment is at 7% and the larger investment is at 7.6%. Find the two investments if her total annual income is \$2798.
21. Huff invested part of \$60,000 at 9% and the rest at 12%. If his income from the 9% investment was \$1620 more than that from the 12% investment, how much money was invested at each rate?
22. Bob invested part of \$40,000 at 8.6% and the rest at 11.8%. If his income from the 11.8% investment was \$176 less than that from the 8.6% investment, how much money was invested at each rate?
23. Holly has \$12,000 invested at 7.8%. How much additional money must she invest at 10.6% so that her total annual income will equal 9% of the entire investment?
24. Donna has \$8000 invested at 8.4%. How much additional money must she invest at 13% so that her total annual income will equal 12% of the entire investment?
25. Pat made two investments totaling \$12,000. On one investment she made a 25% profit, but on the other she took a 10% loss. If her net gain was \$375, how much money was in each investment?
26. Marg made two investments totaling \$17,000. On one investment she made 18% profit, but on the other she took a 23% loss. If her net loss was \$712, how much money was in each investment?
27. The amount of annual interest earned by \$6000 is \$140 less than that earned by \$8000 at .5% less interest per year. What is the rate of interest on each amount of money?
28. The amount of annual interest earned by \$4000 is \$146 less than that earned by \$7000 at .4% less interest per year. What is the rate of interest on each amount of money?
29. The amount of annual interest earned by \$9000 is \$30 more than that earned by \$8000 at .75% more interest per year. What is the rate of interest on each amount of money?

30. The amount of annual interest earned by \$24,000 is \$360 more than that earned by \$15,000 at 1.2% more interest per year. What is the rate of interest on each amount of money?
31. How many liters of a 60% acid solution must be added to 14 liters of a 10% acid solution to make a 25% acid solution?
32. How many ounces of alcohol must be added to 36 ounces of a 15% iodine solution to make a 9% iodine solution?
33. Riad mixed 70 grams of a 96% silver alloy with 28 grams of a 40% silver alloy. What is the percentage of silver in the mixture?
34. Ramsis mixed 120 pounds of a 60% aluminum alloy with 180 pounds of an 80% aluminum alloy. What is the percentage of aluminum in the mixture?
35. A chemist mixed 40 milliliters of a 30% iodine solution with 60 milliliters of a second iodine solution. What is the percentage of iodine in the second solution, if the mixture is a 21% iodine solution?
36. A man mixed 80 pounds of a 45% copper alloy with 70 pounds of a second copper alloy. What is the percentage of copper in the second alloy, if the mixture contains 66% copper?
37. John mixed a 48% copper alloy with an 80% copper alloy to make a 60% copper alloy. If there are 20 pounds more of the 48% alloy than the 80% alloy, how many pounds are there in the total mixture?
38. Bill mixed a 48% silver alloy with a 72% silver alloy to make a 56% silver alloy. If there are 30 ounces more of the 48% alloy than the 72% alloy, how many ounces are there in the total mixture?
39. A food processing plant wants to make 5600 liters of catsup that is 30% sugar. If they have one catsup that is 24% sugar and one that is 45% sugar, how much of each kind of catsup should be used?
40. A food processing plant wants to make 3600 liters of jam that is 55% sugar. If they have one jam that is 43% sugar and one that is 73% sugar how much of each kind of jam should be used?

Value Problems

EXAMPLE

Doni has \$4.60 in nickels, dimes, and quarters. If she has 44 coins in all, and there are three fewer quarters than dimes, how many coins of each kind does she have?

Solution

Let the number of dimes be x.

The number of quarters is $x - 3$.

The number of nickels is $44 - (x + x - 3) = 47 - 2x$.

Nickels	*Dimes*	*Quarters*
5¢	10¢	25¢
$47 - 2x$	x	$x - 3$

The value of the nickels is $5(47 - 2x)$ cents

The value of the dimes is $10x$ cents

The value of the quarters is $25(x - 3)$ cents

The sum of the values of the coins is equal to the total amount of money:

$$\begin{aligned} 5(47 - 2x) + 10x + 25(x - 3) &= 460 \\ 235 - 10x + 10x + 25x - 75 &= 460 \\ 25x &= 300 \\ x &= 12 \end{aligned}$$

Hence she has 12 dimes, 9 quarters, and 23 nickels.

EXAMPLE A grocer mixes two kinds of coffee beans, one worth \$3.22 a pound and the other \$1.84 a pound. If the mixture weighs 78 pounds and sells for \$2.30 a pound, how many pounds of each kind does he use?

Solution

	First Kind	*Second Kind*	*Mixture*
Weight:	x pounds at \$3.22	$(78 - x)$ pounds at \$1.84	78 pounds at \$2.30
Value:	$\$3.22x$	$\$1.84(78 - x)$	$\$2.30(78)$

The sum of the values of the individual kinds is equal to the value of the mixture:

$$3.22x + 1.84(78 - x) = 2.30(78) \qquad \text{Multiply by 100}$$

$$\begin{aligned} 322x + 184(78 - x) &= 230(78) \\ 322x + 14{,}352 - 184x &= 17{,}940 \\ 138x &= 3588 \\ x &= 26 \end{aligned}$$

He uses 26 pounds of beans at \$3.22 per pound.

He uses 52 pounds of beans at \$1.84 per pound.

Exercise 3.4C

1. Sarah has \$2.85 in 5¢ and 10¢ coins. If she has 40 coins in all, how many coins of each kind does she have?
2. Janine has \$8.70 in 10¢ and 25¢ coins. If she has 60 coins in all, how many coins of each kind does she have?
3. Audrey has 12 more nickels than quarters. If their value is \$10.20 how many coins of each kind does she have?
4. Marcie has 15 more dimes than nickels. If their value is \$5.70 how many coins of each kind does she have?
5. Don bought \$12 worth of 17¢ and 22¢ stamps. If he bought 60 stamps in all, how many stamps of each kind did he buy?

6. Donna bought $10.10 worth of 17¢ and 22¢ stamps. If she bought 50 stamps in all, how many stamps of each kind did she buy?
7. Sophie bought $8.01 worth of 5¢, 17¢, and 22¢ stamps, for a total of 45 stamps. If she bought twice as many 17¢ stamps as 5¢ stamps, how many stamps of each kind did she buy?
8. Brian bought $13 worth of 5¢, 17¢, and 22¢ stamps, for a total of 71 stamps. If he bought three times as many 22¢ stamps as 17¢ stamps, how many stamps of each kind did he buy?
9. David bought $12.60 worth of 5¢, 17¢, and 22¢ stamps, for a total of 76 stamps. If the 17¢ stamps are twelve more than the 5¢ stamps, how many stamps of each kind did he buy?
10. Mark bought $12.01 worth of 5¢, 17¢, and 22¢ stamps, for a total of 70 stamps. If the 22¢ stamps are five less than three times the 5¢ stamps, how many stamps of each kind did he buy?
11. Nadia has $6 in 5¢, 10¢, and 25¢ coins. If she has 42 coins in all, and there are sixteen more dimes than nickels, how many coins of each kind does she have?
12. Diane has $5.55 in 10¢, 25¢, and 50¢ coins. If she has 24 coins in all, and there are seven more 25¢ coins than twice the 50¢ coins, how many coins of each kind does she have?
13. Debra has $4.40 in 5¢, 10¢, and 25¢ coins. If she has 34 coins in all, and there are eight less 5¢ coins than four times the 10¢ coins, how many coins of each kind does she have?
14. Maria has $14 in 10¢, 25¢, and 50¢ coins. If she has 57 coins in all, and there are six less 25¢ coins than five times the 50¢ coins, how many coins of each kind does she have?
15. A grocer mixes two kinds of nuts, one worth $2.40 per pound and the other $3.60 per pound. If the mixture weighs 120 pounds and is worth $2.80 a pound, how many pounds of each kind of nut does he use?
16. A grocer mixes two kinds of coffee beans, one worth $1.80 a pound and the other $3 a pound. If the mixture weighs 240 pounds and sells for $2.52 a pound, how many pounds of each kind of coffee does he use?
17. How many pounds of a tea worth $3.60 a pound must be mixed with 28 pounds of tea worth $5.25 a pound to make a mixture of tea that sells for $4.65 a pound?
18. How many pounds of candy worth 96¢ a pound must be mixed with 100 pounds of candy worth $1.80 a pound to make a mixture of candy that sells for $1.56 a pound?
19. The receipts from the sale of 48,000 tickets for a football game totaled $348,000. The tickets were sold at $10.50, $7.50, and $4.25, and there were three times as many $7.50 tickets sold as the $10.50 tickets. How many tickets of each kind were sold?
20. The receipts from the sale of 16,000 tickets for a basketball game totaled $88,600. The tickets were sold at $8.75, $5.50, and $3.25, and there were 800 more tickets of the $5.50 price sold than four times the $8.75 tickets. How many tickets of each kind were sold?

Motion Problems

The distance traveled in miles equals the product of the speed, in miles per hour (mph), and the time, in hours. In symbols,

$$d = rt$$

EXAMPLE Two cars, which are 464 miles apart and whose speeds differ by 8 miles per hour, are moving toward each other. They will meet in 4 hours. What is the speed of each car?

Solution

	First Car	*Second Car*
Speed:	x mph	$(x + 8)$ mph
Time:	4 hours	4 hours
Distance:	$4x$ miles	$4(x + 8)$ miles

The sum of the distances traveled equals 464 miles.

$$4x + 4(x + 8) = 464$$
$$4x + 4x + 32 = 464$$
$$8x = 432$$
$$x = 54$$

Speed of the first car is 54 miles per hour.

Speed of the second car is 62 miles per hour.

EXAMPLE A jet plane flying at a speed of 800 mph is to overtake another plane that has a head start of $2\frac{1}{2}$ hours and is flying at a speed of 550 mph. How long will it take the jet to overtake the other plane?

Solution

	Jet	*Plane*
Speed:	800 mph	550 mph
Time:	t hours	$\left(t + 2\frac{1}{2}\right)$ hours

The distance traveled by the jet is equal to the distance traveled by the plane:

$$800t = 500\left(t + 2\frac{1}{2}\right)$$
$$800t = 550t + 1375$$
$$250t = 1375$$
$$t = 5\frac{1}{2}$$

Time required for the jet to overtake the plane is $5\frac{1}{2}$ hours.

Exercise 3.4D

1. Two cars, which are 400 miles apart and whose speeds differ by 4 mph, are moving toward each other. They will meet in 4 hours. What is the speed of each car?
2. Two cars start from the same place and travel in opposite directions. The first car averages 48 mph and the second car 54 mph. In how many hours will the two cars be 357 miles apart?
3. Jim and Jack start at the same spot and walk in opposite directions around a lake whose shoreline is 9 miles long. Jim walks one-tenth of a mile per hour faster than Jack. If they meet in 2 hours, how fast does each walk?
4. A jet plane flying at a speed of 800 mph is to overtake another plane that has a 3-hour head start and is flying at a speed of 600 mph. How far from the starting point will the jet overtake the other plane?
5. A car is being driven at a speed of 55 mph. A motorcycle starts 3.5 hours later at a speed of 66 mph to overtake the car. In how many hours will the motorcycle overtake the car?
6. Norma drove a car out into the country at a speed of 30 mph and back at a speed of 45 mph. Her round trip took $2\frac{1}{2}$ hours. How far did Norma go?
7. Terry drove her car 36 miles, the first 18 miles at an average speed of 54 mph and the second 18 miles at an average speed of 36 mph. What was her average speed for the total distance traveled?
8. Liz drove her car for 4 hours at a certain speed. When fog set in, it forced her to reduce her speed by 40 miles per hour for the remainder of the trip. If the total distance traveled was 260 miles and it took Liz 5 hours to make the trip, how long would it have taken her to drive the distance if the weather had remained clear?
9. Rick went to the city, 32 miles away, on a bus and rode his bicycle back home. The bus traveled three times as fast as the bicycle, and the round trip took 2 hours and 40 minutes. How fast was Rick riding his bicycle?
10. Jeff had a dinner engagement 72 miles away. He drove his car at an average speed of 30 mph in town and 60 mph on the freeway. If the trip took 1.5 hours, how far did Jeff drive in town?
11. An officer of an aircraft carrier missed his ship. He passed over the dock in a plane 20 minutes after the ship left. The carrier averages 40 mph while the plane averages 240 mph. If it took 6 minutes for the plane to land after it overtook the carrier, how far was the carrier from the dock when the officer stepped out of the plane?
12. A mechanized army division moved in a column at 24 mph. A messenger rode from the rear of the column to the front of the column, and then returned to the rear in 18 minutes. If the messenger traveled at a speed of 36 mph, find the length of the column.

Temperature Problems

Three scales used to measure temperature are the Fahrenheit scale, the Celsius (centigrade) scale, and the Kelvin scale.

On the **Fahrenheit scale**, the freezing point of water is calibrated as 32, and the boiling point of water is calibrated as 212. There are 180 equal divisions between the freezing temperature of water and its boiling point. Each division is called a degree Fahrenheit, denoted by F.

On the **Celsius scale**, the freezing point of water is calibrated as 0, and the boiling point of water is calibrated as 100. There are 100 equal divisions between the freezing point of water and its boiling point. Each division is called a degree Celsius, denoted by C.

The **Kelvin scale** has the same divisions as the Celsius scale except that the freezing point of water on the Kelvin scale is 273. Each division is denoted by K.

Note that 180 divisions of the Fahrenheit scale are equivalent to 100 divisions of the Celsius scale. Each degree Fahrenheit equals $\frac{5}{9}$ of a degree Celsius. Each degree Celsius equals $\frac{9}{5}$ of a degree Fahrenheit.

Since a reading of 32 on the Fahrenheit scale is the same as a 0 reading on the Celsius scale, we have the following:

1. Given the Fahrenheit reading, subtract 32 from it and multiply by $\frac{5}{9}$ to get the Celsius reading:

$$C = \frac{5}{9}(F - 32)$$

2. Multiply the Celsius reading by $\frac{9}{5}$ and add 32 to get the Fahrenheit reading:

$$F = \frac{9}{5}C + 32$$

3. Add 273 to the Celsius reading to get the Kelvin reading:

$$K = C + 273$$

EXAMPLE

The normal human body temperature is 37 C. What is it on the Fahrenheit scale?

Solution

$$F = \frac{9}{5}C + 32 = \frac{9}{5}(37) + 32$$

$$= 66.6 + 32 = 98.6$$

EXAMPLE

Normal room temperature is 70 F. What is it on the Celsius scale?

Solution

$$C = \frac{5}{9}(F - 32) = \frac{5}{9}(70 - 32)$$

$$= \frac{5}{9}(38)$$

$$= 21.11$$

EXAMPLE

Find the corresponding Kelvin temperature reading of an 86 F reading.

Solution

In order to find the Kelvin reading, we first find the Celsius reading:

$$C = \frac{5}{9}(86 - 32) = \frac{5}{9}(54)$$

$$= 30$$

$$K = C + 273 = 30 + 273$$

$$= 303$$

Exercise 3.4E

Change the following Fahrenheit readings to corresponding Celsius readings and Kelvin readings:

1. 68 F
2. 95 F
3. 149 F
4. 239 F
5. −22 F
6. 5 F

Change the following Celsius readings to corresponding Fahrenheit readings:

7. 30 C
8. 41 C
9. −50 C
10. −273 C
11. 125 C
12. 600 C

Change the following Kelvin readings to corresponding Fahrenheit readings:

13. 288 K
14. 508 K
15. 700 K
16. 100 K

Lever Problems

When a uniform bar with negligible weight is balanced on a support, with weights on both sides of the support, it is called a **lever** (see Figure 3.1). The support, called the **fulcrum**, is denoted by F. The weights are denoted by $W_1, W_2, W_3, \ldots,$ (the first weight, second weight, and so on).

The distance from a weight to the fulcrum is called the **arm of the weight**. The length of the arm of the first weight is denoted by L_1, the length of the arm of the second weight by L_2, and so on.

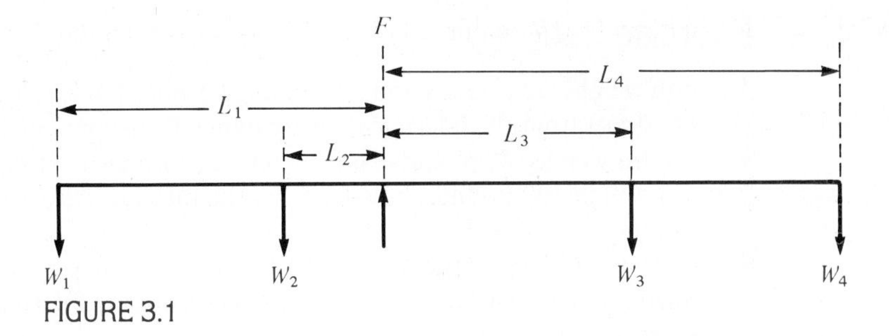

FIGURE 3.1

In order for the lever to balance, the sum of the products of the weights on one side of the fulcrum and their arms must equal the sum of the products of the weights on the other side of the fulcrum and their arms:

$$W_1L_1 + W_2L_2 = W_3L_3 + W_4L_4$$

Some machines for which the lever law applies are the teeterboard, the nutcracker, the scissors, and the balance.

EXAMPLE

A bar with negligible weight is in balance when a 120-pound weight is placed on one side 8 feet from the fulcrum, and 40- and 160-pound weights are placed 1 foot apart on the other side of the fulcrum, with the 40-pound weight closer to the fulcrum. How far from the fulcrum is the 40-pound weight?

Solution

Let the 40-pound weight be x feet from the fulcrum.

The 160-pound weight will be $(x + 1)$ feet from the fulcrum. See Figure 3.2.

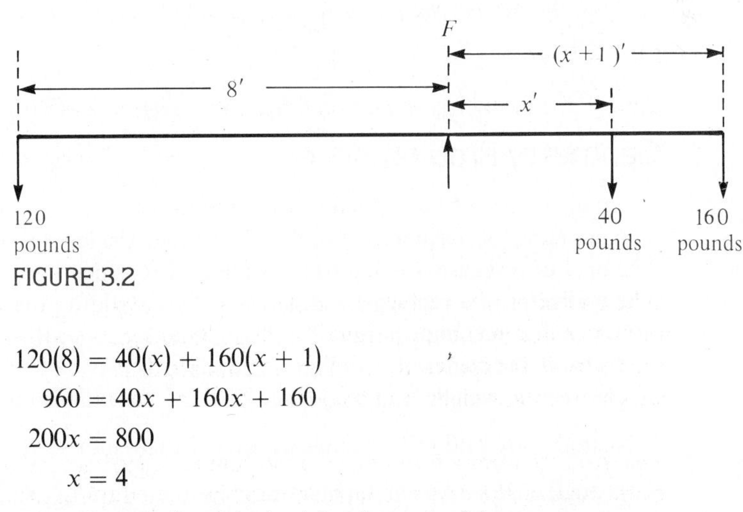

FIGURE 3.2

$$\begin{aligned} 120(8) &= 40(x) + 160(x + 1) \\ 960 &= 40x + 160x + 160 \\ 200x &= 800 \\ x &= 4 \end{aligned}$$

The 40-pound weight is 4 feet from the fulcrum.

Exercise 3.4F

1. Bob weighs 72 pounds and sits on a teeterboard 6 feet from the fulcrum. If Bill sits 8 feet from the fulcrum to just balance Bob, how much does Bill weigh?
2. Sandra weighs 48 pounds and sits on a teeterboard 4 feet from the fulcrum. If Ann weighs 32 pounds, how far from the fulcrum must Ann sit to balance Sandra?
3. Kristen and Doug together weigh 216 pounds. They balance a teeterboard when Kristen is 4 feet and Doug is 5 feet from the fulcrum. Find their weights.
4. A teeterboard 11 feet long is balanced by Greg, who weighs 180 pounds and Jeff who weighs 150 pounds. Find the distance of each man from the fulcrum.
5. A bar with negligible weight is in balance when a 160-pound weight is placed 9 feet from the fulcrum on one side, and 80-and 120-pound weights are placed 2 feet apart on the other side of the fulcrum, with the 80-pound weight closer to the fulcrum. How far from the fulcrum is the 80-pound weight?
6. A bar with negligible weight is balanced when an 88-pound weight is placed 8 feet from the fulcrum on one side, and 32- and 96-pound weights are placed 2 feet apart on the other side of the fulcrum, with the 32-pound weight closer to the fulcrum. How far from the fulcrum is the 32-pound weight?
7. A bar with negligible weight is in balance when a 148-pound weight is placed 10 feet from the fulcrum on one side, and two weights that differ by 20 pounds are placed on the other side, with the lighter weight 6 feet from the fulcrum, and the heavier weight 10 feet from the fulcrum. How much does each weight weigh?
8. A bar with negligible weight is in balance when a 240-pound weight is placed on one side 8 feet from the fulcrum, and two weights that differ by 80 pounds are placed on the other side, so that the lighter weight is 6 feet from the fulcrum, and the heavier weight is 4 feet from the fulcrum. How much does each weight weigh?

Geometry Problems

From geometry we have the following relations:
The **perimeter of a square** is equal to four times the length of its side.
The **area of a square** is equal to the square of its side.
The **perimeter of a rectangle** is equal to twice its width plus twice its length.
The **area of a rectangle** is equal to the product of its width and its length.
The **sum of the angles of a triangle** is equal to 180°.
The **area of a triangle** is equal to one-half the base times the height.

Two angles are said to be **complementary** if their sum is 90°.

Two angles are said to be **supplementary** if their sum is 180°.

EXAMPLE

The length of a rectangle is 4 feet less than twice its width. The perimeter of the rectangle is 100 feet. Find the area of the rectangle.

Solution

Width	*Length*
x feet	$(2x - 4)$ feet

$$2(x) + 2(2x - 4) = 100$$
$$2x + 4x - 8 = 100$$
$$6x = 108$$
$$x = 18$$

The width of the rectangle = 18 feet

The length of the rectangle = $2(18) - 4 = 32$ feet

The area of the rectangle = $18(32) = 576$ square feet

EXAMPLE

The length of a painting is 6 inches less than three times its width. If the frame is 3 inches wide and its area is 432 square inches, find the dimensions of the painting alone.

Solution

	Width of Painting	*Length of Painting*	*Area of Painting*
Without frame:	x inches	$(3x - 6)$ inches	$x(3x - 6)$ square inches
Including frame:	$(x + 6)$ inches	$3x$ inches	$3x(x + 6)$ square inches

The area of the painting including the frame minus the area of the painting without the frame is equal to the area of the frame:

$$3x(x + 6) - x(3x - 6) = 432$$
$$3x^2 + 18x - 3x^2 + 6x = 432$$
$$24x = 432$$
$$x = 18$$

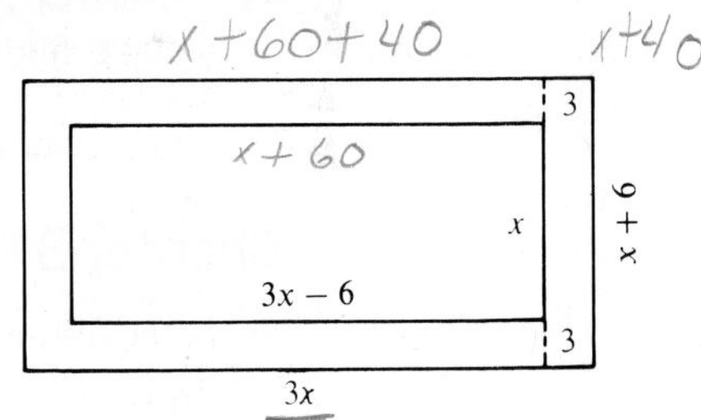

Width of painting is 18 inches.

Length of painting is 48 inches.

Exercise 3.4G

1. The length of a rectangle is 4 inches more than its width. The perimeter of the rectangle is 36 inches. Find the dimensions of the rectangle.
2. The length of a rectangle is 5 inches more than its width. The perimeter of the rectangle is 62 inches. Find the dimensions of the rectangle.
3. The length of a rectangle is 20 feet less than twice its width. The perimeter of the rectangle is 152 feet. Find the area of the rectangle.

4. The length of a rectangle is 16 feet less than three times its width. The perimeter of the rectangle is 128 feet. Find the area of the rectangle.
5. The length of a rectangle is 6 feet more than twice its width. The perimeter of the rectangle is 84 feet. Find the area of the rectangle.
6. The length of a rectangle is 9 feet more than three times its width. The perimeter of the rectangle is 106 feet. Find the area of the rectangle.
7. If two opposite sides of a square are each increased by 3 inches, and the other two sides are each decreased by 4 inches, the area is decreased by 23 square inches. Find the side of the square.
8. If two opposite sides of a square are each increased by 6 inches, and the other two sides are each decreased by 3 inches, the area is increased by 33 square inches. Find the side of the square.
9. The length of a picture without its border is 5 inches less than twice its width. If the border is 1 inch wide and its area is 66 square inches, what are the dimensions of the picture alone?
10. The length of a building is 40 feet less than twice its width. The sidewalk around the building is 10 feet wide and its area is 4400 square feet. What are the dimensions of the building?
11. One angle of a triangle is 30° less than three times a second angle. The third angle is 10° more than the second angle. How many degrees are in each angle?
12. One angle of a triangle is 5° more than twice a second angle. The third angle is 5° less than three times the second angle. How many degrees are in each angle?
13. One of two complementary angles is 12° less than twice the other angle. Find the two angles.
14. If one of two supplementary angles is 8° more than three times the other angle, find the two angles.

Chapter 3 Review

Solve the following equations:

1. $3x + 6 - x = 4x - 3$
2. $5x + 7 + x = 3x - 2$
3. $2x - 9 - x = 1 - 2x$
4. $x - 10 - 6x = x + 5$
5. $6x - 2 = 11 - x + 8$
6. $10x - 5 = 32 + 3x - 9$
7. $15 - 10x = 8 - 14x + 7$
8. $6 - 2x = 4 - 3x + 12$
9. $\frac{x}{2} + \frac{3}{4} = \frac{2}{3} + \frac{5x}{6}$
10. $\frac{x}{4} + \frac{5}{3} = \frac{7}{2} - \frac{2x}{3}$
11. $\frac{7x}{6} - \frac{5}{8} = \frac{2x}{3} - \frac{7}{4}$
12. $\frac{4x}{9} - 1 = \frac{x}{6} - \frac{1}{6}$
13. $\frac{x}{3} + 2 = \frac{7x}{6} + \frac{1}{3}$
14. $\frac{2x}{3} - 2 = x - \frac{7}{3}$

15. $\frac{x}{4} - \frac{1}{2} = \frac{x}{3} + \frac{1}{6}$

16. $\frac{3x}{7} - \frac{1}{4} = \frac{x}{2} + \frac{2}{7}$

17. $5(x - 4) + 2(x + 3) = 0$

18. $2(x + 3) + 7(x - 3) = 0$

19. $3(x + 4) + 2(x - 5) = 7$

20. $4(2x - 1) + 3(x - 4) = 6$

21. $3(x - 5) - (x - 8) = 3$

22. $2(x + 6) - 3(x + 4) = 1$

23. $2(3x - 1) - 5(2x + 3) = 3$

24. $4(3x + 2) - 7(2x + 1) = 11$

25. $8(x + 2) - 7(3x - 2) = 4$

26. $6(x - 1) - 4(x - 4) = 7$

27. $3 + 2(5x - 1) = 4(3x - 1)$

28. $9 + 6(3x - 7) = 7(2x - 5)$

29. $7 - 4(x + 3) = 5(x - 1)$

30. $11 - 8(x - 2) = 3(9 - x)$

31. $(x + 2)(x - 1) - x(x + 3) = 2$

32. $(x - 3)(x + 1) - x(x + 2) = 1$

33. $x(2x - 1) - (x + 1)(2x + 3) = 3$

34. $3x(x - 4) - (x - 3)(3x - 2) = 1$

35. $2x(3x - 4) - (2x - 1)(3x + 1) = 8$

36. $(2x + 3)^2 - (x + 5)(4x - 3) = 4$

37. $(2x - 1)^2 - 4(x - 1)(x - 2) = 2$

38. $(3x + 2)^2 - 9(x + 1)(x - 1) = 1$

39. $(4x + 5)(x - 2) - (2x - 3)^2 = 5$

40. $(2x - 3)(3x - 1) - 6(x - 1)^2 = 0$

41. $\frac{x - 7}{5} + \frac{8 - x}{4} = 1$

42. $\frac{3x - 2}{6} + \frac{3 - x}{3} = 1$

43. $\frac{2x - 5}{2} + \frac{4 - x}{3} = \frac{1}{3}$

44. $\frac{5x - 3}{9} - \frac{x - 1}{2} = \frac{1}{3}$

45. $\frac{x + 2}{3} - \frac{x - 1}{4} = 1$

46. $\frac{4x - 1}{5} - \frac{2x - 3}{3} = \frac{3}{5}$

47. $\frac{3}{4}(x + 5) + \frac{5}{2}(2x - 1) = -\frac{9}{2}$

48. $\frac{3}{5}(x + 2) + \frac{2}{3}(x - 4) = -\frac{1}{5}$

49. $\frac{2}{3}(x - 1) + \frac{5}{6}(x - 2) = \frac{13}{6}$

50. $\frac{7}{8}(x - 1) - \frac{2}{3}(x - 2) + \frac{1}{6} = 0$

51. $\frac{2}{9}(2x - 1) - \frac{1}{4}(x - 1) + \frac{1}{6} = 0$

52. $\frac{5}{7}(x + 2) - \frac{3}{4}(x + 3) + \frac{1}{2} = 0$

53. Given $y = x - 1$, solve $2x - 3y = 2$ for x.

54. Given $y = 2x - 3$, solve $3x - 4y = 7$ for x.

55. Given $y = x + 2$, solve $\frac{3}{4}x - \frac{5}{6}y = -2$ for x.

56. Given $y = 2x - 1$, solve $\frac{2}{3}x - \frac{1}{4}y = 1$ for x.

57. Given $y = 2x - 5$, solve $2x^2 - xy + 5 = 0$ for x.

58. Given $y = 3x - 2$, solve $3x^2 - xy + 4 = 0$ for x.

59. The sum of two numbers is 86. Five times the smaller number is 14 more than three times the larger number. Find the two numbers.

60. The sum of two numbers is 75. Eight times the smaller number is 11 less than five times the larger number. Find the two numbers.

61. The sum of three numbers is 74. The second number is 4 more than the first number and the third number is 6 less than twice the first number. Find the numbers.
62. Find three consecutive even integers such that three times the sum of the second and third integers is 6 less than seven times the first integer.
63. Find two numbers whose difference is 6, and the difference of their squares is 4 less than 11 times the larger number.
64. The sum of the digits of a three-digit number is 18. The units digit is 1 more than the tens digit. If the number is 6 less than 48 times the units digit, find the number.
65. The sum of the digits of a three-digit number is 14. The tens digit is 3 more than the hundreds digit. If the number is 5 more than 16 times the units digit, find the number.
66. The cost of a tape deck is \$840. What is the selling price if the markup is 40% of the selling price of the tape deck?
67. The selling price of a dryer is \$325. What is the cost of the dryer if the markup is 30% of the cost?
68. The owner of a business increased his sales by \$4800 a month when he rented a computer. Assuming that he made a 60% profit on his additional cost, what was the cost of renting the computer per month?
69. A field yielded crops valued at \$2000 an acre. By adding fertilizer at a cost of \$400 an acre, the farmer increased the yield by 28%. What percent profit did this give the farmer on his additional cost?
70. A grocer bought 600 pounds of oranges. He marked up the price 60% and sold 520 pounds. The other 80 pounds rotted. His profit was \$46.40 Find the cost and the selling price per pound of oranges.
X 71. Carol has \$15,000 invested at 5.52%. How much additional money must she invest at 7.2% so that her total annual income will equal 6.5% of the entire investment?
72. The amount of annual interest earned by \$14,000 is \$40 more than that earned by \$12,000 at .75% more interest per year. What is the rate of interest on each amount of money?
73. Brenda has \$5.90 in 5¢, 10¢, and 25¢ coins. If she has 44 coins in all and there are 6 more dimes than nickels, how many coins of each kind does she have?
74. Rae bought \$10.38 worth of 5¢, 17¢, and 22¢ stamps for a total of 60 stamps. If the 22¢ stamps are eight more than twice the 5¢ stamps, how many stamps of each kind did she buy?
75. A butcher mixed two kinds of ground meat, one worth \$1.29 a pound and the other 69¢ a pound. If the mixture weighed 600 pounds and sold for 93¢ a pound, how many pounds of each kind of ground meat made up the mixture?
76. A movie house sold 1200 tickets for a total of \$2940. If the tickets were sold for \$2 and \$3.50 how many tickets of each kind were sold?
77. A man drove a car 57 miles. He drove at an average speed of 36 mph in town and 54 mph on the freeway. If the trip took 1 hour and 10 minutes, how far did he drive in town?

78. A bar with negligible weight is in balance when a 185-pound weight is placed on one side 12 feet from the fulcrum, and two weights that differ by 60 pounds are placed on the other side such that the heavier weight is 9 feet from the fulcrum, and the lighter weight is 5 feet from the fulcrum. How much does each weight weigh?

79. The length of a building is 60 feet more than its width. The sidewalk around the building is 20 feet wide and its area is 15,200 square feet. What are the dimensions of the building?

80. In a biological study of cell growth, it has been found that the velocity (speed) of growth of cells is given by $v = \dfrac{kn}{t}$, when n is the number of generations formed and t is the time in minutes. If for a certain type of cell, it was found that the velocity of growth was 6 when 200 generations formed in 25 minutes, find the velocity of cell growth, when 600 generations formed in 40 minutes.

81. From Problem 80, determine the number of generations formed in 1 hour, when the velocity of growth is 5.

82. From Problem 80, find the time required for 1000 generations to form if the velocity of growth is 10.

83. Economic studies show that the change in the quantity of certain commodities that buyers will purchase is given by the relation $q_1 - q_2 = -k(p_1 - p_2)$ where $p_1 - p_2$ is the change in price. It was found that when the price of meat was \$1.50 a pound, buyers purchased 10,000 pounds, but they bought only 8000 pounds when the price was \$1.60 a pound. Under these conditions, how many pounds of meat would be purchased if the price were \$1.54 a pound?

84. Under the conditions in Problem 83, how many pounds of meat would be purchased if the price were \$1.39 a pound?

85. Under the conditions in Problem 83, what is the price of meat per pound if 15,000 pounds were sold?

chapter 4

Linear Inequalities and Absolute Values in One Variable

4.1 Definitions and notation

4.2 Properties of the order relations

4.3 Solution of linear inequalities in one variable

4.4 Solution of systems of linear inequalities in one variable

4.5 Absolute values

4.6 Solution of linear equations involving absolute values

4.7 Solution of linear inequalities involving absolute values

4.1 Definitions and Notation

When the real numbers a and b are represented by points on a number line, one of the following relationships holds:

1. When the graph of a lies to the right of the graph of b, then a is **greater than** b, denoted by $a > b$.
2. When a and b represent the same point, then a **equals** b, denoted by $a = b$.
3. When the graph of a lies to the left of the graph of b, then a is **less than** b, denoted by $a < b$.

Note that the statements $a > b$ and $b < a$ are equivalent.

The statement $10 > 4$ means that, if we subtract 4 from 10 we get a positive number, $10 - 4 = +6$.

The statement $-3 > -8$ means that if we subtract -8 from -3 we get a positive number, $(-3) - (-8) = -3 + 8 = +5$.

DEFINITION

For $a, b \in R$, $a > b$ means that $a - b$ is a positive number.

If $a - b$ is a positive number we can write

$$a - b = k \qquad \text{where} \qquad a, b, k \in R, k > 0$$

Since $a - b = k$ and $a = b + k$ are equivalent statements,

$$a > b \qquad \text{means} \qquad a = b + k \qquad \text{where} \qquad k > 0$$

If a **is greater than or equal to** b, the notation $a \geq b$ is used.
Thus $a \geq b$ means either $a > b$ or $a = b$.
If a **is less than or equal to** b, the notation $a \leq b$ is used.
Thus $a \leq b$ means either $a < b$ or $a = b$.

DEFINITION

The relations $>, <, \geq, \leq$ are called **order relations.**

To represent $\{x \mid x < 2, x \in R\}$ graphically, draw a ray from the point whose coordinate is 2 in the negative direction. Place a hollow circle at the point to denote that the point is not included on the ray (Figure 4.1).

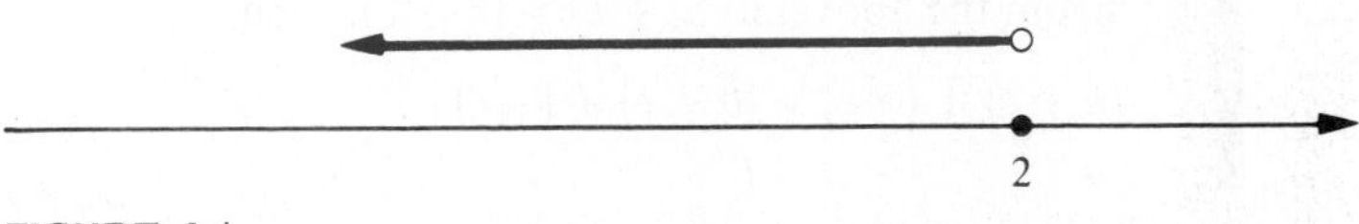

FIGURE 4.1

To represent $\{x \mid x \geq 7, x \in R\}$ graphically, draw a ray from the point whose coordinate is 7 in the positive direction. Place a solid dot at the point to denote that the point is included on the ray (Figure 4.2).

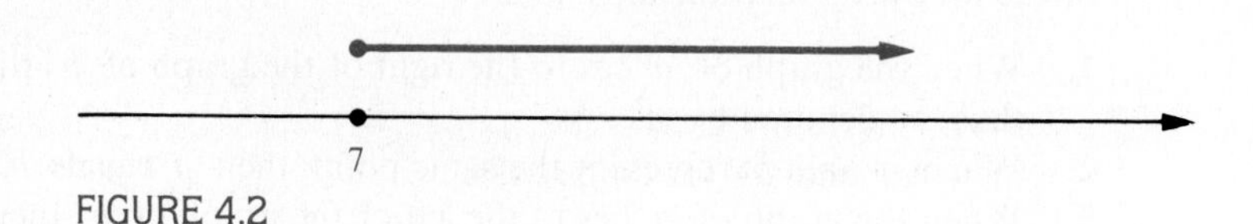

FIGURE 4.2

4.2 Properties of the Order Relations

Since $a > b$ and $b < a$ are equivalent statements, from the theorems below for the "greater than" relation we can derive similar theorems for the "less than" relation.

THEOREM 1 Let $a, b, c \in R$; if $a > b$ and $b > c$, then $a > c$.
See Appendix E for the proof.

EXAMPLE $-5 > -8$ and $-8 > -15$

Hence $\quad -5 > -15$

THEOREM 2 Let $a, b, c, d \in R$; if $a > b$ and $c > d$, then $a + c > b + d$.
See Appendix E for the proof.

EXAMPLE

	$7 > 3$	and	$-10 > -15$
	$7 + (-10) = -3$	and	$3 + (-15) = -12$
Since	$-3 > -12$	then	$7 + (-10) > 3 + (-15)$

THEOREM 3 Let $a, b, c \in R$; if $a > b$, then $a + c > b + c$.

Proof $a > b$ means $a = b + k$ where $k \in R, k > 0$.
Add c to both sides of the equation $a = b + k$:

$$a + c = b + k + c$$

$$(a + c) = (b + c) + k$$

Hence $\quad a + c > b + c$

From the above, if $(a + c) > (b + c)$, then

$$(a + c) + (-c) > (b + c) + (-c)$$

or $\qquad a > b$

EXAMPLE

$$8 > 5$$

$$8 + (-11) = -3 \quad \text{and} \quad 5 + (-11) = -6$$

Since $-3 > -6$ then $8 + (-11) > 5 + (-11)$

THEOREM 4

Let $a, b, c, d \in R$, $a, b, c, d > 0$; if $a > b$ and $c > d$, then $ac > bd$. See Appendix E for the proof.

EXAMPLE

$$5 > 2 \quad \text{and} \quad 6 > 4$$

$$5(6) = 30 \quad \text{and} \quad 2(4) = 8$$

Since $30 > 8$ then $5(6) > 2(4)$

THEOREM 5

Let $a, b, c \in R$, $c > 0$; if $a > b$, then $ac > bc$.

Proof

$a > b$ means $a = b + k$ where $k \in R$, $k > 0$.
Multiply both sides of the equation $a = b + k$ by c:

$$ac = (b + k)c = bc + kc$$

Since $k > 0$ and $c > 0$, it follows that $kc > 0$.
Therefore, $ac > bc$.
From above, if $ac > bc$ with $c > 0$, then

$$ac\left(\frac{1}{c}\right) > bc\left(\frac{1}{c}\right)$$

or $\quad a > b$

EXAMPLE

$$9 > -4 \quad \text{and} \quad 6 > 0$$

$$9(6) = 54 \quad \text{and} \quad -4(6) = -24$$

Since $54 > -24$ then $9(6) > -4(6)$

COROLLARY

Let $a, b \in R$; if $a > b$ where $ab > 0$, then $\frac{1}{a} < \frac{1}{b}$.

See Appendix E for the proof.

EXAMPLE

$10 > 2$ and $10(2) > 0$

Hence $\frac{1}{10} < \frac{1}{2}$

THEOREM 6 Let $a, b, c \in R$, $c < 0$; if $a > b$, then $ac < bc$.

Proof $a > b$ means $a = b + k$ where $k \in R$, $k > 0$.
Multiply both sides of the equation $a = b + k$ by c where $c < 0$:

$$\begin{aligned} ac &= (b + k)c \\ &= bc + kc \end{aligned}$$

or $\quad ac - kc = bc$

or $\quad ac + (-kc) = bc$

Since $k > 0$ and $c < 0$, $kc < 0$ or $-kc > 0$.
Thus $\quad bc > ac \quad$ or $\quad ac < bc$
From above, if $ac < bc$ with $c < 0$, then

$$ac\left(\frac{1}{c}\right) > bc\left(\frac{1}{c}\right)$$

or $\quad a > b$

EXAMPLE

$$20 > 14 \quad \text{and} \quad -3 < 0$$
$$20(-3) = -60 \quad \text{and} \quad 14(-3) = -42$$

Since $\quad -60 < -42 \quad$ then $\quad 20(-3) < 14(-3)$

COROLLARY Let $a, b \in R$; if $a > b$, then $-a < -b$.
The proof follows from Theorem 6 if we take $c = -1$.

EXAMPLE $15 > 8$
Hence $\quad -15 < -8$

4.3 Solution of Linear Inequalities in One Variable

The following are examples of statements of order of two algebraic expressions:

1. $2x < 2x + 7$
2. $x^2 + y^2 \geq 0$
3. $5x < -10$
4. $3x - 7 \geq 2x$
5. $2x + 5 \leq 2(x + 1)$
6. $3(x + y) > 3x + 3y + 1$

Statements 1 and 2 are true for all real values of the variables involved.
Such statements are called **absolute statements**.

Statements 3 and 4 are true for some, but not all, real values of the variable involved.
Statement 3 is true when $x < -2$. Statement 4 is true when $x \geq 7$.
Such statements are called **conditional inequalities**, or simply, **inequalities**.

Statements 5 and 6 are not true for any real number.

The set of all numbers that satisfies the inequality is called the **solution set** of the inequality.

Note A linear equation in one variable has one element in its solution set. The solution set of the equation $2x - 3 = 5$ is $\{4\}$.
A linear inequality in one variable has more than one element in its solution set. The solution set of the inequality $2x - 3 > 5$ is $\{x \mid x > 4\}$, that is, all the real numbers that are greater than 4.

DEFINITION

Two inequalities are said to be **equivalent** if they have the same solution set.

In order to find the solution set of an inequality, we must first state some theorems.

THEOREM 1

If $P, Q,$ and T are polynomials in the same variable and $P > Q$ is an inequality, then $P > Q$ and $P + T > Q + T$ are equivalent.

Theorem 1 shows that we can add a polynomial to both sides of the inequality and obtain an equivalent inequality.

THEOREM 2

If P and Q are polynomials in the same variable, $a > 0$, $a \in R$, and if $P > Q$ is an equality, then $P > Q$ and $aP > aQ$ are equivalent.

Theorem 2 shows that if we multiply both sides of an inequality by a positive real number, we obtain an equivalent inequality.

THEOREM 3

If P and Q are polynomials in the same variable, $a < 0$, $a \in R$, and if $P > Q$ is an inequality, then $P > Q$ and $aP < aQ$ are equivalent.

Theorem 3 shows that if we multiply both sides of an equality by a negative number, an equivalent inequality results with the direction of the order relation reversed.

In order to find the solution set of a linear inequality in one variable, apply the previous theorems to obtain an equivalent inequality of the form

$x > a$ whose solution set is $\{x \mid x > a, x \in R\}$

or of the form

$x < b$ whose solution set is $\{x \mid x < b, x \in R\}$

Note When both sides of the inequality contain terms that have the variable as a factor and terms that do not have the variable as a factor, form an equivalent inequality that has all the terms with the variable as a factor on one side and the terms not having the variable on the other side.
This can be accomplished by adding the additive inverses (negatives) of the terms to both sides of the inequality.

EXAMPLE Find the solution set of the inequality $6x - 3 \le 5x - 2$.

Solution Adding $(3 - 5x)$ to both sides of the inequality, we have

$$6x - 3 + 3 - 5x \le 5x - 2 + 3 - 5x$$
$$x \le 1$$

Hence the solution set is $\{x \mid x \le 1, x \in R\}$.

Note When the inequality involves grouping symbols, first perform the operations that these symbols designate.

EXAMPLE Find the solution set of the inequality $2(3x + 1) - 3(x - 2) \ge 2x + 5$.

Solution Apply the distributive law to remove the parentheses:

$$6x + 2 - 3x + 6 \ge 2x + 5$$

Combining like terms, we get

$$3x + 8 \ge 2x + 5$$

Add $(-8 - 2x)$ to both sides of the inequality:

$$3x + 8 - 8 - 2x \ge 2x + 5 - 8 - 2x$$
$$x \ge -3$$

Hence the solution set is $\{x \mid x \ge -3, x \in R\}$.

EXAMPLE Find the solution set of the inequality $(3x - 1)^2 - 3(x + 4)(3x - 7) > 15$.

Solution

$$(3x - 1)^2 - 3(x + 4)(3x - 7) > 15$$
$$(9x^2 - 6x + 1) - 3(3x^2 + 5x - 28) > 15$$

or
$$9x^2 - 6x + 1 - 9x^2 - 15x + 84 > 15$$

Hence
$$-21x + 85 > 15$$

Adding -85 to both sides of the inequality, we get

$$-21x > -70$$

Multiply both sides of the inequality by $-\frac{1}{21}$ and reverse the direction of the order relation:

$$x < \frac{10}{3}$$

Hence the solution set is $\left\{x \middle| x < \frac{10}{3}, x \in R\right\}$.

EXAMPLE

Find the solution set of the inequality

$$\frac{3x+1}{2} - \frac{7x+3}{6} < \frac{3-x}{3}$$

Solution

Multiply both sides of the inequality by 6:

$$3(3x+1) - (7x+3) < 2(3-x)$$
$$9x + 3 - 7x - 3 < 6 - 2x$$
$$2x < 6 - 2x$$

Adding $2x$ to both sides of the inequality, we get

$$4x < 6$$

Multiply both sides of the inequality by $\frac{1}{4}$:

$$x < \frac{6}{4} \quad \text{that is} \quad x < \frac{3}{2}$$

The solution set is $\left\{x \middle| x < \frac{3}{2}, x \in R\right\}$.

EXAMPLE

List the elements in the set $\{x \mid 10 - 3(x+1) < 3(2-x), x \in R\}$.

Solution

Consider the statement:

$$10 - 3(x+1) < 3(2-x)$$
$$10 - 3x - 3 < 6 - 3x$$
$$-3x + 3x < 6 + 3 - 10$$
$$0x < -1$$

Since $0x < -1$ is not true for any real value of x, we have

$\{x \mid 10 - 3(x+1) < 3(2-x), x \in R\} = \varnothing$

EXAMPLE List the elements in the set $\{x \mid (x+3)(x-1)-5<(x+1)^2, x \in R\}$.

Solution Consider the statement:

$$(x+3)(x-1)-5<(x+1)^2$$
$$x^2+2x-3-5<x^2+2x+1$$
$$x^2+2x-x^2-2x<1+3+5$$
$$0x<9$$

Since $0x<9$ is true for all real values of x, we have

$$\{x \mid (x+3)(x-1)-5<(x+1)^2, x \in R\}=\{x \mid x \in R\}$$

Exercise 4.3

Find the solution set of each of the following inequalities:

1. $4x-x+6<2x+7$
2. $6x-1+2x<4+7x$
3. $3-x+5x \geq 3x-2$
4. $2x+12+6x \geq 7x+10$
5. $3x-8+2x \leq 4x-10$
6. $2x+15-7x \leq 8-6x$
7. $6x-9x-2>3-4x$
8. $10x+20-x>8x+24$
9. $8+23x-1<3x-3$
10. $6+5x-2<4-5x$
11. $7x+6-2x \geq 2-3x$
12. $10+9x-3 \geq 4-3x$
13. $13x+5-18x>4-x$
14. $6x+7-4x>7x-3$
15. $20-3x-4<6x-11$
16. $12-5x-5<2x-7$
17. $2(7x-8)+7(2-x) \leq 26$
18. $4(3-x)-3(4-x) \leq 0$
19. $5-3(4x+3) \geq 4(x-1)$
20. $7(x-3)-6(2x-3) \geq 2$
21. $4(2x-3)>11-3(7x-2)$
22. $5(x-2)-4(2x-1)>3$
23. $(x+2)(x-3)<x(x-4)$
24. $(x+3)(x+1)<x(x+6)$
25. $(x+3)(x-4)-x(x+2)>-3$
26. $(x-1)(x-5)-x(x-2)>8$
27. $(2x-1)(x-2)-2x(x-1) \geq 6$
28. $(3x+1)(x+2)-3x(x-2) \geq -10$
29. $(2x+3)(2x-1)-(4x-1)(x+1) \leq 2$
30. $(2x+1)(3x-1)-(x-3)(6x+1) \leq -4$
31. $(3x-2)(x+4)-3(x-2)(x+2)<0$
32. $(4x-3)(3x+1)-6(2x+5)(x-1)<4$
33. $\frac{2x}{3}-\frac{1}{6}>\frac{3x}{2}+\frac{2}{3}$
34. $\frac{3x}{5}-\frac{1}{15}>\frac{5x}{9}+\frac{1}{5}$
35. $\frac{3x}{4}-\frac{2x}{3}<\frac{2}{9}-\frac{7}{18}$
36. $\frac{2x}{9}-\frac{1}{4}>\frac{3x}{8}+\frac{1}{18}$

37. $\frac{2x}{7} - \frac{19}{21} > -\frac{x}{9} - \frac{1}{9}$

38. $\frac{7x}{3} - \frac{6}{5} < \frac{12x}{5} - \frac{2}{3}$

39. $\frac{2x-1}{4} + \frac{4x+1}{3} < 1$

40. $\frac{3x-2}{3} + \frac{9x+1}{7} \geq 1$

41. $\frac{3x-1}{2} - \frac{2x+3}{3} \geq 1$

42. $\frac{5x+2}{7} - \frac{x+1}{4} \leq \frac{1}{2}$

43. $\frac{2x-1}{5} - \frac{2}{3} \leq \frac{x-8}{6}$

44. $\frac{2x+13}{3} > 2 - \frac{4-x}{4}$

45. $\frac{3(x+4)}{2} > 14 - \frac{2(x-1)}{3}$

46. $\frac{5(x-9)}{6} + \frac{3(x-1)}{4} < -13$

47. $\frac{7(2x-5)}{9} + \frac{5(x-4)}{3} < -17$

48. $\frac{3(x-3)}{4} > 5 - \frac{2(x-2)}{5}$

List the elements in the following sets given that $x \in R$.

49. $\left\{x \,\middle|\, \frac{2}{3}x - x < \frac{1}{3}(2-x)\right\}$

50. $\left\{x \,\middle|\, \frac{3}{4}x + 8 > x - \frac{1}{8}(2x-3)\right\}$

51. $\{x \mid 7 - 2(9-x) > 2(x-15)\}$

52. $\{x \mid 11(2x-1) \leq 7x - 5(4-3x)\}$

53. $\{x \mid 4x(x+3) - (2x+1)^2 < 8(x-1)\}$

54. $\{x \mid (8x-1)(2x+3) - (4x+1)^2 \leq 7(2x-1)\}$

55. $\left\{x \,\middle|\, 2x - \frac{3-x}{4} > 3x - \frac{3x-1}{4}\right\}$

56. $\left\{x \,\middle|\, \frac{3x-2}{4} - \frac{2x-1}{3} > \frac{x}{12} - \frac{1}{4}\right\}$

57. $\left\{x \,\middle|\, \frac{x+5}{9} - \frac{2x-7}{4} \leq 3 - \frac{7x}{18}\right\}$

58. $\{x \mid (4x-3)^2 < 25(x^2-1) - (3x+4)^2\}$

4.4 Solution of Systems of Linear Inequalities in One Variable

Sometimes it is necessary to find the common solution, or solution set, of two or more inequalities, called a **system of inequalities**. The solution set of a system of inequalities is thus the intersection of the solution set of each inequality in the system.

EXAMPLE

Find the solution set of the following system:

$$9(2x + 1) + 8(x - 6) \geq 0$$
$$2(2 - x) - (5 - 4x) < 13$$

Solution

We first find the solution set of each inequality:

$$9(2x + 1) + 8(x - 6) \geq 0$$
$$18x + 9 + 8x - 48 \geq 0$$
$$26x \geq 39$$
$$x \geq \frac{3}{2}$$

The solution set is

$$\left\{x \,\middle|\, x \geq \frac{3}{2}\right\}$$

$$2(2 - x) - (5 - 4x) < 13$$
$$4 - 2x - 5 + 4x < 13$$
$$x < 7$$

The solution set is

$$\{x \mid x < 7\}$$

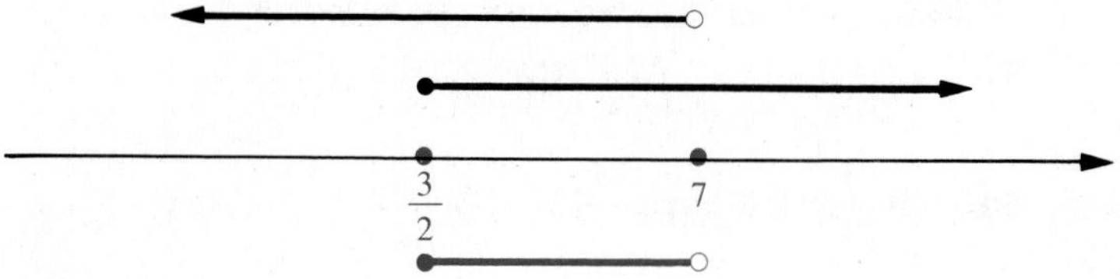

FIGURE 4.3

The solution set of the system is (Figure 4.3)

$$\left\{x \,\middle|\, x \geq \frac{3}{2}\right\} \cap \{x \mid x < 7\} = \left\{x \,\middle|\, \frac{3}{2} \leq x < 7\right\}$$

The system $-3 < x - 2$ and $x - 2 < 5$ can be written as $-3 < x - 2 < 5$. When there are no terms in x in the first and third members of the statement, as in $-3 < x - 2 < 5$, the solution set can be found by adding 2 to each member to get $-1 < x < 7$.
Thus the solution set is $\{x \mid -1 < x < 7\}$.

EXAMPLE

Find the values of x that satisfy $7 < 3 - 2x < 15$.

Solution

Adding -3 to each member we get $4 < -2x < 12$.

Multiplying each member by $-\frac{1}{2}$ and reversing the direction of the order relations, we get

$$-2 > x > -6 \qquad \text{or} \qquad -6 < x < -2$$

The solution set is $\{x \mid -6 < x < -2\}$.

When each member of the statement has terms containing x as a factor, write the statement as a system of inequalities, and find the solution set of the system.

EXAMPLE Find the solution set of $2x - 5 < 3x + 6 < 5x - 3$.

Solution The statement is equivalent to the system:

$$2x - 5 < 3x + 6 \quad \text{and} \quad 3x + 6 < 5x - 3$$

$$x > -11 \quad \text{and} \quad x > \frac{9}{2}$$

The solution set is $\left\{x \middle| x > \frac{9}{2}\right\}$.

Exercise 4.4

Find the solution set of each of the following systems:

1. $2x + 5 < x + 7$, $3x - 6 > 2x - 9$

2. $4x + 3 > 3x - 2$, $7x - 4 < 6x + 3$

3. $5x - 4 \geq x - 8$, $3x + 5 < x + 7$

4. $3x - 2 > 8 - 2x$, $4x + 3 \leq 2x + 11$

5. $2x + 5 > x - 1$, $5x + 1 > x + 9$

6. $3x + 2 \geq x + 4$, $x - 3 > 5 - x$

7. $2x - 1 < 2 - x$, $3x + 4 \leq 2x + 1$

8. $4x - 7 < 3 - x$, $7x - 1 \leq x - 1$

9. $5x - 2 \geq x + 10$, $2x + 5 \geq 3x + 2$

10. $3x + 4 \leq x - 1$, $4x + 3 \geq 2x - 2$

11. $7x - 6 \leq 3x - 2$, $1 - 2x \leq 3x - 4$

12. $8x + 7 \geq 1 - x$, $9 - 5x \geq 11 - 2x$

13. $3x - 7 < 8 - x$, $4x - 2 > x + 13$

14. $2x - 9 < 5x - 3$, $x - 1 > 7x + 17$

15. $x + 4 \leq 2x - 2$, $3x - 1 < x + 11$

16. $4x - 3 \geq 3x + 1$, $7x - 1 < 5x + 7$

17. $2(5x - 3) + 7(2 - 3x) < -13$, $7 - 3(x + 1) \geq 2(x - 8)$

18. $5(1 + 6x) + 9(3 + x) + 7 < 0$, $x - 2(x + 3) < 5x - 3(x - 4)$

19. $2 - 2(7 - 2x) < 3(3 - x)$, $3(4x + 3) > 7 - 4(x - 2)$

20. $3(7x - 2) < 11 - 4(2x - 3)$, $5(3 - 8x) \leq 16 - 7(4x - 5)$

21. $7(2x - 3) + 2(3x - 1) > 17$, $11 - 13(2 - x) > 5(3x - 7)$

22. $3(2x + 7) - 4(2 - x) \leq 3$, $5(3 - x) - 6(3 - 2x) \geq 18$

Find the values of x that satisfy each of the following:

23. $-1 < x - 2 < 7$

24. $-3 < x - 1 < 4$

25. $2 < x - 4 < 5$

26. $-5 < x - 5 < 5$

27. $-3 < x + 3 \leq 11$

28. $2 < x + 2 \leq 3$

29. $0 < x + 7 \leq 2$

30. $-6 < x + 3 \leq 0$

31. $-8 \leq 3x - 5 < 10$

32. $7 \leq 2x - 3 < 15$

33. $2 \leq 4 - x \leq 9$

34. $-2 \leq 5 - x \leq 4$

35. $-1 \le 2 - x \le 3$

36. $-9 \le 6 - 5x \le 16$

37. $-13 < 3 - 8x < 3$

38. $8 < -4 - 3x < 20$

39. $-2 < 7 - 3x < 13$

40. $-4 < 5 - 6x < 8$

41. $2x - 1 < 3x + 7 < x + 9$

42. $x + 3 < 3x - 1 < 2x + 3$

43. $3x - 5 < x - 1 < 4x + 11$

44. $4x - 2 < x + 4 < 3x + 10$

45. $2x - 3 < 3x + 1 < 4x - 5$

46. $x + 1 < 2x + 3 < 3x - 2$

47. $3x + 1 < 2x - 1 < x + 3$

48. $5x - 2 < 3x - 4 < 2x + 1$

4.5 Absolute Values

DEFINITION

The **absolute value** of a number $a \in R$, denoted by $|a|$, is either $+a$ or $-a$, whichever is positive, and zero if $a = 0$.

The definition may be written as

$$|a| = \begin{cases} a & \text{if } a \ge 0 \\ -a & \text{if } a < 0 \end{cases}$$

Notes

1. Since $-0 = 0$, $|0| = |-0| = 0$.
2. If $a \in R$, $a \neq 0$, $|a| > 0$.
3. If $a, b \in R$, and $a = b$, then $|a| = |b|$.
4. If $a, b \in R$, and $|a| = |b|$, then $a = b$ or $a = -b$.

EXAMPLES

1. $|10| = 10$

2. $|-3| = -(-3) = 3$

3. $|7 - 12| = |-5| = -(-5) = 5$

Note that the absolute value of any real number is either zero or a positive number, never a negative number. That is, $|a| \ge 0$ for all $a \in R$.

Properties of the Absolute Values of Real Numbers

The following theorems give some of the properties of the absolute values of real numbers.

THEOREM 1

If $a \in R$, then $|a| = |-a|$.
See Appendix E for the proof.

THEOREM 2 If $a \in R$, then $|a| \geq a$; also, $|a| \geq -a$.
See Appendix E for the proof.

THEOREM 3 If $a, b \in R$, then $|ab| = |a| \cdot |b|$.
See Appendix E for the proof.

THEOREM 4 If $a, b \in R$, $b \neq 0$, then $\left|\frac{a}{b}\right| = \frac{|a|}{|b|}$.
See Appendix E for the proof.

THEOREM 5 When $a, b \in R$, $b > 0$, then $|a| \leq b$ if and only if $-b \leq a \leq b$.

Proof

First: We have to prove that if $|a| \leq b$, then $-b \leq a \leq b$.

If $a \geq 0$, then $|a| = a \leq b$,

so $\quad 0 \leq a \leq b \qquad (1)$

If $a < 0$, then $|a| = -a \leq b$,

or $\quad -b \leq a < 0. \qquad (2)$

From (1) and (2) we have $-b \leq a \leq b$.

Second: We have to prove that if $-b \leq a \leq b$, then $|a| \leq b$.

If $-b \leq a < 0$, then $b \geq -a = |a| > 0$,

or $\quad |a| \leq b$

If $0 \leq a \leq b$, then $0 \leq |a| = a \leq b$,

or $\quad |a| \leq b$

THEOREM 6 When $a, b \in R$, if $|a| > b$, then $a < -b$ or $a > b$.

Proof

If $a < 0$, then $|a| = -a > b$, or $a < -b$.
If $a > 0$, then $|a| = a > b$.

4.6 Solution of Linear Equations Involving Absolute Values

When we have the absolute value of a quantity involving a variable such as $|x - 3|$, that quantity, $x - 3$ could be

1. greater than or equal to zero
2. less than zero

When $x - 3$ is greater than or equal to zero, that is, $x - 3 \geq 0$, then

$$|x - 3| = x - 3$$

When $x - 3$ is less than zero, that is, $x - 3 < 0$, then

$$|x - 3| = -(x - 3) = -x + 3$$

The following examples illustrate how to solve a linear equation in one variable involving absolute value.

EXAMPLE Solve the equation $|x - 3| = 4$.

Solution To find the solution set of the equation, we have to consider the two cases:

First: When $x - 3 \geq 0$, that is, $x \geq 3$,

$$|x - 3| = x - 3$$

The equation now becomes

$$|x - 3| = x - 3 = 4 \qquad \text{that is,} \qquad x = 7$$

The solution set is the intersection of the solution sets of

$$x \geq 3 \qquad \text{and} \qquad x = 7$$

The solution set is $\{7\}$.

Second: When $x - 3 < 0$, that is, $x < 3$,

$$|x - 3| = -(x - 3) = -x + 3$$

The equation now becomes

$$|x - 3| = -x + 3 = 4 \qquad \text{that is,} \qquad x = -1$$

The solution set is the intersection of the solution sets of

$$x < 3 \qquad \text{and} \qquad x = -1$$

The solution set is $\{-1\}$.

The solution set of $|x - 3| = 4$ is the union of the solution sets in the two cases.

Hence the solution set is $\{-1, 7\}$.

EXAMPLE Find the solution set of $|2 - 3x| = 10$.

Solution *First:* When $2 - 3x \geq 0$, that is, $x \leq \frac{2}{3}$,

$$|2 - 3x| = 2 - 3x$$

The equation now becomes

$$|2 - 3x| = 2 - 3x = 10 \qquad \text{that is,} \qquad x = -\frac{8}{3}$$

The solution set is the intersection of the solution sets of

$$x \leq \frac{2}{3} \quad \text{and} \quad x = -\frac{8}{3}$$

The solution set is $\left\{-\frac{8}{3}\right\}$.

Second: When $2 - 3x < 0$, that is, $x > \frac{2}{3}$,

$$|2 - 3x| = -(2 - 3x) = -2 + 3x$$

The equation now becomes

$$|2 - 3x| = -2 + 3x = 10 \qquad \text{that is,} \qquad x = 4$$

The solution set is the intersection of the solution sets of

$$x > \frac{2}{3} \quad \text{and} \quad x = 4$$

The solution set is $\{4\}$.

The solution set of $|2 - 3x| = 10$ is the union of the solution sets in the two cases.

Hence the solution set is $\left\{-\frac{8}{3}, 4\right\}$.

EXAMPLE

Find the solution set of $|7x - 1| = -8$.

Solution

Since the absolute value of any real number is never negative, the solution set is $\varnothing$.

EXAMPLE

Find the solution set of $|2x - 3| = x + 5$.

Solution

First: When $2x - 3 \geq 0$, that is, $x \geq \frac{3}{2}$,

$$|2x - 3| = 2x - 3$$

The equation now becomes

$$2x - 3 = x + 5 \qquad \text{that is,} \qquad x = 8$$

The solution set is the intersection of the solution sets of

$$x \geq \frac{3}{2} \quad \text{and} \quad x = 8$$

The solution set is $\{8\}$.

Second: When $2x - 3 < 0$, that is, $x < \frac{3}{2}$,

$|2x - 3| = -(2x - 3) = -2x + 3$

The equation now becomes

$$-2x + 3 = x + 5 \qquad \text{that is,} \qquad x = -\frac{2}{3}$$

The solution set is the intersection of the solution sets of

$$x < \frac{3}{2} \qquad \text{and} \qquad x = -\frac{2}{3}$$

The solution set is $\left\{-\frac{2}{3}\right\}$.

The solution set of $|2x - 3| = x + 5$ is the union of the solution sets in the two cases.

Hence the solution set is $\left\{-\frac{2}{3}, 8\right\}$.

EXAMPLE

Find the solution set of $|3x + 1| = 4x + 3$.

Solution

First: When $3x + 1 \geq 0$, that is, $x \geq -\frac{1}{3}$,

$|3x + 1| = 3x + 1$

The equation now becomes

$$3x + 1 = 4x + 3 \qquad \text{that is,} \qquad x = -2$$

The solution set is the intersection of the solution sets of

$$x \geq -\frac{1}{3} \qquad \text{and} \qquad x = -2$$

The solution set is $\varnothing$.

Second: When $3x + 1 < 0$, that is, $x < -\frac{1}{3}$,

$|3x + 1| = -(3x + 1) = -3x - 1$

The equation now becomes

$$-3x - 1 = 4x + 3 \qquad \text{that is,} \qquad x = -\frac{4}{7}$$

The solution set is the intersection of the solution sets of

$$x < -\frac{1}{3} \quad \text{and} \quad x = -\frac{4}{7}$$

The solution set is $\left\{-\frac{4}{7}\right\}$.

The solution set of $|3x + 1| = 4x + 3$ is the union of the solution sets in the two cases.

Hence the solution set is $\left\{-\frac{4}{7}\right\}$.

EXAMPLE

Find the solution set of $|1 - 2x| = x - 7$.

Solution

First: When $1 - 2x \geq 0$, that is, $x \leq \frac{1}{2}$,

$$|1 - 2x| = 1 - 2x$$

The equation now becomes

$$1 - 2x = x - 7 \quad \text{that is,} \quad x = \frac{8}{3}$$

The solution set is the intersection of the solution sets of

$$x \leq \frac{1}{2} \quad \text{and} \quad x = \frac{8}{3}$$

The solution set is $\varnothing$.

Second: When $1 - 2x < 0$, that is, $x > \frac{1}{2}$,

$$|1 - 2x| = -(1 - 2x) = -1 + 2x$$

The equation now becomes

$$-1 + 2x = x - 7 \quad \text{that is,} \quad x = -6$$

The solution set is the intersection of the solution sets of

$$x > \frac{1}{2} \quad \text{and} \quad x = -6$$

The solution set is $\varnothing$.

The solution set of $|1 - 2x| = x - 7$ is the union of the solution sets in the two cases.

Hence the solution set is $\varnothing$.

EXAMPLE Find the solution set of $|2x - 7| = 7 - 2x$.

Solution

First: When $2x - 7 \geq 0$, that is, $x \geq \frac{7}{2}$,

$|2x - 7| = 2x - 7$

The equation now becomes

$2x - 7 = 7 - 2x$ that is, $x = \frac{7}{2}$

The solution set is the intersection of the solution sets of

$x \geq \frac{7}{2}$ and $x = \frac{7}{2}$

The solution set is $\left\{\frac{7}{2}\right\}$.

Second: When $2x - 7 < 0$, that is, $x < \frac{7}{2}$,

$|2x - 7| = -(2x - 7) = -2x + 7$

The equation now becomes

$-2x + 7 = 7 - 2x$ that is, $0x = 0$

which is true for all $x \in R$.

The solution set is the intersection of the solution sets of

$x < \frac{7}{2}$ and $0x = 0$

The solution set is $\left\{x \,\middle|\, x < \frac{7}{2}\right\}$.

The solution set of $|2x - 7| = 7 - 2x$ is the union of the solution sets in the two cases.

Hence the solution set is $\left\{x \,\middle|\, x \leq \frac{7}{2}\right\}$.

EXAMPLE Find the solution set of $|3x - 1| = |x + 3|$.

Solution If $a, b \in R$, and $|a| = |b|$, then $a = b$ or $a = -b$.

First: When $3x - 1 = x + 3$

$$2x = 4$$

$$x = 2$$

The solution set is $\{2\}$.

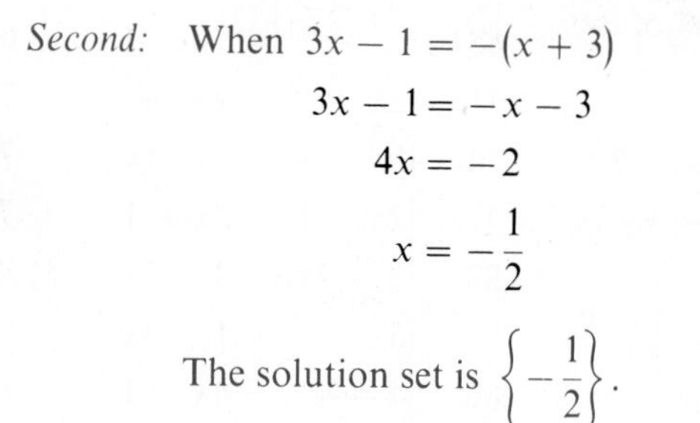

Second: When $3x - 1 = -(x + 3)$

$$3x - 1 = -x - 3$$
$$4x = -2$$
$$x = -\frac{1}{2}$$

The solution set is $\left\{-\frac{1}{2}\right\}$.

The solution set of $|3x - 1| = |x + 3|$ is the union of the solution sets in the two cases.

Hence the solution set is $\left\{-\frac{1}{2}, 2\right\}$.

Exercise 4.6

Find the solution sets of the following equations:

1. $|x| = 1$ **2.** $|x| = 2$ **3.** $|x| = 6$ **4.** $|x| = 3$

5. $|x| = -5$ **6.** $|2x| = 4$ **7.** $|2x| = -2$ **8.** $|3x| = 6$

9. $|2x| = 3$ **10.** $|3x| = 2$ **11.** $|2x| = |-4|$ **12.** $|3x| = |-9|$

13. $|5x| = |-10|$ **14.** $|6x| = |-24|$ **15.** $|x - 7| = 0$

16. $|x - 4| = 0$ **17.** $|x + 3| = 0$ **18.** $|x + 1| = 0$

19. $|4x + 3| = 0$ **20.** $|5x + 4| = 0$ **21.** $|x - 3| = 1$

22. $|x - 1| = 5$ **23.** $|x - 2| = 2$ **24.** $|x - 5| = 5$

25. $|x + 4| = 4$ **26.** $|x + 6| = 6$ **27.** $|2x - 1| = 3$

28. $|2x - 3| = 5$ **29.** $|3x - 1| = 2$ **30.** $|3x - 4| = 4$

31. $|3x - 2| = 7$ **32.** $|5x - 6| = 6$ **33.** $|2x + 1| = 5$

34. $|1 - x| = 7$ **35.** $|2 - x| = 2$ **36.** $|3 - x| = 3$

37. $|4 - x| = 8$ **38.** $|5 - 2x| = 1$ **39.** $|3 - 2x| = 10$

40. $|5 - 3x| = 2$ **41.** $|7 - 4x| = 1$ **42.** $|1 - 5x| = 4$

43. $|9 - 7x| = 5$ **44.** $|8 - 3x| = 7$ **45.** $|9 - 4x| = 3$

46. $|8 - 5x| = 2$ **47.** $|4 - 7x| = 3$ **48.** $|1 - 2x| = 9$

49. $|2x - 5| = x + 8$ **50.** $|2x + 3| = x + 6$ **51.** $|3x - 2| = x + 4$

52. $|4x - 3| = x + 6$ **53.** $|6x + 1| = 4x + 3$ **54.** $|7x - 1| = 3x + 5$

55. $|3 - 8x| = x + 1$ **56.** $|3 - 4x| = 2x + 1$ **57.** $|8 - 5x| = x + 2$

58. $|x + 2| = 2x + 3$ **59.** $|x + 1| = 2x + 7$ **60.** $|2x + 1| = 3x + 2$

61. $|2x + 3| = 4x + 5$ **62.** $|3x + 2| = 5x + 6$ **63.** $|4x + 1| = 6x + 7$

64. $|4 - x| = 2x - 3$ **65.** $|7 - x| = 4x - 7$ **66.** $|x - 4| = 3x + 5$

67. $|x - 8| = 4x + 1$ **68.** $|2x - 3| = 5x + 3$ **69.** $|3x - 1| = 7x + 1$
70. $|2 - 7x| = x - 3$ **71.** $|2x + 1| = x - 4$ **72.** $|3x + 2| = 2x - 7$
73. $|4x - 5| = 2x - 3$ **74.** $|5x + 1| = 2x - 9$ **75.** $|8 - 3x| = x - 5$
76. $|x - 3| = x - 3$ **77.** $|2x - 1| = 2x - 1$ **78.** $|3x - 2| = 3x - 2$
79. $|2 - x| = 2 - x$ **80.** $|3 - 2x| = 3 - 2x$ **81.** $|4 - 3x| = 4 - 3x$
82. $|x - 1| = 1 - x$ **83.** $|3x - 1| = 1 - 3x$ **84.** $|2x - 5| = 5 - 2x$
85. $|2 - 5x| = 5x - 2$ **86.** $|1 - 4x| = 4x - 1$ **87.** $|3 - 2x| = 2x - 3$
88. $|x - 3| = |2x + 3|$ **89.** $|2x - 1| = |3x + 1|$
90. $|x + 2| = |3x - 2|$ **91.** $|2x + 5| = |3x - 5|$
92. $|x + 6| = |2x - 3|$ **93.** $|3x + 4| = |x - 2|$

4.7 Solution of Linear Inequalities Involving Absolute Values

The solution of a linear inequality involving absolute values can be found when we apply Theorem 5 or Theorem 6.

EXAMPLE Find the solution set of $|x + 4| < 7$.

Solution From Theorem 5, the given inequality is equivalent to

$$-7 < x + 4 < 7$$

Adding (-4) to each member, we get

$$-11 < x < 3$$

The solution set is $\{x \mid -11 < x < 3, x \in R\}$.

EXAMPLE Find the solution set of $|2x - 1| > 11$.

Solution From Theorem 6, the given inequality is equivalent to

$$2x - 1 < -11 \quad \text{or} \quad 2x - 1 > 11$$
$$2x < -10 \quad \text{or} \quad 2x > 12$$
$$x < -5 \quad \text{or} \quad x > 6$$

The solution set is $\{x \mid x < -5 \text{ or } x > 6, x \in R\}$.

EXAMPLE Find the solution set of $|4x - 5| < 2x + 6$.

Solution The given inequality is equivalent to

$$-2x - 6 < 4x - 5 < 2x + 6$$

The above statement is equivalent to the system:

$$-2x - 6 < 4x - 5 \quad \text{and} \quad 4x - 5 < 2x + 6$$

$$6x > -1 \quad \text{and} \quad 2x < 11$$

$$x > -\frac{1}{6} \quad \text{and} \quad x < \frac{11}{2}$$

See Figure 4.4.

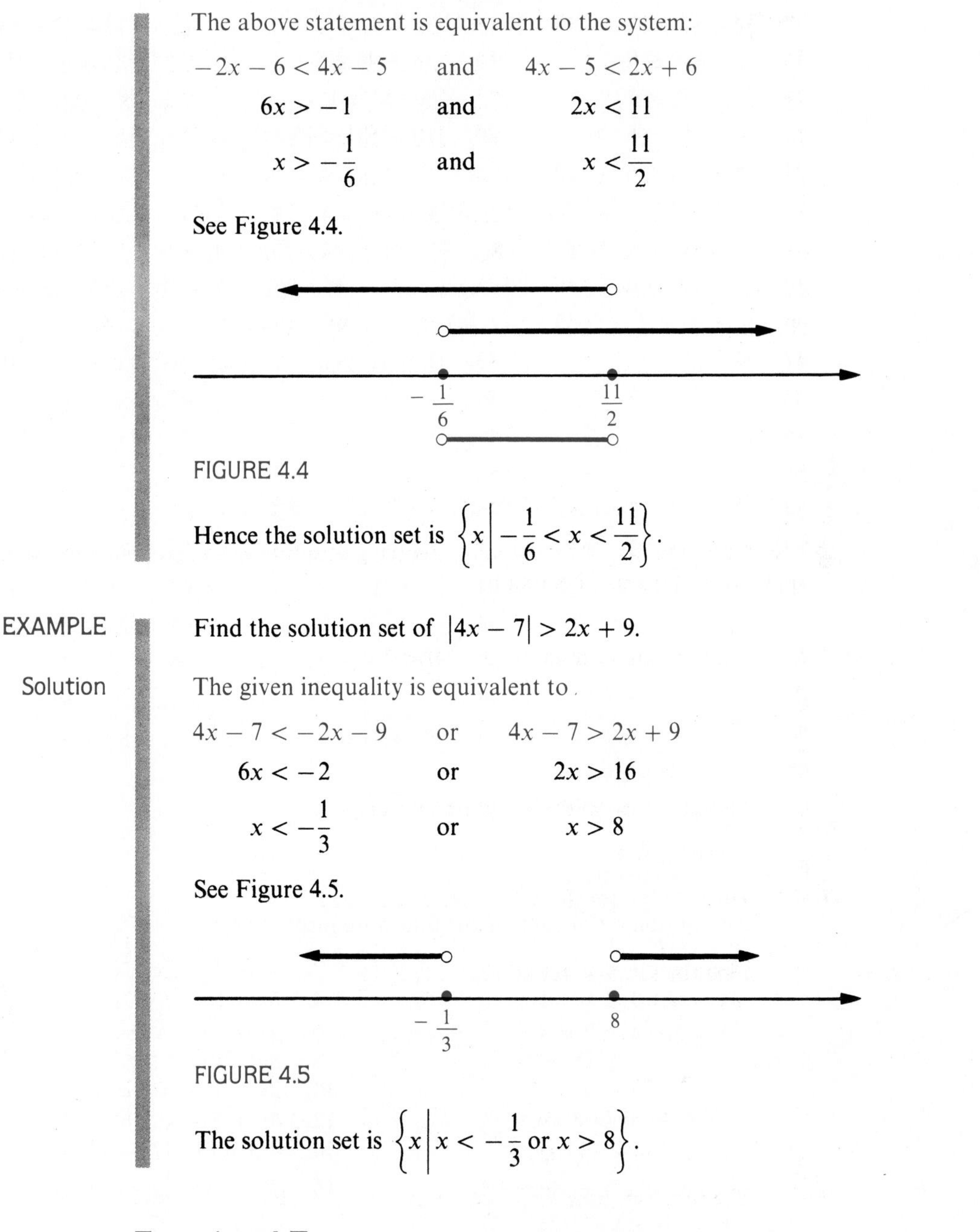

FIGURE 4.4

Hence the solution set is $\left\{x \,\middle|\, -\frac{1}{6} < x < \frac{11}{2}\right\}$.

EXAMPLE

Find the solution set of $|4x - 7| > 2x + 9$.

Solution

The given inequality is equivalent to

$$4x - 7 < -2x - 9 \quad \text{or} \quad 4x - 7 > 2x + 9$$

$$6x < -2 \quad \text{or} \quad 2x > 16$$

$$x < -\frac{1}{3} \quad \text{or} \quad x > 8$$

See Figure 4.5.

$-\frac{1}{3}$ 8

FIGURE 4.5

The solution set is $\left\{x \,\middle|\, x < -\frac{1}{3} \text{ or } x > 8\right\}$.

Exercise 4.7

Find the solution sets of the following inequalities:

1. $|x| < 1$ **2.** $|x| < 2$ **3.** $|2x| \leq 3$ **4.** $|3x| \leq 2$

5. $|x - 1| \leq 1$ **6.** $|x - 3| \leq 2$ **7.** $|x - 4| < 5$ **8.** $|2x - 1| < 2$

9. $|3x - 1| < 8$ **10.** $|4x - 3| < 9$ **11.** $|x + 2| \le 7$ **12.** $|x + 4| \le 1$
13. $|2x + 3| \le 9$ **14.** $|3x + 4| \le 5$ **15.** $|2x + 7| < 3$
16. $|5x + 2| < 12$ **17.** $|8 - x| < 1$ **18.** $|4 - x| < 3$
19. $|1 - 2x| < 5$ **20.** $|10 - 3x| < 13$ **21.** $|5 - 2x| \le 11$
22. $|6 - 5x| < 6$ **23.** $|x| > 1$ **24.** $|x| > 3$ **25.** $|3x| \ge 5$
26. $|7x| \ge 4$ **27.** $|x - 3| \ge 1$ **28.** $|x - 2| \ge 2$ **29.** $|2x - 1| > 3$
30. $|3x - 2| > 1$ **31.** $|3x - 4| > 7$ **32.** $|5x - 4| > 6$ **33.** $|2x - 5| > 0$
34. $|5x - 3| > 0$ **35.** $|4x + 1| \ge 0$ **36.** $|3x + 7| \ge 0$ **37.** $|x + 4| \ge 2$
38. $|x + 3| > 7$ **39.** $|2x + 7| > 1$ **40.** $|3x + 8| > 4$ **41.** $|1 - x| > 3$
42. $|4 - x| > 1$ **43.** $|1 - 2x| > 3$ **44.** $|2 - 3x| \ge 10$
45. $|3 - 5x| \ge 7$ **46.** $|x + 1| < 3x - 2$ **47.** $|x + 4| < 2x - 5$
48. $|x + 5| < 2x + 7$ **49.** $|x - 3| < 3x - 1$ **50.** $|x - 1| < 4x - 5$
51. $|x - 2| < 3x - 8$ **52.** $|2x - 5| < x + 3$ **53.** $|3x - 1| < x + 5$
54. $|4x - 3| < 2x + 11$ **55.** $|3x - 7| < x + 3$ **56.** $|5 - 2x| < 6 - x$
57. $|9 - 4x| < 3 - x$ **58.** $|5x - 7| < x - 5$ **59.** $|7x - 3| < 4x - 9$
60. $|3x - 10| < 2x - 7$ **61.** $|x - 1| > 3x - 4$ **62.** $|2x - 1| > 4x - 5$
63. $|2x - 5| > 6x - 1$ **64.** $|4x - 3| > 7x + 6$ **65.** $|2x + 3| > 4x + 7$
66. $|3x - 8| > 6x - 4$ **67.** $|7x - 2| > 3x + 8$ **68.** $|3x - 2| > x + 6$
69. $|4x - 1| > x + 5$ **70.** $|8x - 1| > 2x + 9$ **71.** $|5x - 9| > 3x + 1$
72. $|9x - 5| > 5x + 7$ **73.** $|3x + 1| > x - 8$ **74.** $|8x + 3| > 5x - 9$
75. $|6x + 11| > 3x - 4$

Chapter 4 Review

Find the solution set of each of the following inequalities:

1. $3x + 4 - x < 2 + x$ **2.** $7x - 5 + 2x < 6 - 2x$
3. $4x - x + 1 \le x - 5$ **4.** $6x + 8 - x \le 2x - 1$
5. $5x - 3 - x > 6 + x$ **6.** $9x + 11 + x \ge 5x - 4$
7. $x + 4 - 2x \ge 18 - 8x$ **8.** $8 + 3x - 2 \ge 6 - 3x$
9. $x + 8 - 7x < 3 - x$ **10.** $2x - 1 - 10x < 6 - x$
11. $2x - 7 - 5x > 8x + 4$ **12.** $4x + 5 - x \ge 6x - 4$
13. $3(x - 1) < 4 - (1 - x)$ **14.** $4(3 - x) < 7 + 3(2 - x)$
15. $5(8x - 3) \le 3 - 2(4x - 3)$ **16.** $7(x - 1) - 2(x + 1) > 2(x + 9)$
17. $2(x - 5) > 3 + 3(2x - 3)$
18. $6x(x - 3) > 55 + (2x - 1)(3x + 5)$
19. $8x + 5x(x - 3) < (5x - 3)(x + 1)$ **20.** $(x - 3)(x - 4) - x(x - 5) < 4$
21. $(x - 5)(3x - 2) - 3x(x - 7) > 2$ **22.** $(x + 4)(2x - 1) - 2x(x + 6) > 6$
23. $(2x - 3)^2 - (2x + 1)^2 \le 0$ **24.** $(3x + 1)^2 - (3x + 2)^2 \ge 3$

25. $(4x - 1)(x + 6) - (2x + 1)(2x - 3) \leq 15$

26. $(3x - 4)(x + 3) - 3(x + 1)(x - 2) \leq 0$

27. $\frac{2x + 1}{5} - \frac{x - 8}{6} < \frac{5}{6}$

28. $\frac{x + 3}{3} > 2 + \frac{3 - x}{4}$

29. $\frac{x + 1}{3} - \frac{x - 2}{7} \geq \frac{x}{2}$

30. $\frac{x - 3}{2} - \frac{4x - 1}{6} \leq \frac{2}{3}$

31. $\frac{3x + 2}{4} - \frac{x - 5}{6} \leq \frac{1}{6}$

32. $\frac{3x + 1}{4} - \frac{7x + 2}{8} > \frac{3}{4}$

33. $\frac{2x - 1}{3} - \frac{3x + 2}{9} > \frac{4}{9}$

34. $\frac{x + 2}{3} - \frac{3x - 1}{7} \geq \frac{1}{7}$

Find the values of x that satisfy each of the following:

35. $-2 < x - 4 < 1$

36. $3 < x - 5 < 4$

37. $1 \leq x + 6 < 7$

38. $-5 \leq x + 1 < 0$

39. $-3 < 2x - 1 \leq 5$

40. $-1 < 3x - 2 \leq 7$

41. $4 < 7x + 4 < 18$

42. $1 < 6x + 7 < 19$

43. $-4 \leq 2 - 3x \leq 8$

44. $-5 \leq 1 - 2x \leq 9$

45. $3 < 3 - 7x < 10$

46. $1 < 5 - 4x < 7$

47. $3x - 4 < 4x + 9 < x + 12$

48. $x - 7 < 3x + 1 < x + 3$

49. $2x - 1 < 3x + 5 < x + 7$

50. $2x - 9 < 4x + 1 < 3x + 4$

51. $2x + 3 < 3x + 7 < 5x - 6$

52. $x + 8 < 2x + 3 < 3x - 5$

53. $7x - 3 < 5x - 4 < 2x + 1$

54. $4x - 5 < 2x - 7 < x + 3$

Solve the following pairs of inequalities simultaneously:

55. $7x - 3 > 5 + 3x$, $2x + 1 < x + 7$

56. $8x + 4 \geq 7x - 1$, $9x - 10 < 4 + 2x$

57. $3x + 22 > 4 - 3x$, $2x - 5 \geq 3x - 6$

58. $6x + 1 \leq 7x + 2$, $5x - 3 \geq 8x - 9$

59. $4x - 2 > 3x - 4$, $7x + 1 \geq 2x + 6$

60. $6x - 5 < 3x + 4$, $2x + 3 \geq 9x + 10$

61. $5x - 4 < 4x - 6$, $x + 5 \geq 2x + 11$

62. $7x - 6 > x - 6$, $2x - 5 < 4x + 1$

63. $4x + 5 \geq 6x - 3$, $3x - 1 \geq x + 7$

64. $9x - 3 \leq 7x + 1$, $2x + 1 \geq 5x - 5$

65. $8x - 9 \geq x - 16$, $4x - 1 \geq 9x + 4$

66. $2x - 7 \leq 7x + 8$, $3x + 22 \leq 1 - 4x$

Find the solution sets of the following equations:

67. $|x| = 4$

68. $|5x| = 2$

69. $|6x| = -6$

70. $|7x| = -14$

71. $|34x| = |-51|$

72. $|2x| = |-1|$

73. $|x - 3| = 5$

74. $|x - 1| = 4$

75. $|x - 5| = 1$

76. $|x + 1| = 2$

77. $|x + 3| = 3$

78. $|x + 4| = 7$

79. $|2x - 1| = 5$

80. $|2x - 7| = 3$

81. $|3x - 11| = 4$

82. $|5 - x| = 3$

83. $|7 - 2x| = 7$

84. $|4 - 3x| = 14$

85. $|6 - 5x| = 19$

86. $|2x - 5| = x - 1$

87. $|3x - 2| = x + 6$

88. $|5x + 1| = x + 9$

89. $|5x - 8| = x + 4$

90. $|3x - 4| = x + 6$

91. $|x + 6| = 7x + 2$
92. $|3x + 4| = 5x + 1$
93. $|2x + 7| = 4x + 5$
94. $|3x - 4| = 5x - 9$
95. $|2x - 1| = 4x - 7$
96. $|3x - 11| = 6x - 13$
97. $|5x - 8| = 2x - 5$
98. $|3x - 10| = x - 6$
99. $|2x - 9| = x - 8$
100. $|x - 1| = x - 1$
101. $|2x - 3| = 2x - 3$
102. $|5x - 2| = 5x - 2$
103. $|3x - 5| = 3x - 5$
104. $|3x + 4| = 3x + 4$
105. $|2x + 1| = 2x + 1$
106. $|x - 2| = 2 - x$
107. $|6 - x| = x - 6$
108. $|7 - 2x| = 2x - 7$

Find the solution sets of the following inequalities:

109. $|x| > 2$
110. $|x| > 4$
111. $|3x| \leq 9$
112. $|4x| \leq 2$
113. $|x + 1| < 1$
114. $|x + 3| < 5$
115. $|x - 2| > 3$
116. $|x - 4| > 1$
117. $|2x + 5| \leq 3$
118. $|3x + 2| \leq 7$
119. $|2x - 1| \leq 1$
120. $|3x - 5| < 4$
121. $|2x + 1| \geq 5$
122. $|3x + 11| \geq 1$
123. $|2x - 7| > 7$
124. $|3x - 1| > 2$
125. $|13 - 7x| \leq 15$
126. $|11 - 3x| \leq 1$
127. $|1 - 2x| > 9$
128. $|3 - 4x| \geq 27$
129. $|5 - 7x| \geq 9$
130. $|x + 6| < 2x - 1$
131. $|2x - 3| < 4x - 7$
132. $|2x + 5| < 3x - 11$
133. $|2x - 1| < 3x - 8$
134. $|2x - 9| < x - 1$
135. $|3x - 8| < x + 7$
136. $|7 - 3x| < 8 - x$
137. $|1 - 4x| < 7 - 2x$
138. $|5x - 2| < x - 4$
139. $|3x + 8| < 2x - 1$
140. $|4x - 9| < x - 3$
141. $|7x - 6| < 3x - 8$
142. $|x - 4| > 2x - 7$
143. $|3x - 2| > 5x - 1$
144. $|4x - 1| > 6x - 3$
145. $|2x - 3| > 7 - x$
146. $|5x - 7| > 2x + 5$
147. $|7x - 2| > x + 10$
148. $|3x + 1| > x - 6$
149. $|2x + 9| > x - 4$
150. $|7x + 1| > x - 8$

Cumulative Review

Chapter 1 Perform the indicated operations and simplify:

1. $6 + 2(-3) - 4(5)$

2. $7 + 3(-4) + 5(-9)$

3. $8 - 3(-2) - 6(-3)$

4. $9 - 4(-3) - 7(4)$

5. $10 - 4(7 - 3) + 3(8 - 6)$

6. $14 - 4(9 - 6) + 8(4 - 1)$

7. $15 + 3(3 - 8) - 6(5 - 7)$

8. $9 + 6(4 - 11) - 3(8 - 13)$

9. $16 \times 4 \div 8 - 9 \div (-3)$

10. $12 \times 3 \div 9 - 10 \div (-5)$

11. $24 \div 3 \times 2 + 27 \div (-3)$

12. $54 \div 9 \times 3 + 36 \div (-4)$

13. $\frac{2}{3} - \frac{5}{4} + \frac{1}{6}$

14. $\frac{7}{8} - \frac{5}{6} + \frac{11}{12}$

15. $\frac{5}{4} - \frac{7}{6} - \frac{3}{8}$

16. $\frac{8}{9} - \frac{3}{4} - \frac{1}{6}$

17. $\frac{7}{12} - \frac{11}{18} - \frac{5}{9}$

18. $\frac{4}{7} + \frac{1}{6} - \frac{9}{14}$

19. $\frac{11}{15} + \frac{10}{9} - \frac{7}{5}$

20. $\frac{3}{8} + \frac{5}{9} - \frac{13}{18}$

21. $\frac{48}{91} \times \frac{26}{35} \times \frac{49}{64}$

22. $\frac{27}{68} \times \frac{119}{18} \times \frac{16}{21}$

23. $\frac{32}{51} \times \frac{81}{80} \div \frac{63}{68}$

24. $\frac{35}{36} \times \frac{16}{45} \div \frac{28}{27}$

25. $\frac{45}{38} \div \frac{40}{57} \times \frac{64}{81}$

26. $\frac{32}{33} \div \frac{128}{63} \times \frac{22}{21}$

27. $\frac{40}{27} \div \frac{16}{81} \div \frac{15}{8}$

28. $\frac{18}{49} \div \frac{27}{56} \div \frac{32}{35}$

29. $\frac{69}{68} \div \left(\frac{46}{51} \times \frac{27}{16}\right)$

30. $\frac{152}{99} \div \left(\frac{128}{81} \times \frac{171}{44}\right)$

31. $\frac{5}{4} + \frac{7}{9} \times \frac{3}{14}$

32. $\frac{5}{6} - \frac{3}{22} \times \frac{11}{4}$

33. $\frac{8}{9} - \frac{4}{5} \times \frac{15}{16}$

34. $\frac{7}{3} - \frac{9}{8} \div \frac{3}{2}$

35. $\frac{4}{9} - \frac{1}{3} \div \frac{4}{13}$

36. $\frac{7}{8} - \frac{15}{4} \div \frac{9}{2}$

37. $\frac{3}{4} + \frac{4}{11}\left(\frac{5}{6} - \frac{3}{8}\right)$

38. $\frac{2}{3} + \frac{3}{2}\left(\frac{7}{6} - \frac{5}{4}\right)$

39. $\frac{5}{9} - \frac{17}{8} \div \left(\frac{5}{6} - \frac{9}{4}\right)$

40. $\frac{1}{6} + \frac{2}{27} \div \left(\frac{8}{9} - \frac{11}{12}\right)$

Write each of the following common fractions in decimal form:

41. $\frac{10}{9}$
42. $\frac{8}{11}$
43. $\frac{13}{12}$
44. $\frac{22}{15}$
45. $\frac{14}{33}$
46. $\frac{40}{33}$
47. $\frac{9}{37}$
48. $\frac{25}{37}$

Chapter 2 Remove the grouping symbols and combine like terms:

49. $6 - 3(2x - 5) + 2(x - 7)$
50. $4x + [8 - (3x - 1)]$
51. $9x + [2x - (3 + x)]$
52. $5x - [7 - 2(7 - 3x)]$
53. $7x + [x - (3x - 2)] - [8 - 2(3 - x)]$
54. $x - 2[3x + (4 - x)] - [9 - 3(x - 1)]$
55. $12 + \{x - 3[y + (x - 3) - 2(y - 5)]\}$
56. $7 - 4\{x - 2[x - (y - 3) + 2(y - 4)] - 5\}$

Evaluate the following expressions when $a = 2$, $b = 4$, $c = -1$, and $d = -3$:

57. $ac^6 - bd$
58. $c^3(ad - bc)^2$
59. $(b + c)^2(4d + a^2b)$
60. $(a^2 + bc)^{10}(a^3b - d^4)^5$
61. $\frac{3a^2 + bd}{a - bc^2}$
62. $\frac{2b^2 + cd}{2b^2 - cd}$

Perform the indicated operations and simplify:

63. $4x^3y^2(yz^2)(-3xz^3)$
64. $3xy^2(-4x^2y)(-2yz^2)$
65. $(-2xy^2)^3$
66. $(-2^2x^2y^3)^3$
67. $(-3x^2y)^4$
68. $(-2^3xy^3)^2$
69. $(2^2xy^3)^2(3x^2y)^3$
70. $(-2^3x^2y^2)^2(y^2z)^3$
71. $(-xy)^3(2x^2yz)^2(-4xz^3)$
72. $(-3xy^2)^2(2xy^3)(-x^2z^3)^3$
73. $[3a(x - 1)]^2[2a^2(x - 1)^2]^3$
74. $[-a^2(x + 2)]^3[a(x + 2)^3]^2$
75. $(2^2x^2)^3 - x^2(-3x)^4$
76. $(3x)^3(-x^2)^3 + 7x(-x^4)^2$
77. $(x - 1)(2x^2 + 2x - 3)$
78. $(x - 2)(3x^2 + 6x - 1)$
79. $(x^2 + x - 1)^2$
80. $(x^2 - x + 2)^2$
81. $(x - 1)(x - 3) - x(x - 4)$
82. $(x + 3)(x - 2) - x(x + 1)$
83. $(2x + 3)(x - 2) - 2(x - 1)(x + 3)$
84. $(3x + 1)(x + 5) - 3(x - 1)^2$
85. $\frac{24x^4y^3}{-9x^2y^6}$
86. $\frac{-12x^2y^8}{36x^6y^4}$
87. $\left(\frac{2x^3yz^2}{6x^4y^3}\right)^3$

88. $\left(\frac{14xy^2z^4}{21x^3y^2z^2}\right)^4$

89. $\frac{6^4}{9^3}$

90. $\frac{18^3}{12^4}$

91. $\frac{(4a^2b^3)^3}{(8a^4b^2)^2}$

92. $\frac{(-4a^2b^3)^2}{(2ab^2)^3}$

93. $\frac{(15a^2b^3c^4)^3}{(10a^2bc^2)^4}$

94. $\frac{(12a^3b^4c^2)^4}{(9a^2b^4c^3)^3}$

95. $4x^8 \div (x^3) - x(3x^2)^2$

96. $9x^9 \div (x^2)^3 - 4x^2(2x - 1)$

97. $7x^6y^8 \div (xy^2)^3 - 12x^7y^4 \div (3x^2y)^2$

98. $16x^5y^{10} \div (-2xy^3)^3 - 6x^6y^5 \div (-xy)^4$

99. $(2x^4 - x^3 + 2x^2 + 3x - 2) \div (x^2 - x + 2)$

100. $(2x^4 - 5x^3 - 2x^2 - x - 6) \div (2x^2 - x + 2)$

101. $(x^4 + 3x^3 - 3x^2 - 8x + 7) \div (x^2 + 2x - 3)$

102. $(3x^4 - 2x^3 + 9x^2 + 5x - 6) \div (3x^2 + x - 2)$

103. $(9x^4 + 14x^2 + 9) \div (3x^2 - 2x + 3)$

104. $(4x^4 - x^2 + 8x - 16) \div (2x^2 - x + 4)$

105. $(2x^5 - 5x^3 + 16x^2 - 13x + 3) \div (2x^2 + 4x - 3)$

106. $(18x^5 - 11x^3 + 17x^2 + 7x - 3) \div (6x^2 + 2x - 1)$

107. $(2x^4 - 3x^3y - 5x^2y^2 - 4xy^3 - 2y^4) \div (x^2 - 2xy - 2y^2)$

108. $(6x^4 - x^3y - 15x^2y^2 - 3xy^3 + 4y^4) \div (3x^2 + xy - y^2)$

109. $(2x^4 - 25x^2y^2 + 27xy^3 - 4y^4) \div (x^2 + 3xy - 4y^2)$

110. $(3x^4 - 14x^3y + 13x^2y^2 - 2y^4) \div (x^2 - 4xy + 2y^2)$

Chapter 3 Solve the following equations:

111. $2(x - 1) + 3(x + 2) = -1$

112. $6(x + 1) + 2(x + 3) = 5x$

113. $4(x - 3) - 3(2x - 1) = 1$

114. $5(2x - 1) - 2(3x - 2) = 3$

115. $(x - 1)(x + 2) - x(x - 3) = 0$

116. $(x - 3)(x + 1) - x(x - 4) = 0$

117. $(2x + 1)(x + 3) - 2x(x + 2) = 3$

118. $(3x - 1)(x + 2) - 3x(x - 4) = 7x$

119. $(2x - 1)(3x + 2) - (6x + 1)(x - 2) = 15$

120. $(4x - 5)(x - 1) - 2(2x - 1)(x + 3) = -8$

121. $\frac{2x}{3} + \frac{1}{6} = \frac{7x}{8} - \frac{1}{4}$

122. $\frac{3x}{4} - \frac{2}{9} = \frac{2x}{3} - \frac{7}{18}$

123. $\frac{2x}{9} - \frac{1}{18} = \frac{3x}{8} + \frac{1}{4}$

124. $\frac{x}{3} - \frac{9}{8} = \frac{11x}{16} - \frac{5}{12}$

125. $1\frac{1}{3}x + 2\frac{2}{3} = 1\frac{3}{4}x + \frac{1}{6}$

126. $2\frac{2}{5}x + 1\frac{1}{5} = 2\frac{1}{3}x + \frac{2}{3}$

127. $\dfrac{9x+1}{7} + \dfrac{3x-2}{3} = 1$

128. $\dfrac{x-4}{3} - \dfrac{2x+3}{4} = \dfrac{1}{12}$

129. $\dfrac{4x-3}{9} - \dfrac{x-1}{4} = \dfrac{1}{12}$

130. $\dfrac{x-3}{5} - \dfrac{2-7x}{4} = \dfrac{1}{5}$

131. $\dfrac{3}{2}(x+4) + \dfrac{2}{3}(x-1) = 14$

132. $\dfrac{2}{3}(x+1) + \dfrac{3}{4}(x-1) = \dfrac{4}{3}$

133. $\dfrac{7}{4}(x-1) - \dfrac{8}{9}(2x-1) = -\dfrac{5}{6}$

134. $\dfrac{3}{8}(2x-3) - \dfrac{4}{3}(x-2) = \dfrac{2}{3}$

135. $.06(60{,}000 - x) - .08x = 520$

136. $.25(30{,}000 - x) - .2x = 3000$

137. $.15(15{,}000 - x) - .08x = 1330$

138. $.08x - .03(21{,}000 - x) = 800$

139. $.25x - .1(12{,}000 - x) = 375$

140. $.05x - .06(x - 20{,}000) = 1080$

List the elements in each of the following sets:

141. $\{x \mid 3 - 4(2x-1) = 2(3-4x)\}$

142. $\{x \mid 2 + 3(4x+1) = 6(2x+5)\}$

143. $\{x \mid 4 + 5(3x+1) = 3(5x+3)\}$

144. $\{x \mid 2 - (6x-1) = 3(1-2x)\}$

145. One number is 18 more than another number. Find the two numbers if 3 times the smaller number is 1 more than twice the larger number.

146. Find three consecutive odd integers such that 5 times the sum of the second and third integers is 9 more than 11 times the first integer.

147. Find two numbers whose difference is 6 and the difference of their squares is 13 less than 11 times the larger number.

148. The tens digit of a two-digit number is 4 more than the units digit. If the number is 7 less than 8 times the sum of the digits, find the number.

149. The sum of the digits of a three-digit number is 17. The units digit is 2 more than the tens digit. If the number is 4 more than 106 times the hundreds digit, find the number.

150. The sum of the digits of a three-digit number is 20. The tens digit is 1 more than the hundreds digit. If the units and hundreds digits are interchanged, the number is decreased by 495. Find the original number.

151. The discount on a watch was \$75.25 based on a 35% rate of discount. What was the regular price of the watch?

152. A personal computer was sold for \$269.50 after a 23% reduction sale. What was the regular price of the computer?

153. The selling price of a camera is \$177.80 What is the cost if markup is 27% of the cost?

154. Two sums of money totaling \$46,000 earn respectively 7.6% and 6.8% interest per year. Find the two amounts if together they earn \$3344.

155. Emily has \$14,000 invested at 7.4%. How much must she invest at 9.75% so that the interest from both investments makes her income \$3181?

156. The amount of annual interest earned by \$9000 is \$74 more than that earned by \$7000 at .6% more interest per year. What is the rate of interest on each amount?

157. How many gallons of a 5.4% acid solution must be added to 21 gallons of a 7.8% acid solution to produce a 6.8% acid solution?

158. A man mixed 120 pounds of a 26% zinc alloy with 70 pounds of a 45% zinc alloy. What is the percentage of zinc in the mixture?

159. A man mixed 120 pounds of a 60% copper alloy with 90 pounds of a second copper alloy. What is the percentage of copper in the second alloy if the mixture is 69% copper?

160. A woman has $4.55 in 5¢, 10¢, and 25¢ coins. If she has 40 coins in all and there are 6 more 10¢ coins than 25¢ coins, how many coins of each kind does she have?

161. Helen bought $8.60 worth of 5¢, 17¢, and 22¢ stamps for a total of 48 stamps. If the 17¢ stamps are 4 more than the 5¢ stamps, how many stamps of each kind did she buy?

162. The receipts from the sale of 56,000 tickets for a football game totaled $519,000. The tickets were sold at $7.50, $10.50, and $14. Five times as many $7.50 tickets were sold as $14 tickets. How many tickets of each kind were sold?

163. Tom had an appointment 57 miles away. He drove his car at an average speed of 30 mph in town and 54 mph on the freeway. If the trip took 1 hour and 14 minutes, how far did Tom drive in town?

164. Find the corresponding Celsius and Kelvin readings of a 50 F reading.

165. Find the corresponding Fahrenheit reading of a 25 C reading.

166. Andy weighs 57 pounds and sits on a teeterboard 10 feet from the fulcrum. If Albert weighs 95 pounds, how far from the fulcrum must he sit to just balance Andy?

167. A bar with negligible weight is in balance when a 124-pound weight is placed 9 feet from the fulcrum on one side and 48- and 65-pound weights are placed 5 feet apart on the other side of the fulcrum, with the 48-pound weight closer to the fulcrum. How far from the fulcrum is the 48-pound weight?

168. If two opposite sides of a square are each increased by 5 inches and the other two sides are each decreased by 2 inches, the area is increased by 17 square inches. Find the side of the square.

169. The length of a building is 40 feet less than twice its width. The sidewalk around the building is 12 feet wide and its area is 8256 square feet. What are the dimensions of the building?

170. The second angle of a triangle is 4° more than the first angle. The third angle is 6° less than 1.5 times the first angle. How many degrees are in each angle?

Chapter 4 Find the solution set of each of the following inequalities:

171. $4x - 7 > 2x + 5$

172. $3x + 9 > x + 3$

173. $8x - 2 < 5x + 7$

174. $6x - 5 < 4x - 3$

175. $7 - 3(2x - 1) \geq 4$

176. $5 - 2(x + 5) \geq 1$

177. $9 - 5(x - 2) \leq 4$

178. $4 - 3(x + 2) \leq 10$

179. $3(x + 1) + 7 < 2(3x - 1)$

180. $2(x - 3) - 5 < 3(x - 4)$

181. $8(x-2)-1>5(3x-2)$

182. $3(4x-1)-14>7(2x-3)$

183. $(x+3)(x-1)-x(x-2)<5$

184. $(x-4)(x+2)-x(x+1)<1$

185. $(2x+1)(x-2)-2x(x+4)\geq 9$

186. $(3x-2)(2x-1)-x(6x+1)\geq 6$

187. $\frac{2x}{3}-\frac{2}{3}>\frac{7x}{6}-\frac{1}{2}$

188. $\frac{3x}{8}-\frac{5}{6}>\frac{x}{2}-\frac{1}{3}$

189. $\frac{x}{6}-\frac{7}{12}<\frac{1}{4}+\frac{x}{3}$

190. $\frac{x}{9}-\frac{1}{4}<\frac{x}{6}-\frac{5}{12}$

191. $\frac{2x+1}{2}-\frac{x-2}{3}<-\frac{1}{6}$

192. $\frac{x+1}{3}-\frac{2x-1}{4}<\frac{1}{12}$

193. $\frac{3(x+2)}{5}-\frac{x-3}{4}\geq\frac{8}{5}$

194. $\frac{2(x+1)}{3}-\frac{3(x-1)}{8}\geq-\frac{5}{12}$

List the elements in each of the following sets given that $x \in R$:

195. $\{x \mid 2(x+3)-3(x-2)>5-x\}$

196. $\{x \mid 7(x+1)-5(x-4)>2x+3\}$

197. $\{x \mid 3(2x-5)-5(x-1)<x+1\}$

198. $\{x \mid 4(x+2)-7(x+3)<1-3x\}$

199. $\{x \mid 6(x-3)-5(x-2)<x-9\}$

200. $\{x \mid 2(2x-3)-3(x+4)<x-20\}$

201. $\{x \mid 4(x+1)-3(x+7)>x-3\}$

202. $\{x \mid 6(3x-1)-7(2x+1)>4x+1\}$

Find the solution set of each of the following systems:

203. $3x-2<2x+3$
$7x-4>3x+4$

204. $4x-1>2x-5$
$x+7>3x-1$

205. $2x-3\leq 7x+12$
$3x+4<10-3x$

206. $x-4\leq 5x+24$
$4x+9<2x+5$

207. $7-2x\geq x-11$
$6-3x<x-6$

208. $7x-3\leq x-3$
$3x-10<5x-2$

209. $3x+4>x+6$
$2x+1<5x-8$

210. $4x+5>2x+1$
$x+2<3x-2$

211. $3x+1<5x+4$
$6x-7\geq 2x-3$

212. $4x-3<5x-4$
$8x-9\geq 4x+1$

213. $3(x-1)<x-5$
$4(x-4)<x-1$

214. $5(x-3)<x+1$
$3(x+2)>5x+2$

215. $2(x-5)\geq 9x+4$
$7(x-3)<3(x+1)$

216. $4(x+2)\leq x-1$
$3(x+3)>7x+5$

217. $4(2x-3)\geq x+9$
$5(x+3)\geq 2(4x+3)$

218. $6(x-1)\leq 7x-5$
$2(3x-4)\geq 9x-5$

219. $3(x+2)\leq 7x-10$
$7(x-1)<3x+1$

220. $4(2x+3)\leq 3x-13$
$3(4x+7)\geq 4x-3$

Find the values of x that satisfy each of the following:

221. $2 < x + 4 < 9$
222. $-2 < x - 3 < 4$
223. $-1 < 2x - 5 < 7$
224. $4 < 3x + 7 < 10$
225. $-3 < 4 - x < 2$
226. $2 < 5 - x < 9$
227. $5 < 3 - 2x < 13$
228. $7 < 3 - 4x < 11$
229. $4x - 5 < 2x - 1 < 5x + 8$
230. $3x - 7 < x + 1 < 4x - 2$
231. $x - 3 < 2x + 1 < 3x - 5$
232. $2x - 7 < 3x + 2 \leq 5x - 4$
233. $5x - 1 < 4x + 3 \leq 3x + 1$
234. $3x - 4 \leq 2x + 1 < x - 6$
235. $2x + 3 < 4x - 1 < 3x + 3$
236. $4x + 1 < 6x - 5 < 3x + 1$

Find the solution set of each of the following equations:

237. $|x - 1| = 6$
238. $|x - 2| = 5$
239. $|x + 3| = 8$
240. $|x + 4| = 7$
241. $|x + 7| = 0$
242. $|x - 9| = 0$
243. $|2x - 1| = 5$
244. $|3x - 2| = 10$
245. $|4x + 5| = 11$
246. $|3x + 7| = 13$
247. $|4 - 3x| = 16$
248. $|7 - 2x| = 1$
249. $|2x - 1| = x + 4$
250. $|4x - 3| = x + 6$
251. $|1 - 7x| = 3x + 4$
252. $|8 - 5x| = x + 4$
253. $|2x + 1| = 3x + 2$
254. $|3x + 4| = 5x + 6$
255. $|4x + 3| = 6x + 7$
256. $|5x + 2| = 7x + 4$
257. $|3x + 7| = x + 1$
258. $|5x + 6| = 2x - 3$
259. $|3x - 1| = 3x - 1$
260. $|4x - 3| = 4x - 3$
261. $|4 - 3x| = 3x - 4$
262. $|6 - 5x| = 5x - 6$

Find the solution set of each of the following inequalities:

263. $|6x + 1| < 13$
264. $|4x + 9| < 1$
265. $|3x + 10| < 2$
266. $|3x - 4| < 5$
267. $|2x - 5| < 1$
268. $|5x - 2| < 7$
269. $|7 - 4x| < 3$
270. $|9 - 7x| < 5$
271. $|1 - 3x| < 11$
272. $|3x + 5| > 10$
273. $|5x + 1| > 4$
274. $|2x + 6| > 3$
275. $|5x - 6| > 4$
276. $|3x - 1| > 8$
277. $|11 - 3x| > 1$
278. $|4 - 7x| > 18$
279. $|4x + 3| > 0$
280. $|2x - 7| > 0$
281. $|4x - 1| < 2x + 5$
282. $|3x - 3| < x + 8$
283. $|7 - 2x| < 8 - x$
284. $|6 - 4x| < 6 - x$
285. $|2x + 4| < 3x - 2$
286. $|4x + 5| < 5x + 7$
287. $|8x - 1| < 3x - 11$
288. $|4x - 1| < 2x - 15$
289. $|4x - 3| > x + 3$
290. $|5x - 4| > x + 6$
291. $|7x - 3| > 2x + 7$
292. $|2x - 5| > 6x - 7$
293. $|3x + 1| > 5x + 9$
294. $|3x - 2| > 5x - 6$
295. $|2x + 3| > x - 1$
296. $|5x + 7| > x - 4$

chapter 5

Factoring Polynomials

5.1 Factors common to all terms

5.2 Factoring a binomial

5.3 Factoring a trinomial

5.4 Factoring by completing the square

5.5 Factoring four-term polynomials

When numbers are multiplied together each of the numbers multiplied to get the product is called a **factor**. Sometimes it is desirable to write a polynomial as the product of certain of its factors. This operation is called **factoring**. Here, we are interested in factoring polynomials with integral coefficients.

A polynomial is said to be **factored completely** if it is expressed as the product of polynomials with integral coefficients and no one of the factors can still be written as the product of two polynomials with integral coefficients.

Following is a discussion of factoring some special polynomials.

5.1 Factors Common to All Terms

The **greatest common factor (GCF)** of a set of integers is defined as the greatest integer that divides each element of that set of integers.
The GCF can be obtained as follows:

1. Factor the integers into their prime factors.
2. Write the factors in the exponent form.
3. Take the common bases each to its lowest exponent.

EXAMPLE Find the GCF of $\{144, 216, 360\}$.

Solution

$144 = 2^4 \cdot 3^2$

$216 = 2^3 \cdot 3^3$

$360 = 2^3 \cdot 3^2 \cdot 5$

The common bases are 2 and 3.

The least exponent of 2 is 3 and of 3 is 2.

Hence the GCF $= 2^3 \cdot 3^2 = 72$.

The greatest common factor of a set of monomials can be found by taking the product of the GCF of the coefficients of the monomials and the common literal bases, each to its lowest exponent.

EXAMPLE Find the GCF of $48x^3y^2$, $72x^4y$, and $120x^2y^3z$.

Solution

$$48x^3y^2 = 2^4 \cdot 3x^3y^2$$
$$72x^4y = 2^3 \cdot 3^2x^4y$$
$$120x^2y^3z = 2^3 \cdot 3 \cdot 5x^2y^3z$$

The common bases are 2, 3, x, and y.

The least exponent of 2 is 3, of 3 is 1, of x is 2, and of y is 1.

Hence the GCF $= 2^3 \cdot 3^1x^2y^1 = 24x^2y$.

EXAMPLE Find the GCF of $4a^3(x-3)^2, 8a^2(x-3)^3, 12a^4(x-3)$.

Solution
$$4a^3(x-3)^2 = 2^2a^3(x-3)^2$$
$$8a^2(x-3)^3 = 2^3a^2(x-3)^3$$
$$12a^4(x-3) = 2^2 \cdot 3a^4(x-3)$$
The GCF $= 2^2a^2(x-3)$.

Note Since $(b-a) = -(a-b)$, the greatest common factor of $c(a-b)$ and $d(b-a)$ is either $(a-b)$, or $(b-a)$.

When the terms of a polynomial have a common factor, the distributive law,

$$ab_1 + ab_2 + \cdots + ab_n = a(b_1 + b_2 + \cdots + b_n)$$

is used to factor the polynomial. One factor is the greatest common factor of all the terms of the polynomial. The other factor is the entire quotient, obtained by dividing each term of the polynomial by the common factor:

$$\begin{aligned} ab_1 + ab_2 + \cdots + ab_n &= a\left(\frac{ab_1}{a} + \frac{ab_2}{a} + \cdots + \frac{ab_n}{a}\right) \\ &= a(b_1 + b_2 + \cdots + b_n) \end{aligned}$$

EXAMPLE Factor $18x^5y - 24x^3y^2 + 36x^4y^3$.

Solution The greatest common factor is $6x^3y$:

$$\begin{aligned} 18x^5y - 24x^3y^2 + 36x^4y^3 &= 6x^3y\left(\frac{18x^5y}{6x^3y} - \frac{24x^3y^2}{6x^3y} + \frac{36x^4y^3}{6x^3y}\right) \\ &= 6x^3y(3x^2 - 4y + 6xy^2) \end{aligned}$$

EXAMPLE Factor $16x^2(x-2)^2 - 24x(x-2)^3$.

Solution The greatest common factor is $8x(x-2)^2$:

$$\begin{aligned} 16x^2(x-2)^2 - 24x(x-2)^3 &= 8x(x-2)^2\left[\frac{16x^2(x-2)^2}{8x(x-2)^2} - \frac{24x(x-2)^3}{8x(x-2)^2}\right] \\ &= 8x(x-2)^2[2x - 3(x-2)] \\ &= 8x(x-2)^2(2x - 3x + 6) \\ &= 8x(x-2)^2(6-x) \end{aligned}$$

EXAMPLE Factor $12x(x-3)^2 + 8x^2(3-x)$.

Solution
$$\begin{aligned} 12x(x-3)^2 + 8x^2(3-x) &= 12x(x-3)^2 - 8x^2(x-3) \\ &= 4x(x-3)[3(x-3) - 2x] \\ &= 4x(x-3)(x-9) \end{aligned}$$

Exercise 5.1

Find the greatest common factor of each of the following:

1. 6, 9, 12 **2.** 12, 16, 18 **3.** 18, 24, 30 **4.** 14, 28, 42
5. 10, 15, 20 **6.** 16, 24, 32 **7.** 24, 36, 60 **8.** 26, 39, 52
9. 36, 48, 72 **10.** 32, 48, 60 **11.** 45, 54, 63 **12.** 48, 72, 120

13. $x^2y^2,\ x^3y,\ x^4y^3$
14. $x^5y^2,\ x^2y^3,\ x^2y^1z$
15. $20xy^2z^2,\ 15x^2y^3,\ 10xy^4z^3$
16. $12x^3y^2,\ 24x^3yz,\ 18x^3y^3$
17. $15(x+1),\ 6(x+1)^2$
18. $36(x+4),\ 48(4+x)$
19. $x(x+3)^2,\ x^2(3+x)$
20. $12x^3(x+2),\ 6x^2(2+x)^2$
21. $76(x-1)^2,\ 57(x-1)$
22. $24(2x-1),\ 36(2x-1)^2$
23. $x(x-2),\ x^2(2-x)$
24. $6x^2(x-3),\ 9x(3-x)$
25. $(x+2)(x+4),\ (6+x)(4+x)$
26. $(x+1)^2,\ (x+1)(x-1)$
27. $(x-2)(x+3),\ (x-2)(x-3)$
28. $(x+2)(x-5),\ (3+x)(5-x)$
29. $(x-3)(2x+1),\ (3-x)(2x+3)$
30. $(2x-1)(x+1),\ (1-2x)(4+x)$

Factor the following polynomials:

31. $7x+7$ **32.** $8x+4$ **33.** $18x+27$ **34.** $4x-8$
35. $15x-5$ **36.** $8-16x$ **37.** $9-3x$ **38.** $18-12x$
39. $6x^2+3x$ **40.** $9x-18x^2$ **41.** $3xy-6y$ **42.** $5y-10xy$
43. $8xy+12x$ **44.** $xy+x^2$ **45.** x^2y+y^2x **46.** $3x^2-6x^2y$
47. $21y^2-14xy$ **48.** $16xy^2+24y^3$ **49.** $18xy-24x^2y^2$
50. $12x^4-12x^3$ **51.** $27x^3-18x^2$ **52.** $14x^4+21x^3$
53. $2x^2y^2+y^3x$ **54.** $12x^5-24x^4$ **55.** $6x^2-12x+18$
56. $8x^2+4x-8$ **57.** x^3-x^2+x **58.** $2x^3+x^2+3x$
59. $6x^3+9x^2-3x$ **60.** $14x^4-21x^3+28x^2$

61. $3(2x+1)+x(2x+1)$
62. $5(3x+1)+x(3x+1)$
63. $(x+6)^2+(x+6)$
64. $(x+2)+(x+2)^2$
65. $4(x+1)^2+8(x+1)$
66. $(x+5)^2+(x+5)^3$
67. $(2x+1)^3+2(2x+1)^2$
68. $10(x+3)^2+5(x+3)^3$
69. $2(x-4)^2+6(x-4)$
70. $4(x-1)^2+12(x-1)$
71. $8x(x-3)-12y(x-3)$
72. $x^2(x+2)^2-x^3(x+2)$
73. $3x^2(3x+1)-6x(3x+1)^2$
74. $7x(4x+1)^2-14x^2(4x+1)$
75. $(x+3)(x-2)-2(x+5)(x-2)$
76. $(x+4)(x-1)-3(x+3)(x-1)$
77. $(x+1)(x-4)-3(x-3)(x-4)$
78. $2(x-2)(x-3)-(x-3)(x-4)$
79. $(x+1)^2-2(x+1)(x+2)$
80. $(x-1)^2-3(x+3)(x-1)$
81. $2(x+2)^2-(x+2)(x-1)$
82. $3(x+4)^2-(x-2)(x+4)$
83. $x(x-2)+5(2-x)$
84. $3(x-3)+x(3-x)$
85. $x(x-4)+y(4-x)$
86. $2x(x-y)+y(y-x)$
87. $4x(x-7)-2y(7-x)$
88. $3x(x-1)-6y(1-x)$
89. $x^2(2x-1)-x(1-2x)$
90. $6x(3x-1)-12x^2(1-3x)$

5.2 Factoring a Binomial

The methods of factoring polynomials will be presented according to the number of terms in the polynomial to be factored.

A monomial is already in factored form, so the first type of polynomial to be considered for factoring is a binomial. Here we shall discuss factoring three types of binomials.

Squares and Square Roots

The **squares** of the numbers $2,\ 3^2,\ x,\ y^2$, and x^2y
are, respectively, $2^2,\ 3^4,\ x^2,\ y^4$, and x^4y^2

The $2, 3^2, x, y^2$, and x^2y are called the **square roots** of $2^2, 3^4, x^2, y^4$, and x^4y^2, respectively.

The square root of a number x is denoted by $\sqrt[2]{x}$. The $\sqrt{\ }$ is called a **radical sign**, the 2 is called the **index**, and the x is called the **radicand**. When there is no index written, the index 2 is implied.

Although the square of both $(+5)$ and (-5) is 25, when we talk about the square root of 25, we will mean the positive number 5 and not the negative number (-5).

DEFINITION

> A number is said to be a **perfect square** if its square root is a rational number.

The square root of a specific number can be found by factoring the number into its prime factors, writing it in the exponent form, and then taking each base to one-half of its original exponent (when we square a number we multiply its exponent by 2).

EXAMPLES

1. $\sqrt{324} = \sqrt{2^2 \cdot 3^4} = 2^1 \cdot 3^2 = 18$

2. $\sqrt{784} = \sqrt{2^4 \cdot 7^2} = 2^2 \cdot 7^1 = 28$

DEFINITION

> When a is a literal number and $n \in N$, we define $\sqrt{a^{2n}}$ as $(\sqrt{a})^{2n} = a^n$. If the exponent is not divisible by 2, the number is not a perfect square.

EXAMPLES

1. $\sqrt{x^4} = x^2$

2. $\sqrt{x^2y^4} = xy^2$

3. $\sqrt{36x^6} = \sqrt{2^2 \cdot 3^2x^6} = 2 \cdot 3x^3 = 6x^3$

The numbers 2, 3, 5, 7, 8, 10, etc., are not perfect square numbers. This means that there are no rational numbers whose squares are 2, 3, 5, etc.

DEFINITION

The square roots of numbers that are not perfect squares are called **irrational numbers**.

Difference of Two Squares

The product of the two factors $(x + y)(x - y)$ is $x^2 - y^2$, the difference of two perfect square terms. The factors of the difference of two squares are the sum and difference of the respective square roots of the two squares.

EXAMPLE Factor $4x^2 - 25$.

Solution The square root of $4x^2$ is $2x$ and of 25 is 5.

Hence $4x^2 - 25 = (2x + 5)(2x - 5)$

Note Remember to factor the polynomial completely.

EXAMPLE Factor completely $x^4 - 16y^4$.

Solution

$$\begin{aligned} x^4 - 16y^4 &= (x^2 + 4y^2)(x^2 - 4y^2) \\ &= (x^2 + 4y^2)(x + 2y)(x - 2y) \end{aligned}$$

Note Before checking to see if the binomial is a difference of two squares, check for a common factor. That is always the first operation to be performed.

EXAMPLE Factor completely $3x^2y^2 - 3x^4$.

Solution

$$\begin{aligned} 3x^2y^2 - 3x^4 &= 3x^2(y^2 - x^2) \\ &= 3x^2(y + x)(y - x) \end{aligned}$$

EXAMPLE Factor $x^2 - 9(y - 2)^2$.

Solution

$$x^2 - 9(y - 2)^2 = [x + 3(y - 2)][x - 3(y - 2)]$$
$$= (x + 3y - 6)(x - 3y + 6)$$

Note $(a + b)(a - b) = (a - b)(a + b)$

Exercise 5.2A

Factor completely:

1. $x^2 - 4$
2. $x^2 - 9$
3. $x^2 - 25$
4. $x^2 - 49$
5. $x^2 - 81$
6. $x^2 - 144$
7. $x^2 - 225$
8. $x^2 + 16$
9. $x^2 + 36$
10. $1 - x^2$
11. $16 - x^2$
12. $64 - x^2$
13. $169 - x^2$
14. $4x^2 - 1$
15. $36x^2 - 1$
16. $9x^2 - 4$
17. $9x^2 - 25$
18. $4x^2 - 121$
19. $16x^2 - 9$
20. $25x^2 - 16$
21. $4 - x^2y^2$
22. $25 - 16x^2$
23. $x^2 - y^2z^2$
24. $4x^2 - 9y^2$
25. $4 - x^4$
26. $x^4 - 9y^2$
27. $3x^2 - 12y^2$
28. $36x^2 - 4y^2$
29. $9x^2 - 36y^2$
30. $7x^2y^2 - 28z^2$
31. $x^3 - x$
32. $3x - 3x^3$
33. $x^4 - x^2$
34. $2x^4 - 18x^2y^2$
35. $5xy^2 - 20x^3$
36. $4x^3 - 36xy^2$
37. $x^4 - y^4$
38. $x^4 - 81y^4$
39. $16x^4 - 625y^4$
40. $81x^4 - 16y^4$
41. $x^8 - 256y^4$
42. $64 - 4x^4$
43. $3x^4 - 48y^4z^4$
44. $162x^4 - 32$
45. $144x^4 - 729$
46. $5x - 80x^5$
47. $243x^6 - 3x^2$
48. $4x^6 - 4x^2y^4$
49. $(x + y)^2 - 4$
50. $(2x - y)^2 - 9$
51. $(x - 2y)^2 - 36$
52. $(3x - y)^2 - 49$
53. $(x + 2y)^2 - z^2$
54. $(3x + 1)^2 - 4y^2$
55. $5(x - y)^2 - 45$
56. $3(x + y)^2 - 12$
57. $3x^2(x - y)^2 - 12z^2$
58. $b^3(x + 2y)^2 - 4b^5$
59. $x^2 - (y + 1)^2$
60. $4x^2 - (y + 2)^2$
61. $9x^2 - (y + 3)^2$
62. $16x^2 - (y + 4)^2$
63. $x^2 - (y - 1)^2$
64. $25x^2 - (y - 2)^2$
65. $x^4 - (2y - 1)^2$
66. $49x^4 - (3y - 1)^2$
67. $x^2 - 4(y - 3)^2$
68. $4x^2 - 9(y - 4)^2$
69. $9x^2 - 16(y - 6)^2$
70. $(x + y)^2 - (x + 1)^2$
71. $(x - y)^2 - (x + 2)^2$
72. $(x - 2y)^2 - (x + 3)^2$
73. $(2x + y)^2 - (x + 4)^2$
74. $(2x - y)^2 - (x - 2)^2$
75. $(x - 3y)^2 - (y - 1)^2$
76. $(x - y)^2 - 4(y - 4)^2$
77. $(x + 4y)^2 - 9(y - 2)^2$
78. $(x - 3y)^3 + (3y - x)$
79. $(x - 2y)^3 + 4(2y - x)$
80. $(2x - y)^3 + 9(y - 2x)$
81. $(4x - y)^3 + 16(y - 4x)$
82. $9(3x - y)^3 + 4(y - 3x)$
83. $8(x - 2y)^3 + 18(2y - x)$

Cubes and Cube Roots

The **cubes** of $2,\ a,\ b^2$, and c^3d^5
are, respectively, $2^3,\ a^3,\ b^6$, and c^9d^{15}.

The $2, a, b^2$, and c^3d^5 are called the **cube roots** of $2^3, a^3, b^6$, and c^9d^{15}, respectively.

The cube root of a number x is denoted by $\sqrt[3]{x}$.

The cube root of a specific number can be found by factoring the number to its prime factors, writing it in the exponent form, and then taking each base to one-third of its original exponent (when we cube a number we multiply its exponent by 3).

EXAMPLES

1. $\sqrt[3]{27} = \sqrt[3]{3^3} = 3$
2. $\sqrt[3]{1728} = \sqrt[3]{2^6 \cdot 3^3} = 2^2 \cdot 3 = 12$

The cube root of any literal number raised to a power is equal to the literal number raised to one-third of its power.

EXAMPLES

1. $\sqrt[3]{a^6} = a^2$
2. $\sqrt[3]{a^9b^{12}} = a^3b^4$

Note When the exponent of the literal number is not divisible by 3, the literal number is not a perfect cube. The cube root of a nonperfect cube is an **irrational number**.

Sum of Two Cubes

Consider the following products:

$$(a + b)(a^2 - ab + b^2) = a^3 + b^3$$
$$(x + 2y)(x^2 - 2xy + 4y^2) = x^3 + 8y^3$$
$$(2 + 3a)(4 - 6a + 9a^2) = 8 + 27a^3$$
$$(3x^2 + 4y^2)(9x^4 - 12x^2y^2 + 16y^4) = 27x^6 + 64y^6$$

Each product is the sum of two perfect cube terms.

The first factor is the sum of the respective cube roots of the two cube terms:

cube root

$$a^3 + 27b^3 = (a + 3b)(\text{second factor})$$

cube root

The second factor consists of three terms and can be arrived at easily from the first factor. The terms of the second factor are

1. The square of the first term in the first factor
2. The negative of the product of the two terms in the first factor
3. The square of the second term in the first factor

negative of the product

$$(a + 3b)(a^2 - 3ab + 9b^2)$$

square

square

EXAMPLES

1. $8x^3 + 1 = (2x + 1)(4x^2 - 2x + 1)$
2. $64 + b^3 = (4 + b)(16 - 4b + b^2)$
3. $(a + b)^3 + c^3 = [(a + b) + c][(a + b)^2 - c(a + b) + c^2]$
4. $54a^3 + 16b^3 = 2(27a^3 + 8b^3)$
   $= 2(3a + 2b)(9a^2 - 6ab + 4b^2)$
5. $x^6 + y^6 = (x^2 + y^2)(x^4 - x^2y^2 + y^4)$

Difference of Two Cubes

Consider the following products:

$$(a - b)(a^2 + ab + b^2) = a^3 - b^3$$
$$(2a - b)(4a^2 + 2ab + b^2) = 8a^3 - b^3$$
$$(5a - 3)(25a^2 + 15a + 9) = 125a^3 - 27$$

Each product is the difference of two perfect cube terms.

The first factor is the difference of the respective cube roots of the two cube terms.

The second factor consists of three terms and can be arrived at easily from the first factor. The terms of the second factor are

1. The square of the first term in the first factor
2. The negative of the product of the two terms in the first factor
3. The square of the second term in the first factor

EXAMPLES

1. $a^3 - 64 = (a - 4)(a^2 + 4a + 16)$
2. $27x^3 - 1 = (3x - 1)(9x^2 + 3x + 1)$

3. $16x^3 - 250y^3 = 2(8x^3 - 125y^3)$
$= 2(2x - 5y)(4x^2 + 10xy + 25y^2)$

4. $8x^3 - (3y - 1)^3 = [2x - (3y - 1)][4x^2 + 2x(3y - 1) + (3y - 1)^2]$

Note When the polynomial can be factored as the difference of either two squares or two cubes, it should be factored as the difference of two squares.

EXAMPLE

$$x^6 - y^6 = (x^3 + y^3)(x^3 - y^3)$$
$$= (x + y)(x^2 - xy + y^2)(x - y)(x^2 + xy + y^2)$$

Exercise 5.2B

Factor completely:

1. $x^3 + 1$ **2.** $x^3 + 8$ **3.** $x^3 + 27$ **4.** $x^3 + 64$
5. $x^3 + 216$ **6.** $x^3 - 1$ **7.** $x^3 - 8$ **8.** $x^3 - 125$
9. $1 - 8x^3$ **10.** $1 - 27x^3$ **11.** $27 - x^3$ **12.** $x^3 - 8y^3$
13. $64x^3 - y^3$ **14.** $8x^3 - 27y^3$ **15.** $3x^3 + 24$ **16.** $4x^3 + 108$
17. $4x^3 + 32y^3$ **18.** $3x^3 - 81$ **19.** $16 - 2x^3$ **20.** $4x^3 - 32$
21. $250x^3 - 2$ **22.** $7x^3 - 56y^3$ **23.** $16x^3 + 54y^3$
24. $3x^3y^6 + 81$ **25.** $81x^3 + 24y^3$ **26.** $54x^3 - 16$
27. $x^4 + 8x$ **28.** $x^3y + y^4$ **29.** $x^4y^2 - xy^5$
30. $54x^4 - 2x$ **31.** $54x^4 + 2xy^3$ **32.** $40x^5 + 5x^2$
33. $x^6 + y^3$ **34.** $2x^6 + 16y^3$ **35.** $x^3 - x^6$ **36.** $x^3y^3 - y^6$
37. $x^6 + 1$ **38.** $64x^6 + 1$ **39.** $x^6 + 8y^6$ **40.** $27x^6 + y^6$
41. $x^8 + x^2y^6$ **42.** $x^6 - 1$ **43.** $64 - x^6$ **44.** $64x^6 - 1$
45. $x^6 - 729$ **46.** $256 - 4x^6$ **47.** $x^2y^6 - x^8$ **48.** $x^9 - x^3$
49. $(x + 1)^3 + y^3$ **50.** $(x + 2)^3 + y^3$ **51.** $(x - 2)^3 + 8y^3$
52. $(x - 1)^3 - 8y^3$ **53.** $(x - 4)^3 - 64y^3$ **54.** $8(x + 3)^3 - y^3$
55. $27(x - 2)^3 - y^3$ **56.** $x^3 + (y + 2)^3$ **57.** $x^3 + (y - 3)^3$
58. $x^3 + 8(y - 1)^3$ **59.** $x^3 + 27(y - 2)^3$ **60.** $x^3 - (y + 2)^3$
61. $27x^3 - (2y + 1)^3$ **62.** $8x^3 - 27(y + 1)^3$ **63.** $x^3 - 64(y - 2)^3$
64. $4x^3 - 32(y - 3)^3$ **65.** $x^4 - x(y - 1)^3$ **66.** $x^3y^3 - y^3(2y - 1)^3$
67. $(x + y)^3 + (y - 1)^3$ **68.** $(x - y)^3 - (y + 2)^3$
69. $(x - 2y)^3 - 8(y - 2)^3$ **70.** $(x + 3y)^3 - 27(y - 3)^3$

5.3 Factoring a Trinomial

Factoring trinomials is divided into two cases:

1. When the trinomial is of the form $x^2 + bx + c,\ b,\ c \in I$
2. When the trinomial is of the form $ax^2 + bx + c,\ a, b,\ c \in I,\ a \neq 1$

Trinomials of the Form $x^2 + bx + c$, $b, c \in I$

Consider the following products:

$$(x + m)(x + n) = x^2 + (m + n)x + mn$$

$$(x - m)(x - n) = x^2 + (-m - n)x + mn$$

$$(x + m)(x - n) = x^2 + (m - n)x - mn$$

$$(x - m)(x + n) = x^2 + (-m + n)x - mn$$

We note the following relations between the products and their factors:

1. The first term in each factor is the square root of the square term in the trinomial.
2. The product of the second terms of the factors is the third term in the trinomial.
3. The sum of the second terms, the signed numbers, is the coefficient of the middle term in the trinomial.

Notes

1. To find the second terms in the factors, look for two signed numbers whose product is the third term in the trinomial, and whose sum is the coefficient of the middle term in the trinomial.
2. When the sign of the third term in the trinomial is plus, the two signed numbers have like signs, which are the same as the sign of the middle term in the trinomial.
3. When the sign of the third term in the trinomial is minus, the two signed numbers have different signs, and the larger one numerically has the sign of the middle term in the trinomial.

EXAMPLE Factor $x^2 + 6x + 8$.

Solution The first term of each factor is $\sqrt{x^2} = x$.

Hence $\quad x^2 + 6x + 8 = (x \quad)(x \quad)$

Since the sign of the last term $(+8)$ is plus, the two signed numbers in the factors have like signs.

Since the sign of the middle term $(+6x)$ is plus, the two signed numbers are positive:

$$x^2 + 6x + 8 = (x + \quad)(x + \quad)$$

We look for two natural numbers whose product is 8 and whose sum is 6. The two numbers are 2 and 4.

Hence $\quad x^2 + 6x + 8 = (x + 2)(x + 4)$

EXAMPLE Factor $x^2 - 11x + 24$.

Solution The first term of each factor is $\sqrt{x^2} = x$.

Hence $x^2 - 11x + 24 = (x \quad)(x \quad)$

Since the sign of the last term $(+24)$ is plus, the two signed numbers in the factors have like signs.

Since the sign of the middle term $(-11x)$ is minus, the two signed numbers are negative:

$x^2 - 11x + 24 = (x - \quad)(x - \quad)$

We look for two natural numbers whose product is 24 and whose sum is 11. The two numbers are 3 and 8.

Hence $x^2 - 11x + 24 = (x - 3)(x - 8)$

EXAMPLE Factor $x^2 - 4x - 12$.

Solution $x^2 - 4x - 12 = (x \quad)(x \quad)$

Since the sign of the last term (-12) is minus, the two numbers in the factors have different signs:

$x^2 - 4x - 12 = (x + \quad)(x - \quad)$

Since the sign of the middle term $(-4x)$ is minus, the numerically larger number has the negative sign:

$x^2 - 4x - 12 = (x + \text{smaller number})(x - \text{larger number})$

We look for two natural numbers whose product is 12 and whose difference is 4. The two numbers are 2 and 6.

Hence $x^2 - 4x - 12 = (x + 2)(x - 6)$

EXAMPLE Factor $x^2 + 16x - 36$.

Solution $x^2 + 16x - 36 = (x \quad)(x \quad)$

Since the sign of the last term (-36) is minus, the two numbers in the factors have different signs:

$x^2 + 16x - 36 = (x + \quad)(x - \quad)$

Since the sign of the middle term $(+16x)$ is plus, the numerically larger number has the plus sign:

$x^2 + 16x - 36 = (x + \text{larger number})(x - \text{smaller number})$

We look for two natural numbers whose product is 36 and whose difference is 16. The two numbers are 2 and 18.

Hence $x^2 + 16x - 36 = (x + 18)(x - 2)$

EXAMPLE Factor $y^4 - 3y^2 - 10$.

Solution The first term in each factor is $\sqrt{y^4} = y^2$.

Hence $y^4 - 3y^2 - 10 = (y^2 - 5)(y^2 + 2)$

EXAMPLE Factor $y^4 - 29y^2 + 100$ completely.

Solution

$$\begin{aligned} y^4 - 29y^2 + 100 &= (y^2 - 4)(y^2 - 25) \\ &= (y + 2)(y - 2)(y + 5)(y - 5) \end{aligned}$$

EXAMPLE Factor $4x^2 + 28xy - 32y^2$.

Solution

$$\begin{aligned} 4x^2 + 28xy - 32y^2 &= 4(x^2 + 7xy - 8y^2) \\ &= 4(x + 8y)(x - y) \end{aligned}$$

EXAMPLE Factor $(a - 2b)^2 + (a - 2b) - 12$.

Solution $(a - 2b)^2 + (a - 2b) - 12$ is of the form $x^2 + x - 12$, whose factors are $(x - 3)(x + 4)$.

Hence

$$\begin{aligned} (a - 2b)^2 + (a - 2b) - 12 &= [(a - 2b) - 3][(a - 2b) + 4] \\ &= (a - 2b - 3)(a - 2b + 4) \end{aligned}$$

Note When the third term of the trinomial is a large number and its factors are not obvious, write the number as the product of its prime factors. Then you can make products of factors using combinations of the primes.

Exercise 5.3A

Factor completely:

1. $x^2 + 5x + 6$	**2.** $x^2 + 7x + 10$	**3.** $x^2 + 4x + 4$
4. $x^2 + 8x + 12$	**5.** $x^2 + 11x + 24$	**6.** $x^2 + 9x + 18$
7. $x^2 + 13x + 12$	**8.** $x^2 + 12x + 32$	**9.** $x^2 + 11x + 18$
10. $x^2 + 14x + 24$	**11.** $x^2 - 3x + 2$	**12.** $x^2 - 4x + 4$
13. $x^2 - 8x + 15$	**14.** $x^2 - 10x + 16$	**15.** $x^2 - 9x + 20$
16. $x^2 - 13x + 30$	**17.** $x^2 - 18x + 45$	**18.** $x^2 - 12x + 35$
19. $x^2 - 14x + 45$	**20.** $x^2 - 15x + 56$	**21.** $x^2 + x - 2$
22. $x^2 + x - 12$	**23.** $x^2 + 3x - 4$	**24.** $x^2 + 6x - 16$
25. $x^2 + 9x - 36$	**26.** $x^2 + 5x - 24$	**27.** $x^2 + 2x - 35$
28. $x^2 + 2x - 24$	**29.** $x^2 + 3x - 54$	**30.** $x^2 + x - 72$
31. $x^2 - 2x - 3$	**32.** $x^2 - x - 30$	**33.** $x^2 - 4x - 12$
34. $x^2 - 5x - 24$	**35.** $x^2 - 4x - 5$	**36.** $x^2 - 5x - 36$
37. $x^2 - 7x - 30$	**38.** $x^2 - 6x - 72$	**39.** $x^2 - 9x - 10$

40. $x^2 - 2x - 63$ **41.** $x^2 + 2x + 24$ **42.** $x^2 + 5x + 14$
43. $x^2 - 3x + 10$ **44.** $x^2 - 8x + 20$ **45.** $x^2 + 6x - 8$
46. $x^2 + 10x - 16$ **47.** $x^2 - 4x - 3$ **48.** $x^2 - 7x - 12$
49. $x^2 + 7x + 12$ **50.** $x^2 + 3x + 2$ **51.** $x^2 + 8x + 15$
52. $x^2 - 4x + 3$ **53.** $x^2 - 6x + 5$ **54.** $x^2 - 9x + 8$
55. $x^2 + 2x - 8$ **56.** $x^2 + x - 6$ **57.** $x^2 + 5x - 14$
58. $x^2 - 2x - 24$ **59.** $x^2 - 3x - 10$ **60.** $x^2 - 3x - 28$
61. $x^2 + 10x + 24$ **62.** $x^2 + 10x + 16$ **63.** $x^2 + 12x + 20$
64. $x^2 + 13x + 40$ **65.** $x^2 - 11x + 30$ **66.** $x^2 - 21x + 20$
67. $x^2 - 20x + 36$ **68.** $x^2 - 14x + 49$ **69.** $x^2 + 8x - 9$
70. $x^2 + 12x - 28$ **71.** $x^2 + x - 42$ **72.** $x^2 + 2x - 48$
73. $x^2 - 6x - 27$ **74.** $x^2 - 4x - 45$ **75.** $x^2 - 8x - 20$
76. $x^2 - 7x - 60$ **77.** $x^2 + 4x + 12$ **78.** $x^2 - 5x + 14$
79. $x^2 + 8x - 12$ **80.** $x^2 - 9x - 18$ **81.** $x^2 - 10x - 16$
82. $x^2 + 18 + 19x$ **83.** $x^2 + 36 + 15x$ **84.** $x^2 + 36 + 12x$
85. $x^2 + 42 - 13x$ **86.** $x^2 + 60 - 17x$ **87.** $x^2 + 72 - 27x$
88. $x^2 - 50 + 5x$ **89.** $6x - 72 + x^2$ **90.** $2x - 63 + x^2$
91. $x^2 - 80 - 2x$ **92.** $x^2 - 36 - 16x$ **93.** $x^2 - 63 - 18x$
94. $x^2 + 2xy + y^2$ **95.** $x^2 + 7xy + 6y^2$ **96.** $x^2 + 10xy + 9y^2$
97. $x^2 + 10xy + 25y^2$ **98.** $x^2 - 6xy + 8y^2$ **99.** $x^2 - 10xy + 21y^2$
100. $x^2 - 11xy + 18y^2$ **101.** $x^2 - 11xy + 28y^2$ **102.** $x^2 + 4xy - 12y^2$
103. $x^2 + 6xy - 7y^2$ **104.** $x^2 + 10xy - 24y^2$ **105.** $x^2 + 3xy - 40y^2$
106. $x^2 - 2xy - 15y^2$ **107.** $x^2 - 3xy - 18y^2$ **108.** $x^2 - 7xy - 18y^2$
109. $x^2 - 13xy - 30y^2$ **110.** $3x^2 + 15x + 12$ **111.** $2x^2 + 16x + 32$
112. $ax^2 + 12ax + 27a$ **113.** $b^2x^2 + 16b^2x + 48b^2$
114. $4x^2 - 28x + 40$ **115.** $3x^2 - 33x + 72$
116. $x^4 + 16x^3 + 28x^2$ **117.** $x^2y^2 - 17xy + 72$
118. $x^2y - 24xy + 63y$ **119.** $b^4x^2 - 19b^3x + 70b^2$
120. $6x^2 - 108 + 18x$ **121.** $4x^2 - 80 + 4x$
122. $5x^2 - 100 + 5x$ **123.** $4x^3 - 32x^2 + x^4$
124. $x^2y^2 - 60 + 11xy$ **125.** $x^4 + 11x^2 - 80$
126. $x^4y^2 - x^2y - 42$ **127.** $ax^3 - 12ax^2 - 45ax$
128. $ax^4 - 6ax^3 - 91ax^2$ **129.** $x^2y^2 - 21xy^3 - 72y^4$
130. $x^4 + x^2 - 2$ **131.** $x^4 - 3x^2 + 2$ **132.** $x^4 - 2x^2 - 8$
133. $x^4 - x^2 - 12$ **134.** $x^4 - 9x^2 + 20$ **135.** $x^4 - 8x^2 - 9$
136. $x^4 - 7x^2 - 18$ **137.** $x^4 - 28x^2 + 75$ **138.** $x^4 - 5x^2 + 4$
139. $x^4 - 2x^2 + 1$ **140.** $x^4 - 10x^2 + 9$ **141.** $x^4 - 26x^2 + 25$
142. $x^4 - 13x^2 + 36$ **143.** $x^4 - 29x^2 + 100$ **144.** $x^6 + 3x^3 + 2$
145. $x^6 - 2x^3 - 3$ **146.** $x^6 + 6x^3 - 16$ **147.** $x^6 + 28x^3 + 27$
148. $x^6 + 9x^3 + 8$ **149.** $x^6 + 2x^3 - 3$ **150.** $x^6 - 5x^3 + 4$
151. $(x - y)^2 - 3(x - y) + 2$ **152.** $(2x + y)^2 + (2x + y) - 6$
153. $(x + y)^2 - 6(x + y) + 5$ **154.** $(x - 2y)^2 + 6(x - 2y) + 8$
155. $(x - y)^2 + 5(x - y) + 6$ **156.** $(2x - y)^2 + 2(2x - y) - 3$
157. $(x + 3y)^2 - 6(x + 3y) - 7$ **158.** $(x + y)^2 + 4(x + y) - 12$
159. $(2x + y)^2 + 3(2x + y) - 10$ **160.** $(x + 2y)^2 - 8(x + 2y) + 15$

Trinomials of the Form $ax^2 + bx + c$, $a, b, c \in I$, $a \neq 1$

Consider the product $(3x + 3)(x + 2) = 3x^2 + 9x + 6$.

The first factor on the left contains the common factor 3:

$$3x + 3 = 3(x + 1)$$

Also, the expanded product contains the common factor 3:

$$3x^2 + 9x + 6 = 3(x^2 + 3x + 2)$$

In general, if a factor of a product contains a common factor, then the expanded product also will contain that common factor.

On the other hand, if no factor in a product, such as $(x + 3)(2x - 1)$, contains a common factor, then the expanded product, $2x^2 + 5x - 3$, will have no common factor.

Conversely, if the terms of a product do not have a common factor, then neither will any of its factors.

In order to learn how to factor a trinomial of the form $ax^2 + bx + c$, let us first look at how we multiply two factors together to get a product of this form.
We multiply $(2x + 3)(4x - 5)$ as follows:

$$\begin{array}{rrr}
2x & + & 3 \\
4x & - & 5 \\
\hline
8x^2 & + 12x & \\
 & - 10x & - 15 \\
\hline
8x^2 & + \ 2x & - 15
\end{array}$$

Let us go over the same multiplication again, as shown in Figure 5.1.

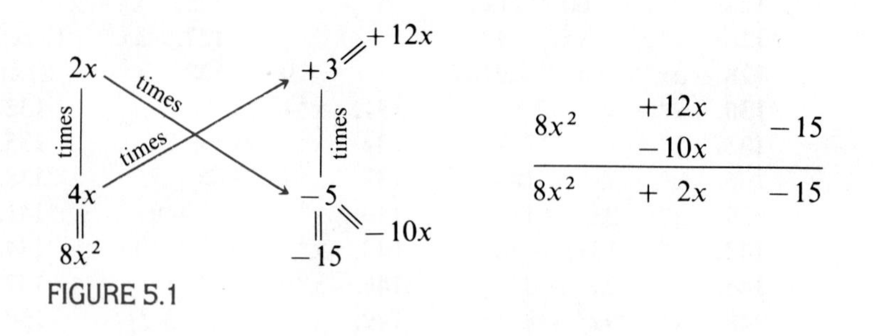

FIGURE 5.1

The crossed arrows [scissors symbol] we will refer to as **scissors**.

On the left side of the scissors $\begin{smallmatrix} 2x \\ 4x \end{smallmatrix}$ [scissors symbol] are factors of $8x^2$, the first term of the trinomial.

On the right side of the scissors $\begin{matrix} +3 \\ -5 \end{matrix}$ are factors of -15, the third term of the trinomial.

The sum of the products in the direction of the arrows

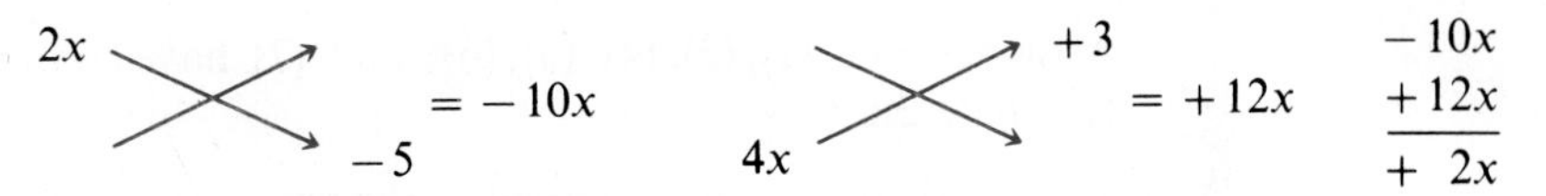

is the middle term of the trinomial.

The following example illustrates how to use the scissors in factoring a trinomial $ax^2 + bx + c,\ a \neq 1,\ a, b,\ c \in I$.

EXAMPLE Factor $6x^2 - 5x - 6$.

Solution Find all possible pairs of factors whose product is the first term of the trinomial; each factor must contain the square root of the literal number. Write these factors at the left side of the scissors:

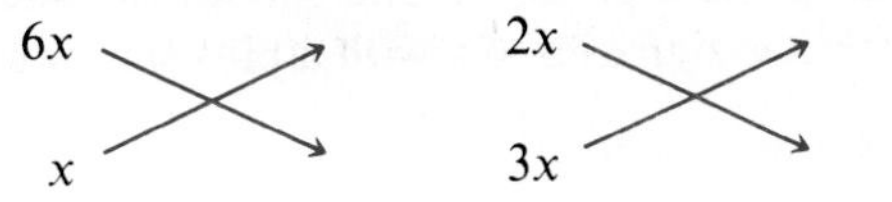

Find all possible pairs of factors whose product is the third term of the trinomial, disregarding the signs, and write them at the right side of the scissors.

Write all possible arrangements using the factors of the first term and the factors of the third term:

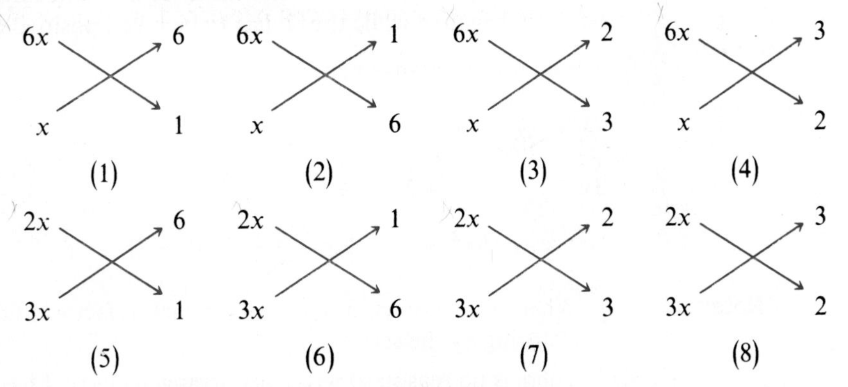

The eight scissors shown above give all possible arrangements of the factors of the first term of the trinomial and the factors of the third term of the trinomial.

The terms on top of the scissors form the first factor of the product, and the terms on the bottom of the scissors form the second factor of the product.

Since there is no common factor in the trinomial, there should not be a common factor between the terms at the top of the scissors or a common factor between the terms at the bottom of the scissors. If there is a common factor between the terms at the top or the terms at the bottom in an arrangement, that arrangement cannot be the correct one.

Arrangements (1), (3), (4), (5), (6), and (7) have common factors, and we eliminate them.

The candidates are now limited to the two arrangements:

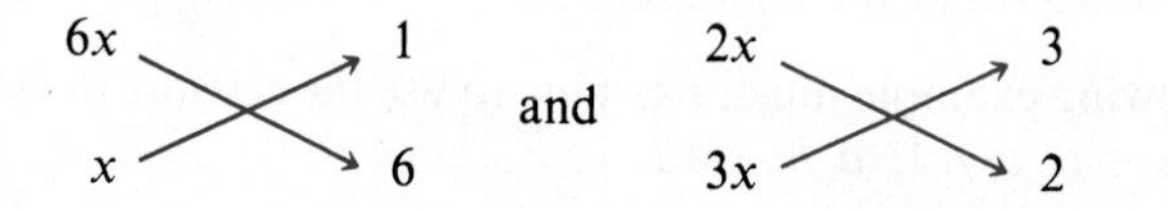

The middle term of the trinomial, which is the sum of the products in the direction of the arrows, will indicate which arrangement is the correct one.

Since the first arrangement gives x and $36x$ for the middle term, which cannot give a sum of $-5x$, the first arrangement is not the correct one.

The second arrangement gives $9x$ and $4x$ for the middle term. By taking the $9x$ with a minus sign and $4x$ with a plus sign, we get $-9x + 4x = -5x$.

Hence the correct arrangement is

$2x$ ⟶ 3

$3x$ ⟶ 2

The factors of the first term of the trinomial are always taken positive. Thus in order to arrive at $-9x$, the 3 at the right side of the scissors has to be taken negative while the 2 has to be taken positive to arrive at $+4x$.

The complete arrangement is

$2x$ ⟶ -3

$3x$ ⟶ $+2$

Hence $\quad 6x^2 - 5x - 6 = (2x - 3)(3x + 2)$

Notes

1. When the trinomial has a common factor, factor it first before you attempt factoring by the scissors.
2. There is no reason to write any arrangement that has a common factor between the top terms or a common factor between the bottom terms.
3. When the coefficient of the first term or the third term of the trinomial is a large number, write the number as the product of its prime factors and form products of factors using combination of the primes.

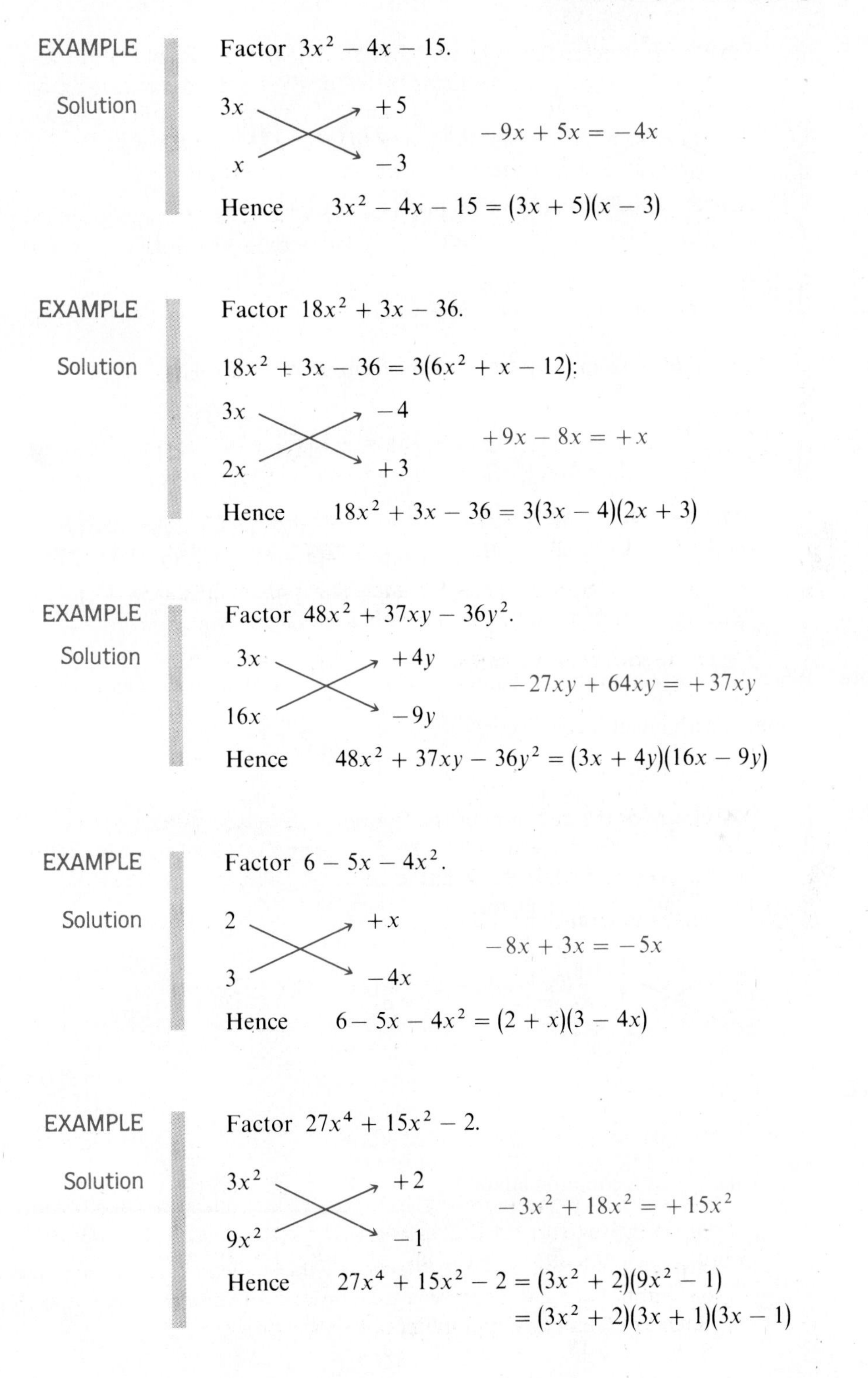

EXAMPLE Factor $3x^2 - 4x - 15$.

Solution

$$\begin{matrix} 3x & \searrow\nearrow & +5 \\ x & \nearrow\searrow & -3 \end{matrix} \qquad -9x + 5x = -4x$$

Hence $\quad 3x^2 - 4x - 15 = (3x + 5)(x - 3)$

EXAMPLE Factor $18x^2 + 3x - 36$.

Solution $18x^2 + 3x - 36 = 3(6x^2 + x - 12)$:

$$\begin{matrix} 3x & \searrow\nearrow & -4 \\ 2x & \nearrow\searrow & +3 \end{matrix} \qquad +9x - 8x = +x$$

Hence $\quad 18x^2 + 3x - 36 = 3(3x - 4)(2x + 3)$

EXAMPLE Factor $48x^2 + 37xy - 36y^2$.

Solution

$$\begin{matrix} 3x & \searrow\nearrow & +4y \\ 16x & \nearrow\searrow & -9y \end{matrix} \qquad -27xy + 64xy = +37xy$$

Hence $\quad 48x^2 + 37xy - 36y^2 = (3x + 4y)(16x - 9y)$

EXAMPLE Factor $6 - 5x - 4x^2$.

Solution

$$\begin{matrix} 2 & \searrow\nearrow & +x \\ 3 & \nearrow\searrow & -4x \end{matrix} \qquad -8x + 3x = -5x$$

Hence $\quad 6 - 5x - 4x^2 = (2 + x)(3 - 4x)$

EXAMPLE Factor $27x^4 + 15x^2 - 2$.

Solution

$$\begin{matrix} 3x^2 & \searrow\nearrow & +2 \\ 9x^2 & \nearrow\searrow & -1 \end{matrix} \qquad -3x^2 + 18x^2 = +15x^2$$

Hence
$$\begin{aligned} 27x^4 + 15x^2 - 2 &= (3x^2 + 2)(9x^2 - 1) \\ &= (3x^2 + 2)(3x + 1)(3x - 1) \end{aligned}$$

EXAMPLE Factor $8x^6 - 19x^3 - 27$.

Solution

$$\begin{array}{lcl} 8x^3 & \searrow\nearrow & -27 \\ x^3 & & +1 \end{array} \qquad +8x^3 - 27x^3 = -19x^3$$

Hence
$$\begin{aligned} 8x^6 - 19x^3 - 27 &= (8x^3 - 27)(x^3 + 1) \\ &= (2x - 3)(4x^2 + 6x + 9)(x + 1)(x^2 - x + 1) \end{aligned}$$

EXAMPLE Factor $20(a - b)^2 - 23(a - b) + 6$.

Solution $20(a - b)^2 - 23(a - b) + 6$ is of the form $20x^2 - 23x + 6$:

$$\begin{array}{lcl} 4x & \searrow\nearrow & -3 \\ 5x & & -2 \end{array} \qquad -8x - 15x = -23x$$

$20x^2 - 23x + 6 = (4x - 3)(5x - 2)$

Hence
$$\begin{aligned} 20(a - b)^2 - 23(a - b) + 6 &= [4(a - b) - 3][5(a - b) - 2] \\ &= (4a - 4b - 3)(5a - 5b - 2) \end{aligned}$$

Note When a, b, or c are rational numbers, taking $\dfrac{1}{\text{LCD}}$ as a common factor yields a trinomial with integral coefficients.

EXAMPLE Factor $\frac{1}{3}x^2 - \frac{1}{2}x - \frac{1}{3}$.

Solution Take $\frac{1}{6}$ as a common factor:

$$\frac{1}{3}x^2 - \frac{1}{2}x - \frac{1}{3} = \frac{1}{6}(2x^2 - 3x - 2) = \frac{1}{6}(2x + 1)(x - 2)$$

EXAMPLE Factor $\frac{1}{2}x^2 - \frac{1}{3}x - \frac{4}{9}$.

Solution Take $\frac{1}{18}$ as a common factor:

$$\begin{aligned} \frac{1}{2}x^2 - \frac{1}{3}x - \frac{4}{9} &= \frac{1}{18}(9x^2 - 6x - 8) \\ &= \frac{1}{18}(3x + 2)(3x - 4) \end{aligned}$$

Exercise 5.3B

Factor completely:

1. $2x^2 + 3x + 1$
2. $2x^2 + 7x + 3$
3. $4x^2 + 4x + 1$
4. $2x^2 + 5x + 3$
5. $2x^2 + 13x + 21$
6. $3x^2 + 16x + 21$
7. $6x^2 + 7x + 2$
8. $2x^2 + 17x + 8$
9. $4x^2 + 9x + 2$
10. $6x^2 + 11x + 4$
11. $2x^2 - 5x + 2$
12. $2x^2 - 9x + 4$
13. $4x^2 - 8x + 3$
14. $2x^2 - 7x + 6$
15. $3x^2 - 4x + 1$
16. $6x^2 - 5x + 1$
17. $2x^2 - 13x + 6$
18. $4x^2 - 5x + 1$
19. $4x^2 - 13x + 3$
20. $8x^2 - 14x + 5$
21. $2x^2 + x - 1$
22. $2x^2 + 5x - 3$
23. $2x^2 + x - 3$
24. $3x^2 + 2x - 1$
25. $3x^2 + 8x - 3$
26. $3x^2 + x - 2$
27. $2x^2 + 11x - 6$
28. $4x^2 + 3x - 1$
29. $4x^2 + 11x - 3$
30. $2x^2 + 5x - 12$
31. $2x^2 - 3x - 2$
32. $2x^2 - 7x - 4$
33. $2x^2 - 11x - 6$
34. $2x^2 - x - 6$
35. $3x^2 - 5x - 2$
36. $3x^2 - 11x - 4$
37. $3x^2 - 4x - 4$
38. $6x^2 - 5x - 4$
39. $12x^2 - x - 1$
40. $12x^2 - 5x - 3$
41. $2x^2 + 9x + 9$
42. $2x^2 + 15x + 18$
43. $4x^2 + 11x + 6$
44. $4x^2 - 9x + 2$
45. $4x^2 - 15x + 9$
46. $10x^2 - 11x + 3$
47. $4x^2 + 5x - 6$
48. $4x^2 + 11x - 3$
49. $10x^2 + 3x - 1$
50. $4x^2 - 4x - 3$
51. $8x^2 - 2x - 1$
52. $8x^2 - 2x - 3$
53. $10x^2 + 13x + 4$
54. $12x^2 + 8x + 1$
55. $12x^2 + 20x + 7$
56. $12x^2 - 16x + 5$
57. $4x^2 - 12x + 5$
58. $4x^2 - 20x + 9$
59. $12x^2 + 8x - 7$
60. $4x^2 + 12x - 7$
61. $6x^2 + 11x - 7$
62. $4x^2 - 7x - 2$
63. $4x^2 - 9x - 9$
64. $10x^2 - x - 2$
65. $4x^2 + 17x - 18$
66. $9x^2 + 18x - 8$
67. $6x^2 - x + 12$
68. $4x^2 - 5x + 6$
69. $8x^2 - 22x - 15$
70. $6x^2 - 19x - 3$
71. $6x^2 + 23x + 10$
72. $8x^2 + 22x + 9$
73. $3x^2 + 11x + 6$
74. $3x^2 + 19x + 6$
75. $8x^2 - 18x + 7$
76. $3x^2 - 16x + 5$
77. $3x^2 - 14x + 8$
78. $4x^2 - 12x + 9$
79. $8x^2 + 14x - 9$
80. $3x^2 + 7x - 6$
81. $4x^2 + 4x - 15$
82. $4x^2 + 12x - 27$
83. $10x^2 - x - 3$
84. $10x^2 - 7x - 6$
85. $12x^2 - 4x - 1$
86. $12x^2 - 4x - 5$
87. $4x^2y^2 + 16xy + 15$
88. $9x^2 + 24xy + 16y^2$
89. $9x^2 + 31xy + 12y^2$
90. $24x^2 + 25xy + 6y^2$
91. $4x^2 - 20xy + 21y^2$
92. $24x^2 - 17xy + 3y^2$
93. $6x^2y^4 - 17xy^2 + 12$
94. $9x^2 - 12xy^2 + 4y^4$
95. $12x^2 - 17xy + 6y^2$
96. $9x^2y^2 + 3xy - 2$
97. $9x^2 + 12xy^2 - 5y^4$
98. $24x^2 + 7xy^3 - 5y^6$
99. $6x^2y^4 + xy^2 - 12$
100. $9x^2 + 23xy - 12y^2$
101. $4x^2y^2 - 16xy - 9$
102. $6x^2 - 13xy - 8y^2$
103. $9x^2 - 6xy - 8y^2$
104. $4x^2 - 13xy^2 - 12y^4$
105. $4x^2 - 8xy^2 - 21y^4$
106. $36x^2 + 24x + 4$
107. $27x^2 + 45x + 12$
108. $24x^2y^2 + 11xy^2 + y^2$
109. $24x^2 + 52x + 24$
110. $36x^2 + 33x + 6$
111. $6x^3 - 19x^2 + 3x$
112. $6x^2y - 37xy + 6y$
113. $18x^2 - 18x + 4$
114. $27x^2 - 54x + 15$

115. $8x^4 - 14x^3 + 3x^2$
116. $18x^2 + 6x - 4$
117. $9x^4 + 12x^3 - 5x^2$
118. $24x^3 + 7x^2 - 5x$
119. $16x^2 - 32x - 20$
120. $40x^2 - 50x - 35$
121. $3x^4 - 14x^3 - 5x^2$
122. $6x^2y^2 - xy^2 - 2y^2$
123. $4 + 23x - 6x^2$
124. $9 + 6x - 8x^2$
125. $3 + 16x - 12x^2$
126. $6 + 19x - 36x^2$
127. $24 + 11x - 18x^2$
128. $36 + 5x - 24x^2$
129. $3 - x - 24x^2$
130. $6 - 5x - 6x^2$
131. $6 - x - 12x^2$
132. $8 - 2x - 21x^2$
133. $2x^4 - 3x^2 + 1$
134. $2x^4 - 7x^2 - 4$
135. $3x^4 - 26x^2 - 9$
136. $9x^4 + 35x^2 - 4$
137. $16x^4 + 15x^2 - 1$
138. $4x^4 + 7x^2 - 36$
139. $9x^4 + 23x^2 - 12$
140. $25x^4 + 49x^2 - 2$
141. $2x^4 - 73x^2 + 36$
142. $4x^4 - 5x^2 + 1$
143. $9x^4 - 37x^2 + 4$
144. $16x^4 - 65x^2 + 4$
145. $2x^6 + 3x^3 + 1$
146. $4x^6 + 35x^3 + 24$
147. $8x^6 - 31x^3 - 4$
148. $6x^6 - 49x^3 + 8$
149. $2x^6 + 51x^3 - 81$
150. $3x^6 - 80x^3 - 27$
151. $27x^6 + 53x^3 - 2$
152. $64x^6 - 191x^3 - 3$
153. $64x^6 + 127x^3 - 2$
154. $8x^6 + 7x^3 - 1$
155. $8x^6 - 7x^3 - 1$
156. $8x^6 + 9x^3 + 1$
157. $3(x + y)^2 + 2(x + y) - 5$
158. $6(x - y)^2 - (x - y) - 2$
159. $2(2x + y)^2 - 5(2x + y) - 3$
160. $2(2x - y)^2 - (2x - y) - 6$
161. $6(x - 2y)^2 - 13(x - 2y) + 6$
162. $5(x + 2y)^2 + 12(x + 2y) - 9$
163. $8(3x + y)^2 + 6(3x + y) - 9$
164. $6(x - 3y)^2 - 7(x - 3y) + 2$
165. $6(x + y)^2 - (x + y) - 12$
166. $4(x + y)^2 - 5(x + y) - 6$
167. $9(x - y)^2 + 18(x - y) + 8$
168. $8(x - y)^2 - 22(x - y) + 15$
169. $4(x + 2y)^2 + 17(x + 2y) + 18$
170. $6(x - 2y)^2 - 19(x - 2y) + 3$
171. $\frac{1}{2}x^2 - \frac{9}{4}x + 1$
172. $\frac{2}{3}x^2 + \frac{5}{6}x - 1$
173. $x^2 + \frac{4}{3}x - \frac{5}{9}$
174. $2x^2 - \frac{2}{3}x - \frac{1}{6}$
175. $\frac{1}{3}x^2 + x - \frac{9}{4}$
176. $\frac{2}{3}x^2 - 2x + \frac{3}{2}$
177. $\frac{1}{3}x^2 + \frac{1}{6}x - \frac{1}{2}$
178. $\frac{3}{4}x^2 - \frac{1}{2}x - \frac{2}{3}$
179. $\frac{1}{4}x^2 - \frac{1}{3}x - \frac{1}{3}$
180. $\frac{2}{9}x^2 - \frac{5}{6}x + \frac{1}{2}$

5.4 Factoring by Completing the Square

This method applies to special types of binomials and trinomials that cannot be factored by the previous methods but that can be expressed as the difference of two squares.

The quantities $(a + b)^2 = a^2 + 2ab + b^2$ and $(a - b)^2 = a^2 - 2ab + b^2$ are perfect squares. Given $a^2 + b^2$, the term $2ab$ may be added to or subtracted from $a^2 + b^2$

to result in a perfect square trinomial. The term $2ab$ is twice the product of the square roots of a^2 and b^2. The principle of completing the square is used in factoring binomials of the form $a^{4m} + 4b^{4n}$, $m, n \in N$, and trinomials of the form

$$r^2a^{4m} + sa^{2m}b^{2n} + t^2b^{4n}, m, n \in N$$

Consider the binomial $a^{4m} + 4b^{4n}$, $m, n \in N$. The term that makes the binomial a perfect square is $4a^{2m}b^{2n}$. Add $+4a^{2m}b^{2n} - 4a^{2m}b^{2n}$ to the binomial. Note that for all numerical values of a and b this expression is zero. Therefore the addition of this expression does not affect the value of the binomial:

$$a^{4m} + 4b^{4n} = a^{4m} + 4b^{4n} + 4a^{2m}b^{2n} - 4a^{2m}b^{2n}$$

Group $+4a^{2m}b^{2n}$ with $a^{4m} + 4b^{4n}$ and factor:

$$\begin{aligned} a^{4m} + 4b^{4n} &= (a^{4m} + 4a^{2m}b^{2n} + 4b^{4n}) - 4a^{2m}b^{2n} \\ &= (a^{2m} + 2b^{2n})^2 - 4a^{2m}b^{2n} \end{aligned}$$

Factor as the difference of two squares:

$$\begin{aligned} &= [(a^{2m} + 2b^{2n}) + 2a^m b^n][(a^{2m} + 2b^{2n}) - 2a^m b^n] \\ &= (a^{2m} + 2a^m b^n + 2b^{2n})(a^{2m} - 2a^m b^n + 2b^{2n}) \end{aligned}$$

Note Although $a^{2m} + 2a^m b^n + 2b^{2n}$ and $a^{2m} - 2a^m b^n + 2b^{2n}$ are trinomials, they are not factorable.

EXAMPLE Factor $a^4 + 4$.

Solution Add $(4a^2 - 4a^2)$:

$$\begin{aligned} a^4 + 4 &= a^4 + 4 + 4a^2 - 4a^2 \\ &= (a^4 + 4a^2 + 4) - 4a^2 \\ &= (a^2 + 2)^2 - 4a^2 \\ &= [(a^2 + 2) + 2a][(a^2 + 2) - 2a] \\ &= (a^2 + 2a + 2)(a^2 - 2a + 2) \end{aligned}$$

EXAMPLE Factor $x^4 + 324y^4$.

Solution Add $(36x^2y^2 - 36x^2y^2)$:

$$\begin{aligned} x^4 + 324y^4 &= x^4 + 324y^4 + 36x^2y^2 - 36x^2y^2 \\ &= (x^4 + 36x^2y^2 + 324y^4) - 36x^2y^2 \\ &= (x^2 + 18y^2)^2 - 36x^2y^2 \\ &= [(x^2 + 18y^2) + 6xy][(x^2 + 18y^2) - 6xy] \\ &= (x^2 + 6xy + 18y^2)(x^2 - 6xy + 18y^2) \end{aligned}$$

Consider the trinomial $r^2a^{4m} + sa^{2m}b^{2n} + t^2b^{4n}$, $m, n \in N$. The term that makes $r^2a^{4m} + t^2b^{4n}$ a perfect square is $2tra^{2m}b^{2n}$.

Add $+2tra^{2m}b^{2n} - 2tra^{2m}b^{2n}$ to the trinomial:

$$r^2a^{4m} + sa^{2m}b^{2n} + t^2b^{4n} = r^2a^{4m} + sa^{2m}b^{2n} + t^2b^{4n} + 2tra^{2m}b^{2n} - 2tra^{2m}b^{2n}$$

Combine either $+2tra^{2m}b^{2n}$ or $-2tra^{2m}b^{2n}$ with $sa^{2m}b^{2n}$, whichever one makes the sum of the form $-p^2a^{2m}b^{2n}$; combine the other term with $r^2a^{4m} + t^2b^{4n}$; then factor.

Notes

1. If neither $2tra^{2m}b^{2n}$ nor $-2tra^{2m}b^{2n}$ when combined with $sa^{2m}b^{2n}$ yields a term of the form $-p^2a^{2m}b^{2n}$, the original trinomial cannot be factored.

   Consider, $x^4 + 5x^2 + 16$; the term that when combined with $x^4 + 16$ forms a square trinomial is $+8x^2$ or $-8x^2$. Since neither $+8x^2$ nor $-8x^2$ when combined with $5x^2$ yields a negative square term, $x^4 + 5x^2 + 16$ cannot be factored.

2. If either $2tra^{2m}b^{2n}$ or $-2tra^{2m}b^{2n}$ when combined with $sa^{2m}b^{2n}$ yields a term of the form $-p^2a^{2m}b^{2n}$, the original polynomial can be factored by the method of factoring trinomials. If the polynomial is factored by the method of completing the square, the result can still be factored.

EXAMPLE

Factor $x^4 + 3x^2 + 4$.

Solution

Add $+4x^2 - 4x^2$ to the trinomial:

$$\begin{aligned} x^4 + 3x^2 + 4 &= x^4 + 3x^2 + 4 + 4x^2 - 4x^2 \\ &= (x^4 + 4x^2 + 4) + (3x^2 - 4x^2) \\ &= (x^2 + 2)^2 - x^2 \\ &= [(x^2 + 2) + x][(x^2 + 2) - x] \\ &= (x^2 + x + 2)(x^2 - x + 2) \end{aligned}$$

EXAMPLE

Factor $x^4 - 7x^2 + 9$.

Solution

Add $(+6x^2 - 6x^2)$ to the trinomial:

$$\begin{aligned} x^4 - 7x^2 + 9 &= x^4 - 7x^2 + 9 + 6x^2 - 6x^2 \\ &= (x^4 - 6x^2 + 9) + (-7x^2 + 6x^2) \\ &= (x^2 - 3)^2 - x^2 \\ &= [(x^2 - 3) + x][(x^2 - 3) - x] \\ &= (x^2 + x - 3)(x^2 - x - 3) \end{aligned}$$

EXAMPLE

Factor $x^4 - 13x^2 + 36$.

Solution

Add $(+12x^2 - 12x^2)$ to the trinomial:

$$x^4 - 13x^2 + 36 = x^4 - 13x^2 + 36 + 12x^2 - 12x^2$$

Note that $-13x^2 + 12x^2 = -x^2$ and $-13x^2 - 12x^2 = -25x^2$; that is, we can combine $-13x^2$ with either $+12x^2$ or $-12x^2$:

$$\begin{aligned} x^4 - 13x^2 + 36 &= (x^4 + 12x^2 + 36) + (-13x^2 - 12x^2) \\ &= (x^2 + 6)^2 - 25x^2 \\ &= [(x^2 + 6) + 5x][(x^2 + 6) - 5x] \\ &= (x^2 + 5x + 6)(x^2 - 5x + 6) \end{aligned}$$

The last two factors may be factored further:

$$= (x + 3)(x + 2)(x - 3)(x - 2)$$

Notes

1. $x^4 - 13x^2 + 36 = (x^2 - 9)(x^2 - 4)$
   $= (x + 3)(x - 3)(x + 2)(x - 2)$
2. We arrive at the same factors if we combine $-13x^2$ with $+12x^2$

Exercise 5.4

Factor completely:

1. $4x^4 + 1$
2. $x^4 + 324$
3. $x^8 + 4$
4. $4x^8 + 1$
5. $x^4 + 64$
6. $x^4 + 2500$
7. $x^4 + x^2 + 1$
8. $x^4 + 5x^2 + 9$
9. $x^4 + 4x^2 + 16$
10. $x^4 + x^2 + 25$
11. $x^4 + 7x^2 + 16$
12. $x^4 - 15x^2 + 9$
13. $x^4 - 12x^2 + 4$
14. $x^4 - 8x^2 + 4$
15. $x^4 - 6x^2 + 1$
16. $x^4 - 7x^2 + 1$
17. $x^4 - 9x^2 + 16$
18. $x^4 - 21x^2 + 4$
19. $x^4 - 6x^2 + 25$
20. $x^4 - 23x^2 + 1$
21. $x^4 - x^2 + 16$
22. $x^4 - 24x^2 + 16$
23. $x^4 - 33x^2 + 16$
24. $x^4 - 5x^2 + 4$
25. $x^4 - 10x^2 + 9$
26. $x^4 - 29x^2 + 100$
27. $9x^4 + 2x^2 + 1$
28. $9x^4 + 8x^2 + 4$
29. $16x^4 + 7x^2 + 4$
30. $16x^4 + 4x^2 + 1$
31. $9x^4 - 22x^2 + 1$
32. $4x^4 - 8x^2 + 1$
33. $4x^4 - 12x^2 + 1$
34. $4x^4 - 16x^2 + 9$
35. $4x^4 - 4x^2 + 9$
36. $9x^4 - 21x^2 + 4$
37. $16x^4 - 41x^2 + 4$
38. $16x^4 - 24x^2 + 1$
39. $25x^4 - 6x^2 + 1$
40. $4x^4 - 5x^2 + 25$
41. $25x^4 - 16x^2 + 4$
42. $25x^4 - 6x^2 + 9$
43. $9x^4 - 36x^2 + 25$
44. $25x^4 - 39x^2 + 9$
45. $x^8 + x^4 + 1$

5.5 Factoring Four-Term Polynomials

The methods of factoring polynomials with more than three terms are called **factoring by grouping**. There are two types of four-term polynomials that can be factored.

In the first type, the terms are grouped with three terms in one group and the fourth term as the other group.

In the second type, the terms are grouped in pairs.

Grouping as Three and One

The polynomial $(x + y)^2 - z^2$ can be factored as the difference of two squares. When $(x + y)^2 - z^2$ is expanded, we get

$$(x + y)^2 - z^2 = x^2 + 2xy + y^2 - z^2$$

Note that disregarding the signs, three out of four terms, x^2, y^2, z^2, are square terms.
The fourth term, $2xy$, equals $2\sqrt{x^2}\sqrt{y^2}$. This fourth term and the two related square terms form one group which, when factored, results in a square quantity:

$$x^2 + 2xy + y^2 = (x + y)^2$$

EXAMPLE Factor $x^2 - y^2 + 4z^2 - 4xz$.

Solution There are three square terms: x^2, y^2, and $4z^2$.

The fourth term is $4xz = 2\sqrt{x^2}\sqrt{4z^2}$.

The two square terms related to $4xz$ are x^2 and $4z^2$.

Hence $x^2, 4xz,$ and $4z^2$ form one group:

$$\begin{aligned} x^2 - y^2 + 4z^2 - 4xz &= (x^2 - 4xz + 4z^2) - y^2 \\ &= (x - 2z)^2 - y^2 \\ &= [(x - 2z) + y][(x - 2z) - y] \\ &= (x - 2z + y)(x - 2z - y) \end{aligned}$$

EXAMPLE Factor $9x^2 - y^2 - 25z^2 + 10yz$.

Solution

$$\begin{aligned} 9x^2 - y^2 - 25z^2 + 10yz &= 9x^2 - (y^2 - 10yz + 25z^2) \\ &= 9x^2 - (y - 5z)^2 \\ &= [3x + (y - 5z)][3x - (y - 5z)] \\ &= (3x + y - 5z)(3x - y + 5z) \end{aligned}$$

EXAMPLE Factor $x^2 - y^2 - 9 - 6y$.

Solution

$$\begin{aligned} x^2 - y^2 - 9 - 6y &= x^2 - (y^2 + 6y + 9) \\ &= x^2 - (y + 3)^2 \\ &= [x + (y + 3)][x - (y + 3)] \\ &= (x + y + 3)(x - y - 3) \end{aligned}$$

Grouping in Pairs

When the four terms cannot be grouped as three and one, group them in pairs. The following examples illustrate the principle behind grouping in pairs.

EXAMPLE Factor $x^3 + x^2 + 2x + 2$.

Solution Group the first two terms in one group and the last two terms in a second group:

$$\begin{aligned} x^3 + x^2 + 2x + 2 &= (x^3 + x^2) + (2x + 2) \\ &= x^2(x + 1) + 2(x + 1) \end{aligned}$$

Now we have a common factor $(x + 1)$:

$$x^3 + x^2 + 2x + 2 = (x + 1)(x^2 + 2)$$

EXAMPLE Factor $ax + ay + bx + by$.

Solution

$$\begin{aligned} ax + ay + bx + by &= (ax + ay) + (bx + by) \\ &= a(x + y) + b(x + y) \\ &= (x + y)(a + b) \end{aligned}$$

Note A different grouping is possible in some problems, but remember that the final factors will be the same, except for their order.

The example above could also be grouped as

$$\begin{aligned} ax + ay + bx + by &= (ax + bx) + (ay + by) \\ &= x(a + b) + y(a + b) \\ &= (a + b)(x + y) \end{aligned}$$

EXAMPLE Factor $12ax - 20bx - 9ay + 15by$.

Solution

$$12ax - 20bx - 9ay + 15by = (12ax - 20bx) - (9ay - 15by)$$

When you enclose $-9ay + 15by$ in parentheses preceded by a minus sign, you get $-(9ay - 15by)$:

$$\begin{aligned} 12ax - 20bx - 9ay + 15by &= 4x(3a - 5b) - 3y(3a - 5b) \\ &= (3a - 5b)(4x - 3y) \end{aligned}$$

Note If there is no common factor, group the terms differently.

EXAMPLE Factor $x^3 + x^2 - 2x - 8$.

Solution Grouping the first two terms in one group and the last two terms in a second group does not yield a common factor:

$$\begin{aligned} x^3 + x^2 - 2x - 8 &= (x^3 + x^2) - (2x + 8) \\ &= x^2(x + 1) - 2(x + 4) \end{aligned}$$

Since there is no common factor, we try another grouping:

$$\begin{aligned} x^3 + x^2 - 2x - 8 &= (x^3 - 8) + (x^2 - 2x) \\ &= (x - 2)(x^2 + 2x + 4) + x(x - 2) \\ &= (x - 2)[(x^2 + 2x + 4) + x] \\ &= (x - 2)(x^2 + 2x + 4 + x) \\ &= (x - 2)(x^2 + 3x + 4) \end{aligned}$$

Note When there are two cubes in the polynomial, try grouping them together.

EXAMPLE Factor $27x^3 - 9x^2 + y^2 + y^3$.

Solution

$$\begin{aligned} 27x^3 - 9x^2 + y^2 + y^3 &= (27x^3 + y^3) - (9x^2 - y^2) \\ &= (3x + y)(9x^2 - 3xy + y^2) - (3x + y)(3x - y) \\ &= (3x + y)[(9x^2 - 3xy + y^2) - (3x - y)] \\ &= (3x + y)(9x^2 - 3xy + y^2 - 3x + y) \end{aligned}$$

EXAMPLE Factor $8x^3 + 2x - y^3 - y$.

Solution

$$\begin{aligned} 8x^3 + 2x - y^3 - y &= (8x^3 - y^3) + (2x - y) \\ &= (2x - y)(4x^2 + 2xy + y^2) + (2x - y) \\ &= (2x - y)[(4x^2 + 2xy + y^2) + 1] \\ &= (2x - y)(4x^2 + 2xy + y^2 + 1) \end{aligned}$$

Note When factoring out a common factor, the second factor is the result of dividing every term of the polynomial by that common factor.

Exercise 5.5

Factor completely:

1. $x^2 + 2xy + y^2 - z^2$
2. $x^2 - 2xy + y^2 - z^2$
3. $4x^2 - 4xy + y^2 - 4z^2$
4. $x^2 + 4xy + 4y^2 - 16z^2$
5. $x^2 - y^2 + 4x + 4$
6. $y^2 - 4x^2 + 2y + 1$
7. $y^2 - 9x^2 + 6y + 9$
8. $x^2 - 4 + 4y^2 - 4xy$

9. $4x^2 + 4xy - 25 + y^2$
10. $9x^2 + y^2 - 6xy - 36$
11. $4x^2 + 4y^2 + 8xy - 25$
12. $9x^2 - 4 + 9y^2 - 18xy$
13. $2x^2 + 2y^2 - 18 - 4xy$
14. $4x^2 - 1 + 24xy + 36y^2$
15. $x^3 - 16x + 2x^2y + xy^2$
16. $4x^2y - 9y - 4xy^2 + y^3$
17. $x^2 - y^2 - z^2 - 2yz$
18. $4x^2 - 4z^2 - y^2 - 4yz$
19. $9x^2 - 9 - y^2 - 6y$
20. $x^2 - 4y^2 - 16y - 16$
21. $25x^2 - 9y^2 - 9 - 18y$
22. $1 - x^2 - y^2 - 2xy$
23. $4 - 4xy - x^2 - 4y^2$
24. $16 - y^2 - 4x^2 - 4xy$
25. $9 - 4x^2 - 8xy - 4y^2$
26. $49 - 81x^2 - 54xy - 9y^2$
27. $2x^2 - 2y^2 - 16y - 32$
28. $3x^4 - 12y^2 - 3z^2 - 12yz$
29. $x^3 - x - xy^2 - 2xy$
30. $16 - 4x^4 - 4y^2 - 8x^2y$
31. $y^2 - 4z^2 - x^2 + 4xz$
32. $4x^2 - y^2 - z^4 + 2yz^2$
33. $9x^4 - 9y^2 - 4 + 12y$
34. $4x^2 - 9y^2 - 81 + 54y$
35. $16x^2 - y^4 - 16 + 8y^2$
36. $36y^2 - x^4 - 1 + 2x^2$
37. $1 - x^4 - 4y^2 + 4x^2y$
38. $4 - 4x^2 - y^4 + 4xy^2$
39. $9 - 4y^2 - 36x^2 + 24xy$
40. $1 - 16x^2 - 16y^2 + 32xy$
41. $25 - x^2 + 6xy - 9y^2$
42. $4 - 25x^2 - 25y^2 + 50xy$
43. $6xz + 3y^2 - 3x^2 - 3z^2$
44. $8 + 36xy - 18x^2 - 18y^2$
45. $4 + 64xy - 16x^2 - 64y^2$
46. $30xy - 45x^2 + 45 - 5y^2$
47. $2x^2y + 4x - x^3 - xy^2$
48. $36x^2 - x^2y^2 - x^4 + 2x^3y$
49. $3x^2 - xy - 6x + 2y$
50. $x^2 + y + x + xy$
51. $x^2 - 3y - 3x + xy$
52. $2x^2 + 5x - 2xy^2 - 5y^2$
53. $6xy - 3yz - 14x + 7z$
54. $8xy + 3z - 8xz - 3y$
55. $27x^2 + 12x - 18xy - 8y$
56. $2ax + 3by + 2bx + 3ay$
57. $14ax + 7by - 14ay - 7bx$
58. $6a^2x^2 + 3a^2y + 6b^2x^2 + 3b^2y$
59. $12x^3 - 4x^2y - 4y^3 + 12xy^2$
60. $40ax - 45bx + 24ay - 27by$
61. $x^3 - x - y^3 + y$
62. $8x^3 + 27y^3 + 2x + 3y$
63. $x^3 + 5y - 125y^3 - x$
64. $3x + 27x^3 - 2y - 8y^3$
65. $64x^3 - 4x + y^3 - y$
66. $x^3 - 4x + y^3 - 4y$
67. $8x^3 - 6x + y^3 - 3y$
68. $x^3 + 6y - 6x - y^3$
69. $x^3 - 8y^3 - 6y + 3x$
70. $x^3 + x^2 - 8y^3 - 4y^2$
71. $8x^3 - 4x^2 - 27y^3 + 9y^2$
72. $y^3 + y^2 + 216x^3 - 36x^2$
73. $x^2 - 25y^2 + 125y^3 + x^3$
74. $16y^2 + x^3 - x^2 - 64y^3$
75. $x^3 + 3x^2 - 9x - 27$
76. $x^3 + 8 - 2x^2 - 4x$
77. $4x^3 + 2 - x - 8x^2$
78. $18x^3 - 16 - 32x + 9x^2$
79. $x^4 + x^3 + x + 1$
80. $x^4 - 16 + 2x^3 - 8x$
81. $x^4 - 54 - 2x^3 + 27x$
82. $8x^4 + 24x^3 + x + 3$

Chapter 5 Review

Factor completely:

1. $x^2 - 1$
2. $x^2 - 16$
3. $x^2 - 36$
4. $x^2 - 64$
5. $4x^2 - 9$
6. $25x^2 - 9$
7. $32 - 2x^2y^2$
8. $45x^2 - 5y^2$
9. $75x^2 - 12y^4$

10. $4x^3 - 4x$ **11.** $12x^2 - 3x^4$ **12.** $x^4 - 1$
13. $16x^4 - y^4$ **14.** $81x^4 - y^4$ **15.** $(x + 1)^2 - y^2$
16. $(x + y)^2 - 25$ **17.** $(x - 2)^2 - 4y^2$ **18.** $4(2x - 1)^2 - y^2$
19. $x^2 - (2y + 1)^2$ **20.** $16x^2 - (y + 3)^2$ **21.** $x^3 + 125$
22. $8x^3 + 27$ **23.** $27x^3 + y^3$ **24.** $x^3 - 64$
25. $8 - x^3$ **26.** $x^3 - 27y^3$ **27.** $32x^3 + 4$
28. $81x^4 + 3xy^3$ **29.** $x^6 + x^3y^3$ **30.** $5x^3 - 40y^3$
31. $9x^4 - 72xy^6$ **32.** $4x^6 - 4$ **33.** $x^3 + (y - 1)^3$
34. $x^3 + (y + 3)^3$ **35.** $(2x - 1)^3 + 8y^3$ **36.** $27x^3 - (y - 2)^3$
37. $x^3 - 27(y + 1)^3$ **38.** $4x^3 - 4(y - 4)^3$ **39.** $6(x - 1)^3 - 6y^3$
40. $x^2 + 4x + 3$ **41.** $x^2 + 9x + 14$ **42.** $x^2 + 13x + 30$
43. $x^2 + 18x + 45$ **44.** $x^2 - 2x + 1$ **45.** $x^2 - 5x + 4$
46. $x^2 - 9x + 18$ **47.** $x^2 - 12x + 27$ **48.** $x^2 + 2x - 3$
49. $x^2 + 3x - 10$ **50.** $x^2 + 5x - 6$ **51.** $x^2 + 7x - 18$
52. $x^2 - x - 2$ **53.** $x^2 - x - 20$ **54.** $x^2 - 5x - 14$
55. $x^2 - x - 72$ **56.** $x^2 + 20x + 51$ **57.** $x^2 + 23x + 60$
58. $x^2 + 14x + 40$ **59.** $x^2 - 21x + 54$ **60.** $x^2 - 22x + 57$
61. $x^2 - 23x + 42$ **62.** $x^2 + 6x - 40$ **63.** $x^2 + 11x - 26$
64. $x^2 + 18x - 63$ **65.** $x^2 - 5x - 50$ **66.** $x^2 - 8x - 48$
67. $x^2 - 6x - 40$ **68.** $x^2 + 18x + 56$ **69.** $x^2 + 19x + 60$
70. $x^2 - 16x + 63$ **71.** $x^2 - 14x + 40$ **72.** $x^2 + 2x - 80$
73. $x^2 + 22x - 48$ **74.** $x^2 - 11x - 80$ **75.** $x^2 - 25x - 54$
76. $3x^2 + 7x + 2$ **77.** $12x^2 + 13x + 3$ **78.** $4x^2 + 19x + 12$
79. $6x^2 + 17x + 7$ **80.** $3x^2 - 10x + 3$ **81.** $12x^2 - 19x + 5$
82. $10x^2 - 7x + 1$ **83.** $6x^2 - 19x + 8$ **84.** $2x^2 + 3x - 2$
85. $2x^2 + 7x - 4$ **86.** $2x^2 + x - 6$ **87.** $3x^2 + 5x - 2$
88. $2x^2 + 15x - 8$ **89.** $4x^2 + 7x - 2$ **90.** $6x^2 + 5x - 4$
91. $12x^2 + 5x - 3$ **92.** $2x^2 - x - 1$ **93.** $2x^2 - 5x - 3$
94. $3x^2 - 2x - 1$ **95.** $3x^2 - 8x - 3$ **96.** $3x^2 - x - 2$
97. $2x^2 - 11x - 6$ **98.** $4x^2 - 3x - 1$ **99.** $4x^2 - 11x - 3$
100. $12x^2 - 11x - 5$ **101.** $8x^2 - 6x - 5$ **102.** $2x^2 + 3x - 9$
103. $4x^2 + 9x - 9$ **104.** $10x^2 + x - 2$ **105.** $4x^2 + 8x - 5$
106. $4x^2 - x - 3$ **107.** $10x^2 - 3x - 1$ **108.** $4x^2 - 12x - 7$
109. $6x^2 - 11x - 7$ **110.** $4x^2y^2 + 16xy - 9$ **111.** $6x^2 + 13xy - 8y^2$
112. $8x^2 + 10xy - 7y^2$ **113.** $3x^4 + 14x^2 - 5$ **114.** $6x^2y^2 - 17xy - 10$
115. $8x^4 - 14x^2 - 9$ **116.** $3x^2 - 7xy - 6y^2$ **117.** $4x^2 - 4xy - 15y^2$
118. $8x^2 + 26x - 24$ **119.** $4x^4 + 8x^3 - 21x^2$ **120.** $24x^2 + 92x - 16$
121. $9x^3y + 9x^2y - 4xy$ **122.** $6x^3 - 17x^2 - 3x$ **123.** $6x^2y - 35xy - 6y$
124. $36x^2 - 12x - 8$ **125.** $27x^2 - 36x - 15$ **126.** $9 - 6x - 8x^2$
127. $3 - 16x - 12x^2$ **128.** $10 - 9x - 9x^2$ **129.** $2 - 5x - 12x^2$
130. $12 + x - 6x^2$ **131.** $12 + 23x - 9x^2$ **132.** $3 + 10x - 8x^2$
133. $4 + 21x - 18x^2$ **134.** $4x^4 + 7x^2 - 2$ **135.** $4x^4 - 63x^2 - 16$
136. $2x^4 - 51x^2 + 25$ **137.** $36x^4 + 107x^2 - 3$ **138.** $16x^4 - 17x^2 + 1$
139. $36x^4 - 25x^2 + 4$ **140.** $3x^6 - 4x^3 + 1$ **141.** $24x^6 + 5x^3 - 1$

142. $27x^6 - 53x^3 - 2$
143. $2x^6 + 129x^3 + 64$
144. $3x^6 - 194x^3 + 128$
145. $8x^6 - 9x^3 + 1$
146. $27x^6 + 28x^3 + 1$
147. $27x^6 - 26x^3 - 1$
148. $(x + y)^2 + (x + y) - 2$
149. $(x - y)^2 - 4(x - y) + 3$
150. $(2x + y)^2 + 2(2x + y) - 15$
151. $(3x - y)^2 - 13(3x - y) + 12$
152. $6(x - y)^2 - 17(x - y) - 14$
153. $12(x + y)^2 - 41(x + y) + 24$
154. $24(x - y)^2 + 38(x - y) + 15$
155. $12(x + y)^2 + 47(x + y) - 4$
156. $3(x - 2y)^2 + 5(x - 2y) - 12$
157. $3(2x - y)^2 + 14(2x - y) - 24$
158. $10(x + 2y)^2 - 7(x + 2y) - 12$
159. $21(x - y)^2 - 8(x - y) - 45$
160. $12(x + y)^2 - 11(x + y) - 36$
161. $8(x + y)^2 - (x + y) - 9$
162. $x^4 + 6x^2 + 25$
163. $x^4 + 8x^2 + 36$
164. $x^4 - 7x^2 + 9$
165. $x^4 - 13x^2 + 4$
166. $x^4 - 11x^2 + 25$
167. $x^4 - 39x^2 + 49$
168. $x^4 - 16x^2 + 36$
169. $x^4 - 11x^2 + 49$
170. $x^4 - 53x^2 + 4$
171. $4x^4 - 28x^2 + 9$
172. $4x^4 - x^2 + 36$
173. $4x^4 - 33x^2 + 36$
174. $15ax - 20a + 24b - 18bx$
175. $7ax - 8a + 16b - 14bx$
176. $30ax + 35ay - 56by - 48bx$
177. $18ay - 22az - 33bz + 27by$
178. $24ax + 16ay + 78bx + 52by$
179. $120ax + 48ay - 90bx - 36by$
180. $x^2 + x - 9y^2 - 3y$
181. $4x^2 - 2x - y^2 - y$
182. $x^3 - x^2 + 343y^3 + 49y^2$
183. $125x^3 - 25x^2 - 27y^3 + 9y^2$
184. $x^4 - x^2 - 81y^4 + 9y^2$
185. $x^6 + x^2 + 8y^6 + 2y^2$
186. $x^2 - 36z^2 - 2xy + y^2$
187. $x^2 + 4x - 9y^2 + 4$
188. $25x^2 - 16z^2 + 9y^2 + 30xy$
189. $9 + x^2 - 4y^2 - 6x$
190. $81x^2 - y^2 - 49z^2 - 14yz$
191. $72xy - 72x^2 + 128 - 18y^2$
192. $x^3 - 4x^2 - 4x + 16$
193. $x^4 + 2x^3 + 8x + 16$
194. $3x^4 - x^3 - 81x + 27$
195. $x^4 - x^3 - x + 1$

chapter 6

Algebraic Fractions

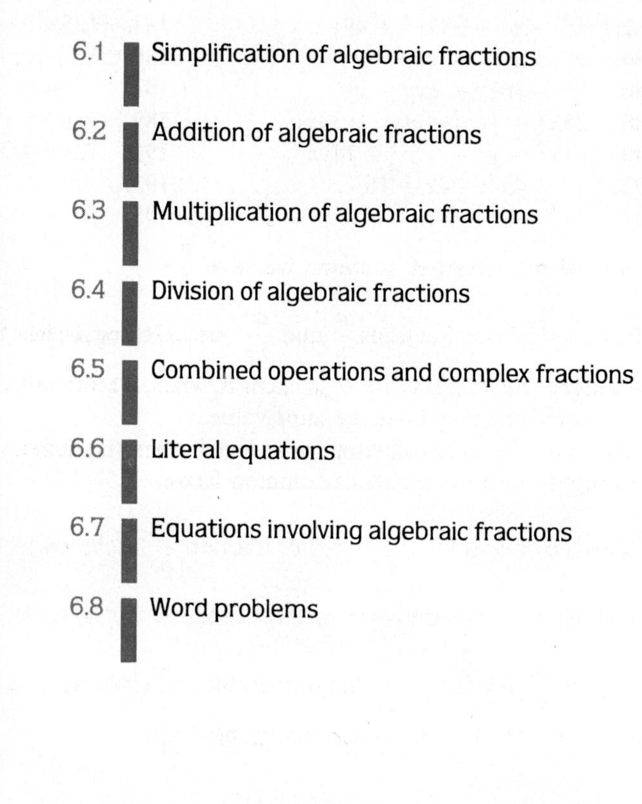

6.1 Simplification of algebraic fractions

6.2 Addition of algebraic fractions

6.3 Multiplication of algebraic fractions

6.4 Division of algebraic fractions

6.5 Combined operations and complex fractions

6.6 Literal equations

6.7 Equations involving algebraic fractions

6.8 Word problems

Algebraic fractions are similar to arithmetic fractions in that they both indicate a division operation. The specific number $\frac{2}{5}$ means $2 \div 5$; the literal number $\frac{a}{b}$ means $a \div b$.

When a specific number is divided by a literal number, $\frac{2}{x}$, or a literal number is divided by another literal number, $\frac{x}{y}$, the result is an **algebraic fraction**. The number x is called the **numerator** of the fraction, and the number y is called the **denominator**.

The notation

$$\frac{a+b}{c} \quad \text{means} \quad (a+b) \div c$$

and

$$\frac{a+b}{c+d} \quad \text{means} \quad (a+b) \div (c+d)$$

Note The literal numbers in the denominators of the algebraic fractions may not be assigned specific values that make the denominator equal to zero, since division by zero is not defined.

6.1 Simplification of Algebraic Fractions

From the properties of fractions, we have $\frac{a}{b} = \frac{ac}{bc}$.

The two algebraic fractions $\frac{a}{b}$ and $\frac{ac}{bc}$ are called **equivalent fractions.**

Two algebraic fractions are equivalent if, when specific values are assigned to their literal numbers, they have the same value.

A fraction is in its **lowest terms**, or **reduced**, when the numerator and the denominator do not possess a common factor.

To reduce or simplify the algebraic fraction $\frac{ac}{bc}$ to its lowest terms, we divide both numerator and denominator by their common factor, c, to get $\frac{a}{b}$. The a and the c in $\frac{ac}{bc}$ are factors of the numerator, not terms as in $a + c$. The b and the c are factors of the denominator, not terms.

The fraction $\frac{a+c}{b+c}$ cannot be reduced to any simpler form; it is not equal to either $\frac{a}{b}$ or $\frac{a+1}{b+1}$. Similarly,

$$\frac{5a+b}{6a} \neq \frac{5+b}{6} \quad \text{but} \quad \frac{5a+b}{6a} = \frac{5a}{6a} + \frac{b}{6a} = \frac{5}{6} + \frac{b}{6a}$$

To reduce to its lowest terms a fraction whose numerator and denominator are monomials, divide both numerator and denominator by their greatest common factor.

EXAMPLE Reduce $\frac{96x^5y^7z^4}{72x^6y^4z^4}$ to its lowest terms.

Solution The greatest common factor of the monomials $96x^5y^7z^4$ and $72x^6y^4z^4$ is $24x^5y^4z^4$.
Dividing both numerator and denominator by $24x^5y^4z^4$, we get

$$\frac{96x^5y^7z^4}{72x^6y^4z^4} = \frac{4y^3}{3x}.$$

To find the **greatest common factor (GCF)** of a set of polynomials, factor the polynomials completely and take all common factors, each to the least exponent to which it appears in any of the given polynomials.

When either the numerator or the denominator or both are polynomials, factor them completely, determine their greatest common factor, and then divide them by that greatest common factor.

EXAMPLE Reduce $\frac{15x^3y - 10x^2y^2}{15x^3y^3}$ to its lowest terms.

Solution

$$\frac{15x^3y - 10x^2y^2}{15x^3y^3} = \frac{5x^2y(3x-2y)}{15x^3y^3}$$

Dividing both numerator and denominator by $5x^2y$, we obtain

$$\frac{15x^3y - 10x^2y^2}{15x^3y^3} = \frac{5x^2y(3x-2y)}{15x^3y^3} = \frac{3x-2y}{3xy^2}$$

EXAMPLE Reduce $\frac{28x^3}{14x^3 + 7x^2}$ to its lowest terms.

Solution

$$\frac{28x^3}{14x^3 + 7x^2} = \frac{28x^3}{7x^2(2x+1)} = \frac{4x}{2x+1}$$

EXAMPLE Reduce $\dfrac{2x^2 + 5x - 3}{6x^2 - 7x + 2}$ to its lowest terms.

Solution Factoring both numerator and denominator, we obtain

$$\frac{2x^2 + 5x - 3}{6x^2 - 7x + 2} = \frac{(x + 3)(2x - 1)}{(2x - 1)(3x - 2)}$$

Dividing both numerator and denominator by their greatest common factor, $(2x - 1)$, we get

$$\frac{2x^2 + 5x - 3}{6x^2 - 7x + 2} = \frac{(x + 3)\overset{1}{\cancel{(2x - 1)}}}{\underset{1}{\cancel{(2x - 1)}}(3x - 2)} = \frac{x + 3}{3x - 2}$$

Note The fraction $\dfrac{x + 3}{3x - 2}$ is in its lowest terms; the numerator and denominator do not possess a common factor.

EXAMPLE Reduce $\dfrac{18x^2 + 3x - 36}{12x^3 - 25x^2 + 12x}$ to its lowest terms.

Solution

$$\frac{18x^2 + 3x - 36}{12x^3 - 25x^2 + 12x} = \frac{3(6x^2 + x - 12)}{x(12x^2 - 25x + 12)}$$

$$= \frac{3\overset{1}{\cancel{(3x - 4)}}(2x + 3)}{x(4x - 3)\underset{1}{\cancel{(3x - 4)}}} = \frac{3(2x + 3)}{x(4x - 3)}$$

EXAMPLE Reduce $\dfrac{4x^2 - 9y^2 + 16x + 16}{4x^2 + 9y^2 - 12xy - 16}$ to its lowest terms.

Solution

$$\frac{4x^2 - 9y^2 + 16x + 16}{4x^2 + 9y^2 - 12xy - 16} = \frac{(4x^2 + 16x + 16) - 9y^2}{(4x^2 - 12xy + 9y^2) - 16} = \frac{(2x + 4)^2 - 9y^2}{(2x - 3y)^2 - 16}$$

$$= \frac{[(2x + 4) + 3y][(2x + 4) - 3y]}{[(2x - 3y) + 4][(2x - 3y) - 4]}$$

$$= \frac{(2x + 3y + 4)\overset{1}{\cancel{(2x - 3y + 4)}}}{\underset{1}{\cancel{(2x - 3y + 4)}}(2x - 3y - 4)}$$

$$= \frac{2x + 3y + 4}{2x - 3y - 4}$$

Note From the commutative and distributive laws,

$$a - b = -b + a = -(b - a)$$

Hence $\dfrac{a - b}{b - a} = \dfrac{-(b - a)}{(b - a)} = -1$

However, the fraction $\dfrac{a + b}{a - b}$ cannot be reduced to any simpler form, since $a + b$ can never be written as a multiple of $a - b$.

EXAMPLE Reduce $\dfrac{6x^2 + 23x - 4}{3 - 16x - 12x^2}$.

Solution

$$\frac{6x^2 + 23x - 4}{3 - 16x - 12x^2} = \frac{(6x - 1)(x + 4)}{(3 + 2x)(1 - 6x)}$$

$$= \frac{-\cancel{(1 - 6x)}^{1}(x + 4)}{(3 + 2x)\cancel{(1 - 6x)}_{1}} = -\frac{x + 4}{3 + 2x}$$

Exercise 6.1

Reduce each of the following fractions to their lowest terms:

1. $\dfrac{x^3}{x^5}$ 2. $\dfrac{x^6}{x^4}$ 3. $\dfrac{x^{10}}{x^5}$ 4. $\dfrac{36x^2}{48x^7}$

5. $\dfrac{57x^8}{38x^2}$ 6. $\dfrac{x^3y^4}{x^5y^3}$ 7. $\dfrac{24x^7}{36x^4y}$ 8. $\dfrac{-28x^3y^2}{21x^3y}$

9. $\dfrac{25xy^2z}{-15x^2y^3}$ 10. $\dfrac{-8x^9y^6}{-12x^3y^2}$ 11. $\dfrac{-24x^4y^6}{-32x^3y^7}$ 12. $\left(\dfrac{xy^3}{x^2y^2}\right)^3$

13. $\left(\dfrac{10x^2y}{15x^3y^3}\right)^4$ 14. $\left(\dfrac{-x^2y^4z^2}{x^3yz^2}\right)^4$ 15. $\left(\dfrac{x^3y^6z^3}{-x^5y^7z}\right)^3$ 16. $\dfrac{(-x)^3}{(-x)^4}$

17. $\dfrac{(-x^2)^5}{(-x^3)^4}$ 18. $\dfrac{(-6x^3)^3}{(-9x^2)^2}$ 19. $\dfrac{(10x^2y)^3}{(15x^2y^3)^4}$

20. $\dfrac{5x^3(x + y)^3}{10x^4(x + y)^2}$ 21. $\dfrac{14x^5(x - y)^4}{42x^3(x - y)^5}$ 22. $\dfrac{x^7(x + 2y)}{x^6(2y + x)}$

23. $\dfrac{(2x + 1)(3x - 2)}{(3 - x)(1 + 2x)}$ 24. $\dfrac{(x + 2)(x - 2)}{(4 + x)(2 - x)}$ 25. $\dfrac{(x - 4)(3x - 1)}{(1 - 3x)(5 + x)}$

26. $\dfrac{(x-4)^2}{(6-x)(4-x)}$

27. $\dfrac{(2x-1)^2}{(1+x)(1-2x)}$

28. $\dfrac{(x-3)(x+3)}{(3-x)^2}$

29. $\dfrac{14x^2+7x}{7x}$

30. $\dfrac{6x^2-12x}{12x}$

31. $\dfrac{2x^2-4x}{2x^2}$

32. $\dfrac{12x^2+18x}{6x^2}$

33. $\dfrac{18x^2-18x}{24x^2}$

34. $\dfrac{4x}{8x^2-6x}$

35. $\dfrac{xy}{xy+x}$

36. $\dfrac{35x^3}{14x^3+28x^2}$

37. $\dfrac{12x^3}{6x^4+6x^3}$

38. $\dfrac{8x^3+16x^2}{4x^4+16x^3}$

39. $\dfrac{27x^2-81x}{27x^2+9x^3}$

40. $\dfrac{18x^3+27x^4}{54x^5+36x^4}$

41. $\dfrac{2x^3+2x^2y}{3x^4+3x^3y}$

42. $\dfrac{3x^2-6x}{2x^2-x^3}$

43. $\dfrac{x^2+4}{x^2-4}$

44. $\dfrac{x^2-9}{4x+12}$

45. $\dfrac{4x^2-y^2}{6x+3y}$

46. $\dfrac{3x^3-x^2y}{9x^2-y^2}$

47. $\dfrac{(2x-3)^2}{4x^2-9}$

48. $\dfrac{x^2-4}{(x+2)^2}$

49. $\dfrac{9x^2-1}{1-3x}$

50. $\dfrac{36x^2-1}{(1-6x)^2}$

51. $\dfrac{x^3+8}{x+2}$

52. $\dfrac{3x-1}{27x^3-1}$

53. $\dfrac{4x^2-1}{8x^3-1}$

54. $\dfrac{x^2+y^2}{x^3+y^3}$

55. $\dfrac{x^2-x-2}{x^2+x-6}$

56. $\dfrac{x^2-x-6}{x^2-4x-12}$

57. $\dfrac{x^2-5x+6}{x^2-4x+3}$

58. $\dfrac{x^2+7x+12}{x^2+5x+6}$

59. $\dfrac{x^2+2x-24}{x^2-8x+16}$

60. $\dfrac{x^2+6x-16}{x^2+5x-24}$

61. $\dfrac{x^2-4x-32}{2x^2+10x-48}$

62. $\dfrac{x^2-6x+9}{3x^2-21x+36}$

63. $\dfrac{x^2-10x-24}{x^3+9x^2+14x}$

64. $\dfrac{x^3+3x^2-10x}{2x^2+2x-40}$

65. $\dfrac{x^2-11x+28}{x^2-4x-21}$

66. $\dfrac{x^2-5xy+6y^2}{x^2+4xy-12y^2}$

67. $\dfrac{x^2+7xy-8y^2}{x^2+11xy-12y^2}$

68. $\dfrac{x^2+xy-12y^2}{x^2-xy-6y^2}$

69. $\dfrac{2x^2+x-1}{2x^2-3x+1}$

70. $\dfrac{2x^2-7x-4}{2x^2-5x-12}$

71. $\dfrac{2x^2+3x-9}{2x^2+5x-12}$

72. $\dfrac{4x^2-11x+6}{4x^2+21x-18}$

73. $\dfrac{3x^2-7x+2}{3x^2-10x+3}$

74. $\dfrac{4x^2+23x-6}{4x^2+7x-2}$

75. $\dfrac{8x^2+10x+3}{12x^2+17x+6}$

76. $\dfrac{9x^2-6x-8}{9x^2-18x+8}$

77. $\dfrac{12x^2-2x-4}{12x^2-17x+6}$

78. $\dfrac{6x^2+7x-3}{24x^2+30x-9}$

79. $\dfrac{9x^2+23x-12}{9x^3+14x^2-8x}$

80. $\dfrac{6x^3 - 7x^2 - 5x}{6x^2 + 7x + 2}$

81. $\dfrac{24x^2 - 5x - 1}{1 + 6x - 16x^2}$

82. $\dfrac{6 - 7x - 3x^2}{3x^2 + 16x - 12}$

83. $\dfrac{12 - 5x - 2x^2}{2x^2 - 7x + 6}$

84. $\dfrac{6 - 35x - 6x^2}{12x^2 - 8x + 1}$

85. $\dfrac{5 - 14x - 24x^2}{4x^2 + 31x - 8}$

86. $\dfrac{6 + x - 12x^2}{12x^2 - 5x - 3}$

87. $\dfrac{x^4 - 7x^2 + 9}{x^4 + x^3 + 3x - 9}$

88. $\dfrac{x^4 + 4}{x^4 - 2x^3 + 4x - 4}$

89. $\dfrac{6x^2 - 15xy + 14x - 35y}{8x^2 - 2xy - 45y^2}$

90. $\dfrac{6y^2 - 27xy - 16y + 72x}{36x^2 + 19xy - 6y^2}$

91. $\dfrac{x^2 - 9y^2 - 2x - 6y}{x^2 - 9y^2 - 4x + 4}$

92. $\dfrac{4x^2 - y^2 + 9 - 12x}{4x^2 - y^2 - 9 - 6y}$

93. $\dfrac{x^2 + y^2 + 2xy - 4}{x^2 - y^2 + 4x + 4}$

94. $\dfrac{x^2 + 4y^2 - 4xy - 9}{4y^2 - x^2 + 12y + 9}$

95. $\dfrac{x^2 - 8x + 16}{x^3 - 12x^2 + 48x - 64}$

96. $\dfrac{64x^3 + 144x^2 + 108x + 27}{16x^2 + 24x + 9}$

6.2 Addition of Algebraic Fractions

Addition of algebraic fractions is similar to addition of fractions in arithmetic. We will start by discussing adding algebraic fractions with like denominators, and then extend the discussion to addition of algebraic fractions with unlike denominators.

Fractions with Like Denominators

Addition of fractions with like denominators is defined by the relation

$$\frac{a}{c} + \frac{b}{c} = \frac{a + b}{c}$$

This shows that the sum of two fractions with the same denominator is a fraction whose numerator is the sum of the numerators and whose denominator is the common denominator.

Notes

1. To avoid making mistakes in adding the numerators, it is advisable to enclose the numerators in parentheses, apply the distributive law, and then combine.
2. After combining the two fractions into one fraction, combine like terms, and then reduce the new fraction to its lowest terms.

EXAMPLE Combine $\frac{2x+1}{4x^2} - \frac{1-2x}{4x^2}$.

Solution

$$\frac{2x+1}{4x^2} - \frac{1-2x}{4x^2} = \frac{(2x+1)-(1-2x)}{4x^2} = \frac{2x+1-1+2x}{4x^2}$$

$$= \frac{4x}{4x^2} = \frac{1}{x}$$

EXAMPLE Combine $\frac{x^2}{x^2+x} - \frac{x}{x^2+x}$.

Solution

$$\frac{x^2}{x^2+x} - \frac{x}{x^2+x} = \frac{x^2-x}{x^2+x} = \frac{x(x-1)}{x(x+1)}$$

$$= \frac{x-1}{x+1}$$

EXAMPLE Combine $\frac{7x^2}{9x^2-4} - \frac{6x-2x^2}{9x^2-4}$.

Solution

$$\frac{7x^2}{9x^2-4} - \frac{6x-2x^2}{9x^2-4} = \frac{7x^2-(6x-2x^2)}{9x^2-4} = \frac{7x^2-6x+2x^2}{9x^2-4}$$

$$= \frac{9x^2-6x}{9x^2-4} = \frac{3x(3x-2)}{(3x+2)(3x-2)}$$

$$= \frac{3x}{3x+2}$$

EXAMPLE Combine $\frac{x^2+3x}{3x^2+x-2} - \frac{7x^2-x}{3x^2+x-2}$.

Solution

$$\frac{x^2+3x}{3x^2+x-2} - \frac{7x^2-x}{3x^2+x-2} = \frac{(x^2+3x)-(7x^2-x)}{3x^2+x-2}$$

$$= \frac{x^2+3x-7x^2+x}{3x^2+x-2}$$

$$= \frac{4x-6x^2}{3x^2+x-2} = \frac{2x(2-3x)}{(3x-2)(x+1)}$$

$$= \frac{-2x(3x-2)}{(3x-2)(x+1)}$$

$$= -\frac{2x}{x+1}$$

Remark The rule for combining two fractions can be extended to any number of fractions:

$$\frac{a_1}{c} + \frac{a_2}{c} + \frac{a_3}{c} + \cdots + \frac{a_n}{c} = \frac{a_1 + a_2}{c} + \frac{a_3}{c} + \cdots + \frac{a_n}{c}$$

$$\vdots$$

$$= \frac{a_1 + a_2 + a_3 + \cdots + a_n}{c}$$

EXAMPLE Combine $\dfrac{5x^2 + 4}{3x^2 + 10x - 8} - \dfrac{3x^2 - x + 1}{3x^2 + 10x - 8} + \dfrac{x^2 - 5}{3x^2 + 10x - 8}$.

Solution

$$\frac{5x^2 + 4}{3x^2 + 10x - 8} - \frac{3x^2 - x + 1}{3x^2 + 10x - 8} + \frac{x^2 - 5}{3x^2 + 10x - 8}$$

$$= \frac{(5x^2 + 4) - (3x^2 - x + 1) + (x^2 - 5)}{3x^2 + 10x - 8}$$

$$= \frac{5x^2 + 4 - 3x^2 + x - 1 + x^2 - 5}{3x^2 + 10x - 8}$$

$$= \frac{3x^2 + x - 2}{3x^2 + 10x - 8} = \frac{(3x - 2)(x + 1)}{(3x - 2)(x + 4)} = \frac{x + 1}{x + 4}$$

Exercise 6.2A

Combine each of the following fractions and simplify:

1. $\dfrac{5}{x} - \dfrac{3}{x} + \dfrac{1}{x}$
2. $\dfrac{9}{2x} - \dfrac{5}{2x} - \dfrac{3}{2x}$
3. $\dfrac{3}{x^2} + \dfrac{4}{x^2} - \dfrac{2}{x^2}$
4. $\dfrac{10}{x^2} - \dfrac{7}{x^2} - \dfrac{4}{x^2}$
5. $\dfrac{x + 1}{2x^2} + \dfrac{x - 1}{2x^2}$
6. $\dfrac{4x - 3}{3x^2} + \dfrac{2x + 3}{3x^2}$
7. $\dfrac{5x + 2}{4x^2} - \dfrac{x - 6}{4x^2}$
8. $\dfrac{3x - 2}{3x^2} - \dfrac{6x - 8}{3x^2}$
9. $\dfrac{6x - 5}{5x + 4} + \dfrac{9 - x}{5x + 4}$
10. $\dfrac{3x + 1}{x - 2} + \dfrac{x - 9}{x - 2}$
11. $\dfrac{x + 4}{x + 7} + \dfrac{x + 10}{x + 7}$
12. $\dfrac{3x}{2x + 3} + \dfrac{x + 6}{2x + 3}$
13. $\dfrac{2x - 5}{3x - 1} + \dfrac{7x + 2}{3x - 1}$
14. $\dfrac{7x - 2}{2x + 1} + \dfrac{5 - x}{2x + 1}$
15. $\dfrac{5x - 2}{7x - 3} - \dfrac{1 - 2x}{7x - 3}$
16. $\dfrac{7x + 4}{4x - 3} - \dfrac{3x + 7}{4x - 3}$
17. $\dfrac{5x + 2}{2x - 3} - \dfrac{x + 8}{2x - 3}$
18. $\dfrac{4x - 3}{x - 4} - \dfrac{x + 9}{x - 4}$
19. $\dfrac{2x - 7}{x + 5} - \dfrac{x - 12}{x + 5}$
20. $\dfrac{5x + 11}{3x + 2} - \dfrac{5 - 4x}{3x + 2}$
21. $\dfrac{4x - 5}{6x - 8} + \dfrac{5x - 7}{6x - 8}$

22. $\frac{3x-2}{6x-9}+\frac{x-4}{6x-9}$

23. $\frac{2x+1}{x^2-x}+\frac{x-4}{x^2-x}$

24. $\frac{x^2+2}{2x^2+3x}-\frac{2-3x}{2x^2+3x}$

25. $\frac{x^2+3}{2x^2+4x}-\frac{3-x}{2x^2+4x}$

26. $\frac{2x^2-5}{3x^2-9x}-\frac{6x-5}{3x^2-9x}$

27. $\frac{2x}{27x^3-1}+\frac{x-1}{27x^3-1}$

28. $\frac{x^2}{x^3+8}-\frac{2x-4}{x^3+8}$

29. $\frac{x^2+3x}{x^2+5x+6}+\frac{6-x^2}{x^2+5x+6}$

30. $\frac{x^2+2x+1}{x^2+5x+4}+\frac{3+2x-x^2}{x^2+5x+4}$

31. $\frac{x^2-8}{2x^2-5x-3}+\frac{2-x}{2x^2-5x-3}$

32. $\frac{x^2+x}{2x^2+7x+6}+\frac{x^2-3}{2x^2+7x+6}$

33. $\frac{x^2+4x}{x^2-x-6}-\frac{3x+12}{x^2-x-6}$

34. $\frac{x^2+2x}{x^2+2x-8}-\frac{3x+20}{x^2+2x-8}$

35. $\frac{3x^2-5x}{3x^2-10x-8}-\frac{2x+6}{3x^2-10x-8}$

36. $\frac{2x^2+x}{4x^2+8x+3}-\frac{6x+12}{4x^2+8x+3}$

37. $\frac{x+1}{2x^2+7x-4}-\frac{2x^2+1}{2x^2+7x-4}$

38. $\frac{6x-5}{4x^2-11x-3}-\frac{2x^2-5}{4x^2-11x-3}$

39. $\frac{3x^2-2}{3x^2-x-4}-\frac{2x+1}{3x^2-x-4}+\frac{4x-5}{3x^2-x-4}$

40. $\frac{2x^2-1}{2x^2-7x+6}-\frac{x^2-3x}{2x^2-7x+6}-\frac{2x+5}{2x^2-7x+6}$

41. $\frac{x^2+3x}{x^2-x-12}+\frac{2x^2+9x}{x^2-x-12}-\frac{2x-3}{x^2-x-12}$

42. $\frac{3x^2+x-2}{x^2+4x-12}+\frac{x^2+11x-3}{x^2+4x-12}-\frac{x^2-4x+7}{x^2+4x-12}$

43. $\frac{x^2+5x}{x^2-y^2+6x+9}-\frac{3x+2xy}{x^2-y^2+6x+9}+\frac{x+xy}{x^2-y^2+6x+9}$

44. $\frac{5x-y}{4x^2-16+y^2+4xy}+\frac{3+2y}{4x^2-16+y^2+4xy}-\frac{3x+7}{4x^2-16+y^2+4xy}$

45. $\frac{x^2+3}{x^2-y^2-2y-1}-\frac{2xy+x}{x^2-y^2-2y-1}+\frac{xy-3}{x^2-y^2-2y-1}$

46. $\frac{4x^2-xy}{9x^2-1-y^2+2y}+\frac{2x^2-x}{9x^2-1-y^2+2y}-\frac{xy-3x}{9x^2-1-y^2+2y}$

Least Common Multiple of Polynomials

To find the least common multiple (LCM) of a set of numbers, factor the numbers into their prime factors and write them in exponent form. Take all the prime numbers, each to the highest exponent.

DEFINITION

> A polynomial P is the **least common multiple (LCM)** of a set of polynomials if
>
> **1.** Each polynomial in the set divides P.
>
> **2.** Any polynomial divisible by all the polynomials in the set is also divisible by P.

To find the LCM of a set of polynomials, factor the polynomials completely and take all distinct factors, each to the largest exponent to which it appears in any of the given polynomials.

EXAMPLE Find the LCM of x^3y, x^2y^3, and xy^5z.

Solution The literal factors are x, y, and z.
The greatest power of x is 3, of y is 5, and of z is 1.
Hence $\quad \text{LCM} = x^3y^5z$

EXAMPLE Find the LCM of $36x^2, 48x^3y$, and $60y^2$.

Solution $36 = 2^2 \cdot 3^2$
$48 = 2^4 \cdot 3$
$60 = 2^2 \cdot 3 \cdot 5$
The LCM of the specific numbers $= 2^4 \cdot 3^2 \cdot 5 = 720$.
The LCM of the monomials $= 720x^3y^2$.

EXAMPLE Find the LCM of $x(x+1), x^2(x-2)$, and $(x+1)(x-2)$.

Solution The distinct factors are $x, (x+1)$, and $(x-2)$.
The greatest power of x is 2, of $(x+1)$ is 1, and of $(x-2)$ is 1.
Hence $\quad \text{LCM} = x^2(x+1)(x-2)$

Note The LCM of $(x-2)$ and $(x-6)$ is $(x-2)(x-6)$.

EXAMPLE

Find the LCM of $x^3 - x^2$ and $x^2 + x - 2$.

Solution

First factor each polynomial completely:

$$x^3 - x^2 = x^2(x-1)$$
$$x^2 + x - 2 = (x-1)(x+2)$$

Hence $\quad \text{LCM} = x^2(x-1)(x+2)$

EXAMPLE

Find the LCM of $9x^2 + 6x + 1$ and $6x^2 + 11x + 3$.

Solution

$$9x^2 + 6x + 1 = (3x+1)^2$$
$$6x^2 + 11x + 3 = (3x+1)(2x+3)$$

Hence $\quad \text{LCM} = (3x+1)^2(2x+3)$

EXAMPLE

Find the LCM of $3x^2 + 4x + 1$, $9x^2 - 1$, and $1 - 2x - 3x^2$.

Solution

$$3x^2 + 4x + 1 = (3x+1)(x+1)$$
$$9x^2 - 1 = (3x+1)(3x-1)$$
$$1 - 2x - 3x^2 = (1-3x)(1+x)$$

Since $(1-3x) = -(3x-1)$, either we can write $(1-3x)$ as $-(3x-1)$, or we can write $(3x-1)$ as $-(1-3x)$.

Remember that $1 + x = x + 1$.

Hence
$$3x^2 + 4x + 1 = (3x+1)(x+1)$$
$$9x^2 - 1 = (3x+1)(3x-1)$$
$$1 - 2x - 3x^2 = -(3x-1)(x+1)$$

Thus $\quad \text{LCM} = (3x+1)(x+1)(3x-1)$

Exercise 6.2B

Find the least common multiple of each of the following:

1. 6, 8, 10 **2.** 8, 9, 12 **3.** 9, 15, 24 **4.** 12, 16, 18

5. 14, 21, 35 **6.** 24, 32, 36 **7.** $2x$, $3x$, x^2 **8.** $4x^2$, $6x$, $8x^3$

9. $3x$, $4xy$, $5y^2$ **10.** x^2, xy^2, y^3 **11.** xy^3, x^2y, x^4

12. $2x^2y$, $5xy^3$, $6y^4$ **13.** x^3y^2, x^4y, y^3z **14.** xy^4, x^2z^3, $-y^2z$

15. $x(x+1)$, $2x(x+1)$, $x^2(x+1)$ **16.** $9(x-2)$, $x^2(x-2)$, $x(x-2)^2$

17. $x(x+1)$, $(x+1)(x-2)$, $x^2(x-2)$ **18.** $x(x-3)$, $(x-3)(x+1)$, $(x+1)^2$

19. $(x+3)(x-1)$, $(x+3)^2$, $(x-1)^2$ **20.** $(x-4)^2$, $(x-4)(x+2)$, $(x+2)^2$

21. $(x-3)(x+1)$, $(x+1)(x-4)$, $(x-3)(x-4)$

22. $(2x+1)(x-1)$, $(x-1)(2x+3)$, $(2x+1)(2x+3)$

23. $(x+2)(x-6)$, $(6-x)(3+x)$, $(x+2)(x+3)$

24. $(2x - 1)(x + 4),\ (x + 4)(x + 8),\ (8 + x)(1 - 2x)$
25. $(x + 2)(x - 1),\ (3 + x)(1 - x),\ (x + 2)(x + 3)$
26. $(x + 1)(3x - 1),\ (1 - 3x)(4 + x),\ (x + 4)(x + 1)$
27. $8x + 8,\ 12x + 12,\ 4x + 4$
28. $6x + 12,\ 15x + 30,\ 3x^2 + 6x$
29. $4x - 12,\ x^2 - 3x,\ 6x - 18$
30. $x^2 - x,\ 6 - 6x,\ 9x - 9$
31. $x^2 + 4,\ x + 2,\ x^2 + 4x + 4$
32. $x^2 - 6x + 9,\ x^2 + 9,\ 3 - x$
33. $x^2 + 3x,\ x^3 - 2x^2,\ x^2 + x - 6$
34. $2x^2 + 2x,\ x^3 - x^2,\ x^2 - 1$
35. $9x^2 - 3x,\ 8x^2 - 4x,\ 6x^2 - 5x + 1$
36. $x^2 - 2x - 8,\ x^2 - x - 6,\ x^2 - 7x + 12$
37. $x^2 + x - 6,\ x^2 + 4x - 12,\ x^2 + 9x + 18$
38. $x^2 - 5x + 6,\ x^2 + 2x - 8,\ x^2 + x - 12$
39. $3x^2 - x - 2,\ 4 - 3x - x^2,\ 3x^2 + 14x + 8$
40. $2x^2 - 9x + 4,\ 3 - 2x - 8x^2,\ 4x^2 - 13x - 12$
41. $x^2 - 5x + 6,\ x^2 + x - 6,\ 9 - x^2$
42. $2x^2 + 7x - 4,\ 4 + 9x + 2x^2,\ 1 - 4x^2$
43. $x^2 + x + 1,\ x^2 - 2x + 1,\ 1 - x^3$
44. $x^3 + 4x,\ 16 - x^4,\ x^2 - 2x$
45. $x^3 + 27,\ x^2 - 6x + 9,\ 9 - x^2$
46. $x^3 - 8,\ 4 - x^2,\ x^2 + 2x + 4$

Fractions with Unlike Denominators

Fractions can be added only when their denominators are alike. When the denominators are not alike, we must find the least common multiple of the denominators, called the **least common denominator (LCD)**. Change each fraction to an equivalent fraction having the LCD as a denominator by the rule

$$\frac{a}{b} = \frac{ac}{bc}$$

and then combine. The sum of algebraic fractions having unlike denominators is, therefore, a fraction whose numerator is the sum of the numerators of the equivalent fractions and whose denominator is the LCD. The final fraction must be reduced to lowest terms.

EXAMPLE Combine $\frac{2}{3x} - \frac{4}{x^2} + \frac{6}{5x}$.

Solution The LCD $= 15x^2$.

We write equivalent fractions with denominators $15x^2$ and then combine:

$$\begin{aligned}\frac{2}{3x} - \frac{4}{x^2} + \frac{6}{5x} &= \frac{2(5x)}{3x(5x)} - \frac{4(15)}{x^2(15)} + \frac{6(3x)}{5x(3x)} \\ &= \frac{10x}{15x^2} - \frac{60}{15x^2} + \frac{18x}{15x^2} \\ &= \frac{10x - 60 + 18x}{15x^2} = \frac{28x - 60}{15x^2}\end{aligned}$$

EXAMPLE

Combine $\frac{x+1}{x-2}+\frac{3}{x+3}-\frac{4x-3}{x^2+x-6}$.

Solution

We first factor the denominators of the fractions:

$$\begin{aligned} x-2 &= (x-2) \\ x+3 &= (x+3) \\ x^2+x-6 &= (x-2)(x+3) \end{aligned}$$

The LCD $= (x-2)(x+3)$.

Writing equivalent fractions with denominators $(x-2)(x+3)$ and then combining, we get

$$\frac{x+1}{x-2}+\frac{3}{x+3}-\frac{4x-3}{x^2+x-6}=\frac{x+1}{x-2}+\frac{3}{x+3}-\frac{4x-3}{(x-2)(x+3)}$$

$$=\frac{(x+1)(x+3)}{(x-2)(x+3)}+\frac{3(x-2)}{(x+3)(x-2)}-\frac{4x-3}{(x-2)(x+3)}$$

$$=\frac{(x+1)(x+3)+3(x-2)-(4x-3)}{(x-2)(x+3)}$$

$$=\frac{x^2+4x+3+3x-6-4x+3}{(x-2)(x+3)}$$

$$=\frac{x^2+3x}{(x-2)(x+3)}$$

$$=\frac{x(x+3)}{(x-2)(x+3)}$$

$$=\frac{x}{x-2}$$

EXAMPLE

Combine and simplify:

$$\frac{7x-22}{x^2-6x+8}-\frac{x+26}{x^2+3x-10}+\frac{17-2x}{x^2+x-20}$$

Solution

$$\frac{7x-22}{x^2-6x+8}-\frac{x+26}{x^2+3x-10}+\frac{17-2x}{x^2+x-20}$$

$$=\frac{7x-22}{(x-2)(x-4)}-\frac{x+26}{(x+5)(x-2)}+\frac{17-2x}{(x+5)(x-4)}$$

The LCD $= (x-2)(x-4)(x+5)$.

Instead of writing equivalent fractions with denominators equal to the least common denominator, and then combining the numerators of the fractions, we write one fraction with the LCD as the denominator. Divide the least common denominator by the denominator of the first fraction, and then multiply the resulting quotient by the numerator of that fraction to get the first expression of the numerator. Repeat the process for each fraction, connecting them by the signs of the corresponding fractions.

$$\otimes \quad \frac{(7x-22)}{(x-2)(x-4)} - \frac{(x+26)}{(x+5)(x-2)} + \frac{(17-2x)}{(x+5)(x-4)}$$

$$\div \quad = \frac{(x+5)(7x-22)-(x-4)(x+26)+(x-2)(17-2x)}{(x-2)(x-4)(x+5)}$$

$$= \frac{(7x^2+13x-110)-(x^2+22x-104)+(21x-34-2x^2)}{(x-2)(x-4)(x+5)}$$

$$= \frac{7x^2+13x-110-x^2-22x+104+21x-34-2x^2}{(x-2)(x-4)(x+5)}$$

$$= \frac{4x^2+12x-40}{(x-2)(x-4)(x+5)} = \frac{4(x+5)(x-2)}{(x-2)(x-4)(x+5)} = \frac{4}{x-4}$$

EXAMPLE

Combine and simplify $\dfrac{x^2-2x-2}{2x^2-13x+6} - \dfrac{x-20}{6+5x-x^2} - \dfrac{7x-2}{2x^2+x-1}$.

Solution

$$\frac{x^2-2x-2}{2x^2-13x+6} - \frac{x-20}{6+5x-x^2} - \frac{7x-2}{2x^2+x-1}$$

$$= \frac{x^2-2x-2}{(2x-1)(x-6)} - \frac{x-20}{(1+x)(6-x)} - \frac{7x-2}{(2x-1)(x+1)}$$

Take the LCD $= (2x-1)(x-6)(x+1)$

$$= \frac{x^2-2x-2}{(2x-1)(x-6)} - \frac{x-20}{-(x+1)(x-6)} - \frac{7x-2}{(2x-1)(x+1)}$$

$$= \frac{x^2-2x-2}{(2x-1)(x-6)} + \frac{x-20}{(x+1)(x-6)} - \frac{7x-2}{(2x-1)(x+1)}$$

$$= \frac{(x+1)(x^2-2x-2)+(2x-1)(x-20)-(x-6)(7x-2)}{(2x-1)(x-6)(x+1)}$$

$$= \frac{(x^3-x^2-4x-2)+(2x^2-41x+20)-(7x^2-44x+12)}{(2x-1)(x-6)(x+1)}$$

$$= \frac{x^3-x^2-4x-2+2x^2-41x+20-7x^2+44x-12}{(2x-1)(x-6)(x+1)}$$

$$= \frac{x^3-6x^2-x+6}{(2x-1)(x-6)(x+1)} = \frac{(x^3-6x^2)-(x-6)}{(2x-1)(x-6)(x+1)}$$

$$= \frac{x^2(x-6)-(x-6)}{(2x-1)(x-6)(x+1)} = \frac{(x-6)(x^2-1)}{(2x-1)(x-6)(x+1)}$$

$$= \frac{(x-6)(x+1)(x-1)}{(2x-1)(x-6)(x+1)} = \frac{x-1}{2x-1}$$

Exercise 6.2C

Combine into a single fraction and simplify:

1. $\frac{2}{3}+\frac{1}{4}-\frac{5}{6}$
2. $\frac{3}{5}-\frac{1}{6}-\frac{1}{4}$
3. $\frac{3}{4}-\frac{1}{6}+\frac{7}{8}$
4. $\frac{7}{8}-\frac{5}{9}+\frac{7}{12}$
5. $\frac{5}{x}+\frac{3}{2x}-\frac{4}{3x}$
6. $\frac{6}{5x}-\frac{2}{3x}+\frac{4}{x}$
7. $\frac{1}{x}-\frac{3}{2x}+\frac{2}{5x}$
8. $\frac{3}{7x}-\frac{5}{14x}-\frac{8}{21x}$
9. $\frac{2}{x^2}-\frac{3}{2x}+\frac{7}{4x}$
10. $\frac{4}{x^2}-\frac{2}{3x^2}+\frac{1}{2x}$
11. $\frac{4}{5x^2}-\frac{3}{4x^2}+\frac{2}{3x}$
12. $\frac{6}{x}-\frac{4}{3x}-\frac{2}{x^2}+\frac{5}{2x^2}$
13. $\frac{1}{x}-\frac{2}{y}+\frac{3}{2x}-\frac{5}{2y}$
14. $\frac{4}{3x}+\frac{5}{y}-\frac{3}{2x}-\frac{3}{2y}$
15. $\frac{3}{x}-\frac{2}{y}-\frac{7}{4x}+\frac{4}{3y}$
16. $\frac{1}{2x}+\frac{2}{3x}-\frac{5}{4y}-\frac{7}{6y}$
17. $\frac{2x+1}{4x}+\frac{x+1}{3x}$
18. $\frac{x-2}{6x}+\frac{2x-1}{8x}$
19. $\frac{3x-2}{9x}+\frac{x+1}{12x}$
20. $\frac{x+4}{5x}+\frac{2x-3}{2x}$
21. $\frac{x-4}{2x}-\frac{x+2}{3x}$
22. $\frac{2x-3}{4x}-\frac{x+3}{6x}$
23. $\frac{3x-1}{7x}-\frac{x-2}{14x}$
24. $\frac{2x-5}{10x}-\frac{3x-4}{15x}$
25. $\frac{x}{x-1}+\frac{2}{x-2}$
26. $\frac{2x}{2x-1}+\frac{3}{x-3}$
27. $\frac{2x}{x+2}+\frac{x}{x-2}$
28. $\frac{x}{2x-3}-\frac{1}{x+2}$
29. $\frac{2x}{2x-1}-\frac{1}{x+1}$
30. $\frac{x}{x-4}-\frac{x-1}{x+3}$
31. $\frac{x}{x+4}-\frac{x-3}{x-2}$
32. $\frac{x-1}{3x-2}-\frac{2x-3}{6x+1}$
33. $\frac{4}{x-4}+\frac{1}{x+2}+\frac{x+8}{x^2-2x-8}$
34. $\frac{2}{x+1}+\frac{3}{x-1}+\frac{1-3x}{x^2-1}$
35. $\frac{3}{2x-1}+\frac{4}{2x+1}+\frac{3-10x}{4x^2-1}$
36. $\frac{4}{x+2}+\frac{2}{x-2}-\frac{x+6}{x^2-4}$
37. $\frac{2}{x+3}+\frac{3}{x-2}-\frac{2x+11}{x^2+x-6}$
38. $\frac{2}{3x+1}-\frac{3}{x-1}+\frac{10x+2}{3x^2-2x-1}$
39. $\frac{3x+5}{x^2+4x+3}+\frac{5-x}{x^2+2x-3}$
40. $\frac{3x}{x^2+x-2}+\frac{2}{x^2-4x+3}$
41. $\frac{3x-8}{x^2-5x+6}+\frac{x+2}{x^2-6x+8}$
42. $\frac{x+7}{x^2+5x+4}+\frac{3x+10}{x^2+6x+8}$

43. $\dfrac{x+13}{x^2+5x-6} - \dfrac{2x+4}{12-4x-x^2}$

44. $\dfrac{x+3}{2x^2-9x+4} - \dfrac{3x+2}{3-5x-2x^2}$

45. $\dfrac{4x+4}{4x^2+8x+3} - \dfrac{x+3}{2+3x-2x^2}$

46. $\dfrac{2x+2}{3x^2-10x+3} - \dfrac{4x+1}{2-5x-3x^2}$

47. $\dfrac{4x+2}{x^2+x-12} - \dfrac{3x+8}{x^2+6x+8}$

48. $\dfrac{9x+12}{2x^2+7x+3} - \dfrac{7x-7}{6x^2-x-2}$

49. $\dfrac{7x-7}{3x^2-11x+6} - \dfrac{4x+2}{3x^2+10x-8}$

50. $\dfrac{17x-13}{4x^2-7x+3} - \dfrac{5x-1}{4x^2+5x-6}$

51. $\dfrac{x+2}{x^2-5x+4} + \dfrac{x+10}{x^2-x-12} + \dfrac{2x+2}{x^2+2x-3}$

52. $\dfrac{x+7}{x^2-4x-5} + \dfrac{x+13}{x^2-x-20} + \dfrac{2x+5}{x^2+5x+4}$

53. $\dfrac{x+2}{x^2-6x+8} + \dfrac{2x-5}{x^2-7x+12} + \dfrac{3x-8}{x^2-5x+6}$

54. $\dfrac{x+1}{x^2+5x+6} + \dfrac{x-11}{x^2-x-12} + \dfrac{2x-2}{x^2-2x-8}$

55. $\dfrac{x+5}{2x^2+5x+2} + \dfrac{2x-1}{x^2-x-6} + \dfrac{x-10}{2x^2-5x-3}$

56. $\dfrac{x-2}{2x^2-3x+1} + \dfrac{2x+13}{2x^2+5x-3} + \dfrac{2x+2}{x^2+2x-3}$

57. $\dfrac{3x}{x^2+x-2} + \dfrac{2x-4}{x^2-4x+3} - \dfrac{3x-4}{x^2-x-6}$

58. $\dfrac{2x+1}{x^2+x-20} + \dfrac{3x+8}{x^2+3x-10} - \dfrac{3x-10}{x^2-6x+8}$

59. $\dfrac{3x-3}{x^2-x-2} - \dfrac{x+5}{x^2+4x+3} - \dfrac{5}{x^2+x-6}$

60. $\dfrac{5x-4}{2x^2-11x-6} + \dfrac{3x+4}{2x^2+7x+3} - \dfrac{3x}{x^2-3x-18}$

61. $\dfrac{5x-7}{2x^2-5x+2} + \dfrac{6x+6}{2x^2+7x-4} - \dfrac{3x}{x^2+2x-8}$

62. $\dfrac{4x+5}{3x^2+17x-6} + \dfrac{2x+7}{x^2+7x+6} - \dfrac{4x}{3x^2+2x-1}$

6.3 Multiplication of Algebraic Fractions

The product of the two fractions $\frac{a}{b}$ and $\frac{c}{d}$ is defined as $\frac{ac}{bd}$; that is,

$$\frac{a}{b} \times \frac{c}{d} = \frac{ac}{bd}$$

Thus the product of two fractions is a fraction whose numerator is the product of the numerators and whose denominator is the product of the denominators. In general,

$$\frac{a_1}{b_1} \cdot \frac{a_2}{b_2} \cdots \frac{a_n}{b_n} = \frac{a_1 a_2 \cdots a_n}{b_1 b_2 \cdots b_n}$$

Note Always reduce the resulting fraction to lowest terms.

EXAMPLE Find the product of $\frac{56a^4b^2}{27xy^4}$ and $\frac{9x^3y^2}{35a^5b^2}$.

Solution

$$\frac{56a^4b^2}{27xy^4} \cdot \frac{9x^3y^2}{35a^5b^2} = \frac{56 \cdot 9a^4b^2x^3y^2}{27 \cdot 35xy^4a^5b^2}$$

$$= \frac{8x^2}{15ay^2}$$

Note It is easier to reduce $\frac{56 \cdot 9}{27 \cdot 35}$ than $\frac{504}{945}$, which is the result of the products. That is, numbers should not be multiplied together until the fraction has been simplified.

EXAMPLE Perform the indicated multiplications and simplify

$$\left(\frac{4a^2b^3}{6x^4y^3}\right)^3 \cdot \left(\frac{3x^2y}{ab^3}\right)^4$$

Solution

$$\left(\frac{4a^2b^3}{6x^4y^3}\right)^3 \cdot \left(\frac{3x^2y}{ab^3}\right)^4 = \left(\frac{2^2a^2b^3}{2 \cdot 3x^4y^3}\right)^3 \cdot \left(\frac{3x^2y}{ab^3}\right)^4$$

$$= \frac{2^6a^6b^9}{2^3 \cdot 3^3x^{12}y^9} \cdot \frac{3^4x^8y^4}{a^4b^{12}}$$

$$= \frac{2^6 \cdot 3^4a^6b^9x^8y^4}{2^3 \cdot 3^3a^4b^{12}x^{12}y^9}$$

$$= \frac{2^3 \cdot 3a^2}{b^3x^4y^5}$$

$$= \frac{24a^2}{b^3x^4y^5}$$

EXAMPLE

Perform the indicated operations and reduce $\dfrac{(6x^3yz^3)^3}{(9x^2yz^2)^2} \cdot \dfrac{(-x^3y)^2}{(2xz)^4}$.

Solution

$$\frac{(6x^3yz^3)^3}{(9x^2yz^2)^2} \cdot \frac{(-x^3y)^2}{(2xz)^4} = \frac{(2 \cdot 3x^3yz^3)^3}{(3^2x^2yz^2)^2} \cdot \frac{(-x^3y)^2}{(2xz)^4}$$

$$= \frac{2^3 \cdot 3^3x^9y^3z^9}{3^4x^4y^2z^4} \cdot \frac{x^6y^2}{2^4x^4z^4}$$

$$= \frac{2^3 \cdot 3^3x^{15}y^5z^9}{2^4 \cdot 3^4x^8y^2z^8} = \frac{x^7y^3z}{2 \cdot 3} = \frac{x^7y^3z}{6}$$

To multiply fractions whose numerators or denominators are polynomials, first factor the polynomials completely. Consider the fractions as just one fraction and divide the numerators and denominators by their greatest common factor to get an equivalent fraction in lowest terms.

EXAMPLE

Simplify $\dfrac{3x-12}{2x^2-7x+6} \cdot \dfrac{6x^2-5x-6}{3x^2-10x-8}$.

Solution

$$\frac{3x-12}{2x^2-7x+6} \cdot \frac{6x^2-5x-6}{3x^2-10x-8} = \frac{3(x-4)}{(x-2)(2x-3)} \cdot \frac{(2x-3)(3x+2)}{(x-4)(3x+2)} = \frac{3}{x-2}$$

EXAMPLE

Simplify $\dfrac{27-8x^3}{12x^2-19x+4} \cdot \dfrac{3x^2+2x-8}{4x^3+6x^2+9x} \cdot \dfrac{4x^2-x}{4x^2-4x-3}$.

Solution

$$\frac{27-8x^3}{12x^2-19x+4} \cdot \frac{3x^2+2x-8}{4x^3+6x^2+9x} \cdot \frac{4x^2-x}{4x^2-4x-3}$$

$$= \frac{(3-2x)(9+6x+4x^2)}{(3x-4)(4x-1)} \cdot \frac{(3x-4)(x+2)}{x(4x^2+6x+9)} \cdot \frac{x(4x-1)}{(2x-3)(2x+1)}$$

$$= -\frac{x+2}{2x+1}$$

Exercise 6.3

Perform the following multiplications and simplify:

1. $\dfrac{16}{27} \cdot \dfrac{39}{40} \cdot \dfrac{15}{78}$

2. $\dfrac{34}{21} \cdot \dfrac{32}{51} \cdot \dfrac{63}{128}$

3. $\dfrac{36}{77} \cdot \dfrac{33}{32} \cdot \dfrac{56}{27}$

4. $\dfrac{48}{95} \cdot \dfrac{57}{42} \cdot \dfrac{35}{64}$

5. $\dfrac{x^4}{12} \cdot \dfrac{15}{x^3} \cdot \dfrac{x}{10}$

6. $\dfrac{18x^2}{14} \cdot \dfrac{28}{3x^3} \cdot \dfrac{x}{4}$

7. $\dfrac{36x^2}{7y} \cdot \dfrac{y^3}{3x^4} \cdot \dfrac{x}{4y^2}$

8. $\dfrac{6a^4b}{9x^3y^2} \cdot \dfrac{15x^2y^3}{10a^3b^5}$

9. $\dfrac{64a^3b^4}{9x^2y^3} \cdot \dfrac{27x^5y}{16a^6b^2}$

10. $\dfrac{6^2a^4b^{10}}{9^3x^8y^9} \cdot \dfrac{12^4x^4y^{12}}{16^2a^6b^5}$

11. $\dfrac{a^2b}{xy^3} \cdot \dfrac{x^4y}{ab^5} \cdot \dfrac{ab^2}{x^2z}$

12. $\dfrac{a^2b^3}{2xy^4} \cdot \dfrac{3xy^2}{ab^4} \cdot \dfrac{b^2}{y}$

13. $\left(\dfrac{-x^2}{y^2}\right)^4\left(\dfrac{y^3}{-x}\right)^3$

14. $\left(\dfrac{3x^2}{y^3}\right)^3\left(\dfrac{-y}{6x}\right)^2$

15. $\left(\dfrac{5x}{3y}\right)^3\left(\dfrac{-y}{5x}\right)^2$

16. $\left(\dfrac{x^2}{3y}\right)^4\left(\dfrac{9y}{-x^2}\right)^3$

17. $\left(\dfrac{4a^2b^4}{9xy^3}\right)^3 \cdot \left(\dfrac{3x^2y^2}{4ab^3}\right)^4$

18. $\left(\dfrac{3a^5b^2}{2x^2y^4}\right)^3 \cdot \left(\dfrac{4x^3y^6}{9a^8b^4}\right)^2$

19. $\dfrac{(2x^2y^3)^2}{(3xy^3)^3} \cdot \dfrac{(-x^2y^2)^3}{(4x^3y^2)^2}$

20. $\dfrac{(6xz)^3}{(9x^2y)^4} \cdot \dfrac{(10x^3yz)^3}{(25xy^2)^2}$

21. $\dfrac{(14x^2)^4}{(21xy^2)^3} \cdot \dfrac{(9xy^4)^3}{(-2x^2)^5}$

22. $\dfrac{(6xy^2)^3}{(12x^2y^2)^2} \cdot \dfrac{(18x^2z)^4}{(-27x^3y)^3}$

23. $\dfrac{3x^2 - 6x}{x^3 + 3x^2} \cdot \dfrac{x^2 - 4x}{6x - 12}$

24. $\dfrac{10x^2 - 15x}{9x + 36} \cdot \dfrac{3x^2 + 3x}{2x^3 - 3x^2}$

25. $\dfrac{12x + 4}{14x - 7} \cdot \dfrac{8x^2 - 2x}{24x^2 + 8x}$

26. $\dfrac{3x^3 + 5x^2}{12x + 6} \cdot \dfrac{2x^2 + x}{x^4 - 2x^3}$

27. $\dfrac{x^2 - x - 2}{x^3y} \cdot \dfrac{x^2y^2}{x^2 - 4}$

28. $\dfrac{x^2 + 2x - 3}{4xy^4} \cdot \dfrac{6x^3y}{x^2 + 9x + 18}$

29. $\dfrac{x^3 + 8}{x^2 - x - 6} \cdot \dfrac{x^2 - 2x - 3}{x^2 - 2x + 4}$

30. $\dfrac{x^3 - 1}{x^2 + 3x - 4} \cdot \dfrac{x^2 - 16}{x^2 + x + 1}$

31. $\dfrac{x^3 - 27}{x^2 - 6x + 9} \cdot \dfrac{x^2 - 9}{x^2 + 3x + 9}$

32. $\dfrac{x^3 + 64}{x^2 + 9x + 20} \cdot \dfrac{x^2 + 10x + 25}{x^2 - 4x + 16}$

33. $\dfrac{x^2 + 5x - 6}{x^2 - 3x + 2} \cdot \dfrac{x^2 - 6x + 8}{x^2 - x - 12}$

34. $\dfrac{x^2 + 5x + 6}{x^2 + 7x + 12} \cdot \dfrac{x^2 + x - 12}{x^2 - 4x - 12}$

35. $\dfrac{x^2 - 5x - 24}{x^2 + x - 6} \cdot \dfrac{x^2 + 2x - 8}{x^2 - 7x - 8}$

36. $\dfrac{x^2 - 4x + 4}{x^2 - 8x + 12} \cdot \dfrac{x^2 - 3x - 18}{x^2 + 4x - 12}$

37. $\dfrac{x^2 + 2x - 24}{x^2 - x - 20} \cdot \dfrac{x^2 - 5x - 36}{x^2 - 3x - 54}$

38. $\dfrac{x^2 - 11x - 12}{x^2 - 10x - 24} \cdot \dfrac{x^2 + 8x + 12}{x^2 - 3x - 4}$

39. $\dfrac{2x^2 - 5x - 3}{3x^2 + 5x - 2} \cdot \dfrac{3x^2 + 11x - 4}{2x^2 + 9x + 4}$

40. $\dfrac{3x^2 - 4x - 4}{3x^2 + 14x + 8} \cdot \dfrac{3x^2 + 5x - 12}{3x^2 - 10x + 8}$

41. $\dfrac{3x^2 + 10x - 8}{3x^2 - 11x + 6} \cdot \dfrac{4x^2 + 11x - 3}{4x^2 + 15x - 4}$

42. $\dfrac{6x^2 + 11x - 2}{6x^2 - 37x + 6} \cdot \dfrac{4x^2 - 21x - 18}{4x^2 + 11x + 6}$

43. $\dfrac{2x^2 - x - 10}{3x^2 + 4x - 4} \cdot \dfrac{12x^2 - 11x + 2}{2x^2 - 11x + 15}$

44. $\dfrac{6x^2 - 5x - 6}{24x^2 + 13x - 2} \cdot \dfrac{8x^2 + 7x - 1}{2x^2 - 5x + 3}$

45. $\dfrac{6 - 23x - 4x^2}{3x^2 + 10x - 48} \cdot \dfrac{3x^2 + x - 24}{8x^2 - 6x + 1}$

46. $\dfrac{9 - 3x - 2x^2}{2x^2 + 11x - 21} \cdot \dfrac{2x^2 + 9x - 35}{2x^2 + 11x + 15}$

47. $\dfrac{6x^2 + 13x + 2}{8 - 14x - 9x^2} \cdot \dfrac{9x^2 + 23x - 12}{6x^2 - 23x - 4}$

48. $\dfrac{24 - 29x - 4x^2}{4x^2 + 21x - 18} \cdot \dfrac{12 - 16x - 3x^2}{6x^2 - 7x + 2}$

49. $\dfrac{2x^2 + 7x - 9}{3x^2 + 5x - 12} \cdot \dfrac{2x^2 + 5x - 3}{2x^2 + x - 36} \cdot \dfrac{3x^2 - 16x + 16}{2x^2 + x - 3}$

50. $\dfrac{4x^2 + 11x + 6}{4x^2 - 21x - 18} \cdot \dfrac{4x^2 - 27x + 18}{2x^2 - x - 10} \cdot \dfrac{2x^2 - 3x - 5}{4x^2 - 7x + 3}$

51. $\dfrac{x^2 - 9 + y^2 - 2xy}{(x - y)^2 + (x - y) - 6} \cdot \dfrac{x^2 - y^2 - 4x + 4}{(x - y)^2 - 2(x - y) - 3}$

52. $\dfrac{(x + 2y)^2 + 3(x + 2y) - 4}{x^2 - 1 + 4y^2 + 4xy} \cdot \dfrac{x^2 + 1 - 4y^2 + 2x}{(x + 2y)^2 + (x + 2y) - 12}$

53. $\dfrac{x^4 - 5x^2 - 36}{x^4 - 14x^2 - 32} \cdot \dfrac{x^3 - 8 - 4x^2 + 2x}{x^3 + 12 + 3x^2 + 4x}$

54. $\dfrac{4x^4 - 21x^2 + 27}{9x^4 - 49x^2 + 20} \cdot \dfrac{3x^3 + 2x^2 - 15x - 10}{2x^3 - 3x^2 - 6x + 9}$

6.4 Division of Algebraic Fractions

From the definition of division of fractions, we have

$$\frac{a}{b} \div \frac{c}{d} = \frac{a}{b} \cdot \frac{d}{c}$$

The above result shows how to transform division of fractions into multiplication of fractions.

The fractions $\dfrac{c}{d}$ and $\dfrac{d}{c}$ are called **multiplicative inverses** or **reciprocals**.

Notes

1. The reciprocal of the expression $a + b$ is $\dfrac{1}{a + b}$ not $\dfrac{1}{a} + \dfrac{1}{b}$.
2. The reciprocal of $\dfrac{1}{a} + \dfrac{1}{b}$ is $\dfrac{1}{\frac{1}{a} + \frac{1}{b}}$, or simplified $\dfrac{ab}{b + a}$:

$$\frac{1}{\frac{1}{a} + \frac{1}{b}} = \frac{1}{\frac{1}{a} + \frac{1}{b}} \cdot \frac{ab}{ab} = \frac{ab}{b + a}$$

EXAMPLE

Simplify $\dfrac{10x^2}{9y} \div \dfrac{4x^3}{27y^2}$.

Solution

$$\frac{10x^2}{9y} \div \frac{4x^3}{27y^2} = \frac{10x^2}{9y} \cdot \frac{27y^2}{4x^3} = \frac{15y}{2x}$$

EXAMPLE Simplify $\left(\frac{4x^2y^3}{9ab^2}\right)^3 \div \left(\frac{8x^5y^4}{6a^2b^4}\right)^2$.

Solution

$$\left(\frac{4x^2y^3}{9ab^2}\right)^3 \div \left(\frac{8x^5y^4}{6a^2b^4}\right)^2 = \left(\frac{2^2x^2y^3}{3^2ab^2}\right)^3 \div \left(\frac{2^3x^5y^4}{2\cdot 3a^2b^4}\right)^2$$
$$= \frac{2^6x^6y^9}{3^6a^3b^6}\cdot\frac{2^2\cdot 3^2a^4b^8}{2^6x^{10}y^8}$$
$$= \frac{4yab^2}{81x^4}$$

Note $\frac{a}{b} \div \frac{c}{d}\cdot\frac{e}{f} = \frac{a}{b}\cdot\frac{d}{c}\cdot\frac{e}{f} = \frac{ade}{bcf}$

EXAMPLE Simplify $\frac{9x^4y^7}{28a^3b^6} \div \frac{x^3y^5}{a^2b^3}\cdot\frac{49a^2b}{3xy}$.

Solution

$$\frac{9x^4y^7}{28a^3b^6} \div \frac{x^3y^5}{a^2b^3}\cdot\frac{49a^2b}{3xy} = \frac{9x^4y^7}{28a^3b^6}\cdot\frac{a^2b^3}{x^3y^5}\cdot\frac{49a^2b}{3xy} = \frac{21ya}{4b^2}$$

Note $\frac{a}{b} \div \left(\frac{c}{d}\cdot\frac{e}{f}\right) = \frac{a}{b} \div \frac{ce}{df} = \frac{a}{b}\cdot\frac{df}{ce} = \frac{adf}{bce}$

EXAMPLE Simplify $\frac{x^3y^8}{a^2b^6} \div \left(\frac{x^4y^2}{a^3b^2}\cdot\frac{ab^3}{x^2y}\right)$.

Solution

$$\frac{x^3y^8}{a^2b^6} \div \left(\frac{x^4y^2}{a^3b^2}\cdot\frac{ab^3}{x^2y}\right) = \frac{x^3y^8}{a^2b^6} \div \frac{x^4y^2\cdot ab^3}{a^3b^2\cdot x^2y}$$
$$= \frac{x^3y^8}{a^2b^6}\cdot\frac{a^3b^2\cdot x^2y}{x^4y^2\cdot ab^3} = \frac{xy^7}{b^7}$$

EXAMPLE Simplify $\frac{6x^2+5x-6}{12x^2-11x+2} \div \frac{2x^2+11x+12}{4x^2+7x-2}$.

Solution As in multiplication of fractions, we factor the numerators and the denominators:

$$\frac{6x^2+5x-6}{12x^2-11x+2} \div \frac{2x^2+11x+12}{4x^2+7x-2} = \frac{(2x+3)(3x-2)}{(3x-2)(4x-1)} \div \frac{(2x+3)(x+4)}{(4x-1)(x+2)}$$
$$= \frac{(2x+3)(3x-2)}{(3x-2)(4x-1)}\cdot\frac{(4x-1)(x+2)}{(2x+3)(x+4)}$$
$$= \frac{x+2}{x+4}$$

EXAMPLE

Perform the indicated operations and simplify:

$$\frac{3x^2 - 14x - 5}{2x^2 - x - 3} \div \frac{3x^2 - 23x - 8}{2x^2 - 11x + 12} \cdot \frac{3x^2 - x - 4}{x^2 - 9x + 20}$$

Solution

$$\frac{3x^2 - 14x - 5}{2x^2 - x - 3} \div \frac{3x^2 - 23x - 8}{2x^2 - 11x + 12} \cdot \frac{3x^2 - x - 4}{x^2 - 9x + 20}$$

$$= \frac{(3x+1)(x-5)}{(2x-3)(x+1)} \div \frac{(3x+1)(x-8)}{(2x-3)(x-4)} \cdot \frac{(3x-4)(x+1)}{(x-4)(x-5)}$$

$$= \frac{(3x+1)(x-5)}{(2x-3)(x+1)} \cdot \frac{(2x-3)(x-4)}{(3x+1)(x-8)} \cdot \frac{(3x-4)(x+1)}{(x-4)(x-5)} = \frac{3x-4}{x-8}$$

Exercise 6.4

Perform the indicated operations and simplify each of the following:

1. $\frac{18}{35} \div \frac{27}{49}$
2. $\frac{48}{34} \div \frac{64}{68}$
3. $\frac{63}{44} \div \frac{56}{33} \cdot \frac{32}{81}$
4. $\frac{57}{45} \div \frac{38}{36} \cdot \frac{30}{108}$
5. $\frac{45}{36} \div \left(\frac{12}{56} \cdot \frac{35}{72}\right)$
6. $\frac{32}{49} \div \left(\frac{48}{343} \cdot \frac{28}{243}\right)$
7. $\frac{12x^4}{35y^2} \div \frac{3x}{14y^3}$
8. $\frac{15x^2}{4y^3} \div \frac{5x^2}{8y^2}$
9. $\frac{4x^3y}{9a^2b^3} \div \frac{8x^2y}{3ab^2}$
10. $\frac{10x^2y^3}{7a^3b} \div \frac{5x^3y^2}{21a^2b^2}$
11. $\left(\frac{x^2}{y^3}\right)^4 \div \left(\frac{x^3}{2y^2}\right)^3$
12. $\left(\frac{ax}{b^2y}\right)^3 \div \left(\frac{a^2x}{b^3y^2}\right)^2$
13. $\left(\frac{2xy^3}{3a^2b}\right)^3 \div \left(\frac{4xy^2}{3ab^2}\right)^2$
14. $\left(\frac{3x^3y^2}{4a^2b}\right)^3 \div \left(\frac{9x^5y^3}{8a^4b}\right)^2$
15. $\frac{a^2b^3}{x^2y} \cdot \frac{xy^3}{ab^2} \div \frac{a^3b}{xy^2}$
16. $\frac{xy^3}{a^2b} \cdot \frac{x^2y}{ab^3} \div \frac{x^3y^2}{a^2b^5}$
17. $\frac{7x^2}{5y^4} \div \frac{21x}{10y^3} \cdot \frac{3y}{x^3}$
18. $\frac{15x^3}{14y^2} \div \frac{20x^2y}{21x} \cdot \frac{16y^5}{9x}$
19. $\frac{xy^3}{a^2b} \div \frac{x^2y^4}{a^4b^3} \cdot \frac{x^3y}{ab}$
20. $\frac{a^3x^2}{b^2y^3} \div \frac{a^2x}{b^5y^2} \cdot \frac{a}{b^3y}$
21. $\frac{xy^5}{a^2b^3} \div \left(\frac{x^2y^3}{a^4b^2} \cdot \frac{ab^3}{x^3y}\right)$
22. $\frac{a^2x^6}{b^2y^3} \div \left(\frac{ax^3}{b^3y^2} \cdot \frac{b^2y}{ax^2}\right)$
23. $\frac{a^2b^6}{x^3y^4} \div \left(\frac{ab^4}{x^2y^3} \cdot \frac{a^2b^2}{xy^2}\right)$
24. $\frac{x^2y^3}{a^4b^3} \div \left(\frac{a^3b}{x^4y^2} \cdot \frac{x^3y}{a^2b^4}\right)$

25. $\dfrac{2x^3y - 4x^2y}{x^3y} \div \dfrac{4x^4y - 8x^3y}{x^4y}$

26. $\dfrac{9x^2y - 9xy^2}{x^2y} \div \dfrac{6x^4 - 6x^3y}{x^4}$

27. $\dfrac{3x^2 + 9x}{(-x)^2} \div \dfrac{4x^3 + 12x^2}{-x^2}$

28. $\dfrac{(-2x)^4}{x^4 + 2x^2} \div \dfrac{-8x^3}{x^3 + 2x}$

29. $\dfrac{x^2 + 3x + 2}{x^2 + 4x + 4} \div \dfrac{x^2 - 1}{x^2 + x - 2}$

30. $\dfrac{x^2 + 2x - 3}{x^2 - 4x + 3} \div \dfrac{x^2 - x - 12}{x^2 - 5x + 6}$

31. $\dfrac{x^2 - x - 2}{x^2 + 4x + 3} \div \dfrac{x^2 - 4x + 4}{x^2 + x - 6}$

32. $\dfrac{x^2 + 2x - 8}{x^2 - 2x - 8} \div \dfrac{x^2 + 10x + 24}{x^2 + 2x - 24}$

33. $\dfrac{x^2 - x - 6}{x^2 + 5x - 24} \div \dfrac{x^2 - 4}{x^2 + 6x - 16}$

34. $\dfrac{x^2 + 7x + 12}{x^2 + 5x + 4} \div \dfrac{x^2 - 5x - 24}{x^2 - 7x - 8}$

35. $\dfrac{x^2 - 10x + 24}{x^2 - 2x - 24} \div \dfrac{x^2 + x - 20}{x^2 + 9x + 20}$

36. $\dfrac{x^2 - 7x - 18}{x^2 - 4x - 12} \div \dfrac{x^2 - 6x - 27}{x^2 - 3x - 18}$

37. $\dfrac{x^2 + 10x - 24}{x^2 - 7x + 10} \div \dfrac{x^2 + 14x + 24}{x^2 - 3x - 10}$

38. $\dfrac{x^2 - 4x - 21}{x^2 + 8x + 15} \div \dfrac{x^2 + 3x - 4}{x^2 + 4x - 5}$

39. $\dfrac{x^2 + 9xy + 8y^2}{x^2 + 7xy - 8y^2} \div \dfrac{x^2 - y^2}{x^2 + 5xy - 6y^2}$

40. $\dfrac{2x^2 - 9x + 9}{2x^2 + 3x - 9} \div \dfrac{2x^2 + 3x - 2}{2x^2 + 5x - 3}$

41. $\dfrac{3x^2 + 17x - 6}{3x^2 - 13x + 4} \div \dfrac{2x^2 + 7x + 6}{2x^2 - 5x - 12}$

42. $\dfrac{2x^2 + 11x + 12}{3x^2 - 5x - 2} \div \dfrac{2x^2 - x - 6}{3x^2 - 11x - 4}$

43. $\dfrac{3x^2 - 2x - 8}{3x^2 + 10x + 8} \div \dfrac{2x^2 - 9x + 10}{2x^2 + 7x + 6}$

44. $\dfrac{4x^2 + 25x + 6}{4x^2 + 11x - 3} \div \dfrac{x^2 + 8x + 12}{4x^2 + 7x - 2}$

45. $\dfrac{6x^2 + x - 12}{6x^2 - 17x + 12} \div \dfrac{9x^2 - 4}{6x^2 - 13x + 6}$

46. $\dfrac{3x^2 - 8x - 16}{2x^2 + 5x - 12} \div \dfrac{3x^2 + 13x + 12}{9 - 3x - 2x^2}$

47. $\dfrac{4x^2 + 19x - 5}{6x^2 - 11x - 2} \div \dfrac{6 - 19x - 20x^2}{5x^2 - 4x - 12}$

48. $\dfrac{4x^2 - 12x - 7}{24 - 14x - 3x^2} \div \dfrac{8x^2 + 10x + 3}{6x^2 - 5x - 4}$

49. $\dfrac{x^3 - y^3}{x^2 - y^2} \div \dfrac{x^2 + xy + y^2}{x^2 - xy - 2y^2}$

50. $\dfrac{x^6 + y^6}{x^4 + 4x^2y^2 + 3y^4} \div \dfrac{x^4 + 3x^2y^2 + 2y^4}{x^4 + 5x^2y^2 + 6y^4}$

51. $\dfrac{3x^2 - 20x + 12}{x^2 + 4x - 32} \cdot \dfrac{x^2 + 2x - 24}{4x^2 + 21x - 18} \div \dfrac{x^2 - 8x + 12}{4x^2 + 29x - 24}$

52. $\dfrac{3x^2 + 5x + 2}{3x^2 + 11x + 6} \cdot \dfrac{2x^2 - 3x + 1}{2x^2 - 15x - 8} \div \dfrac{x^2 - 1}{x^2 - 5x - 24}$

53. $\dfrac{3x^2 + 2x - 8}{2x^2 + 5x + 2} \div \dfrac{3x^2 + 8x - 16}{2x^2 + 5x - 12} \cdot \dfrac{6x^2 - x - 2}{2x^2 - 7x + 6}$

54. $\dfrac{2x^2 - x - 6}{2x^2 - 7x - 4} \div \dfrac{2x^2 + 9x + 9}{3x^2 - 10x - 8} \cdot \dfrac{2x^2 + 7x + 3}{3x^2 - 7x - 6}$

55. $\dfrac{24x^2 - 2x - 15}{24x^2 - 37x - 72} \div \left[\dfrac{18x^2 + 9x - 20}{8x^2 - 87x - 108} \cdot \dfrac{4x^2 - 45x - 36}{9x^2 - 12x - 32}\right]$

56. $\dfrac{12x^2 - 9xy + 8x - 6y}{8x^2 + 48xy - x - 6y} \div \left[\dfrac{12x^2 - x - 6}{x^2 + x - 36y^2 + 6y} \cdot \dfrac{64x^3 - 27y^3}{32x^2 - 28x + 3}\right]$

6.5 Combined Operations and Complex Fractions

In the previous sections, we discussed the addition and subtraction of fractions as well as their multiplication and division. In all cases, the final answer was one fraction in its simplest form. Here we shall confront the four operations in one problem and still require the final answer to be one fraction in its simplest form.

When there are no symbols of grouping in the problem, multiplications and divisions are performed first in the order in which they appear in the problem. Only after all the multiplications and divisions have been done are the additions and subtractions performed.

EXAMPLE

Perform the indicated operations and simplify:

$$\frac{3}{2x - 1} - \frac{4x + 12}{3x^2 + 13x + 4} \div \frac{2x^2 + 7x + 3}{2x^2 + 9x + 4}$$

Solution

$$\frac{3}{2x - 1} - \frac{4x + 12}{3x^2 + 13x + 4} \div \frac{2x^2 + 7x + 3}{2x^2 + 9x + 4}$$

$$= \frac{3}{2x - 1} - \frac{4(x + 3)}{(x + 4)(3x + 1)} \div \frac{(x + 3)(2x + 1)}{(x + 4)(2x + 1)}$$

$$= \frac{3}{2x - 1} - \frac{4(x + 3)}{(x + 4)(3x + 1)} \cdot \frac{(x + 4)(2x + 1)}{(x + 3)(2x + 1)}$$

$$= \frac{3}{2x - 1} - \frac{4}{3x + 1} = \frac{3(3x + 1) - 4(2x - 1)}{(2x - 1)(3x + 1)}$$

$$= \frac{9x + 3 - 8x + 4}{(2x - 1)(3x + 1)} = \frac{x + 7}{(2x - 1)(3x + 1)}$$

When there are symbols of grouping, as in the problem

$$\left(\frac{x}{4} - \frac{9}{x}\right)\left(\frac{x}{x + 6} - \frac{x}{x - 6}\right)$$

we have the choice of performing the multiplication first or of combining the terms within parentheses. Combining the terms within parentheses is easier, as illustrated in the following examples.

EXAMPLE

Perform the indicated operations and simplify:

$$\left(\frac{x}{4}-\frac{9}{x}\right)\left(\frac{x}{x+6}-\frac{x}{x-6}\right)$$

Solution

$$\left(\frac{x}{4}-\frac{9}{x}\right)\left(\frac{x}{x+6}-\frac{x}{x-6}\right)=\frac{x^2-36}{4x}\cdot\frac{x(x-6)-x(x+6)}{(x+6)(x-6)}$$

$$=\frac{(x+6)(x-6)}{4x}\cdot\frac{x^2-6x-x^2-6x}{(x+6)(x-6)}$$

$$=\frac{(x+6)(x-6)}{4x}\cdot\frac{-12x}{(x+6)(x-6)}=-3$$

EXAMPLE

Perform the indicated operations and simplify:

$$\left(2x+1-\frac{5}{x-1}\right)\left(x+3+\frac{4}{x-2}\right)$$

Solution

$$\left(2x+1-\frac{5}{x-1}\right)\left(x+3+\frac{4}{x-2}\right)$$

$$=\left(\frac{2x}{1}+\frac{1}{1}-\frac{5}{x-1}\right)\left(\frac{x}{1}+\frac{3}{1}+\frac{4}{x-2}\right)$$

$$=\frac{2x(x-1)+(x-1)-5}{(x-1)}\cdot\frac{x(x-2)+3(x-2)+4}{(x-2)}$$

$$=\frac{2x^2-2x+x-1-5}{(x-1)}\cdot\frac{x^2-2x+3x-6+4}{(x-2)}$$

$$=\frac{2x^2-x-6}{(x-1)}\cdot\frac{x^2+x-2}{(x-2)}$$

$$=\frac{(2x+3)(x-2)}{(x-1)}\cdot\frac{(x+2)(x-1)}{(x-2)}$$

$$=(2x+3)(x+2)$$

EXAMPLE

Perform the indicated operations and simplify:

$$\left(\frac{x}{4}-\frac{2}{x^2}\right)\div\left(\frac{1}{2}-\frac{1}{x}\right)$$

Solution

$$\left(\frac{x}{4}-\frac{2}{x^2}\right)\div\left(\frac{1}{2}-\frac{1}{x}\right)=\frac{x^3-8}{4x^2}\div\frac{x-2}{2x}$$

$$=\frac{(x-2)(x^2+2x+4)}{4x^2}\cdot\frac{2x}{(x-2)}$$

$$=\frac{x^2+2x+4}{2x}$$

Note Since $(a+b) \div (c+d)$ can be written as $\dfrac{a+b}{c+d}$, we can write

$$\left(3+\frac{11}{x}-\frac{4}{x^2}\right) \div \left(6+\frac{7}{x}-\frac{3}{x^2}\right)$$

in the form $\dfrac{3+\dfrac{11}{x}-\dfrac{4}{x^2}}{6+\dfrac{7}{x}-\dfrac{3}{x^2}}$, which is a **complex fraction.**

Given a complex fraction, we can either simplify the problem as it is, in fraction form, or write it in division form and simplify. Sometimes a complex fraction can be easily simplified by multiplying both numerator and denominator by the least common multiple of all the denominators involved.

EXAMPLE Simplify $\dfrac{3+\dfrac{11}{x}-\dfrac{4}{x^2}}{6+\dfrac{7}{x}-\dfrac{3}{x^2}}$.

Solution The LCM of the denominators is x^2:

$$\frac{3+\dfrac{11}{x}-\dfrac{4}{x^2}}{6+\dfrac{7}{x}-\dfrac{3}{x^2}} = \frac{\dfrac{x^2}{1}\left(\dfrac{3}{1}+\dfrac{11}{x}-\dfrac{4}{x^2}\right)}{\dfrac{x^2}{1}\left(\dfrac{6}{1}+\dfrac{7}{x}-\dfrac{3}{x^2}\right)} = \frac{3x^2+11x-4}{6x^2+7x-3}$$

$$= \frac{(3x-1)(x+4)}{(3x-1)(2x+3)} = \frac{x+4}{2x+3}$$

EXAMPLE Simplify $\dfrac{x-3}{x-2-\dfrac{4}{x+1}}$.

Solution The LCM of the denominators is $x+1$:

$$\frac{x-3}{x-2-\dfrac{4}{x+1}} = \frac{(x+1)(x-3)}{\dfrac{x+1}{1}\left(\dfrac{x-2}{1}-\dfrac{4}{x+1}\right)} = \frac{(x+1)(x-3)}{(x+1)(x-2)-4}$$

$$= \frac{(x+1)(x-3)}{x^2-x-2-4} = \frac{(x+1)(x-3)}{x^2-x-6}$$

$$= \frac{(x+1)(x-3)}{(x-3)(x+2)} = \frac{x+1}{x+2}$$

EXAMPLE

Simplify $\dfrac{x-2-\dfrac{7}{x+4}}{x-5+\dfrac{14}{x+4}}$.

Solution

The LCM of the denominators is $x + 4$:

$$\frac{x-2-\dfrac{7}{x+4}}{x-5+\dfrac{14}{x+4}} = \frac{\dfrac{x+4}{1}\left(\dfrac{x-2}{1}-\dfrac{7}{x+4}\right)}{\dfrac{x+4}{1}\left(\dfrac{x-5}{1}+\dfrac{14}{x+4}\right)}$$

$$= \frac{(x+4)(x-2)-7}{(x+4)(x-5)-14}$$

$$= \frac{x^2+2x-8-7}{x^2-x-20+14}$$

$$= \frac{x^2+2x-15}{x^2-x-6}$$

$$= \frac{(x+5)(x-3)}{(x-3)(x+2)}$$

$$= \frac{x+5}{x+2}$$

EXAMPLE

Simplify $\dfrac{\dfrac{x+1}{2x-1}-\dfrac{x-1}{2x+1}}{\dfrac{x+1}{2x-1}+\dfrac{x-1}{2x+1}}$.

Solution

The LCM of the denominators is $(2x - 1)(2x + 1)$:

$$\frac{\dfrac{x+1}{2x-1}-\dfrac{x-1}{2x+1}}{\dfrac{x+1}{2x-1}+\dfrac{x-1}{2x+1}} = \frac{\dfrac{(2x-1)(2x+1)}{1}\left(\dfrac{x+1}{2x-1}-\dfrac{x-1}{2x+1}\right)}{\dfrac{(2x-1)(2x+1)}{1}\left(\dfrac{x+1}{2x-1}+\dfrac{x-1}{2x+1}\right)}$$

$$= \frac{(2x+1)(x+1)-(2x-1)(x-1)}{(2x+1)(x+1)+(2x-1)(x-1)}$$

$$= \frac{2x^2+3x+1-2x^2+3x-1}{2x^2+3x+1+2x^2-3x+1}$$

$$= \frac{6x}{4x^2+2} = \frac{6x}{2(2x^2+1)}$$

$$= \frac{3x}{2x^2+1}$$

Exercise 6.5

Perform the indicated operations and simplify:

1. $\dfrac{3}{x-2}+\dfrac{4x-2}{x^2-3x-4}\cdot\dfrac{x^2-x-12}{2x^2+5x-3}$

2. $\dfrac{4}{2x-1}-\dfrac{3x+6}{3x^2-x-2}\cdot\dfrac{4x^2-3x-1}{4x^2+9x+2}$

3. $\dfrac{3x^2+3x}{4x^2+11x-3}\cdot\dfrac{3x^2+5x-12}{3x^2-x-4}-\dfrac{2x}{3x+1}$

4. $\dfrac{4}{x+5}-\dfrac{6x-15}{3x^2+8x-3}\div\dfrac{6x^2-11x-10}{9x^2+3x-2}$

5. $\dfrac{3}{2x-3}-\dfrac{2x-8}{3x^2+10x-8}\div\dfrac{x^2-6x+8}{x^2+2x-8}$

6. $\dfrac{6x-3}{12x^2-25x+12}\div\dfrac{6x^2+5x-4}{9x^2-16}+\dfrac{2}{x-4}$

7. $\left(1+\dfrac{1}{x}\right)\dfrac{6x}{x^2-1}$

8. $\left(3+\dfrac{1}{x}\right)\dfrac{9}{9x^2-1}$

9. $\left(2-\dfrac{3}{x}\right)\dfrac{x^2}{4x^2-9}$

10. $\left(\dfrac{x}{2}-\dfrac{2}{x}\right)\dfrac{4x}{x^2-4}$

11. $\left(\dfrac{x}{3}-\dfrac{3}{x}\right)\dfrac{6x}{x+3}$

12. $\left(6+\dfrac{30}{x}\right)\dfrac{x}{x^2-25}$

13. $\left(\dfrac{1}{x}+\dfrac{2}{x^2}\right)\left(\dfrac{1}{x+2}+\dfrac{1}{x-2}\right)$

14. $\left(\dfrac{2}{x}+\dfrac{1}{x^2}\right)\left(\dfrac{1}{3x-1}+\dfrac{1}{2x+1}\right)$

15. $\left(\dfrac{1}{3}-\dfrac{1}{x}\right)\left(\dfrac{4}{x+4}+\dfrac{3}{x-3}\right)$

16. $\left(\dfrac{2}{3}-\dfrac{1}{2x}\right)\left(\dfrac{1}{x+2}-\dfrac{4}{4x-3}\right)$

17. $\left(x+\dfrac{6}{x+5}\right)\left(x-\dfrac{15}{x+2}\right)$

18. $\left(x-\dfrac{3}{x+2}\right)\left(x+\dfrac{2}{x+3}\right)$

19. $\left(x-\dfrac{28}{x-3}\right)\left(x-\dfrac{21}{x+4}\right)$

20. $\left(x+3-\dfrac{1}{x+3}\right)\left(x+\dfrac{3}{x+4}\right)$

21. $\left(2x-8-\dfrac{x+10}{3x+1}\right)\left(x-6-\dfrac{x-6}{3x+2}\right)$

22. $\left(2x-4-\dfrac{x-12}{3x+4}\right)\left(3x-2-\dfrac{10}{2x+1}\right)$

23. $\left(1+\dfrac{1}{x}\right)\div\left(1-\dfrac{1}{x^2}\right)$

24. $\left(1-\dfrac{4}{x}\right)\div\left(1-\dfrac{16}{x^2}\right)$

25. $\left(3+\dfrac{1}{x}\right)\div\left(9-\dfrac{1}{x^2}\right)$

26. $\left(1+\dfrac{1}{x^3}\right)\div\left(1+\dfrac{1}{x}\right)$

27. $\left(1 - \frac{8}{x^3}\right) \div \left(1 - \frac{2}{x}\right)$

28. $\left(x + 1 - \frac{4}{x - 2}\right) \div \left(x - \frac{3}{x - 2}\right)$

29. $\left(x + 1 - \frac{5}{x - 3}\right) \div \left(x - \frac{4}{x - 3}\right)$

30. $\left(x + \frac{3}{x + 4}\right) \div \left(x - 5 + \frac{18}{x + 4}\right)$

31. $\left(x - \frac{15}{x + 2}\right) \div \left(x - 1 - \frac{10}{x + 2}\right)$

32. $\left(2x + \frac{x - 1}{2x + 1}\right) \div \left(2x - \frac{x + 3}{2x + 1}\right)$

33. $\left(x - 1 - \frac{5}{x + 3}\right) \div \left(x - 4 - \frac{24}{x + 1}\right)$

34. $\left(x - 2 - \frac{22}{x + 7}\right) \div \left(x - 3 + \frac{2}{x - 6}\right)$

35. $\dfrac{\frac{5}{3} - \frac{3}{2}}{\frac{3}{4} - \frac{2}{3}}$

36. $\dfrac{\frac{5}{2} - \frac{3}{5}}{\frac{7}{10} + \frac{1}{4}}$

37. $\dfrac{\frac{4}{3} - \frac{1}{7}}{\frac{5}{7} - \frac{2}{3}}$

38. $\dfrac{\frac{7}{6} - \frac{5}{2}}{\frac{4}{5} - \frac{4}{3}}$

39. $\dfrac{\frac{5}{6} - \frac{4}{9}}{\frac{3}{4} - \frac{1}{6}}$

40. $\dfrac{\frac{7}{8} - \frac{5}{12}}{\frac{1}{6} - \frac{11}{16}}$

41. $\dfrac{\frac{1}{x^2} - \frac{1}{4}}{\frac{1}{x} + \frac{1}{2}}$

42. $\dfrac{\frac{9}{x^2} - 1}{\frac{3}{x} + 1}$

43. $\dfrac{\frac{2}{x} - 1}{\frac{4}{x^2} - 1}$

44. $\dfrac{\frac{4}{x} - \frac{1}{3}}{\frac{16}{x^2} - \frac{1}{9}}$

45. $\dfrac{\frac{1}{x^3} + 1}{\frac{1}{x} + 1}$

46. $\dfrac{\frac{8}{x^3} - \frac{1}{8}}{\frac{2}{x} - \frac{1}{2}}$

47. $\dfrac{1 + \frac{2}{x} - \frac{3}{x^2}}{1 + \frac{5}{x} + \frac{6}{x^2}}$

48. $\dfrac{1 + \frac{1}{x} - \frac{12}{x^2}}{1 + \frac{2}{x} - \frac{8}{x^2}}$

49. $\dfrac{1 + \frac{2}{x} - \frac{24}{x^2}}{\frac{12}{x^2} + \frac{1}{x} - 1}$

50. $\dfrac{24 - \frac{5}{x} - \frac{36}{x^2}}{\frac{36}{x^2} + \frac{23}{x} - 8}$

51. $\dfrac{x + 2}{x + 3 - \frac{1}{x + 3}}$

52. $\dfrac{x - 4}{x - 1 - \frac{9}{x - 1}}$

53. $\dfrac{x + 3}{x + 10 + \frac{35}{x - 2}}$

54. $\dfrac{2x + 1}{2x - 1 - \frac{2}{4x + 1}}$

55. $\dfrac{3x + 2}{3x + 4 + \frac{6}{3x - 1}}$

56. $\dfrac{2x - 3}{4x - 9 + \frac{12}{2x + 1}}$

57. $\dfrac{x + \frac{8}{x + 6}}{x + 1 + \frac{6}{x + 6}}$

58. $\dfrac{x - \frac{16}{x - 6}}{x + 1 - \frac{18}{x - 6}}$

59. $\dfrac{x + 3 + \frac{10}{2x - 3}}{x - \frac{2}{2x - 3}}$

60. $\dfrac{x - 1 - \dfrac{4}{3x - 2}}{x - \dfrac{1}{3x - 2}}$

61. $\dfrac{x + 3 + \dfrac{5}{3x + 1}}{x - \dfrac{4}{3x + 1}}$

62. $\dfrac{x + 2 - \dfrac{6}{x + 1}}{x + 1 - \dfrac{9}{x + 1}}$

63. $\dfrac{x - \dfrac{6}{x - 1}}{x + 5 + \dfrac{9}{x - 1}}$

64. $\dfrac{x - 6 + \dfrac{14}{x + 3}}{x - 5 + \dfrac{12}{x + 3}}$

65. $\dfrac{x + 9 + \dfrac{42}{x - 4}}{x + 8 + \dfrac{35}{x - 4}}$

66. $\dfrac{x + 4 + \dfrac{8}{x - 5}}{x + 6 + \dfrac{24}{x - 5}}$

67. $\dfrac{x - 9 + \dfrac{33}{x + 5}}{x - 10 + \dfrac{44}{x + 5}}$

68. $\dfrac{2x - 1 + \dfrac{1}{3x + 2}}{x + 1 - \dfrac{4}{3x + 2}}$

69. $\dfrac{x - \dfrac{6}{x + 1}}{x + 2 - \dfrac{4}{x - 1}}$

70. $\dfrac{x - \dfrac{8}{x + 2}}{x + 5 + \dfrac{7}{x - 3}}$

71. $\dfrac{x + 2 - \dfrac{18}{x - 5}}{x - 5 - \dfrac{20}{x + 3}}$

72. $\dfrac{2x + 1 - \dfrac{9}{x + 2}}{2x - 1 - \dfrac{60}{x - 4}}$

73. $\dfrac{\dfrac{x + 2}{x - 2} + \dfrac{x - 2}{x + 2}}{\dfrac{x + 2}{x - 2} - \dfrac{x - 2}{x + 2}}$

74. $\dfrac{\dfrac{2x - 1}{2x + 1} - \dfrac{2x + 1}{2x - 1}}{\dfrac{2x - 1}{2x + 1} + \dfrac{2x + 1}{2x - 1}}$

75. $\left(x + \dfrac{3}{x + 4}\right)\left(x + 1 - \dfrac{9}{x + 1}\right) \div \left(x - 1 - \dfrac{4}{x + 2}\right)$

76. $\left(x + 2 + \dfrac{4}{x - 3}\right)\left(x + 1 - \dfrac{4}{x - 2}\right) \div \left(x + 4 + \dfrac{6}{x - 1}\right)$

77. $\left(x - 5 + \dfrac{7}{x + 3}\right) \div \left(x - \dfrac{24}{x - 2}\right) \cdot \left(x + 5 + \dfrac{2}{x + 2}\right)$

78. $\left(x + 4 - \dfrac{9}{2x + 1}\right) \div \left(2x + 9 + \dfrac{35}{x - 4}\right) \cdot \left(2x - 3 + \dfrac{18}{x + 5}\right)$

6.6 Literal Equations

Some equations, called **literal equations**, involve more than one literal number. We can solve for one of the literals, called the **variable**, in terms of the other literals. Thus by assigning values for those literals we obtain corresponding values for the variable.

To find the solution set of a literal equation, form an equivalent equation with all the terms that have the variable as a factor on one side of the equation and those terms that do not have the variable as a factor on the other side. Factor the variable from the terms that have the variable as a factor and then divide both sides of the equation by the coefficient of the variable.

EXAMPLE Solve the following equation for x:

$$a(x - a) = 2(x - 2)$$

Solution

$$a(x - a) = 2(x - 2)$$
$$ax - a^2 = 2x - 4$$
$$ax - 2x = a^2 - 4$$
$$x(a - 2) = a^2 - 4$$

If $a - 2 \neq 0$, that is, $a \neq 2$, we can divide both sides of the equation by $(a - 2)$ to get

$$x = \frac{a^2 - 4}{a - 2} = \frac{(a + 2)(a - 2)}{(a - 2)}$$
$$= a + 2$$

Hence the solution set is $\{a + 2 \mid a \neq 2\}$.

Note For any value of $a \neq 2$ the value of x is $a + 2$. When $a = 2$ the equation becomes an identity, that is, a statement that is true for all values of x.

EXAMPLE Solve the following equation for x:

$$a(x - 3) = 1 - x$$

Solution

$$a(x - 3) = 1 - x$$
$$ax - 3a = 1 - x$$
$$ax + x = 3a + 1$$
$$x(a + 1) = 3a + 1$$

If $a + 1 \neq 0$, that is, $a \neq -1$, we can divide both sides of the equation by $(a + 1)$ to get

$$x = \frac{3a + 1}{a + 1}$$

Hence the solution set is $\left\{\frac{3a + 1}{a + 1} \,\middle|\, a \neq -1\right\}$.

Note When $a = -1$ we have a false statement.

Formulas are rules expressed in symbols or literal numbers. They are used in many fields of study. Formulas may be considered special types of literal equations. Many problems require the solution of a formula for one of the letters involved.

EXAMPLE

The focal length f of a thin lens is given by $\frac{1}{f} = \frac{1}{d_0} + \frac{1}{d_i}$, where d_0 is the distance between the object and the lens and d_i is the distance between the image and the lens. Solve for f and d_i.

Solution

$$\frac{1}{f} = \frac{1}{d_0} + \frac{1}{d_i} = \frac{d_i + d_0}{d_0 d_i}$$

Hence $$f = \frac{d_0 d_i}{d_i + d_0}$$

$$\frac{1}{d_i} = \frac{1}{f} - \frac{1}{d_0} = \frac{d_0 - f}{f d_0}$$

Hence $$d_i = \frac{f d_0}{d_0 - f}$$

Exercise 6.6

Solve the following equations for x:

1. $2x + y = 3$ **2.** $3x + 2y = 5$ **3.** $2x - y = 4$
4. $5x - 3y = 7$ **5.** $3x - 4y + 8 = 0$ **6.** $2x - 5y + 11 = 0$
7. $4x + 3y + 10 = 0$ **8.** $7x + 2y + 6 = 0$ **9.** $y - 3x = 6$
10. $3y - 4x = 4$ **11.** $2y - 5x + 3 = 0$ **12.** $7y - 3x + 7 = 0$
13. $ax - a = 2$ **14.** $3ax - 4a = 1$ **15.** $2a - 2ax = 3$
16. $4a - ax = -2$ **17.** $ax - 3a = b$ **18.** $ax + by = c$
19. $bx - y = 3a$ **20.** $4ax = 3 + ax$ **21.** $2ax = 5 - 3ax$
22. $4ax = 6 + 7ax$ **23.** $ax = 2a - 3x$ **24.** $ax = a + 2x$
25. $2x - 3a = 3ax$ **26.** $2a - x = 2ax$ **27.** $4a - x = 4ax$
28. $bx = 2 - ax$ **29.** $bx = 3 - 2ax$ **30.** $2bx + 4 = 3ax$
31. $a - ax = 2bx$ **32.** $2a + bx = ax$ **33.** $3b - ax = 3bx$
34. $3a + 2x = 2ax + 3$ **35.** $2ax + 3b = 3x + 2ab$
36. $ax - 3b = ab - 3x$ **37.** $4ax + 8a = x + 2$
38. $2x + 8 = 3ax + 12a$ **39.** $4a^2 + x = 2ax + 1$
40. $3ax + 1 = 9a^2 - x$ **41.** $ax - 1 = a^3 - x$ **42.** $a^3 + x = ax + 1$
43. $b(1 - 6x) = 2ax$ **44.** $a(x - 3) = 3 - x$ **45.** $3(x - 5) = a(5 - x)$
46. $a(x - 3) = 3(x - 3)$ **47.** $2a(x - 2) = x - 2$ **48.** $a(x - 4) = 4 - x$
49. $a(3x - 1) = 2(1 - 3x)$ **50.** $a(a - x) - 1 = x$
51. $a(a - 2x) - 4 = 4x$ **52.** $8 - a(x - a^2) = 2x$
53. $x - 2a(x - 4a^2) = 1$ **54.** $a^2(x - a) = 9(x + 3)$
55. $a(x - a^2) = 2b(4b^2 - x)$ **56.** $3a(x - 9a^2) = b(x - b^2)$
57. $(x - 2a)^2 - (x + 3b)^2 = 0$ **58.** $a(x - 4) = a^2 - 3x + 3$

59. $a(x - a) = 3a - 4 - 4x$

60. $2a^2 + x = ax + a + 1$

61. $a(x + 7) = 3(a^2 + x - 2)$

62. $a(2x - 2a - 5) = 3(x - 4)$

63. $a(3x - 6a + 1) = -x - 1$

Solve the following formulas for the indicated letters:

64. $d = rt$, solve for r

65. $A = \frac{1}{2}bh$, solve for h

66. $I = Prt$, solve for t

67. $F = \frac{km_1m_2}{d^2}$, solve for m_1

68. $\frac{P_1V_1}{T_1} = \frac{P_2V_2}{T_2}$, solve for V_1 and T_1

69. $C = \frac{5}{9}(F - 32)$, solve for F

70. $a = \frac{v_f - v_0}{t}$, solve for v_0

71. $A = P + Prt$, solve for r and P

72. $\frac{1}{R} = \frac{1}{R_1} + \frac{1}{R_2}$, solve for R and R_1

73. $S_n = \frac{n}{2}(a_1 + a_n)$, solve for n and a_1

74. $A = \frac{1}{2}h(b_1 + b_2)$, solve for h and b_1

75. $a_n = a_1 + (n - 1)d$, solve for d and n

76. $S_n = \frac{n}{2}[2a_1 + (n - 1)d]$, solve for a_1 and d

77. $S_n = \frac{a_nr - a_1}{r - 1}$, solve for a_1 and r

78. $f' = f\left(\frac{v + v_0}{v - v_s}\right)$, solve for v_0 and v

6.7 Equations Involving Algebraic Fractions

When an equation involves fractions, it can be put in a simpler form if both sides of the equation are multiplied by the LCD of all the fractions in the equation.

When an equation is multiplied by the LCD (which is a polynomial in the variable), the resulting equation may not be equivalent to the original equation. The resulting equation may have a solution set with elements that do not satisfy the original equation. In all such cases, the elements in the solution set must be checked in the original equation.

The values of the variable that do not satisfy the original equation are called **extraneous roots**.

EXAMPLE

Solve the equation $\dfrac{2x+5}{2x^2-5x-3} - \dfrac{2x-3}{8x^2+2x-1} = \dfrac{3x+10}{4x^2-13x+3}$.

Solution

$$\frac{2x+5}{(2x+1)(x-3)} - \frac{2x-3}{(4x-1)(2x+1)} = \frac{3x+10}{(4x-1)(x-3)}$$

The LCD $= (2x+1)(x-3)(4x-1)$.

Multiplying both sides of the equation by the LCD, we get

$$(4x-1)(2x+5) - (x-3)(2x-3) = (2x+1)(3x+10)$$
$$(8x^2+18x-5) - (2x^2-9x+9) = (6x^2+23x+10)$$
$$8x^2+18x-5-2x^2+9x-9 = 6x^2+23x+10$$
$$4x = 24$$
$$x = 6$$

To check, substitute 6 for x in the original equation:

Left Side	*Right Side*
$= \dfrac{12+5}{72-30-3} - \dfrac{12-3}{288+12-1}$	$= \dfrac{18+10}{144-78+3}$
$= \dfrac{17}{39} - \dfrac{9}{299}$	$= \dfrac{28}{69}$
$= \dfrac{391-27}{897}$	$= \dfrac{28}{69}$
$= \dfrac{364}{897}$	$= \dfrac{28}{69}$
$= \dfrac{28}{69}$	$= \dfrac{28}{69}$

The solution set is $\{6\}$.

EXAMPLE

Solve the equation $\dfrac{x}{3x-2} - \dfrac{8}{2x+3} = \dfrac{2x^2-24x+18}{6x^2+5x-6}$.

Solution

$$\frac{x}{(3x-2)} - \frac{8}{(2x+3)} = \frac{2x^2-24x+18}{(3x-2)(2x+3)}$$

Multiplying both sides of the equation by $(3x-2)(2x+3)$, we obtain

$$x(2x+3) - 8(3x-2) = 2x^2-24x+18$$
$$2x^2+3x-24x+16 = 2x^2-24x+18$$
$$3x = 2$$
$$x = \frac{2}{3}$$

Substituting $\frac{2}{3}$ for x in the original equation, we find that the denominator of the first fraction becomes zero.
Since division by zero is not defined, the solution set of the equation is $\varnothing$.

EXAMPLE

Find the solution set of $\frac{3}{x-4} - \frac{5}{2x+1} = \frac{x+23}{2x^2-7x-4}$.

Solution

$$\frac{3}{x-4} - \frac{5}{2x+1} = \frac{x+23}{(2x+1)(x-4)}$$

Multiplying both sides of the equation by $(x-4)(2x+1)$, we get

$$3(2x+1) - 5(x-4) = x + 23$$
$$6x + 3 - 5x + 20 = x + 23$$
$$0x = 0$$

Since $0x = 0$ is true for all real values of x, the original equation is true for all real values of x except when $2x + 1 = 0$, or $x - 4 = 0$.

Thus the solution set is $\left\{x \,\middle|\, x \in R, x \neq -\frac{1}{2} \text{ or } x \neq 4\right\}$.

Exercise 6.7

Solve the following equations for x:

1. $\frac{1}{x} + \frac{1}{2x} = \frac{3}{2}$

2. $\frac{1}{3x} + \frac{1}{2x} = \frac{1}{4}$

3. $\frac{2}{3x} - \frac{1}{2x} = \frac{3}{4}$

4. $\frac{1}{x} - \frac{3}{2x} = \frac{1}{5}$

5. $\frac{3}{8x} - \frac{1}{2x} = \frac{1}{x^2}$

6. $\frac{1}{3x} - \frac{1}{4x} = \frac{1}{x^2}$

7. $\frac{1}{2x} + \frac{3}{4x} = \frac{5}{2x^2}$

8. $\frac{2}{3x} - \frac{1}{2x} = \frac{3}{4x^2}$

9. $\frac{5}{x+2} = 3$

10. $\frac{2}{x-1} = 1$

11. $\frac{7}{3x-2} = 4$

12. $\frac{2a}{x+a} = 1$

13. $\frac{3a}{x-a} = 4$

14. $\frac{2a}{2x-3a} = 1$

15. $\frac{4a}{3x+a} = 3$

16. $\frac{x}{x+5} = 2$

17. $\frac{3x}{5x-1} = 2$

18. $\frac{2x}{3x+2} = -3$

19. $\frac{3x}{4x-3} = -2$

20. $\frac{2}{x-3} = \frac{1}{x+2}$

21. $\frac{4}{x+1} = \frac{3}{x-2}$

22. $\frac{3}{x+4} = \frac{2}{2x+1}$

23. $\frac{5}{3x-2}-\frac{1}{x-4}=0$

24. $\frac{7}{3x-4}-\frac{3}{x-3}=0$

25. $\frac{4}{2x-3}-\frac{7}{3x-5}=0$

26. $\frac{7}{2x-3}+\frac{1}{2x-2}=\frac{3}{x-1}$

27. $\frac{1}{2x-1}+\frac{3}{12x-8}=\frac{2}{3x-2}$

28. $\frac{2}{x+1}-\frac{3}{x+2}=\frac{1}{3x+3}$

29. $\frac{3}{x+3}-\frac{1}{x-2}=\frac{5}{2x+6}$

30. $\frac{2}{x-1}+\frac{4}{x+3}=\frac{3x+11}{x^2+2x-3}$

31. $\frac{6}{x+3}+\frac{20}{x^2+x-6}=\frac{5}{x-2}$

32. $\frac{7}{x+2}+\frac{2}{x+3}=\frac{1}{x^2+5x+6}$

33. $\frac{4}{x-2}+\frac{x}{x+1}=\frac{x^2-2}{x^2-x-2}$

34. $\frac{x}{x+2}-\frac{x^2+1}{x^2-2x-8}=\frac{3}{x-4}$

35. $\frac{x}{2x+1}-\frac{2}{x-3}=\frac{x^2+5}{2x^2-5x-3}$

36. $\frac{2}{3x+1}-\frac{15}{6x^2-x-1}=\frac{3}{2x-1}$

37. $\frac{9}{3x-1}-\frac{5-x}{3x^2-4x+1}=\frac{4}{x-1}$

38. $\frac{3}{x+1}+\frac{2}{x+2}=\frac{5x+4}{x^2+3x+2}$

39. $\frac{5}{x-2}+\frac{2}{x+4}=\frac{3x}{x^2+2x-8}$

40. $\frac{4}{x+6}+\frac{1}{x-3}=\frac{9}{x^2+3x-18}$

41. $\frac{x}{x-3}-\frac{2x^2+9}{2x^2-3x-9}=\frac{1}{2x+3}$

42. $\frac{2}{x+6}+\frac{1}{x+1}=\frac{3x+8}{x^2+7x+6}$

43. $\frac{x}{2x-3}+\frac{1}{x-3}=\frac{x^2-x-3}{2x^2-9x+9}$

44. $\frac{3}{2x-1}+\frac{1}{x+4}=\frac{5x+11}{2x^2+7x-4}$

45. $\frac{3}{x+2}-\frac{2x-20}{3x^2+4x-4}=\frac{7}{3x-2}$

46. $\frac{2}{x^2+3x+2}+\frac{1}{x^2+5x+6}=\frac{5}{x^2+4x+3}$

47. $\frac{3}{x^2+x-2}+\frac{2}{x^2-4}=\frac{1}{x^2-3x+2}$

48. $\frac{x}{x^2-1}+\frac{3}{x^2-2x-3}=\frac{x}{x^2-4x+3}$

49. $\frac{2}{x^2-x-6}+\frac{x+1}{x^2+x-12}=\frac{x}{x^2+6x+8}$

50. $\frac{x-1}{2x^2-3x-2}-\frac{x}{2x^2+7x+3}=\frac{4}{x^2+x-6}$

51. $\dfrac{x+1}{2x^2+7x-4} - \dfrac{x}{2x^2-7x+3} = \dfrac{1}{x^2+x-12}$

52. $\dfrac{3x}{6x^2-7x-3} - \dfrac{x-2}{2x^2-5x+3} = \dfrac{3}{3x^2-2x-1}$

53. $\dfrac{2x}{6x^2+7x-3} - \dfrac{x-3}{3x^2+11x-4} = \dfrac{5}{2x^2+11x+12}$

54. $\dfrac{3x-2}{x^2-12x+20} + \dfrac{4x+3}{x^2+6x-16} = \dfrac{7x+11}{x^2-2x-80}$

55. $\dfrac{2x-5}{x^2+5x-36} - \dfrac{x-6}{x^2+3x-28} = \dfrac{x+8}{x^2+16x+63}$

56. $\dfrac{2x-1}{x^2+4x-5} + \dfrac{x-2}{x^2-10x+9} = \dfrac{3x-12}{x^2-4x-45}$

57. $\dfrac{3x-1}{18x^2+3x-28} - \dfrac{4x}{24x^2+23x-12} = \dfrac{3}{48x^2-74x+21}$

58. $\dfrac{x-2}{4x^2-29x+30} - \dfrac{x+1}{20x^2-13x-15} = \dfrac{x+2}{5x^2-27x-18}$

59. $\dfrac{x+8}{x^2+x-6} + \dfrac{x-8}{x^2-x-2} = \dfrac{2x+8}{x^2+4x+3}$

60. $\dfrac{3x-1}{x^2-1} + \dfrac{1}{x^2-3x+2} = \dfrac{3x-3}{x^2-x-2}$

61. $\dfrac{2x+3}{x^2+3x+2} + \dfrac{6}{x^2-x-6} = \dfrac{2x-2}{x^2-2x-3}$

62. $\dfrac{x+2}{6x^2-x-1} + \dfrac{3x-5}{2x^2+5x-3} = \dfrac{5x-2}{3x^2+10x+3}$

6.8 Word Problems

The following are illustrations of certain verbal phrases and questions and their algebraic equivalents.

1. The denominator of a fraction exceeds its numerator by 5. What is the fraction?

Let the numerator be x. The denominator is $x + 5$.

The fraction is $\dfrac{x}{x+5}$.

2. What is the fraction whose numerator is 2 less than its denominator?

Let the denominator be x. The numerator is $x - 2$.

The fraction is $\frac{x-2}{x}$.

3. If 72 is divided by x, the quotient is 5 and the remainder is 7:

$$\frac{72}{x} = 5 + \frac{7}{x}$$

4. If a man can do a job in 10 hours, what part of the job can he do in 3 hours?

Job Time	*Time Worked*
10 hours	3 hours

He can do $\frac{3}{10}$ of the job.

5. If a man can do a job in x hours, what part of the job can he do (a) in 1 hour? (b) in 12 hours?

	Job Time	*Time Worked*
(a)	x hours	1 hour
(b)	x hours	12 hours

(a) He can do $\frac{1}{x}$ of the job.

(b) He can do $\frac{12}{x}$ of the job.

6. If Art can do a job in 72 hours, and Bruce can do the same job in 96 hours, what part of the job can be done by the two men working together for x hours?

Job Time for Art	*Job Time for Bruce*	*Time Worked*
72 hours	96 hours	x hours

Art can do $\frac{x}{72}$ of the job.

Bruce can do $\frac{x}{96}$ of the job.

Working together they can do $\frac{x}{72} + \frac{x}{96}$ of the job.

7. If water from one pipe can fill a pool in 30 hours and water from another pipe can fill the same pool in x hours, what part of the pool will be filled in 11 hours if both pipes are open at the same time?

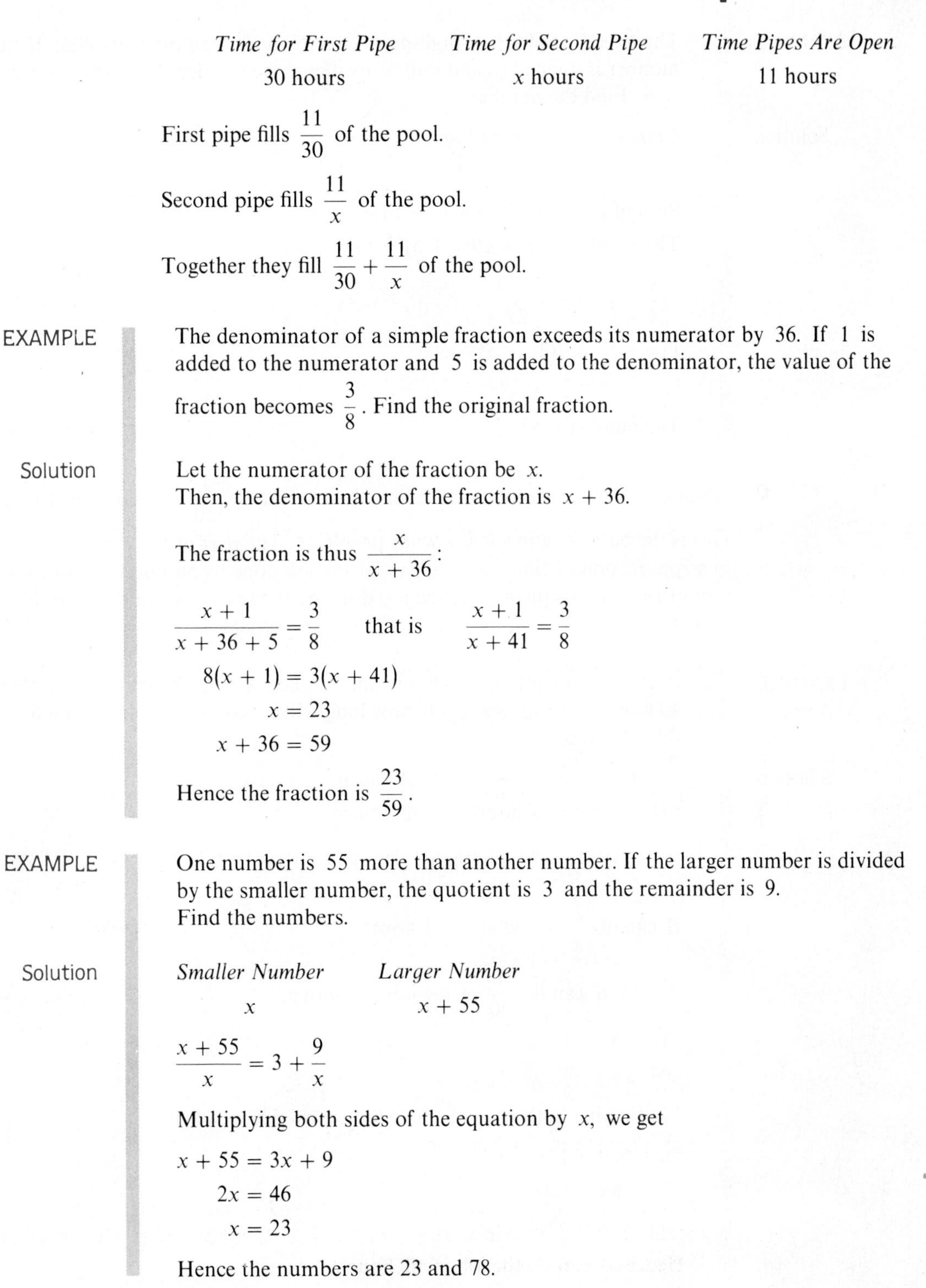

Time for First Pipe	*Time for Second Pipe*	*Time Pipes Are Open*
30 hours	x hours	11 hours

First pipe fills $\frac{11}{30}$ of the pool.

Second pipe fills $\frac{11}{x}$ of the pool.

Together they fill $\frac{11}{30} + \frac{11}{x}$ of the pool.

EXAMPLE

The denominator of a simple fraction exceeds its numerator by 36. If 1 is added to the numerator and 5 is added to the denominator, the value of the fraction becomes $\frac{3}{8}$. Find the original fraction.

Solution

Let the numerator of the fraction be x.
Then, the denominator of the fraction is $x + 36$.

The fraction is thus $\frac{x}{x + 36}$:

$$\frac{x + 1}{x + 36 + 5} = \frac{3}{8} \quad \text{that is} \quad \frac{x + 1}{x + 41} = \frac{3}{8}$$

$$8(x + 1) = 3(x + 41)$$

$$x = 23$$

$$x + 36 = 59$$

Hence the fraction is $\frac{23}{59}$.

EXAMPLE

One number is 55 more than another number. If the larger number is divided by the smaller number, the quotient is 3 and the remainder is 9.
Find the numbers.

Solution

Smaller Number	*Larger Number*
x	$x + 55$

$$\frac{x + 55}{x} = 3 + \frac{9}{x}$$

Multiplying both sides of the equation by x, we get

$$x + 55 = 3x + 9$$

$$2x = 46$$

$$x = 23$$

Hence the numbers are 23 and 78.

EXAMPLE

The tens digit of a two-digit number is 5 more than the units digit. If the number is divided by the sum of its digits, the quotient is 7 and the remainder is 6. Find the number.

Solution

Units Digit	*Tens Digit*
x	$x + 5$

Sum of the digits is $x + (x + 5) = 2x + 5$.

The number is $x + 10(x + 5) = 11x + 50$.

$$\frac{11x + 50}{2x + 5} = 7 + \frac{6}{2x + 5}$$

$$11x + 50 = 7(2x + 5) + 6$$

$$x = 3$$

The number is 83.

When a man can do a job in 10 hours, then he can do $\frac{1}{10}$ of the job in 1 hour. This is the basic idea in solving work problems. The part of the job done by a man in a specific unit of time plus the part of the job done by another man in the same unit of time equals the part of the job done by the two men working together the same unit of time.

EXAMPLE

If A can do a job in 104 hours and it takes A and B working together 40 hours to do the same job, how long will it take B working alone to do the same job?

Solution

A	B	A and B
104 hours	x hours	40 hours

A can do $\frac{1}{104}$ of the job in 1 hour.

B can do $\frac{1}{x}$ of the job in 1 hour.

A and B can do $\frac{1}{40}$ of the job in 1 hour:

$$\frac{1}{104} + \frac{1}{x} = \frac{1}{40}$$

Multiplying both sides of the equation by $520x$, we get

$$5x + 520 = 13x$$

$$8x = 520$$

$$x = 65$$

Hence B can do the job in 65 hours.

Exercise 6.8

1. What number must be added to both the numerator and the denominator of the fraction $\frac{17}{29}$ to give a fraction equal to $\frac{3}{4}$?
2. What number must be added to both the numerator and the denominator of the fraction $\frac{21}{47}$ to give a fraction equal to $\frac{2}{3}$?
3. What number must be subtracted from both the numerator and the denominator of the fraction $\frac{87}{137}$ to give a fraction equal to $\frac{3}{5}$?
4. What number must be subtracted from both the numerator and the denominator of the fraction $\frac{145}{183}$ to give a fraction equal to $\frac{5}{7}$?
5. What number must be subtracted from the numerator and added to the denominator of the fraction $\frac{73}{97}$ to give a fraction equal to $\frac{6}{11}$?
6. What number must be subtracted from the numerator and added to the denominator of the fraction $\frac{81}{139}$ to give a fraction equal to $\frac{2}{9}$?
7. The denominator of a simple fraction is 3 more than its numerator. If 1 is added to the numerator and 2 is added to the denominator, the value of the fraction becomes $\frac{2}{3}$. Find the original fraction.
8. The denominator of a simple fraction is 4 more than its numerator. If 3 is added to the numerator and 7 is added to the denominator, the value of the fraction becomes $\frac{3}{5}$. Find the original fraction.
9. The denominator of a fraction exceeds its numerator by 5. If 1 is subtracted from the numerator and 2 is added to the denominator, the value of the fraction becomes $\frac{1}{5}$. Find the original fraction.
10. The denominator of a fraction exceeds its numerator by 8. If 3 is added to the numerator and 1 is subtracted from the denominator, the value of the fraction becomes $\frac{3}{4}$. Find the original fraction.
11. The numerator of a fraction is 41 less than its denominator. If 3 is added to the numerator and 4 is added to the denominator, the value of the fraction becomes $\frac{1}{3}$. Find the original fraction.
12. The numerator of a fraction is 63 less than its denominator. If 2 is added to the numerator and 1 is subtracted from the denominator, the value of the fraction becomes $\frac{3}{8}$. Find the original fraction.

13. One number is 46 more than another number. If the larger number is divided by the smaller number, the quotient is 4 and the remainder is 7. Find the two numbers.
14. One number is 112 more than another number. If the larger number is divided by the smaller number, the quotient is 6 and the remainder is 17. Find the two numbers.
15. One number is 88 more than another number. If the larger number is divided by the smaller number, the quotient is 3 and the remainder is 2. Find the numbers.
16. One number is 205 more than another number. If the larger number is divided by the smaller number, the quotient is 6 and the remainder is 15. Find the numbers.
17. The units digit of a two-digit number exceeds its tens digit by 4. If the number is divided by the sum of its digits, the quotient is 4 and the remainder is 3. Find the number.
18. The tens digit of a two-digit number exceeds its units digit by 3. If the number is divided by the sum of its digits, the quotient is 7 and the remainder is 3. Find the number.
19. The units digit of a two-digit number is 3 less than its tens digit. If the number is divided by twice its units digit, the quotient is 9 and the remainder is 2. Find the number.
20. The sum of the digits of a three-digit number is 11, and the units digit is 3 times its hundreds digit. If the number is divided by twice its units digit, the quotient is 19 and the remainder is 8. Find the number.
21. The sum of the digits of a three-digit number is 15, and the tens digit is twice its hundreds digit. If the number is divided by its units digit, the quotient is 27 and the remainder is 6. Find the number.
22. The sum of the digits of a three-digit number is 13, and the units digit is twice its tens digit. If the number is divided by its units digit, the quotient is 18 and the remainder is 4. Find the number.
23. If A can do a job in 30 hours and B can do the same job in 70 hours, how long will it take the two of them working together to do the job?
24. If A can do a job in 78 hours and it takes A and B working together 42 hours to do the same job, how long will it take B working alone to do the job?
25. If A can do a job in 112 hours and it takes A and B working together 63 hours to do the same job, how long will it take B working alone to do the job?
26. It takes A twice as long as it takes B to do a job. Together they finish the job in 6 hours. How long does it take each separately to finish the job?
27. It takes A $\frac{2}{3}$ as long as it takes B to do a job. If A and B working together can do the job in 12 hours, how long will it take each alone to do the job?
28. It takes A $\frac{3}{5}$ as long as it takes B to do a job. If A and B working together can do the job in 15 hours, how long will it take each alone to do the job?

29. A tank can be filled by one pipe in 21 minutes and by another pipe in 28 minutes. How long will it take both pipes open at the same time to fill the tank?

30. A tank can be filled by one pipe in 9 minutes and drained by another pipe in 12 minutes. How long will it take to fill the tank if the two pipes are open at the same time?

31. A supply pipe can fill a tank in 8 minutes. How long will it take a drainage pipe to drain the tank, if when both the supply and drainage pipes are open at the same time, the tank will be filled in 24 minutes?

32. It takes one pipe $\frac{3}{5}$ as long as it takes another pipe to fill a tank. If the two pipes together fill the tank in 45 minutes, how long does it take each pipe to fill the tank?

33. It takes one pipe $\frac{1}{3}$ as long as it takes another pipe to fill a tank. If the two pipes together fill the tank in 3 minutes, how long does it take each pipe alone to fill the tank?

34. A man drove 8 miles in town in the same time that he drove 15 miles on the freeway. His freeway speed was 28 mph faster than in town. What was his average speed in town?

35. A man drove 6 miles in town in the same time that he drove 9 miles on the freeway. His freeway speed was 18 mph faster than in town. What was his average speed on the freeway?

36. It took a man the same time to drive 48 miles as it took him to fly 560 miles. The average speed of the plane was 12 mph slower than 12 times the speed of the car. What was the average speed of the plane?

Chapter 6 Review

Perform the indicated operations and simplify:

1. $\dfrac{21^2x^4y^3}{14^2x^2y^6}$

2. $\dfrac{6^3x^8y^3}{9^2x^4y^9}$

3. $\dfrac{(4x^2yz^3)^3}{(8x^3y^2z)^2}$

4. $\dfrac{(12xy^4z^2)^4}{(18x^2y^4z^2)^3}$

5. $\dfrac{x^4 - x^2}{x^3 - x^6}$

6. $\dfrac{x^2 - 3x - 4}{8 + 2x - x^2}$

7. $\dfrac{4x^2 + 4x - 3}{4 - 5x - 6x^2}$

8. $\dfrac{x^2 - 9 - 2xy + y^2}{(x - y)^2 + (x - y) - 6}$

9. $\dfrac{2}{x} - \dfrac{3}{4x} + \dfrac{5}{6x}$

10. $\dfrac{3}{x^2} - \dfrac{2}{3x} - \dfrac{7}{4x^2}$

11. $\dfrac{3x - 4}{4x} + \dfrac{2x + 3}{3x}$

12. $\dfrac{x + 1}{3x - 2} - \dfrac{x + 4}{12x - 8}$

13. $\dfrac{x}{x - 2} - \dfrac{x}{x + 3}$

14. $\dfrac{x}{x + 2} - \dfrac{4}{x + 4}$

15. $\dfrac{2}{x - 2} + \dfrac{1}{x - 1} + \dfrac{1}{x^2 - 3x + 2}$

16. $\dfrac{3}{x + 3} + \dfrac{1}{x - 3} - \dfrac{6}{x^2 - 9}$

17. $\dfrac{2}{x+1}+\dfrac{3}{x-4}-\dfrac{x-9}{x^2-3x-4}$

18. $\dfrac{3}{2x+1}+\dfrac{2}{3x-1}-\dfrac{7x-4}{6x^2+x-1}$

19. $\dfrac{x-7}{x^2+x-6}+\dfrac{2x-1}{x^2-x-2}$

20. $\dfrac{4x+3}{2x^2+3x-2}+\dfrac{x+5}{x^2+x-2}$

21. $\dfrac{2x+6}{3x^2-10x-8}+\dfrac{4x-1}{3x^2-7x-6}$

22. $\dfrac{5x}{6x^2+x-1}+\dfrac{x+2}{2x^2-x-1}$

23. $\dfrac{3x-1}{x^2-1}-\dfrac{3}{x^2-x-2}+\dfrac{1}{x^2-3x+2}$

24. $\dfrac{x+4}{x^2+5x+6}+\dfrac{x-8}{x^2-x-6}+\dfrac{2x}{x^2-9}$

25. $\dfrac{x-6}{x^2-6x+8}+\dfrac{x+8}{x^2+x-6}+\dfrac{2x-1}{x^2-x-12}$

26. $\dfrac{2x-9}{x^2+5x-6}+\dfrac{x+2}{x^2+10x+24}+\dfrac{2x+3}{x^2+3x-4}$

27. $\dfrac{x-10}{2x^2-5x-3}+\dfrac{2x-3}{4x^2-1}+\dfrac{3x-4}{2x^2-7x+3}$

28. $\dfrac{x-5}{6x^2-11x+3}+\dfrac{4x+1}{3x^2+5x-2}-\dfrac{x-5}{2x^2+x-6}$

29. $\dfrac{4x-1}{3x^2-7x-6}+\dfrac{5x+5}{6x^2+13x+6}-\dfrac{3x}{2x^2-3x-9}$

30. $\dfrac{4x+5}{6x^2-7x-3}-\dfrac{2x+5}{3x^2-11x-4}+\dfrac{x+1}{2x^2-11x+12}$

31. $\dfrac{20xy^5z^3}{39a^5bc^4}\cdot\dfrac{26a^4bc}{15x^2y^5z}$

32. $\dfrac{x^4y^6z^5}{a^5b^2c^6}\cdot\dfrac{a^4bc^6}{x^3y^2z^7}$

33. $\dfrac{x^3y^6z^2}{a^4b^6c}\div\dfrac{x^2y^3z^2}{a^2b^3c^4}$

34. $\dfrac{x^{10}y^8z^3}{a^9b^6c}\div\dfrac{x^5y^4z^6}{a^3b^3c^4}$

35. $\left(\dfrac{6x^2yz}{a^4b^2c}\right)^3\left(\dfrac{a^4b^3c^2}{9xy^2z^3}\right)^2$

36. $\left(\dfrac{xy^3z}{8a^3b^4}\right)^3\left(\dfrac{4a^3b^2c}{xy^2z}\right)^4$

37. $\dfrac{2x^2-6x}{2x^3+x^2}\cdot\dfrac{4x^2-1}{x^2-9}$

38. $\dfrac{x^2-1}{x^2+2x-3}\cdot\dfrac{x^2-4}{x^2-x-2}$

39. $\dfrac{x^3+1}{x^2-2x-3}\cdot\dfrac{x^2-4x+3}{x^2-x+1}$

40. $\dfrac{2x^2-7x-4}{x^2-6x+8}\cdot\dfrac{x^2-8x+12}{8x^3+1}$

41. $\dfrac{6x^2+13x-8}{10x^2-9x+2}\cdot\dfrac{5x^2+8x-4}{3x^2+11x+8}$

42. $\dfrac{3x^2-7x-6}{3x^2+4x-4}\cdot\dfrac{3x^2+7x-6}{3x^2-10x-8}$

43. $\dfrac{2x^2-5x-12}{6x^2-25x+4}\cdot\dfrac{6x^2-19x+3}{4x^2+8x+3}$

44. $\dfrac{8x^2-6x-9}{6x^2-5x-6}\cdot\dfrac{3x^2-4x-4}{12x^2+x-6}$

45. $\left(\frac{5x^3y^2}{ab^2c^3}\right)^4 \div \left(\frac{10x^4y}{ab^2c^3}\right)^3$

46. $\left(\frac{xy^3z^2}{12a^4b^2c^3}\right)^2 \div \left(\frac{xy^2z^3}{6ab^2c^2}\right)^3$

47. $\frac{x^2 - x - 6}{x^2 + 6x + 9} \div \frac{x^2 + 6x + 8}{x^2 + 7x + 12}$

48. $\frac{x^2 - 4x + 4}{x^2 - 3x + 2} \div \frac{x^2 + 3x - 18}{x^2 + 5x - 6}$

49. $\frac{x^2 + x - 12}{x^2 - 6x + 9} \div \frac{x^2 + 5x + 4}{x^2 - 2x - 3}$

50. $\frac{x^2 - 10x + 24}{x^2 + x - 20} \div \frac{x^2 - 4x - 12}{x^2 + 3x - 10}$

51. $\frac{6x^2 - 7x - 5}{3x^2 + 4x - 15} \div \frac{6x^2 + x - 1}{3x^2 - 10x + 3}$

52. $\frac{6 - x - 2x^2}{2x^2 - x - 3} \div \frac{2x^2 + 7x + 6}{2x^2 - 5x - 12}$

53. $\frac{3 - 11x - 4x^2}{4x^2 + 23x - 6} \div \frac{3x^2 + 11x + 6}{9x^2 + 9x + 2}$

54. $\frac{12 - x - 6x^2}{8x^2 + 18x + 9} \div \frac{6x^2 - 5x - 4}{4x^2 - 5x - 6}$

55. $\frac{x^2 - 2x - 3}{2x^2 + 7x + 3} \cdot \frac{x^2 + x - 6}{x^2 - 16} \div \frac{x^2 - x - 2}{x^2 - 2x - 8}$

56. $\frac{2x^2 + 11x + 12}{6x^2 - 7x + 2} \cdot \frac{2x^2 + 5x - 3}{x^2 + 3x - 4} \div \frac{x^2 + 9x + 18}{6x^2 - 13x + 6}$

57. $\frac{4x^2 - 21x + 5}{6x^2 - 17x + 12} \div \frac{3x^2 - 4x - 4}{6x^2 - 5x - 6} \cdot \frac{6x^2 - 5x - 4}{4x^2 + 11x - 3}$

58. $\frac{12x^2 + x - 6}{x^2 - 4x + 3} \div \frac{3x^2 + 4x - 4}{3x^2 + x - 4} \cdot \frac{4x^2 - 11x - 3}{4x^2 - 13x - 12}$

59. $\frac{3}{x - 4} - \frac{8x - 6}{2x^2 - 7x + 3} \cdot \frac{6x^2 - 5x + 1}{12x^2 - 13x + 3}$

60. $\frac{5}{2x - 3} - \frac{12x + 9}{6x^2 + 11x + 3} \cdot \frac{2x^2 - 5x - 12}{4x^2 - 13x - 12}$

61. $\frac{6}{3x - 2} - \frac{15x - 5}{12x^2 - 31x + 20} \div \frac{3x^2 - 10x + 3}{3x^2 - 13x + 12}$

62. $\frac{x}{x - 3} - \frac{12x^2 - 2x}{6x^2 - 13x + 6} \div \frac{6x^2 + 17x - 3}{3x^2 + 7x - 6}$

63. $\left(x + 4 + \frac{6}{x - 1}\right)\left(x - \frac{3}{x + 2}\right)$

64. $\left(x - \frac{4}{x - 3}\right)\left(x - 2 - \frac{4}{x + 1}\right)$

65. $\left(x - 3 + \frac{6}{x + 4}\right)\left(x + 5 + \frac{6}{x - 2}\right)$

66. $\left(x - 6 + \frac{14}{x + 3}\right)\left(x + 9 + \frac{42}{x - 4}\right)$

67. $\left(2x - 5 + \frac{26}{x + 6}\right) \div \left(x + 2 - \frac{5}{2x + 1}\right)$

68. $\left(x + 4 + \frac{14}{2x - 3}\right) \div \left(x + 2 + \frac{3}{2x - 1}\right)$

69. $\left(2x + \dfrac{x-2}{2x+3}\right) \div \left(4x - 9 + \dfrac{26}{x+3}\right)$

70. $\left(x + 2 + \dfrac{4}{3x-1}\right) \div \left(3x + 8 - \dfrac{20}{x+4}\right)$

71. $\dfrac{x-4}{x - 3 - \dfrac{1}{x-3}}$

72. $\dfrac{2x-1}{x - 6 - \dfrac{11}{2x-3}}$

73. $\dfrac{x - 4 + \dfrac{16}{x+6}}{x + \dfrac{8}{x+6}}$

74. $\dfrac{2x - \dfrac{x-2}{3x+4}}{2x - 4 - \dfrac{x-10}{3x+4}}$

75. $\dfrac{x - 1 - \dfrac{x}{4x-3}}{x + 3 - \dfrac{x-4}{4x-3}}$

76. $\dfrac{2x + 1 - \dfrac{2}{3x+1}}{2x + 3 - \dfrac{5}{3x+1}}$

77. $\dfrac{x + 1 - \dfrac{2}{3x-2}}{3x - 2 + \dfrac{10}{x+3}}$

78. $\dfrac{x + 2 - \dfrac{16}{x-4}}{x + 1 - \dfrac{28}{x-2}}$

79. $\left(x - 8 + \dfrac{35}{x+4}\right) \div \left(x - 8 + \dfrac{30}{x+3}\right) \cdot \left(x + 3 - \dfrac{5}{x-1}\right)$

80. $\left(x + 1 + \dfrac{x-2}{4x-1}\right) \div \left(2x - 1 - \dfrac{2x-1}{3x+2}\right) \cdot \left(6x - 8 - \dfrac{x-23}{2x+3}\right)$

Solve the following equations for x:

81. $3x = 5a - ax$

82. $4ax = 6a + x$

83. $ax - 4a = 2x - 8$

84. $3ax + 15a = 5 + x$

85. $2ax + 1 = 4a^2 - x$

86. $ax - a^3 = 8 - 2x$

87. $3a(x - a) = ab - 2b(b - x)$

88. $a(x + 17b) = 4a^2 + 4b(x + b)$

89. $a(2x - a) = a - 6(x + 1)$

90. $2a(3x - a) = 3b(b - x) + 7ab$

91. $\dfrac{4}{x-3} - \dfrac{3}{x-4} = 0$

92. $\dfrac{2}{2x-3} - \dfrac{5}{6x-5} = 0$

93. $\dfrac{3}{x-1} - \dfrac{2}{x+2} = \dfrac{7}{x^2+x-2}$

94. $\dfrac{4}{3x-1} - \dfrac{5}{6x^2+x-1} = \dfrac{3}{2x+1}$

95. $\dfrac{5}{2x+3} - \dfrac{12}{2x^2+9x+9} = \dfrac{4}{x+3}$

96. $\dfrac{4}{x-2} - \dfrac{7}{2x-3} = \dfrac{4}{2x^2-7x+6}$

97. $\dfrac{3}{2x-5} + \dfrac{1}{x+3} = \dfrac{5x+4}{2x^2+x-15}$

98. $\dfrac{2}{x+4} - \dfrac{3}{2x-1} = \dfrac{x-14}{2x^2+7x-4}$

99. $\dfrac{x}{x^2-3x+2} + \dfrac{4}{x^2-5x+6} = \dfrac{x}{x^2-4x+3}$

100. $\dfrac{3}{x^2+4x+3} + \dfrac{x}{x^2+x-6} = \dfrac{x-1}{x^2-x-2}$

101. $\dfrac{x+2}{x^2+x-6}+\dfrac{2x+8}{x^2-9}=\dfrac{3x-7}{x^2-5x+6}$

102. $\dfrac{3x+3}{x^2+6x+8}+\dfrac{x-2}{x^2+3x+2}=\dfrac{4x-1}{x^2+5x+4}$

103. $\dfrac{4x+9}{4x^2-4x-15}+\dfrac{x}{2x^2-x-6}=\dfrac{3x-8}{2x^2-9x+10}$

104. $\dfrac{2x+1}{6x^2-13x+6}+\dfrac{5x-1}{3x^2+7x-6}=\dfrac{4x-2}{2x^2+3x-9}$

105. What number must be added to the numerator and the denominator of the fraction $\frac{10}{43}$ to give a fraction equal to $\frac{2}{5}$?

106. What number must be subtracted from the numerator and the denominator of the fraction $\frac{77}{97}$ to give a fraction equal to $\frac{3}{4}$?

107. The denominator of a simple fraction is 4 more than its numerator. If 1 is added to the numerator and 3 is added to the denominator, the value of the fraction becomes $\frac{2}{3}$. Find the original fraction.

108. The numerator of a simple fraction is 24 less than its denominator. If 5 is added to the numerator and 11 is subtracted from the denominator, the value of the fraction becomes $\frac{6}{7}$. Find the original fraction.

109. One number is 141 more than another number. If the larger number is divided by the smaller number, the quotient is 4 and the remainder is 3. Find the two numbers.

110. The units digit of a two-digit number is 6 more than the tens digit. If the number is divided by 3 times the units digit, the quotient is 1 and the remainder is 4. Find the number.

111. If A can do a job in 35 hours and B can do the same job in 14 hours, how long will it take the two of them working together to do the job?

112. It takes A $\frac{4}{5}$ as long as it takes B to do a job. If A and B working together can do the job in 20 hours, how long will it take each alone to do the job?

113. It takes one pipe $\frac{2}{3}$ as long as it takes another pipe to fill a tank. If the two pipes together fill the tank in 6 minutes, how long does it take each pipe alone to fill the tank?

114. A supply pipe can fill a tank in 35 minutes. How long will it take a drainage pipe to drain the tank, if, when both the supply and drainage pipes are open at the same time, the tank will be filled in 84 minutes?

chapter 7

Exponents and Applications

7.1 Positive fractional exponents

7.2 Zero and negative exponents

The purpose of this chapter is to extend the scope of the rules of exponents and to study some of their applications in algebra.

The following theorems of exponents were introduced in Chapter 2.
When $a, b \in R$, $a \neq 0, b \neq 0$, and $m, n \in N$, we have the following theorems.

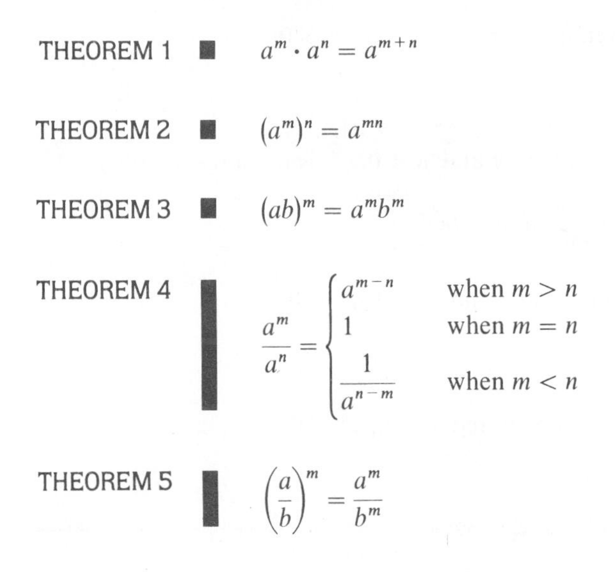

THEOREM 1 ■ $a^m \cdot a^n = a^{m+n}$

THEOREM 2 ■ $(a^m)^n = a^{mn}$

THEOREM 3 ■ $(ab)^m = a^m b^m$

THEOREM 4

$$\frac{a^m}{a^n} = \begin{cases} a^{m-n} & \text{when } m > n \\ 1 & \text{when } m = n \\ \dfrac{1}{a^{n-m}} & \text{when } m < n \end{cases}$$

THEOREM 5

$$\left(\frac{a}{b}\right)^m = \frac{a^m}{b^m}$$

7.1 Positive Fractional Exponents

In order for Theorem 2 to hold for positive fractional exponents we must have this definition:

DEFINITION

When $a \in R$ and $m, n \in N$ we define

$$(a^m)^{\frac{1}{n}} = \left(a^{\frac{1}{n}}\right)^m = a^{\frac{m}{n}}$$

From the definition we have $\left(a^{\frac{1}{n}}\right)^n = a^{\frac{n}{n}} = a$

When m is an even number, a^m is a positive number if a is either a positive number or a negative number; for example,

$$(+2)^6 = 64 \qquad \text{and} \qquad (-2)^6 = 64$$

When m is an odd number, a^m is a positive number when a is a positive number, and a negative number when a is a negative number; for example,

$$(+2)^5 = 32 \qquad \text{and} \qquad (-2)^5 = -32$$

DEFINITION

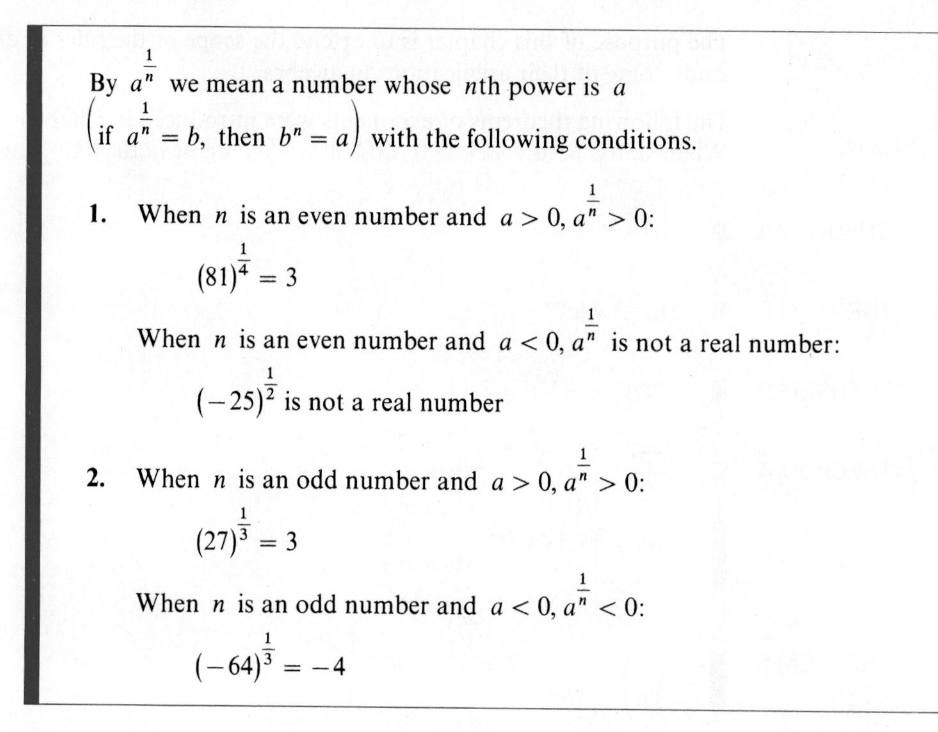

By $a^{\frac{1}{n}}$ we mean a number whose nth power is a $\left(\text{if } a^{\frac{1}{n}} = b, \text{ then } b^n = a\right)$ with the following conditions.

1. When n is an even number and $a > 0$, $a^{\frac{1}{n}} > 0$:

$$(81)^{\frac{1}{4}} = 3$$

When n is an even number and $a < 0$, $a^{\frac{1}{n}}$ is not a real number:

$$(-25)^{\frac{1}{2}} \text{ is not a real number}$$

2. When n is an odd number and $a > 0$, $a^{\frac{1}{n}} > 0$:

$$(27)^{\frac{1}{3}} = 3$$

When n is an odd number and $a < 0$, $a^{\frac{1}{n}} < 0$:

$$(-64)^{\frac{1}{3}} = -4$$

DEFINITION

For $a \in R$ and $m, n \in N$, whenever $a^{\frac{1}{n}}$ is defined, we define $a^{\frac{m}{n}}$ as $\left(a^{\frac{1}{n}}\right)^m$.

According to these definitions, Theorems 1–3 at the beginning of this chapter are valid when $a > 0, b > 0$ and m, n are positive fractional exponents. See Appendix E for the proofs.

The following are direct applications of the theorems.

1. $a^2 \cdot a^{\frac{1}{3}} = a^{2+\frac{1}{3}} = a^{\frac{7}{3}}$

2. $2^{\frac{3}{2}} \cdot 2^{\frac{5}{2}} = 2^{\frac{3}{2}+\frac{5}{2}} = 2^4$

3. $(3^4)^{\frac{3}{2}} = 3^{4 \cdot \frac{3}{2}} = 3^6$

4. $\left(a^{\frac{2}{3}}\right)^6 = a^{\frac{2}{3} \cdot 6} = a^4$

5. $\left(2^{\frac{2}{3}}\right)^{\frac{3}{4}} = 2^{\frac{2}{3} \cdot \frac{3}{4}} = 2^{\frac{1}{2}}$

6. $(2a)^{\frac{2}{3}} = 2^{\frac{2}{3}} a^{\frac{2}{3}}$

Note When $a, b \in R$, $a, b > 0$, and $p, q, r, s, u, v \in N$, then by use of Theorems 3 and 2, we have

$$\left(a^{\frac{p}{q}} b^{\frac{r}{s}}\right)^{\frac{u}{v}} = a^{\frac{pu}{qv}} b^{\frac{ru}{sv}}$$

EXAMPLE Multiply $\left(3a^{\frac{1}{2}}b^{\frac{1}{3}}\right)$ and $\left(2ab^{\frac{2}{3}}\right)$.

Solution

$$\left(3a^{\frac{1}{2}}b^{\frac{1}{3}}\right)\cdot\left(2ab^{\frac{2}{3}}\right) = (3 \times 2)\left(a^{\frac{1}{2}} \cdot a^{1}\right)\left(b^{\frac{1}{3}} \cdot b^{\frac{2}{3}}\right)$$
$$= 6a^{\frac{1}{2}+1}b^{\frac{1}{3}+\frac{2}{3}}$$
$$= 6a^{\frac{3}{2}}b$$

EXAMPLE Evaluate $(32)^{\frac{3}{5}}$.

Solution $(32)^{\frac{3}{5}} = (2^5)^{\frac{3}{5}} = 2^{5\cdot\frac{3}{5}} = 2^3 = 8$

EXAMPLE Simplify $\left(2a^{\frac{2}{3}}b^{\frac{1}{2}}\right)^6$.

Solution

$$\left(2^1a^{\frac{2}{3}}b^{\frac{1}{2}}\right)^6 = 2^{1\cdot 6}a^{\frac{2}{3}\cdot 6}b^{\frac{1}{2}\cdot 6}$$
$$= 2^6a^4b^3 = 64a^4b^3$$

EXAMPLE Multiply $\left(-3a^{\frac{3}{2}}b^{\frac{2}{3}}\right)^6$ and $\left(a^{\frac{1}{4}}b^{\frac{1}{2}}\right)^{12}$.

Solution

$$\left(-3a^{\frac{3}{2}}b^{\frac{2}{3}}\right)^6\cdot\left(a^{\frac{1}{4}}b^{\frac{1}{2}}\right)^{12} = \left[(-3)^6a^{\frac{3}{2}\cdot 6}b^{\frac{2}{3}\cdot 6}\right]\left[a^{\frac{1}{4}\cdot 12}b^{\frac{1}{2}\cdot 12}\right]$$
$$= (729a^9b^4)(a^3b^6)$$
$$= 729a^{12}b^{10}$$

EXAMPLE Multiply $\left(16a^{\frac{2}{9}}b^{\frac{2}{3}}\right)^{\frac{3}{4}}$ and $\left(27a^{\frac{5}{4}}b^{\frac{9}{4}}\right)^{\frac{2}{3}}$.

Solution

$$\left(16a^{\frac{2}{9}}b^{\frac{2}{3}}\right)^{\frac{3}{4}}\left(27a^{\frac{5}{4}}b^{\frac{9}{4}}\right)^{\frac{2}{3}} = \left(2^4a^{\frac{2}{9}}b^{\frac{2}{3}}\right)^{\frac{3}{4}}\left(3^3a^{\frac{5}{4}}b^{\frac{9}{4}}\right)^{\frac{2}{3}}$$
$$= \left(2^3a^{\frac{1}{6}}b^{\frac{1}{2}}\right)\left(3^2a^{\frac{5}{6}}b^{\frac{3}{2}}\right) = 72ab^2$$

EXAMPLE Perform the indicated multiplication $a^{\frac{1}{3}}b^{\frac{1}{2}}\left(a^{\frac{2}{3}}b - ab^{\frac{1}{2}}\right)$.

Solution

$$a^{\frac{1}{3}}b^{\frac{1}{2}}\left(a^{\frac{2}{3}}b - ab^{\frac{1}{2}}\right) = \left(a^{\frac{1}{3}}b^{\frac{1}{2}}\right)\left(a^{\frac{2}{3}}b\right) + \left(a^{\frac{1}{3}}b^{\frac{1}{2}}\right)\left(-ab^{\frac{1}{2}}\right)$$
$$= \left(a^{\frac{1}{3}}\cdot a^{\frac{2}{3}}\right)\left(b^{\frac{1}{2}}\cdot b\right) - \left(a^{\frac{1}{3}}\cdot a\right)\left(b^{\frac{1}{2}}\cdot b^{\frac{1}{2}}\right)$$
$$= ab^{\frac{3}{2}} - a^{\frac{4}{3}}b$$

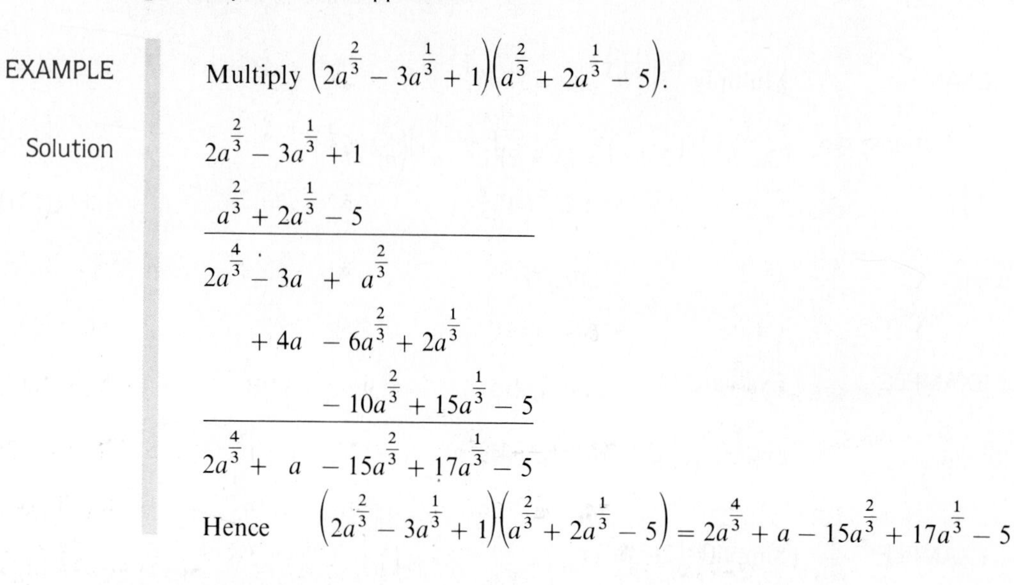

EXAMPLE Multiply $\left(2a^{\frac{2}{3}} - 3a^{\frac{1}{3}} + 1\right)\left(a^{\frac{2}{3}} + 2a^{\frac{1}{3}} - 5\right)$.

Solution

$$\begin{array}{rrrrr}
 & & 2a^{\frac{2}{3}} & - 3a^{\frac{1}{3}} & + 1 \\
 & & a^{\frac{2}{3}} & + 2a^{\frac{1}{3}} & - 5 \\
\hline
2a^{\frac{4}{3}} & - 3a & + a^{\frac{2}{3}} & & \\
 & + 4a & - 6a^{\frac{2}{3}} & + 2a^{\frac{1}{3}} & \\
 & & - 10a^{\frac{2}{3}} & + 15a^{\frac{1}{3}} & - 5 \\
\hline
2a^{\frac{4}{3}} & + a & - 15a^{\frac{2}{3}} & + 17a^{\frac{1}{3}} & - 5
\end{array}$$

Hence $\left(2a^{\frac{2}{3}} - 3a^{\frac{1}{3}} + 1\right)\left(a^{\frac{2}{3}} + 2a^{\frac{1}{3}} - 5\right) = 2a^{\frac{4}{3}} + a - 15a^{\frac{2}{3}} + 17a^{\frac{1}{3}} - 5$

Exercise 7.1A

Perform the indicated operations and simplify given that $m, n, k \in N$:

1. $2^{\frac{1}{2}} \cdot 2^{\frac{3}{2}}$ **2.** $3^{\frac{1}{4}} \cdot 3^{\frac{3}{4}}$ **3.** $5^{\frac{1}{3}} \cdot 5^{\frac{5}{3}}$ **4.** $7^{\frac{1}{6}} \cdot 7^{\frac{5}{6}}$

5. $2^{\frac{1}{2}} \cdot 2^{\frac{1}{3}}$ **6.** $3^{\frac{3}{8}} \cdot 3^{\frac{1}{4}}$ **7.** $2^2 \cdot 2^{\frac{1}{2}}$ **8.** $3 \cdot 3^{\frac{2}{3}}$

9. $8 \cdot 2^{\frac{1}{2}}$ **10.** $9 \cdot 3^{\frac{1}{3}}$ **11.** $x^{\frac{3}{5}} \cdot x^{\frac{4}{5}}$ **12.** $x^{\frac{2}{7}} \cdot x^{\frac{5}{7}}$

13. $x^{\frac{1}{9}} \cdot x^{\frac{5}{9}}$ **14.** $x^3 \cdot x^{\frac{1}{2}}$ **15.** $x^{\frac{3}{2}} \cdot x$ **16.** $x^{\frac{2}{3}} \cdot x^{\frac{1}{2}}$

17. $x^{\frac{1}{6}} \cdot x^{\frac{4}{3}}$ **18.** $x^{\frac{1}{8}} \cdot x^{\frac{3}{4}}$ **19.** $3a^2\left(a^{\frac{1}{2}}b\right)$ **20.** $5a\left(2a^{\frac{1}{3}}b^{\frac{1}{2}}\right)$

21. $a^{\frac{1}{2}}\left(a^{\frac{3}{2}}b^{\frac{1}{2}}\right)$ **22.** $a^{\frac{1}{3}}\left(a^{\frac{1}{3}}b^{\frac{2}{3}}\right)$ **23.** $ab\left(a^{\frac{1}{2}}b^{\frac{1}{3}}\right)$

24. $ab^{\frac{1}{2}}\left(-2ab^{\frac{3}{2}}\right)$ **25.** $2a^{\frac{1}{2}}b^{\frac{1}{3}} \cdot 3a^{\frac{1}{2}}b^{\frac{2}{3}}$ **26.** $\left(6a^{\frac{3}{2}}b^{\frac{9}{2}}\right)\left(-8a^{\frac{7}{2}}b^{\frac{5}{2}}\right)$

27. $a^{\frac{8}{3}}b^{\frac{7}{12}} \cdot a^{\frac{4}{3}}b^{\frac{1}{12}}$ **28.** $\left(2a^{\frac{1}{4}}b^{\frac{1}{2}}\right)\left(-3a^{\frac{1}{2}}b^{\frac{1}{4}}\right)$ **29.** $5x^{\frac{1}{2}}y^{\frac{3}{2}} \cdot x^{\frac{1}{3}}y^{\frac{1}{4}}$

30. $2c^{\frac{1}{6}}d^{\frac{1}{2}} \cdot c^{\frac{1}{3}}d^{\frac{2}{3}}$ **31.** $3a^{\frac{3}{4}}b^{\frac{5}{6}} \cdot 2a^{\frac{7}{4}}b^{\frac{1}{3}}$ **32.** $a^{\frac{1}{8}}b^{\frac{1}{3}}c^{\frac{1}{5}} \cdot a^{\frac{3}{8}}b^{\frac{2}{3}}c^{\frac{9}{5}}$

33. $\left(-a^2b^{\frac{1}{3}}\right)\left(-2^2a^{\frac{1}{2}}b^{\frac{1}{4}}\right)$ **34.** $5^{\frac{1}{2}}a^{\frac{1}{3}}b^{\frac{1}{6}} \cdot 5^{\frac{3}{2}}a^{\frac{2}{3}}b^{\frac{7}{12}}$

35. $\left(2^{\frac{1}{3}}\right)^3$ **36.** $\left(3^{\frac{1}{2}}\right)^4$ **37.** $\left(5^{\frac{2}{3}}\right)^6$ **38.** $\left(2^{\frac{3}{4}}\right)^8$

39. $\left(x^{\frac{1}{2}}\right)^4$ **40.** $\left(x^{\frac{1}{3}}\right)^6$ **41.** $\left(x^{\frac{2}{3}}\right)^9$ **42.** $\left(x^{\frac{3}{4}}\right)^{12}$

43. $(x^4)^{\frac{3}{2}}$ **44.** $(x^3)^{\frac{2}{3}}$ **45.** $(x^8)^{\frac{3}{4}}$ **46.** $(x^{12})^{\frac{1}{6}}$

47. $(x^4)^{\frac{n}{2}}$ **48.** $(x^9)^{\frac{n}{3}}$ **49.** $\left(x^{\frac{2}{3}}\right)^{\frac{9}{4}}$ **50.** $\left(x^{\frac{3}{2}}\right)^{\frac{4}{3}}$

51. $\left(x^{\frac{3}{5}}\right)^{\frac{5}{9}}$ **52.** $\left(x^{\frac{7}{8}}\right)^{\frac{6}{7}}$ **53.** $\left(x^{\frac{n}{2}}\right)^{\frac{6}{n}}$ **54.** $\left(x^{\frac{n}{6}}\right)^{\frac{9}{n}}$

55. $(4)^{\frac{1}{2}}$ **56.** $(49)^{\frac{1}{2}}$ **57.** $(36)^{\frac{1}{2}}$ **58.** $(121)^{\frac{1}{2}}$

59. $(27)^{\frac{1}{3}}$ **60.** $(64)^{\frac{1}{3}}$ **61.** $(125)^{\frac{1}{3}}$ **62.** $(216)^{\frac{1}{3}}$

63. $(16)^{\frac{1}{4}}$ **64.** $(81)^{\frac{1}{4}}$ **65.** $(8)^{\frac{2}{3}}$ **66.** $(9)^{\frac{3}{2}}$

67. $(16)^{\frac{5}{4}}$ **68.** $(243)^{\frac{2}{5}}$ **69.** $(64)^{\frac{5}{6}}$ **70.** $(1000)^{\frac{2}{3}}$

71. $(-36)^{\frac{1}{2}}$ **72.** $(-49)^{\frac{1}{2}}$ **73.** $(-8)^{\frac{1}{3}}$ **74.** $(-27)^{\frac{1}{3}}$

75. $4^{\frac{1}{2}} \cdot 2^2$ **76.** $4^{\frac{1}{2}} \cdot 9^{\frac{1}{2}}$ **77.** $8^{\frac{2}{3}} \cdot 9^{\frac{3}{2}}$ **78.** $16^{\frac{3}{4}} \cdot 25^{\frac{3}{2}}$

79. $\left(2a^{\frac{1}{2}}b^{\frac{1}{4}}\right)^2\left(a^{\frac{1}{3}}b^{\frac{1}{2}}\right)^6$ **80.** $\left(x^{\frac{3}{4}}y^{\frac{1}{8}}\right)^8\left(x^{\frac{1}{3}}y^{\frac{1}{6}}\right)^{12}$ **81.** $\left(3a^{\frac{3}{2}}b^{\frac{3}{4}}\right)^4\left(a^{\frac{2}{3}}b^{\frac{2}{5}}\right)^{15}$

82. $\left(a^{\frac{2}{3}}b^{\frac{1}{6}}\right)^6\left(a^{\frac{3}{4}}b^{\frac{1}{2}}\right)^4$ **83.** $(a^9b^6)^{\frac{1}{3}}(a^2b^{10})^{\frac{1}{2}}$ **84.** $(81a^4b^8)^{\frac{1}{4}}(8a^6b^3)^{\frac{1}{3}}$

85. $(49a^2b^4)^{\frac{1}{2}}(125a^3b^3)^{\frac{2}{3}}$ **86.** $(4a^4b^6)^{\frac{3}{2}}(216a^3b^3)^{\frac{1}{3}}$

87. $(64a^6b^{12})^{\frac{5}{6}}(243a^5b^{10})^{\frac{2}{5}}$ **88.** $(36a^4b^8)^{\frac{1}{2}}(25a^2b^4)^{\frac{3}{2}}$

89. $\left(a^{\frac{2}{3}}b^{\frac{5}{6}}\right)^{\frac{3}{10}}\left(a^{\frac{2}{5}}b^{\frac{2}{3}}\right)^{\frac{3}{4}}$ **90.** $\left(a^{\frac{3}{8}}b^{\frac{1}{4}}\right)^{\frac{4}{3}}\left(a^{\frac{2}{7}}b^{\frac{8}{21}}\right)^{\frac{7}{4}}$ **91.** $\left(a^{\frac{4}{9}}b^{\frac{5}{3}}\right)^{\frac{3}{2}}\left(a^{\frac{3}{2}}b^{\frac{9}{4}}\right)^{\frac{2}{9}}$

92. $\left(a^{\frac{7}{6}}b^{\frac{5}{8}}\right)^{\frac{8}{7}}\left(a^{\frac{4}{3}}b^{\frac{4}{7}}\right)^{\frac{1}{2}}$ **93.** $\left(8^{\frac{1}{3}}a^{\frac{1}{2}}b^{\frac{1}{3}}\right)^{\frac{6}{5}}\left(4^{\frac{2}{3}}a^{\frac{2}{3}}b\right)^{\frac{3}{5}}$ **94.** $\left(4a^{\frac{5}{6}}b^{\frac{2}{9}}\right)^{\frac{3}{4}}\left(16ab^{\frac{4}{3}}\right)^{\frac{3}{8}}$

95. $\left(a^{\frac{3}{m}}b^{\frac{2}{n}}\right)^{mn}\left(a^{\frac{2}{n}}b^{\frac{3}{m}}\right)^{2mn}$ **96.** $\left(a^{\frac{m}{2n}}b^{\frac{n}{3m}}\right)^{6mn}\left(a^{\frac{k}{3}}b^{\frac{k}{4}}\right)^{\frac{12}{k}}$

97. $a\left(a^{\frac{1}{2}} - 1\right)$ **98.** $a^2\left(2a^{\frac{1}{2}} + 3\right)$ **99.** $a^2\left(a^{\frac{1}{3}} + 2\right)$

100. $a\left(3a^{\frac{2}{3}} - 2\right)$ **101.** $a^{\frac{1}{4}}(2a - 3)$ **102.** $a^{\frac{3}{4}}(a^2 + 1)$

103. $a^{\frac{1}{2}}\left(a^{\frac{1}{2}} + b^{\frac{1}{2}}\right)$ **104.** $2a^{\frac{1}{4}}\left(3a^{\frac{1}{4}} - b^{\frac{1}{4}}\right)$

105. $a^{\frac{2}{3}}b^{\frac{2}{3}}\left(a^{\frac{1}{3}} - b^{\frac{1}{3}}\right)$ **106.** $ab^{\frac{3}{2}}\left(a^{\frac{1}{2}}b + ab^{\frac{1}{2}}\right)$

107. $3a^{\frac{1}{6}}b^{\frac{1}{3}}\left(2a^{\frac{1}{3}}b^{\frac{1}{6}} - 4a^{\frac{2}{3}}b^{\frac{1}{2}}\right)$ **108.** $a^{\frac{2}{3}}\left(a^{\frac{4}{3}} - a^{\frac{2}{3}} + 4\right)$

109. $a^{\frac{3}{4}}\left(a^2 - 2a^{\frac{7}{4}} + a^{\frac{3}{2}}\right)$ **110.** $a^{\frac{1}{2}}b^{\frac{1}{2}}\left(a^{\frac{1}{2}} - a^{\frac{1}{4}}b^{\frac{1}{4}} + b^{\frac{1}{2}}\right)$

111. $(2x + 1)\left(x^{\frac{1}{2}} + 2\right)$ **112.** $(x^2 - 2)\left(x^{\frac{1}{2}} + 3\right)$

113. $\left(x^{\frac{1}{2}} + y^{\frac{1}{2}}\right)\left(x^{\frac{1}{2}} - y^{\frac{1}{2}}\right)$ **114.** $\left(2x^{\frac{1}{2}} + 3y^{\frac{1}{3}}\right)\left(3x^{\frac{1}{2}} - y^{\frac{1}{3}}\right)$

115. $\left(5x^{\frac{2}{3}} - y^{\frac{1}{4}}\right)\left(4x^{\frac{2}{3}} - 2y^{\frac{1}{4}}\right)$

116. $\left(a^{\frac{1}{2}} + b^{\frac{1}{2}}\right)^2$

117. $\left(a^{\frac{1}{2}} - 2b^{\frac{1}{2}}\right)^2$

118. $\left(2a - 3b^{\frac{1}{3}}\right)^2$

119. $\left(x^{\frac{1}{3}} + y^{\frac{1}{3}}\right)\left(x^{\frac{2}{3}} - x^{\frac{1}{3}}y^{\frac{1}{3}} + y^{\frac{2}{3}}\right)$

120. $\left(2x^{\frac{1}{3}} - y^{\frac{1}{3}}\right)\left(4x^{\frac{2}{3}} + 2x^{\frac{1}{3}}y^{\frac{1}{3}} + y^{\frac{2}{3}}\right)$

121. $\left(2x^{\frac{2}{3}} - 3y^{\frac{2}{3}}\right)\left(4x^{\frac{4}{3}} + 6x^{\frac{2}{3}}y^{\frac{2}{3}} + 9y^{\frac{4}{3}}\right)$

122. $\left(x^{\frac{1}{4}} - y^{\frac{1}{4}}\right)\left(x^{\frac{3}{4}} + x^{\frac{1}{2}}y^{\frac{1}{4}} + x^{\frac{1}{4}}y^{\frac{1}{2}} + y^{\frac{3}{4}}\right)$

123. $\left(x^{\frac{1}{2}} - x^{\frac{1}{4}} + 3\right)\left(x^{\frac{1}{2}} + x^{\frac{1}{4}} + 3\right)$

124. $\left(x^{\frac{1}{2}} - 2x^{\frac{1}{4}} + 3\right)\left(x^{\frac{1}{2}} + 2x^{\frac{1}{4}} - 3\right)$

125. $\left(x^{\frac{2}{3}} + 3x^{\frac{1}{3}} + 1\right)\left(x^{\frac{2}{3}} + 3x^{\frac{1}{3}} - 1\right)$

126. $\left(x^{\frac{2}{3}} - 3x^{\frac{1}{3}} - 2\right)\left(x^{\frac{2}{3}} + 3x^{\frac{1}{3}} + 2\right)$

127. $\left(3x^{\frac{2}{3}} - x^{\frac{1}{3}} - 2\right)\left(x^{\frac{2}{3}} + x^{\frac{1}{3}} - 1\right)$

128. $\left(x^{\frac{1}{3}} + y^{\frac{1}{3}}\right)^3$

129. $\left(x^{\frac{1}{6}} + 1\right)^3$

130. $\left(2x^{\frac{2}{3}} - 3y^{\frac{2}{3}}\right)^3$

According to the definitions on page 216, Theorems 4 and 5 page 215 are valid when $a > 0$, $b > 0$, and m, n are positive fractional exponents. See Appendix E for the proofs.

The following are direct applications of the theorems:

1. $\dfrac{a^{\frac{2}{3}}}{a^{\frac{1}{2}}} = a^{\frac{2}{3}-\frac{1}{2}} = a^{\frac{1}{6}}$

2. $\dfrac{a^{\frac{1}{8}}}{a^{\frac{3}{4}}} = \dfrac{1}{a^{\frac{3}{4}-\frac{1}{8}}} = \dfrac{1}{a^{\frac{5}{8}}}$

3. $\left(\dfrac{2}{x}\right)^{\frac{1}{3}} = \dfrac{2^{\frac{1}{3}}}{x^{\frac{1}{3}}}$

EXAMPLE

Simplify $\dfrac{a^{\frac{7}{3}}b^{\frac{1}{2}}c^{\frac{4}{5}}}{a^{\frac{1}{3}}b^{\frac{3}{2}}c^{\frac{1}{5}}}$

Solution

$$\frac{a^{\frac{7}{3}}b^{\frac{1}{2}}c^{\frac{4}{5}}}{a^{\frac{1}{3}}b^{\frac{3}{2}}c^{\frac{1}{5}}} = \frac{a^{\frac{7}{3}}}{a^{\frac{1}{3}}} \cdot \frac{b^{\frac{1}{2}}}{b^{\frac{3}{2}}} \cdot \frac{c^{\frac{4}{5}}}{c^{\frac{1}{5}}}$$

$$= a^{\frac{7}{3}-\frac{1}{3}} \cdot \frac{1}{b^{\frac{3}{2}-\frac{1}{2}}} \cdot c^{\frac{4}{5}-\frac{1}{5}}$$

$$= a^2 \cdot \frac{1}{b} \cdot c^{\frac{3}{5}} = \frac{a^2c^{\frac{3}{5}}}{b}$$

EXAMPLE

Simplify $\dfrac{\left(a^{\frac{3}{4}}b^{\frac{5}{6}}\right)^{12}}{\left(a^{\frac{2}{3}}b^{\frac{4}{9}}\right)^{9}}$.

Solution

$$\frac{\left(a^{\frac{3}{4}}b^{\frac{5}{6}}\right)^{12}}{\left(a^{\frac{2}{3}}b^{\frac{4}{9}}\right)^{9}} = \frac{a^9b^{10}}{a^6b^4} = a^3b^6$$

EXAMPLE

Simplify $\dfrac{(27a^3b^9)^{\frac{2}{3}}}{(16a^8b^{16})^{\frac{1}{4}}}$.

Solution

$$\frac{(27a^3b^9)^{\frac{2}{3}}}{(16a^8b^{16})^{\frac{1}{4}}} = \frac{(3^3a^3b^9)^{\frac{2}{3}}}{(2^4a^8b^{16})^{\frac{1}{4}}}$$

$$= \frac{3^2a^2b^6}{2a^2b^4}$$

$$= \frac{9b^2}{2}$$

EXAMPLE

Divide $\left(27a^{\frac{1}{3}}b^{\frac{2}{3}} - 36a^{\frac{3}{2}}b^{\frac{1}{2}}\right)$ by $9a^{\frac{1}{3}}b^{\frac{1}{2}}$.

Solution

$$\frac{27a^{\frac{1}{3}}b^{\frac{2}{3}} - 36a^{\frac{3}{2}}b^{\frac{1}{2}}}{9a^{\frac{1}{3}}b^{\frac{1}{2}}} = \frac{27a^{\frac{1}{3}}b^{\frac{2}{3}}}{9a^{\frac{1}{3}}b^{\frac{1}{2}}} + \frac{-36a^{\frac{3}{2}}b^{\frac{1}{2}}}{9a^{\frac{1}{3}}b^{\frac{1}{2}}}$$

$$= 3b^{\frac{2}{3}-\frac{1}{2}} - 4a^{\frac{3}{2}-\frac{1}{3}}$$

$$= 3b^{\frac{1}{6}} - 4a^{\frac{7}{6}}$$

When we divide two algebraic expressions with fractional exponents, we arrange the terms of the dividend according to the decreasing exponents of one of the literals. The smallest increment in which the exponents of the divisor decrease indicates the missing powers in the dividend.

Arrange the terms of the dividend in decreasing steps of that smallest increment. Where an exponent is missing, add a term with a zero coefficient. Arrange the terms of the divisor also according to the decreasing exponents of the same literal used in arranging the terms of the dividend.

EXAMPLE

Divide $(x^2 + x + 1)$ by $\left(x + x^{\frac{1}{2}} + 1\right)$.

Solution

Since the exponents of x in the divisor are in steps of $\frac{1}{2}$, arrange the dividend in decreasing powers of x, in steps of $\frac{1}{2}$. Include terms with zero coefficients for the missing terms:

$$\begin{array}{r|l}
 & x - x^{\frac{1}{2}} + 1 \\
\hline
x + x^{\frac{1}{2}} + 1 & x^2 + 0x^{\frac{3}{2}} + x + 0x^{\frac{1}{2}} + 1 \\
 & \ominus\ \ \ominus\ \ \ \ominus \\
 & +x^2 + x^{\frac{3}{2}} + x \\
\cline{2-2}
 & -x^{\frac{3}{2}} + 0x^{\frac{1}{2}} + 1 \\
 & \oplus\ \ \ \oplus\ \ \oplus \\
 & -x^{\frac{3}{2}} - x - x^{\frac{1}{2}} \\
\cline{2-2}
 & +x + x^{\frac{1}{2}} + 1 \\
 & \ominus\ \ \ominus\ \ \ \ominus \\
 & +x + x^{\frac{1}{2}} + 1 \\
\cline{2-2}
 & 0
\end{array}$$

Hence

$$\frac{x^2 + x + 1}{x + x^{\frac{1}{2}} + 1} = x - x^{\frac{1}{2}} + 1$$

Exercise 7.1B

Perform the indicated operations and simplify:

1. $\dfrac{3}{3^{\frac{1}{4}}}$ **2.** $\dfrac{5}{5^{\frac{2}{3}}}$ **3.** $\dfrac{7^2}{7^{\frac{1}{3}}}$ **4.** $\dfrac{3^{\frac{1}{2}}}{3^{\frac{1}{4}}}$

5. $\dfrac{2^{\frac{2}{3}}}{2^{\frac{1}{2}}}$ **6.** $\dfrac{2^{\frac{1}{2}}}{2^{\frac{1}{3}}}$ **7.** $\dfrac{2^{\frac{1}{2}}}{2}$ **8.** $\dfrac{5}{5^{\frac{3}{2}}}$

9. $\dfrac{3^{\frac{1}{4}}}{3^{\frac{3}{8}}}$ **10.** $\dfrac{4}{2^{\frac{1}{3}}}$ **11.** $\dfrac{3^{\frac{1}{2}}}{9}$ **12.** $\dfrac{x^2}{x^{\frac{1}{2}}}$

13. $\dfrac{x^{\frac{2}{3}}}{x^{\frac{1}{3}}}$

14. $\dfrac{x^{\frac{5}{6}}}{x^{\frac{2}{3}}}$

15. $\dfrac{x^{\frac{3}{2}}}{x^{\frac{1}{6}}}$

16. $\dfrac{x^{\frac{2}{5}}}{x}$

17. $\dfrac{x^{\frac{1}{4}}}{x^{\frac{3}{4}}}$

18. $\dfrac{x^{\frac{3}{4}}}{x^{\frac{7}{8}}}$

19. $\dfrac{15a^{\frac{5}{3}}b^{\frac{1}{3}}}{20a^{\frac{2}{3}}b^{\frac{4}{3}}}$

20. $\dfrac{12a^{\frac{2}{3}}b^{\frac{3}{2}}}{36a^{\frac{8}{3}}b^{\frac{7}{2}}}$

21. $\dfrac{21a^{\frac{2}{3}}b^{\frac{1}{8}}}{14a^{\frac{1}{6}}b^{\frac{5}{8}}}$

22. $\dfrac{24a^{\frac{1}{2}}b^{\frac{7}{8}}}{32a^{\frac{7}{6}}b^{\frac{3}{8}}}$

23. $\dfrac{a^{\frac{3}{2}}b^{2}c^{\frac{2}{3}}}{a^{3}b^{\frac{5}{3}}c^{2}}$

24. $\dfrac{ab^{\frac{1}{6}}c^{\frac{3}{5}}}{a^{\frac{5}{8}}bc^{2}}$

25. $\left(\dfrac{a^{2}}{b^{3}}\right)^{\frac{1}{6}}$

26. $\left(\dfrac{a^{6}}{b^{8}}\right)^{\frac{1}{12}}$

27. $\left(\dfrac{a^{\frac{5}{6}}}{b^{\frac{10}{3}}}\right)^{\frac{3}{5}}$

28. $\left(\dfrac{a^{\frac{9}{4}}}{b^{\frac{3}{4}}}\right)^{\frac{2}{3}}$

29. $\left(\dfrac{a^{\frac{1}{2}}}{a^{\frac{1}{4}}}\right)^{3}$

30. $\left(\dfrac{a^{\frac{3}{4}}}{a^{\frac{1}{2}}}\right)^{2}$

31. $\left(\dfrac{a^{\frac{1}{3}}}{a^{\frac{5}{6}}}\right)^{3}$

32. $\left(\dfrac{a^{\frac{4}{3}}}{a^{\frac{2}{5}}}\right)^{\frac{1}{2}}$

33. $\left(\dfrac{a^{\frac{3}{8}}}{a^{\frac{3}{4}}}\right)^{\frac{2}{9}}$

34. $\left(\dfrac{a^{\frac{6}{7}}}{a^{\frac{3}{14}}}\right)^{\frac{7}{3}}$

35. $\left(\dfrac{a^{\frac{1}{3}}b^{\frac{5}{6}}}{a^{\frac{1}{4}}b^{\frac{3}{4}}}\right)^{12}$

36. $\left(\dfrac{a^{\frac{3}{2}}b^{\frac{1}{4}}}{a^{\frac{3}{4}}b^{\frac{5}{8}}}\right)^{4}$

37. $\left(\dfrac{a^{\frac{2}{7}}b^{\frac{3}{8}}}{a^{\frac{4}{7}}b^{\frac{1}{4}}}\right)^{14}$

38. $\left(\dfrac{a^{\frac{3}{2}}b^{\frac{3}{4}}}{a^{\frac{1}{6}}b^{\frac{1}{3}}}\right)^{6}$

39. $\left(\dfrac{a^{\frac{3}{4}}b^{\frac{1}{4}}}{a^{\frac{1}{2}}b^{\frac{3}{8}}}\right)^{\frac{4}{3}}$

40. $\left(\dfrac{a^{\frac{1}{3}}b^{\frac{3}{4}}}{a^{\frac{1}{6}}b^{\frac{1}{3}}}\right)^{\frac{6}{5}}$

41. $\dfrac{4^{\frac{1}{3}}}{2^{\frac{2}{3}}}$

42. $\dfrac{9^{\frac{1}{6}}}{3^{\frac{1}{3}}}$

43. $\dfrac{4}{8^{\frac{1}{2}}}$

44. $\dfrac{8^{\frac{2}{3}}}{16^{\frac{1}{4}}}$

45. $\dfrac{27^{\frac{3}{2}}}{81^{\frac{1}{8}}}$

46. $\dfrac{32^{\frac{2}{5}}}{8^{\frac{4}{3}}}$

47. $\dfrac{\left(2a^{\frac{1}{3}}b^{\frac{1}{6}}\right)^{6}}{\left(a^{\frac{1}{2}}b^{\frac{1}{4}}\right)^{8}}$

48. $\dfrac{\left(a^{\frac{3}{4}}b^{\frac{5}{2}}\right)^{4}}{\left(a^{\frac{2}{3}}b^{\frac{5}{6}}\right)^{6}}$

49. $\dfrac{\left(a^{\frac{3}{7}}b^{\frac{2}{7}}\right)^{7}}{\left(a^{\frac{1}{2}}b\right)^{4}}$

50. $\dfrac{\left(x^{\frac{1}{2}}y^{\frac{3}{2}}\right)^{2}}{\left(x^{\frac{3}{4}}y^{\frac{2}{3}}\right)^{12}}$

51. $\dfrac{(64a^{3}b^{6})^{\frac{1}{6}}}{(128a^{7}b^{14})^{\frac{3}{7}}}$

52. $\dfrac{(9a^{2}b^{4}c^{6})^{\frac{1}{2}}}{(8a^{3}b^{6}c^{9})^{\frac{1}{3}}}$

53. $\dfrac{(25a^{4}b^{2}c^{2})^{\frac{3}{2}}}{(64a^{6}b^{12}c^{6})^{\frac{5}{6}}}$

54. $\dfrac{(16a^{8}b^{12}c^{8})^{\frac{1}{4}}}{(8a^{3}b^{3}c^{12})^{\frac{2}{3}}}$

55. $\dfrac{\left(x^{\frac{2}{3}}y^{\frac{2}{5}}\right)^{\frac{3}{4}}}{\left(x^{\frac{3}{4}}y^{\frac{3}{5}}\right)^{\frac{5}{6}}}$

56. $\dfrac{\left(x^{\frac{7}{3}}y^{\frac{3}{2}}\right)^{\frac{4}{7}}}{\left(x^{\frac{5}{9}}y^{\frac{5}{7}}\right)^{\frac{3}{10}}}$

57. $\dfrac{\left(x^{\frac{4}{3}}y^{\frac{2}{3}}\right)^{\frac{3}{2}}}{\left(x^{2}y^{\frac{3}{5}}\right)^{\frac{5}{3}}}$

58. $\dfrac{\left(x^{\frac{3}{2}}y^{\frac{5}{2}}\right)^{\frac{4}{3}}}{\left(x^{\frac{4}{3}}y^{\frac{5}{6}}\right)^{\frac{6}{5}}}$

59. $\dfrac{x^{2}-x^{3}}{x^{\frac{3}{2}}}$

60. $\dfrac{2x^{\frac{5}{4}}+x}{x^{\frac{3}{4}}}$

61. $\dfrac{10x^{\frac{4}{3}} - 5x}{5x^{\frac{2}{3}}}$

62. $\dfrac{6x^{\frac{2}{3}} - 4x^{\frac{1}{2}}y}{2x^{\frac{1}{3}}}$

63. $\dfrac{14bc^{\frac{1}{3}} - 21b^{\frac{1}{2}}c}{-7b^{\frac{1}{2}}c^{\frac{1}{3}}}$

64. $\dfrac{2y^{\frac{1}{4}} - 4y^{\frac{3}{8}} + 6y^{\frac{1}{2}}}{-2y^{\frac{1}{4}}}$

65. $\dfrac{72b^{\frac{5}{6}} - 84b^{\frac{2}{3}} + 48b^{\frac{1}{2}}}{12b^{\frac{1}{2}}}$

66. $\dfrac{8a^{\frac{6}{5}} - 6a^{\frac{4}{5}} - 2a^{\frac{2}{5}}}{-2a^{\frac{2}{5}}}$

67. $\dfrac{x^{\frac{4}{9}} - 5x^{\frac{1}{3}} + 6x^{\frac{2}{9}}}{x^{\frac{2}{9}}}$

68. $\left(a^2b^{\frac{5}{2}} + 3a^{\frac{7}{3}}b^2 - a^{\frac{8}{3}}b^{\frac{3}{2}}\right) \div \left(a^{\frac{1}{3}}b^{\frac{1}{2}}\right)$

69. $\left(6x + 11x^{\frac{1}{2}} - 10\right) \div \left(3x^{\frac{1}{2}} - 2\right)$

70. $\left(8x - 18x^{\frac{1}{2}} + 9\right) \div \left(4x^{\frac{1}{2}} - 3\right)$

71. $\left(8x + 2x^{\frac{1}{2}}y^{\frac{1}{2}} - 15y\right) \div \left(2x^{\frac{1}{2}} + 3y^{\frac{1}{2}}\right)$

72. $\left(x + x^{\frac{1}{2}}y^{\frac{1}{2}} - 6y\right) \div \left(x^{\frac{1}{2}} + 3y^{\frac{1}{2}}\right)$

73. $\left(x^{\frac{1}{2}} - y^{\frac{1}{2}}\right) \div \left(x^{\frac{1}{4}} - y^{\frac{1}{4}}\right)$

74. $\left(x^{\frac{2}{3}} - y^{\frac{2}{3}}\right) \div \left(x^{\frac{1}{3}} + y^{\frac{1}{3}}\right)$

75. $\left(4x^{\frac{3}{2}} - 9y^{\frac{3}{2}}\right) \div \left(2x^{\frac{3}{4}} + 3y^{\frac{3}{4}}\right)$

76. $\left(25x^{\frac{1}{3}} - y^{\frac{1}{3}}\right) \div \left(5x^{\frac{1}{6}} - y^{\frac{1}{6}}\right)$

77. $(x^2 + 5x + 9) \div \left(x - x^{\frac{1}{2}} + 3\right)$

78. $(x^2 - 10x + 9) \div \left(x - 2x^{\frac{1}{2}} - 3\right)$

79. $(x^2 - 12x + 4) \div \left(x - 4x^{\frac{1}{2}} + 2\right)$

80. $(4x^2 - 4x + 9) \div \left(2x - 4x^{\frac{1}{2}} + 3\right)$

81. $(9x^2 - 21x + 4) \div \left(3x - 3x^{\frac{1}{2}} - 2\right)$

82. $\left(x^2 - 6x^{\frac{3}{2}} + 9x - 4\right) \div \left(x - 3x^{\frac{1}{2}} - 2\right)$

83. $\left(12x^2 + 2x^{\frac{3}{2}} + 9x + 13x^{\frac{1}{2}} - 15\right) \div \left(3x - x^{\frac{1}{2}} + 5\right)$

84. $\dfrac{x - 10x^{\frac{1}{2}} + 9x^{\frac{1}{4}} - 2}{x^{\frac{1}{2}} - 3x^{\frac{1}{4}} + 1}$

85. $\dfrac{x^{\frac{3}{2}} - x + 4x^{\frac{3}{4}} + 4}{x^{\frac{3}{4}} - x^{\frac{1}{2}} + 2}$

86. $\dfrac{x^2 - 7x + 5x^{\frac{2}{3}} + 10x^{\frac{1}{3}} + 12}{x^{\frac{2}{3}} + 2x^{\frac{1}{3}} + 4}$

87. $\dfrac{6x^2 + 3x^{\frac{5}{3}} + 9x^{\frac{4}{3}} + 2x - 5x^{\frac{2}{3}} - 9}{2x^{\frac{2}{3}} + x^{\frac{1}{3}} + 3}$

88. $(x + y) \div \left(x^{\frac{1}{3}} + y^{\frac{1}{3}}\right)$

89. $(x^2 - y^2) \div \left(x^{\frac{1}{2}} + y^{\frac{1}{2}}\right)$

7.2 Zero and Negative Exponents

For the first and second parts of Theorem 4 for exponents (page 215) to be consistent, we must have, for $n = m$ and $a \neq 0$,

$$a^{m-m} = 1 \qquad \text{or} \qquad a^0 = 1$$

So we define

$$\text{if} \quad a \neq 0, \quad a^0 = 1$$

When $a = 0$, we have 0^0, which is indeterminate.

According to this definition, it can be shown that Theorems 1–5 for exponents are valid when a zero exponent occurs.
See Appendix E for the proofs.

EXAMPLES

1. $1^0 = 1$ **2.** $(-25)^0 = 1$ **3.** $\left(a^2 b^{\frac{1}{2}} c^{\frac{2}{3}}\right)^0 = 1$

Notes

1. $3a^0 = 3(1) = 3$
2. If $a \neq -b$, $(a + b)^0 = 1$
3. $a^0 + b^0 = 1 + 1 = 2$

Again, for the first and third parts of Theorem 4 for exponents to be consistent, we must have, when $m = 0$ and $a \neq 0$,

$$a^{0-n} = \frac{1}{a^{n-0}}$$

So we define

$$\text{if} \quad a \neq 0 \quad a^{-n} = \frac{1}{a^n}$$

EXAMPLES

1. $2^{-3} = \frac{1}{2^3} = \frac{1}{8}$ **2.** $-10^{-3} = -\frac{1}{10^3} = -\frac{1}{1000}$

3. $a^{-5} = \frac{1}{a^5}$ **4.** $a^{-\frac{1}{2}} = \frac{1}{a^{\frac{1}{2}}}$

According to the definition of negative exponents, $a \neq 0$, $a^{-n} = \frac{1}{a^n}$, Theorems 1–5 (page 215) for exponents are still valid.
See Appendix E for the proofs.

Note that when $n \in Q$, $1^n = 1$.

Remark Theorems 1–5 are true when $a > 0, b > 0$, and m, n are rational numbers.

Now we can write Theorem 4 as

$$\frac{a^m}{a^n} = a^{m-n}$$

The following are direct applications of the theorems:

1. $a^{-3} \cdot a^5 = a^{-3+5} = a^2$

2. $a^{-2} \cdot a^{-4} = a^{-2-4} = a^{-6} = \dfrac{1}{a^6}$

3. $(a^3)^{-4} = a^{3(-4)} = a^{-12} = \dfrac{1}{a^{12}}$

4. $(a^{-2})^3 = a^{(-2)(3)} = a^{-6} = \dfrac{1}{a^6}$

5. $(a^{-3})^{-5} = a^{(-3)(-5)} = a^{15}$

6. $(ab)^{-4} = a^{-4}b^{-4} = \dfrac{1}{a^4b^4}$

7. $\dfrac{a^2}{a^{-5}} = a^{2-(-5)} = a^7$

8. $\dfrac{a^{-3}}{a} = \dfrac{1}{a^{1-(-3)}} = \dfrac{1}{a^4}$

9. $\left(\dfrac{a}{b}\right)^{-3} = \dfrac{a^{-3}}{b^{-3}} = \dfrac{b^3}{a^3}$

10. $.007 = \dfrac{7}{1000} = \dfrac{7}{10^3} = 7 \times 10^{-3}$

Notes

1. $-a^{-n} = -\dfrac{1}{a^n}$

2. $\dfrac{1}{a^{-n}} = \dfrac{1}{\dfrac{1}{a^n}} = a^n$

3. $\left(\dfrac{a}{b}\right)^{-n} = \dfrac{a^{-n}}{b^{-n}} = \dfrac{b^n}{a^n} = \left(\dfrac{b}{a}\right)^n$

4. $(a+b)^{-n} = \dfrac{1}{(a+b)^n}, \quad a \neq -b$

5. $a^{-n} + b^{-n} = \dfrac{1}{a^n} + \dfrac{1}{b^n} = \dfrac{b^n + a^n}{a^nb^n}$

LCD $= a^n b^n$

EXAMPLE Express a^2b^{-3} with positive exponents.

Solution $a^2b^{-3} = a^2 \cdot \dfrac{1}{b^3} = \dfrac{a^2}{b^3}$

EXAMPLE Multiply x^2y^{-3} by $x^{-3}y^5$ and write the answer with positive exponents.

Solution

$$(x^2y^{-3})(x^{-3}y^5) = (x^2 \cdot x^{-3})(y^{-3} \cdot y^5)$$

$$= x^{2-3}y^{-3+5} = x^{-1}y^2 = \frac{y^2}{x}$$

EXAMPLE Simplify $(2a^{-1}b^2)^2$ and write the answer with positive exponents.

Solution $(2a^{-1}b^2)^2 = 2^2a^{-2}b^4 = 2^2 \cdot \dfrac{1}{a^2} \cdot b^4 = \dfrac{4b^4}{a^2}$

EXAMPLE Simplify $(3a^2b^{-3})^{-2}$ and write the answer with positive exponents.

Solution
$$(3a^2b^{-3})^{-2} = 3^{-2}a^{-4}b^6 = \frac{1}{3^2}\cdot\frac{1}{a^4}\cdot b^6 = \frac{b^6}{9a^4}$$

EXAMPLE Simplify $(2ab^{-2}c)^2(3^{-1}b^2c^3)^{-3}$ and write the answer with positive exponents.

Solution
$$\begin{aligned}(2ab^{-2}c)^2(3^{-1}b^2c^3)^{-3} &= (2^2a^2b^{-4}c^2)(3^3b^{-6}c^{-9})\\ &= (2^2\cdot 3^3)(a^2)(b^{-4}\cdot b^{-6})(c^2\cdot c^{-9})\\ &= 108a^2b^{-10}c^{-7} = \frac{108a^2}{b^{10}c^7}\end{aligned}$$

EXAMPLE Multiply $(a^{-2} + 2a^{-1}b^{-1} + 4b^{-2})$ by $(a^{-1} - 2b^{-1})$ and write the answer with positive exponents.

Solution
$$\begin{array}{l} a^{-2} + 2a^{-1}b^{-1} + 4b^{-2} \\ a^{-1} - 2b^{-1} \\ \hline a^{-3} + 2a^{-2}b^{-1} + 4a^{-1}b^{-2} \\ \qquad - 2a^{-2}b^{-1} - 4a^{-1}b^{-2} - 8b^{-3} \\ \hline a^{-3} \qquad\qquad\qquad\qquad - 8b^{-3} \end{array}$$

Hence
$$(a^{-2} + 2a^{-1}b^{-1} + 4b^{-2})\cdot(a^{-1} - 2b^{-1}) = a^{-3} - 8b^{-3} = \frac{1}{a^3} - \frac{8}{b^3}$$

EXAMPLE Simplify $\dfrac{a^{-2}b^2c}{a^3b^{-4}c^2}$ and write the answer with positive exponents.

Solution
$$\begin{aligned}\frac{a^{-2}b^2c}{a^3b^{-4}c^2} &= \frac{a^{-2}}{a^3}\cdot\frac{b^2}{b^{-4}}\cdot\frac{c}{c^2}\\ &= \frac{1}{a^3\cdot a^2}\cdot\frac{b^2\cdot b^4}{1}\cdot\frac{1}{c^2\cdot c^{-1}}\\ &= \frac{1}{a^5}\cdot b^6\cdot\frac{1}{c} = \frac{b^6}{a^5c}\end{aligned}$$

Note $\dfrac{ab^{-3}}{c^{-2}d^4} = \dfrac{ac^2}{b^3d^4}$; that is, when we have a factor in the numerator and we write it in the denominator, or a factor in the denominator and we write it in the numerator, we take that factor to the negative of its exponent.

EXAMPLE

Simplify $\left(\dfrac{2a^3b^{-2}c^{-1}}{3a^{-2}b^{-2}c^3}\right)^3$ and write the answer with positive exponents.

Solution

$$\left(\frac{2a^3b^{-2}c^{-1}}{3a^{-2}b^{-2}c^3}\right)^3 = \left(\frac{2}{3}\cdot\frac{a^3}{a^{-2}}\cdot\frac{b^{-2}}{b^{-2}}\cdot\frac{c^{-1}}{c^3}\right)^3$$
$$= \left(\frac{2}{3}\cdot\frac{a^5}{1}\cdot\frac{b^0}{1}\cdot\frac{1}{c^4}\right)^3$$
$$= \left(\frac{2a^5}{3c^4}\right)^3 = \frac{2^3a^{15}}{3^3c^{12}} = \frac{8a^{15}}{27c^{12}}$$

EXAMPLE

Simplify $\dfrac{(2a^2b^{-3}c^{-2})^3}{(a^{-1}b^2c^2)^{-2}}$ and write the answer with positive exponents.

Solution

$$\frac{(2a^2b^{-3}c^{-2})^3}{(a^{-1}b^2c^2)^{-2}} = \frac{2^3a^6b^{-9}c^{-6}}{a^2b^{-4}c^{-4}} = \frac{8a^4}{b^5c^2}$$

EXAMPLE

Simplify $\left(\dfrac{5^2a^2b^{-4}c^{-6}}{2^4x^{-2}y^{-2}z^6}\right)^{-\frac{3}{2}}\cdot\left(\dfrac{2a^{-3}b^2c^{-1}}{x^3y^2z^{-2}}\right)^{-1}$ and express the answer with positive exponents.

Solution

$$\left(\frac{5^2a^2b^{-4}c^{-6}}{2^4x^{-2}y^{-2}z^6}\right)^{-\frac{3}{2}}\cdot\left(\frac{2a^{-3}b^2c^{-1}}{x^3y^2z^{-2}}\right)^{-1} = \frac{5^{-3}a^{-3}b^6c^9}{2^{-6}x^3y^3z^{-9}}\cdot\frac{2^{-1}a^3b^{-2}c}{x^{-3}y^{-2}z^2}$$
$$= \frac{5^{-3}\cdot 2^{-1}}{2^{-6}}\cdot\frac{(a^{-3}\cdot a^3)(b^6\cdot b^{-2})(c^9\cdot c)}{(x^3\cdot x^{-3})(y^3\cdot y^{-2})(z^{-9}\cdot z^2)}$$
$$= \frac{2^5}{5^3}\cdot\frac{a^0b^4c^{10}}{x^0yz^{-7}} = \frac{32b^4c^{10}z^7}{125y}$$

EXAMPLE

Simplify $\dfrac{4a^{-1}+3b^{-2}}{a^{-1}-2b^{-2}}$ and write the answer with positive exponents.

Solution

$$\frac{4a^{-1}+3b^{-2}}{a^{-1}-2b^{-2}} = \frac{\dfrac{4}{a}+\dfrac{3}{b^2}}{\dfrac{1}{a}-\dfrac{2}{b^2}} = \frac{\dfrac{4b^2+3a}{ab^2}}{\dfrac{b^2-2a}{ab^2}}$$
$$= \frac{4b^2+3a}{ab^2}\cdot\frac{ab^2}{b^2-2a}$$
$$= \frac{4b^2+3a}{b^2-2a}$$

The last result can be accomplished easily by multiplying both numerator and denominator by ab^2:

$$\frac{4a^{-1}+3b^{-2}}{a^{-1}-2b^{-2}} = \frac{ab^2(4a^{-1}+3b^{-2})}{ab^2(a^{-1}-2b^{-2})}$$

$$= \frac{4a^0b^2+3ab^0}{a^0b^2-2ab^0} = \frac{4b^2+3a}{b^2-2a}$$

EXAMPLE

Write $\dfrac{3(2x+1)^{\frac{1}{2}}-(3x-4)(2x+1)^{-\frac{1}{2}}}{(2x+1)}$ as one fraction with positive exponents.

Solution

Multiply both numerator and denominator by $(2x+1)^{\frac{1}{2}}$:

$$\frac{\left[3(2x+1)^{\frac{1}{2}}-(3x-4)(2x+1)^{-\frac{1}{2}}\right](2x+1)^{\frac{1}{2}}}{(2x+1)(2x+1)^{\frac{1}{2}}} = \frac{3(2x+1)-(3x-4)(2x+1)^0}{(2x+1)^{\frac{3}{2}}}$$

$$= \frac{3(2x+1)-(3x-4)}{(2x+1)^{\frac{3}{2}}}$$

$$= \frac{3x+7}{(2x+1)^{\frac{3}{2}}}$$

EXAMPLE

Write $\dfrac{\frac{1}{3}(2x-3)^{\frac{1}{3}}(x+4)^{-\frac{2}{3}}-\frac{2}{3}(x+4)^{\frac{1}{3}}(2x-3)^{-\frac{2}{3}}}{(2x-3)^{\frac{2}{3}}}$ as one fraction with positive exponents.

Solution

Multiply both numerator and denominator by $3(x+4)^{\frac{2}{3}}(2x-3)^{\frac{2}{3}}$:

$$\frac{\left[\frac{1}{3}(2x-3)^{\frac{1}{3}}(x+4)^{-\frac{2}{3}}-\frac{2}{3}(x+4)^{\frac{1}{3}}(2x-3)^{-\frac{2}{3}}\right]\cdot 3(x+4)^{\frac{2}{3}}(2x-3)^{\frac{2}{3}}}{(2x-3)^{\frac{2}{3}}\cdot 3(x+4)^{\frac{2}{3}}(2x-3)^{\frac{2}{3}}}$$

$$= \frac{(2x-3)(x+4)^0-2(x+4)(2x-3)^0}{3(2x-3)^{\frac{4}{3}}(x+4)^{\frac{2}{3}}}$$

$$= \frac{(2x-3)-2(x+4)}{3(2x-3)^{\frac{4}{3}}(x+4)^{\frac{2}{3}}} = -\frac{11}{3(2x-3)^{\frac{4}{3}}(x+4)^{\frac{2}{3}}}$$

Any positive number in decimal notation can be written as the product of a number between 1 and 10 and a power of 10. For example:

1. $32.5 = 3.25 \times 10^1$

2. $738.6 = 7.386 \times 100 = 7.386 \times 10^2$

3. $6.78 = 6.78 \times 10^0$

4. $.967 = \frac{9.67}{10} = 9.67 \times 10^{-1}$

5. $.064 = \frac{6.4}{100} = \frac{6.4}{10^2} = 6.4 \times 10^{-2}$

6. $.008 = \frac{8.0}{1000} = \frac{8.0}{10^3} = 8.0 \times 10^{-3}$

The decimal point is always placed after the leftmost digit. This notation is called the **scientific notation for a number.**

Exercise 7.2

Simplify the following and write the answers with positive exponents, given that all literal numbers are not equal to zero and $n \in N$:

1. 2^0 **2.** -3^0 **3.** $(-5)^0$ **4.** $(-1)^0$

5. $\left(\frac{1}{2}\right)^0$ **6.** $\left(\frac{2}{3}\right)^0$ **7.** 0^0 **8.** 2^0a^2

9. $3a^0$ **10.** $(a^0)^4$ **11.** $(3a^0)^2$ **12.** $(a^2)^0$

13. $(-7a^3)^0$ **14.** $\left(a^{\frac{1}{2}}\right)^0$ **15.** $(a+b)^0, a \neq -b$

16. $2(a+b)^0, a \neq -b$ **17.** $5(a-b)^0, a \neq b$ **18.** $a^0 \cdot a^4$

19. $\frac{a^5}{a^0}$ **20.** $\frac{a^0}{a^2}$ **21.** $3a^0(2-a)$

22. $(a^0+1)^2$ **23.** $(2+a^0)^3$ **24.** $(b^0-1)^{10}$

25. $(2b^0-1)^8$ **26.** $(a+b^0)^2$ **27.** $(2a+b^0)^2$

28. $(a-2b^0)^2$ **29.** $(3a-2b^0)^2$ **30.** 2^{-1}

31. $2^{-1} \cdot 3$ **32.** $3 \cdot 5^{-1}$ **33.** $3a^{-1}$ **34.** $5^{-1}a$

35. a^2b^{-1} **36.** $a^{-3}b$ **37.** $2^{-2}a^{-3}$ **38.** $2^{-1}x^{-2}y$

39. $x^{-5}y^{-2}$ **40.** $2x^{-3}y^{-2}$ **41.** $3^{-1}xy^{-1}$ **42.** $2^{-1}+3^{-2}$

43. $3^{-1}-5^{-2}$ **44.** $x^{-4}+y^{-4}$ **45.** $x^{-2}-y^{-1}$

46. $3x^{-1}+y^{-3}$ **47.** $2x^{-1}-3y^{-2}$ **48.** $(1+3^{-1})^2$

49. $(3+2^{-1})^2$ **50.** $(2^{-1}+3^{-1})^2$ **51.** $(2^{-1}-4^{-1})^3$

52. $(3^{-1}-5^{-1})^2$ **53.** $2^4 \cdot 2^{-4}$ **54.** $2^3 \cdot 2^{-2}$

55. $3^2 \cdot 3^{-4}$ **56.** $2^{-1} \cdot 2^{-3}$ **57.** $3^{-2} \cdot 3^{-3}$

58. $x^2 \cdot x^{-4}$ **59.** $x^5 \cdot x^{-2}$ **60.** $x^{-2} \cdot x^{-3}$

61. $x^{-2n} \cdot x^{3n}$ **62.** $x^n \cdot x^{-2n}$ **63.** $x^{-n} \cdot x^{-2n}$

64. $2x^{-2}y \cdot y^{-3}$ **65.** $(4^{-1}x^2y^{-1})(6x^{-3}y^2)$

66. $(-3^{-1}x^2y^{-3})(2x^{-4}y^3)$ **67.** $(-3^{-2}x^{-1}y^2)(2x^{-2}y^{-1})$

68. $(-2^{-2}x^{-3}y^{-1})(5^{-2}xy^{2})$

69. $(-2^{-3}xy^{-2})(3x^{-3}y^{2})$

70. $(-4x^{-2}y^{3})(3^{-1}x^{3}y^{-4})$

71. $(2^{-2}x^{3}y^{-1})(-5^{2}x^{-1}y^{4})$

72. $(2a^{-3}b^{2}c)(3ab^{2}c^{-2})$

73. $(2^{-1}a^{-2}b^{-3}c^{2})(2^{2}a^{-1}b^{4}c^{-3})$

74. $(-3^{-1}a^{-2}b^{2})(2^{2}a^{-1}b^{3}c^{-1})$

75. $(-6a^{2}b^{-3})(2^{-2}a^{2}c^{-3})$

76. $(2^{-3})^{2}$

77. $(3^{-2})^{2}$

78. $(2^{2})^{-2}$

79. $(3^{-1})^{-2}$

80. $(x^{2})^{-3}$

81. $(x^{-3})^{3}$

82. $(x^{-2})^{-2}$

83. $(x^{-3})^{-4}$

84. $(x^{-2})^{n}$

85. $(x^{3})^{-n}$

86. $(x^{-2n})^{n}$

87. $(x^{-n})^{-2n}$

88. $(3^{2}xy^{-3})^{2}$

89. $(-3^{-1}x^{-2}y)^{2}$

90. $(5x^{-1}y^{2})^{-2}$

91. $(-4x^{2}y^{-3})^{-1}$

92. $(2^{3}x^{-6}y^{9})^{-\frac{2}{3}}$

93. $(4x^{-1}y^{-2})^{-\frac{1}{2}}$

94. $(27x^{-3}y)^{-\frac{1}{3}}$

95. $(16x^{4}y^{-6})^{-\frac{1}{4}}$

96. $(xy^{-2}z^{-1})^{2}(-2x^{-1}z^{2})^{-3}$

97. $(x^{-2}yz^{2})^{-2}(3^{-2}x^{-1}y^{2})^{3}$

98. $(2^{5}x^{-1}y^{-3})^{\frac{1}{2}}(2x^{-3}y)^{-\frac{1}{2}}$

99. $(3^{4}x^{3}y^{2})^{-\frac{1}{3}}(3x^{-3}y^{-1})^{\frac{1}{3}}$

100. $x^{-1}y^{-1}(3xy^{-2}+2x^{2}y^{-1}-xy)$

101. $x^{\frac{3}{2}}y^{\frac{3}{2}}\left(x^{-1}y^{-1}+x^{-\frac{3}{2}}y^{-\frac{3}{2}}-x^{-2}y^{-2}\right)$

102. $x^{-\frac{1}{2}}y^{-\frac{1}{2}}\left(2x^{-\frac{3}{2}}y^{-\frac{3}{2}}-3x^{-1}y^{-1}+x^{-\frac{1}{2}}y^{-\frac{1}{2}}\right)$

103. $(x^{-1}+y^{-1})(x^{-1}-y^{-1})$

104. $(2x^{-1}-3y^{-2})(3x^{-1}+4y^{-2})$

105. $(x^{-1}-3)(2x^{-2}-x^{-1}-1)$

106. $(2x^{-1}+1)(3x^{-2}+x^{-1}-4)$

107. $(x^{-1}-4)(3x^{-2}+x^{-1}-2)$

108. $(x^{-1}-y^{-1})(x^{-2}+x^{-1}y^{-1}+y^{-2})$

109. $(2x^{-1}+y^{-1})^{2}$

110. $(3x^{-1}-2y^{-2})^{2}$

111. $\left(x^{-\frac{1}{2}}+4y^{-\frac{1}{2}}\right)^{2}$

112. $\left(x^{-\frac{1}{3}}+y^{-\frac{1}{3}}\right)^{3}$

113. $\left(x^{-\frac{2}{3}}-y^{-\frac{2}{3}}\right)^{3}$

114. $\dfrac{1}{3^{-1}}$

115. $\left(\dfrac{1}{2}\right)^{-1}$

116. $\left(\dfrac{2}{3}\right)^{-1}$

117. $\dfrac{2^{-1}}{3^{-3}}$

118. $\dfrac{2^{-4}}{2^{2}}$

119. $\dfrac{2^{3}}{2^{-2}}$

120. $\dfrac{3^{-4}}{3^{-6}}$

121. $\dfrac{2^{-5}}{2^{-2}}$

122. $\dfrac{1}{2^{-3}}+\dfrac{1}{2^{-1}}$

123. $\dfrac{1}{3^{-1}}+\dfrac{1}{3^{-2}}$

124. $\dfrac{1}{2^{-2}}-\dfrac{1}{3^{-1}}$

125. $\dfrac{1}{3^{-3}}-\dfrac{1}{2^{-4}}$

126. $\dfrac{3}{a^{-2}}$

127. $\dfrac{a^{2}}{3b^{-2}}$

128. $\dfrac{4^{-1}}{a^{-1}}$

129. $\dfrac{2a^{-2}}{3b^{-1}}$

130. $\dfrac{x^{2}}{x^{-3}}$

131. $\dfrac{x^{-2}}{x^{4}}$

132. $\dfrac{x^{-3}}{x^{-5}}$

133. $\dfrac{x^{-6}}{x^{-2}}$

134. $\dfrac{xy^{-2}}{x^{-2}y^{2}}$

135. $\dfrac{x^{-2}y^{-1}}{x^{-5}y^{-3}}$

136. $\dfrac{x^{-4}y^{-3}}{x^{-1}y^{-2}}$

137. $\dfrac{x^{-8}y^{-1}}{x^{-2}y^{-5}}$

138. $\dfrac{3xy^{-4}}{2^{-2}x^{-3}y}$

139. $\dfrac{6x^2y^{-3}}{2^{-1}x^{-3}y}$ **140.** $\dfrac{2^{-3}x^{-2}y^2}{6^{-1}x^4y^{-6}}$ **141.** $\dfrac{6^{-2}x^{-8}y^7}{-18^{-1}x^{-4}y^{-2}}$

142. $\dfrac{-4^{-2}xy^{-2}}{12^{-2}x^{-5}y^{-1}}$ **143.** $\dfrac{9^{-2}x^{-3}y^{-5}}{36^{-1}x^{-6}y^2}$ **144.** $\left(\dfrac{14x^{-2}}{21x^{-4}}\right)^{-3}$

145. $\left(\dfrac{x^{-5}y^2}{x^{-2}y^{-6}}\right)^{-2}$ **146.** $\left(\dfrac{x^{-3}y^{-2}}{x^{-4}y^3}\right)^{-3}$ **147.** $\left(\dfrac{2x^{-2}y^2}{3xy^{-3}}\right)^{-1}$

148. $\left(\dfrac{2x^{-1}y^{-3}}{3x^2y^{-5}}\right)^{-2}$ **149.** $\left(\dfrac{3x^{-1}y^{-1}}{2x^{-3}y^2}\right)^{-3}$ **150.** $\left(\dfrac{8x^{-5}y}{12x^{-2}y^{-2}}\right)^{-2}$

151. $\left(\dfrac{4x^{-2}y^{-3}}{6x^{-1}y^{-4}}\right)^{-2}$ **152.** $\left(\dfrac{3^{-1}x^{-2}y^{-5}}{9^{-1}x^{-1}y^{-3}}\right)^{-2}$ **153.** $\left(\dfrac{x^{-10}y^8z^{-4}}{x^{-12}y^{-2}z^2}\right)^{-\frac{1}{2}}$

154. $\left(\dfrac{16x^8y^{-4}}{81x^{-4}y^4}\right)^{-\frac{1}{4}}$ **155.** $\left(\dfrac{27x^{-3}y^{-6}}{8^{-1}x^6y^{-12}}\right)^{-\frac{1}{3}}$ **156.** $\dfrac{(2x^{-2}y^2z^{-1})^{-2}}{(3xy^{-2}z^{-1})^{-3}}$

157. $\dfrac{(4x^{-1}y^{-2}z)^{-4}}{(2x^2y^{-1}z^{-3})^{-3}}$ **158.** $\dfrac{(6^{-1}x^{-3}y^{-1}z^{-2})^{-3}}{(12^{-1}x^{-2}yz^{-2})^{-4}}$

159. $\dfrac{(14^{-1}x^{-2}yz^{-3})^{-5}}{(21^{-1}x^{-3}y^{-1}z^{-2})^{-4}}$ **160.** $\left(\dfrac{x^{-2}y^2}{-a^{-3}b}\right)\left(\dfrac{x^{-2}y^{-1}}{a^{-2}b^2}\right)^{-2}$

161. $\left(\dfrac{6a^3b^{-3}}{x^2y^{-1}}\right)^{-2}\left(\dfrac{x^{-3}y^2}{4a^4b^{-2}}\right)^{-3}$ **162.** $\left(\dfrac{2^2a^{-6}b^{-2}}{x^{-2}y^4}\right)^{-\frac{1}{2}}\left(\dfrac{xy^{-1}}{a^{-3}b^2}\right)^{-1}$

163. $\left(\dfrac{2^{-3}a^6b^{-9}}{x^3y^{-6}}\right)^{-\frac{1}{3}}\left(\dfrac{6x^{-1}y^2}{27a^{-2}b^{-3}}\right)^{-1}$ **164.** $(2x^{-1} + y^{-1})^{-1}$

165. $(3x^{-1} - y^{-2})^{-1}$ **166.** $(x^{-2} - 5y^{-2})^{-1}$ **167.** $\dfrac{x^{-1} + y^{-1}}{(xy)^{-1}}$

168. $\dfrac{x^{-1} - 3y^{-1}}{(y - 3x)^{-1}}$ **169.** $\dfrac{(x^{-2} + y^{-1})^{-1}}{(x^2 + y)^{-1}}$ **170.** $\dfrac{2x^{-1} - 3y^{-1}}{2x^{-1} + 3y^{-1}}$

171. $\dfrac{27x^{-2} + 9x^{-3}}{27x^{-2} + 81x^{-1}}$ **172.** $\dfrac{x^{-2} - y^{-2}}{x^{-4} - y^{-4}}$ **173.** $\dfrac{4 - x^{-2}}{2x^{-1} + x^{-2}}$

174. $\dfrac{x^{-2} - 9}{3x^{-2} - 9x^{-1}}$ **175.** $\dfrac{1 + x^{-1} - 6x^{-2}}{2 + x^{-1} - 15x^{-2}}$ **176.** $\dfrac{2 + x^{-1} - x^{-2}}{3 + 4x^{-1} + x^{-2}}$

177. $\dfrac{1 + 2x^{-1} - 8x^{-2}}{1 + x^{-1} - 12x^{-2}}$ **178.** $\dfrac{2 + 7x^{-1} - 4x^{-2}}{2 + 11x^{-1} - 6x^{-2}}$ **179.** $\dfrac{x^{-3} + 1}{x^{-1} + 1}$

180. $\dfrac{x^{-3} - 1}{x^{-1} - 1}$ **181.** $\dfrac{64x^{-6} + y^{-6}}{4x^{-2} + y^{-2}}$

182. $(x^{-1} - 2)^{-1}(1 - 2x) + (x^{-1} + 2)^{-1}(1 + 2x)$

183. $(1 + x^{-1})^{-1}(1 - x^{-1}) + (x + 1)^{-1}$

184. $[(x^2 + 2x^{-2})^2 - 4]^{-1}(x^4 + 4x^{-4})$

185. $[(x - 3x^{-1})^2 + 6]^{-1}(x^3 + 9x^{-1})$

186. $[(3x^{-1} + x)^2 - 6]^{-1}(x^2 + 9x^{-2})$

Write each of the following as one fraction with positive exponents, and simplify:

187. $\dfrac{(x-1)^{\frac{1}{2}} - \frac{1}{2}x(x-1)^{-\frac{1}{2}}}{(x-1)}$

188. $\dfrac{(x-3)^{\frac{1}{2}} - \frac{1}{2}(x+3)(x-3)^{-\frac{1}{2}}}{(x-3)}$

189. $\dfrac{(x+2)^{\frac{1}{3}} - \frac{1}{3}x(x+2)^{-\frac{2}{3}}}{(x+2)^{\frac{2}{3}}}$

190. $\dfrac{3(x-1)^{\frac{1}{3}} - x(x-1)^{-\frac{2}{3}}}{(x-1)^{\frac{2}{3}}}$

191. $\dfrac{(x+4)^{\frac{1}{3}} - \frac{1}{3}(x+3)(x+4)^{-\frac{2}{3}}}{(x+4)^{\frac{2}{3}}}$

192. $\dfrac{(x^2+1)^{\frac{1}{2}} - x^2(x^2+1)^{-\frac{1}{2}}}{(x^2+1)}$

193. $\dfrac{x^2(x^2+3)^{-\frac{1}{2}} - (x^2+3)^{\frac{1}{2}}}{x^2}$

194. $\dfrac{2x(4-x^2)^{\frac{1}{2}} + x^3(4-x^2)^{-\frac{1}{2}}}{(4-x^2)}$

195. $\dfrac{2x(3x^2-1)^{\frac{1}{2}} - 3x^3(3x^2-1)^{-\frac{1}{2}}}{(3x^2-1)}$

196. $\dfrac{2x(x^2+1)^{\frac{1}{3}} - 2x^3(x^2+1)^{-\frac{2}{3}}}{(x^2+1)^{\frac{2}{3}}}$

197. $\dfrac{\frac{1}{3}x^2(x-1)^{-\frac{2}{3}} - x(x-1)^{\frac{1}{3}}}{x^4}$

198. $\dfrac{\frac{1}{2}x^{\frac{1}{2}}(x+2)^{-\frac{1}{2}} - \frac{1}{2}x^{-\frac{1}{2}}(x+2)^{\frac{1}{2}}}{x}$

199. $\dfrac{\frac{1}{2}x^{-\frac{1}{2}}(x+4)^{\frac{1}{2}} - \frac{1}{2}x^{\frac{1}{2}}(x+4)^{-\frac{1}{2}}}{(x+4)}$

200. $\dfrac{\frac{1}{3}x^{-\frac{2}{3}}(x+8)^{\frac{1}{3}} - \frac{1}{3}x^{\frac{1}{3}}(x+8)^{-\frac{2}{3}}}{(x+8)^{\frac{2}{3}}}$

201. $\dfrac{\frac{2}{3}x^{-\frac{1}{3}}(x^2+1)^{\frac{1}{3}} - \frac{2}{3}x^{\frac{5}{3}}(x^2+1)^{-\frac{2}{3}}}{(x^2+1)^{\frac{2}{3}}}$

202. $\dfrac{\frac{1}{2}(x-1)^{\frac{1}{2}}(x+1)^{-\frac{1}{2}} - \frac{1}{2}(x+1)^{\frac{1}{2}}(x-1)^{-\frac{1}{2}}}{(x-1)}$

203. $\dfrac{\frac{1}{2}(x+4)^{\frac{1}{2}}(x-2)^{-\frac{1}{2}} - \frac{1}{2}(x-2)^{\frac{1}{2}}(x+4)^{-\frac{1}{2}}}{(x+4)}$

204. $\dfrac{\frac{1}{2}(x-3)^{\frac{1}{2}}(x+5)^{-\frac{1}{2}} - \frac{1}{2}(x+5)^{\frac{1}{2}}(x-3)^{-\frac{1}{2}}}{(x-3)}$

205. $\dfrac{x(x^2-3)^{\frac{1}{2}}(x^2+1)^{-\frac{1}{2}} - x(x^2+1)^{\frac{1}{2}}(x^2-3)^{-\frac{1}{2}}}{(x^2-3)}$

206. $\dfrac{\frac{1}{3}(x+4)^{\frac{1}{3}}(x+2)^{-\frac{2}{3}} - \frac{1}{3}(x+2)^{\frac{1}{3}}(x+4)^{-\frac{2}{3}}}{(x+4)^{\frac{2}{3}}}$

Write the following numbers in scientific notation:

207. 27 **208.** 354 **209.** 4500 **210.** 7.9
211. 1.3 **212.** .84 **213.** .921 **214.** .032
215. .074 **216.** .006 **217.** .0049 **218.** .00012

Chapter 7 Review

Evaluate each of the following:

1. $(625)^{\frac{1}{4}}$ **2.** $(225)^{\frac{1}{2}}$ **3.** $(27)^{\frac{2}{3}}$ **4.** $(32)^{\frac{4}{5}}$

5. $(16)^{\frac{3}{2}}$ **6.** $(81)^{\frac{3}{4}}$ **7.** $(4)^{\frac{5}{2}}$ **8.** $(36)^{\frac{3}{2}}$

Perform the indicated operations and simplify:

9. $a^2 \cdot a^{\frac{1}{3}}$ **10.** $a^{\frac{1}{3}} \cdot a^{\frac{1}{6}}$ **11.** $a^{\frac{1}{4}} \cdot a^{\frac{1}{2}}$ **12.** $a^{\frac{3}{4}} \cdot a^{\frac{1}{8}}$

13. $(a^5)^{\frac{3}{5}}$ **14.** $(a^6)^{\frac{2}{3}}$ **15.** $\left(-a^{\frac{1}{3}}\right)^6$ **16.** $\left(-a^{\frac{3}{2}}\right)^4$

17. $\left(-a^{\frac{2}{3}}\right)^3$ **18.** $\left(a^{\frac{3}{4}}\right)^2$ **19.** $\left(a^{\frac{3}{8}}\right)^{\frac{8}{9}}$ **20.** $\left(a^{\frac{2}{3}}\right)^{\frac{3}{4}}$

21. $a^3b^{\frac{3}{2}} \cdot a^{\frac{1}{2}}b^2$ **22.** $a^{\frac{1}{2}}b^{\frac{3}{4}} \cdot a^{\frac{3}{2}}b^{\frac{1}{4}}$ **23.** $a^{\frac{3}{7}}b^{\frac{2}{5}} \cdot a^{\frac{4}{7}}b^{\frac{3}{5}}$

24. $a^{\frac{1}{3}}b^{\frac{1}{6}} \cdot a^{\frac{1}{9}}b^{\frac{7}{12}}$ **25.** $a^{\frac{1}{3}}b^{\frac{2}{3}}c \cdot a^{\frac{5}{3}}b^{\frac{7}{3}}c^3$ **26.** $a^{\frac{5}{6}}b^{\frac{4}{7}}c^{\frac{1}{8}} \cdot a^{\frac{1}{2}}b^{\frac{10}{7}}c^{\frac{3}{4}}$

27. $2a^{\frac{1}{2}}b^{\frac{2}{5}}c \cdot 3a^{\frac{1}{3}}b^{\frac{1}{4}}c^{\frac{1}{6}}$ **28.** $\left(3a^{\frac{1}{2}}b^{\frac{3}{4}}\right)^4\left(a^{\frac{1}{3}}b^{\frac{1}{6}}c\right)^3$ **29.** $\left(2a^{\frac{1}{6}}b^{\frac{1}{4}}\right)^2\left(a^{\frac{2}{3}}b^{\frac{1}{2}}\right)^6$

30. $\left(a^{\frac{1}{6}}b^{\frac{1}{4}}c^{\frac{1}{3}}\right)^{12}\left(a^{\frac{1}{3}}b^{\frac{3}{4}}c^{\frac{5}{6}}\right)^6$ **31.** $\left(16^{\frac{1}{8}}a^{\frac{5}{8}}b^{\frac{1}{2}}\right)^4\left(32^{\frac{2}{5}}a^{\frac{1}{3}}b^{\frac{7}{6}}\right)^3$

32. $(16a^8b^4)^{\frac{1}{4}}(27b^6c^3)^{\frac{1}{3}}$ **33.** $(25x^4y^2)^{\frac{1}{4}}(x^3y^6z^9)^{\frac{1}{6}}$

34. $(8a^3b^6)^{\frac{1}{3}}(81a^8b^4)^{\frac{3}{4}}$ **35.** $(25a^8b^2)^{\frac{3}{2}}(64a^6b^{12})^{\frac{1}{6}}$

36. $\left(45a^{\frac{8}{5}}b^{\frac{2}{9}}\right)^{\frac{3}{2}}\left(20a^{\frac{6}{5}}b^{\frac{4}{3}}\right)^{\frac{1}{2}}$ **37.** $a^3\left(-4a^{\frac{1}{4}}\right) + 2a^2\left(a^{\frac{5}{4}}\right)$

38. $a\left(2a^{\frac{2}{3}}\right) - a^{\frac{1}{3}}\left(3a^{\frac{4}{3}}\right)$

39. $a^2\left(a^{\frac{3}{2}}\right) - a^3\left(a^{\frac{1}{2}}\right)$

40. $2a^2\left(-a^{\frac{1}{2}}\right) - a\left(a^{\frac{1}{2}}\right)$

41. $x\left(2x^{\frac{3}{2}}\right) - 3x^2\left(-x^{\frac{1}{2}}\right)$

42. $\left(a^{\frac{2}{3}}\right)^6\left(a^{\frac{1}{4}}\right)^4 - \left(a^{\frac{3}{5}}\right)^5\left(a^{\frac{1}{3}}\right)^6$

43. $\left(3x^{\frac{1}{2}} - 2\right)\left(5x^{\frac{1}{2}} + 1\right)$

44. $\left(2x^{\frac{1}{4}} - 3y^{\frac{1}{2}}\right)\left(3x^{\frac{1}{4}} + y^{\frac{1}{2}}\right)$

45. $\left(4x^{\frac{1}{6}} + y^{\frac{1}{3}}\right)\left(6x^{\frac{1}{6}} - y^{\frac{1}{3}}\right)$

46. $\left(x^{\frac{1}{2}} - x^{\frac{1}{4}} + 2\right)\left(x^{\frac{1}{2}} - x^{\frac{1}{4}} - 2\right)$

47. $\left(x^{\frac{1}{3}} + 2x^{\frac{1}{6}} - 4\right)\left(x^{\frac{1}{3}} - 2x^{\frac{1}{6}} - 4\right)$

48. $\left(x^{\frac{1}{2}} + 2y^{\frac{1}{2}}\right)^2 - \left(2x^{\frac{1}{2}} - y^{\frac{1}{2}}\right)^2$

49. $a^{\frac{1}{6}} \div a^{\frac{1}{3}}$

50. $a^{\frac{5}{8}} \div a^{\frac{1}{4}}$

51. $a^{\frac{4}{9}} \div a^{\frac{2}{3}}$

52. $a^{\frac{7}{3}} \div a$

53. $\left(a^{\frac{5}{2}} \div b^{\frac{2}{3}}\right)^6$

54. $\left(a^{\frac{7}{4}} \div b^{\frac{3}{8}}\right)^8$

55. $(a^9 \div b^6)^{\frac{2}{3}}$

56. $(a^8 \div b^{12})^{\frac{3}{4}}$

57. $\left(a^{\frac{3}{5}} \div b^{\frac{6}{5}}\right)^{\frac{5}{12}}$

58. $\left(a^{\frac{4}{9}} \div b^{\frac{2}{3}}\right)^{\frac{3}{4}}$

59. $\left(a^{\frac{3}{7}}b^{\frac{4}{5}}\right) \div \left(a^{\frac{2}{7}}b^{\frac{3}{5}}\right)$

60. $\left(a^{\frac{3}{4}}b^{\frac{1}{6}}\right) \div \left(a^{\frac{3}{8}}b^{\frac{1}{12}}\right)$

61. $\left(6a^{\frac{2}{3}}b^{\frac{1}{2}}c\right) \div \left(15a^{\frac{5}{6}}b^{\frac{3}{4}}c^3\right)$

62. $\dfrac{a^{\frac{3}{8}}b^{\frac{2}{5}}c^{\frac{4}{3}}}{a^{\frac{3}{4}}b^{\frac{7}{5}}c^{\frac{1}{3}}}$

63. $\dfrac{(36a^4b^2)^{\frac{3}{2}}}{(64a^6b^3)^{\frac{2}{3}}}$

64. $\dfrac{(32a^5b^{10})^{\frac{3}{5}}}{(16a^8b^4)^{\frac{3}{4}}}$

65. $\dfrac{(a^4b^3c)^{\frac{2}{3}}}{(a^2b^3c^5)^{\frac{1}{3}}}$

66. $\dfrac{\left(2a^{\frac{2}{3}}b^{\frac{5}{6}}\right)^6}{\left(4a^{\frac{3}{2}}b^{\frac{1}{2}}\right)^4}$

67. $\dfrac{\left(3a^{\frac{3}{4}}b^{\frac{1}{3}}\right)^{12}}{\left(6a^{\frac{5}{3}}b^{\frac{1}{2}}\right)^6}$

68. $\dfrac{\left(a^{\frac{2}{3}}b^{\frac{3}{4}}c^{\frac{1}{2}}\right)^6}{\left(a^{\frac{1}{6}}b^{\frac{3}{8}}c^{\frac{5}{12}}\right)^{12}}$

69. $\left(3x - 2x^{\frac{1}{2}}y^{\frac{1}{2}} - 8y\right) \div \left(x^{\frac{1}{2}} - 2y^{\frac{1}{2}}\right)$

70. $\left(12x^{\frac{2}{3}} - 16x^{\frac{1}{3}}y^{\frac{1}{3}} - 3y^{\frac{2}{3}}\right) \div \left(6x^{\frac{1}{3}} + y^{\frac{1}{3}}\right)$

71. $\left(x^2 + x^{\frac{3}{2}} - 5x + 3x^{\frac{1}{2}}\right) \div \left(x^{\frac{1}{2}} - 1\right)$

72. $\left(2x^2 - 5x^{\frac{4}{3}} + 3x^{\frac{2}{3}} - 2\right) \div \left(x^{\frac{2}{3}} - 2\right)$

73. $\left(x + 4x^{\frac{1}{2}} + 16\right) \div \left(x^{\frac{1}{2}} - 2x^{\frac{1}{4}} + 4\right)$

74. $\left(x^2 - x^{\frac{3}{2}} - 8x + 11x^{\frac{1}{2}} - 3\right) \div \left(x + 2x^{\frac{1}{2}} - 3\right)$

Simplify and write the answers with positive exponents:

75. $(3a^{-1}b^2c^{-3})(3^{-2}a^{-2}b^{-1}c^2)$

76. $(4^{-1}a^2b^{-4}c^{-3})(8^{-2}a^{-1}b^2c^4)$

77. $(8^{-1}a^{-1}bc^{-2})^{-1}(4^{-2}a^3b^{-5}c^{-1})^2$

78. $(10a^2b^{-3}c^{-4})^{-2}(15^{-1}a^{-6}b^{-1}c^5)^{-3}$

79. $(3a^{-2}b^{-1}c)^{-3}(2^{-2}ab^{-3}c^{2})^{2}$

80. $(4a^{-6}b^{4}c^{-2})^{\frac{1}{2}}(27a^{-3}b^{-6}c^{9})^{-\frac{1}{3}}$

81. $(8a^{-3}b^{3}c^{-6})^{-\frac{1}{3}}(a^{-4}b^{8}c^{-8})^{\frac{3}{4}}$

82. $(9a^{4}b^{-1}c^{-2})^{\frac{3}{2}}(36a^{-2}bc^{-4})^{-\frac{5}{2}}$

83. $(8a^{-3}b^{6}c^{3})^{-\frac{2}{3}}(81^{-1}a^{2}b^{-4}c^{2})^{-\frac{1}{2}}$

84. $(25a^{-4}b^{2})^{-\frac{1}{6}}(625a^{-6}b^{9}c^{-3})^{\frac{1}{3}}$

85. $(3x^{-1}-y^{-1})(x^{-1}+2y^{-1})$

86. $(a^{-1}+b^{-2})(a^{-2}-a^{-1}b^{-2}+a^{-2}b^{-4})$

87. $(3^{-1}x^{-1}-2y^{-1})^{2}$

88. $\left(4x^{-\frac{1}{2}}-y^{-\frac{1}{2}}\right)^{2}$

89. $\left(a^{-\frac{4}{3}}-b^{-\frac{4}{3}}\right)^{3}$

90. $\dfrac{14^{-2}a^{-3}b^{2}}{21^{-1}a^{-2}b^{-3}}$

91. $\left(\dfrac{4^{-2}a^{2}b^{-7}}{6^{-1}a^{-4}b^{9}}\right)^{-2}$

92. $\left(\dfrac{a^{-6}b^{2}c^{-2}}{a^{-3}b^{-1}c^{-4}}\right)^{-3}$

93. $\dfrac{(6^{-2}a^{2}b^{-3})^{2}}{(12a^{-2}b^{2})^{-3}}$

94. $\dfrac{(72a^{-4}b^{-3}c)^{-2}}{(36^{2}a^{2}b^{-4}c^{-2})^{-\frac{3}{2}}}$

95. $\dfrac{(125a^{7}b^{-5})^{-\frac{1}{3}}}{(64a^{2}b^{8})^{-\frac{1}{6}}}$

96. $\left[\dfrac{(16)^{-\frac{3n}{4}}(4)^{\frac{n}{2}}}{(64)^{\frac{n}{6}}(32)^{\frac{n}{5}}}\right]^{\frac{1}{n}}$

97. $\left[\dfrac{(27)^{\frac{2}{3n}}(81)^{-\frac{3}{4n}}}{(729)^{\frac{5}{6n}}(9)^{-\frac{1}{n}}}\right]^{\frac{n}{2}}$

98. $\left[\dfrac{(32)^{-\frac{n}{5}}(8)^{\frac{2n}{3}}}{(128)^{\frac{3n}{7}}(16)^{-\frac{n}{2}}}\right]^{-\frac{3}{n}}$

99. $\left[\dfrac{(81)^{-\frac{3n}{4}}(27)^{\frac{2n}{3}}}{(243)^{-\frac{n}{5}}(9)^{\frac{7n}{2}}}\right]^{\frac{2}{n}}$

100. $\dfrac{6x^{-2}-3x^{-1}}{8x^{-3}-4x^{-2}}$

101. $\dfrac{4x^{-2}-9}{2x^{-2}-3x^{-1}}$

102. $\dfrac{2x^{-2}+5x^{-1}-3}{3x^{-2}+8x^{-1}-3}$

103. $\dfrac{25x^{-2}-1}{15x^{-2}+17x^{-1}-4}$

104. $\dfrac{4x^{-2}-15x^{-1}-4}{2x^{-2}-7x^{-1}-4}$

105. $\dfrac{9x^{-2}+3x^{-1}-2}{3x^{-2}+14x^{-1}+8}$

106. $[(x+x^{-1})^{2}-2]^{-1}(x^{2}+x^{-2})$

107. $[(3x-2x^{-1})^{2}+12]^{-1}(9x^{2}+4x^{-2})$

Write each of the following as one fraction with positive exponents and simplify:

108. $\dfrac{(x+1)^{\frac{1}{2}}-\frac{1}{2}x(x+1)^{-\frac{1}{2}}}{(x+1)}$

109. $\dfrac{(x-2)^{\frac{1}{2}}-\frac{1}{2}x(x-2)^{-\frac{1}{2}}}{(x-2)}$

110. $\dfrac{(2x-1)^{\frac{1}{3}}-\frac{2}{3}x(2x-1)^{-\frac{2}{3}}}{(2x-1)^{\frac{2}{3}}}$

111. $\dfrac{(4x-3)^{\frac{1}{3}}-\frac{4}{3}x(4x-3)^{-\frac{2}{3}}}{(4x-3)^{\frac{2}{3}}}$

112. $\dfrac{(x^{2}-2)^{\frac{1}{2}}-x^{2}(x^{2}-2)^{-\frac{1}{2}}}{(x^{2}-2)}$

113. $\dfrac{(x^{2}+3)^{\frac{1}{2}}-x^{2}(x^{2}+3)^{-\frac{1}{2}}}{(x^{2}+3)}$

chapter 8

Radical Expressions

8.1 Definitions and notation

8.2 Standard form of radicals

8.3 Combination of radical expressions

8.4 Multiplication of radical expressions

8.5 Division of radical expressions

8.6 Equations involving radical expressions

8.1 Definitions and Notation

The nth powers of $3, a, b^2$, and c^3 are respectively $3^n, a^n, b^{2n}$, and c^{3n}.

When n is an even number, the nth power of a positive or a negative number is a positive number. For example,

$$(+3)^4 = 81 \quad \text{and} \quad (-3)^4 = 81$$

When n is an odd number, the nth power of a positive number is a positive number, and the nth power of a negative number is a negative number. For example,

$$(+2)^5 = 32 \quad \text{and} \quad (-2)^5 = -32$$

DEFINITION

The **nth root** of a number a is denoted by $\sqrt[n]{a}$. It is a number whose nth power is a, that is, $(\sqrt[n]{a})^n = a$, with the following conditions:

1. When n is an even number and $a > 0$, $\sqrt[n]{a}$ is a positive number, called the **principal root**.
   When n is an even number and $a < 0$, $\sqrt[n]{a}$ is not a real number.

2. When n is an odd number and $a > 0$, $\sqrt[n]{a}$ is a positive number.
   When n is an odd number and $a < 0$, $\sqrt[n]{a}$ is a negative number.

The n in $\sqrt[n]{a}$ (always a natural number greater than 1) is called the **index** or the **order of the radical**, and a is called the **radicand**. When there is no indicated index, as in $\sqrt{a}$, the index 2 is implied and is read "the **square root** of a." When the index is 3, as in $\sqrt[3]{a}$, it is read "the **cube root** of a."

The expression $\sqrt[n]{a^m}$ is defined as $(\sqrt[n]{a})^m$, providing $\sqrt[n]{a}$ is defined.

EXAMPLES

1. $\sqrt{16} = \sqrt{(4)^2} = (\sqrt{4})^2 = 4$

2. $\sqrt{-25}$ is not a real number

3. $\sqrt[3]{8} = \sqrt[3]{(2)^3} = (\sqrt[3]{2})^3 = 2$

4. $\sqrt[3]{-216} = \sqrt[3]{(-6)^3} = (\sqrt[3]{-6})^3 = -6$

From the definition of the radical and the definition of fractional exponents, for $a \in R$ and $m, n \in N$, we have

$$\sqrt[n]{a} = a^{\frac{1}{n}} \quad \text{and} \quad \sqrt[n]{a^m} = a^{\frac{m}{n}}$$

providing $\sqrt[n]{a}$ and $a^{\frac{1}{n}}$ are defined.

The relations just discussed enable us to express radicals as fractional exponents and fractional exponents as radicals.

EXAMPLES

1. $\sqrt[3]{6} = 6^{\frac{1}{3}}$
2. $\sqrt[5]{3^2} = 3^{\frac{2}{5}}$
3. $\sqrt{x-5} = (x-5)^{\frac{1}{2}}$
4. $x^{\frac{2}{3}} = \sqrt[3]{x^2}$
5. $2x^{\frac{5}{7}} = 2\sqrt[7]{x^5}$
6. $x^{\frac{2}{3}}y^{\frac{1}{4}} = \sqrt[3]{x^2}\,\sqrt[4]{y}$

When the value of a radical expression is a rational number, we say it is a **perfect root**. Since $\sqrt[n]{a^{nk}} = a^k$, a radical expression is a perfect root if the radicand can be expressed as a product of factors, each to an exponent that is an integral multiple of the radical index.

The value of the radical is obtained by forming the product of the factors, where the exponent of each factor is its original exponent divided by the radical index.

EXAMPLES

1. $\sqrt{729} = \sqrt{3^6} = 3^{\frac{6}{2}} = 3^3 = 27$
2. $\sqrt[3]{64x^9y^{15}} = \sqrt[3]{2^6x^9y^{15}} = 2^{\frac{6}{3}}x^{\frac{9}{3}}y^{\frac{15}{3}} = 2^2x^3y^5 = 4x^3y^5$

Nonperfect roots such as $\sqrt{2}$, $\sqrt[3]{2}$, $\sqrt{3}$, $\sqrt[4]{5}$, $\sqrt[5]{4}$, $1+\sqrt{2}$, and $5-\sqrt[3]{9}$ are **irrational numbers**.

Notes

1. Since $a^{\frac{m}{n}} = a^{\frac{mk}{nk}}$ for all $a \in R$, $a > 0$, and $m, n \in N$, $k \in Q$, $k > 0$, we have $\sqrt[n]{a^m} = \sqrt[nk]{a^{mk}}$ provided nk and $mk \in N$:

   $\sqrt[3]{a} = \sqrt[6]{a^2}$

   $\sqrt[4]{a^2} = \sqrt{a}$

2. $1^n = 1$ and $\sqrt[n]{1} = 1$

8.2 Standard Form of Radicals

THEOREM If $a, b \in R$, $a > 0$, $b > 0$, and $n \in N$, then $\sqrt[n]{ab} = \sqrt[n]{a}\,\sqrt[n]{b}$.

Proof $\sqrt[n]{ab} = (ab)^{\frac{1}{n}} = a^{\frac{1}{n}}b^{\frac{1}{n}} = \sqrt[n]{a}\,\sqrt[n]{b}$

EXAMPLES

1. $\sqrt[3]{32} = \sqrt[3]{2^5} = \sqrt[3]{2^3 \cdot 2^2}$
   $= \sqrt[3]{2^3}\,\sqrt[3]{2^2} = 2\sqrt[3]{4}$
2. $\sqrt{9a^5b} = \sqrt{3^2a^5b} = \sqrt{3^2(a^4 \cdot a)b}$
   $= \sqrt{3^2a^4}\,\sqrt{ab} = 3a^2\sqrt{ab}$

simplicant

The expression $3a^2\sqrt{ab}$ is called the **standard form** of the expression $\sqrt{9a^5b}$.

A radical expression is said to be in standard form if the following are true:

1. The radicand is positive.
2. The radical index is as small as possible.
3. The exponent of each factor of the radicand is a natural number less than the radical index.
4. There are no fractions in the radicand.
5. There are no radicals in the denominator of a fraction.

By **simplifying a radical** expression, we mean putting the radical expression in standard form.

We will now discuss the first three conditions necessary for a radical to be in standard form. The last two conditions are discussed in Section 8.5.

When the radicand is negative we have from the definition of radicals:

1. When n is even and $a > 0$, $\sqrt[n]{-a}$ is not a real number.
2. When n is odd and $a > 0$, $\sqrt[n]{-a} = -\sqrt[n]{a}$.

EXAMPLES 1. $\sqrt[3]{-4} = -\sqrt[3]{4}$ 2. $\sqrt[5]{-ab^2} = -\sqrt[5]{ab^2}$

When the radical index and the exponents of all the factors in the radicand have a common factor, divide both the radical index and the exponents of the factors of the radicand by their common factor.
That is, apply $\sqrt[nk]{a^{mk}} = \sqrt[n]{a^m}$ to obtain the smallest possible radical index.

EXAMPLE $\sqrt[6]{a^4b^2} = \sqrt[3]{a^2b}$

When the exponents of some factors of the radicand are greater than the radical index, but not an integral multiple of it, write each of these factors as a product of two factors: one factor with an exponent that is an integral multiple of the radical index, and the other factor with an exponent that is less than the radical index.

EXAMPLE $\sqrt[3]{a^7} = \sqrt[3]{a^6 \cdot a}$

Then apply the theorem $\sqrt[n]{ab} = \sqrt[n]{a}\,\sqrt[n]{b}$. Write the factors with exponents that are integral multiples of the index under one radical, thus obtaining a perfect root; write the other factors with exponents less than the radical index under the other radical.

EXAMPLE $\sqrt[3]{a^7} = \sqrt[3]{a^6 \cdot a} = \sqrt[3]{a^6}\,\sqrt[3]{a} = a^2\sqrt[3]{a}$

The cases when there are fractions in the radicand and radicals in the denominator of a fraction are discussed later (Section 8.5).

EXAMPLE Put $\sqrt{a^3b^2c^5}$ in standard form.

Solution
$$\begin{aligned}\sqrt{a^3b^2c^5} &= \sqrt{(a^2 \cdot a)b^2(c^4 \cdot c)}\\ &= \sqrt{a^2b^2c^4}\,\sqrt{ac}\\ &= abc^2\sqrt{ac}\end{aligned}$$

EXAMPLE Simplify $\sqrt[3]{-81a^4b^6}$.

Solution
$$\begin{aligned}\sqrt[3]{-81a^4b^6} &= -\sqrt[3]{3^4a^4b^6} = -\sqrt[3]{(3^3 \cdot 3)(a^3 \cdot a)b^6}\\ &= -\sqrt[3]{3^3a^3b^6}\,\sqrt[3]{3a}\\ &= -3ab^2\sqrt[3]{3a}\end{aligned}$$

EXAMPLE Simplify $\sqrt[4]{a^8b^{10}c^6}$.

Solution
$$\begin{aligned}\sqrt[4]{a^8b^{10}c^6} &= \sqrt[4]{a^8(b^8 \cdot b^2)(c^4 \cdot c^2)}\\ &= \sqrt[4]{a^8b^8c^4}\,\sqrt[4]{b^2c^2}\\ &= a^2b^2c\sqrt{bc}\end{aligned}$$

EXAMPLE Simplify $\sqrt[3n]{a^{4n}b^{11n}}$, given that $n \in N$.

Solution
$$\sqrt[3n]{a^{4n}b^{11n}} = \sqrt[3]{a^4b^{11}} = ab^3\sqrt[3]{ab^2}$$

EXAMPLE Simplify $\sqrt[n]{3^{2n}a^{n+1}b^{n+2}}$, given that $n > 2, n \in N$.

Solution
$$\begin{aligned}\sqrt[n]{3^{2n}a^{n+1}b^{n+2}} &= \sqrt[n]{3^{2n}(a^n \cdot a)(b^n \cdot b^2)}\\ &= \sqrt[n]{3^{2n}a^nb^n}\,\sqrt[n]{ab^2}\\ &= 3^2ab\sqrt[n]{ab^2}\\ &= 9ab\sqrt[n]{ab^2}\end{aligned}$$

EXAMPLE Put $\sqrt[3]{8x^3y - 8x^3}$ in standard form.

Solution
$$\sqrt[3]{8x^3y - 8x^3} = \sqrt[3]{8x^3(y-1)} = 2x\sqrt[3]{y-1}$$

Notes

1. $\sqrt[n]{a^nb^n} = ab$
2. $\sqrt[n]{(a+b)^n} = (a+b)$
3. $\sqrt[n]{a^n + b^n} \neq a + b$ since $(a+b)^n \neq a^n + b^n$

Exercises 8.1–8.2

Put the following radicals in standard form given that $n > 2, n \in N$:

1. $\sqrt{9}$	**2.** $\sqrt{36}$	**3.** $\sqrt{-64}$	**4.** $\sqrt{-81}$
5. $\sqrt[3]{8}$	**6.** $\sqrt[3]{27}$	**7.** $\sqrt[3]{-125}$	**8.** $\sqrt[3]{-216}$
9. $\sqrt[4]{16}$	**10.** $\sqrt[4]{81}$	**11.** $\sqrt[5]{32}$	**12.** $\sqrt[6]{64}$
13. $\sqrt[n]{2^n}$	**14.** $\sqrt[n]{3^{2n}}$	**15.** $\sqrt{x^2}$	**16.** $\sqrt{x^6y^4}$
17. $\sqrt{x^8y^2}$	**18.** $\sqrt{4x^{10}}$	**19.** $\sqrt{(x+2)^2}$	**20.** $\sqrt{(x-1)^2}$
21. $\sqrt[3]{x^3y^6}$	**22.** $\sqrt[3]{-x^6y^3}$	**23.** $\sqrt[3]{x^{12}y^9}$	**24.** $\sqrt[3]{(x-3)^3}$
25. $\sqrt[4]{x^{16}}$	**26.** $\sqrt[5]{x^{10}}$	**27.** $\sqrt[5]{-x^{25}}$	**28.** $\sqrt[3n]{x^{9n}}$
29. $\sqrt[4]{25}$	**30.** $\sqrt[4]{9}$	**31.** $\sqrt[4]{121}$	**32.** $\sqrt[6]{9}$
33. $\sqrt[6]{49}$	**34.** $\sqrt[6]{8}$	**35.** $\sqrt[6]{125}$	**36.** $\sqrt{8}$
37. $\sqrt{12}$	**38.** $5\sqrt{20}$	**39.** $\sqrt{24}$	**40.** $2\sqrt{28}$
41. $3\sqrt{32}$	**42.** $\sqrt{45}$	**43.** $\sqrt{48}$	**44.** $\sqrt{54}$
45. $2\sqrt{60}$	**46.** $-3\sqrt{63}$	**47.** $-\sqrt{72}$	**48.** $-2\sqrt{75}$
49. $-3\sqrt{80}$	**50.** $\sqrt{96}$	**51.** $\sqrt{108}$	**52.** $\sqrt{128}$
53. $\sqrt{16+9}$	**54.** $\sqrt{25+9}$	**55.** $\sqrt{49-9}$	**56.** $\sqrt{169-25}$
57. $\sqrt[3]{32}$	**58.** $\sqrt[3]{-24}$	**59.** $\sqrt[3]{-54}$	**60.** $\sqrt[3]{-72}$
61. $\sqrt[3]{81}$	**62.** $\sqrt[3]{128}$	**63.** $\sqrt[3]{250}$	**64.** $\sqrt[3]{432}$
65. $\sqrt[3]{8+27}$	**66.** $\sqrt[3]{125+8}$	**67.** $\sqrt[3]{27-8}$	**68.** $\sqrt[3]{27-125}$
69. $\sqrt[4]{32}$	**70.** $\sqrt[4]{64}$	**71.** $\sqrt[4]{128}$	**72.** $\sqrt[4]{144}$
73. $\sqrt[4]{162}$	**74.** $\sqrt[5]{-64}$	**75.** $\sqrt[5]{-96}$	**76.** $\sqrt[5]{128}$
77. $\sqrt[6]{256}$	**78.** $\sqrt[6]{192}$	**79.** $\sqrt[6]{576}$	**80.** $\sqrt{x^3}$
81. $\sqrt{x^5}$	**82.** $\sqrt{4x^7}$	**83.** $\sqrt{2x^9}$	**84.** $\sqrt{16x^4y}$
85. $2\sqrt{x^3y^2}$	**86.** $x\sqrt{x^6y^3}$	**87.** $\sqrt{6xy^8}$	**88.** $\sqrt{12x^3y}$
89. $\sqrt{9x^3y^5}$	**90.** $\sqrt{x^5y^9}$	**91.** $\sqrt{x^3y^7z^2}$	**92.** $\sqrt{8x^4y^7}$

93. $\sqrt{27x^5y^6}$	**94.** $\sqrt{(2x+1)^3}$	**95.** $\sqrt{x^2(x+3)}$
96. $\sqrt{x^3(x-2)^2}$	**97.** $\sqrt{x^4(x+1)^5}$	**98.** $\sqrt{4y^4(x^2-y^2)}$

99. $\sqrt{x^2+4}$	**100.** $\sqrt{x^3+8}$	**101.** $\sqrt[3]{-2x^9}$	**102.** $\sqrt[3]{16x^5}$
103. $\sqrt[3]{8x^4y^6}$	**104.** $\sqrt[3]{27x^6y^4}$	**105.** $\sqrt[3]{-x^6y^3z}$	**106.** $\sqrt[3]{64x^3y}$

107. $\sqrt[3]{-x^8y^6z^3}$	**108.** $\sqrt[3]{x^3-1}$	**109.** $\sqrt[3]{8x^3+27}$
110. $\sqrt[4]{x^5}$	**111.** $\sqrt[4]{x^6y}$	**112.** $\sqrt[4]{x^{10}y^2}$
113. $\sqrt[4]{32x^4y^5}$	**114.** $\sqrt[4]{729x^2y^6}$	**115.** $\sqrt[4]{x^4+y^4}$
116. $\sqrt[5]{x^7y^3}$	**117.** $\sqrt[5]{-x^{10}y^6}$	**118.** $\sqrt[6]{x^7y}$
119. $\sqrt[6]{x^6y^8}$	**120.** $\sqrt[6]{x^9y^{15}}$	**121.** $\sqrt[n]{5^{3n+1}}$
122. $\sqrt[2n]{2^{4n+1}}$	**123.** $\sqrt[3n]{2^{6n+3}}$	**124.** $\sqrt[3n]{7^{8n}}$
125. $\sqrt[5n]{2^{8n}}$	**126.** $\sqrt[n]{x^{2n}y^{3n+2}}$	**127.** $\sqrt[n]{x^{n+2}y^{2n+1}}$

128. $\sqrt[2n]{x^{2n+1}y^{4n}}$ **129.** $\sqrt[3n]{x^{6n}y^{5n+1}}$ **130.** $\sqrt[6n]{x^{9n}y^{15n}}$

131. $\sqrt[4n]{x^{8n}y^{10n}}$ **132.** $\sqrt[4n]{x^{4n+4}y^{8n+8}}$ **133.** $\sqrt[5n]{x^{10n+5}y^{15n+10}}$

134. $\sqrt{4x^2 + 4y^2}$ **135.** $\sqrt{x^2 - x^2y^2}$ **136.** $\sqrt[3]{27x^3 - 27y^3}$

137. $\sqrt[3]{x^3y + x^3}$ **138.** $\sqrt[4]{x^8 + x^4}$ **139.** $\sqrt[4]{81x^2 - 81y^2}$

8.3 Combination of Radical Expressions

DEFINITION

Radical expressions are said to be **similar** when they have the same radical index and the same radicand.

EXAMPLES

1. The radical expressions $3\sqrt{2}$ and $5\sqrt{2}$ are similar.

2. The radical expressions $\sqrt{24}$ and $\sqrt{54}$ can be shown to be similar:

$$\sqrt{24} = \sqrt{2^3 \cdot 3} = 2\sqrt{6}$$

and $$\sqrt{54} = \sqrt{2 \cdot 3^3} = 3\sqrt{6}$$

3. The radical expressions $\sqrt{18}$ and $\sqrt{27}$ are not similar:

$$\sqrt{18} = \sqrt{2 \cdot 3^2} = 3\sqrt{2}$$

and $$\sqrt{27} = \sqrt{3^3} = 3\sqrt{3}$$

Radical expressions can be combined only when they are similar. First we put the radical expressions in standard form and then combine similar radicals using the distributive law.

EXAMPLE

Combine $\sqrt{20} - \sqrt{125} + \sqrt{180}$.

Solution

$$\begin{aligned}\sqrt{20} - \sqrt{125} + \sqrt{180} &= \sqrt{2^2 \cdot 5} - \sqrt{5^3} + \sqrt{2^2 \cdot 3^2 \cdot 5}\\ &= 2\sqrt{5} - 5\sqrt{5} + 2 \cdot 3\sqrt{5}\\ &= 2\sqrt{5} - 5\sqrt{5} + 6\sqrt{5}\\ &= (2 - 5 + 6)\sqrt{5} = 3\sqrt{5}\end{aligned}$$

EXAMPLE

Simplify $\sqrt{48} - \sqrt[3]{16} + \sqrt{108} + \sqrt[3]{54}$ and combine similar radical expressions.

Solution

$$\begin{aligned}\sqrt{48} - \sqrt[3]{16} + \sqrt{108} + \sqrt[3]{54} &= \sqrt{2^4 \cdot 3} - \sqrt[3]{2^4} + \sqrt{2^2 \cdot 3^3} + \sqrt[3]{2 \cdot 3^3}\\ &= 4\sqrt{3} - 2\sqrt[3]{2} + 6\sqrt{3} + 3\sqrt[3]{2} = 10\sqrt{3} + \sqrt[3]{2}\end{aligned}$$

EXAMPLE Put $\sqrt{3a^3} + \sqrt[3]{3a^4} - \sqrt[4]{9a^6}$ in standard form and combine similar radical expressions.

Solution

$$\begin{aligned}\sqrt{3a^3} + \sqrt[3]{3a^4} - \sqrt[4]{9a^6} &= \sqrt{3a^3} + \sqrt[3]{3a^4} - \sqrt[4]{3^2a^6}\\ &= a\sqrt{3a} + a\sqrt[3]{3a} - a\sqrt[4]{3^2a^2}\\ &= a\sqrt{3a} + a\sqrt[3]{3a} - a\sqrt{3a}\\ &= a\sqrt[3]{3a}\end{aligned}$$

EXAMPLE Simplify $\sqrt{25x^2(x+y)} + \sqrt{36y^2(x+y)} - 2\sqrt{(x+y)^3}$ and combine similar radical expressions.

Solution

$$\begin{aligned}&\sqrt{25x^2(x+y)} + \sqrt{36y^2(x+y)} - 2\sqrt{(x+y)^3}\\ &\quad = 5x\sqrt{x+y} + 6y\sqrt{x+y} - 2(x+y)\sqrt{x+y}\\ &\quad = [5x + 6y - 2(x+y)]\sqrt{x+y}\\ &\quad = (3x+4y)\sqrt{x+y}\end{aligned}$$

EXAMPLE Simplify $3\sqrt[n]{x^{2n+1}y^{n+1}} - \frac{x}{y}\sqrt[2n]{x^{2n+2}y^{4n+2}} + \frac{y}{x}\sqrt[n]{2^n x^{3n+1}y}$ and combine similar radical expressions.

Solution

$$\begin{aligned}&3\sqrt[n]{x^{2n+1}y^{n+1}} - \frac{x}{y}\sqrt[2n]{x^{2n+2}y^{4n+2}} + \frac{y}{x}\sqrt[n]{2^n x^{3n+1}y}\\ &\quad = 3x^2y\sqrt[n]{xy} - \frac{x}{y}\cdot\frac{xy^2}{1}\sqrt[2n]{x^2y^2} + \frac{y}{x}\cdot\frac{2x^3}{1}\sqrt[n]{xy}\\ &\quad = 3x^2y\sqrt[n]{xy} - x^2y\sqrt[n]{xy} + 2x^2y\sqrt[n]{xy}\\ &\quad = (3x^2y - x^2y + 2x^2y)\sqrt[n]{xy} = 4x^2y\sqrt[n]{xy}\end{aligned}$$

Exercise 8.3

Simplify and combine similar radical expressions, given that $n \in N$:

1. $4\sqrt{2} + 2\sqrt{2} - \sqrt{2} - 3\sqrt{2}$

2. $5\sqrt{3} - 7\sqrt{3} - 2\sqrt{3} + 4\sqrt{3}$

3. $7\sqrt[3]{5} - 10\sqrt[3]{5} - 4\sqrt[3]{5} + 2\sqrt[3]{5}$

4. $2\sqrt{2} + 3\sqrt[3]{2} - 6\sqrt[3]{2} - \sqrt{2}$

5. $\sqrt{x} - 2\sqrt{x} + 5\sqrt{x} + 3\sqrt{x}$

6. $a\sqrt{x} - 2\sqrt{x} + 3\sqrt{x} - b\sqrt{x}$

7. $3x\sqrt[3]{2} - 6y\sqrt[3]{2} - 3(x-2y)\sqrt[3]{2}$

8. $4x\sqrt[3]{2} - 3x\sqrt{3} + y\sqrt[3]{2} + y\sqrt{3}$

9. $\sqrt{8} - \sqrt{18} + \sqrt{25}$

10. $\sqrt{54} - \sqrt{64} + \sqrt{96}$

11. $\sqrt{12} - \sqrt{75} - \sqrt{48}$

12. $\sqrt{45} - \sqrt{20} + \sqrt{125}$

13. $\sqrt{32} - \sqrt{48} + \sqrt{108}$

14. $\sqrt{28} - \sqrt{72} + \sqrt{98}$

15. $\sqrt[3]{24} + \sqrt[3]{81} - \sqrt[3]{192}$

16. $\sqrt[3]{16} - \sqrt[3]{54} + 2\sqrt[3]{162}$

17. $3\sqrt[3]{32} + \sqrt[3]{108} + \sqrt[3]{-256}$

18. $5\sqrt[3]{56} - 2\sqrt[3]{-189} + \sqrt[3]{875}$

19. $3\sqrt{8} - \sqrt[3]{81} - \sqrt{128} + \sqrt[3]{375}$

20. $\sqrt{128} + \sqrt[3]{54} - \sqrt[3]{128} - \sqrt{50}$

21. $\sqrt[4]{36} - \sqrt{54} + \sqrt{96}$

22. $\sqrt{98} + \sqrt[3]{81} + \sqrt[4]{324} - \sqrt[6]{576}$

23. $\sqrt{320} - \sqrt[3]{1080} - \sqrt[4]{2025} + \sqrt[6]{1600}$

24. $2x\sqrt{x} - 5\sqrt{x^3} + \frac{1}{3x}\sqrt{9x^5}$

25. $\sqrt{36x^3} - \frac{1}{x}\sqrt{16x^5} + \frac{1}{x^2}\sqrt{25x^7}$

26. $x\sqrt{4xy^3} + y\sqrt{9x^3y} - \sqrt{49x^3y^3}$

27. $\frac{1}{x}\sqrt{12x^3} + \frac{3}{x^2}\sqrt{32x^5} + \sqrt{48x} - \sqrt{18x}$

28. $\frac{2x}{y}\sqrt{50y^5} - \frac{3y}{x}\sqrt{18x^4y} + \frac{2}{x^2y}\sqrt{98x^6y^5}$

29. $x\sqrt[3]{8x} - 5\sqrt[3]{-x^4} - \frac{7}{x}\sqrt[3]{x^7}$

30. $3x\sqrt[3]{x^4y} + 2y\sqrt[3]{-xy^4} + 2\sqrt[3]{x^7y}$

31. $2\sqrt[3]{x^5y} + \frac{x}{y}\sqrt[3]{125x^2y^4} + \frac{3}{x}\sqrt[3]{-x^8y}$

32. $\sqrt[3]{16x^4y^2} - \frac{1}{x}\sqrt[3]{-54x^7y^2} + \frac{x}{y}\sqrt[3]{2xy^5}$

33. $\sqrt{36x^3} + 2\sqrt[3]{-54xy^3} + \sqrt[3]{16x^4} - \sqrt{9xy^2}$

34. $\sqrt{128x^3} + \sqrt[3]{-2xy^3} - \sqrt[3]{128x^4} - \sqrt{50xy^2}$

35. $y\sqrt{25x^5y} + x^2\sqrt{9xy^3} + y\sqrt[3]{8x^4y} - x\sqrt[3]{xy^4}$

36. $y\sqrt{48x^3y^2} - x\sqrt[4]{144x^2y^8} + xy\sqrt{108xy^2}$

37. $x\sqrt{75xy^2} - y\sqrt{12x^3} + \sqrt[4]{9x^6y^4}$

38. $3\sqrt{2x^5} - x\sqrt[4]{4x^6} + \sqrt[3]{16x^7}$

39. $\sqrt{75x^3} + \sqrt[4]{729x^6} - \sqrt[3]{24x^4}$

40. $x\sqrt[3]{8xy^4} - \sqrt[6]{x^8y^8} + y\sqrt[3]{x^4y}$

41. $2\sqrt[4]{x^9} - 3x\sqrt[4]{x^5} + x\sqrt[8]{x^{10}}$

42. $\sqrt[5]{xy^{13}} + 3\sqrt[10]{x^2y^6} + \sqrt[5]{-32x^{11}y^3}$

43. $\sqrt{4x+4} + \sqrt{9x+9} - \sqrt{25x+25}$

44. $\sqrt{x^2+y^2} - \sqrt{4x^2+4y^2} - \sqrt{16x^2+16y^2}$

45. $\sqrt{x^2-y^2} - \sqrt{9x^2-9y^2} + \sqrt{36x^2-36y^2}$

46. $\sqrt{16x^3-16x^2y} + \sqrt{9xy^2-9y^3} - \sqrt{(x+y)(x^2-y^2)}$

47. $\sqrt{28x^3(y-1)^2} + y\sqrt{63x(x+1)^2} - 2\sqrt{175x(y-x)^2}$

48. $\sqrt[3]{8x^3y - 8x^3z} + \sqrt[3]{27y^3z - 27y^4} + 3\sqrt[3]{(x+y)^3(y-z)}$

49. $2\sqrt[n]{x^{n+1}y^n} + 5\sqrt[n]{xy^{2n}} - 6\sqrt[n]{x^{2n+1}}$

50. $\sqrt[n]{2^nx^n(x-3y)} - \sqrt[n]{5^ny^n(x-3y)} - \sqrt[n]{2^n(x-3y)^{n+1}}$

51. $\sqrt[2n]{2^nx^{3n}} - \frac{3}{x}\sqrt[2n]{2^{3n}x^{5n}} + x\sqrt[2n]{2^{5n}x^n}$

52. $\sqrt[n]{6^nx^ny} - \frac{1}{y}\sqrt[2n]{3^{2n}x^{2n}y^{2n+2}} + \sqrt[3n]{y^3(x-y)^{3n}}$

53. $\frac{1}{2}\sqrt[n]{2^{n+1}\cdot 3^nx^{2n+1}} + \frac{7}{xy}\sqrt[2n]{2^2x^{2n+2}y^{6n}} - \frac{3}{x}\sqrt[4n]{2^4x^{4n+4}(x^2-y^2)^{4n}}$

8.4 Multiplication of Radical Expressions

When a and b are positive real numbers, k is a rational number, and m, n, mk, and nk are natural numbers, we have the following two rules:

1. $\sqrt[n]{a}\,\sqrt[n]{b} = \sqrt[n]{ab}$ **2.** $\sqrt[n]{a^m} = \sqrt[nk]{a^{mk}}$

These rules are used to multiply radical expressions.

EXAMPLES

1. $\sqrt{3} \cdot \sqrt{7} = \sqrt{3 \cdot 7} = \sqrt{21}$

2. $2\sqrt[3]{5} \cdot 3\sqrt[3]{6} = (2 \cdot 3)\sqrt[3]{5 \cdot 6} = 6\sqrt[3]{30}$

3. $\sqrt[3]{2} \cdot \sqrt[3]{12} = \sqrt[3]{2 \cdot 12} = \sqrt[3]{24} = 2\sqrt[3]{3}$

4. $\sqrt{x-2}\,\sqrt{x^2-4} = \sqrt{(x-2)(x^2-4)}$
$= \sqrt{(x-2)^2(x+2)} = (x-2)\sqrt{x+2}$

5. $\sqrt[n]{32x^3}\,\sqrt[n]{2^{n-2}x^{n-1}} = \sqrt[n]{2^5x^3}\,\sqrt[n]{2^{n-2}x^{n-1}}$ $\quad n > 3$
$= \sqrt[n]{2^{n+3}x^{n+2}}$
$= 2x\sqrt[n]{2^3x^2} = 2x\sqrt[n]{8x^2}$

When the radicals have different indices, we apply the second rule before the first rule.

EXAMPLES

1. $\sqrt{2}\,\sqrt[3]{2^2} = \sqrt[6]{2^3}\,\sqrt[6]{2^4} = \sqrt[6]{2^7} = 2\sqrt[6]{2}$

2. $\sqrt[3]{a^2}\,\sqrt[4]{a^3} = \sqrt[12]{a^8}\,\sqrt[12]{a^9} = \sqrt[12]{a^{17}} = a\sqrt[12]{a^5}$

3. $\sqrt[n]{xy^2}\,\sqrt[3n]{x^{3n-1}y^{9n-6}} = \sqrt[3n]{x^3y^6}\,\sqrt[3n]{x^{3n-1}y^{9n-6}}$
$= \sqrt[3n]{x^{3n+2}y^{9n}} = xy^3\sqrt[3n]{x^2}$

Note The final radical must be in standard form.

In order to multiply one radical by a radical expression of more than one term, we use the distributive law.

EXAMPLE Multiply $5\sqrt[3]{4}(2\sqrt[3]{6} + \sqrt[3]{14})$.

Solution $5\sqrt[3]{4}(2\sqrt[3]{6} + \sqrt[3]{14}) = 10\sqrt[3]{24} + 5\sqrt[3]{56} = 20\sqrt[3]{3} + 10\sqrt[3]{7}$

EXAMPLE Multiply $\sqrt{10xy}(\sqrt{5x} - \sqrt{2y})$.

Solution $\sqrt{10xy}(\sqrt{5x} - \sqrt{2y}) = \sqrt{50x^2y} - \sqrt{20xy^2} = 5x\sqrt{2y} - 2y\sqrt{5x}$

To multiply two radical expressions, each with more than one term, follow the same arrangement used in multiplying polynomials.

EXAMPLE

Multiply $3\sqrt{2} - 4\sqrt{3}$ by $\sqrt{2} - 3\sqrt{3}$ and simplify.

Solution

$$\begin{array}{l} 3\sqrt{2} - 4\sqrt{3} \\ \sqrt{2} - 3\sqrt{3} \\ \hline 3\sqrt{4} - 4\sqrt{6} \\ \qquad - 9\sqrt{6} + 12\sqrt{9} \\ \hline 3\sqrt{4} - 13\sqrt{6} + 12\sqrt{9} \end{array}$$

Hence
$$\begin{aligned}(3\sqrt{2} - 4\sqrt{3})(\sqrt{2} - 3\sqrt{3}) &= 3\sqrt{4} - 13\sqrt{6} + 12\sqrt{9} \\ &= 6 - 13\sqrt{6} + 36 \\ &= 42 - 13\sqrt{6}\end{aligned}$$

EXAMPLE

Multiply $\sqrt{x} + 3\sqrt{y}$ by $2\sqrt{x} - \sqrt{y}$ and simplify.

Solution

$$\begin{array}{l} \sqrt{x} + 3\sqrt{y} \\ 2\sqrt{x} - \sqrt{y} \\ \hline 2\sqrt{x^2} + 6\sqrt{xy} \\ \qquad - \sqrt{xy} - 3\sqrt{y^2} \\ \hline 2\sqrt{x^2} + 5\sqrt{xy} - 3\sqrt{y^2} \end{array}$$

Hence
$$\begin{aligned}(\sqrt{x} + 3\sqrt{y})(2\sqrt{x} - \sqrt{y}) &= 2\sqrt{x^2} + 5\sqrt{xy} - 3\sqrt{y^2} \\ &= 2x + 5\sqrt{xy} - 3y\end{aligned}$$

EXAMPLE

Expand $(\sqrt{2x+1} - 3\sqrt{x-2})^2$ and simplify.

Solution

$$\begin{array}{l} \sqrt{2x+1} - 3\sqrt{x-2} \\ \sqrt{2x+1} - 3\sqrt{x-2} \\ \hline \sqrt{(2x+1)^2} - 3\sqrt{(2x+1)(x-2)} \\ \qquad - 3\sqrt{(2x+1)(x-2)} + 9\sqrt{(x-2)^2} \\ \hline \sqrt{(2x+1)^2} - 6\sqrt{(2x+1)(x-2)} + 9\sqrt{(x-2)^2} \end{array}$$

Hence
$$\begin{aligned}(\sqrt{2x+1} - 3\sqrt{x-2})^2 &= \sqrt{(2x+1)^2} - 6\sqrt{(2x+1)(x-2)} \\ &\qquad + 9\sqrt{(x-2)^2} \\ &= 2x + 1 - 6\sqrt{2x^2 - 3x - 2} + 9(x-2) \\ &= 2x + 1 - 6\sqrt{2x^2 - 3x - 2} + 9x - 18 \\ &= 11x - 17 - 6\sqrt{2x^2 - 3x - 2}\end{aligned}$$

Note $(\sqrt{a} + \sqrt{b})^2 \neq a + b$

$(\sqrt{a} + \sqrt{b})^2 = (\sqrt{a} + \sqrt{b})(\sqrt{a} + \sqrt{b}) = a + 2\sqrt{ab} + b$

Exercise 8.4

Perform the indicated multiplications and simplify, given that $n > 3,\ n \in N$:

1. $\sqrt{2}\,\sqrt{3}$ **2.** $2\sqrt{2}\,\sqrt{5}$ **3.** $\sqrt{2}\,\sqrt{7}$ **4.** $\sqrt{3}\,\sqrt{11}$

5. $\sqrt[3]{2}\,\sqrt[3]{2}$ **6.** $\sqrt[3]{2}\,\sqrt[3]{3}$ **7.** $\sqrt[3]{4}\,\sqrt[3]{3}$ **8.** $\sqrt[3]{6}\,\sqrt[3]{3}$

9. $\sqrt{2}\,\sqrt{2}$ **10.** $\sqrt{3}\,\sqrt{3}$ **11.** $2\sqrt{5}\,\sqrt{5}$ **12.** $5\sqrt{6}\,\sqrt{6}$

13. $\sqrt[3]{2}\,\sqrt[3]{4}$ **14.** $\sqrt[3]{3}\,\sqrt[3]{9}$ **15.** $\sqrt[3]{25}\,\sqrt[3]{5}$ **16.** $\sqrt[3]{7}\,\sqrt[3]{49}$

17. $5\sqrt{2}\,\sqrt{6}$ **18.** $\sqrt{2}\,\sqrt{10}$ **19.** $\sqrt{3}(-\sqrt{21})$

20. $\sqrt{5}(-\sqrt{10})$ **21.** $\sqrt{7}\,\sqrt{14}$ **22.** $\sqrt{5}\,\sqrt{35}$

23. $\sqrt{6}\,\sqrt{15}$ **24.** $\sqrt{6}\,\sqrt{21}$ **25.** $\sqrt{10}\,\sqrt{15}$ **26.** $\sqrt{26}\,\sqrt{39}$

27. $2\sqrt[3]{6}\,\sqrt[3]{4}$ **28.** $\sqrt[3]{9}\,\sqrt[3]{18}$ **29.** $\sqrt[3]{-15}\,\sqrt[3]{25}$

30. $\sqrt[3]{-10}\,\sqrt[3]{-75}$ **31.** $\sqrt[3]{-22}\,\sqrt[3]{121}$ **32.** $\sqrt[3]{14}\,\sqrt[3]{-98}$

33. $\sqrt[4]{6}\,\sqrt[4]{27}$ **34.** $\sqrt[4]{8}\,\sqrt[4]{6}$ **35.** $\sqrt[5]{-8}\,\sqrt[5]{12}$

36. $\sqrt[5]{27}\,\sqrt[5]{-18}$ **37.** $\sqrt[n]{2^{n-1}}\,\sqrt[n]{16}$ **38.** $\sqrt[n]{3^{n+3}}\,\sqrt[n]{3^{n-1}}$

39. $\sqrt{2}\,\sqrt[4]{8}$ **40.** $\sqrt{3}\,\sqrt[4]{27}$ **41.** $\sqrt{6}\,\sqrt[6]{32}$

42. $\sqrt{14}\,\sqrt[3]{49}$ **43.** $\sqrt{15}\,\sqrt[3]{50}$ **44.** $\sqrt[n]{8}\,\sqrt[2n]{2^{2n-3}}$

45. $\sqrt[n]{27}\,\sqrt[2n]{3^{2n-1}}$ **46.** $\sqrt{2}\,\sqrt{x}$ **47.** $\sqrt{2x}\,\sqrt{3y}$

48. $3\sqrt{x}\,\sqrt{2y}$ **49.** $\sqrt{2x}\,\sqrt{2x}$ **50.** $\sqrt{3xy}\,\sqrt{3xy}$

51. $\sqrt{x+1}\,\sqrt{x+1}$ **52.** $\sqrt{x-2}\,\sqrt{x-2}$ **53.** $\sqrt{2x}\,\sqrt{3xy}$

54. $\sqrt{6x}\,\sqrt{15xy}$ **55.** $\sqrt[3]{x}\,\sqrt[3]{2x^2}$ **56.** $\sqrt[3]{9x}\,\sqrt[3]{12x^2}$

57. $\sqrt[3]{4xy^2}\,\sqrt[3]{6x^2}$ **58.** $\sqrt[3]{3x^2y}\,\sqrt[3]{9xy}$ **59.** $\sqrt[5]{x^2}\,\sqrt[5]{x^4}$

60. $\sqrt{2}\,\sqrt{x+2}$ **61.** $\sqrt{x}\,\sqrt{x+1}$ **62.** $\sqrt{2x}\,\sqrt{x-1}$

63. $\sqrt{3}\,\sqrt{6x+9}$ **64.** $\sqrt{5}\,\sqrt{10x-5}$ **65.** $\sqrt{3x}\,\sqrt{3x-3}$

66. $\sqrt{2x}\,\sqrt{2x^2+8x}$ **67.** $\sqrt{x+1}\,\sqrt{x+2}$ **68.** $\sqrt{x-1}\,\sqrt{x+1}$

69. $\sqrt{x+3}\,\sqrt{x-3}$ **70.** $\sqrt{x+1}\,\sqrt{x^2-1}$ **71.** $\sqrt{x-2}\,\sqrt{x^2-4}$

72. $\sqrt{x+1}\,\sqrt{x^3+1}$ **73.** $\sqrt{x-4}\,\sqrt{x^3-64}$

74. $\sqrt[n]{xy^4}\,\sqrt[n]{x^{n-1}y^{2n-3}}$ **75.** $\sqrt[n]{x^5y^3}\,\sqrt[n]{x^{2n-2}y^{n-1}}$

76. $\sqrt{x}\,\sqrt[4]{x^3}$ **77.** $\sqrt[3]{xy^2}\,\sqrt[6]{x^5y^4}$

78. $\sqrt{xy}\,\sqrt[3]{2x^2}$ **79.** $\sqrt[3]{2x^2y}\,\sqrt[4]{x^3y^3}$

80. $\sqrt[n]{x^5y}\,\sqrt[2n]{x^{6n-3}y^{4n-1}}$ **81.** $\sqrt[n]{32x^4y^3}\,\sqrt[2n]{2^{2n-7}x^{2n-5}y^{4n-6}}$

82. $\sqrt{2}(2+\sqrt{2})$ **83.** $\sqrt{3}(\sqrt{3}-1)$ **84.** $\sqrt{5}(3-\sqrt{5})$

85. $\sqrt{3}(4+2\sqrt{3})$ **86.** $\sqrt{2}(\sqrt{2}+2\sqrt{3})$ **87.** $\sqrt{3}(\sqrt{3}-\sqrt{2})$

88. $\sqrt{2}(\sqrt{6}-\sqrt{2})$ **89.** $\sqrt{2}(\sqrt{6}+\sqrt{10})$ **90.** $\sqrt{3}(\sqrt{6}+2\sqrt{3})$

91. $\sqrt{3}(\sqrt{15}-\sqrt{21})$ **92.** $\sqrt{6}(\sqrt{10}+\sqrt{15})$ **93.** $\sqrt{10}(\sqrt{14}-\sqrt{15})$

94. $\sqrt[3]{6}(\sqrt[3]{4}-\sqrt[3]{9})$ **95.** $\sqrt[3]{9}(\sqrt[3]{21}-\sqrt[3]{12})$ **96.** $\sqrt[3]{4}(\sqrt[3]{6}-\sqrt[3]{14})$

97. $\sqrt[3]{15}(\sqrt[3]{9}+\sqrt[3]{75})$ **98.** $\sqrt{2}(\sqrt{2x}+\sqrt{6y})$ **99.** $\sqrt{3}(\sqrt{6x}-\sqrt{12y})$

100. $\sqrt{x}(\sqrt{x}-3)$ **101.** $\sqrt{x}(\sqrt{2x}+1)$ **102.** $\sqrt{x}(\sqrt{2x}+\sqrt{3y})$

103. $\sqrt{2x}(\sqrt{10x}+\sqrt{2y})$ **104.** $\sqrt{3x}(\sqrt{2xy}-\sqrt{3xz})$

105. $\sqrt{xy}(\sqrt{x} - \sqrt{y})$
106. $\sqrt{6xy}(\sqrt{2x} + \sqrt{15xy})$
107. $\sqrt{10xy}(3\sqrt{6x} - \sqrt{5y})$
108. $\sqrt{2x}(\sqrt{x+1} - \sqrt{x-1})$
109. $\sqrt{6x}(\sqrt{2x-2} + \sqrt{3x+3})$
110. $\sqrt{x-3}(\sqrt{x-3} + \sqrt{x+4})$
111. $2\sqrt{x-2}(\sqrt{x-2} + 3\sqrt{x+1})$
112. $\sqrt[3]{4a^2b}(\sqrt[3]{2a^2b^2} + 3\sqrt[3]{6ab})$
113. $\sqrt[3]{9ab^2}(\sqrt[3]{33a^2} - \sqrt[3]{15b^2})$
114. $\sqrt[3]{2ab}(\sqrt[3]{4a^2} - \sqrt[3]{2b^2})$
115. $\sqrt[3]{3ab}(\sqrt[3]{18ab^2} + \sqrt[3]{9a^2b})$
116. $(1 + \sqrt{2})(1 - \sqrt{2})$
117. $(2 - \sqrt{3})(2 + \sqrt{3})$
118. $(4 + \sqrt{5})(4 - \sqrt{5})$
119. $(2 + \sqrt{7})(2 - \sqrt{7})$
120. $(3\sqrt{3} + 2)(3\sqrt{3} - 2)$
121. $(2\sqrt{5} + 3)(2\sqrt{5} - 3)$
122. $(\sqrt{3} + \sqrt{2})(\sqrt{3} - \sqrt{2})$
123. $(\sqrt{5} + \sqrt{6})(\sqrt{5} - \sqrt{6})$
124. $(\sqrt{5} + 2\sqrt{2})(\sqrt{5} - 2\sqrt{2})$
125. $(2\sqrt{3} + \sqrt{2})(2\sqrt{3} - \sqrt{2})$
126. $(\sqrt{2} + 2\sqrt{3})(\sqrt{2} + 3\sqrt{3})$
127. $(\sqrt{3} + 2\sqrt{2})(3\sqrt{3} + \sqrt{2})$
128. $(3\sqrt{2} + \sqrt{5})(\sqrt{2} + 3\sqrt{5})$
129. $(\sqrt{3} + 2\sqrt{5})(\sqrt{3} + 4\sqrt{5})$
130. $(4\sqrt{2} - \sqrt{3})(3\sqrt{2} - \sqrt{3})$
131. $(\sqrt{2} - 2\sqrt{3})(3\sqrt{2} - \sqrt{3})$
132. $(\sqrt{5} + \sqrt{2})^2$
133. $(\sqrt{3} + \sqrt{5})^2$
134. $(2\sqrt{3} - \sqrt{2})^2$
135. $(4\sqrt{2} - \sqrt{7})^2$
136. $(2\sqrt[3]{3} - \sqrt[3]{4})(4\sqrt[3]{9} + 2\sqrt[3]{12} + 2\sqrt[3]{2})$
137. $(\sqrt[3]{6} + \sqrt[3]{5})(\sqrt[3]{36} - \sqrt[3]{30} + \sqrt[3]{25})$
138. $(\sqrt[3]{9} - \sqrt[3]{4})(3\sqrt[3]{3} + \sqrt[3]{36} + 2\sqrt[3]{2})$
139. $(\sqrt{x} + 1)(\sqrt{x} - 1)$
140. $(\sqrt{x} + 2)(\sqrt{x} - 2)$
141. $(\sqrt{x} + 3)(\sqrt{x} - 3)$
142. $(\sqrt{2x} + 1)(\sqrt{2x} - 1)$
143. $(\sqrt{x} + \sqrt{3})(\sqrt{x} - \sqrt{3})$
144. $(\sqrt{2x} + \sqrt{5})(\sqrt{2x} - \sqrt{5})$
145. $(\sqrt{x} + \sqrt{y})(\sqrt{x} - \sqrt{y})$
146. $(\sqrt{2x} + \sqrt{3y})(\sqrt{2x} - \sqrt{3y})$
147. $(\sqrt{x} + \sqrt{y})(2\sqrt{x} + 3\sqrt{y})$
148. $(\sqrt{x} + 4\sqrt{y})(2\sqrt{x} + \sqrt{y})$
149. $(\sqrt{x} + \sqrt{y})^2$
150. $(2\sqrt{x} + \sqrt{2y})^2$
151. $(\sqrt{x} - \sqrt{y})^2$
152. $(\sqrt{x} - \sqrt{3y})^2$
153. $(\sqrt{x+1} + 1)(\sqrt{x+1} - 1)$
154. $(\sqrt{x-1} + 2)(\sqrt{x-1} - 2)$
155. $(\sqrt{x+2} + 3)(\sqrt{x+2} - 3)$
156. $(\sqrt{2x-1} + 4)(\sqrt{2x-1} - 4)$
157. $(\sqrt{x-1} + 1)^2$
158. $(\sqrt{x+2} + 2)^2$
159. $(\sqrt{2x-1} - 2)^2$
160. $(\sqrt{x+3} - 3)^2$
161. $(4 - \sqrt{x-1})^2$
162. $(3 - \sqrt{x-2})^2$
163. $(3 - 4\sqrt{x-3})^2$
164. $(2 - 5\sqrt{x-1})^2$
165. $(\sqrt{x} + \sqrt{x-1})(\sqrt{x} - \sqrt{x-1})$
166. $(\sqrt{2x} + \sqrt{x+3})(\sqrt{2x} - \sqrt{x+3})$
167. $(\sqrt{3x} + \sqrt{x-2})(\sqrt{3x} - \sqrt{x-2})$
168. $(\sqrt{x} + \sqrt{2x-3})(\sqrt{x} - \sqrt{2x-3})$
169. $(3\sqrt{x} + \sqrt{x-1})(\sqrt{x} - 2\sqrt{x-1})$
170. $(\sqrt{x} + 3\sqrt{x+1})(2\sqrt{x} - 3\sqrt{x+1})$
171. $(\sqrt{x} + \sqrt{x-1})^2$
172. $(\sqrt{x} + \sqrt{x+1})^2$
173. $(\sqrt{2x} + 2\sqrt{x+3})^2$
174. $(\sqrt{x} + 4\sqrt{x-4})^2$

175. $(\sqrt{x} - 3\sqrt{x-3})^2$

176. $(\sqrt{2x} - 2\sqrt{3x-2})^2$

177. $(\sqrt{x+1} + \sqrt{x-1})^2$

178. $(\sqrt{x-2} + \sqrt{x+2})^2$

179. $(2\sqrt{x+3} + \sqrt{x-1})^2$

180. $(\sqrt{x+3} - \sqrt{x-4})^2$

181. $(\sqrt{2x+1} - 2\sqrt{2x-1})^2$

182. $(\sqrt{x+1} - 2\sqrt{x-2})^2$

183. $(\sqrt[3]{a} + \sqrt[3]{b})(\sqrt[3]{a^2} - \sqrt[3]{ab} + \sqrt[3]{b^2})$

184. $(\sqrt[3]{3a} - \sqrt[3]{2b})(\sqrt[3]{9a^2} + \sqrt[3]{6ab} + \sqrt[3]{4b^2})$

185. $(\sqrt{10} - \sqrt{2} - \sqrt{6})(\sqrt{10} - \sqrt{2} + \sqrt{6})$

186. $(\sqrt{7} + \sqrt{3} - 2)(\sqrt{7} - \sqrt{3} - 2)$

187. $(\sqrt{15} - \sqrt{5} + \sqrt{3})(\sqrt{15} + \sqrt{5} - \sqrt{3})$

188. $(\sqrt{42} + \sqrt{6} - \sqrt{7})(\sqrt{42} - \sqrt{6} + \sqrt{7})$

8.5 Division of Radical Expressions

THEOREM When $a, b \in R$, $a > 0, b > 0$, and $n \in N$, then $\dfrac{\sqrt[n]{a}}{\sqrt[n]{b}} = \sqrt[n]{\dfrac{a}{b}}$.

Proof

$$\frac{\sqrt[n]{a}}{\sqrt[n]{b}} = \frac{a^{\frac{1}{n}}}{b^{\frac{1}{n}}} = \left(\frac{a}{b}\right)^{\frac{1}{n}} = \sqrt[n]{\frac{a}{b}}$$

Radical expressions can be divided according to this theorem only when the radical indices are the same. For different radical indices, the preliminary step to make them the same must be carried out.

EXAMPLES

1. $\dfrac{\sqrt{21}}{\sqrt{7}} = \sqrt{\dfrac{21}{7}} = \sqrt{3}$

2. $\dfrac{\sqrt[3]{4}}{\sqrt[3]{2}} = \sqrt[3]{\dfrac{4}{2}} = \sqrt[3]{2}$

3. $\dfrac{\sqrt{x^5y^3}}{\sqrt{x^3y^2}} = \sqrt{\dfrac{x^5y^3}{x^3y^2}} = \sqrt{x^2y} = x\sqrt{y}$

4. $\dfrac{\sqrt{2}}{\sqrt[3]{2}} = \dfrac{\sqrt[6]{2^3}}{\sqrt[6]{2^2}} = \sqrt[6]{\dfrac{2^3}{2^2}} = \sqrt[6]{2}$

Sometimes the numerator of a fractional radical is not an exact multiple of the denominator; for example, $\sqrt{\dfrac{3}{2}}$. In order to simplify such a radical, multiply both numerator and denominator of the radicand by the smallest number that will make the denominator a perfect root.

Note The denominator is a perfect root if the exponent of each factor is an integral multiple of the radical index.

To simplify $\sqrt{\frac{3}{2}}$, multiply the numerator and denominator of the radicand by 2:

$$\sqrt{\frac{3}{2}} = \sqrt{\frac{3 \cdot 2}{2 \cdot 2}} = \frac{1}{2}\sqrt{6}$$

Remark When the radical expression is of the form $\frac{a}{b\sqrt[n]{c^m}}$ where $m < n$, multiply the numerator and the denominator by $\sqrt[n]{c^{n-m}}$:

$$\frac{3}{\sqrt{2}} = \frac{3}{\sqrt{2}} \cdot \frac{\sqrt{2}}{\sqrt{2}} = \frac{3\sqrt{2}}{2}$$

$$\frac{2}{\sqrt{12x}} = \frac{2}{\sqrt{2^2 \cdot 3x}} = \frac{2}{2\sqrt{3x}} = \frac{1}{\sqrt{3x}} \cdot \frac{\sqrt{3x}}{\sqrt{3x}} = \frac{\sqrt{3x}}{3x}$$

$$\frac{5}{\sqrt[3]{2}} = \frac{5}{\sqrt[3]{2}} \cdot \frac{\sqrt[3]{2^2}}{\sqrt[3]{2^2}} = \frac{5\sqrt[3]{4}}{2}$$

$$\frac{2}{3\sqrt[7]{a^3}} = \frac{2}{3\sqrt[7]{a^3}} \cdot \frac{\sqrt[7]{a^4}}{\sqrt[7]{a^4}} = \frac{2\sqrt[7]{a^4}}{3a}$$

EXAMPLE Divide $\sqrt{6}$ by $\sqrt{10}$.

Solution

$$\frac{\sqrt{6}}{\sqrt{10}} = \sqrt{\frac{6}{10}} = \sqrt{\frac{3}{5}} = \sqrt{\frac{3 \cdot 5}{5 \cdot 5}} = \sqrt{\frac{15}{5^2}} = \frac{1}{5}\sqrt{15}$$

EXAMPLE Put $\sqrt{\frac{2a^3bc^2}{9xy^3}}$ in standard form.

Solution

$$\sqrt{\frac{2a^3bc^2}{9xy^3}} = \sqrt{\frac{2a^3bc^2}{3^2xy^3}} = \sqrt{\frac{2a^3bc^2 \cdot xy}{3^2xy^3 \cdot xy}}$$

$$= \sqrt{\frac{2a^3bc^2xy}{3^2x^2y^4}}$$

$$= \frac{ac}{3xy^2}\sqrt{2abxy}$$

EXAMPLE

Put $\sqrt[3]{\dfrac{2a^2b^7}{3x^7y^5}}$ in standard form.

Solution

$$\sqrt[3]{\frac{2a^2b^7}{3x^7y^5}} = \sqrt[3]{\frac{2a^2b^7 \cdot 3^2x^2y}{3x^7y^5 \cdot 3^2x^2y}} = \sqrt[3]{\frac{18a^2b^7x^2y}{3^3x^9y^6}}$$

$$= \frac{b^2}{3x^3y^2}\sqrt[3]{18a^2bx^2y}$$

EXAMPLE

Put $\dfrac{\sqrt{x}}{\sqrt{2x^2 - x}}$ in standard form.

Solution

$$\frac{\sqrt{x}}{\sqrt{2x^2 - x}} = \sqrt{\frac{x}{x(2x-1)}} = \sqrt{\frac{1}{(2x-1)} \cdot \frac{(2x-1)}{(2x-1)}}$$

$$= \sqrt{\frac{(2x-1)}{(2x-1)^2}} = \frac{1}{(2x-1)}\sqrt{(2x-1)}$$

EXAMPLE

Put $\dfrac{\sqrt{2a^2b}}{\sqrt[3]{4ab^2}}$ in standard form.

Solution

$$\frac{\sqrt{2a^2b}}{\sqrt[3]{4ab^2}} = \frac{\sqrt{2a^2b}}{\sqrt[3]{2^2ab^2}} = \frac{\sqrt[6]{2^3a^6b^3}}{\sqrt[6]{2^4a^2b^4}} = \sqrt[6]{\frac{2^3a^6b^3}{2^4a^2b^4}}$$

$$= \sqrt[6]{\frac{a^4}{2b}} = \sqrt[6]{\frac{a^4 \cdot 2^5b^5}{2b \cdot 2^5b^5}}$$

$$= \sqrt[6]{\frac{2^5a^4b^5}{2^6b^6}} = \frac{1}{2b}\sqrt[6]{32a^4b^5}$$

EXAMPLE

Simplify $\dfrac{\sqrt[n]{2^{n+5}a^{2n+3}b^2}}{\sqrt[3n]{2^8a^{n+6}b^{2n+4}}}$ given that $n \in N, n > 3$.

Solution

$$\frac{\sqrt[n]{2^{n+5}a^{2n+3}b^2}}{\sqrt[3n]{2^8a^{n+6}b^{2n+4}}} = \frac{\sqrt[3n]{2^{3n+15}a^{6n+9}b^6}}{\sqrt[3n]{2^8a^{n+6}b^{2n+4}}} = \sqrt[3n]{\frac{2^{3n+15}a^{6n+9}b^6}{2^8a^{n+6}b^{2n+4}}}$$

$$= \sqrt[3n]{\frac{2^{3n+7}a^{5n+3}}{b^{2n-2}}} = \sqrt[3n]{\frac{2^{3n+7}a^{5n+3} \cdot b^{n+2}}{b^{2n-2} \cdot b^{n+2}}}$$

$$= \sqrt[3n]{\frac{2^{3n+7}a^{5n+3}b^{n+2}}{b^{3n}}}$$

$$= \frac{2a}{b}\sqrt[3n]{2^7a^{2n+3}b^{n+2}}$$

The definition of addition of fractions

$$\frac{a_1 + a_2 + \cdots + a_n}{a} = \frac{a_1}{a} + \frac{a_2}{a} + \cdots + \frac{a_n}{a}$$

is used to divide a radical expression with more than one term by a one-term radical.

EXAMPLE Divide and simplify $\dfrac{\sqrt{3x} - \sqrt{5y}}{\sqrt{15xy}}$.

Solution

$$\frac{\sqrt{3x} - \sqrt{5y}}{\sqrt{15xy}} = \frac{\sqrt{3x}}{\sqrt{15xy}} - \frac{\sqrt{5y}}{\sqrt{15xy}} = \sqrt{\frac{3x}{15xy}} - \sqrt{\frac{5y}{15xy}}$$

$$= \sqrt{\frac{1}{5y}} - \sqrt{\frac{1}{3x}} = \sqrt{\frac{1 \cdot 5y}{5y \cdot 5y}} - \sqrt{\frac{1 \cdot 3x}{3x \cdot 3x}}$$

$$= \sqrt{\frac{5y}{5^2y^2}} - \sqrt{\frac{3x}{3^2x^2}} = \frac{1}{5y}\sqrt{5y} - \frac{1}{3x}\sqrt{3x}$$

EXAMPLE Divide and simplify $\dfrac{\sqrt[3]{3a} + \sqrt[3]{6b}}{\sqrt[3]{3ab}}$.

Solution

$$\frac{\sqrt[3]{3a} + \sqrt[3]{6b}}{\sqrt[3]{3ab}} = \frac{\sqrt[3]{3a}}{\sqrt[3]{3ab}} + \frac{\sqrt[3]{6b}}{\sqrt[3]{3ab}} = \sqrt[3]{\frac{3a}{3ab}} + \sqrt[3]{\frac{6b}{3ab}}$$

$$= \sqrt[3]{\frac{1}{b}} + \sqrt[3]{\frac{2}{a}} = \sqrt[3]{\frac{1 \cdot b^2}{b \cdot b^2}} + \sqrt[3]{\frac{2 \cdot a^2}{a \cdot a^2}}$$

$$= \sqrt[3]{\frac{b^2}{b^3}} + \sqrt[3]{\frac{2a^2}{a^3}} = \frac{1}{b}\sqrt[3]{b^2} + \frac{1}{a}\sqrt[3]{2a^2}$$

When we multiply the radical expressions $(\sqrt{a} + \sqrt{b})$ and $(\sqrt{a} - \sqrt{b})$, we get the rational expression $(a - b)$. Each of the expressions $(\sqrt{a} + \sqrt{b})$ and $(\sqrt{a} - \sqrt{b})$ is called a **rationalizing factor** of the other.

This relation can be obtained when we factor $a - b$ as the difference of two squares.

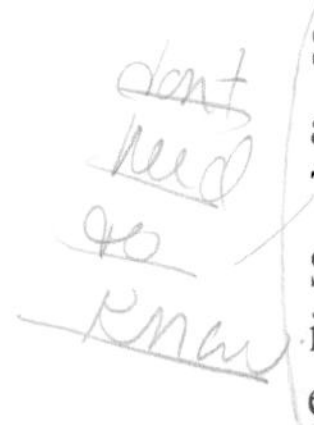

Since $(\sqrt[3]{a} + \sqrt[3]{b})(\sqrt[3]{a^2} - \sqrt[3]{ab} + \sqrt[3]{b^2}) = a + b$, each of the expressions $(\sqrt[3]{a} + \sqrt[3]{b})$ and $(\sqrt[3]{a^2} - \sqrt[3]{ab} + \sqrt[3]{b^2})$ is a rationalizing factor of the other.

This relation can be obtained when we factor $a + b$ as the sum of two cubes.

Similarly, each of the expressions $(\sqrt[3]{a} - \sqrt[3]{b})$ and $(\sqrt[3]{a^2} + \sqrt[3]{ab} + \sqrt[3]{b^2})$ is a rationalizing factor of the other, since their product is the rational expression $(a - b)$.

This relation can be obtained when we factor $a - b$ as the difference of two cubes.

EXAMPLES

1. $(\sqrt{2} - \sqrt{3})$ is a rationalizing factor of $(\sqrt{2} + \sqrt{3})$.
2. $(2 + 3\sqrt{2})$ is a rationalizing factor of $(2 - 3\sqrt{2})$.
3. $(\sqrt[3]{2} - \sqrt[3]{3})$ is a rationalizing factor of $(\sqrt[3]{4} + \sqrt[3]{6} + \sqrt[3]{9})$.
4. $(\sqrt[3]{16} - \sqrt[3]{20} + \sqrt[3]{25})$ is a rationalizing factor of $(\sqrt[3]{4} + \sqrt[3]{5})$.

The proofs are left to the reader.

In order to facilitate the manipulation with a radical expression such as $\dfrac{\sqrt{a} + \sqrt{b}}{\sqrt{c} + \sqrt{d}}$, we change the fraction to an equivalent one with a rational denominator. This can be accomplished by multiplying both numerator and denominator by the rationalizing factor of the denominator, $\sqrt{c} - \sqrt{d}$.
This operation is called **rationalizing the denominator**.

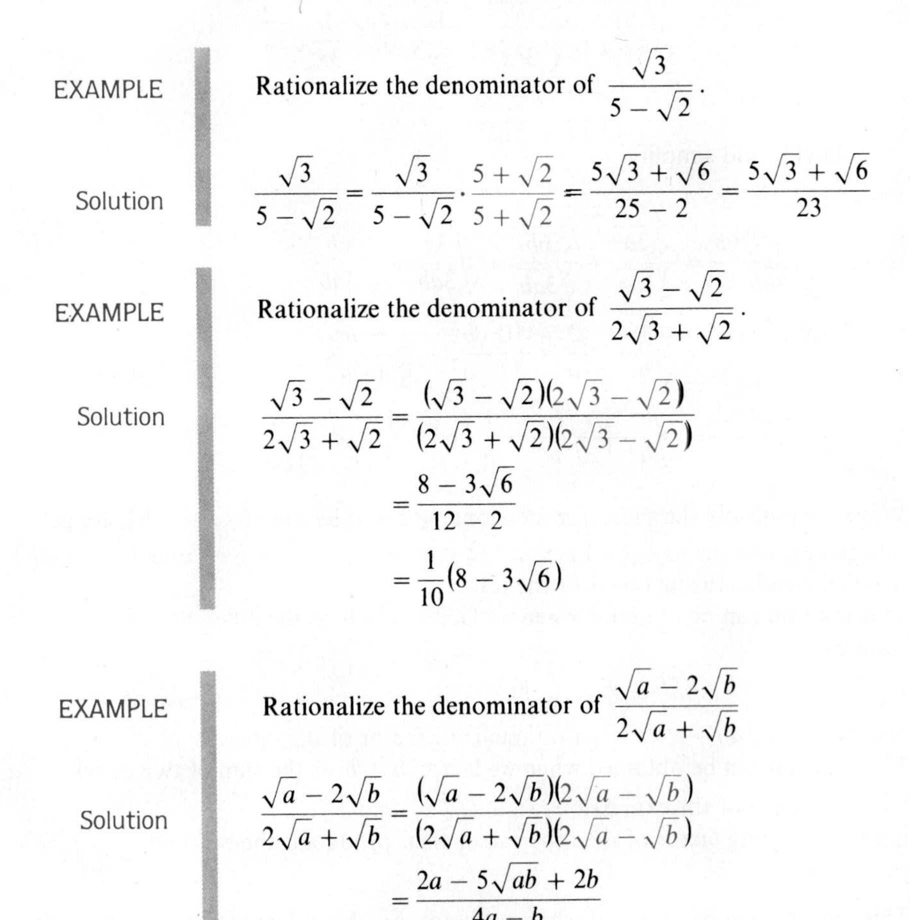

EXAMPLE

Rationalize the denominator of $\dfrac{\sqrt{3}}{5 - \sqrt{2}}$.

Solution

$$\frac{\sqrt{3}}{5 - \sqrt{2}} = \frac{\sqrt{3}}{5 - \sqrt{2}} \cdot \frac{5 + \sqrt{2}}{5 + \sqrt{2}} = \frac{5\sqrt{3} + \sqrt{6}}{25 - 2} = \frac{5\sqrt{3} + \sqrt{6}}{23}$$

EXAMPLE

Rationalize the denominator of $\dfrac{\sqrt{3} - \sqrt{2}}{2\sqrt{3} + \sqrt{2}}$.

Solution

$$\frac{\sqrt{3} - \sqrt{2}}{2\sqrt{3} + \sqrt{2}} = \frac{(\sqrt{3} - \sqrt{2})(2\sqrt{3} - \sqrt{2})}{(2\sqrt{3} + \sqrt{2})(2\sqrt{3} - \sqrt{2})}$$

$$= \frac{8 - 3\sqrt{6}}{12 - 2}$$

$$= \frac{1}{10}(8 - 3\sqrt{6})$$

EXAMPLE

Rationalize the denominator of $\dfrac{\sqrt{a} - 2\sqrt{b}}{2\sqrt{a} + \sqrt{b}}$

Solution

$$\frac{\sqrt{a} - 2\sqrt{b}}{2\sqrt{a} + \sqrt{b}} = \frac{(\sqrt{a} - 2\sqrt{b})(2\sqrt{a} - \sqrt{b})}{(2\sqrt{a} + \sqrt{b})(2\sqrt{a} - \sqrt{b})}$$

$$= \frac{2a - 5\sqrt{ab} + 2b}{4a - b}$$

EXAMPLE Rationalize the denominator of $\dfrac{\sqrt{a}+\sqrt{a-b}}{\sqrt{a}-\sqrt{a-b}}$.

Solution

$$\frac{\sqrt{a}+\sqrt{a-b}}{\sqrt{a}-\sqrt{a-b}} = \frac{\sqrt{a}+\sqrt{a-b}}{\sqrt{a}-\sqrt{a-b}} \cdot \frac{\sqrt{a}+\sqrt{a-b}}{\sqrt{a}+\sqrt{a-b}}$$

$$= \frac{a+2\sqrt{a(a-b)}+(a-b)}{a-(a-b)}$$

$$= \frac{2a-b+2\sqrt{a^2-ab}}{b}$$

EXAMPLE Rationalize the denominator of $\dfrac{\sqrt[3]{15}}{\sqrt[3]{5}-\sqrt[3]{3}}$.

Solution

$$\frac{\sqrt[3]{15}}{\sqrt[3]{5}-\sqrt[3]{3}} = \frac{\sqrt[3]{15}(\sqrt[3]{25}+\sqrt[3]{15}+\sqrt[3]{9})}{(\sqrt[3]{5}-\sqrt[3]{3})(\sqrt[3]{25}+\sqrt[3]{15}+\sqrt[3]{9})}$$

$$= \frac{5\sqrt[3]{3}+\sqrt[3]{225}+3\sqrt[3]{5}}{5-3}$$

$$= \frac{5\sqrt[3]{3}+\sqrt[3]{225}+3\sqrt[3]{5}}{2}$$

EXAMPLE Rationalize the denominator of $\dfrac{\sqrt{14}}{\sqrt{7}-\sqrt{2}+\sqrt{5}}$.

Solution

$$\frac{\sqrt{14}}{\sqrt{7}-\sqrt{2}+\sqrt{5}} = \frac{\sqrt{14}}{\sqrt{7}-(\sqrt{2}-\sqrt{5})}$$

$$= \frac{\sqrt{14}}{\sqrt{7}-(\sqrt{2}-\sqrt{5})} \cdot \frac{\sqrt{7}+(\sqrt{2}-\sqrt{5})}{\sqrt{7}+(\sqrt{2}-\sqrt{5})}$$

$$= \frac{7\sqrt{2}+2\sqrt{7}-\sqrt{70}}{7-(2-2\sqrt{10}+5)} = \frac{7\sqrt{2}+2\sqrt{7}-\sqrt{70}}{2\sqrt{10}}$$

$$= \frac{7\sqrt{2}+2\sqrt{7}-\sqrt{70}}{2\sqrt{10}} \cdot \frac{\sqrt{10}}{\sqrt{10}}$$

$$= \frac{14\sqrt{5}+2\sqrt{70}-10\sqrt{7}}{20}$$

$$= \frac{7\sqrt{5}+\sqrt{70}-5\sqrt{7}}{10}$$

Note Rationalizing the denominator is important since it facilitates two things:

1. For numerical computation, it is far easier to compute $\frac{1}{2}\sqrt{6}$ than $\frac{\sqrt{3}}{\sqrt{2}}$.
2. In algebraic manipulations, similar terms are often revealed.

However, in some instances, it is necessary to rationalize the numerator.

EXAMPLE Rationalize the numerator of $\frac{\sqrt{3}-\sqrt{2}}{3\sqrt{2}+\sqrt{3}}$.

Solution

$$\frac{\sqrt{3}-\sqrt{2}}{3\sqrt{2}+\sqrt{3}} = \frac{(\sqrt{3}-\sqrt{2})(\sqrt{3}+\sqrt{2})}{(3\sqrt{2}+\sqrt{3})(\sqrt{3}+\sqrt{2})} = \frac{3-2}{9+4\sqrt{6}} = \frac{1}{9+4\sqrt{6}}$$

Exercise 8.5

Divide and simplify the following radical expressions, given that $n \in N, n > 3$.

1. $\frac{\sqrt{6}}{\sqrt{2}}$ **2.** $\frac{\sqrt{10}}{\sqrt{5}}$ **3.** $\frac{2\sqrt{15}}{\sqrt{3}}$ **4.** $\frac{\sqrt{42}}{\sqrt{6}}$

5. $\frac{\sqrt{20}}{\sqrt{5}}$ **6.** $\frac{\sqrt{45}}{\sqrt{5}}$ **7.** $\frac{\sqrt{32}}{\sqrt{2}}$ **8.** $\frac{\sqrt{50}}{\sqrt{2}}$

9. $\frac{\sqrt[3]{24}}{\sqrt[3]{6}}$ **10.** $\frac{\sqrt[3]{21}}{\sqrt[3]{7}}$ **11.** $\frac{\sqrt[3]{40}}{\sqrt[3]{5}}$ **12.** $\frac{\sqrt[3]{-54}}{\sqrt[3]{2}}$

13. $\frac{2}{\sqrt{3}}$ **14.** $\frac{4}{\sqrt{2}}$ **15.** $\frac{3}{\sqrt{5}}$ **16.** $\frac{8}{\sqrt{6}}$

17. $\frac{2}{\sqrt{8}}$ **18.** $\frac{10}{\sqrt{12}}$ **19.** $\frac{5}{\sqrt{20}}$ **20.** $\frac{6}{\sqrt{18}}$

21. $\frac{6}{\sqrt{27}}$ **22.** $\frac{21}{\sqrt{28}}$ **23.** $\frac{2}{\sqrt[3]{3}}$ **24.** $\frac{1}{\sqrt[3]{5}}$

25. $\frac{3}{\sqrt[3]{4}}$ **26.** $\frac{2}{\sqrt[3]{9}}$ **27.** $\frac{4}{\sqrt[3]{16}}$ **28.** $\frac{3}{\sqrt[3]{81}}$

29. $\frac{2}{\sqrt[4]{2}}$ **30.** $\frac{5}{\sqrt[4]{3}}$ **31.** $\frac{1}{\sqrt[4]{8}}$ **32.** $\frac{3}{\sqrt[4]{27}}$

33. $\frac{4}{\sqrt[5]{2}}$ **34.** $\frac{6}{\sqrt[5]{3}}$ **35.** $\frac{1}{\sqrt[5]{9}}$ **36.** $\frac{4}{\sqrt[5]{4}}$

37. $\frac{6}{\sqrt[5]{-8}}$ **38.** $\frac{9}{\sqrt[5]{-27}}$ **39.** $\frac{8}{\sqrt[5]{16}}$ **40.** $\frac{2}{\sqrt[6]{32}}$

41. $\frac{\sqrt{5}}{\sqrt{2}}$ 42. $\frac{\sqrt{7}}{\sqrt{3}}$ 43. $\frac{\sqrt{2}}{\sqrt{5}}$ 44. $\frac{\sqrt{5}}{\sqrt{7}}$

45. $\frac{\sqrt{2}}{\sqrt{6}}$ 46. $\frac{\sqrt{3}}{\sqrt{6}}$ 47. $\frac{\sqrt{6}}{\sqrt{30}}$ 48. $\frac{\sqrt{14}}{\sqrt{21}}$

49. $\frac{\sqrt{6}}{\sqrt{15}}$ 50. $\frac{\sqrt{10}}{\sqrt{15}}$ 51. $\frac{\sqrt{24}}{\sqrt{18}}$ 52. $\frac{\sqrt{8}}{\sqrt{63}}$

53. $\frac{\sqrt{12}}{\sqrt{45}}$ 54. $\frac{\sqrt{50}}{\sqrt{27}}$ 55. $\frac{\sqrt[3]{6}}{\sqrt[3]{12}}$ 56. $\frac{\sqrt[3]{7}}{\sqrt[3]{21}}$

57. $\frac{\sqrt[3]{9}}{\sqrt[3]{36}}$ 58. $\frac{\sqrt[3]{25}}{\sqrt[3]{15}}$ 59. $\frac{\sqrt[3]{14}}{\sqrt[3]{10}}$ 60. $\frac{\sqrt[4]{2}}{\sqrt[4]{72}}$

61. $\frac{\sqrt[4]{3}}{\sqrt[4]{108}}$ 62. $\frac{\sqrt[4]{5}}{\sqrt[4]{1125}}$ 63. $\frac{\sqrt{2}}{\sqrt[3]{3}}$ 64. $\frac{\sqrt{2}}{\sqrt[3]{4}}$

65. $\frac{\sqrt[3]{3}}{\sqrt{2}}$ 66. $\frac{\sqrt{3}}{\sqrt[4]{27}}$ 67. $\frac{\sqrt{3}}{\sqrt[4]{243}}$ 68. $\frac{\sqrt{2}}{\sqrt[6]{32}}$

69. $\frac{\sqrt[3]{4}}{\sqrt[6]{32}}$ 70. $\frac{2}{\sqrt{x}}$ 71. $\frac{3}{\sqrt{3x}}$ 72. $\frac{4}{\sqrt{5x}}$

73. $\frac{6}{\sqrt{9x}}$ 74. $\frac{\sqrt{x}}{\sqrt{2}}$ 75. $\frac{\sqrt{3x}}{\sqrt{6}}$ 76. $\frac{\sqrt{5x}}{\sqrt{10}}$

77. $\frac{\sqrt{3}}{\sqrt{x}}$ 78. $\frac{\sqrt{5}}{\sqrt{2x}}$ 79. $\frac{\sqrt{2x}}{\sqrt{3x}}$ 80. $\frac{\sqrt{3x}}{\sqrt{15x}}$

81. $\frac{\sqrt{2x}}{\sqrt{3y}}$ 82. $\frac{\sqrt{3x^3y^2}}{\sqrt{6xy^3}}$ 83. $\frac{\sqrt{12xy^3}}{\sqrt{27x^3y^6}}$ 84. $\frac{\sqrt{2xy^2}}{\sqrt{3a^2b}}$

85. $\frac{\sqrt{4x^3}}{\sqrt{27a^3b^2}}$ 86. $\frac{\sqrt{6xy^5}}{\sqrt{7a^4b^3}}$ 87. $\frac{\sqrt{12x^2y}}{\sqrt{15ab^4}}$ 88. $\frac{3}{\sqrt{x+3}}$

89. $\frac{2}{\sqrt{x-2}}$ 90. $\frac{6}{\sqrt{x+4}}$ 91. $\frac{\sqrt{2}}{\sqrt{x-1}}$ 92. $\frac{\sqrt{3}}{\sqrt{3x+3}}$

93. $\frac{\sqrt{5}}{\sqrt{10x+5}}$ 94. $\frac{\sqrt{x}}{\sqrt{x+1}}$ 95. $\frac{\sqrt{2x}}{\sqrt{2x-1}}$ 96. $\frac{\sqrt{3x}}{\sqrt{x^2+2x}}$

97. $\frac{\sqrt{2x}}{\sqrt{x^2+3x}}$ 98. $\frac{\sqrt{x-2}}{\sqrt{x^2-4}}$ 99. $\frac{\sqrt{x+3}}{\sqrt{x^2-9}}$ 100. $\frac{\sqrt{x+1}}{\sqrt{x-1}}$

101. $\frac{\sqrt{x+2}}{\sqrt{x+3}}$ 102. $\frac{2}{\sqrt[3]{xy^2}}$ 103. $\frac{3}{\sqrt[3]{9xy}}$ 104. $\frac{x}{\sqrt[4]{xy^2}}$

105. $\frac{y}{\sqrt[4]{2x^2y^3}}$ 106. $\frac{\sqrt[3]{2x}}{\sqrt[3]{6xy}}$ 107. $\frac{\sqrt[3]{3y}}{\sqrt[3]{9xy}}$ 108. $\frac{\sqrt[3]{4x^2y}}{\sqrt[3]{2x^2y^2}}$

109. $\dfrac{\sqrt[3]{12xy^2}}{\sqrt[3]{18x^2y^2}}$

110. $\dfrac{\sqrt[3]{81a^6b^7}}{\sqrt[3]{625x^8y^{11}}}$

111. $\dfrac{\sqrt[4]{a^5b^8}}{\sqrt[4]{x^3y^9}}$

112. $\dfrac{\sqrt[4]{3a^7b^{10}}}{\sqrt[4]{8x^4y^6}}$

113. $\dfrac{\sqrt[5]{x^2y^{10}}}{\sqrt[5]{81a^5b^{12}}}$

114. $\dfrac{\sqrt[3]{b^4c}}{\sqrt[3]{a^2(b+c)^4}}$

115. $\dfrac{\sqrt[5]{a^6(a-b)}}{\sqrt[5]{c^8(a+b)^9}}$

116. $\dfrac{\sqrt[4]{a^9b^6}}{\sqrt[4]{c^7(a^4+b^4)}}$

117. $\dfrac{\sqrt[3]{b^3c^7}}{\sqrt[3]{a^5(b^2+c^2)}}$

118. $\dfrac{\sqrt{xy^3}}{\sqrt[3]{x^2y^7}}$

119. $\dfrac{\sqrt[3]{4x^2y^4}}{\sqrt{2xy^3}}$

120. $\dfrac{\sqrt[n]{2^{n+2}x^3}}{\sqrt[2n]{2^{n+4}x^{n+6}}}$

121. $\dfrac{\sqrt[4n]{2^8x^{3n-2}}}{\sqrt[2n]{2^{3n+4}x^{2n-1}}}$

122. $\dfrac{\sqrt[n]{2^{n+5}x^{2n+3}}}{\sqrt[3n]{256x^{n+6}}}$

123. $\dfrac{\sqrt{8}+\sqrt{18}}{\sqrt{2}}$

124. $\dfrac{\sqrt{12}+\sqrt{18}}{\sqrt{3}}$

125. $\dfrac{6-\sqrt{14}}{\sqrt{2}}$

126. $\dfrac{9-\sqrt{15}}{\sqrt{3}}$

127. $\dfrac{5-\sqrt{30}}{\sqrt{5}}$

128. $\dfrac{2+\sqrt{21}}{\sqrt{7}}$

129. $\dfrac{2\sqrt{3}-3\sqrt{2}}{\sqrt{6}}$

130. $\dfrac{\sqrt{6}+\sqrt{5}}{\sqrt{15}}$

131. $\dfrac{\sqrt{15}-\sqrt{35}}{\sqrt{21}}$

132. $\dfrac{\sqrt{10}-\sqrt{21}}{\sqrt{35}}$

133. $\dfrac{2+\sqrt{x}}{\sqrt{x}}$

134. $\dfrac{4-\sqrt{x}}{\sqrt{2x}}$

135. $\dfrac{\sqrt{x}+\sqrt{y}}{\sqrt{2xy}}$

136. $\dfrac{\sqrt{2x}-\sqrt{3y}}{\sqrt{6xy}}$

137. $\dfrac{\sqrt{6x}+\sqrt{10y}}{\sqrt{15xy}}$

138. $\dfrac{\sqrt{5x}+\sqrt{6y}}{\sqrt{30xy}}$

139. $\dfrac{\sqrt{6x}+\sqrt{21y}}{\sqrt{14xy}}$

140. $\dfrac{\sqrt[3]{2x}+\sqrt[3]{3y}}{\sqrt[3]{6xy}}$

141. $\dfrac{\sqrt[3]{4x^2y}+\sqrt[3]{12xy^2}}{\sqrt[3]{6x^2y^2}}$

142. $\dfrac{\sqrt[3]{3x^2y}-\sqrt[3]{18x^5y^4}}{\sqrt[3]{12x^2y^5}}$

- Rationalize the denominator:

143. $\dfrac{1}{4+\sqrt{5}}$

144. $\dfrac{3}{2-\sqrt{3}}$

145. $\dfrac{1}{4-\sqrt{7}}$

146. $\dfrac{\sqrt{2}}{2+\sqrt{3}}$

147. $\dfrac{\sqrt{3}}{1-\sqrt{3}}$

148. $\dfrac{\sqrt{3}}{\sqrt{3}+\sqrt{2}}$

149. $\dfrac{3\sqrt{5}}{\sqrt{5}+\sqrt{2}}$

150. $\dfrac{2\sqrt{7}}{\sqrt{7}-\sqrt{5}}$

151. $\dfrac{\sqrt{15}}{\sqrt{5}-\sqrt{3}}$

152. $\dfrac{5-2\sqrt{2}}{\sqrt{5}+\sqrt{2}}$

153. $\dfrac{2-\sqrt{3}}{\sqrt{3}-\sqrt{2}}$

154. $\dfrac{\sqrt{3}-\sqrt{2}}{\sqrt{3}+\sqrt{2}}$

155. $\dfrac{\sqrt{5}+\sqrt{2}}{\sqrt{5}-\sqrt{2}}$

156. $\dfrac{\sqrt{6}+\sqrt{2}}{2\sqrt{6}+\sqrt{2}}$

157. $\dfrac{\sqrt{6}-\sqrt{7}}{2\sqrt{6}+\sqrt{7}}$

158. $\dfrac{\sqrt{x}}{\sqrt{x}+\sqrt{y}}$

159. $\dfrac{2+\sqrt{x}}{\sqrt{x}-\sqrt{y}}$

160. $\dfrac{\sqrt{x}-\sqrt{y}}{\sqrt{2x}-\sqrt{y}}$

161. $\dfrac{\sqrt{3x}-\sqrt{2y}}{\sqrt{3x}+\sqrt{2y}}$

162. $\dfrac{\sqrt{x+4}}{\sqrt{x+4}-\sqrt{x}}$

163. $\dfrac{\sqrt{x}+\sqrt{x-9}}{\sqrt{x}-\sqrt{x-9}}$

164. $\dfrac{\sqrt{x+5}-\sqrt{x}}{\sqrt{x+5}+\sqrt{x}}$

165. $\dfrac{\sqrt{x+y}+\sqrt{x}}{\sqrt{x+y}-\sqrt{x}}$

166. $\dfrac{\sqrt{y}-\sqrt{x-y}}{\sqrt{y}+\sqrt{x-y}}$

167. $\dfrac{\sqrt[3]{4}}{\sqrt[3]{2}-1}$

168. $\dfrac{\sqrt[3]{12}}{\sqrt[3]{9}+2}$

169. $\dfrac{\sqrt[3]{6}}{\sqrt[3]{4}+\sqrt[3]{2}}$

170. $\dfrac{\sqrt[3]{9}}{\sqrt[3]{6}-\sqrt[3]{4}}$

171. $\dfrac{\sqrt[3]{100}}{\sqrt[3]{25}+\sqrt[3]{10}+\sqrt[3]{4}}$

172. $\dfrac{\sqrt[3]{7}-\sqrt[3]{3}}{\sqrt[3]{49}-\sqrt[3]{21}+\sqrt[3]{9}}$

173. $\dfrac{\sqrt{6}}{\sqrt{3}+\sqrt{2}+1}$

174. $\dfrac{\sqrt{10}}{\sqrt{6}+\sqrt{5}-1}$

175. $\dfrac{\sqrt{14}}{\sqrt{7}+\sqrt{6}-1}$

176. $\dfrac{3\sqrt{6}+2}{\sqrt{11}+\sqrt{6}+3}$

177. $\dfrac{\sqrt{42}}{\sqrt{10}-\sqrt{7}+\sqrt{3}}$

178. $\dfrac{\sqrt{3}}{\sqrt{3}+\sqrt{2}-1}$

179. $\dfrac{\sqrt{35}}{\sqrt{14}+\sqrt{7}+\sqrt{5}}$

180. $\dfrac{\sqrt{6}}{\sqrt{6}+\sqrt{3}+2\sqrt{2}}$

181. $\dfrac{\sqrt{11}}{\sqrt{21}-\sqrt{11}+\sqrt{10}}$

Rationalize the numerator:

182. $\dfrac{1+\sqrt{3}}{\sqrt{3}-\sqrt{2}}$

183. $\dfrac{3-\sqrt{2}}{\sqrt{3}-2\sqrt{2}}$

184. $\dfrac{\sqrt{6}-\sqrt{2}}{\sqrt{3}+1}$

185. $\dfrac{\sqrt{3}-3\sqrt{2}}{\sqrt{3}+\sqrt{2}}$

186. $\dfrac{\sqrt{5}+\sqrt{3}}{\sqrt{5}-2\sqrt{3}}$

187. $\dfrac{\sqrt{5}-\sqrt{2}}{2\sqrt{5}-\sqrt{2}}$

188. $\dfrac{\sqrt{10}+\sqrt{6}}{2\sqrt{10}-\sqrt{6}}$

189. $\dfrac{\sqrt{x}-\sqrt{y}}{\sqrt{x}+\sqrt{y}}$

190. $\dfrac{\sqrt{x}-2\sqrt{y}}{\sqrt{x}+3\sqrt{y}}$

191. $\dfrac{\sqrt{2x}+\sqrt{3y}}{\sqrt{3x}-\sqrt{2y}}$

192. $\dfrac{x+\sqrt{x^2+y^2}}{x-\sqrt{x^2+y^2}}$

193. $\dfrac{\sqrt{x}+\sqrt{x-6}}{\sqrt{x}-\sqrt{x-6}}$

8.6 Equations Involving Radical Expressions

When an equation involves a variable under one or more radicals, we eliminate the radicals one at a time. Form an equivalent equation in which one radical expression forms one side of the equation. Raise both sides of the equation to the power of the index of the radical, and simplify.

Repeat the process if the equation still contains a radical. Remember, only the principal roots are considered. All elements of the resulting solution set must be checked in the original equation.

EXAMPLE Solve $\sqrt{x+2} = 3$.

Solution Square both sides of the equation:

$$(\sqrt{x+2})^2 = (3)^2$$
$$x + 2 = 9$$
$$x = 7$$

Substitute 7 for x in the original equation:

Left Side	*Right Side*
$= \sqrt{(7)+2}$	$= 3$
$= \sqrt{9}$	$= 3$
$= 3$	$= 3$

Hence the solution set is $\{7\}$.

EXAMPLE Solve $\sqrt{x-3} + \sqrt{x} = 3$.

Solution

$$\sqrt{x-3} + \sqrt{x} = 3$$
$$\sqrt{x-3} = 3 - \sqrt{x}$$

Square both sides of the equation:

$$(\sqrt{x-3})^2 = (3 - \sqrt{x})^2$$
$$x - 3 = 9 - 6\sqrt{x} + x$$
$$6\sqrt{x} = 12$$
$$\sqrt{x} = 2$$
$$(\sqrt{x})^2 = 2^2$$
$$x = 4$$

Substitute 4 for x in the original equation:

Left Side	*Right Side*
$= \sqrt{(4)-3} + \sqrt{4}$	$= 3$
$= \sqrt{1} + \sqrt{4}$	$= 3$
$= 1 + 2$	$= 3$
$= 3$	$= 3$

Hence the solution set is $\{4\}$.

Note $(3 - \sqrt{x})^2 = (3 - \sqrt{x})(3 - \sqrt{x}) = 9 - 6\sqrt{x} + x$

EXAMPLE Solve $\sqrt{4x+17}+\sqrt{x+7}=\sqrt{x+2}$.

Solution Square both sides of the equation:

$$(\sqrt{4x+17}+\sqrt{x+7})^2=(\sqrt{x+2})^2$$
$$4x+17+2\sqrt{(4x+17)(x+7)}+x+7=x+2$$
$$2\sqrt{4x^2+45x+119}=-4x-22$$
$$\sqrt{4x^2+45x+119}=-2x-11$$
$$(\sqrt{4x^2+45x+119})^2=(-2x-11)^2$$
$$4x^2+45x+119=4x^2+44x+121$$
$$x=2$$

Substitute 2 for x in the original equation:

Left Side	*Right Side*
$=\sqrt{4(2)+17}+\sqrt{2+7}$	$=\sqrt{2+2}$
$=\sqrt{25}+\sqrt{9}$	$=\sqrt{4}$
$=5+3$	$=2$
$=8$	$=2$

Hence 2 is not a root.

The solution set is $\varnothing$.

Exercise 8.6

Solve the following equations:

1. $\sqrt{x+1}=2$
2. $\sqrt{x-1}=3$
3. $\sqrt{2x+1}=4$
4. $\sqrt{2x-3}=7$
5. $\sqrt{3x-3}=6$
6. $\sqrt{3x-5}=4$
7. $\sqrt{1-7x}=6$
8. $\sqrt{1-4x}=3$
9. $\sqrt{6x-2}=-2$
10. $\sqrt{9x-2}=-5$
11. $\sqrt[3]{3x-1}=2$
12. $\sqrt[3]{5x+2}=3$
13. $\sqrt[3]{6x-26}=4$
14. $\sqrt[3]{x-4}=-2$
15. $\sqrt[3]{7x-6}=-3$
16. $\sqrt{x-5}+\sqrt{x}=5$
17. $\sqrt{x+3}+\sqrt{x-2}=5$
18. $\sqrt{x+4}+\sqrt{x-3}=7$
19. $\sqrt{4x-3}+2\sqrt{x+2}=11$
20. $\sqrt{x+10}+3=\sqrt{x+37}$
21. $\sqrt{x+1}-\sqrt{x-2}=1$
22. $\sqrt{5x+9}-\sqrt{5x-4}=1$
23. $\sqrt{x+3}-2=\sqrt{x-5}$
24. $\sqrt{6x+22}-\sqrt{6x+3}=1$
25. $\sqrt{x-4}+\sqrt{x-1}=\sqrt{4x-11}$
26. $\sqrt{x-3}+\sqrt{x+5}=2\sqrt{x}$
27. $\sqrt{x-4}+\sqrt{x+4}=2\sqrt{x-1}$
28. $\sqrt{4x+13}+\sqrt{x+7}=3\sqrt{x+4}$
29. $\sqrt{x+6}+\sqrt{x+14}=2\sqrt{x+9}$
30. $\sqrt{x+2}+2\sqrt{x-1}=\sqrt{9x-2}$
31. $4\sqrt{x+8}-\sqrt{4x+25}=\sqrt{4x+41}$
32. $5\sqrt{x+4}-\sqrt{4x+21}=\sqrt{9x+31}$

33. $\sqrt{x+7} - 2\sqrt{x-8} = \sqrt{x-5}$

34. $\sqrt{9x+12} - 2\sqrt{x+5} = \sqrt{x-4}$

35. $\sqrt{x+1} - 2\sqrt{x+6} = \sqrt{x+13}$

36. $\sqrt{x+3} - 3\sqrt{x+11} = 2\sqrt{x+18}$

37. $\sqrt{x-1} - 3\sqrt{x+5} = \sqrt{4x+38}$

38. $\sqrt{x+13} - 3\sqrt{x-8} = \sqrt{4x-47}$

39. $\sqrt{x+4} - 2\sqrt{x+10} = \sqrt{x+20}$

40. $\dfrac{x+3}{\sqrt{x+1}} = \sqrt{x+6}$

41. $\dfrac{x+1}{\sqrt{x+4}} = \sqrt{x-1}$

42. $\dfrac{1-x}{\sqrt{x+5}} = \sqrt{x+1}$

43. $\dfrac{6}{\sqrt{x+2}} = \sqrt{x+2} + \sqrt{x-1}$

44. $\dfrac{10}{\sqrt{x-3}} = \sqrt{x-3} + \sqrt{x+2}$

45. $\dfrac{4}{\sqrt{x+7}} = \sqrt{x+7} + \sqrt{x+15}$

Chapter 8 Review

Put in standard form and combine similar radicals, given that $n \in N, n > 3$.

1. $\sqrt{12} - \sqrt{27} + \sqrt{75}$

2. $\sqrt{18} + \sqrt{72} - \sqrt{98}$

3. $\sqrt{32} + \sqrt{50} - \sqrt{128}$

4. $\sqrt{20} - \sqrt{45} + \sqrt{180}$

5. $\sqrt[3]{40} - \sqrt[3]{-135} - \sqrt[3]{320}$

6. $\sqrt[3]{72} - \sqrt[3]{243} - \sqrt[3]{-576}$

7. $\sqrt{54} + \sqrt[4]{36} - \sqrt[6]{196}$

8. $\sqrt{147} - \sqrt[4]{144} + \sqrt[6]{27}$

9. $\sqrt{150} + \sqrt[4]{576} - \sqrt[6]{216}$

10. $\sqrt{75} - \sqrt[4]{729} + \sqrt[6]{1728}$

11. $x\sqrt{4x^3y} - y\sqrt{9xy^3} + \sqrt{16x^5y^5}$

12. $\dfrac{1}{x}\sqrt{12x^4y^3} - \dfrac{1}{y}\sqrt{27x^2y^5} - \sqrt{48x^2y^3}$

13. $\dfrac{1}{y}\sqrt{2x^5y^4} + \dfrac{1}{x}\sqrt{8x^7y^2} - \dfrac{1}{xy}\sqrt{18x^7y^4}$

14. $\sqrt{4x+8} + \sqrt{x^3+2x^2} - \sqrt{(x+2)^3}$

15. $\dfrac{x}{2y}\sqrt[3]{16x^3y^5} + \dfrac{1}{3x}\sqrt[3]{54x^9y^2} + \dfrac{2}{x^2y^2}\sqrt[3]{-2x^{12}y^8}$

16. $\dfrac{1}{x}\sqrt{4x^5y^3} - \dfrac{1}{y}\sqrt{36x^3y^5} + \sqrt[4]{x^6y^6}$

17. $\dfrac{1}{x}\sqrt[3]{16x^5y} + \dfrac{3}{y}\sqrt[6]{4x^4y^8} + \sqrt[9]{8x^6y^3}$

18. $y\sqrt{18x^3y} - x\sqrt{8xy^3} - \dfrac{1}{y}\sqrt[6]{8x^9y^{15}}$

19. $\sqrt[n]{3^nx^{2n+1}y^{3n}} - \sqrt[n]{5^nx^ny^{2n+2}} + \sqrt[n]{2^{2n}x^{n+2}y^{4n+1}}$

20. $\sqrt[2n]{x^{4n+6}y^{6n+2}} + \sqrt[3n]{x^{6n+9}y^{9n+3}} - 6\sqrt[n]{x^{2n+3}y^{3n+1}}$

Perform the indicated operations and simplify, given that $n \in N, n > 5$.

21. $3\sqrt{7}\sqrt{14}$ **22.** $\sqrt{5}\sqrt{10}$ **23.** $\sqrt{3}\sqrt{21}$

24. $\sqrt{5}\sqrt{30}$ **25.** $\sqrt[3]{-10}\sqrt[3]{50}$ **26.** $\sqrt{10}\sqrt[3]{75}$

27. $\sqrt[3]{12}\sqrt[4]{216}$ **28.** $\sqrt{2x}\sqrt{6x}$ **29.** $\sqrt{3x}\sqrt{6x}$

30. $\sqrt{3x^2y^3}\sqrt{12xy^5}$ **31.** $\sqrt{14xy^2}\sqrt{21x^2y^4}$ **32.** $\sqrt{13x^3y^2}\sqrt{26xy^3}$

33. $\sqrt{7}\sqrt{7x+7}$ **34.** $\sqrt{x}\sqrt{xy+x}$ **35.** $\sqrt{2x}\sqrt{2x-4}$

36. $\sqrt{3x+1}\sqrt{9x^2-1}$ **37.** $\sqrt{2x-1}\sqrt{4x^2-1}$

38. $\sqrt{2x+3}\sqrt{8x^3+27}$ **39.** $\sqrt[3]{-3x}\sqrt[3]{18x^2}$

40. $\sqrt[3]{10x^2y^3}\sqrt[3]{25xy^2}$ **41.** $\sqrt[3]{21x^7y^5}\sqrt[3]{49x^4y^2}$

42. $\sqrt{6xy}\sqrt[3]{4x^2y}$ **43.** $\sqrt[3]{2x^3y}\sqrt[4]{32x^2y^6}$

44. $\sqrt[n]{x^{n-2}y^3}\sqrt[n]{x^{2n+3}y^{n+1}}$ **45.** $\sqrt[n]{2^5x^{n+1}y^{n-3}}\sqrt[n]{2^{n-3}x^{2n-1}y^{n+3}}$

46. $\sqrt[n]{27x^{n-4}y^{n+7}}\sqrt[2n]{3^{5n-1}x^{4n+10}y^{2n-5}}$

47. $\sqrt{10}(\sqrt{35}-\sqrt{6})$ **48.** $\sqrt{22}(\sqrt{6}+\sqrt{55})$ **49.** $\sqrt{6x}(\sqrt{3x}-\sqrt{2})$

50. $\sqrt{7x}(\sqrt{14x}+\sqrt{35})$ **51.** $(\sqrt{7}+3)(\sqrt{7}-2)$ **52.** $(\sqrt{5}-2)(\sqrt{5}+4)$

53. $(\sqrt{3}+\sqrt{5})(\sqrt{3}-\sqrt{5})$ **54.** $(2\sqrt{2}+\sqrt{7})^2$

55. $(\sqrt{5}-\sqrt{2})^2$ **56.** $(\sqrt{2x}+1)(\sqrt{2x}-1)$

57. $(\sqrt{3x}-\sqrt{2})(\sqrt{3x}+\sqrt{2})$ **58.** $(\sqrt{x}-3\sqrt{2y})(2\sqrt{x}+\sqrt{2y})$

59. $(\sqrt{3-x}+2)^2$ **60.** $(3\sqrt{x-1}+2)^2$ **61.** $(2\sqrt{x+1}-3)^2$

62. $(4-3\sqrt{x-3})^2$ **63.** $(\sqrt{x+3}+\sqrt{x-1})^2$

64. $(\sqrt{2x+1}+\sqrt{x-2})^2$ **65.** $(\sqrt{x+4}-\sqrt{x-2})^2$

66. $(\sqrt{3x-1}+2\sqrt{x+2})^2$ **67.** $(\sqrt{x+5}+3\sqrt{x-5})^2$

68. $(\sqrt{x+4}-3\sqrt{x-4})^2$ **69.** $(\sqrt{2x-3}-2\sqrt{x+3})^2$

70. $(\sqrt[3]{4}+\sqrt[3]{2})(\sqrt[3]{16}-\sqrt[3]{8}+\sqrt[3]{4})$ **71.** $(\sqrt[3]{9}+2)(\sqrt[3]{81}-\sqrt[3]{72}+4)$

72. $(\sqrt[3]{x}-3)(\sqrt[3]{x^2}+3\sqrt[3]{x}+9)$ **73.** $(\sqrt[3]{x}-2\sqrt[3]{y})(\sqrt[3]{x^2}+2\sqrt[3]{xy}+4\sqrt[3]{y^2})$

74. $(\sqrt{5}+\sqrt{3}-1)(\sqrt{5}-\sqrt{3}-1)$ **75.** $(\sqrt{6}+\sqrt{3}-\sqrt{2})(\sqrt{6}-\sqrt{3}+\sqrt{2})$

76. $\sqrt{\dfrac{54x^3y^2}{125a^5b^4}}$ **77.** $\sqrt[3]{\dfrac{64x^3y^7}{27ab^8}}$ **78.** $\sqrt[4]{\dfrac{4x^6(y+1)^2}{9a^8b^{14}}}$ **79.** $\dfrac{\sqrt{18x^3y^2}}{\sqrt[3]{36x^2y^5}}$

80. $\dfrac{\sqrt[4]{15x^5y^3}}{\sqrt[3]{15x^2y^7}}$ **81.** $\dfrac{\sqrt[n]{2^{n-1}x^{2n-3}y^{n+5}}}{\sqrt[2n]{2^{n+2}x^{3n-1}y^{2n+3}}}$ **82.** $\sqrt[n]{\dfrac{16^n-8^n}{4^n-2^n}}$

83. $\sqrt[n]{\dfrac{27^n+9^n}{3^{6n}+3^{5n}}}$ **84.** $\sqrt[n]{\dfrac{64^n+16^n}{32^n+8^n}}$ **85.** $\sqrt[n]{\dfrac{18^n+9^n}{6^n+3^n}}$

86. $\sqrt[n]{\dfrac{24^n-8^n}{12^n-4^n}}$ **87.** $\sqrt[n]{\dfrac{24^n-6^n}{8^n-2^n}}$ **88.** $\sqrt[n]{\dfrac{2^{3n+2}-2^{3n+1}}{2^{5n+1}-2^{5n}}}$

89. $\sqrt[n]{\dfrac{5^{4n+4}+5^{4n+2}}{5^{6n+1}+5^{6n-1}}}$ **90.** $\dfrac{2\sqrt{2}-\sqrt{3}}{3\sqrt{2}+\sqrt{3}}$ **91.** $\dfrac{\sqrt{7}-\sqrt{3}}{\sqrt{7}+2\sqrt{3}}$

92. $\dfrac{\sqrt{3x}-\sqrt{x}}{\sqrt{3x}+2\sqrt{x}}$ **93.** $\dfrac{\sqrt{x+3}-\sqrt{x}}{\sqrt{x+3}+\sqrt{x}}$ **94.** $\dfrac{\sqrt{x+y}+\sqrt{y}}{\sqrt{x+y}-\sqrt{y}}$

95. $\dfrac{\sqrt{x^2+y^2}+\sqrt{x^2-y^2}}{\sqrt{x^2+y^2}-\sqrt{x^2-y^2}}$ **96.** $\dfrac{2}{\sqrt[3]{3}-1}$

97. $\dfrac{\sqrt[3]{2}}{\sqrt[3]{3}+\sqrt[3]{2}}$ **98.** $\dfrac{\sqrt[3]{5}+\sqrt[3]{3}}{\sqrt[3]{25}+\sqrt[3]{15}+\sqrt[3]{9}}$

99. $\dfrac{\sqrt[3]{3}-2\sqrt[3]{2}}{\sqrt[3]{9}+2\sqrt[3]{6}+4\sqrt[3]{4}}$ **100.** $\dfrac{\sqrt{2}}{\sqrt{2}+\sqrt{3}+\sqrt{5}}$

101. $\dfrac{\sqrt{3}-\sqrt{2}}{\sqrt{6}-\sqrt{3}+\sqrt{2}}$ **102.** $\dfrac{\sqrt{5}}{\sqrt{3}-\sqrt{2}+1}$

Write equivalent radicals with 1 as a coefficient and reduce the radicand:

103. $5\sqrt{2}$ **104.** $2\sqrt{3}$ **105.** $3\sqrt{5}$ **106.** $\dfrac{1}{5}\sqrt{35}$

107. $\dfrac{1}{2}\sqrt[3]{2}$ **108.** $\dfrac{1}{3}\sqrt[3]{3}$ **109.** $\dfrac{2}{3}\sqrt[3]{\dfrac{21}{2}}$ **110.** $\dfrac{2}{3}\sqrt[3]{\dfrac{3}{2}}$

111. $2\sqrt[4]{x}$ **112.** $x\sqrt[4]{3}$ **113.** $\dfrac{x}{2y}\sqrt{6xy}$ **114.** $\dfrac{2x}{y}\sqrt{\dfrac{y}{x}}$

115. $\dfrac{x^2y}{3}\sqrt{\dfrac{6}{xy}}$ **116.** $\dfrac{y^3}{x^2}\sqrt{\dfrac{x^3}{y^5}}$ **117.** $\dfrac{2x}{5y^2}\sqrt{\dfrac{5y}{2x}}$ **118.** $\dfrac{3y^2}{2x}\sqrt{\dfrac{4x^2}{27y^5}}$

119. $(x+1)\sqrt{\dfrac{3}{x^2-1}}$ **120.** $\dfrac{2x}{3y}\sqrt[3]{\dfrac{3y^2}{4x}}$

121. $\dfrac{5x}{6y^2}\sqrt[3]{\dfrac{9y^5}{25x^4}}$ **122.** $\dfrac{2x^2}{7y}\sqrt[3]{\dfrac{147y^2}{4x^3}}$

Solve the following equations:

123. $\sqrt{x+2}+\sqrt{x-3}=5$ **124.** $\sqrt{x+8}-\sqrt{x+4}=2$

125. $\sqrt{x+3}+\sqrt{x-2}=\sqrt{4x+1}$ **126.** $\sqrt{x-2}+\sqrt{x+1}=\sqrt{4x-3}$

127. $\sqrt{x-7}-2\sqrt{x-1}=\sqrt{x+9}$ **128.** $\sqrt{x+4}-3\sqrt{x-1}=\sqrt{4x-11}$

129. $\dfrac{x+4}{\sqrt{x+2}}=\sqrt{x+7}$ **130.** $\dfrac{x-2}{\sqrt{x+1}}=\sqrt{x-4}$

131. $\dfrac{14}{\sqrt{x-3}}=\sqrt{x-3}+\sqrt{x+18}$ **132.** $\dfrac{12}{\sqrt{x+8}}=\sqrt{x+20}+\sqrt{x+8}$

Cumulative Review

Chapter 5 Factor completely:

1. $(x + y)^2 + (x + y)$
2. $(x + y)^2 - 2(x + y)$
3. $(x - y)^2 - 4(x - y)$
4. $(x - y)^2 - 7(x - y)$
5. $(x - y)^2 + 2(y - x)$
6. $(x - y)^2 - 3(y - x)$
7. $(x - 3)^3 + 2(3 - x)^2$
8. $(x - 1)^3 - 6(1 - x)^2$
9. $(x - y)^3 - 4(y - x)^2$
10. $(x - y)^3 + 5(y - x)^2$
11. $4x^2 - 1$
12. $9x^2 - 49$
13. $16x^4 - x^2$
14. $x^4 - 25x^2$
15. $x^4 - 1$
16. $16x^4 - 81$
17. $(x - 1)^2 - y^2$
18. $(x + 1)^2 - 4y^2$
19. $4x^2 - (y + 1)^2$
20. $9x^2 - (y - 1)^2$
21. $8x^3 + 1$
22. $27x^4 + x$
23. $81x^4 + 3x$
24. $x^6 + 64$
25. $27x^3 - 1$
26. $27x^4 - 8x$
27. $x^6 - 1$
28. $x^3 - 8(y - 1)^3$
29. $x^2 + 6x + 8$
30. $x^2 + 8x + 16$
31. $x^2 + 16x + 63$
32. $x^2 + 12xy + 20y^2$
33. $x^2 - 8x + 15$
34. $x^2 - 13x + 30$
35. $x^2 + 2x - 80$
36. $x^2 - 15x + 36$
37. $x^2 + 2x - 35$
38. $x^2 - 11x + 18$
39. $x^2 - 13x + 36$
40. $x^2 - 19x + 48$
41. $x^2 - 3x - 18$
42. $x^2 - 4x - 21$
43. $x^2 - 6x - 72$
44. $x^2 - 10xy - 24y^2$
45. $x^4 + 3x^2 - 4$
46. $x^4 - 4x^2 + 3$
47. $x^4 - 5x^2 + 4$
48. $x^4 - 8x^2 + 16$
49. $6x^2 + 7x + 2$
50. $2x^2 + 15x + 18$
51. $9x^2 + 12x + 4$
52. $9x^2 + 39x + 12$
53. $6x^2 - 11x + 4$
54. $2x^2 - 13x + 15$
55. $3x^4 - 8x^3 + 4x^2$
56. $9x^2 - 6x + 8$
57. $3x^2 + 7x - 6$
58. $9x^2 + 9x - 4$
59. $12x^2 + 5x - 2$
60. $12x^2 + 19x - 18$
61. $3x^2 - 13x - 10$
62. $6x^2 - 7x - 20$
63. $3x^2 - 7x - 6$
64. $3x^2 - 16x - 12$
65. $6 - 5x - 4x^2$
66. $12 + 5x - 2x^2$
67. $3 + x - 4x^2$
68. $21 - 5x - 6x^2$
69. $2x^4 - 5x^2 - 12$
70. $27x^4 - 30x^2 + 8$
71. $2x^6 - 13x^3 - 24$
72. $3x^6 - x^3 - 4$
73. $6(x - y)^2 - 25(x - y) + 4$
74. $6(x + y)^2 - 5(x + y) - 6$
75. $81x^4 + 4$
76. $625x^4 + 324$
77. $x^4 - 11x^2 + 1$
78. $x^4 - 20x^2 + 4$
79. $4x^4 - 29x^2 + 1$
80. $4x^4 + 8x^2 + 9$
81. $9x^4 + 3x^2 + 4$
82. $16x^4 + 20x^2 + 9$
83. $x^2 - 9 + 4y^2 - 4xy$
84. $4x^2 - 4 + 9y^2 - 12xy$
85. $x^2 - z^2 + 6xy + 9y^2$
86. $4x^2 - 25z^2 + 4xy + y^2$
87. $4x^2 - 20x - 9y^2 + 25$
88. $x^2 - 4x - y^2 + 4$
89. $x^2 - 16 - y^2 - 8y$
90. $4x^2 - 36 - y^2 - 12y$
91. $9x^2 + 4y - 4y^2 - 1$
92. $16x^2 + 6y - y^2 - 9$

93. $3x^2 + 4x - 4y^2 - 3xy^2$

94. $3x^2 - xy - 9x + 3y$

95. $21x^2 + 12x - 14xy - 8y$

96. $4x^3 - 4x^2y - 4y^3 + 4xy^2$

97. $x^2 - 3x - 4y^2 + 6y$

98. $9x^2 - 12x - y^2 - 4y$

99. $x^3 + 2x^2 - y^3 - 2y^2$

100. $x^3 - 3x^2 + 8y^3 + 12y^2$

Chapter 6 Reduce the following fractions to their lowest terms:

101. $\dfrac{80x^8y^6z^4}{64x^4y^9z^5}$

102. $\dfrac{36x^4y^6z^3}{48x^5y^8z^3}$

103. $\left(\dfrac{x^3y^6}{x^4y^2}\right)^3$

104. $\left(\dfrac{2x^7y}{x^5y^4}\right)^4$

105. $\dfrac{(28x^5y^6)^2}{(21x^4y^9)^3}$

106. $\dfrac{(15x^3y)^3}{(25x^3y^2)^2}$

107. $\dfrac{(x-2)^5}{(2-x)^3}$

108. $\dfrac{(x-1)^4}{(1-x)^2}$

109. $\dfrac{4x^2 - 9}{2x^2 - 11x + 12}$

110. $\dfrac{3x^2 + 4x + 1}{2x^2 + x - 1}$

111. $\dfrac{6x^2 + x - 12}{6x^2 - 23x + 20}$

112. $\dfrac{x^2 + 7x - 8}{6 - 5x - x^2}$

113. $\dfrac{3 - 4x - 4x^2}{6x^2 + x - 2}$

114. $\dfrac{x^4 + 4x^2 - 12}{3x^4 + 14x^2 - 24}$

115. $\dfrac{2(x+y)^2 + 5(x+y) - 12}{2(x+y)^2 + (x+y) - 6}$

116. $\dfrac{x^2 + 2xy + y^2 - 9}{(x+y)^2 - (x+y) - 12}$

Combine into a single fraction and simplify:

117. $\dfrac{3x}{x^2 - 2x - 8} + \dfrac{4}{x^2 - 4}$

118. $\dfrac{3x + 4}{x^2 + 3x + 2} + \dfrac{6}{x^2 + x - 2}$

119. $\dfrac{3x + 6}{x^2 + x - 20} - \dfrac{x}{x^2 - 6x + 8}$

120. $\dfrac{7x + 14}{3x^2 + 10x - 8} - \dfrac{x + 6}{3x^2 + x - 2}$

121. $\dfrac{2x + 4}{x^2 + 4x + 3} + \dfrac{2x + 5}{x^2 - x - 2} - \dfrac{1}{x + 3}$

122. $\dfrac{4x - 5}{x^2 + x - 12} + \dfrac{9}{18 - 3x - x^2} + \dfrac{2}{x^2 + 10x + 24}$

123. $\dfrac{4}{4x^2 + 4x - 3} - \dfrac{5x + 5}{3 - x - 2x^2} + \dfrac{x}{2x^2 - 3x + 1}$

124. $\dfrac{2x - 2}{x^2 - 2x - 8} + \dfrac{2x - 9}{x^2 - 7x + 12} + \dfrac{x + 7}{x^2 - x - 6}$

125. $\dfrac{2x - 1}{x^2 - x - 2} - \dfrac{3x - 1}{x^2 - 2x - 3} + \dfrac{3x + 7}{x^2 - 5x + 6}$

126. $\dfrac{x - 1}{2x^2 + 5x + 2} - \dfrac{x - 5}{2x^2 + x - 6} + \dfrac{8x}{4x^2 - 4x - 3}$

127. $\dfrac{4x - 5}{3x^2 - 11x + 6} - \dfrac{x - 3}{6x^2 - x - 2} - \dfrac{x + 4}{2x^2 - 5x - 3}$

128. $\dfrac{2x-10}{6x^2-5x-4}-\dfrac{4x-14}{3x^2+5x-12}+\dfrac{7x+11}{2x^2+7x+3}$

Perform the indicated operations and simplify:

129. $\dfrac{4x^2+13x+3}{3x^2+11x+6}\cdot\dfrac{3x^2+14x+8}{3x^2+13x+4}$

130. $\dfrac{4x^2+4x+1}{6x^2+x-1}\cdot\dfrac{9x^2+9x-4}{8x^2+10x+3}$

131. $\dfrac{12x^3+41x^2+24x}{27x^2-18x-24}\cdot\dfrac{18x^2-36x+16}{12x^3+x^2-6x}$

132. $\dfrac{9x^2-6xy+y^2}{12x^2+13xy-4y^2}\cdot\dfrac{6x^2+5xy-6y^2}{6x^2+7xy-3y^2}$

133. $\dfrac{30+x-x^2}{x^2-2x-48}\cdot\dfrac{x^2-12x+32}{x^2-10x+24}$

134. $\dfrac{x^2-2x-35}{x^2-x-42}\cdot\dfrac{x^2-14x+48}{40+3x-x^2}$

135. $\dfrac{x^4+x^2-2}{x^2+3x+2}\cdot\dfrac{x^2+x-6}{x^4-x^2-12}$

136. $\dfrac{x^4-7x^2+12}{x^2-5x+6}\cdot\dfrac{x^2+4x+3}{x^4-10x^2+9}$

137. $\dfrac{3(x-y)^2-11(x-y)-4}{6(x-y)^2-7(x-y)-3}\cdot\dfrac{2(x-y)^2+5(x-y)-12}{(x-y)^2-16}$

138. $\dfrac{8x^2+10x-3}{12x^2-23x-24}\cdot\dfrac{12x^2-41x+24}{12x^2-25x+12}\cdot\dfrac{4x^2+27x+18}{2x^2+15x+18}$

139. $\dfrac{50x^4y}{16x^2}\div\dfrac{25x^3}{32y}\cdot\dfrac{x}{y^3}$

140. $\dfrac{a^3x^2}{b^2y}\div\dfrac{a^2x^3}{3by^2}\cdot\dfrac{ay}{bx}$

141. $\dfrac{x^2+8x+7}{x^2-6x+5}\div\dfrac{x^2+5x-14}{x^2-7x+10}$

142. $\dfrac{x^2-12x+35}{x^2+3x-40}\div\dfrac{x^2+5x+4}{x^2+12x+32}$

143. $\dfrac{6x^2-23x-18}{6x^2+13x+6}\div\dfrac{4x^2-20x+9}{6x^2+5x-6}$

144. $\dfrac{2x^2-4x-48}{x^3+10x^2+24x}\div\dfrac{4x^2-12x-72}{x^3-5x^2-24x}$

145. $\dfrac{x^2-10x+16}{x^2-7x-8}\div\dfrac{x^2-8x-20}{10+9x-x^2}$

146. $\dfrac{20x^2+7x-6}{16x^2+24x+9}\div\dfrac{10x^2-39x+14}{28+13x-6x^2}$

147. $\dfrac{3(x+y)^2-11(x+y)-4}{(x+y)^2-2(x+y)-8}\div\dfrac{3(x+y)^2-17(x+y)-6}{2(x+y)^2+(x+y)-6}$

148. $\dfrac{3(x-y)^2+10(x-y)-8}{2(x-y)^2+5(x-y)-12}\div\dfrac{9(x-y)^2-4}{2(x-y)^2+3(x-y)-9}$

149. $\dfrac{6x^2-19x-36}{2x^2-17x+36}\div\dfrac{6x^2-23x-18}{4x^2-19x+12}\cdot\dfrac{12x^2-35x+18}{12x^2-11x-36}$

150. $\dfrac{3x^2 - 11x - 42}{3x^2 - 20x + 12} \div \dfrac{6x^2 + 5x - 21}{8x^2 - 26x + 21} \cdot \dfrac{3x^2 + 16x - 12}{4x^2 + 25x - 56}$

151. $\dfrac{3}{x - 2} - \dfrac{8x - 4}{2x^2 - 5x - 3} \cdot \dfrac{x^2 - x - 6}{2x^2 + 3x - 2}$

152. $\dfrac{x}{2x - 3} + \dfrac{3x + 12}{x^2 - 4x - 12} \div \dfrac{x^2 + 8x + 16}{x^2 + 6x + 8}$

153. $\left(x - \dfrac{1}{2x + 1}\right)\left(x - \dfrac{1}{2x - 1}\right)$

154. $\left(x + 4 + \dfrac{5}{x - 2}\right)\left(x - \dfrac{x}{x - 1}\right)$

155. $\left(2x + 5 + \dfrac{9}{x - 2}\right)\left(x - 1 - \dfrac{3}{x + 1}\right)$

156. $\left(3x - 1 + \dfrac{10}{x + 4}\right)\left(x - 3 - \dfrac{7}{x + 3}\right)$

157. $\left(2 - \dfrac{3}{2x + 1}\right) \div \left(7 - \dfrac{9}{2x + 1}\right)$

158. $\left(x - \dfrac{4}{3x + 4}\right) \div \left(x + \dfrac{2}{3x - 5}\right)$

159. $\left(x - 6 - \dfrac{10}{x + 3}\right) \div \left(x + 2 - \dfrac{2}{x + 3}\right)$

160. $\left(x + 1 - \dfrac{8}{x - 1}\right) \div \left(x + 7 + \dfrac{16}{x - 1}\right)$

161. $\dfrac{x + 4}{x + 1 - \dfrac{6}{x + 2}}$

162. $\dfrac{3x + 2}{3x + 8 + \dfrac{26}{2x - 3}}$

163. $\dfrac{x + 1 - \dfrac{9}{2x - 1}}{x + 2 - \dfrac{12}{2x - 1}}$

164. $\dfrac{2x - 3 + \dfrac{2}{3x + 2}}{x + 1 - \dfrac{14}{3x + 2}}$

165. $\dfrac{x + 5 + \dfrac{7}{x - 3}}{2x + 3 - \dfrac{15}{x + 1}}$

166. $\dfrac{x - 9 + \dfrac{21}{x + 1}}{x - 11 + \dfrac{50}{x + 4}}$

Solve the following equations for x:

167. $ax = 2 - 5x$

168. $2ax = 3x - 1$

169. $ax - 2a = 4x - 8$

170. $ax + 3 = -3x - a$

171. $ax - a + 2 = a^2 - 2x$

172. $2ax + 1 = 4a^2 + x$

173. $2ax + a + 6 = 2a^2 - 3x$

174. $ax - 8 = a^3 - 2x$

175. $\dfrac{x + 5}{x^2 + 3x + 2} + \dfrac{x - 1}{x^2 - x - 6} = \dfrac{2x - 1}{x^2 - 2x - 3}$

176. $\dfrac{3x-4}{4x^2-21x+5}-\dfrac{x-1}{2x^2-13x+15}=\dfrac{2x+1}{8x^2-14x+3}$

177. $\dfrac{2x-1}{x^2+4x-5}+\dfrac{x-2}{x^2-10x+9}=\dfrac{3x-12}{x^2-4x-45}$

178. $\dfrac{4x-3}{x^2+5x+6}-\dfrac{2x+3}{x^2-x-6}=\dfrac{2x-7}{x^2-9}$

179. $\dfrac{8x-1}{6x^2-7x-3}-\dfrac{2x-7}{2x^2-11x+12}=\dfrac{x-5}{3x^2-11x-4}$

180. $\dfrac{x^2+1}{x^2-3x-18}-\dfrac{3x+1}{x^2-5x-24}=1$

181. $\dfrac{3x-2}{12x^2+5x-28}+\dfrac{2x+5}{8x^2+2x-21}=\dfrac{3x-5}{6x^2-17x+12}$

182. $\dfrac{2x+1}{4x^2-4x+1}+\dfrac{x}{2x^2+7x-4}=\dfrac{x}{x^2+x-12}$

183. $\dfrac{x+3}{x^2+3x+2}+\dfrac{x+9}{x^2-2x-3}=\dfrac{2x+9}{x^2-x-6}$

184. $\dfrac{x+7}{x^2+2x-3}+\dfrac{x+1}{x^2-3x+2}=\dfrac{2x+11}{x^2+x-6}$

185. $\dfrac{x-6}{x^2-6x+8}+\dfrac{3}{2x^2-9x+4}=\dfrac{2x+2}{2x^2-5x+2}$

186. $\dfrac{x-2}{6x^2+x-1}+\dfrac{3x-2}{2x^2+9x+4}=\dfrac{5x-1}{3x^2+11x-4}$

187. What number must be added to the numerator and subtracted from the denominator of the fraction $\dfrac{53}{112}$ to give a fraction equal to $\dfrac{4}{7}$?

188. The denominator of a simple fraction exceeds its numerator by 38. If 7 is added to the numerator and 13 is subtracted from the denominator, the value of the fraction becomes $\dfrac{3}{4}$. Find the original fraction.

189. One number is 118 more than another number. If the larger number is divided by the smaller number, the quotient is 3 and the remainder is 24. Find the two numbers.

190. The tens digit of a two-digit number exceeds its units digit by 3. If the number is divided by the sum of its digits the quotient is 6 and the remainder is 8. Find the number.

191. The sum of the digits of a three-digit number is 17, and the tens digit is 2 less than its hundreds digit. If the number is divided by its units digit, the quotient is 92 and the remainder is 3. Find the number.

192. If A can do a job in 35 hours and B can do the same job in 14 hours, how long will it take the two of them working together to do the job?

193. If A can do a job in 110 hours and it takes A and B working together 60 hours to do the same job, how long will it take B working alone to do the job?

194. It takes A $\frac{4}{5}$ as long as it takes B to do a job. If A and B working together can do the job in 20 hours, how long will it take each working alone to do the job?

195. It takes one pipe $\frac{2}{3}$ as long as it takes another pipe to fill a tank. If the two pipes together fill the tank in 6 minutes, how long does it take each pipe alone to fill the tank?

196. It took Jim the same time to drive 30 miles as it took him to fly 378 miles. The average speed of the plane was 20 mph slower than 13 times the speed of the car. What was the average speed of the plane?

Chapter 7 Perform the indicated operations and simplify:

197. $x^{\frac{5}{3}}y^{\frac{1}{6}} \cdot x^{\frac{1}{3}}y^{\frac{1}{2}}$

198. $x^{\frac{3}{4}}y^{\frac{2}{3}} \cdot x^{\frac{1}{8}}y^{\frac{1}{2}}$

199. $(x^2y^3)^{\frac{1}{4}}(x^4y^2)^{\frac{1}{8}}$

200. $(x^6y^4)^{\frac{1}{3}}(x^3y^2)^{\frac{1}{2}}$

201. $\left(x^2y^{\frac{3}{2}}\right)^2\left(x^{\frac{1}{8}}y^{\frac{3}{4}}\right)^4$

202. $\left(x^{\frac{5}{6}}y^{\frac{1}{3}}\right)^3\left(x^{\frac{3}{8}}y^{\frac{1}{2}}\right)^4$

203. $\left(x^{\frac{1}{2}}y^{\frac{2}{3}}\right)^{\frac{6}{5}}\left(x^{\frac{2}{3}}y\right)^{\frac{3}{5}}$

204. $(x^3y^6)^{\frac{1}{6}}\left(x^{\frac{5}{4}}y^{\frac{3}{2}}\right)^{\frac{4}{3}}$

205. $(3x-1)\left(x^{\frac{1}{2}}+2\right)$

206. $(x+3)\left(x^{\frac{1}{3}}-1\right)$

207. $\left(2x^{\frac{1}{2}}+3\right)\left(x^{\frac{1}{2}}-4\right)$

208. $\left(4x^{\frac{1}{4}}-1\right)\left(3x^{\frac{1}{4}}+2\right)$

209. $\left(2x^{\frac{2}{3}}+x^{\frac{1}{3}}+1\right)\left(x^{\frac{2}{3}}-2x^{\frac{1}{3}}-1\right)$

210. $\left(3x^{\frac{2}{3}}-2x^{\frac{1}{3}}+1\right)\left(x^{\frac{2}{3}}+3x^{\frac{1}{3}}-2\right)$

211. $\dfrac{x^{\frac{2}{5}}y^{\frac{4}{3}}}{xy^{\frac{1}{3}}}$

212. $\dfrac{x^{\frac{1}{2}}y^{\frac{7}{3}}z^{\frac{1}{5}}}{x^{\frac{3}{2}}y^{\frac{1}{3}}z^{\frac{3}{5}}}$

213. $\left(\dfrac{x^{\frac{1}{3}}y^{\frac{4}{3}}}{x^{\frac{1}{4}}y^{\frac{5}{6}}}\right)^6$

214. $\left(\dfrac{x^{\frac{3}{4}}y^{\frac{5}{4}}}{x^{\frac{7}{8}}y^{\frac{3}{2}}}\right)^4$

215. $\dfrac{\left(x^{\frac{3}{2}}y^{\frac{1}{4}}\right)^4}{\left(x^{\frac{5}{6}}y^{\frac{1}{3}}\right)^6}$

216. $\dfrac{\left(x^{\frac{5}{8}}y^{\frac{3}{4}}\right)^8}{\left(x^{\frac{2}{3}}y^{\frac{3}{2}}\right)^6}$

217. $\dfrac{\left(x^{\frac{2}{3}}y^{\frac{8}{3}}\right)^{\frac{3}{4}}}{\left(x^{\frac{2}{5}}y^3\right)^{\frac{5}{6}}}$

218. $\dfrac{\left(x^{\frac{6}{5}}y^{\frac{3}{4}}\right)^{\frac{5}{6}}}{\left(x^{\frac{7}{2}}y^{\frac{3}{8}}\right)^{\frac{4}{3}}}$

219. $\left(6x+x^{\frac{1}{2}}-2\right) \div \left(3x^{\frac{1}{2}}+2\right)$

220. $\left(3x+10x^{\frac{1}{2}}-8\right) \div \left(x^{\frac{1}{2}}+4\right)$

221. $\left(x-x^{\frac{3}{4}}-3x^{\frac{1}{2}}+5x^{\frac{1}{4}}-2\right) \div \left(x^{\frac{1}{2}}+x^{\frac{1}{4}}-2\right)$

222. $\left(6x^2-5x^{\frac{3}{2}}+2x-1\right) \div \left(2x-x^{\frac{1}{2}}+1\right)$

223. $\left(x^2+4x^{\frac{3}{2}}+4x-9\right) \div \left(x+2x^{\frac{1}{2}}+3\right)$

224. $\left(2x^{\frac{3}{2}} + x^{\frac{5}{4}} - x + x^{\frac{3}{4}} - 5x^{\frac{1}{2}} - 6\right) \div \left(x^{\frac{3}{4}} + x^{\frac{1}{2}} + 2\right)$

225. $\left(9x^2 - 4x + 4x^{\frac{1}{2}} - 1\right) \div \left(3x + 2x^{\frac{1}{2}} - 1\right)$

226. $\left(x^2 - 9x + 12x^{\frac{1}{2}} - 4\right) \div \left(x + 3x^{\frac{1}{2}} - 2\right)$

227. $(4x^2 - 13x + 9) \div \left(2x + x^{\frac{1}{2}} - 3\right)$

228. $(x^2 - 6x + 1) \div \left(x - 2x^{\frac{1}{2}} - 1\right)$

Simplify the following, and write the answers with positive exponents:

229. $2x^2yz^{-1} \cdot x^{-1}y^{-3}z^2$

230. $4^{-2}x^{-4}y^{-1} \cdot 3x^{-1}y^2$

231. $2^3x^{-1}y^3 \cdot 2^{-2}x^3y^{-3}$

232. $2^{-4}x^{-3}y^3 \cdot 2^5xy^{-2}$

233. $(x^3y^{-2})^2(x^2y^3)^{-2}$

234. $(2^{-2}x^2y^{-3})^{-2}(2^{-3}x^{-1}y^2)^2$

235. $(2^3x^{-1}y^{-3})^{\frac{1}{2}}(2x^{-3}y^{-1})^{-\frac{1}{2}}$

236. $(3^{-2}x^{-2}y^3)^{-\frac{1}{3}}(3x^4y^{-6})^{\frac{1}{3}}$

237. $(3x^{-1} + 1)(x^{-1} - 4)$

238. $(2x^{-1} + 3y^{-2})(x^{-1} - 2y^{-2})$

239. $(3x^{-1} + y^{-1})(9x^{-2} - 3x^{-1}y^{-1} + y^{-2})$

240. $(2x^{-1} - 3y^{-1})(4x^{-2} + 6x^{-1}y^{-1} + 9y^{-2})$

241. $\dfrac{x^{-3}y^{-4}}{x^2y^{-6}}$

242. $\dfrac{x^{-4}y^{-3}}{x^{-2}y^{-6}}$

243. $\dfrac{(x^3y^{-2})^{-1}}{(x^{-2}y^{-1})^2}$

244. $\dfrac{(2x^2y^{-2})^{-3}}{(2^{-4}x^{-4}y^5)^{-1}}$

245. $\dfrac{(x^{-1}y^4z^{-3})^3}{(x^3y^{-4}z^{-3})^{-2}}$

246. $\dfrac{(x^3y^{-2}z^{-1})^{-4}}{(x^{-4}y^{-1}z^3)^3}$

247. $\dfrac{2x^{-1} + y^{-2}}{3x^{-1} - y^{-2}}$

248. $\dfrac{3x^{-1} - 6x^{-2}}{1 - 4x^{-2}}$

249. $\dfrac{3x^{-1} + 1}{9x^{-2} - 1}$

250. $\dfrac{3x^{-2} + x^{-1}}{3x^{-2} - 2x^{-1} - 1}$

Write each of the following as one fraction with positive exponents and simplify:

251. $\dfrac{(2x - 1)^{\frac{1}{2}} - x(2x - 1)^{-\frac{1}{2}}}{(2x - 1)}$

252. $\dfrac{(x + 1)^{\frac{1}{2}} - \frac{1}{2}x(x + 1)^{-\frac{1}{2}}}{(x + 1)}$

253. $\dfrac{2x(x - 2)^{\frac{1}{3}} - \frac{1}{3}x^2(x - 2)^{-\frac{2}{3}}}{(x - 2)^{\frac{2}{3}}}$

254. $\dfrac{\frac{1}{2}x^{-\frac{1}{2}}(x^2 + 1)^{\frac{1}{3}} - \frac{2}{3}x^{\frac{3}{2}}(x^2 + 1)^{-\frac{2}{3}}}{(x^2 + 1)^{\frac{2}{3}}}$

Chapter 8 Put the following radicals in standard form given that $n \in N, n > 2$:

255. $\sqrt{108}$ **256.** $\sqrt{243}$ **257.** $\sqrt[3]{-40}$ **258.** $\sqrt[4]{144}$

259. $\sqrt{48x^2y^3z}$ **260.** $\sqrt{16x^5y^2z^4}$ **261.** $\sqrt{12x^6yz^5}$

262. $\sqrt{18x^3y^4z^3}$ **263.** $\sqrt[3]{x^3y^8z^6}$ **264.** $\sqrt[3]{8x^3y^4}$

265. $\sqrt[3]{-16x^4y^6}$ **266.** $\sqrt[3]{-54x^5y^7}$ **267.** $\sqrt[4]{16x^6y^2}$

268. $\sqrt[4]{9x^2y^8}$ **269.** $\sqrt[6]{64x^6y^4}$ **270.** $\sqrt[6]{25x^8y^6}$

271. $\sqrt[n]{2^nx^{n+1}y^{3n}}$ **272.** $\sqrt[n]{3^{n+1}x^{2n+1}y^{n+2}}$

273. $\sqrt[2n]{x^{4n+1}y^{6n+3}}$ **274.** $\sqrt[3n]{x^{6n}y^{7n+1}}$

Simplify and combine similar radical expressions:

275. $\sqrt{24} + \sqrt{36} + \sqrt{96}$ **276.** $\sqrt{49} - \sqrt{27} - \sqrt{48}$

277. $\sqrt{98} + \sqrt{128} - \sqrt{121}$ **278.** $\sqrt[3]{128} + \sqrt{18} - \sqrt{98} - \sqrt[3]{16}$

279. $\sqrt{108} - \sqrt[3]{-4} - \sqrt[3]{32} - \sqrt{75}$ **280.** $\sqrt[3]{16} - \sqrt[3]{-54} + \sqrt[6]{256}$

281. $2x\sqrt{xy^2} - 3y\sqrt{x^3} + 4\sqrt{x^3y^2}$ **282.** $3x\sqrt{x^3y} + 2y\sqrt{xy^3} + 2\sqrt{x^5y}$

283. $\sqrt{x^3 + x^2y} + \sqrt{xy^2 + y^3} + \sqrt{(x+y)^3}$

284. $\sqrt{24x^2} + \sqrt[3]{24x^4} - \sqrt{54x^2} + \sqrt[3]{81x^4}$

285. $\sqrt[4]{16x^2} + \sqrt[4]{81x^2} - \sqrt{36x}$ **286.** $\sqrt{75x^3} + \sqrt[3]{-24x^4} + \sqrt[4]{729x^6}$

287. $\sqrt[6]{x^8y^8} - x\sqrt[3]{8xy^4} - y\sqrt[3]{27x^4y}$ **288.** $\sqrt{9x^3y^3} + x\sqrt[4]{16x^2y^6} - y\sqrt[6]{x^9y^3}$

Perform the indicated multiplications and simplify:

289. $\sqrt{14}\sqrt{21}$ **290.** $\sqrt{26}\sqrt{39}$ **291.** $\sqrt[3]{4}\sqrt[3]{-16}$ **292.** $\sqrt[3]{-12}\sqrt[3]{9}$

293. $\sqrt[3]{-15}\sqrt[3]{-18}$ **294.** $\sqrt[3]{-10}\sqrt[3]{-25}$ **295.** $\sqrt[4]{27}\sqrt[4]{15}$

296. $\sqrt{6}\sqrt[4]{24}$ **297.** $\sqrt{10}\sqrt[3]{25}$ **298.** $\sqrt{x}\sqrt{xy + x}$

299. $\sqrt{x+2}\sqrt{x^2-4}$ **300.** $\sqrt[3]{4xy^2}\sqrt[3]{-6y}$ **301.** $\sqrt[4]{8x^2}\sqrt[4]{10x^3}$

302. $\sqrt[5]{4x^2}\sqrt[5]{8x^4}$ **303.** $\sqrt{2x}\sqrt[4]{4x^3}$ **304.** $\sqrt{3x}\sqrt[3]{9x^2}$

305. $\sqrt{14}(\sqrt{6} + \sqrt{21})$ **306.** $\sqrt{6}(\sqrt{15} - \sqrt{22})$

307. $(3\sqrt{2} - 4\sqrt{3})(\sqrt{2} - 3\sqrt{3})$ **308.** $(\sqrt{5} + 2\sqrt{7})(3\sqrt{5} + \sqrt{7})$

309. $(2x + \sqrt{y})(2x - \sqrt{y})$ **310.** $(\sqrt{2x} + \sqrt{y})(\sqrt{2x} + 3\sqrt{y})$

311. $(\sqrt{x-3} - 2)^2$ **312.** $(\sqrt{x} + \sqrt{x+1})^2$

313. $(\sqrt{x+2} - \sqrt{x-2})^2$ **314.** $(\sqrt{x+1} + \sqrt{2x-3})^2$

315. $(\sqrt{x+3} + 2\sqrt{x+2})^2$ **316.** $(\sqrt{2x+1} - 3\sqrt{x-2})^2$

Divide and simplify the following radical expressions:

317. $9 \div \sqrt{27}$ **318.** $7 \div \sqrt{12}$ **319.** $2 \div \sqrt[3]{5}$ **320.** $3 \div \sqrt[3]{49}$

321. $6 \div \sqrt[4]{3}$ **322.** $1 \div \sqrt[4]{8}$ **323.** $1 \div \sqrt[5]{8}$ **324.** $1 \div \sqrt[5]{81}$

325. $\sqrt{3} \div \sqrt{2}$ **326.** $\sqrt{5} \div \sqrt{3}$ **327.** $3\sqrt{2} \div \sqrt{15}$

328. $\sqrt{27} \div \sqrt{32}$ **329.** $\sqrt[3]{5} \div \sqrt[3]{10}$ **330.** $\sqrt{2} \div \sqrt[4]{8}$

331. $\sqrt{3} \div \sqrt[3]{9}$

332. $\sqrt{5} \div \sqrt[3]{5}$

333. $\sqrt{18a^4} \div \sqrt{20x^2y^5}$

334. $\sqrt{8a^2b^3} \div \sqrt{3x^3y^2}$

335. $\sqrt[3]{2x^3y^4} \div \sqrt[3]{16a^6b^5}$

336. $\sqrt[3]{5x^4y^6} \div \sqrt[3]{15a^3b^7}$

337. $\sqrt[4]{2a^4b} \div \sqrt[4]{32x^3y^5}$

338. $\sqrt[4]{a^6b^2} \div \sqrt[4]{x^6y^{10}}$

Rationalize the denominator:

339. $\dfrac{2}{1+\sqrt{2}}$

340. $\dfrac{1}{1-\sqrt{3}}$

341. $\dfrac{\sqrt{3}}{2-\sqrt{3}}$

342. $\dfrac{\sqrt{7}}{1+\sqrt{7}}$

343. $\dfrac{\sqrt{2}}{\sqrt{2}+\sqrt{3}}$

344. $\dfrac{\sqrt{14}}{\sqrt{7}-\sqrt{2}}$

345. $\dfrac{1+\sqrt{2}}{1-\sqrt{2}}$

346. $\dfrac{\sqrt{2}+\sqrt{3}}{\sqrt{3}-\sqrt{2}}$

347. $\dfrac{2\sqrt{3}+3\sqrt{2}}{4\sqrt{3}+\sqrt{2}}$

348. $\dfrac{\sqrt{15}-\sqrt{6}}{4\sqrt{5}+\sqrt{3}}$

Solve the following equations:

349. $\sqrt{3x+1} = 2$

350. $\sqrt{4x-3} = 3$

351. $\sqrt[3]{2x+1} = 3$

352. $\sqrt[3]{x-1} = 2$

353. $\sqrt{x+7} + \sqrt{x-5} = 6$

354. $\sqrt{x+5} + \sqrt{x-4} = 9$

355. $\sqrt{x+2} + \sqrt{x-6} = 4$

356. $\sqrt{x+4} + \sqrt{x-3} = 7$

357. $\sqrt{x-3} + \sqrt{x+2} = \sqrt{4x-3}$

358. $\sqrt{x-1} + \sqrt{x+6} = \sqrt{4x+9}$

359. $\sqrt{x-8} + \sqrt{x+4} = 2\sqrt{x-3}$

360. $\sqrt{x-15} + \sqrt{x+12} = \sqrt{4x-15}$

chapter 9

Linear Equations and Inequalities in Two Variables

9.1 Ordered pairs of numbers

9.2 Rectangular or Cartesian coordinates

9.3 Distance between two points

9.4 One linear equation in two variables

9.5 Slope of a line

9.6 Equations of lines

9.7 Two linear equations in two variables

9.8 Solution of systems of two linear equations in two variables

9.9 Systems of two linear equations in two variables involving grouping symbols and fractions

9.10 Fractional equations that can be made linear

9.11 Word problems

9.12 Graphs of linear inequalities in two variables

9.13 Graphical solution of a system of linear inequalities in two variables

Linear equations and inequalities in one variable and their solution were treated in Chapters 3 and 4. In this chapter we shall discuss linear equations and inequalities in two variables and systems of such equations and inequalities.

9.1 Ordered Pairs of Numbers

When you are given the combination to open a lock, such as 23L, 15R, you are dealing with what is called an **ordered pair of numbers**. It is important to know which number to use first and which second so that you can open the lock. The first number is called the **first component**, or **first coordinate**, and the second number is the **second component**, or **second coordinate**, of the pair. The ordered pair whose coordinates are a and b is denoted by (a, b).
Note that if $a \neq b$ then $(a, b) \neq (b, a)$.

Any set of ordered pairs of numbers is called a **relation**. The set whose elements are the first coordinates (or components) of the pairs is called the **domain of the relation**. The set whose elements are the second coordinates (components) of the pairs is called the **range of the relation**.

DEFINITION

> A **function** is a set of ordered pairs of numbers (x, y) such that for every x there is a unique y.

The set $\{(0, 1), (1, 2), (2, 3), (3, 4)\}$ is a function.
The set $\{(1, 4), (2, 4), (3, 7), (4, 10)\}$ is a function.
The set $\{(2, 3), (2, 4), (3, 5), (4, 6)\}$ is a relation but not a function.

The set whose elements are the first coordinates of the pairs is called the **domain of the function**. The set whose elements are the second coordinates of the pairs is called the **range of the function**.

Consider the function $\{(0, 4), (1, 7), (2, 10), (3, 13), (4, 16), \ldots\}$.
The domain of the function is $\{0, 1, 2, 3, 4, \ldots\}$.
The range of the function is $\{4, 7, 10, 13, 16, \ldots\}$.

9.2 Rectangular or Cartesian Coordinates

To establish the relationship between ordered pairs of real numbers and points on a plane, construct two perpendicular number lines, one horizontal and the other vertical, as shown in Figure 9.1.

Note Two lines are said to be perpendicular when they form a 90° angle.

The horizontal number line is called the ***x*-axis** and the vertical number line is called the ***y*-axis**. Let the two number lines intersect at their origins. The positive numbers

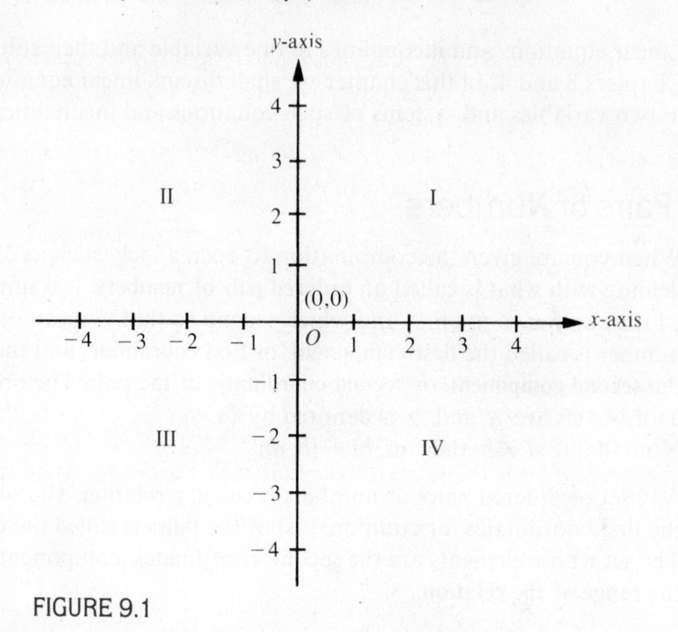

FIGURE 9.1

on the horizontal line are to the right of its origin, and on the vertical line above its origin.

The horizontal and vertical lines are called **coordinate axes**, and their point of intersection is called the **origin**. The whole system is called a **rectangular coordinate system** or a **Cartesian coordinate system**. The two axes separate the plane into four regions called **quadrants**. The upper right quadrant is referred to as the **first quadrant**, the upper left as the **second quadrant**, the lower left as the **third quadrant**, and the lower right as the **fourth quadrant**.

Any point P in the plane can be associated with an ordered pair of real numbers, usually denoted by (x, y), as shown in Figure 9.2. The components x and y of the pair (x, y) are called the **coordinates** of the point P.

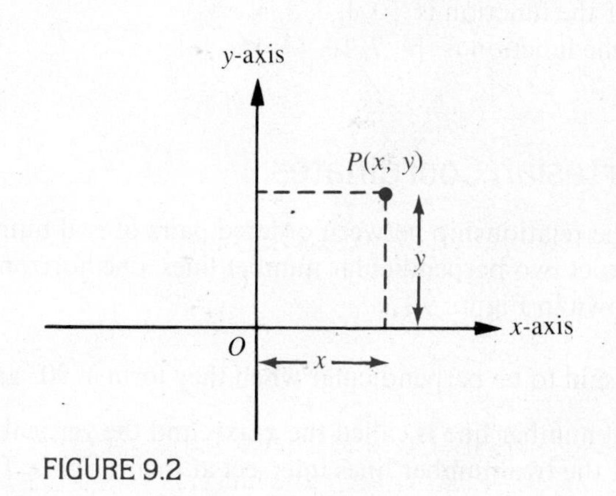

FIGURE 9.2

The first coordinate x is called the **abscissa** or ***x*-coordinate** of the point P. The second coordinate y is called the **ordinate** or ***y*-coordinate** of the point P. The abscissa of a point describes the number of units to the right or left of the origin. The ordinate of a point describes the number of units above or below the origin. The notation $P(x, y)$ is used to denote the point P whose coordinates are (x, y).

The coordinates of a given point in the plane can be found by dropping perpendiculars to the coordinate axes. The coordinate of the point of intersection of the perpendicular on the x-axis is the abscissa of the point. The coordinate of the point of intersection of the perpendicular on the y-axis is the ordinate of the point.

To locate a point P whose coordinates are (a, b), draw a vertical line through the point whose coordinate is a on the x-axis and a horizontal line through the point whose coordinate is b on the y-axis (see Figure 9.3). The point of intersection of these two lines is the point P corresponding to (a, b), or the **graph** of the ordered pair (a, b).

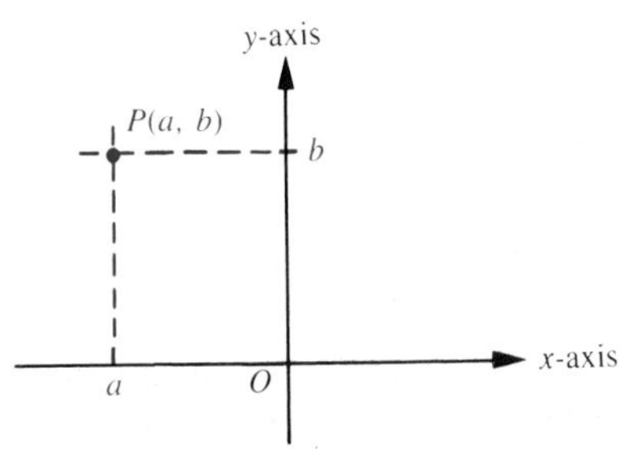

FIGURE 9.3

EXAMPLE

Locate a point P whose coordinates are $(-3, 2)$.

Solution

Draw a vertical line through the point whose coordinate is -3 on the x-axis, and a horizontal line through the point whose coordinate is 2 on the y-axis, as shown in Figure 9.4.

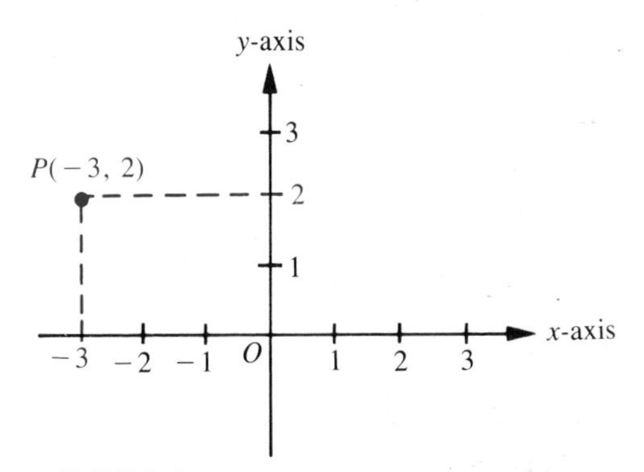

FIGURE 9.4

The point of intersection of these two lines is the point P whose coordinates are $(-3, 2)$. P lies in the second quadrant.

EXAMPLE Locate the point P whose coordinates are $(2, -1)$.

Solution Draw a vertical line through the point whose coordinate is 2 on the x-axis, and a horizontal line through the point whose coordinate is -1 on the y-axis, as shown in Figure 9.5.

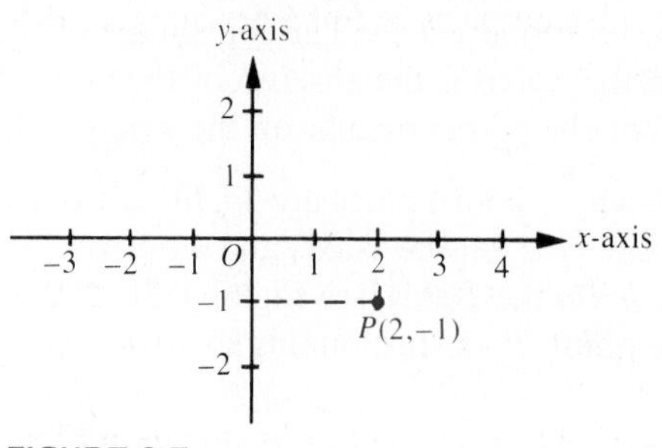

FIGURE 9.5

The point of intersection of these two lines is the point P whose coordinates are $(2, -1)$. P lies in the fourth quadrant.

When the coordinates of the origin are $(0, 0)$, we have the following:

1. All the points on the x-axis have zero as ordinate.
2. All the points on the y-axis have zero as abscissa.
3. All the points in the first quadrant have both coordinates positive.
4. All the points in the second quadrant have negative abscissas and positive ordinates.
5. All the points in the third quadrant have both coordinates negative.
6. All the points in the fourth quadrant have positive abscissas and negative ordinates.

Exercises 9.1–9.2

Determine which of the following relations are functions and which are not. For each function, find its domain and range:

1. $\{(-1, 1), (-1, -2), (0, -3), (1, -4)\}$
2. $\{(-5, -5), (-3, -3), (-1, -1), (0, 0)\}$
3. $\{(1, 3), (2, 4), (3, 4), (4, 5)\}$
4. $\{(1, 5), (2, 10), (3, 15), (7, 20)\}$
5. $\{(3, 1), (4, 1), (5, 1), (6, 1)\}$
6. $\{(1, 3), (2, 6), (3, 9), (4, 12), \ldots\}$
7. $\{(1, 5), (2, 9), (3, 13), (4, 17), \ldots\}$
8. $\{(1, 1), (2, 4), (3, 9), (4, 16), \ldots\}$
9. $\{(0, 0\}, (1, 6), (2, 12), (3, 18), \ldots\}$
10. $\{(1, 3), (2, 5), (3, 7), (4, 9), \ldots\}$

State in which quadrant the graph of each of the following ordered pairs is located, assuming that the coordinates of the origin are $(0, 0)$:

11. $(2, 3)$ **12.** $(4, 1)$ **13.** $(2, -6)$ **14.** $(8, -2)$
15. $(-1, -1)$ **16.** $(-5, -7)$ **17.** $(-2, 9)$ **18.** $(-4, 20)$

Graph the following ordered pairs of numbers on one set of axes and label each point with its coordinates:

19. $(1, 2)$ **20.** $(3, 1)$ **21.** $(2, 0)$ **22.** $(7, 0)$
23. $(-1, 4)$ **24.** $(-3, 3)$ **25.** $(0, 1)$ **26.** $(0, 2)$
27. $(-2, -2)$ **28.** $(-4, -1)$ **29.** $(0, -6)$ **30.** $(0, -4)$
31. $(-3, 0)$ **32.** $(-7, 0)$ **33.** $(4, -3)$ **34.** $(6, -2)$

Give the coordinates of the following points shown in Figure 9.6:

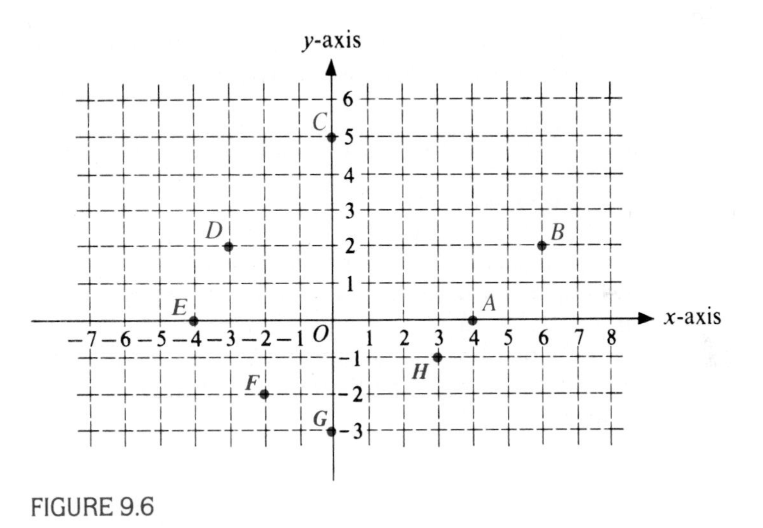

FIGURE 9.6

35. A **36.** B **37.** C **38.** D
39. E **40.** F **41.** G **42.** H

43. Graph the ordered pairs $(1, 2)$ and $(3, -2)$ and connect them with a straight line. What are the coordinates of the points of intersection of the line with the coordinate axes?

44. Graph the ordered pairs $\left(\frac{3}{2}, 6\right)$ and $\left(-\frac{3}{2}, 2\right)$ and connect them with a straight line. What are the coordinates of the points of intersection of the line with the coordinate axes?

45. Graph the ordered pairs (2, 2) and (−2, 6) and connect them with a straight line. On the same set of axes graph the ordered pairs (3, 2) and (−1, 4) and connect them with a straight line. Find the coordinates of the point of intersection of the two lines.

46. Graph the ordered pairs (7, −2) and (−2, 4) and connect them with a straight line. On the same set of axes graph the ordered pairs (2, 5) and (−2, −7) and connect them with a straight line. Find the coordinates of the point of intersection of the two lines.

9.3 Distance Between Two Points

The absolute value of $x_2 - x_1$, denoted by $|x_2 - x_1|$, is the distance between x_2 and x_1 when graphed on an axis.

Given two points in the plane, we can join them by a segment of a line and thus measure the distance between the two points. Algebraically, we can find the distance between two given points as follows:

First: When the two points lie on a horizontal line (a line parallel to the x-axis), the distance between the two points is the absolute value of the difference of their abscissas.
The distance between the two points $P_1(x_1, y_1)$ and $P_2(x_2, y_1)$ is $|x_2 - x_1|$.
See Figure 9.7.

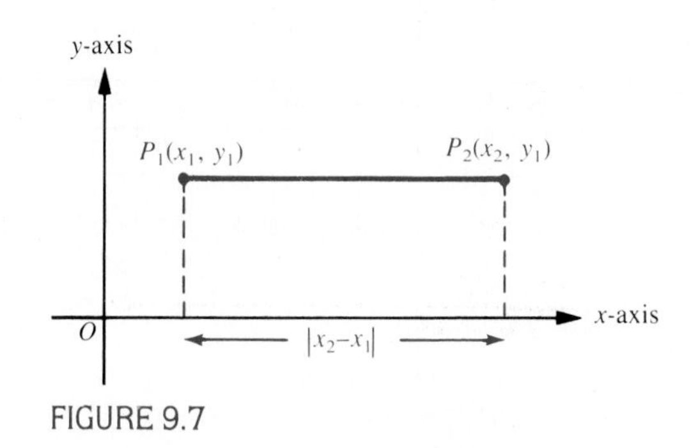

FIGURE 9.7

EXAMPLE

Find the distance between the two points $P_1(-1, 2)$ and $P_2(4, 2)$.

Solution

Here $x_1 = -1$ and $x_2 = 4$.
The distance between the two points P_1 and P_2 is

$$|x_2 - x_1| = |(4) - (-1)| = |4 + 1| = 5$$

See Figure 9.8.

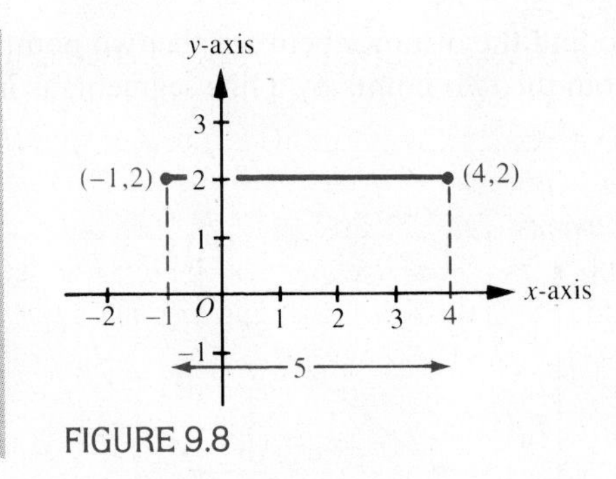

FIGURE 9.8

Second: When the two points lie on a vertical line (a line parallel to the y-axis), the distance between the two points is the absolute value of the difference of their ordinates.

The distance between the two points $P_1(x_1, y_1)$ and $P_2(x_1, y_2)$ is $|y_2 - y_1|$.

See Figure 9.9.

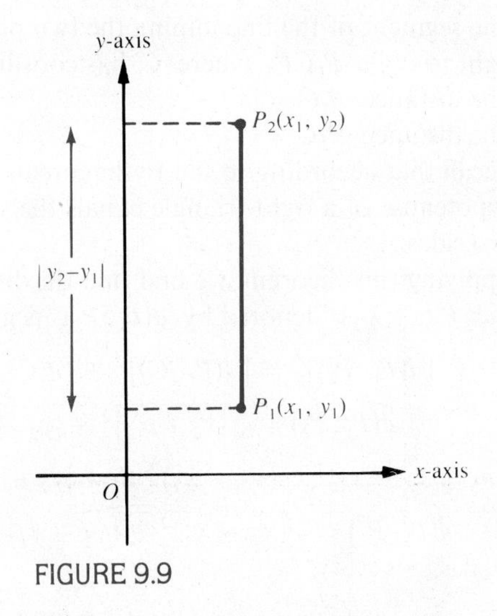

FIGURE 9.9

EXAMPLE Find the distance between the two points $P_1(3, -2)$ and $P_2(3, 5)$.

Solution Here $y_1 = -2$ and $y_2 = 5$.

The distance between the two points P_1 and P_2 is

$|y_2 - y_1| = |(5) - (-2)| = |5 + 2| = 7$

Note $|a - b| = |b - a|$

Third: In general, to find the distance between the two points $P_1(x_1, y_1)$ and $P_2(x_2, y_2)$, join the two points by a line segment, as in Figure 9.10.

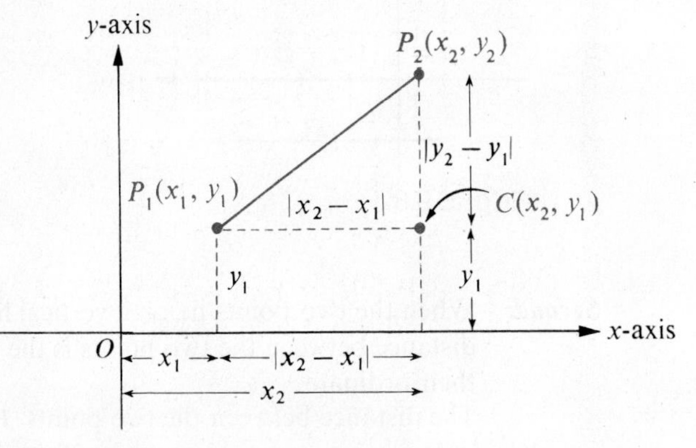

FIGURE 9.10

The segment of the line joining the two points is the hypotenuse of the right triangle P_1CP_2 where C has coordinates (x_2, y_1).
The distance $P_1C = |x_2 - x_1|$.
The distance $P_2C = |y_2 - y_1|$.
Recall that according to the Pythagorean Theorem, the square of the hypotenuse of a right triangle equals the sum of the squares of the other two sides.
Applying this theorem, we find that the distance between $P_1(x_1, y_1)$ and $P_2(x_2, y_2)$, denoted by $d(P_1, P_2)$, is given by

$$[d(P_1, P_2)]^2 = [d(P_1, C)]^2 + [d(C, P_2)]^2$$

or $$d(P_1, P_2) = \sqrt{|x_2 - x_1|^2 + |y_2 - y_1|^2}$$

Since $|x_2 - x_1|^2 = (x_2 - x_1)^2$ and $|y_2 - y_1|^2 = (y_2 - y_1)^2$,

$$d(P_1, P_2) = \sqrt{(x_2 - x_1)^2 + (y_2 - y_1)^2}$$

EXAMPLE Find the distance between the two points $P_1(-2, 6)$ and $P_2(3, -4)$.

Solution Here $x_1 = -2$, $y_1 = 6$, $x_2 = 3$, and $y_2 = -4$:

$$\begin{aligned} d(P_1, P_2) &= \sqrt{[(3) - (-2)]^2 + [(-4) - (6)]^2} \\ &= \sqrt{(5)^2 + (-10)^2} \\ &= \sqrt{25 + 100} \\ &= \sqrt{125} \\ &= 5\sqrt{5} \end{aligned}$$

EXAMPLE

Show that the points $A(2, -1)$, $B(3, 4)$, and $C(-7, 6)$ are vertices of a right triangle.

Solution

$$\begin{aligned} d(A, B) &= \sqrt{(3-2)^2 + [4-(-1)]^2} \\ &= \sqrt{(1)^2 + (5)^2} = \sqrt{26} \\ d(B, C) &= \sqrt{(-7-3)^2 + (6-4)^2} \\ &= \sqrt{(-10)^2 + (2)^2} = \sqrt{104} \\ d(C, A) &= \sqrt{[2-(-7)]^2 + (-1-6)^2} \\ &= \sqrt{(9)^2 + (-7)^2} = \sqrt{81+49} = \sqrt{130} \end{aligned}$$

Since $[d(A, B)]^2 + [d(B, C)]^2 = 26 + 104 = 130 = [d(C, A)]^2$, the triangle ABC is a right triangle.

DEFINITION

A set of points is said to be **collinear** if all the points lie on a straight line.

EXAMPLE

Show that the points $A(-3, -7)$, $B(-1, -3)$, and $C(0, -1)$ are collinear.

Solution

$$\begin{aligned} d(A, B) &= \sqrt{[-1-(-3)]^2 + [-3-(-7)]^2} \\ &= \sqrt{(2)^2 + (4)^2} = \sqrt{20} = 2\sqrt{5} \\ d(B, C) &= \sqrt{[0-(-1)]^2 + [-1-(-3)]^2} \\ &= \sqrt{(1)^2 + (2)^2} = \sqrt{5} \\ d(C, A) &= \sqrt{(-3-0)^2 + [-7-(-1)]^2} \\ &= \sqrt{(-3)^2 + (-6)^2} = \sqrt{45} = 3\sqrt{5} \\ d(A, B) + d(B, C) &= 2\sqrt{5} + \sqrt{5} = 3\sqrt{5} = d(C, A) \end{aligned}$$

Hence points A, B, and C are collinear.

Exercise 9.3

Find the distances between the following points:

1. $A(0, 0)$, $B(4, 3)$

2. $A(0, 0)$, $B(-2, 2)$

3. $A(0, 0)$, $B(-1, -4)$

4. $A(0, 0)$, $B(2, -6)$

5. $A(3, 0)$, $B(0, 1)$

6. $A(5, 0)$, $B(0, -2)$

7. $A(-2, 0)$, $B(0, 3)$

8. $A(0, -4)$, $B(0, 8)$

9. $A(0, 6)$, $B(0, -3)$

10. $A(7, 0)$, $B(-4, 0)$

11. $A(2, 0)$, $B(-5, 0)$

12. $A(1, 1)$, $B(3, 3)$

13. $A(3, 1)$, $B(4, 2)$

14. $A(2, 1)$, $B(6, -3)$

15. $A(8, 3)$, $B(3, -9)$

16. $A(2, 2)$, $B(-1, -2)$

17. $A(-3, 4)$, $B(-2, -5)$

18. $A\left(\frac{2}{3}, -\frac{1}{2}\right)$, $B\left(2, \frac{3}{2}\right)$

19. $A\left(\frac{5}{2}, \frac{1}{3}\right)$, $B\left(\frac{1}{2}, \frac{5}{3}\right)$

20. $A\left(\frac{1}{3}, \frac{3}{5}\right)$, $B\left(\frac{7}{3}, \frac{9}{5}\right)$

21. $A(2, -\sqrt{3})$, $B(6, 3\sqrt{3})$

22. $A(2, 5)$, $B(9, 5 - 4\sqrt{2})$

23. $A(2\sqrt{3}, \sqrt{3})$, $B(-3, 6)$

24. $A(\sqrt{2}, -4 - \sqrt{2})$, $B(3, -1)$

25. Find n so that the distance between $A(n, 1)$ and $B(4, 5)$ is $n + 3$.

26. Find n so that the distance between $A(n, 3)$ and $B(7, 12)$ is $n + 5$.

27. Find n so that the distance between $A(n, 2)$ and $B(3, 1)$ is $n + 2$.

28. Find n so that the distance between $A(n, -2)$ and $B(6, 1)$ is $n - 3$.

Show whether the following points are vertices of a right triangle or not:

29. $A(0, 0)$, $B(2, 5)$, $C(-5, 2)$

30. $A(0, 0)$, $B(2, 3)$, $C(3, -2)$

31. $A(0, 2)$, $B(-2, 1)$, $C(0, -3)$

32. $A(5, 0)$, $B(0, 5)$, $C(2, -1)$

33. $A(1, 1)$, $B(-1, 2)$, $C(-1, -3)$

34. $A(3, 2)$, $B(-2, -3)$, $C(-3, -2)$

35. $A(0, 9)$, $B(-4, 1)$, $C(-1, 0)$

36. $A(1, 0)$, $B(-3, -6)$, $C(0, -7)$

37. $A(-8, 1)$, $B(6, 3)$, $C(4, -2)$

38. $A(2, 4)$, $B(5, 1)$, $C(11, 11)$

39. $A(-1, 3)$, $B(2, 4)$, $C(3, 1)$

40. $A(1, -2)$, $B(5, 10)$, $C(9, 2)$

41. $A(1, 3)$, $B(3, -5)$, $C(5, 4)$

42. $A(-4, 7)$, $B(2, 3)$, $C(4, 6)$

43. $A(-1, 11)$, $B(1, 3)$, $C(3, -5)$

44. $A(1, 9)$, $B(3, 1)$, $C(7, 3)$

45. $A\left(2, -\frac{2}{3}\right)$, $B\left(-5, -\frac{16}{3}\right)$, $C(9, 4)$

46. $A\left(-\frac{3}{2}, 10\right)$, $B\left(\frac{9}{2}, 3\right)$, $C\left(\frac{5}{2}, \frac{7}{2}\right)$

47. $A\left(-\frac{7}{2}, \frac{3}{2}\right)$, $B\left(\frac{1}{2}, -\frac{1}{2}\right)$, $C\left(\frac{9}{2}, \frac{15}{2}\right)$

48. $A\left(-\frac{7}{3}, \frac{1}{3}\right)$, $B\left(\frac{2}{3}, -\frac{5}{3}\right)$, $C\left(\frac{20}{3}, \frac{22}{3}\right)$

Show whether the following points are collinear or not:

49. $A(0, 4)$, $B(2, 2)$, $C(4, 0)$

50. $A(0, -3)$, $B(3, 0)$, $C(2, -1)$

51. $A(0, 2)$, $B(4, 0)$, $C(2, 1)$

52. $A(0, -2)$, $B(1, -1)$, $C(2, 0)$

53. $A(2, 2)$, $B(1, 4)$, $C(-1, 8)$

54. $A(1, -2)$, $B(0, -4)$, $C(-1, -6)$

55. $A(0, -5)$, $B(2, 0)$, $C(1, -3)$

56. $A(0, 3)$, $B(1, 1)$, $C(2, -3)$

57. $A(5, -2)$, $B(3, -3)$, $C(-3, -8)$

58. $A(4, 2)$, $B(7, -4)$, $C(5, 1)$

59. $A(-3, 1)$, $B(4, -1)$, $C(11, -3)$

60. $A(3, 6)$, $B(0, 2)$, $C(-6, -6)$

61. $A(2, -9)$, $B(8, 3)$, $C(10, 7)$

62. $A(3, -2)$, $B(1, -4)$, $C(9, 4)$

63. $A(-1, 1)$, $B(3, -1)$, $C(-5, 2)$

64. $A(-2, 3)$, $B(1, 1)$, $C(7, -2)$

65. $A(2, 9)$, $B\left(-1, \frac{3}{2}\right)$, $C\left(-3, -\frac{7}{2}\right)$

66. $A\left(-\frac{2}{3}, 2\right)$, $B\left(\frac{2}{3}, \frac{6}{5}\right)$, $C\left(\frac{5}{6}, \frac{11}{10}\right)$

9.4 One Linear Equation in Two Variables

The general form of a linear equation in two variables x and y is

$$Ax + By = C$$

where $A, B, C \in R$, and A and B are not both zero.

The elements of the solution set of a linear equation in two variables are the ordered pairs of real numbers (x, y) that satisfy the equation.
The solution set of the equation $Ax + By = C$ is

$$\{(x, y) | Ax + By = C\}$$

In order to find some of the elements in the solution set, assign arbitrary values to x and calculate the corresponding values of y. Since we can assign any real number to x, we have an infinite number of ordered pairs in the solution set. Some of the elements of the solution set of the equation $2x - y = 2$ are

$$(-1, -4), \quad (0, -2), \quad (1, 0), \quad (3, 4)$$

If we take a Cartesian coordinate system on a plane and plot these ordered pairs, we obtain the result shown in Figure 9.11.

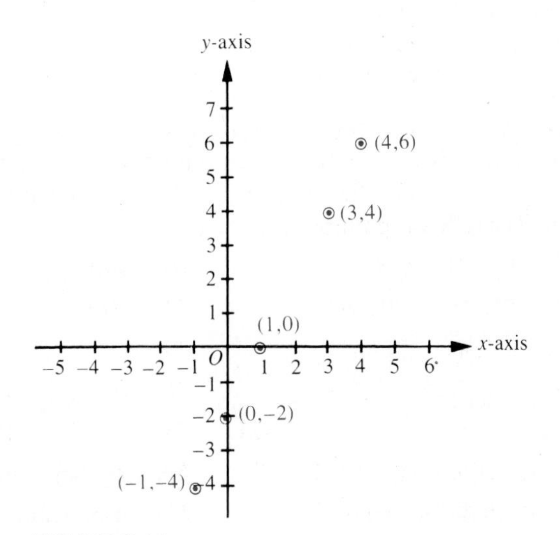

FIGURE 9.11

If we connect these points by a smooth curve, we find that they lie along a straight line. This straight line is called the **graph of the linear equation** $2x - y = 2$.

To simplify the graphing operation, tabulate some elements of the solution set as shown in Figure 9.12.
The arrows at the ends of the graph indicate that the line continues indefinitely in both directions.

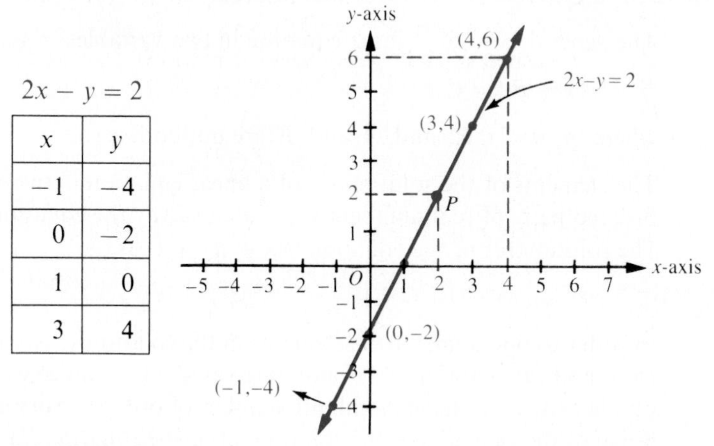

$2x - y = 2$

x	y
−1	−4
0	−2
1	0
3	4

FIGURE 9.12

The graph of any ordered pair of numbers that satisfies the equation, such as $(4, 6)$, lies on the straight line. Also, if we pick a point P on this straight line, the ordered pair of numbers formed from the coordinates of the point P, $(2, 2)$, satisfy the equation:

$$2x - y = 2(2) - (2) = 4 - 2 = 2$$

The graph of any linear equation $Ax + By = C$ where $A, B, C \in R$, and A and B are not both zero is a straight line. The graph of any ordered pair of numbers that satisfies the equation lies on the straight line. Also the coordinates of any point lying on the line satisfy the equation.

Notes

1. Although two points determine a unique line, you should find at least three points as a check.
2. When graphing an equation and the values you obtain for the variables are fractions with p and q as denominators, take every pq divisions on the graph paper to represent one unit.

Remark The solution set of the equation $2x - y = 2$ is a set of ordered pairs of numbers in which no two distinct ordered pairs have the same first coordinates. For every value of x, there is only one value of y. Thus the solution set of the equation is a function. In general, the solution set of a linear equation $Ax + By = C, B \neq 0$, is a function.

EXAMPLE Plot the line whose equation is $2x + y = 6$.

Solution Make a table of three ordered pairs of real numbers that satisfy the equation $2x + y = 6$ and plot the points representing these ordered pairs.
Join these points by a straight line.
The graph of the line is shown in Figure 9.13.

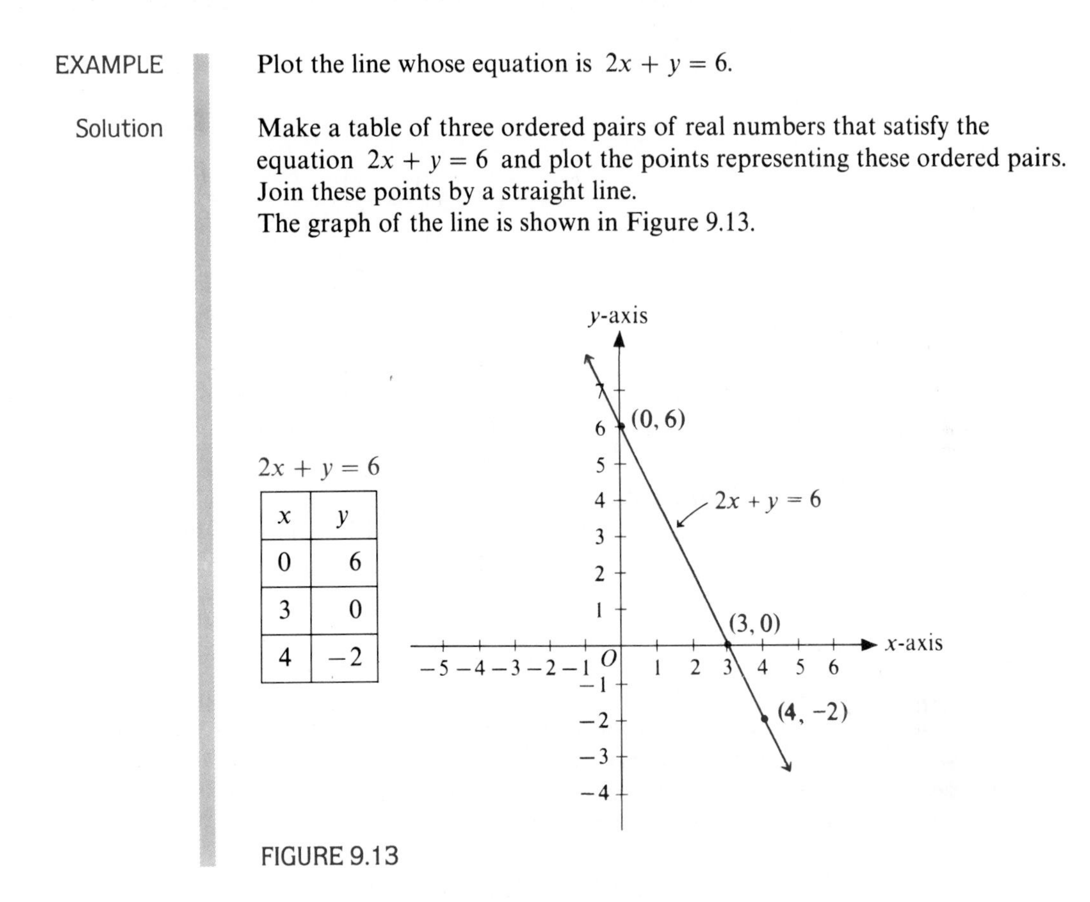

$2x + y = 6$

x	y
0	6
3	0
4	−2

FIGURE 9.13

The equation $By = C$ is equivalent to the equation $0x + By = C$.

Thus for all values of x, we have $y = \frac{C}{B}$.

Hence $By = C$ represents a horizontal line.
The solution set of the equation $By = C$ is a function.

EXAMPLE Plot the line whose equation is $y + 2 = 0$.

Solution The equation $y + 2 = 0$ is equivalent to the equation $0x + y = -2$.
Make a table of three elements of the solution set of the equation and plot the points representing them.
Join these points by a straight line.
Figure 9.14 is the graph of the line.

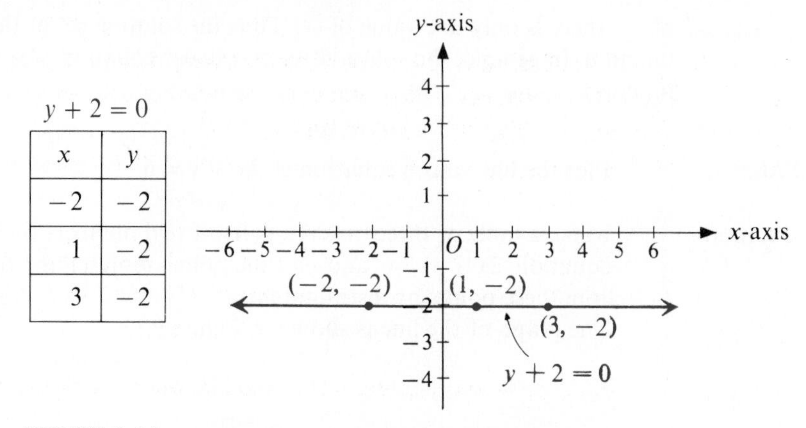

$y + 2 = 0$

x	y
-2	-2
1	-2
3	-2

FIGURE 9.14

The equation $Ax = C$ is equivalent to the equation $Ax + 0y = C$.

Thus for all values of y, we get $x = \dfrac{C}{A}$.

Hence $Ax = C$ represents a vertical line.

The solution set of the equation $Ax = C$ is a relation but not a function.

EXAMPLE Plot the graph of the equation $2x = 3$.

Solution The equation $2x = 3$ is equivalent to the equation $2x + 0y = 3$.
Make a table of three elements of the solution set of the equation and plot the points representing them.
Join these points by a straight line.
Figure 9.15 is the graph of the equation.

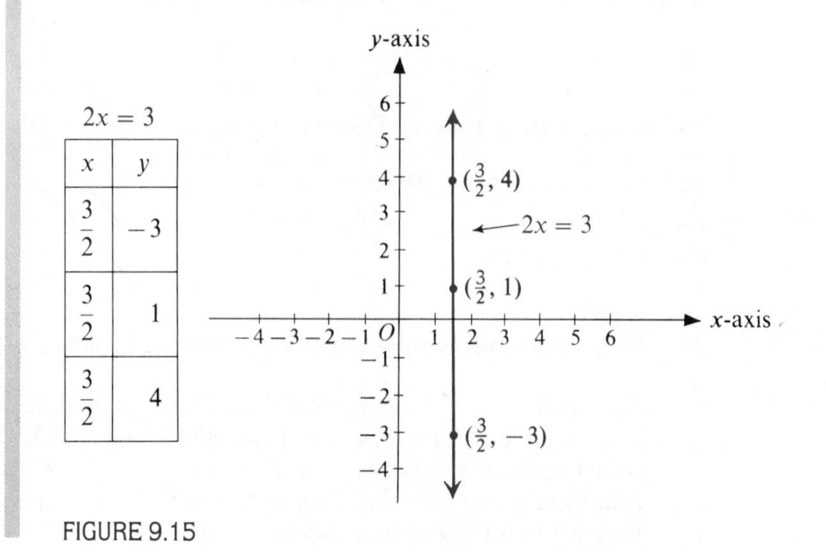

$2x = 3$

x	y
$\frac{3}{2}$	-3
$\frac{3}{2}$	1
$\frac{3}{2}$	4

FIGURE 9.15

DEFINITION

> The abscissa of the point of intersection of the line with the x-axis is called the **x-intercept**. The ordinate of the point of intersection of the line with the y-axis is called the **y-intercept**.

Notes

1. The x-intercept of a line is the value of x when $y = 0$.
2. The y-intercept of a line is the value of y when $x = 0$.

EXAMPLE

Find the x- and y-intercepts of the line whose equation is $4x - 5y = 7$.

Solution

When $y = 0$ we have $4x = 7$ that is, $x = \frac{7}{4}$

When $x = 0$ we have $-5y = 7$ that is, $y = -\frac{7}{5}$

Hence the x-intercept is $\frac{7}{4}$.

The y-intercept is $-\frac{7}{5}$.

Exercise 9.4

Graph the lines represented by the following equations:

1. $x + y = 1$ **2.** $x + y = 3$ **3.** $x - y = 1$
4. $x - y = 2$ **5.** $x - y = 0$ **6.** $x - 2y = 0$
7. $x + 2y = 3$ **8.** $2x - y = 3$ **9.** $x + 3y = 4$
10. $3x + y = 5$ **11.** $x - 2y = -1$ **12.** $x - 3y = -4$
13. $2x - y = -2$ **14.** $3x - y = -6$ **15.** $x = 2$
16. $x = -3$ **17.** $2x = 3$ **18.** $3x = -4$
19. $y = 3$ **20.** $y = -1$ **21.** $2y = 5$
22. $3y = -2$ **23.** $2x + 3y = 6$ **24.** $3x + 2y = 5$

Find the x- and y-intercepts of the lines represented by the equations:

25. $x + y = 3$ **26.** $x + y = -1$ **27.** $2x + y = 4$
28. $4x + y = -2$ **29.** $x + 3y = 6$ **30.** $x + 5y = -1$
31. $2x + 3y = 8$ **32.** $3x + 4y = 7$ **33.** $8x + 3y = 11$
34. $4x + 5y = 9$ **35.** $3x - 2y = -1$ **36.** $5x - 3y = -2$
37. $2x = 3$ **38.** $5x = -2$ **39.** $2y = 9$ **40.** $3y = -7$

9.5 Slope of a Line

Let $A(x_1, y_1)$ and $B(x_2, y_2)$ be two points on the line L. From the point A draw a horizontal line, and from the point B draw a vertical line. Call the point of intersection of these two lines C. The coordinates of the point C are (x_2, y_1), as shown in Figure 9.16.

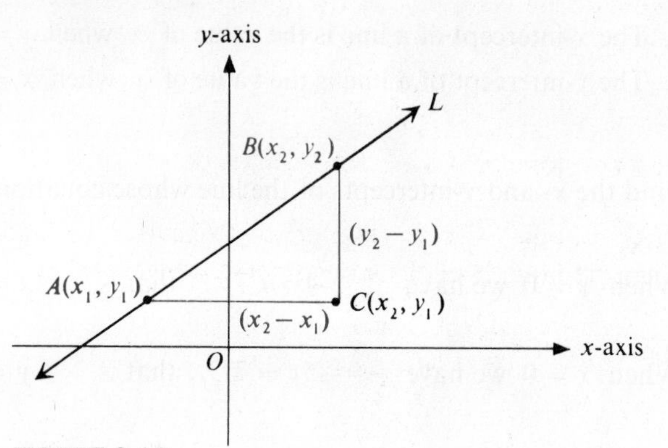

FIGURE 9.16

The directed distance from A to C is $(x_2 - x_1)$.
The directed distance from C to B is $(y_2 - y_1)$.

The quotient $\dfrac{y_2 - y_1}{x_2 - x_1}$, if $x_2 \neq x_1$, is called the **slope of the line**.

Note When $x_2 = x_1$, the slope of the line is not defined.

THEOREM The slope of a line is independent of the pairs of points selected.

Proof From geometry, the two triangles ACB and AED in Figure 9.17 are similar.

Hence $$\frac{CB}{AC} = \frac{ED}{AE}$$

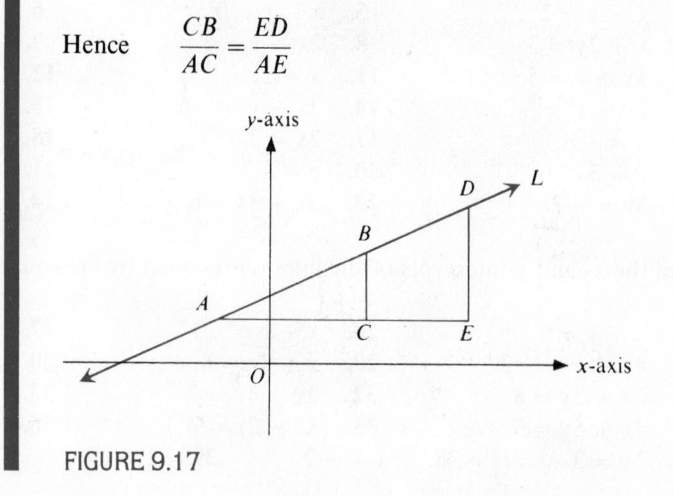

FIGURE 9.17

That is, the slope of the line calculated with respect to the two points A and B is the same as that calculated with respect to the two points A and D.

Note Given the coordinates of two points on a line, the slope of that line can be calculated by dividing the difference between the ordinate of the second point and the ordinate of the first point by the difference between the abscissa of the second point and the abscissa of the first point.

EXAMPLE Find the slope of the line through the points $A(1, -1)$ and $B(7, 3)$.

Solution When we take A as the first point, we have $x_1 = 1$ and $y_1 = -1$. When we take B as the second point, we have $x_2 = 7$ and $y_2 = 3$.

$$\text{The slope of the line} = \frac{y_2 - y_1}{x_2 - x_1}$$

$$= \frac{(3) - (-1)}{(7) - (1)} = \frac{3 + 1}{7 - 1} = \frac{4}{6} = \frac{2}{3}$$

Note The coordinates of a point on a line are an ordered pair of numbers that satisfies the equation of the line.

EXAMPLE Find the slope of the line whose equation is $3x + 4y = 6$.

Solution Find the coordinates of any two points on the line. Consider, for example, the two points $P_1(-2, 3)$ and $P_2(2, 0)$.

$$\text{The slope of the line} = \frac{(0) - (3)}{(2) - (-2)} = -\frac{3}{4}$$

Notes

1. The equation $By = C$ is equivalent to the equation $0x + By = C$.

   Consider the ordered pairs $\left(1, \frac{C}{B}\right)$ and $\left(2, \frac{C}{B}\right)$ that satisfy the equation.

$$\text{The slope of the line} = \frac{\frac{C}{B} - \frac{C}{B}}{2 - 1} = \frac{0}{1} = 0$$

   Hence the slope of a line whose equation is of the form $By = C$, a horizontal line is 0.

2. The equation $Ax = C$ is equivalent to the equation $Ax + 0y = C$. Consider the ordered pairs $\left(\frac{C}{A}, 1\right)$ and $\left(\frac{C}{A}, 2\right)$ that satisfy the equation.

$$\text{The slope of the line} = \frac{2 - 1}{\frac{C}{A} - \frac{C}{A}} = \frac{1}{0}$$

which is not defined.
Hence the slope of a line whose equation is of the form $Ax = C$, a vertical line is not defined.

THEOREM

The slope of the line whose equation is $y = mx + b$ is m.

Proof

Consider the two points with coordinates $(0, b)$ and $(1, m + b)$.

$$\text{The slope of the line} = \frac{(m + b) - b}{1 - 0}$$

$$= \frac{m + b - b}{1} = m$$

Notes

1. When the equation of the line is written in the form $y = mx + b$, then the slope of the line is m, the coefficient of x.
2. When the equation of the line is in the general form, $Ax + By = C$, $B \neq 0$, then

$$y = -\frac{A}{B}x + \frac{C}{B} \quad \text{and} \quad \text{slope } m = -\frac{A}{B}$$

EXAMPLE

Find the slope of the line whose equation is $4y - 3x = 8$.

Solution

The equation $4y - 3x = 8$ is equivalent to the equation

$$y = \frac{3}{4}x + 2$$

The slope of the line $= \frac{3}{4}$.

EXAMPLE

Find the slope of the line whose equation is $2x + 5y = 4$.

Solution

The equation $2x + 5y = 4$ is equivalent to the equation

$$y = -\frac{2}{5}x + \frac{4}{5}$$

The slope of the line $= -\frac{2}{5}$.

Given an equation of a line, the slope of the line can be calculated by one of two ways:

1. Find the coordinates of two points on the line. Substitute in the relation

$$\frac{y_2 - y_1}{x_2 - x_1}, \quad x_2 \neq x_1$$

2. Put the equation of the line in the form $y = mx + b$.
   The coefficient of x is the slope of the line.

Note When the slope of the segment of the line through the points A and B is equal to the slope of the segment of the line through the points B and C, the three points A, B, and C are collinear.

EXAMPLE By the use of slopes determine whether or not the points

$$P_1\left(3, -\frac{1}{5}\right) \qquad P_2\left(\frac{3}{2}, -2\right) \qquad \text{and} \qquad P_3(9, 7)$$

are collinear.

Solution The slope of the line joining P_1 and P_2 is $\dfrac{-2-\left(-\frac{1}{5}\right)}{\frac{3}{2}-3} = \dfrac{6}{5}$.

The slope of the line joining P_2 and P_3 is $\dfrac{7-(-2)}{9-\frac{3}{2}} = \dfrac{6}{5}$.

The slope of P_1P_2 = slope of P_2P_3.
Hence the three points are collinear.

EXAMPLE Determine n so that the three points $P_1(-3, 4)$, $P_2(n, -1)$, and $P_3(12, -6)$ are collinear.

Solution Slope of the line $P_2P_1 = \dfrac{-1-4}{n-(-3)} = \dfrac{-5}{n+3}$.

Slope of the line $P_1P_3 = \dfrac{-6-4}{12-(-3)} = -\dfrac{2}{3}$.

For the three points to be collinear, the two slopes have to be equal.

That is,

$$\frac{-5}{n+3} = -\frac{2}{3}$$

$$2n + 6 = 15$$

$$n = \frac{9}{2}$$

Exercise 9.5

Find the slopes of the lines through the given points:

1. $A(0, 0)$, $B(2, 2)$ **2.** $A(0, 0)$, $B(-3, 1)$ **3.** $A(0, 2)$, $B(-4, 0)$
4. $A(0, -3)$, $B(2, 0)$ **5.** $A(4, 1)$, $B(5, 4)$ **6.** $A(2, 5)$, $B(1, 7)$
7. $A(2, 6)$, $B(4, -4)$ **8.** $A(3, 1)$, $B(5, -5)$
9. $A(-1, -3)$, $B(3, -4)$ **10.** $A(2, -1)$, $B(5, -1)$ – horizontal
11. $A(4, 3)$, $B(-7, 3)$ **12.** $A(2, -2)$, $B(5, -2)$
13. $A(4, 3)$, $B(4, -7)$ **14.** $A(-3, 2)$, $B(-3, 8)$
15. $A(-2, -7)$, $B(-2, 6)$ **16.** $A(-5, 1)$, $B(2, 11)$
17. $A(3, -2)$, $B(-7, 8)$ **18.** $A(-3, -1)$, $B(-2, 4)$

Find the slopes of the lines represented by the following equations in two ways:

19. $x - y = 0$ **20.** $x - 3y = 0$ **21.** $3x - y = 0$ **22.** $3x - 4y = 0$
23. $x + y = 0$ **24.** $x + 2y = 0$ **25.** $2x + y = 0$ **26.** $3x + 2y = 0$
27. $x - y = 1$ **28.** $2x - y = 3$ **29.** $x - 3y = 4$ **30.** $2x - 5y = 7$
31. $x + y = 6$ **32.** $x + 2y = 2$ **33.** $2x + 3y = 5$
34. $3x + 7y = 10$ **35.** $4x + 3y = 7$ **36.** $5x + 3y = 8$
37. $4x + 9y = 13$ **38.** $6x + 8y = 3$ **39.** $3x + 2 = 0$
40. $2x - 3 = 0$ **41.** $5y + 3 = 0$ **42.** $7y - 4 = 0$
43. $2y - 3x = 6$ **44.** $3y - 2x = 12$ **45.** $4y - x = 5$
46. $3y - 4x = 7$ **47.** $3x - 5y = -2$ **48.** $4x - 3y = -1$
49. $7x - 3y = -4$ **50.** $2x - 7y = -5$ **51.** $y - \sqrt{2}x = 4$
52. $\sqrt{3}x + 2y = 5$ **53.** $6x + \sqrt{3}y = 2$ **54.** $3x + \sqrt{5}y = 8$

By the use of slopes determine whether or not the given points are collinear:

55. $A(2, 1)$, $B(4, 2)$, $C(8, 4)$ **56.** $A(-2, -7)$, $B(2, 1)$, $C(5, 7)$
57. $A(-1, 10)$, $B(2, 4)$, $C(6, -4)$ **58.** $A(-3, 16)$, $B(0, 7)$, $C(2, 1)$
59. $A(-4, -5)$, $B(-1, -1)$, $C(2, 3)$ **60.** $A(-4, -3)$, $B(-1, 1)$, $C(2, 5)$
61. $A(-4, -5)$, $B(3, 2)$, $C(9, 6)$ **62.** $A(3, -5)$, $B(5, 1)$, $C(10, 7)$
63. $A(-5, 11)$, $B(1, 2)$, $C(9, -8)$ **64.** $A(-4, -2)$, $B(1, 1)$, $C(4, 3)$

Find the real number n for which all three points lie on the same line:

65. $A(-16, 4)$, $B(4, 0)$, $C(n, 1)$ **66.** $A(-5, n)$, $B(-3, 4)$, $C(-1, -2)$
67. $A(n, -1)$, $B(-4, -5)$, $C(11, 7)$ **68.** $A(-2, -15)$, $B(1, 6)$, $C(3, n)$

69. $A(-12, 9)$, $B\left(-4, \frac{11}{3}\right)$, $C(n, -1)$

70. $A(-4, 7)$, $B\left(\frac{7}{2}, -3\right)$, $C(n, -7)$

71. $A(n, 0)$, $B\left(\frac{9}{4}, 2\right)$, $C\left(\frac{15}{2}, 8\right)$ **72.** $A(-3, 3)$, $B(2, 0)$, $C\left(\frac{13}{3}, n\right)$

9.6 Equations of Lines

A linear equation in two variables represents a line. Given the equation, we can find coordinates of points on the line, the x- and y-intercepts, and the slope of the line. Now, given some of this information about the line, we will discuss how to find the equation of the line.

Equation of a Line Given Two Points

Given the equation of a line, we were able to find the coordinates of two points on the line, and thus determine the line. Since two distinct points determine a line uniquely, we now are going to find the equation of a line given two points on it.

Assume that the two given points are $P_1(x_1, y_1)$ and $P_2(x_2, y_2)$. Let the point $P(x, y)$ be a generic point on the line, different from the two points P_1 and P_2, as shown in Figure 9.18.

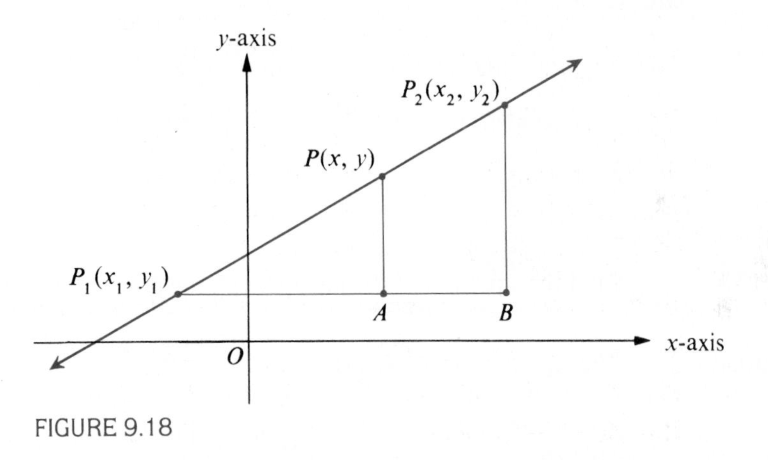

FIGURE 9.18

The slope of the line calculated with respect to the two points $P_1(x_1, y_1)$ and $P(x, y)$ is $\dfrac{y - y_1}{x - x_1}$.

The slope of the line calculated with respect to the two points $P_1(x_1, y_1)$ and $P_2(x_2, y_2)$ is $\dfrac{y_2 - y_1}{x_2 - x_1}$, $x_2 \neq x_1$.

Since the slope of a line is the same for all points on the line, we have

$$\frac{y - y_1}{x - x_1} = \frac{y_2 - y_1}{x_2 - x_1}$$

which is the equation of a line given two points.

Notes

1. When $x_2 = x_1 = a$, the slope of the line is not defined. The line is parallel to the y-axis and its equation is $x = a$. The equation of the y-axis is $x = 0$.
2. When $y_2 = y_1 = b$, the line is parallel to the x-axis and its equation is $y = b$. The equation of the x-axis is $y = 0$.

EXAMPLE

Find the equation of the line through the points $A(2, -3)$ and $B(4, 7)$.

Solution

The equation of the line is $\dfrac{y-(-3)}{x-(2)} = \dfrac{7-(-3)}{4-(2)}$.

That is, $\dfrac{y+3}{x-2} = 5$ or $5x - y = 13$

Equation of a Line Given One Point $P_1\ (x_1, y_1)$ and Slope m

For any point, $P(x, y) \neq P_1(x_1, y_1)$ on the line, the slope $= \dfrac{y - y_1}{x - x_1}$.

Hence the equation of a line given one point and the slope is

$$\frac{y - y_1}{x - x_1} = m$$

EXAMPLE

Find the equation of the line through the point $A(1, -3)$ with slope -2.

Solution

The equation of the line is $\dfrac{y-(-3)}{x-(1)} = -2$.

That is, $\dfrac{y+3}{x-1} = -2$ or $2x + y = -1$

Equation of a Line Given the Intercepts

If a and b are respectively the x- and y-intercepts, and both are not zero, then the two points whose coordinates are $(a, 0), (0, b)$ are on the line.
Hence the equation of the line is

$$\frac{y - 0}{x - a} = \frac{b - 0}{0 - a}$$

That is, $\dfrac{y}{x-a} = \dfrac{b}{-a}$ or $bx + ay = ab$

Dividing both sides of the equation by ab we get

$$\frac{x}{a} + \frac{y}{b} = 1$$

which is the intercept form of the line.

Note If the line passes through the origin, it cannot be written in intercept form.

EXAMPLE Find the equation of the line whose x-intercept is 3 and y-intercept is -5.

Solution The equation of the line is

$$\frac{x}{3} + \frac{y}{-5} = 1 \quad \text{or} \quad 5x - 3y = 15$$

Note When two lines are parallel, their slopes are equal.

EXAMPLE Find the equation of the line through the point $A(1, -5)$ and parallel to the line whose equation is $3x - 2y = 7$.

Solution The slope of the line represented by $3x - 2y = 7$ is $\frac{3}{2}$.

The equation of the required line is

$$\frac{y-(-5)}{x-(1)} = \frac{3}{2} \quad \text{or} \quad 3x - 2y = 13$$

EXAMPLE Find the equation of the line through the point $A(-1, 4)$ and parallel to the line through the points $B(3, 2)$ and $C(-6, 5)$.

Solution The slope of the line through the points B and C is

$$\frac{(5)-(2)}{(-6)-(3)} = \frac{3}{-9} = -\frac{1}{3}$$

The equation of the required line is

$$\frac{y-(4)}{x-(-1)} = -\frac{1}{3} \quad \text{or} \quad x + 3y = 11$$

Exercise 9.6

Find the equation of the line through the given points:

1. $A(0, 0)$, $B(1, 1)$

2. $A(0, 0)$, $B(-2, 1)$

3. $A(0, 0)$, $B(-1, -3)$

4. $A(0, 1)$, $B(2, 0)$

5. $A(0, 2)$, $B(5, 0)$
6. $A(0, 4)$, $B(-3, 0)$
7. $A(1, -1)$, $B(3, 0)$
8. $A(2, 2)$, $B(3, 5)$
9. $A(-2, -1)$, $B(2, 7)$
10. $A(-5, -4)$, $B(10, 6)$
11. $A(1, 3)$, $B(-4, -7)$
12. $A(-3, -4)$, $B(6, 2)$
13. $A(1, -2)$, $B(5, -2)$
14. $A(-3, 4)$, $B(8, 4)$
15. $A(-1, 3)$, $B(7, 3)$
16. $A(2, 3)$, $B(2, 5)$
17. $A(-1, 4)$, $B(-1, 7)$
18. $A(2, -6)$, $B(2, 9)$
19. $A(\sqrt{3}, -1)$, $B(-2\sqrt{3}, 8)$
20. $A(6, \sqrt{2})$, $B(-4, -4\sqrt{2})$

21. Find the value of k for which the line through the points $A(k, 1)$ and $B(2k, 5)$ has a slope of 4.

22. Find the value of k for which the line through the points $A(3, -k)$ and $B(2, -3k)$ has a slope of 3.

Find the equation of the line through the given point with the prescribed slope:

23. $A(5, 2)$; 0
24. $A(2, 6)$; 0
25. $A(3, -2)$; 0
26. $A(-6, -3)$; 0
27. $A(4, 5)$; 0
28. $A(2, 3)$; 5
29. $A(3, 1)$; 1
30. $A(2, 7)$; 3
31. $A(4, 3)$; -2
32. $A(6, 2)$; -3
33. $A(-2, 4)$; -7
34. $A(-1, -2)$; -4
35. $A(-3, 17)$; -5
36. $A(-1, 5)$; -8
37. $A(2, -1)$; $\frac{4}{3}$
38. $A\left(-\frac{7}{2}, -3\right)$; $\frac{2}{5}$
39. $A(1, 3)$; $-\frac{3}{2}$
40. $A(2, 0)$; $-\frac{2}{3}$
41. $A(1, -3)$; $-\frac{1}{7}$
42. $A(\sqrt{2}, \sqrt{3})$; $-\frac{2\sqrt{6}}{3}$
43. $A(2\sqrt{2}, \sqrt{5})$; $\frac{\sqrt{10}}{2}$

Find the equation of the line with the given x- and y-intercepts.

44. 2; 3
45. 3; 5
46. -2; 1
47. -4; 7
48. 5; -8
49. 6; -3
50. -4; -6
51. -1; -5
52. $\frac{1}{2}$; $\frac{2}{3}$
53. $\frac{2}{7}$; $\frac{3}{5}$
54. $-\frac{1}{3}$; $\frac{1}{4}$
55. $-\frac{3}{5}$; $\frac{2}{3}$
56. $\frac{3}{2}$; $-\frac{4}{5}$
57. $\frac{3}{4}$; $-\frac{5}{6}$
58. $-\frac{3}{2}$; $-\frac{7}{4}$
59. $-\frac{5}{2}$; $-\frac{3}{8}$
60. $\sqrt{2}$; 1
61. 2; $\sqrt{3}$
62. $\sqrt{2}$; $\sqrt{3}$
63. $\sqrt{10}$; $\sqrt{6}$

64. Find the equation of the line through the point $A(1, 1)$ and parallel to the line whose equation is $4x - 3y = 6$.

65. Find the equation of the line through the point $A(7, 2)$ and parallel to the line whose equation is $x - 3y = 4$.

66. Find the equation of the line through the point $A(3, -1)$ and parallel to the line whose equation is $4x - 5y = 9$.
67. Find the equation of the line through the point $A(1, -1)$ and parallel to the line whose equation is $2x + 3y = 5$.
68. Find the equation of the line through the point $A(-2, 2)$ and parallel to the line whose equation is $2x + 5y = 8$.
69. Find the equation of the line through the point $A(-4, 2)$ and parallel to the line whose equation is $3x + 7y = 11$.
70. Find the equation of the line through the point $A(-3, 2)$ and parallel to the line through the points $B(2, 1)$ and $C(4, -5)$.
71. Find the equation of the line through the point $A(-3, -2)$ and parallel to the line through the points $B(1, -1)$ and $C(7, 3)$.
72. Find the equation of the line through the point $A(2, -1)$ and parallel to the line through the points $B(-2, 3)$ and $C(2, -2)$.
73. Find the equation of the line through the point $A(-2, -5)$ and parallel to the line through the points $B(1, -1)$ and $C(-2, -3)$.
74. Find the equation of the line through the point $A(1, 5)$ and parallel to the line through the points $B\left(-1, \frac{5}{2}\right)$ and $C(2, 0)$.
75. Find the equation of the line through the point $A(4, 3)$ and parallel to the line through the points $B\left(-1, -\frac{1}{2}\right)$ and $C(1, 3)$.
76. Find the equation of the line with x-intercept 1 and parallel to the line whose equation is $7x - 9y = 15$.
77. Find the equation of the line with x-intercept $\frac{1}{2}$ and parallel to the line whose equation is $6x - 5y = 19$.
78. Find the equation of the line with y-intercept -1 and parallel to the line whose equation is $4x - y = 11$.
79. Find the equation of the line with y-intercept -2 and parallel to the line whose equation is $7x + 8y = 10$.
80. Find the equation of the line parallel to the x-axis and through the point $A(2, 5)$.
81. Find the equation of the line parallel to the x-axis and through the point $A(-2, -3)$.
82. Find the equation of the line parallel to the y-axis and through the point $A(-1, 6)$.
83. Find the equation of the line parallel to the y-axis and through the point $A(8, -2)$.
84. Find the equation of the line through the point $A(2, 2)$ and parallel to the line whose x- and y-intercepts are $\frac{1}{3}; \frac{1}{2}$.
85. Find the equation of the line through the point $(3, 7)$ and parallel to the line whose x- and y-intercepts are $2; -5$.

9.7 Two Linear Equations in Two Variables

The elements of the solution set of a linear equation $ax + by = c$ are an infinite number of ordered pairs (x, y) that can be represented graphically by a straight line. Sometimes it is required to find the common solution, or solution set, of two or more equations, called a **system of equations**. The solution set of a system of equations is thus the intersection of the solution set of each equation in the system.

DEFINITION

> The **solution set of two equations in two variables** is the set of all ordered pairs of numbers that are common solutions to the two equations. It is the intersection of the solution set of one of the equations with the solution set of the other equation.

The solution set of the system $a_1x + b_1y = c_1$ and $a_2x + b_2y = c_2$ is

$$\{(x, y) \mid a_1x + b_1y = c_1\} \cap \{(x, y) \mid a_2x + b_2y = c_2\}$$

Graphical Interpretations

When the graphs of the two equations $a_1x + b_1y = c_1$ and $a_2x + b_2y = c_2$ are drawn using one set of axes, one of the following possibilities arises:

1. The two lines will **coincide**. In this case the two equations are equivalent. That is,

$$a_1 = ka_2, \ b_1 = kb_2, \text{ and } c_1 = kc_2 \qquad \text{or} \qquad \frac{a_1}{a_2} = \frac{b_1}{b_2} = \frac{c_1}{c_2}$$

The solution set of the system is the solution set of either one of the equations. The two equations are said to be **consistent and dependent**.

2. The two lines will be **parallel** but do not coincide. In this case the slopes of the two lines are equal. That is,

$$-\frac{a_1}{b_1} = -\frac{a_2}{b_2} \qquad \text{or} \qquad \frac{a_1}{a_2} = \frac{b_1}{b_2} \neq \frac{c_1}{c_2}$$

The solution set of the system is the null set.
The two equations are said to be **inconsistent**.

3. The two lines will **intersect** at exactly one point. In this case the slopes are not equal. That is,

$$-\frac{a_1}{b_1} \neq -\frac{a_2}{b_2} \qquad \text{or} \qquad \frac{a_1}{a_2} \neq \frac{b_1}{b_2}$$

The solution set of the system is the ordered pair formed from the coordinates of the point of intersection.
The two equations are said to be **consistent and independent**.

Notes

1. When the slopes of two lines are equal, the two lines are either parallel or coincident.
2. The two lines represented by the equations $x = a_1$ and $x = a_2, a_1 \neq a_2$, are parallel but do not coincide.
   The two lines are parallel to the y-axis.
3. The two lines represented by the equations $y = b_1$ and $y = b_2, b_1 \neq b_2$, are parallel but do not coincide.
   The two lines are parallel to the x-axis.
4. The two lines represented by the equations $x = a$ and $y = b$ intersect at exactly one point.
   The coordinates of the point of intersection are (a, b).

EXAMPLES

Determine whether the lines represented by each of the following systems of equations intersect at exactly one point, are parallel but do not coincide, or are coincident. Describe each system as consistent and independent, inconsistent, or consistent and dependent.

1. $3x + 4y = 6$
$6x - 8y = 5$

2. $8x - 2y = 4$
$4x - y = 2$

3. $3x - 9y = 8$
$2x - 6y = 7$

Solutions

1. $3x + 4y = 6 \quad a_1 = 3 \quad b_1 = 4 \quad c_1 = 6$
$6x - 8y = 5 \quad a_2 = 6 \quad b_2 = -8 \quad c_2 = 5$

$$\frac{a_1}{a_2} = \frac{1}{2} \qquad \frac{b_1}{b_2} = -\frac{1}{2} \qquad \frac{c_1}{c_2} = \frac{6}{5}$$

Since $\frac{a_1}{a_2} \neq \frac{b_1}{b_2}$, the two lines intersect at exactly one point.

The system is consistent and independent.

2. $8x - 2y = 4 \quad a_1 = 8 \quad b_1 = -2 \quad c_1 = 4$
$4x - y = 2 \quad a_2 = 4 \quad b_2 = -1 \quad c_2 = 2$

$$\frac{a_1}{a_2} = 2 \qquad \frac{b_1}{b_2} = 2 \qquad \frac{c_1}{c_2} = 2$$

Since $\frac{a_1}{a_2} = \frac{b_1}{b_2} = \frac{c_1}{c_2}$, the two lines are coincident.

The system is consistent and dependent.

3. $3x - 9y = 8 \quad a_1 = 3 \quad b_1 = -9 \quad c_1 = 8$
$2x - 6y = 7 \quad a_2 = 2 \quad b_2 = -6 \quad c_2 = 7$

$$\frac{a_1}{a_2} = \frac{3}{2} \qquad \frac{b_1}{b_2} = \frac{3}{2} \qquad \frac{c_1}{c_2} = \frac{8}{7}$$

Since $\frac{a_1}{a_2} = \frac{b_1}{b_2} \neq \frac{c_1}{c_2}$, the two lines are parallel but do not coincide.

The system is inconsistent.

Exercise 9.7

Determine whether the lines represented by each of the following systems of equations intersect at exactly one point, are parallel but do not coincide, or are coincident.

Describe each system as consistent and independent, inconsistent, or consistent and dependent.

1. $3x - 5y = 4$, $2x - 4y = 7$

2. $x - 2y = 0$, $2x - y = 6$

3. $2x + 3y = 8$, $3x - y = 1$

4. $5x - 7y = 4$, $10x + 14y = 8$

5. $x - 2y = 3$, $2x + 4y = 6$

6. $x + 2y = 3$, $3x + 4y = 6$

7. $2x - 3y = 6$, $4x - 6y = 11$

8. $2x - 4y = 9$, $5x - 10y = 6$

9. $7x + 14y = 8$, $3x + 6y = 10$

10. $6x - 9y = 1$, $8x - 12y = 3$

11. $4x - 6y = 5$, $14x - 21y = 15$

12. $9x - 3y = 2$, $12x - 4y = 7$

13. $x + 6y = 8$, $2x + 12y = 16$

14. $3x + y = 4$, $21x + 7y = 28$

15. $3x - 2y = 4$, $9x - 6y = 12$

16. $2x - y = -1$, $8x - 4y = -4$

17. $6x - 9y = 1$, $8x - 12y = \frac{4}{3}$

18. $6x + 4y = 5$, $9x + 6y = \frac{15}{2}$

19. $x = 3$, $2x + y = 1$

20. $x = -2$, $x - 2y = 4$

21. $3x = 5$, $3x + y = 4$

22. $y = 4$, $3x - 2y = 5$

23. $y = -3$, $x + 5y = 7$

24. $2y = 7$, $x + 2y = 9$

25. $x = 1$, $x = 8$

26. $x = -2$, $x = 7$

27. $2x = 3$, $3x = -5$

28. $y = 3$, $y = 6$

29. $y = -1$, $y = 10$

30. $3y = 2$, $5y = -7$

31. $x = 5$, $y = -2$

32. $x = -8$, $y = 3$

9.8 Solution of Systems of Two Linear Equations in Two Variables

Graphical Solution

To solve two linear equations in two variables graphically, draw the graphs of the two equations on one set of axes. The coordinates of the point of intersection, if it exists, give the ordered pair of numbers that is the solution set of the system.

Note The coordinates of the point of intersection cannot always be read exactly; thus the graphical solution is an approximate one.

EXAMPLE

Graphically find the solution set of the two equations

$x + y = 3$ and $3x - y = 1$

Solution

Plot the two lines representing the two equations on the same set of axes. Drop perpendiculars from the point of intersection of the two lines to the x- and y-axes and find the coordinates of the point (see Figure 9.19). The solution set is $\{(1, 2)\}$.

That is, $\{(x, y) \mid x + y = 3\} \cap \{(x, y) \mid 3x - y = 1\} = \{(1, 2)\}$

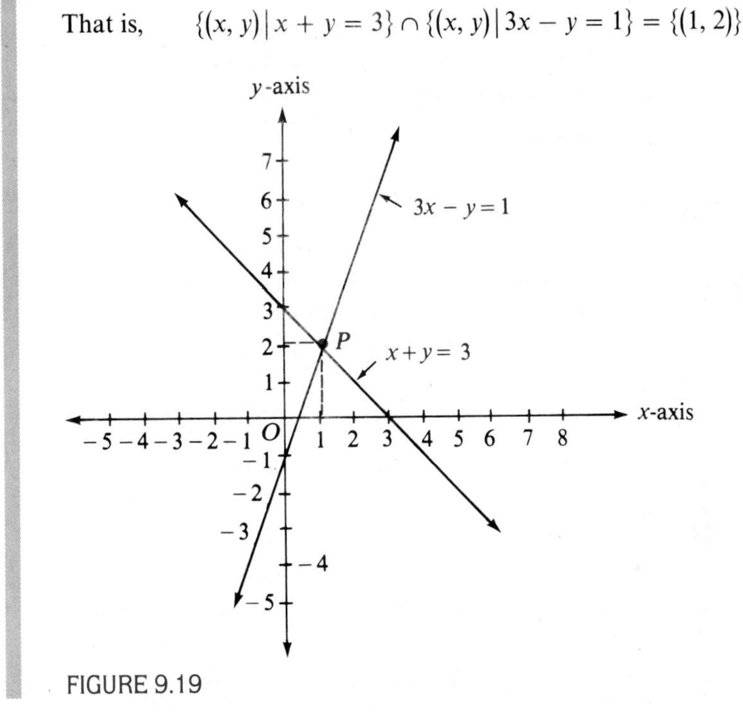

FIGURE 9.19

Exercise 9.8A

Solve the following systems of equations graphically:

1. $x = 1$, $2x + y = 1$

2. $2x = 3$, $2x - y = 1$

3. $y = 2$, $x - y = 2$

4. $y = -2$, $x + y = 1$

5. $x - y = 0$, $x + 2y = 3$

6. $x + y = 0$, $x - 2y = 3$

7. $x + 2y = 0$, $2x - y = 5$

8. $x + y = 3$, $2x + y = 3$

9. $x - 2y = 2$, $2x + y = 4$

10. $x + y = 1$, $x + 3y = 5$

11. $x + 2y = 4$, $x - 3y = -1$

12. $x + y = 4$, $4x - y = 1$

13. $2x - y = 2$, $x + 3y = 8$

14. $x - 2y = 1$, $2x - 3y = 3$

15. $x + y = 2$, $2x + 3y = 3$

16. $2x + y = -1$
$x - 2y = -8$

17. $x - 2y = 1$
$3x + 4y = 8$

18. $2x + 3y = 4$
$6x - y = 2$

19. $x - 3y = 3$
$2x - 6y = 9$

20. $3x - 4y = 6$
$3x - 4y = 12$

21. $x + 2y = 5$
$3x + 6y = 20$

Algebraic Solution

The algebraic solution of two linear equations in two variables gives the exact solution set, not an approximate one as in the case of graphing. There are two methods to solve two linear equations in two variables algebraically: the elimination (or addition) method and the substitution method.

Method of Elimination

To find the solution set of two linear equations in two variables algebraically we transform the given equations to equivalent equations of the form $x = a$ and $y = b$. Hence the solution set is

$$\{(x, y) \mid x = a\} \cap \{(x, y) \mid y = b\} = \{(a, b)\}$$

THEOREM

If (x_1, y_1) is a solution of the equation $a_1x + b_1y + c_1 = 0$ and also a solution of the equation $a_2x + b_2y + c_2 = 0$, then it is a solution of the equation

$$p(a_1x + b_1y + c_1) + q(a_2x + b_2y + c_2) = 0$$

where $p, q \in R$ and p and q are not both zero.

Proof

Given that (x_1, y_1) is a solution of the equation

$$a_1x + b_1y + c_1 = 0 \tag{1}$$

then $a_1x_1 + b_1y_1 + c_1 = 0$

Given that (x_1, y_1) is a solution of the equation

$$a_2x + b_2y + c_2 = 0 \tag{2}$$

then $a_2x_1 + b_2y_1 + c_2 = 0$

Consider the equation

$$p(a_1x + b_1y + c_1) + q(a_2x + b_2y + c_2) = 0 \tag{3}$$

Putting x_1 for x and y_1 for y in Eq. (3), we get

$$p(a_1x_1 + b_1y_1 + c_1) + q(a_2x_1 + b_2y_1 + c_2) = p \cdot 0 + q \cdot 0 = 0$$

Thus if (x_1, y_1) is a solution of Eq. (1) and Eq. (2), it is also a solution of Eq. (3).

The left side of Eq. (3) is said to be a **linear combination** of the left sides of Eqs. (1) and (2).

Since the solution set of Eqs. (1) and (2) is a subset of the solution set of Eq. (3), the system formed by Eqs. (3) and (1), or Eqs. (3) and (2), is equivalent to the system formed by Eqs. (1) and (2).

Equation (3) can be reduced to an equation of the form $rx + t = 0$ (or $r'y + t' = 0$) by choosing p and q in such a manner that the coefficients of y (or x) become additive inverses.
Once the value of x (or y) is found, the value of y (or x) can be determined from the other equation in the system.
Since p and q are chosen so that the coefficient of y is zero, that is, y is eliminated, this is called the **method of elimination**.

EXAMPLE

Using the elimination method, find the solution set of the system

$$2x + 3y - 4 = 0 \quad \text{and} \quad 3x - y + 5 = 0$$

Solution

Consider the equation $p(2x + 3y - 4) + q(3x - y + 5) = 0$.
Taking $p = 3$ and $q = -2$, we have

$$3(2x + 3y - 4) + (-2)(3x - y + 5) = 0$$
$$6x + 9y - 12 - 6x + 2y - 10 = 0$$
$$11y = 22$$
$$y = 2$$

Thus the original system is equivalent to the system

$$2x + 3y - 4 = 0 \quad \text{and} \quad y = 2$$

Putting 2 for y in $2x + 3y - 4 = 0$, we get

$$2x + 3(2) - 4 = 0$$
$$2x = -2$$

Hence $\quad x = -1$

The original system is equivalent to the system

$$x = -1 \quad \text{and} \quad y = 2$$

Hence the solution set is

$$\{(x, y) \mid x = -1\} \cap \{(x, y) \mid y = 2\} = \{(-1, 2)\}$$

When the equations are written in the form $ax + by = c$, the technique of solving the two equations by elimination using the previous theorem (page 304) is illustrated by the following example.

EXAMPLE Using the elimination method, find the solution set of the two equations

$3x + y = 7$ and $2x - 3y = 1$

Solution In order to eliminate x, make the coefficients of x in both equations numerically equal to the least common multiple of their coefficients but with opposite signs.

The least common multiple of 3 and 2 is 6:

$$3x + y = 7 \xrightarrow{\times(2)} 6x + 2y = 14$$

$$2x - 3y = 1 \xrightarrow{\times(-3)} -6x + 9y = -3$$

Adding we get $11y = 11$

Hence $y = 1$

The original system is equivalent to the system

$3x + y = 7$ and $y = 1$

Putting 1 for y in $3x + y = 7$, we obtain

$$3x + (1) = 7$$
$$3x = 6$$

Hence $x = 2$

The original system is equivalent to the system

$x = 2$ and $y = 1$

Hence the solution set is

$$\{(x, y) \mid x = 2\} \cap \{(x, y) \mid y = 1\} = \{(2, 1)\}$$

Remarks Adding the two equations in the manner shown in the previous example is another way of writing

$$p(a_1x + b_1y + c_1) + q(a_2x + b_2y + c_2) = 0$$

Note $\{(x, y) \mid 0x + 0y = k, k \neq 0\} = \varnothing$ and
$\{(x, y) \mid 0x + 0y = 0\} = \{(x, y) \mid x, y \in R\}$

EXAMPLE Using the elimination method, find the solution set of

$4x - 7y = 10$ and $8x - 14y = 15$

Solution The least common multiple of the coefficients of x is 8:

$$4x - 7y = 10 \xrightarrow{\times(-2)} -8x + 14y = -20$$

$$8x - 14y = 15 \longrightarrow 8x - 14y = 15$$

Adding, we get $0x + 0y = -5$

The original system is equivalent to the system

$4x - 7y = 10 \quad \text{and} \quad 0x + 0y = -5$

Hence the solution set is

$$\{(x, y) \mid 4x - 7y = 10\} \cap \{(x, y) \mid 0x + 0y = -5\}$$
$$= \{(x, y) \mid 4x - 7y = 10\} \cap \varnothing = \varnothing$$

EXAMPLE Using the elimination method, find the solution set of

$2x + 3y = 6 \quad \text{and} \quad 6x + 9y = 18$

Solution The least common multiple of the coefficients of x is 6.

$$2x + 3y = 6 \xrightarrow{\times(-3)} -6x - 9y = -18$$
$$6x + 9y = 18 \longrightarrow 6x + 9y = 18$$

Adding, we get $0x + 0y = 0$

The original system is equivalent to the system

$2x + 3y = 6 \quad \text{and} \quad 0x + 0y = 0$

Hence the solution set is

$$\{(x, y) \mid 2x + 3y = 6\} \cap \{(x, y) \mid 0x + 0y = 0\}$$
$$= \{(x, y) \mid 2x + 3y = 6\} \cap \{(x, y) \mid x, y \in R\}$$
$$= \{(x, y) \mid 2x + 3y = 6\}$$

Exercise 9.8B

Solve the following systems of equations by elimination:

1. $x + y = 3$, $2x - y = 3$
2. $x + y = 5$, $3x - y = 3$
3. $2x - y = 6$, $3x + y = 4$
4. $x + 3y = 1$, $x + y = -1$
5. $x - 4y = 1$, $x - 2y = -1$
6. $x + 3y = 7$, $x - y = -5$
7. $x - y = 5$, $2x + y = 4$
8. $2x + y = 3$, $x + 3y = 4$
9. $2x - y = 3$, $x + 2y = -1$
10. $2x - y = 3$, $x + 2y = 9$
11. $x + 2y = 8$, $2x - y = 6$
12. $x - 2y = 1$, $2x - 3y = 1$
13. $2x - 3y = 3$, $x + 2y = 5$
14. $x - 3y = 5$, $2x + y = 3$
15. $x + 2y = 3$, $2x + 3y = 3$
16. $x - 4y = 3$, $2x - y = 6$
17. $x - y = 7$, $2x + 3y = -1$
18. $x + 3y = 6$, $3x + 2y = 4$

19. $3x - 4y = 10$
$5x + 3y = 7$

20. $2x + 3y = 5$
$7x + 4y = -2$

21. $5x + 3y = 1$
$9x + 7y = -3$

22. $4x - 3y = 4$
$2x - 6y = 5$

23. $3x + 5y = 3$
$9x - 10y = 4$

24. $9x + 8y = 9$
$3x + 4y = 2$

25. $2x + 3y = 1$
$4x + 6y = 5$

26. $2x + y = 5$
$8x + 4y = 11$

27. $x - 2y = 4$
$2x - 4y = 9$

28. $x - 3y = 2$
$3x - 9y = 4$

29. $5x + 2y = 3$
$10x + 4y = 7$

30. $2x - 7y = 3$
$6x - 21y = 11$

31. $x + 4y = 1$
$4x + 16y = 4$

32. $x + 5y = 2$
$2x + 10y = 4$

33. $x + 3y = -2$
$3x + 9y = -6$

34. $3x - y = -1$
$6x - 2y = -2$

35. $2x + y = 3$
$12x + 6y = 18$

36. $3x - 2y = 7$
$21x - 14y = 49$

Method of Substitution

The solution set of two linear equations in two variables contains ordered pairs of real numbers (x, y) that satisfy both equations. That is, if (x, y) is in the solution set of two linear equations, then (x, y) must be in the solution set of each equation individually.

The method of solving two linear equations in two variables by **substitution** is based on this principle.

In order to find the solution set of two linear equations in two variables by substitution, we do the following:

1. Express one of the variables in terms of the other variable from one equation.
2. Substitute the expression obtained in Step 1 into the other equation to get a linear equation in one variable.
3. Solve the resulting linear equation in one variable for the specific value of that variable.
4. Substitute the solution obtained in Step 3 into the equation obtained in Step 1 to find the specific value of the other variable.

EXAMPLE

Solve the following system of equations by substitution:

$3x - 2y = 4$ and $5x + 4y = 3$

Solution

Express x in terms of y from the first equation:

$$x = \frac{2y + 4}{3}$$

Substitute $\frac{2y + 4}{3}$ for x in the second equation:

$$5\left(\frac{2y+4}{3}\right) + 4y = 3$$

$$\frac{10y+20}{3} + 4y = 3$$

Multiplying both sides of the equation by 3, we obtain

$$10y + 20 + 12y = 9$$

$$22y = -11$$

Hence $\quad y = \frac{-11}{22} = -\frac{1}{2}$

The original system is equivalent to the system

$$x = \frac{2y+4}{3} \quad \text{and} \quad y = -\frac{1}{2}$$

Putting $\left(-\frac{1}{2}\right)$ for y in $x = \frac{2y+4}{3}$, we get

$$x = \frac{2\left(-\frac{1}{2}\right)+4}{3} = \frac{-1+4}{3} = \frac{3}{3} = 1$$

The original system is equivalent to the system

$$x = 1 \quad \text{and} \quad y = -\frac{1}{2}$$

The solution set is

$$\{(x, y) \mid x = 1\} \cap \left\{(x, y) \,\middle|\, y = -\frac{1}{2}\right\} = \left\{\left(1, -\frac{1}{2}\right)\right\}$$

Exercise 9.8C

Using the substitution method, solve the following systems of equations:

1. $x - y = 0$
$2x + y = 6$

2. $x - 2y = 0$
$2x - y = 3$

3. $2x - y = 0$
$3x + y = 10$

4. $x - 3y = 0$
$2x + y = 7$

5. $x + 3y = 0$
$3x + 5y = 4$

6. $x + 2y = 0$
$2x + y = 3$

7. $x + y = 2$
$2x + y = 3$

8. $x - y = 5$
$x - 2y = 7$

9. $x + 2y = 7$
$2x - y = 4$

10. $x + 3y = 1$
$2x + 3y = 5$

11. $x + y = 6$
$2x - y = 3$

12. $x + 3y = 3$
$2x + y = -4$

13. $x + 5y = 1$
$x - 3y = -7$

14. $2x + y = 2$
$3x + 2y = 2$

15. $2x - 3y = -1$
$3x - 2y = 6$

16. $2x + 5y = 7$
$3x + 2y = -6$

17. $2x + 3y = 2$
$3x + 5y = 2$

18. $3x - 4y = 1$
$4x - 3y = -1$

19. $x + 3y = 1$
$3x - 6y = 8$

20. $2x + y = -1$
$4x + 3y = -4$

21. $2x - 9y = 9$
$6x + 3y = 7$

22. $5x + 4y = 5$
$10x - 12y = -5$

23. $3x + 2y = -4$
$9x + 8y = -11$

24. $3x - 4y = 12$
$9x + 16y = 1$

25. $2x + 3y = 1$
$8x + 12y = 4$

26. $x - 2y = 4$
$2x - 4y = 8$

27. $3x + y = -5$
$9x + 3y = -15$

28. $3x + y = 5$
$6x + 2y = 11$

29. $x - 3y = 2$
$7x - 21y = 10$

30. $3x - 6y = 8$
$2x - 4y = 7$

9.9 Systems of Two Linear Equations in Two Variables Involving Grouping Symbols and Fractions

When either or both equations contain grouping symbols, apply the distributive law to remove the grouping symbols. Write equivalent equations of the form $ax + by = c$ and then solve.

EXAMPLE

Solve the following system of equations:

$3(x - 2y) + 2(x + 3) = 4$ and $4(x + y) - 3(x + 2y) = -2$

Solution

$$\begin{aligned} 3(x - 2y) + 2(x + 3) &= 4 \\ 3x - 6y + 2x + 6 &= 4 \\ 5x - 6y &= -2 \end{aligned} \qquad \begin{aligned} 4(x + y) - 3(x + 2y) &= -2 \\ 4x + 4y - 3x - 6y &= -2 \\ x - 2y &= -2 \end{aligned}$$

Now, we solve the system $5x - 6y = -2$ and $x - 2y = -2$:

$$\begin{aligned} 5x - 6y = -2 &\longrightarrow & 5x - 6y &= -2 \\ x - 2y = -2 &\xrightarrow{\times(-3)} & -3x + 6y &= 6 \end{aligned}$$

Adding, we get $2x = 4$

Hence $x = 2$

The system is equivalent to

$x - 2y = -2$ and $x = 2$

Putting 2 for x in $x - 2y = -2$, we get

$(2) - 2y = -2$ that is, $y = 2$

The original system is equivalent to the system

$x = 2$ and $y = 2$

The solution set is

$\{(x, y) \mid x = 2\} \cap \{(x, y) \mid y = 2\} = \{(2, 2)\}$

When a linear equation has fractional coefficients, we can find an equivalent equation with integral coefficients by multiplying both sides of the equation by the least common multiple of the denominators in the equation.

EXAMPLE

Solve the following system of equations:

$$\frac{5}{6}x + \frac{1}{2}y = 3 \qquad \text{and} \qquad \frac{2}{3}x - \frac{3}{4}y = 7$$

Solution

Multiply the first equation by 6, the second equation by 12, and then solve.

$$\begin{array}{lcl} 5x + 3y = 18 & \xrightarrow{\times (3)} & 15x + 9y = 54 \\ 8x - 9y = 84 & \longrightarrow & 8x - 9y = 84 \\ \hline \text{Adding, we get} & & 23x \qquad = 138 \\ \text{Hence} & & x = 6 \end{array}$$

The system is equivalent to

$$5x + 3y = 18 \qquad \text{and} \qquad x = 6$$

Putting 6 for x in $5x + 3y = 18$, we get

$$5(6) + 3y = 18 \qquad \text{that is,} \qquad y = -4$$

The original system is equivalent to

$$x = 6 \qquad \text{and} \qquad y = -4$$

The solution set is

$$\{(x, y) \mid x = 6\} \cap \{(x, y) \mid y = -4\} = \{(6, -4)\}$$

9.10 Fractional Equations That Can Be Made Linear

Often we encounter fractional equations with variables in the denominators. Clearing the fractions leads to equations of higher degree. In some cases a change of variables offers a linear equation.

For example, consider the equation $\dfrac{2}{3x} - \dfrac{3}{4y} = \dfrac{13}{8}$.

Multiplying by the LCM, $24xy$, we get the equation $16y - 18x = 39xy$, which is not linear.

However if we let $u = \dfrac{1}{x}$ and $v = \dfrac{1}{y}$, then substitution yields the equation $\dfrac{2}{3}u - \dfrac{3}{4}v = \dfrac{13}{8}$, which is a linear equation in u and v. Thus we obtain linear equations in u and v that can be solved. After arriving at the values of u and v, we can calculate the values of x and y.

EXAMPLE

Solve the following system of equations:

$$\frac{2}{3x} - \frac{3}{4y} = \frac{13}{8} \quad \text{and} \quad \frac{3}{4x} + \frac{5}{3y} = \frac{37}{12}$$

Solution

Substituting u for $\frac{1}{x}$ and v for $\frac{1}{y}$, we obtain

$$\frac{2}{3}u - \frac{3}{4}v = \frac{13}{8} \qquad (1)$$

$$\frac{3}{4}u + \frac{5}{3}v = \frac{37}{12} \qquad (2)$$

Multiply Eq. (1) by 24 and Eq. (2) by 12 and solve:

$$16u - 18v = 39 \xrightarrow{\times 10} 160u - 180v = 390$$

$$9u + 20v = 37 \xrightarrow{\times 9} 81u + 180v = 333$$

Adding, we get $241u = 723$

Hence $u = 3$

The system is equivalent to

$16u - 18v = 39$ and $u = 3$

Putting 3 for u in $16u - 18v = 39$, we get

$16(3) - 18v = 39$ that is, $v = \frac{1}{2}$

Since $u = \frac{1}{x} = 3$ and $v = \frac{1}{y} = \frac{1}{2}$

we have $x = \frac{1}{3}$ and $y = 2$

Hence the solution set of the original system is

$$\left\{(x, y) \,\middle|\, x = \frac{1}{3}\right\} \cap \{(x, y) \mid y = 2\} = \left\{\left(\frac{1}{3}, 2\right)\right\}$$

EXAMPLE

Solve the following system of equations:

$$\frac{4}{2x + y} + \frac{5}{x - 2y} = -\frac{9}{2} \quad \text{and} \quad \frac{3}{2x + y} - \frac{2}{x - 2y} = \frac{19}{8}$$

Solution

Substituting u for $\frac{1}{2x + y}$ and v for $\frac{1}{x - 2y}$, we obtain

$$4u + 5v = -\frac{9}{2} \quad \text{or} \quad 8u + 10v = -9$$

$$3u - 2v = \frac{19}{8} \quad \text{or} \quad 24u - 16v = 19$$

$$8u + 10v = -9 \xrightarrow{\times(-3)} -24u - 30v = 27$$

$$24u - 16v = 19 \longrightarrow 24u - 16v = 19$$

Adding, we get

$$-46v = 46$$

$$v = -1$$

The system is equivalent to

$$8u + 10v = -9 \quad \text{and} \quad v = -1$$

Putting -1 for v in $8u + 10v = -9$, we obtain

$$8u + 10(-1) = -9 \quad \text{or} \quad u = \frac{1}{8}$$

Hence $$\frac{1}{2x + y} = \frac{1}{8} \quad \text{and} \quad \frac{1}{x - 2y} = -1$$

or $$2x + y = 8 \quad \text{and} \quad x - 2y = -1$$

$$2x + y = 8 \xrightarrow{\times(2)} 4x + 2y = 16$$

$$x - 2y = -1 \longrightarrow x - 2y = -1$$

Adding, we get

$$5x = 15$$

$$x = 3$$

The original system is equivalent to

$$2x + y = 8 \quad \text{and} \quad x = 3$$

Putting 3 for x in $2x + y = 8$, we get $y = 2$.

The original system is equivalent to

$$x = 3 \quad \text{and} \quad y = 2$$

The solution set is $\{(3, 2)\}$.

Exercises 9.9–9.10

Solve the following systems of equations:

1. $x + 4(y + 3) = 5$
$3(x - 1) - 2(y + 2) = 0$

2. $5x - (2y + x) = 0$
$2(3x - 2) + 3(y - 3) = -1$

3. $2x - 3(y + 1) = 8$
$3(x + 2) + 5y = -6$

4. $2x - (3y + 5) = 2$
$5(x - y) + 3(x + 2y) = 2$

5. $(2x - y) - (x + 2y) = 1$
$2(x - 3y) + 3(x - 3) = -1$

6. $3(2x - y) - 2(x - 3y) = 4$
$6(x + y) - 5(2x + y) = -2$

7. $2(3x - y) - 5(x - y) = 5$
$4(2x - 3y) - 7(x - 2y) = 4$

8. $3(3x - 2y) - 4(2x - 3y) = 11$
$6(x - 4y) - 5(x - 5y) = 1$

9. $5(7x - y) - 6(6x - y) = 1$
$4(3x + y) - 3(5x + 2y) = -12$

10. $5(x + 3y) + 2(x - 4y) = 14$
$7(x - y) - 4(2x + y) = 10$

11. $\frac{3x}{4} + \frac{5y}{2} = 2$
$\frac{3x}{2} + \frac{7y}{2} = 1$

12. $\frac{3x}{4} - \frac{y}{2} = -1$
$\frac{x}{4} - \frac{3y}{8} = -2$

13. $\frac{2x}{3} + \frac{3y}{2} = 8$
$\frac{5x}{3} - \frac{3y}{4} = 2$

14. $\frac{5x}{2} + \frac{y}{3} = 4$
$\frac{7x}{2} - \frac{2y}{3} = 9$

15. $\frac{5x}{6} - \frac{2y}{3} = 3$
$\frac{3x}{2} - \frac{5y}{3} = 4$

16. $\frac{x}{3} + \frac{y}{2} = \frac{4}{3}$
$\frac{x}{2} - \frac{y}{3} = -\frac{1}{6}$

17. $\frac{5x}{4} - \frac{y}{3} = \frac{7}{2}$
$\frac{3x}{4} + \frac{2y}{3} = -\frac{1}{2}$

18. $\frac{2x}{3} - \frac{y}{4} = \frac{11}{12}$
$\frac{2x}{5} - \frac{y}{3} = \frac{11}{15}$

19. $\frac{3x}{8} + \frac{2y}{3} = -\frac{5}{4}$
$\frac{3x}{5} + \frac{5y}{3} = -\frac{19}{5}$

20. $\frac{4x}{7} - \frac{3y}{4} = -\frac{17}{7}$
$\frac{8x}{3} - \frac{5y}{2} = -\frac{22}{3}$

21. $\frac{2x}{5} + \frac{3y}{4} = \frac{9}{20}$
$\frac{3x}{7} + \frac{7y}{5} = -\frac{4}{35}$

22. $\frac{x - y}{2} - \frac{x - 2y}{3} = \frac{5}{6}$
$\frac{x + y}{4} - \frac{x - 3y}{3} = \frac{11}{12}$

23. $\frac{2x - y}{3} - \frac{x - y}{4} = \frac{11}{12}$
$\frac{3x - y}{2} - \frac{2x + y}{3} = \frac{5}{2}$

24. $\frac{x - y}{2} - \frac{3x + y}{3} = \frac{1}{6}$
$\frac{x - 2y}{2} - \frac{x + y}{3} = \frac{19}{6}$

25. $\frac{4x - y}{4} - \frac{2x - y}{3} = \frac{1}{12}$
$\frac{2x - y}{3} - \frac{4x + y}{2} = \frac{7}{6}$

26. $\frac{x + 4y}{2} - \frac{x + 2y}{3} = \frac{5}{2}$
$\frac{x + 3y}{3} - \frac{x + y}{2} = \frac{7}{6}$

27. $\frac{3x - y}{4} - \frac{2x - y}{7} = \frac{19}{28}$
$\frac{4x - y}{5} - \frac{x - y}{4} = \frac{9}{20}$

28. $\frac{1}{x} - \frac{2}{y} = 0$
$\frac{4}{x} + \frac{12}{y} = 5$

29. $\frac{6}{x} + \frac{5}{y} = 3$
$\frac{9}{x} + \frac{10}{y} = 5$

30. $\frac{2}{x} + \frac{3}{y} = 13$
$\frac{5}{x} - \frac{4}{y} = -2$

31. $\frac{1}{x} + \frac{2}{y} = 1$
$\frac{3}{x} + \frac{7}{y} = 1$

32. $\frac{7}{3x} - \frac{5}{2y} = -\frac{1}{3}$
$\frac{4}{3x} + \frac{1}{2y} = \frac{11}{3}$

33. $\frac{1}{x} - \frac{1}{3y} = \frac{2}{3}$
$\frac{3}{2x} + \frac{4}{y} = \frac{49}{4}$

34. $\dfrac{2}{3x} + \dfrac{3}{4y} = \dfrac{35}{36}$
$\dfrac{3}{5x} - \dfrac{4}{3y} = -\dfrac{17}{15}$

35. $\dfrac{1}{3x} - \dfrac{1}{2y} = \dfrac{43}{36}$
$\dfrac{3}{8x} + \dfrac{5}{3y} = -2$

36. $\dfrac{5}{2x} + \dfrac{8}{9y} = \dfrac{1}{18}$
$\dfrac{10}{7x} + \dfrac{2}{5y} = \dfrac{3}{35}$

37. $\dfrac{2}{3x} - \dfrac{3}{4y} = \dfrac{7}{12}$
$\dfrac{1}{2x} - \dfrac{3}{5y} = \dfrac{9}{20}$

38. $\dfrac{5}{x+y} + \dfrac{2}{x-y} = \dfrac{11}{3}$
$\dfrac{4}{x+y} - \dfrac{3}{x-y} = -\dfrac{5}{3}$

39. $\dfrac{2}{2x+y} + \dfrac{3}{2x-y} = \dfrac{19}{15}$
$\dfrac{7}{2x+y} - \dfrac{6}{2x-y} = \dfrac{17}{15}$

40. $\dfrac{5}{x-2y} - \dfrac{15}{x+2y} = 6$
$\dfrac{15}{x-2y} - \dfrac{6}{x+2y} = 5$

41. $\dfrac{3}{x+4y} - \dfrac{4}{x+2y} = 7$
$\dfrac{2}{x+4y} + \dfrac{3}{x+2y} = -1$

42. $\dfrac{7}{3x-y} + \dfrac{2}{x-2y} = \dfrac{13}{12}$
$\dfrac{2}{3x-y} - \dfrac{3}{x-2y} = \dfrac{7}{12}$

43. $\dfrac{5}{2x+3y} - \dfrac{2}{2x+y} = \dfrac{1}{4}$
$\dfrac{3}{2x+3y} - \dfrac{4}{2x+y} = -\dfrac{5}{4}$

44. $\dfrac{3}{3x-y} - \dfrac{8}{x-3y} = \dfrac{11}{6}$
$\dfrac{4}{3x-y} - \dfrac{3}{x-3y} = \dfrac{7}{6}$

45. $\dfrac{3}{2x-3y} - \dfrac{2}{2x+3y} = 3$
$\dfrac{4}{2x-3y} + \dfrac{9}{2x+3y} = -\dfrac{23}{3}$

46. $\dfrac{7}{4x+3y} + \dfrac{5}{4x-y} = \dfrac{4}{15}$
$\dfrac{4}{4x+3y} + \dfrac{3}{4x-y} = \dfrac{1}{5}$

47. $\dfrac{6}{3x+2y} - \dfrac{5}{2x-3y} = \dfrac{6}{13}$
$\dfrac{7}{3x+2y} - \dfrac{9}{2x-3y} = -\dfrac{12}{13}$

9.11 Word Problems

Many word problems can be solved easily by using equations in two variables. Represent two of the unknown quantities by two variables. Express the other unknown quantities in terms of the two variables. Translate the word statements into two equations. Solve the equations in the variables, and calculate the unknown quantities. Check your answer in the word problem.

The following examples illustrate some of the types of problems that lead to equations in two variables.

EXAMPLE

Three times one number is 14 less than twice a second number, while 9 times the second number is 12 less than 16 times the first number. Find the two numbers.

Solution

First Number x *Second Number* y

$3x + 14 = 2y$ that is, $3x - 2y = -14$

$9y + 12 = 16x$ that is, $16x - 9y = 12$

$$3x - 2y = -14 \xrightarrow{\times(-9)} -27x + 18y = 126$$

$$16x - 9y = 12 \xrightarrow{\times(2)} 32x - 18y = 24$$

Adding, we get $5x = 150$

Hence $x = 30$

Putting 30 for x, we get $y = 52$.

The first number is 30. The second number is 52.

EXAMPLE

A two-digit number is 3 more than 7 times the sum of its digits. If the digits are interchanged, the result is 2 more than 5 times the tens digit in the original number. Find the original number.

Solution

Units Digit x *Tens Digit* y

The number is $x + 10y$. The sum of the digits is $x + y$:

$$x + 10y - 3 = 7(x + y) \quad (1)$$

When the digits are interchanged, y will be the units digit and x the tens digit. The new number is $y + 10x$:

$$y + 10x - 2 = 5y \quad (2)$$

Simplifying Eqs. (1) and (2) and solving, we get

$$2x - y = -1 \xrightarrow{\times(-2)} -4x + 2y = 2$$

$$5x - 2y = 1 \longrightarrow 5x - 2y = 1$$

Adding, we get $x = 3$

Putting 3 for x, we get $y = 7$.

Hence the number is 73.

EXAMPLE

If 2 is subtracted from the numerator and 7 is added to the denominator of a fraction, its value becomes $\frac{1}{2}$. If 1 is subtracted from the numerator and 13 is added to the denominator, its value becomes $\frac{4}{9}$. Find the fraction.

Solution

Let the fraction be $\frac{x}{y}$:

$$\frac{x-2}{y+7}=\frac{1}{2} \quad (1)$$

$$\frac{x-1}{y+13}=\frac{4}{9} \quad (2)$$

Multiplying Eq. (1) by $2(y+7)$ and Eq. (2) by $9(y+13)$ and simplifying we obtain

$$2x - y = 11 \qquad \text{and} \qquad 9x - 4y = 61$$

The above system is equivalent to

$$x = 17 \qquad \text{and} \qquad y = 23$$

Hence the fraction is $\frac{17}{23}$.

EXAMPLE

A man invested part of his money at 6% and the rest at 8%. The income from both investments totaled \$2740. If he interchanged his investments, his income would have totaled \$2580. How much did he have in each investment?

Solution

\$x at	\$y at
6%	8%

$$6\%x + 8\%y = 2740 \quad (1)$$

$$8\%x + 6\%y = 2580 \quad (2)$$

Simplifying Eqs. (1) and (2) and solving, we get

$$3x + 4y = 137{,}000 \xrightarrow{\times(4)} 12x + 16y = 548{,}000$$

$$4x + 3y = 129{,}000 \xrightarrow{\times(-3)} -12x - 9y = -387{,}000$$

Adding, we get $7y = 161{,}000$

Hence $y = 23{,}000$

Putting 23,000 for y, we get $x = 15{,}000$.

The investments were \$15,000 at 6% and \$23,000 at 8%.

EXAMPLE

If a given quantity of a 30% alcohol solution is added to a certain quantity of a 90% alcohol solution, the mixture is a 54% alcohol solution. If there were 20 parts less of the 90% solution, the mixture would be a 50% alcohol solution. How many parts of each solution are there?

Solution

x parts	*y parts*	*(x + y) parts*
30%	90%	54%

$$30\%x + 90\%y = 54\%(x + y) \quad (1)$$
$$30\%x + 90\%(y - 20) = 50\%(x + y - 20) \quad (2)$$

Simplifying Eqs. (1) and (2) and solving, we get

$$2x - 3y = 0 \longrightarrow 2x - 3y = 0$$
$$x - 2y = -40 \xrightarrow{\times(-2)} -2x + 4y = 80$$

Adding, we get $y = 80$

Putting 80 for y, we get $x = 120$.

There are 120 parts of the 30% solution.

There are 80 parts of the 90% solution.

EXAMPLE

When flying with a tail wind it took a plane 5 hours to go 3300 miles, but with a head wind 10 mph stronger than the tail wind it took the same plane 6 hours. Find the speed of the plane in still air.

Solution

Let the speed of the tail wind be x mph.
The speed of the head wind is $(x + 10)$ mph.

Let the speed of the plane in still air be y mph.
The speed of the plane with the tail wind is $y + x$ mph.
The speed of the plane with the head wind is $y - (x + 10)$ mph:

$$5(y + x) = 3300 \quad (1)$$
$$6[y - (x + 10)] = 3300 \quad (2)$$

Simplifying Eqs. (1) and (2) and solving, we get

$$y + x = 660$$
$$y - x = 560$$

Adding, we get $2y = 1220$

Hence $y = 610$

The speed of the plane in still air is 610 mph.

EXAMPLE

A fulcrum is placed so that weights of 80 pounds and 120 pounds are in balance. When 20 pounds are added to the 80-pound weight, the 120-pound weight must be moved one foot farther from the fulcrum to preserve the balance. Find the original distance between the two weights.

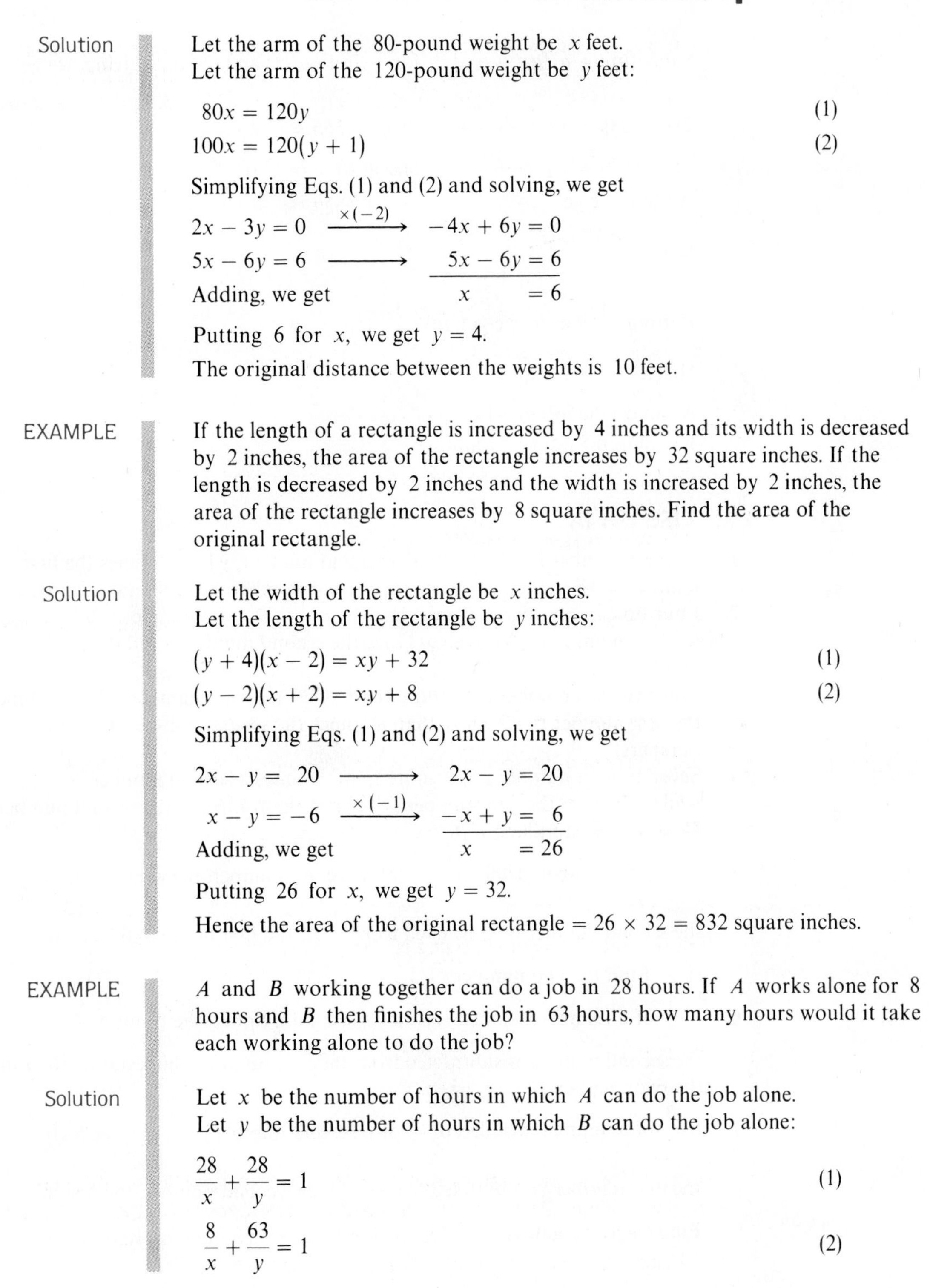

Solution

Let the arm of the 80-pound weight be x feet.
Let the arm of the 120-pound weight be y feet:

$$80x = 120y \quad (1)$$

$$100x = 120(y + 1) \quad (2)$$

Simplifying Eqs. (1) and (2) and solving, we get

$$\begin{array}{lcr} 2x - 3y = 0 & \xrightarrow{\times(-2)} & -4x + 6y = 0 \\ 5x - 6y = 6 & \longrightarrow & 5x - 6y = 6 \\ \hline \text{Adding, we get} & & x \quad = 6 \end{array}$$

Putting 6 for x, we get $y = 4$.

The original distance between the weights is 10 feet.

EXAMPLE

If the length of a rectangle is increased by 4 inches and its width is decreased by 2 inches, the area of the rectangle increases by 32 square inches. If the length is decreased by 2 inches and the width is increased by 2 inches, the area of the rectangle increases by 8 square inches. Find the area of the original rectangle.

Solution

Let the width of the rectangle be x inches.
Let the length of the rectangle be y inches:

$$(y + 4)(x - 2) = xy + 32 \quad (1)$$

$$(y - 2)(x + 2) = xy + 8 \quad (2)$$

Simplifying Eqs. (1) and (2) and solving, we get

$$\begin{array}{lcr} 2x - y = 20 & \longrightarrow & 2x - y = 20 \\ x - y = -6 & \xrightarrow{\times(-1)} & -x + y = 6 \\ \hline \text{Adding, we get} & & x \quad = 26 \end{array}$$

Putting 26 for x, we get $y = 32$.

Hence the area of the original rectangle $= 26 \times 32 = 832$ square inches.

EXAMPLE

A and B working together can do a job in 28 hours. If A works alone for 8 hours and B then finishes the job in 63 hours, how many hours would it take each working alone to do the job?

Solution

Let x be the number of hours in which A can do the job alone.
Let y be the number of hours in which B can do the job alone:

$$\frac{28}{x} + \frac{28}{y} = 1 \quad (1)$$

$$\frac{8}{x} + \frac{63}{y} = 1 \quad (2)$$

Substituting a for $\frac{1}{x}$ and b for $\frac{1}{y}$ in Eqs. (1) and (2) and solving, we get

$$28a + 28b = 1 \xrightarrow{\times(-2)} -56a - 56b = -2$$

$$8a + 63b = 1 \xrightarrow{\times(7)} 56a + 441b = 7$$

Adding, we get $\quad 385b = 5$

Hence $\quad b = \frac{1}{77}$

Putting $\frac{1}{77}$ for b, we get $a = \frac{1}{44}$.

Hence $x = 44$ and $y = 77$.

A can do the job in 44 hours.
B can do the job in 77 hours.

Exercise 9.11

1. Twice a number is 14 less than a second number, while 7 times the first number is 47 more than twice the second number. Find the two numbers.
2. Four times a number is 9 less than 3 times a second number, while 3 times the first number is 1 less than twice the second number. Find the two numbers.
3. Three times a number is 2 more than twice a second number, while 5 times the first number is 22 more than 3 times the second number. Find the two numbers.
4. Seven times a number is 50 more than 3 times a second number, while 12 times the first number is 61 less than 7 times the second number. Find the two numbers.
5. If $\frac{1}{4}$ of a number is added to $\frac{1}{5}$ of a second number, the result is 8. If $\frac{1}{4}$ of the second number is subtracted from $\frac{1}{2}$ of the first number, the result is 3. Find the two numbers.
6. If $\frac{1}{3}$ of a number is added to $\frac{1}{2}$ of a second number, the result is 42. If $\frac{1}{3}$ of the second number is subtracted from the first number, the result is 16. Find the two numbers.
7. If $\frac{3}{2}$ of a number is added to $\frac{2}{3}$ of a second number, the result is 5. If $\frac{9}{8}$ of the first number is subtracted from $\frac{7}{2}$ of the second number, the result is $\frac{21}{4}$. Find the two numbers.

8. If $\frac{2}{5}$ of a number is added to $\frac{5}{3}$ of a second number, the result is $\frac{17}{2}$. If $\frac{5}{2}$ of the second number is subtracted from 4 times the first number, the result is 16. Find the two numbers.

9. The sum of the reciprocals of two numbers is $\frac{49}{6}$ and the difference of their reciprocals is $\frac{5}{6}$. What are the two numbers?

10. The sum of the reciprocals of two numbers is $\frac{82}{35}$ and the difference of their reciprocals is $\frac{2}{35}$. What are the two numbers?

11. A two-digit number is 6 more than 4 times the sum of its digits. If the digits are interchanged, the result is 13 more than 9 times the units digit of the original number. Find the original number.

12. A two-digit number is 7 more than 8 times the sum of its digits. If the digits are interchanged, the result is 3 more than twice the tens digit of the original number. Find the original number.

13. The sum of the digits of a three-digit number is 16. The number is 11 less than 60 times the units digit. If the units and hundreds digits are interchanged, the result is 13 more than 57 times the sum of the digits. Find the original number.

14. The tens digit of a three-digit number is twice the units digit. The number is one more than 57 times the sum of the digits. If the units and hundreds digits are interchanged, the result is 5 less than 36 times the hundreds digit of the original number. Find the original number.

15. If 3 is added to the numerator and 6 is subtracted from the denominator of a fraction, its value becomes $\frac{3}{4}$. If 1 is added to the numerator and 2 is added to the denominator, its value becomes $\frac{1}{4}$. Find the fraction.

16. If 3 is subtracted from the numerator and 1 is added to the denominator of a fraction, its value becomes $\frac{2}{3}$. If 7 is subtracted from the numerator and 5 is added to the denominator, its value becomes $\frac{1}{2}$. Find the fraction.

17. If 6 is subtracted from the numerator and 1 is added to the denominator of a fraction, its value becomes $\frac{1}{2}$. If 4 is added to the numerator and 9 is subtracted from the denominator, its value becomes $\frac{4}{5}$. Find the fraction.

18. If 3 is added to both the numerator and denominator of a fraction, its value becomes $\frac{2}{3}$. If 2 is subtracted from both numerator and denominator, its value becomes $\frac{1}{2}$. What is the fraction?

19. If 1 is subtracted from the numerator and is added to the denominator of a fraction, its value becomes $\frac{1}{2}$. If 3 is added to the numerator and is subtracted from the denominator, its value becomes 2. Find the fraction.

20. If 2 is added to the numerator and 4 is added to the denominator of a fraction, its value becomes $\frac{2}{3}$. If 2 is subtracted from the numerator and 1 is added to the denominator, its value becomes $\frac{1}{2}$. Find the fraction.

21. If 3 is added to the numerator and 5 is added to the denominator of a fraction, its value becomes $\frac{4}{5}$. If 2 is subtracted from both numerator and denominator, its value becomes $\frac{5}{6}$. Find the fraction.

22. The quotient of the difference of two natural numbers divided by their sum equals $\frac{1}{5}$. If the sum of the two numbers is divided by the smaller number, the quotient is 2 and the remainder is 6. Find the two numbers.

23. The quotient of the difference of two natural numbers divided by their sum equals $\frac{4}{9}$. If the sum of the two numbers is divided by 3 times the smaller number, the quotient is 1 and the remainder is 12. Find the two numbers.

24. Twice a number is 3 less than 5 times a second number. If the sum of the two numbers is divided by the second number, the quotient is 4 and the remainder is -6. Find the two numbers.

25. Three times a number is 6 more than twice a second number. If the sum of the second number and twice the first number is divided by 7 less than the first number, the quotient is 4 and the remainder is 4. Find the two numbers.

26. Five times a number is 3 less than 3 times a second number. If 6 times the first number minus the second number is divided by 1 more than the second number, the quotient is 2 and the remainder is 22. Find the two numbers.

27. A man invested part of his money at $5\frac{1}{2}\%$ and the rest at 6%. The income from both investments totaled \$1740. If he interchanged his investments, his income would have totaled \$1710. How much did he have in each investment?

28. The total return from \$15,000 and \$25,000 investments was \$2900. If the investments were interchanged, the return would be \$2700. How much was the rate of return on each investment?

29. The total return from \$10,000 and \$28,000 investments was \$2880. If the investments were interchanged, the return would be \$2250. How much was the rate of return on each investment?
30. If 6 pounds of almonds and 4 pounds of walnuts cost \$5.50, while 4 pounds of almonds and 13 pounds of walnuts cost \$8.73, what is the price of each kind of nut per pound?
31. If 7 pounds of oranges and 4 pounds of apples cost \$3.31, while 3 pounds of oranges and 9 pounds of apples cost \$4.26, what is the price of each per pound?
32. If 4 packages of corn and 9 packages of peas cost \$3.57, but 7 packages of corn and 6 packages of peas cost \$3.42, what is the price of each kind of vegetable per package?
33. If 10 pounds of potatoes and 5 pounds of rice cost \$3.05, but 15 pounds of potatoes and 2 pounds of rice cost \$2.87, what is the price of each per pound?
34. If a 60% acid solution is added to a 40% acid solution, the mixture is a 47.5% acid solution. If there were 20 more gallons of the 40% acid solution, the new mixture would be a 46% acid solution. How many gallons of each solution are there?
35. If a 27% silver alloy is combined with an 18% silver alloy, the mixture would contain 22% silver. If there were 15 more pounds of the 18% silver alloy, the mixture would contain 21% silver. How many pounds of each alloy are there?
36. A jeweler combines a 22-karat gold with a 10-karat gold and obtains 19.6-karat gold. If there were 4 more ounces of the 10-karat gold, he would have 18-karat gold. How many ounces of each kind does he have?
37. A purse contains \$1.55 in nickels and quarters. If the quarters were nickels and the nickels were quarters, the value of the coins would be \$4.15. How many nickels and how many quarters are in the purse?
38. A purse contains \$3.75 in quarters and dimes. If the quarters were dimes and the dimes were quarters, the value of the coins would be \$4.65. How many quarters and how many dimes are in the purse?
39. A man rowed 9 miles up a river in 3 hours and rowed back in $1\frac{1}{2}$ hours. Find the rate of the current and the man's rate of rowing in still water.
40. A power boat travels 20 miles upstream on a river in $2\frac{1}{2}$ hours. At half the previous speed it makes the return trip in 2 hours. Find the rate of flow of the river and the boat's speed in still water.
41. Flying with the wind it took a plane 2 hours and 20 minutes to go 630 miles, while against the wind it took the plane 3 hours. Find the speed of the wind and the speed of the plane in still air.
42. Seven years ago a boy was $\frac{1}{5}$ as old as his father, while 17 years from now he will be $\frac{1}{2}$ as old as his father. Find their present ages.

43. Eighteen years ago a woman was $\frac{1}{2}$ as old as her husband, while 4 years ago she was $\frac{2}{3}$ as old as her husband. Find their ages now.

44. A fulcrum is placed so that weights of 60 pounds and 80 pounds are in balance. When 10 pounds are added to the 60-pound weight, the fulcrum must be moved $\frac{2}{5}$ of a foot farther from the 80 pounds to preserve the balance. Find the length of the lever.

45. A fulcrum is placed so that weights of 36 pounds and 60 pounds are in balance. When 12 pounds are added to the 60-pound weight, the fulcrum must be moved $\frac{1}{2}$ foot farther from the 36 pounds to preserve the balance. Find the length of the lever.

46. A fulcrum is placed so that weights of 80 pounds and 120 pounds are in balance. When 20 pounds are removed from the 120 pounds, the 80 pounds must be moved 1 foot closer to the fulcrum to preserve the balance. Find the original distance between the 80-pound weight and the 120-pound weight.

47. A fulcrum is placed so that weights of 60 pounds and 75 pounds are in balance. When 15 pounds are added to the 75 pounds, the 60 pounds must be moved 2 feet farther from the fulcrum to preserve the balance. Find the original distance between the 60-pound weight and the 75-pound weight.

48. If the length of a rectangle is increased by 2 inches and its width is decreased by 1 inch, the area of the rectangle increases by 3 square inches. If the length is decreased by 4 inches and the width is increased by 3 inches, the area decreases by 11 square inches. Find the area of the original rectangle.

49. If the length of a lot is decreased by 10 feet and the width is increased by 10 feet, the area of the lot increases by 500 square feet. If the length is increased by 10 feet and the width is decreased by 5 feet, the area of the lot decreases by 100 square feet. Find the original area of the lot.

50. A and B working together can do a job in 12 hours. If A works alone for 4 hours and B then finishes the job in 18 hours, how many hours would it take each working alone to do the job?

51. A and B working together can do a job in 20 hours. If A works alone for 8 hours and B then finishes the job in 35 hours, how many hours would it take each working alone to do the job?

52. A farmer plows a field with a team while his son uses a tractor. Working together they can plow the field in 6 days. If the farmer plows alone for 3 days and his son finishes plowing the field in 8 days, how many days would it take each person alone to plow the field?

53. A tank can be filled by two pipes running simultaneously for 45 minutes. If the first pipe was turned on for only 30 minutes, and the second pipe filled the rest of the tank in 72 minutes, how long does it take each pipe separately to fill the tank?

54. An office building with a floor area of 4500 square feet is divided into three offices A, B, and C. The rent per square foot of floor area is \$3 for office A, \$2 for office B, and \$1.50 for office C. The combined rent for offices A and B is $4\frac{1}{2}$ times the rent for office C. The total rent for the building is \$9900. How much is the rent for each office?

55. An office building with a floor area of 48,000 square feet is divided into three offices A, B, and C. The rent per square foot of floor area is \$4 for office A, \$3 for office B, and \$2.50 for office C. The combined rent for offices A and B is twice the rent for office C. The total rent for the building is \$150,000. How much is the rent for each office?

9.12 Graphs of Linear Inequalities in Two Variables

The solution set of a linear inequality in two variables such as $y - x > 5$ is an infinite set of ordered pairs of numbers $\{(x, y) | y - x > 5\}$.

To graph the solution set of the inequality $y - x > 5$ we first consider the linear equation $y - x = 5$.

The graph of the solution set of this equation is a straight line as shown in Figure 9.20.

When $x = 1$ then $y = 6$; that is $(1, 6)$ is an element of the solution set of the equation. Also $(2, 7)$, $(-1, 4)$, and $(-2, 3)$ are elements of the solution set of the equation.

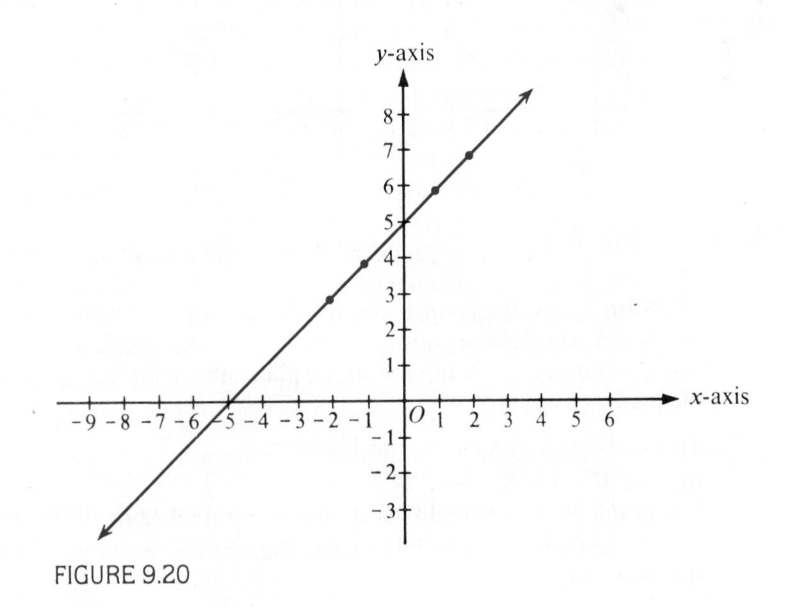

FIGURE 9.20

Now consider the inequality $y - x > 5$.
When $x = 1$, we have $y - 1 > 5$, that is, $y > 6$.

Thus for $x = 1$, any real number y greater than 6 satisfies the inequality.
The coordinates of all the points of the line $x = 1$ above the line $y - x = 5$ are elements of the solution set of the inequality.

When $x = 2$, we have $y - 2 > 5$, that is, $y > 7$.
Thus for $x = 2$, any real number y greater than 7 satisfies the inequality.
The coordinates of all the points of the line $x = 2$ above the line $y - x = 5$ are elements of the solution set of the inequality.

The same holds true for the coordinates of all the points of the lines $x = -1$ and $x = -2$ above the line $y - x = 5$, as shown in Figure 9.21.

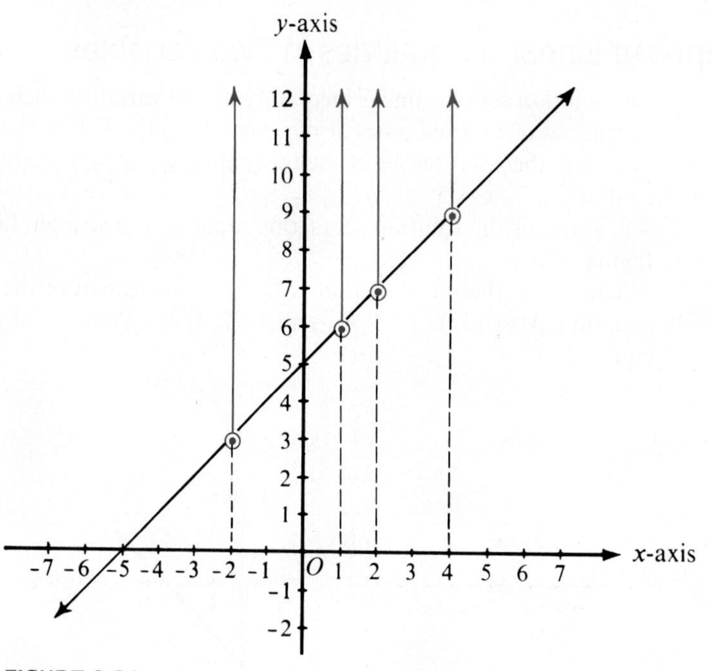

FIGURE 9.21

Thus for $x = a$, the coordinates of all the points of the line $x = a$ above the line $y - x = 5$ are elements of the solution set of the inequality.
The coordinates of each point in the plane above the line $y - x = 5$ satisfy the inequality $y - x > 5$.

Therefore the graphical solution of the inequality $y - x > 5$ is the **half-plane** above the line $y - x = 5$.
The graph of this inequality is shown in Figure 9.22 by the shaded half-plane.
The dashed line $y - x = 5$ indicates that the line is not part of the solution set of the inequality.

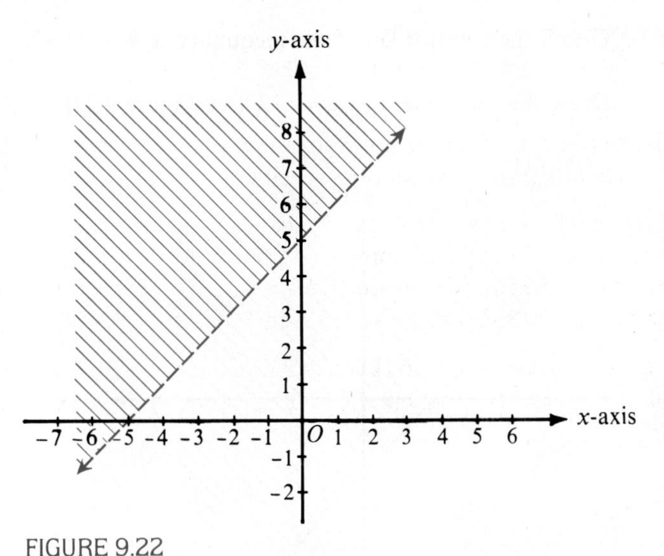

FIGURE 9.22

The graph of the inequality $2x + 3y \leq 6$ is the shaded half-plane under the line $2x + 3y = 6$ shown in Figure 9.23.
The solid line indicates that the line is part of the solution set of the inequality.

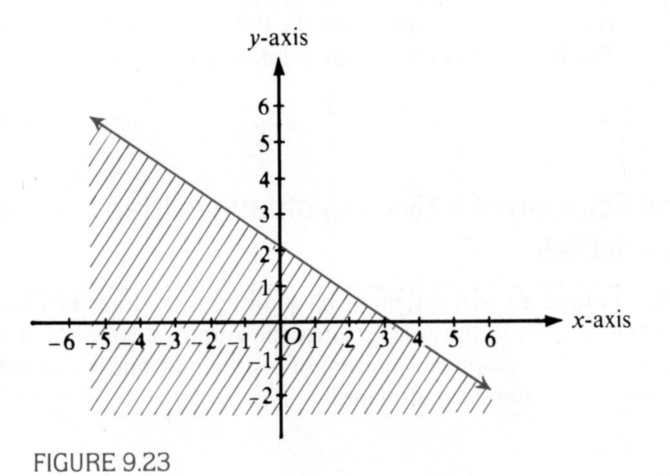

FIGURE 9.23

In order to solve a linear inequality in two variables graphically, replace the order relation by an equal sign. Draw the line representing the equation. Draw a dashed line if the order relation is $>$ or $<$ (the line is not part of the solution set), and draw a solid line if the order relation is $\geq$ or $\leq$ (the line is part of the solution set).

Consider the coordinates of a point not on the line. If the coordinates satisfy the inequality, the half-plane in which the point is located is the solution set of the inequality; otherwise, the solution set is the complementary half-plane.

EXAMPLE

Graph the solution set of the inequality $x + y \geq -2$.

Solution

Draw the solid line $x + y = -2$ (see Figure 9.24).

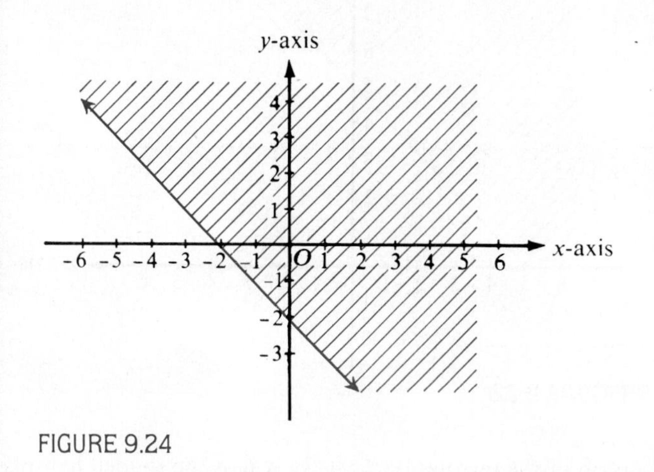

FIGURE 9.24

The point $(0, 0)$ satisfies the inequality, since $0 \geq -2$.
Hence the half-plane above the line is the graph of the inequality.
The line itself is part of the solution set.

9.13 Graphical Solution of a System of Linear Inequalities in Two Variables

The solution set of a system of inequalities is the intersection of the solution set of each inequality in the system. Since the graphical solution of each inequality is a half-plane, the graphical solution of a system of linear inequalities in two variables is the intersection of half-planes.

EXAMPLE

Graph the solution set of the system of inequalities

$$x + y > 2 \qquad \text{and} \qquad x - y > 1$$

Solution

Draw dotted lines representing the graphs of the two linear equations

$$x + y = 2 \qquad \text{and} \qquad x - y = 1$$

Shade the solution set of each inequality.
The double-shaded region of the plane is the solution set of the system (see Figure 9.25).

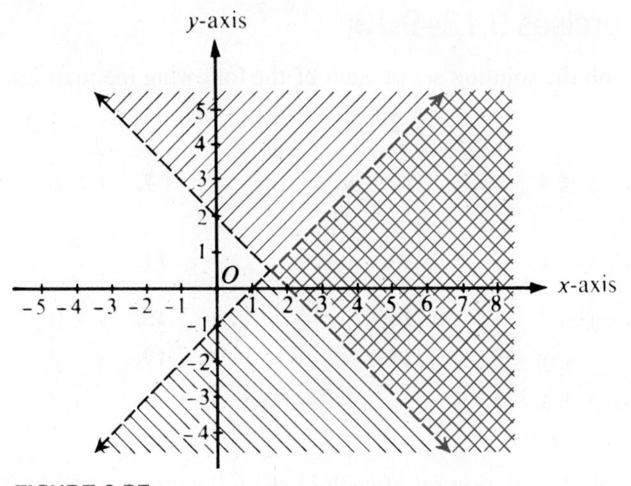

FIGURE 9.25

EXAMPLE

Graph the solution set of the system of inequalities

$y - x \geq 3$ and $y - x < 0$

Solution

Draw a solid line representing the graph of the equation $y - x = 3$ and a dashed line representing the graph of the equation $y - x = 0$.
Shade the solution set of each inequality (see Figure 9.26).
Since the two half-planes do not intersect, the solution set of the system is $\varnothing$.

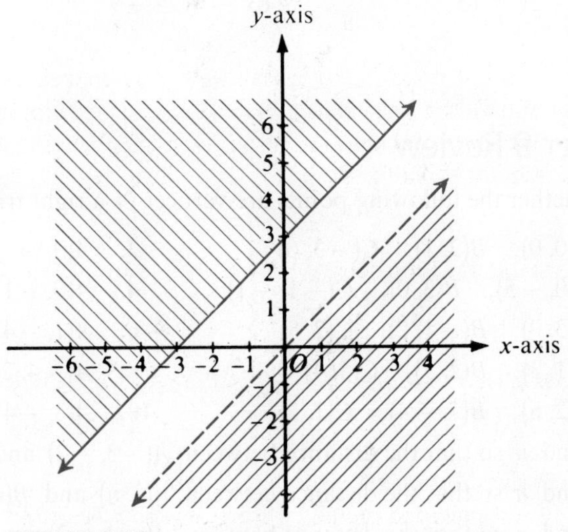

FIGURE 9.26

Exercises 9.12–9.13

Graph the solution set of each of the following inequalities:

1. $x < 2$ **2.** $x < -1$ **3.** $x > -2$ **4.** $x > 3$

5. $x \le 4$ **6.** $x \le \frac{3}{2}$ **7.** $x \ge 0$ **8.** $x \ge -3$

9. $y < 1$ **10.** $y < 5$ **11.** $y > -1$ **12.** $y > \frac{5}{2}$

13. $y \le 2$ **14.** $y \le 3$ **15.** $y \ge 0$ **16.** $y \ge -4$

17. $x + y < 0$ **18.** $x - y < 0$ **19.** $x - 2y > 0$ **20.** $x + 3y > 0$

21. $x + y \le 1$ **22.** $x - 3y \le 4$ **23.** $2x - y \ge 2$

24. $3x + y \ge 3$ **25.** $2x - 3y \le 5$ **26.** $2x + 3y \le 12$

Graph the solution set of each of the following systems of inequalities:

27. $y < 3$, $x > 2$ **28.** $y \ge -2$, $x \le 1$ **29.** $x - y < 0$, $3x - y > 0$ **30.** $x + y > 0$, $x - 2y < 0$

31. $x + y \ge 2$, $x < 4$ **32.** $x - y > 0$, $x + 2y \le 0$ **33.** $x - y > 2$, $y - 2x > 0$ **34.** $3x + y < 4$, $2x + y < 2$

35. $4x + y > 4$, $y - 4x > 4$ **36.** $x + y < 3$, $x - y \ge 3$ **37.** $2x + y \le 3$, $4y - 3x \ge 12$

38. $2y - 3x < 6$, $2y + x > 4$ **39.** $x - y > 3$, $2x + y \le 8$ **40.** $y - x < 6$, $x + y < -2$

41. $3x - y > 3$, $6x - 2y < 9$ **42.** $2x + y < 3$, $4x + 2y > 11$ **43.** $x - 2y > 2$, $2x - 4y > 7$

44. $y - 3x < 3$, $6x - 2y < 15$ **45.** $4x - 3y < 8$, $8x - 6y \ge -9$ **46.** $5x + 3y > 15$, $5x + 3y \le -7$

Chapter 9 Review

Show whether the following points are vertices of a right triangle or not:

1. $A(0, 0)$, $B(2, 1)$, $C(-3, 6)$ **2.** $A(0, 0)$, $B(3, 2)$, $C(4, -6)$

3. $A(0, -5)$, $B(3, 0)$, $C(-1, -1)$ **4.** $A(0, -1)$, $B(-2, 1)$, $C(1, 4)$

5. $A(3, 0)$, $B(-1, 2)$, $C(2, 8)$ **6.** $A(1, -4)$, $B(3, -7)$, $C(4, -2)$

7. $A(1, 3)$, $B(2, 5)$, $C(5, 4)$ **8.** $A(-4, 20)$, $B(-2, 6)$, $C(5, 8)$

9. $A(2, 6)$, $B(7, -4)$, $C(1, 2)$ **10.** $A(3, -4)$, $B(5, 6)$, $C(-1, 3)$

11. Find n so that the distance between $A(-3, -2)$ and $B(5, n)$ is $n + 6$.

12. Find n so that the distance between $A(3, n)$ and $B(6, 11)$ is $n - 2$.

13. Find n so that the distance between $A(-4, -2)$ and $B(2, n)$ is $n + 4$.

14. Find n so that the distance between $A(1, 3)$ and $B(n, 7)$ is $n + 1$.

By the use of slopes show whether the following points are collinear or not:

15. $A(0, -2)$, $B(2, 0)$, $C(3, 1)$
16. $A(1, 0)$, $B(-1, 1)$, $C(3, -1)$
17. $A(0, -4)$, $B(1, -1)$, $C(3, 5)$
18. $A(5, 0)$, $B(-4, 3)$, $C(14, -3)$
19. $A(2, 1)$, $B(-1, 2)$, $C(-4, 3)$
20. $A(2, -2)$, $B(5, 4)$, $C(7, 8)$
21. $A(1, 3)$, $B(-1, -5)$, $C(3, 7)$
22. $A(1, -1)$, $B(-2, -4)$, $C(4, 1)$
23. $A(1, 1)$, $B(6, -3)$, $C(4, -4)$
24. $A(2, 3)$, $B(5, -3)$, $C(-1, 6)$

Find the slopes of the lines represented by the following equations in two ways:

25. $x + 2y = 3$
26. $3x + y = 5$
27. $x + 3y = 4$
28. $2x + 3y = 6$
29. $x - 2y = 1$
30. $3y - x = 4$
31. $y - 2x = 3$
32. $3x - 2y = 6$
33. $4x - 7y = 3$
34. $3x - \sqrt{2}y = 6$
35. $2x + \sqrt{7}y = 3$
36. $\sqrt{5}x + 2y = 1$
37. $x = 0$
38. $x + 8 = 0$
39. $2x + 3 = 0$
40. $y = 0$
41. $y + 2 = 0$
42. $5y - 2 = 0$

Find the real number n for which the three points are collinear:

43. $A(n, 7)$, $B(1, 3)$, $C(9, -5)$
44. $A(n, -8)$, $B(3, 0)$, $C(6, 6)$
45. $A(n, -4)$, $B(3, -3)$, $C(-1, -1)$
46. $A(-4, n)$, $B(2, -1)$, $C(8, 8)$
47. $A(2, n)$, $B(1, 2)$, $C(-1, 8)$
48. $A(9, n)$, $B(3, -2)$, $C(-3, -6)$

Find the equation of the line through the given points:

49. $A(0, 0)$, $B(2, -4)$
50. $A(0, 0)$, $B(3, -2)$
51. $A(-2, -2)$, $B(10, 1)$
52. $A(1, 4)$, $B(3, -2)$
53. $A(-2, -11)$, $B(2, 9)$
54. $A(1, -7)$, $B(3, 1)$
55. $A(-4, 7)$, $B(-2, 1)$
56. $A(2, -1)$, $B(5, -5)$
57. $A(3, 2)$, $B(7, -6)$

Find the equation of the line through the given point with the prescribed slope:

58. $P(2, 1)$; 1
59. $P(2, -3)$; 2
60. $P(1, -1)$; -8
61. $P(-2, 3)$; -4
62. $P(6, 3)$; $\frac{4}{5}$
63. $P(-2, -1)$; $\frac{8}{7}$
64. $P(6, -3)$; $-\frac{2}{3}$
65. $P(1, 1)$; $-\frac{4}{3}$
66. $P(-4, 5)$; $-\frac{5}{6}$

Find the equation of the line with the given x- and y-intercepts:

67. 2; 3
68. 4; -2
69. -12; 6
70. $\frac{1}{7}$; -1
71. $\frac{4}{3}$; -8
72. 4; $\frac{4}{5}$
73. 2; $\frac{6}{5}$
74. -3; $-\frac{4}{3}$
75. $\frac{2}{3}$; $\frac{3}{7}$
76. $-\frac{3}{5}$; $\frac{1}{4}$
77. $\frac{3}{5}$; $-\frac{5}{6}$
78. $-\frac{2}{3}$; $-\frac{5}{3}$

Determine whether the lines represented by each of the following systems of equations intersect at exactly one point, are parallel but do not coincide, or are coincident. Describe each system as consistent and independent, inconsistent, or consistent and dependent:

79. $x - 2y = 6$, $2x - 4y = 5$

80. $3x + y = 7$, $15x + 5y = 9$

81. $2x + y = 8$, $4x - 2y = 5$

82. $x - y = 7$, $3x + 3y = 7$

83. $7x - 2y = 6$, $21x - 6y = 18$

84. $2x - 10y = 2$, $5x - 25y = 5$

Solve the following systems of equations by elimination:

85. $7x - 2y = 1$, $3x + 5y = 18$

86. $3x + y = 8$, $2x - 3y = 9$

87. $4x - 3y = 6$, $3x - 5y = -1$

88. $3x + 2y = 5$, $9x + 4y = 12$

89. $4x - 7y = 8$, $6x - 14y = 11$

90. $9x + 15y = 4$, $6x + 10y = 7$

91. $3x + 6y = 8$, $5x + 10y = 9$

92. $2x - y = -3$, $8x - 4y = -12$

93. $5x - 7y = -2$, $15x - 21y = -6$

Solve the following systems of equations by substitution:

94. $4x + 7y = -2$, $5x + 6y = 3$

95. $9x - 4y = 3$, $6x - 7y = 15$

96. $3x - 4y = -11$, $8x + 9y = 10$

97. $6x - 4y = 1$, $9x - 8y = 0$

98. $5x + 3y = 7$, $10x - 9y = -6$

99. $4x - 3y = -2$, $2x - 9y = 19$

Solve the following systems of equations for x and y:

100. $3(x - 4) - 2(6 - y) = -15$, $7(x + 2) - 3(y + 1) = 9$

101. $2(3x + 2) - (4y - 1) = 1$, $3(3x - y) - 5(2x - y) = 6$

102. $2(2x - 1) - (y - x) = -9$, $3(2y - 3x) - 2(3x - y) = 31$

103. $2(x - y) + 3(2x + y) = 9$, $4(x - 2y) - 5(x + 1) = -14$

104. $8(x - 2y) - 3(x - 4y) = 7$, $3(4x - y) - (10x + y) = -2$

105. $5x - (2y - x) = 10$, $6x - 8(3x - y) = -31$

106. $\frac{4x}{3} + \frac{3y}{8} = 11$, $\frac{5x}{6} - \frac{y}{4} = 3$

107. $\frac{x}{2} + \frac{3y}{4} = -1$, $\frac{3x}{4} - \frac{5y}{2} = 13$

108. $\frac{2x}{3} + \frac{5y}{6} = 6$, $\frac{x}{6} - \frac{7y}{12} = -8$

109. $\frac{3x}{4} + 6y = -\frac{5}{2}$, $\frac{3x}{4} - y = 1$

110. $\frac{2x}{7} - \frac{3y}{4} = -\frac{5}{28}$, $\frac{3x}{4} - \frac{5y}{3} = -\frac{1}{6}$

111. $\frac{2x}{5} + \frac{7y}{4} = -\frac{11}{20}$, $\frac{x}{2} - \frac{5y}{3} = \frac{19}{6}$

112. $\frac{x + y}{2} - \frac{3x - y}{3} = 0$, $\frac{x - y}{2} + \frac{3x + 2y}{7} = 4$

113. $\frac{2x - 3y}{5} - \frac{x + y}{4} = -8$, $\frac{x - 4y}{7} + \frac{3x - y}{8} = -3$

114. $\dfrac{5}{x} - \dfrac{2}{y} = 6$
$\dfrac{2}{x} + \dfrac{3}{y} = 10$

115. $\dfrac{a}{x} + \dfrac{b}{y} = 2ab$
$\dfrac{3a}{x} - \dfrac{2b}{y} = ab$

116. $\dfrac{1}{2x} - \dfrac{1}{3y} = 2$
$\dfrac{3}{2x} + \dfrac{5}{3y} = 22$

117. $\dfrac{2}{3x} + \dfrac{1}{4y} = \dfrac{11}{12}$
$\dfrac{1}{2x} - \dfrac{3}{5y} = -\dfrac{1}{10}$

118. $\dfrac{2}{x} - \dfrac{1}{3y} = 1$
$\dfrac{5}{x} + \dfrac{3}{2y} = \dfrac{19}{2}$

119. $\dfrac{3}{4x} + \dfrac{7}{3y} = 11$
$\dfrac{11}{12x} - \dfrac{5}{9y} = -7$

120. $\dfrac{3}{4x - y} + \dfrac{5}{x + 4y} = -2$
$\dfrac{9}{4x - y} - \dfrac{20}{x + 4y} = 1$

121. $\dfrac{4}{x + y} - \dfrac{3}{x + 2y} = 1$
$\dfrac{2}{x + y} + \dfrac{9}{x + 2y} = 4$

122. $\dfrac{7}{x - y} + \dfrac{5}{2x + y} = 4$
$\dfrac{11}{x - y} - \dfrac{2}{2x + y} = 3$

123. $\dfrac{6}{2x - y} + \dfrac{8}{x + y} = 5$
$\dfrac{5}{2x - y} + \dfrac{6}{x + y} = 4$

124. $\dfrac{5}{3x + y} + \dfrac{4}{2x - 3y} = \dfrac{11}{6}$
$\dfrac{15}{3x + y} - \dfrac{2}{2x - 3y} = \dfrac{5}{6}$

125. $\dfrac{11}{x + 3y} - \dfrac{12}{x - 2y} = 9$
$\dfrac{4}{x + 3y} + \dfrac{18}{x - 2y} = 7$

126. Three times a number is 6 more than twice a second number, while 3 times the second number is 6 more than 4 times the first number. Find the two numbers.

127. A two-digit number is 3 more than 4 times the sum of its digits. If the digits are interchanged, the result is 4 less than 11 times the units digit in the original number. Find the original number.

128. If 3 is subtracted from the numerator and 1 is added to the denominator of a fraction, its value becomes $\dfrac{1}{2}$. If 7 is added to the numerator and 3 is subtracted from the denominator, its value becomes $\dfrac{3}{4}$. Find the fraction.

129. A man invested part of his money at 6% and the rest at 8%. The income from both investments totaled \$2880. If he interchanged his investments, his income would have totaled \$2440. How much did he have in each investment?

130. If a given quantity of a 40% alcohol solution is added to a certain quantity of a 90% alcohol solution, the mixture is a 70% alcohol solution. If there were 10 parts more of the 90% solution, the mixture would be a 71.25% alcohol solution. How many parts of each solution are there?

131. A purse contains \$2.90 in nickels and quarters. If the quarters were nickels and the nickels were quarters, the value of the coins would be \$3.70. How many nickels and how many quarters are in the purse?

132. Flying with a tail wind it took a plane 6 hours to go 3360 miles, while with a head wind 20 miles per hour stronger than the tail wind it took the plane 7 hours. Find the speed of the plane in still air.

133. A fulcrum is placed so that weights of 60 pounds and 100 pounds are in balance. When 20 pounds are added to the 60 pounds, the 100 pounds must be moved 2 feet farther from the fulcrum to preserve the balance. Find the original distance between the 60-pound weight and the 100-pound weight.

134. If the length of a rectangle is increased by 3 inches and its width is decreased by 2 inches, the area of the rectangle decreases by 9 square inches. If the length is decreased by 2 inches and the width is increased by 3 inches, the area of the rectangle increases by 66 square inches. Find the area of the original rectangle.

135. A and B working together can do a job in 15 hours. If A works alone for 9 hours and B then finishes the job in 25 hours, how many hours would it take each working alone to do the job?

Graph the solution set of each of the following systems of inequalities:

136. $y \leq 3$, $y \geq -2$

137. $x \geq -1$, $x \leq 4$

138. $x > 1$, $x + y > 0$

139. $y \leq 1$, $x - 3y < 0$

140. $x - y > 4$, $x + y < 4$

141. $2x - y > 3$, $x - 2y < 3$

142. $x - y > -2$, $2x + y \leq 5$

143. $2x + y \leq 3$, $3x - 2y < -6$

144. $x + 3y < 0$, $x - y \leq -2$

145. $x - y < -1$, $x - y \leq -2$

146. $x + 2y < 3$, $2x + y < 4$

147. $3x + 2y > 12$, $3x + 2y < 6$

chapter 10

Linear Equations in Three Variables

10.1 One linear equation in three variables

10.2 Systems of two linear equations in three variables

10.3 Systems of three linear equations in three variables

10.4 Word problems

In the previous chapter we treated linear equations in two variables. This chapter is an extension of the previous one; here we shall discuss linear equations in three variables and systems of such linear equations.

10.1 One Linear Equation in Three Variables

The standard form of a general linear equation in three variables is

$$ax + by + cz = d \qquad a, b, c, d \in R$$

Elements of the solution set are ordered triples of real numbers (x, y, z). The solution set of the equation $ax + by + cz = d$ is

$$\{(x, y, z) \mid ax + by + cz = d\}$$

In order to find some of the elements of the solution set of a linear equation in three variables, assign arbitrary values to two of the variables and calculate the corresponding value of the third variable. There are an infinite number of ordered triples in the solution set.

EXAMPLE Find three ordered triples that are elements of the solution set of the equation $x - 2y - 3z = 8$.

Solution Putting 1 for x and 1 for y, we get

$$1 - 2(1) - 3z = 8 \qquad \text{or} \qquad z = -3$$

The ordered triple $(1, 1, -3)$ is an element of the solution set.

Putting 0 for y and -3 for z, we obtain

$$x - 2(0) - 3(-3) = 8 \qquad \text{or} \qquad x = -1$$

The ordered triple $(-1, 0, -3)$ is an element of the solution set.

Similarly, if we assign 3 for x and 1 for z, we obtain

$$(3) - 2y - 3(1) = 8 \qquad \text{or} \qquad y = -4$$

The ordered triple $(3, -4, 1)$ is an element of the solution set.

Note An ordered triple of numbers represents a point in three-dimensional space.

The graphical solution of a linear equation in three variables is a plane in three-dimensional space. Since graphing a plane is beyond the scope of this book, we will restrict ourselves to analytic solutions.

Note A point is a **zero-dimensional** figure, a line is a **one-dimensional** figure, and a plane is a **two-dimensional** figure.
The graph of the equation $x = a$ is a point in one-dimensional space.
The equation $x = a$ in two-dimensional space is equivalent to the equation $x + 0y = a$, whose graph is a line.
The equation $x = a$ in three-dimensional space is equivalent to the equation $x + 0y + 0z = a$, whose graph is a plane.

10.2 Systems of Two Linear Equations in Three Variables

The solution set of the system

$$a_1x + b_1y + c_1z = d_1 \quad \text{and} \quad a_2x + b_2y + c_2z = d_2$$

is the intersection of the solution sets of the two equations; that is

$$\{(x, y, z) \mid a_1x + b_1y + c_1z = d_1\} \cap \{(x, y, z) \mid a_2x + b_2y + c_2z = d_2\}$$

Graphical Interpretations

When the graphs of the two equations $a_1x + b_1y + c_1z = d_1$ and $a_2x + b_2y + c_2z = d_2$ are drawn in a three-dimensional space using one set of axes, one of the following possibilities arises:

1. The two planes will coincide. In this case,

$$\frac{a_1}{a_2} = \frac{b_1}{b_2} = \frac{c_1}{c_2} = \frac{d_1}{d_2}$$

The two equations are equivalent and the solution set of the system is the solution set of one of the equations.

2. The two planes will be parallel but do not coincide. In this case,

$$\frac{a_1}{a_2} = \frac{b_1}{b_2} = \frac{c_1}{c_2} \neq \frac{d_1}{d_2}$$

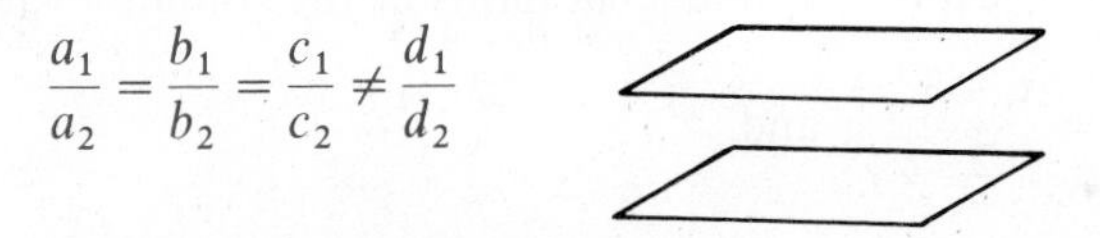

The solution set of the sytem is the null set.

3. The two planes will intersect in exactly one line. In this case,

$$\frac{a_1}{a_2} \neq \frac{b_1}{b_2} \quad \text{or} \quad \frac{a_1}{a_2} \neq \frac{c_1}{c_2}$$

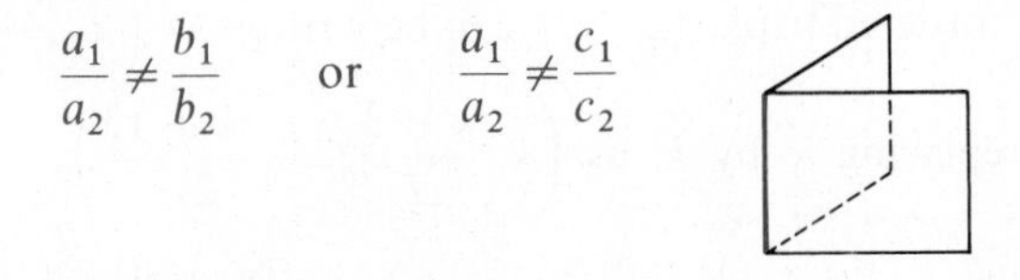

The solution set has an infinite number of ordered triples.

Solution of a System of Two Linear Equations in Three Variables

The solution set of two linear equations in three variables whose graphs intersect in exactly one line is an infinite set of ordered triples.

To find the set of these ordered triples, we derive two equations equivalent to the original system. In one equation one variable is eliminated and in the second equation another variable is eliminated. From these two equations we can express two of the variables in terms of the third (the common variable).

By taking a value for the common variable, the corresponding values of the other two variables can be determined.

EXAMPLE

Find the solution set of the system

$$2x - y + z = 8 \quad \text{and} \quad x + y + z = 5$$

Solution

We shall find expressions for y and z in terms of x. First, eliminate z between the two equations:

$$\begin{array}{lcl} 2x - y + z = 8 & \longrightarrow & 2x - y + z = 8 \\ x + y + z = 5 & \xrightarrow{\times(-1)} & -x - y - z = -5 \\ \hline \text{Adding, we get} & & x - 2y = 3 \end{array}$$

Then eliminate y between the two equations:

$$\begin{array}{ll} & 2x - y + z = 8 \\ & x + y + z = 5 \\ \hline \text{Adding, we get} & 3x + 2z = 13 \end{array}$$

The two equations $x - 2y = 3$ and $3x + 2z = 13$ are equivalent to the two original equations.

Now, express y and z in terms of the common variable x:

$$y = \frac{x - 3}{2} \quad \text{and} \quad z = \frac{-3x + 13}{2}$$

By assigning any real value for x, corresponding values for y and z can be calculated.

The ordered triple (x, y, z) can be written as $\left(x, \dfrac{x-3}{2}, \dfrac{-3x+13}{2}\right)$, or replacing x by k, as $\left(k, \dfrac{k-3}{2}, \dfrac{-3k+13}{2}\right)$.

Thus $\{(x, y, z) \mid 2x - y + z = 8\} \cap \{(x, y, z) \mid x + y + z = 5\}$

$$= \left\{\left(k, \frac{k-3}{2}, \frac{-3k+13}{2}\right) \middle| k \in R\right\}$$

Notes

1. For all values of $k \in R$, the ordered triples $\left(k, \dfrac{k-3}{2}, \dfrac{-3k+13}{2}\right)$ satisfy both equations.
2. We also can find expressions for x and z in terms of y, first by eliminating z and second by eliminating x between the two equations:

$$x = 2y + 3 \quad \text{and} \quad z = -3y + 2$$

Thus the ordered triple (x, y, z) can be written as

$$(2y + 3, y, -3y + 2) \quad \text{or} \quad (2m + 3, m, -3m + 2)$$

The ordered triples $\left(k, \frac{k-3}{2}, \frac{-3k+13}{2}\right)$ and $(2m + 3, m, -3m + 2)$ can be shown to be the same if $2m + 3$ is substituted for k.

EXAMPLE Find the solution set of the system

$$2x + y - z = -3 \quad \text{and} \quad 4x + 3y - 2z = -4$$

Solution Eliminate x between the two equations

$$\begin{aligned} 2x + y - z = -3 &\xrightarrow{\times(-2)} & -4x - 2y + 2z &= 6 \\ 4x + 3y - 2z = -4 &\longrightarrow & 4x + 3y - 2z &= -4 \\ \hline \text{Adding, we get} & & y &= 2 \end{aligned}$$

Eliminate y between the two equations

$$\begin{aligned} 2x + y - z = -3 &\xrightarrow{\times(-3)} & -6x - 3y + 3z &= 9 \\ 4x + 3y - 2z = -4 &\longrightarrow & 4x + 3y - 2z &= -4 \\ \hline \text{Adding, we get} & & -2x + z &= 5 \end{aligned}$$

The system is equivalent to

$$y = 2 \quad \text{and} \quad -2x + z = 5$$

$$\text{or} \quad y = 2 \quad \text{and} \quad z = 2x + 5$$

The ordered triple can be written as

$$(x, y, z) = (x, 2, 2x + 5)$$

or replacing x by k as $(k, 2, 2k + 5)$

Thus $\{(x, y, z) \mid 2x + y - z = -3\} \cap \{(x, y, z) \mid 4x + 3y - 2z = -4\}$
$= \{(k, 2, 2k + 5) \mid k \in R\}$

Exercises 10.1–10.2

Determine whether the planes represented by the following systems of equations are coincident, parallel but do not coincide, or intersect in exactly one line:

1. $x + 2y + 2z = 1$
$4x + 8y + 8z = 11$

2. $x - y - z = 4$
$5x - 5y - 5z = 3$

3. $x + y - 4z = -2$
$2x + 2y + 8z = -4$

4. $x - 2y - 3z = 8$
$3x + 6y + 9z = 24$

5. $x + y - z = 1$
$3x + 3y - 3z = 3$

6. $2x + y - 3z = 2$
$8x + 4y - 12z = 8$

7. $3x + 3y - 6z = 2$
$2x + 2y - 4z = 5$

8. $6x - 3y + 3z = 7$
$10x - 5y + 5z = 1$

9. $3x + 5y - z = 2$
$6x - 10y - 2z = 4$

10. $x + 2y + 4z = 8$
$6x + 12y - 24z = 48$

11. $3x - 6y + 3z = 6$
$7x - 14y + 7z = 14$

12. $2x + 2y - 6z = 8$
$9x + 9y - 27z = 36$

Find the solution set of each of the following systems:

13. $x + y = 1$
$3x - z = 2$

14. $2x + y = 4$
$x - 5z = 8$

15. $2x - 3y = 2$
$y - 2z = 1$

16. $3x + y = 5$
$2y - 3z = -6$

17. $x + z = 2$
$y + 3z = 4$

18. $5x - 2z = 7$
$4y + 3z = 11$

19. $x - 5y = 6$
$x + 3z = 9$

20. $2x - y + z = 7$
$3x - 2y + z = 11$

21. $x + y - 2z = 1$
$x - y + 2z = 1$

22. $x + 3y - 2z = -2$
$2x - 9y + 6z = 31$

23. $3x + y - z = 4$
$3x - y - z = -4$

24. $x + 2y - 2z = 6$
$x - y - 2z = -3$

25. $x - 2y - z = -3$
$2x - 4y + 3z = 4$

26. $x - y + 6z = 2$
$x - y + 3z = 8$

27. $2x + y - 3z = 9$
$2x + 3y - 6z = 14$

28. $x + 2y + 2z = 7$
$2x - 2y - 5z = -1$

29. $x + 3y - 7z = 4$
$3x + 9y - 21z = 12$

30. $2x - 2y + 3z = 5$
$14x - 14y + 21z = 35$

31. $3x - 2y + 5z = 6$
$6x - 4y + 10z = 7$

32. $6x - 9y - 12z = 14$
$4x - 6y - 8z = 25$

33. $x - 3y + z = 23$
$3x + 7y - z = -7$

34. $4x + y + 8z = -6$
$4x + 2y + 11z = 4$

35. $5x - 5y - 12z = -26$
$11x + 5y - 12z = -4$

36. $x - 3y - 2z = 9$
$7x + 6y - 2z = 3$

37. $x + y - 3z = 4$
$3x + 3y - 9z = 10$

38. $x + 2y + 4z = 8$
$5x + 10y + 20z = 12$

39. $x + 12y + 2z = 0$
$x - 2y - 2z = -2$

40. $5x + 3y + 5z = 30$
$10x - 3y + 22z = -9$

10.3 Systems of Three Linear Equations in Three Variables

The solution set of the system

$$a_1x + b_1y + c_1z = d_1$$

$$a_2x + b_2y + c_2z = d_2$$

and $$a_3x + b_3y + c_3z = d_3$$

is the intersection of the solution sets of the three equations, that is

$$\{(x, y, z) \mid a_1x + b_1y + c_1z = d_1\} \cap \{(x, y, z) \mid a_2x + b_2y + c_2z = d_2\}$$
$$\cap \{(x, y, z) \mid a_3x + b_3y + c_3z = d_3\}$$

Graphical Interpretations

The graphical representation of three linear equations in three variables is three planes in three-dimensional space. When all three planes coincide, the three equations are equivalent and their solution set is the solution set of any one of the three. When two of the planes coincide, the system reduces to two linear equations in three variables, which was discussed before. When no two planes coincide, one of the following must be true:

1. The three planes do not possess a common point. The three planes are parallel, two are parallel, or they intersect in pairs.

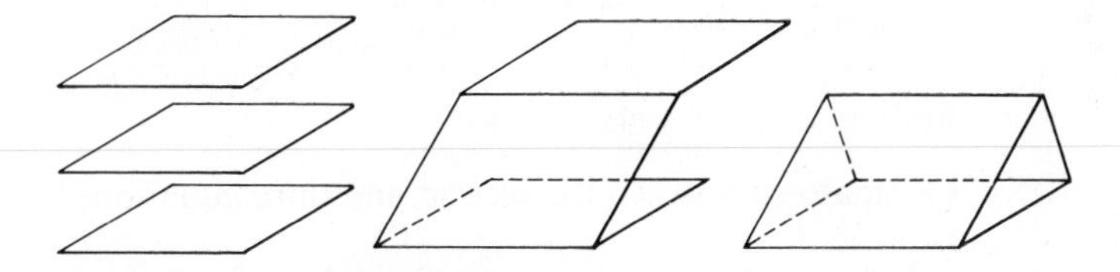

The solution set of the system is the null set.

2. The three planes intersect in one line. One of the equations is a linear combination of the other two.

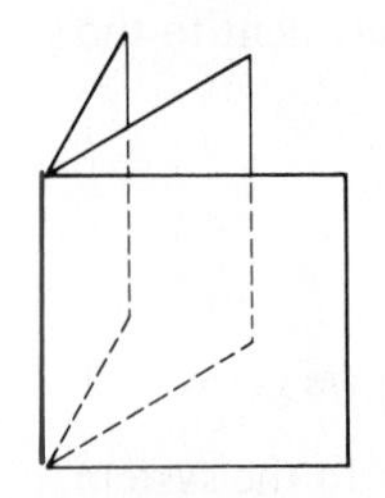

The solution set of the system consists of infinitely many ordered triples as in the case of two linear equations in three variables.

3. The three planes intersect in exactly one point.

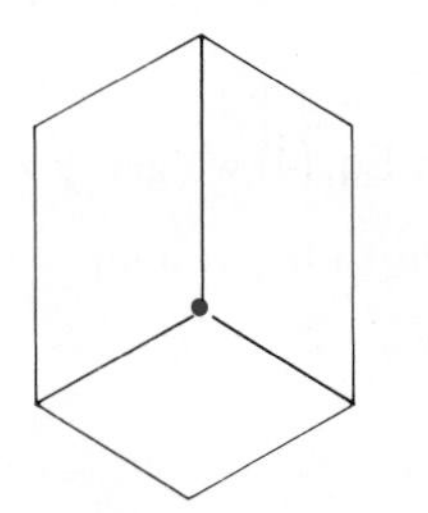

The solution set of the system is the ordered triple formed from the coordinates of the point of intersection.

Solution of Systems of Three Linear Equations in Three Variables

A system of three linear equations in three variables can be solved analytically by either of the methods discussed for systems of two linear equations in two variables. Since the substitution method is complicated, we only illustrate how to solve such a system by the method of elimination.

EXAMPLE Solve the following system:

$$x + y + 2z = 1 \qquad 3x - y - z = 2 \qquad 2x + 3y + 4z = 4$$

Solution Take the equations in pairs and eliminate the same variable. Eliminate y between the first two equations:

$$\begin{array}{rl} & x + y + 2z = 1 \\ & 3x - y - z = 2 \\ \hline \text{Adding, we get} & 4x + z = 3 \end{array}$$

Eliminate y between the second and third equations:

$$\begin{array}{rcl} 3x - y - z = 2 & \xrightarrow{\times (3)} & 9x - 3y - 3z = 6 \\ 2x + 3y + 4z = 4 & \longrightarrow & 2x + 3y + 4z = 4 \\ \hline \text{Adding, we get} & & 11x + z = 10 \end{array}$$

The original system of equations is equivalent to the system

$$x + y + 2z = 1 \tag{1}$$
$$4x + z = 3 \tag{2}$$
$$11x + z = 10 \tag{3}$$

Eliminating z between Eqs. (2) and (3) we get $x = 1$.

Hence the original system is equivalent to the system

$$x + y + 2z = 1 \tag{4}$$
$$4x + z = 3 \tag{5}$$
$$x = 1 \tag{6}$$

Putting 1 for x in Eq. (5) we get $z = -1$.
Putting 1 for x and (-1) for z in Eq. (4) we get $y = 2$.

Thus the original system is equivalent to the system

$$x = 1 \qquad y = 2 \qquad \text{and} \qquad z = -1$$

Hence the solution set is

$$\{(x, y, z) \mid x = 1\} \cap \{(x, y, z) \mid y = 2\} \cap \{(x, y, z) \mid z = -1\} = \{(1, 2, -1)\}$$

EXAMPLE

Solve the following system:

$$2x - y + 3z = 5 \qquad 4x - 2y + 6z = 10 \qquad 6x - 3y + 9z = 15$$

Solution

$$\begin{array}{lcr} 2x - y + 3z = 5 & \xrightarrow{\times(-2)} & -4x + 2y - 6z = -10 \\ 4x - 2y + 6z = 10 & \longrightarrow & 4x - 2y + 6z = 10 \\ \hline \text{Adding, we get} & & 0x + 0y + 0z = 0 \end{array}$$

$$\begin{array}{lcr} 2x - y + 3z = 5 & \xrightarrow{\times(-3)} & -6x + 3y - 9z = -15 \\ 6x - 3y + 9z = 15 & \longrightarrow & 6x - 3y + 9z = 15 \\ \hline \text{Adding, we get} & & 0x + 0y + 0z = 0 \end{array}$$

Hence the original system is equivalent to the system

$$\begin{aligned} 2x - y + 3z &= 5 \\ 0x + 0y + 0z &= 0 \\ 0x + 0y + 0z &= 0 \end{aligned}$$

The solution set is

$$\begin{aligned} &\{(x, y, z) \mid 2x - y + 3z = 5\} \cap \{(x, y, z) \mid 0x + 0y + 0z = 0\} \cap \{(x, y, z) \mid 0x + 0y + 0z = 0\} \\ &= \{(x, y, z) \mid 2x - y + 3z = 5\} \cap \{(x, y, z) \mid x, y, z \in R\} \cap \{(x, y, z) \mid x, y, z \in R\} \\ &= \{(x, y, z) \mid 2x - y + 3z = 5\} \end{aligned}$$

EXAMPLE

Solve the following system:

$$x + 2y - z = 6 \qquad 5x + 10y - 5z = 8 \qquad 7x + 14y - 7z = 10$$

Solution

$$\begin{array}{lcr} x + 2y - z = 6 & \xrightarrow{\times(-5)} & -5x - 10y + 5z = -30 \\ 5x + 10y - 5z = 8 & \longrightarrow & 5x + 10y - 5z = 8 \\ \hline \text{Adding, we get} & & 0x + 0y + 0z = -22 \end{array}$$

The original system is equivalent to the system:

$$\begin{aligned} x + 2y - z &= 6 \\ 0x + 0y + 0z &= -22 \\ 7x + 14y - 7z &= 10 \end{aligned}$$

Hence the solution set is

$$\begin{aligned} &\{(x, y, z) \mid x + 2y - z = 6\} \cap \{(x, y, z) \mid 0x + 0y + 0z = -22\} \cap \{(x, y, z) \mid 7x + 14y - 7z = 10\} \\ &= \{(x, y, z) \mid x + 2y - z = 6\} \cap \varnothing \cap \{(x, y, z) \mid 7x + 14y - 7z = 10\} \\ &= \varnothing \end{aligned}$$

EXAMPLE Solve the following system:

$$x - 2y + z = 7 \qquad 2x + 3y - z = 1 \qquad 5x + 4y - z = 9$$

Solution Eliminate z between the first two equations:

$$\begin{aligned} x - 2y + z &= 7 \\ 2x + 3y - z &= 1 \\ \hline 3x + y \quad &= 8 \end{aligned}$$

Adding, we get $3x + y = 8$

Eliminate z between the first and third equations:

$$\begin{aligned} x - 2y + z &= 7 \\ 5x + 4y - z &= 9 \\ \hline 6x + 2y \quad &= 16 \end{aligned}$$

Adding, we get $6x + 2y = 16$

The original system is equivalent to the system

$$x - 2y + z = 7 \tag{1}$$
$$3x + y = 8 \tag{2}$$
$$6x + 2y = 16 \tag{3}$$

Considering Eqs. (2) and (3), we have

$$\begin{aligned} 3x + y = 8 &\xrightarrow{\times(-2)} & -6x - 2y &= -16 \\ 6x + 2y = 16 &\longrightarrow & 6x + 2y &= 16 \\ \hline \text{Adding, we get} & & 0x + 0y &= 0 \end{aligned}$$

Thus our system is equivalent to

$$\begin{aligned} x - 2y + z &= 7 \\ 3x + y &= 8 \\ 0x + 0y + 0z &= 0 \end{aligned}$$

The solution set of the system is

$$\begin{aligned} &\{(x, y, z) \mid x - 2y + z = 7\} \cap \{(x, y, z) \mid 3x + y = 8\} \\ &\qquad \cap \{(x, y, z) \mid 0x + 0y + 0z = 0\} \\ &= \{(x, y, z) \mid x - 2y + z = 7\} \cap \{(x, y, z) \mid 3x + y = 8\} \\ &\qquad \cap \{(x, y, z) \mid x, y, z \in R\} \\ &= \{(x, y, z) \mid x - 2y + z = 7\} \cap \{(x, y, z) \mid 3x + y = 8\} \end{aligned}$$

Thus the solution set of the system is the intersection of the solution sets of the two equations

$$x - 2y + z = 7 \qquad \text{and} \qquad 3x + y = 8$$

The ordered triple (x, y, z) can be written as $(x, -3x + 8, -7x + 23)$, or $(k, -3k + 8, -7k + 23)$.

Hence the solution set is

$$\{(k, -3k + 8, -7k + 23) \mid k \in R\}$$

Exercise 10.3

Find the solution set of each of the following systems:

1. $3x - y + z = 6$, $2x + 2y - z = 5$, $x - 2y + 2z = 7$

2. $2x + y - z = -1$, $x - 2y + z = 5$, $3x - y + 2z = 8$

3. $3x + 2y - z = 11$, $2x - 3y - 2z = 7$, $x - y + 2z = -5$

4. $x + 4y - z = 5$, $2x + 5y + 8z = 4$, $x - 2y - 3z = -7$

5. $x - 2y - z = 4$, $2x + 3y + z = 3$, $x - y - 2z = -1$

6. $3x + 2y - z = 4$, $x - 5y + 2z = 5$, $4x + 3y - z = 7$

7. $3x + 2y - z = 4$, $4x - y + 3z = 3$, $x + y - 2z = -1$

8. $x + 2y + 2z = -1$, $3x - y - z = -3$, $x + 3y + z = 3$

9. $3x + y - z = 8$, $5x - 2y + 2z = -5$, $2x + 3y + z = 5$

10. $2x + 3y - z = 5$, $3x - 5y - 4z = 5$, $x + y + 2z = 5$

11. $x + y - z = 1$, $2x + y + z = 4$, $x - y - 2z = -2$

12. $x + 2y + z = 4$, $x - 3y - z = -4$, $3x + y + 2z = 3$

13. $x - y + 2z = 5$, $3x + y + z = 9$, $x - 2y - 3z = -6$

14. $x + 3y + 2z = 5$, $2x - y - 2z = 4$, $x + y - z = 2$

15. $x - 3y - 2z + 13 = 0$, $x + 2y + 3z - 7 = 0$, $2x - y + 2z + 5 = 0$

16. $2x + 4y - z = 17$, $x + y - 2z = 11$, $3x - 2y + 3z = -4$

17. $2x - 3y + 3z = 0$, $x + 2y - 3z = 5$, $4x - 5y - 6z = 5$

18. $x - 2y + 2z - 6 = 0$, $3x + 4y - z + 11 = 0$, $2x - 6y + 3z - 13 = 0$

19. $2x + y - z + 3 = 0$, $3x + y + z = 0$, $4x + 3y + 2z - 3 = 0$

20. $2x + y - z - 3 = 0$, $4x + 2y + 3z + 9 = 0$, $6x - 3y - 2z - 12 = 0$

21. $2x - y + 3z = 1$, $4x - 2y + 6z = 2$, $10x - 5y + 15z = 5$

22. $x + 2y - 3z = 2$, $3x + 6y - 9z = 6$, $7x + 14y - 21z = 14$

23. $2x - y + z = 5$, $4x - 2y + 2z = 10$, $8x - 4y + 4z = 20$

24. $x + y - 3z = 1$, $3x + 3y - 9z = 3$, $5x + 5y - 15z = 5$

25. $x - 2y + z = 3$, $2x - 4y + 2z = 6$, $3x - 6y + 3z = 9$

26. $3x - y + z = 2$, $6x - 2y + 2z = 4$, $9x - 3y + 3z = 6$

27. $x - 2y + 3z = 1$, $3x + y + 2z = 3$, $4x - 8y + 12z = 7$

28. $x - 3y + 5z = 6$, $2x - y + z = 8$, $6x - 3y + 3z = 14$

29. $x - 2y - 6z = 3$, $2x + 4y - 12z = 6$, $5x + 10y - 30z = 13$

30. $x - 2y + z = 3$, $4x - 8y + 4z = 7$, $6x - 12y + 6z = 23$

31.
$$\begin{aligned} x - 4y + 2z &= 7 \\ 2x + y - 3z &= 5 \\ 3x + 6y - 8z &= 2 \end{aligned}$$

32.
$$\begin{aligned} 4x - 3y - z &= -1 \\ 3x - 4y + 2z &= -4 \\ x + y - 3z &= 5 \end{aligned}$$

33.
$$\begin{aligned} 3x - 2y + z &= 4 \\ 2x + y - 2z &= 2 \\ 9x + y - 5z &= 10 \end{aligned}$$

34.
$$\begin{aligned} 2x + y - 3z &= 8 \\ 3x + 2y + 5z &= 4 \\ 5x + 3y + 2z &= 12 \end{aligned}$$

35.
$$\begin{aligned} x + y + 3z &= 3 \\ 2x + y + 4z &= 4 \\ 4x + 3y + 10z &= 10 \end{aligned}$$

36.
$$\begin{aligned} 2x - y - z &= 7 \\ x + 2y - 3z &= 1 \\ 3x + y - 4z &= 8 \end{aligned}$$

37.
$$\begin{aligned} x - 3y - 2z &= 9 \\ 2x + y + 3z &= -3 \\ 4x - 5y - z &= 15 \end{aligned}$$

38.
$$\begin{aligned} x - 2y + z &= 8 \\ 2x + y - 3z &= 1 \\ 4x + 7y - 11z &= -13 \end{aligned}$$

39.
$$\begin{aligned} \frac{2}{x} + \frac{1}{y} - \frac{3}{z} &= 13 \\ \frac{1}{x} - \frac{2}{y} + \frac{2}{z} &= -8 \\ \frac{3}{x} + \frac{4}{y} + \frac{7}{z} &= 4 \end{aligned}$$

40.
$$\begin{aligned} \frac{1}{x} + \frac{1}{y} + \frac{3}{z} &= 4 \\ \frac{2}{x} - \frac{1}{y} + \frac{4}{z} &= -2 \\ \frac{3}{x} - \frac{2}{y} - \frac{1}{z} &= 2 \end{aligned}$$

41.
$$\begin{aligned} \frac{2}{x} - \frac{3}{y} + \frac{1}{z} &= -7 \\ \frac{1}{x} + \frac{4}{y} + \frac{3}{z} &= 0 \\ \frac{3}{x} - \frac{1}{y} + \frac{1}{z} &= -2 \end{aligned}$$

42.
$$\begin{aligned} \frac{3}{x} + \frac{1}{y} + \frac{2}{z} &= 5 \\ \frac{2}{x} - \frac{3}{y} + \frac{2}{z} &= -5 \\ \frac{1}{x} + \frac{3}{y} + \frac{1}{z} &= 11 \end{aligned}$$

43.
$$\begin{aligned} \frac{3}{2x} + \frac{1}{3y} - \frac{2}{z} &= \frac{3}{2} \\ \frac{2}{3x} - \frac{3}{2y} + \frac{1}{z} &= \frac{1}{12} \\ \frac{1}{x} - \frac{2}{y} + \frac{5}{z} &= 4 \end{aligned}$$

44.
$$\begin{aligned} \frac{1}{3x} + \frac{1}{2y} - \frac{1}{z} &= -\frac{7}{6} \\ \frac{1}{2x} - \frac{2}{3y} + \frac{3}{z} &= \frac{25}{6} \\ \frac{2}{x} - \frac{3}{y} - \frac{2}{3z} &= \frac{13}{3} \end{aligned}$$

10.4 Word Problems

EXAMPLE

The sum of the digits of a three-digit number is 2 less than 3 times its tens digit. The number is 5 less than 108 times the hundreds digit. If the units and hundreds digits are interchanged, the new number is 1 more than 13 times the sum of the tens and hundreds digits of the original number. Find the original number.

Solution

Units Digit	*Tens Digit*	*Hundreds Digit*
x	y	z

$$x + y + z + 2 = 3y$$

$$x + 10y + 100z + 5 = 108z$$

$$z + 10y + 100x - 1 = 13(y + z)$$

Simplifying the above system we get

$$x - 2y + z = -2$$
$$x + 10y - 8z = -5$$
$$100x - 3y - 12z = 1$$

The above system is equivalent to the system

$x = 1 \qquad y = 5 \qquad z = 7$

Hence the number is 751.

EXAMPLE

Three pounds of peaches, 2 pounds of plums, and 7 pounds of oranges cost \$2.98. Two pounds of peaches, 3 pounds of plums, and 6 pounds of oranges cost \$2.89. Two pounds of peaches, 2 pounds of plums, and 10 pounds of oranges cost \$3.26. What is the price of each fruit per pound?

Solution

	Peaches	*Plums*	*Oranges*
Price per pound:	x ¢	y ¢	z ¢

$$3x + 2y + 7z = 298$$
$$2x + 3y + 6z = 289$$
$$2x + 2y + 10z = 326$$

The above system is equivalent to the system

$x = 29 \qquad y = 39 \qquad z = 19$

Price of peaches is 29¢ per pound.
Price of plums is 39¢ per pound.
Price of oranges is 19¢ per pound.

EXAMPLE

Three machines A, B, and C operating together can do a job in 5 hours. If A operates for 2 hours and B for 5 hours, one-half of the job can be done. If B operates for 3 hours and C for 8 hours, three-fifths of the job can be done. How many hours will it take each machine operating alone to do the job?

Solution

A	B	C	A, B, and C
x hours	y hours	z hours	5 hours

A can do $\frac{1}{x}$ of the job in one hour.

B can do $\frac{1}{y}$ of the job in one hour.

C can do $\frac{1}{z}$ of the job in one hour.

A, B, and C can do $\frac{1}{5}$ of the job in one hour:

$$\frac{1}{x} + \frac{1}{y} + \frac{1}{z} = \frac{1}{5}$$

$$\frac{2}{x} + \frac{5}{y} = \frac{1}{2}$$

$$\frac{3}{y} + \frac{8}{z} = \frac{3}{5}$$

Substituting a for $\frac{1}{x}$, b for $\frac{1}{y}$, and c for $\frac{1}{z}$, we get

$$a + b + c = \frac{1}{5}$$

$$2a + 5b = \frac{1}{2}$$

$$3b + 8c = \frac{3}{5}$$

This system is equivalent to $a = \frac{1}{12}$, $b = \frac{1}{15}$, $c = \frac{1}{20}$.

The original system is equivalent to $x = 12$, $y = 15$, $z = 20$.
A can do the job alone in 12 hours.
B can do the job alone in 15 hours.
C can do the job alone in 20 hours.

Exercise 10.4

1. Three numbers are such that the sum of the second and twice the first is 2 less than twice the third. Three times the second minus the third is 1 more than twice the first. Five times the third is 1 less than 4 times the sum of the first and second. Find the numbers.
2. Three numbers are such that the sum of the first and twice the second is 4 more than twice the third. The sum of the third and twice the first is 10 less than 3 times the second. The sum of the second and third is 6 less than 4 times the first. Find the numbers.
3. Three numbers are such that the sum of the third and twice the first is 1 less than 3 times the second. The sum of the first and 5 times the second is 26 less than 4 times the third. The sum of the second and third is 7 less than 4 times the first. Find the numbers.

4. Three numbers are such that 5 times the first minus the second is 6 more than the third. Three times the third minus twice the second is 10 less than 6 times the first. The sum of the first and twice the third is 4 more than 4 times the second. Find the numbers.
5. The sum of the digits of a three-digit number is 1 less than 6 times its hundreds digit. The number is 23 more than 40 times its tens digit. If the units and hundreds digits are interchanged the new number is 2 more than 60 times the tens digit. Find the original number.
6. The sum of the digits of a three-digit number is 1 less than twice its units digit. The number is 3 more than 105 times its hundreds digit. If the units and hundreds digits are interchanged, the new number is 1 more than 103 times the units digit of the original number. Find the original number.
7. The sum of the digits of a three-digit number is 3 more than twice its tens digit. The number is 4 less than 80 times its units digit. If the hundreds digit is doubled, the new number is 1 more than 125 times its tens digit. Find the original number.
8. The sum of the digits of a three-digit number is 1 more than twice the units digit. The number is 54 times the sum of the units and tens digits. If the tens digit is tripled, the new number is 1 less than 73 times the sum of the tens and hundreds digits of the original number. Find the original number.
9. A, B, and C are packages of three different kinds of vegetables. Three packages of A, 2 of B, and 1 of C, cost \$1.39. Two packages of A, 3 of B, and 2 of C cost \$1.73. Four packages of A, 1 of B, and 3 of C cost \$1.92. What is the price of each per package?
10. A, B, and C are three different kinds of fruits. Four pounds of A, 3 pounds of B, and 1 pound of C cost \$1.34. Two pounds of A, 4 pounds of B, and 3 pounds of C cost \$1.87. Five pounds of A, 2 pounds of B, and 2 pounds of C cost \$1.56. What is the price of each per pound?
11. Three pounds of potatoes, 1 pound of rice, and 2 pounds of macaroni cost \$1.18. Four pounds of potatoes, 2 pounds of rice, and 1 pound of macaroni cost \$1.25. Two pounds of potatoes, 3 pounds of rice, and 2 pounds of macaroni cost \$1.41. What is the price of each per pound?
12. Five pounds of steak, 10 pounds of roast, and 4 pounds of ground meat cost \$24.91. Three pounds of steak, 8 pounds of roast, and 10 pounds of ground meat cost \$24.29. Four pounds of steak, 12 pounds of roast, and 6 pounds of ground meat cost \$27.58. What is the price of each per pound?
13. A, B, and C are three different compounds of copper alloy. Seven pounds of A, 10 pounds of B, and 8 pounds of C form a mixture containing 12.72% copper. Five pounds of A, 6 pounds of B, and 1 pound of C form a mixture containing 11.5% copper. Three pounds of A, 8 pounds of B, and 9 pounds of C form a mixture containing 13.5% copper. What is the percentage of copper in each alloy?
14. A, B, and C are three different kinds of stamps. Thirty stamps of A, 3 of B, and 2 of C cost \$3. Forty stamps of A, 12 of B, and 3 of C cost \$4.85. Fifteen stamps of A, 5 of B, and 4 of C cost \$2.30. What is the price of each kind of stamp?

15. Three men A, B, and C are to do a job. The job can be done when A and B work together for 2 hours and C alone for 1 hour; or by A and B working together for 1 hour and C alone for $5\frac{1}{2}$ hours; or by A working for 3 hours, B for 1 hour, and C for $\frac{1}{2}$ hour. How long will it take each man working alone to do the job?

16. Three men A, B, and C are to do a job. The job can be done when A works for 5 hours, B for 2 hours, and C for 11 hours; or when A works for 4 hours, B for 3 hours, and C for 12 hours; or when A works for 5 hours, B for 1 hour, and C for 19 hours. How long will it take each man working alone to do the job?

17. A tank can be filled by three pipes running simultaneously for 4 hours. If the first and second pipes were turned on for 2 hours, the third pipe would fill the tank in 11 hours. If the second and third pipes were turned on for 6 hours, the first pipe would fill the tank in $1\frac{1}{2}$ hours. How long does it take each pipe separately to fill the tank?

18. A reservoir can be filled by three pipes running simultaneously for 24 hours. If the first and second pipes were turned on for 15 hours, the third pipe would fill the reservoir in 54 hours. If the second and third pipes were turned on for 30 hours, the first pipe would fill the reservoir in 17 hours. How long does it take each pipe separately to fill the tank?

19. A chemist has three mixtures of substances A and B. The first contains 60% of A and 20% of B, the second contains 50% of A and 24% of B, the third contains 40% of A and 40% of B. In what proportion must he mix them in order to obtain a mixture containing 52.5% of A and 26% of B?

20. A chemist has three mixtures of substances A and B. The first contains 80% of A and 10% of B, the second contains 60% of A and 12% of B, the third contains 30% of A and 33% of B. In what proportion must he mix them in order to obtain a mixture containing 65% of A and 16% of B?

21. A farmer has three brands of artificial fertilizer. The first brand contains 5% phosphate and 18% nitrogen, the second brand contains 6% phosphate and 10% nitrogen, the third brand contains 9% phosphate and 7% nitrogen. In what proportion should they be mixed to give a mixture of 6.8% phosphate and 12% nitrogen?

Chapter 10 Review

Find the solution set of each of the following systems:

1. $x + 5z = 1$
$2x + 4y = 3$

2. $2x - z = -1$
$3x - 4y = 8$

3. $3x - y = 6$
$2y + z = 5$

4. $x - 4y = 2$
$2y + z = 7$

5. $4x - 3z = -2$
$y + 6z = 1$

6. $5x + z = 9$
$8y - 2z = 7$

7. $3x + 2y - z = -2$
$2x - 2y + z = -3$

8. $x - y + 3z = 3$
$2x + y - 3z = 3$

9. $3x + 5y - z = -14$
$6x - 3y - 2z = -2$

10. $4x + 2y + z = -2$
$2x + y - z = -7$

11. $2x + y + 4z = 4$
$3x - y + z = 1$

12. $x - 2y + 3z = 1$
$3x + y + 2z = 3$

13. $2x + y - z = -1$
$3x - y + 2z = 8$

14. $2x - y + 3z = 3$
$3x + 2y - z = 6$

15. $x - y + 3z = -4$
$2x + 2y - z = 10$
$3x - y + 2z = 4$

16. $x - 2y + 2z = 1$
$x + 4y - z = -5$
$3x + 2y - z = 1$

17. $2x - 4y + 2z = 5$
$x - 2y + z = 2$
$x + 2y - z = 1$

18. $x - 3y + 2z = 8$
$2x - 4y + 6z = 7$
$3x - 6y + 9z = 15$

19. $3x + 2y - z = 2$
$x + 2y + z = 3$
$5x + 6y + z = 8$

20. $2x + 3y - z = -1$
$x + y + 2z = 4$
$3x + 5y - 4z = -6$

21. $x - 2y + z = 2$
$2x - 4y + 2z = 4$
$3x - 6y + 3z = 6$

22. $2x + y - z = 3$
$4x + 2y - 2z = 6$
$10x + 5y - 5z = 15$

23. $2x - y + 3z = 10$
$2x - 3y - 2z = 3$
$3x + y + 2z = 15$

24. $x - 2y - z = -8$
$2x - y + z = -4$
$x - 5y + 2z = -2$

25. $x + y - 3z = 5$
$2x + y + z = 1$
$6x + 3y + 3z = 4$

26. $3x + 6y - 3z = 5$
$2x + 4y - 2z = 4$
$4x + 8y - 4z = 7$

27. $2x + y - 2z = 3$
$x - 2y + z = -2$
$x + 3y - 3z = 5$

28. $2x - y - z = 2$
$x - 4y + 5z = 10$
$3x + 2y - 7z = -6$

29. $x + y - 2z = -1$
$2x - y + z = 4$
$x + 2y + z = 11$

30. $2x - y - z = 8$
$x - 2y - 3z = 12$
$2x + 3y - 2z = 2$

31. $x + 3y + z = 13$
$x - 2y - 3z = -1$
$2x - 5y + z = 3$

32. $x + 2y + 2z = 10$
$2x - y + 3z = 1$
$2x + 4y - z = 0$

33. $\frac{3}{x} - \frac{2}{y} - \frac{1}{z} = 4$
$\frac{1}{x} + \frac{4}{y} + \frac{2}{z} = -1$
$\frac{2}{x} - \frac{4}{y} - \frac{3}{z} = 6$

34. $\frac{2}{x} - \frac{1}{y} - \frac{1}{z} = 6$
$\frac{1}{x} - \frac{4}{y} - \frac{3}{z} = 4$
$\frac{1}{x} + \frac{3}{y} + \frac{1}{z} = 1$

35. $\frac{1}{x} + \frac{1}{y} - \frac{2}{z} = 0$
$\frac{1}{x} + \frac{2}{y} + \frac{4}{z} = 13$
$\frac{2}{x} + \frac{3}{y} - \frac{2}{z} = 7$

36. $\frac{2}{x} + \frac{1}{y} + \frac{1}{z} = -3$
$\frac{3}{x} + \frac{2}{y} + \frac{1}{z} = -4$
$\frac{4}{x} - \frac{1}{y} + \frac{2}{z} = 3$

chapter 11

Determinants

11.1 Definitions and notation

11.2 Properties of determinants

11.3 Solution of systems of two linear equations in two variables by determinants

11.4 Determinants of order higher than two

11.5 Solution of systems of n linear equations in n variables by determinants

11.1 Definitions and Notation

Determinants are a valuable computational device for solving systems of linear equations.

DEFINITION

A square array of numbers enclosed by vertical lines such as $\begin{vmatrix} a & b \\ c & d \end{vmatrix}$ is called a **determinant** and it stands for the number $ad - bc$.

The numbers in the array are called the **elements of the determinant.** The horizontal lines of elements are called **rows**, and the vertical lines are called **columns**.

When a determinant consists of two rows and two columns, it is called a **second-order determinant**. A determinant with n rows and n columns, $n \in N, n > 1$, is called an ***n*th-order determinant**.

11.2 Properties of Determinants

Although the following properties will be proved for determinants of the second order, they are true for nth-order determinants.

THEOREM 1

The value of a determinant is unchanged if the rows and the columns are interchanged.

Proof

$$\begin{vmatrix} a & c \\ b & d \end{vmatrix} = ad - cb = \begin{vmatrix} a & b \\ c & d \end{vmatrix}$$

Theorem 1 shows that any property of a determinant that holds for rows also holds for columns.

THEOREM 2

The value of a determinant is zero when all the elements of a row (or a column) are zero.

Proof

$$\begin{vmatrix} 0 & 0 \\ a & b \end{vmatrix} = 0(b) - 0(a) = 0$$

THEOREM 3

When two columns (or two rows) of a determinant are identical, the value of the determinant is zero.

Proof

$$\begin{vmatrix} a & a \\ b & b \end{vmatrix} = ab - ab = 0$$

THEOREM 4 The algebraic sign of a determinant is changed when two rows (or two columns) are interchanged.

Proof

$$\begin{vmatrix} c & d \\ a & b \end{vmatrix} = cb - ad = -(ad - bc) = -\begin{vmatrix} a & b \\ c & d \end{vmatrix}$$

THEOREM 5 If each element of a column (or a row) of a determinant is multiplied by a number k, the value of the determinant is multiplied by k.

$$\begin{vmatrix} ka & b \\ kc & d \end{vmatrix} = (ka)d - b(kc) = kad - kbc = k(ad - bc) = k\begin{vmatrix} a & b \\ c & d \end{vmatrix}$$

THEOREM 6 If each element of a row (or a column) of a determinant is expressed as the sum of two numbers, the determinant may be expressed as the sum of two determinants, both of which are identical to the original determinant in all elements except those in the row (or column) expressed as two summands. This row (or column) in one of the determinants consists of one summand from each pair respectively, and this row (or column) in the other determinant consists of the other summand respectively; that is

$$\begin{vmatrix} a_1 + a_2 & b_1 + b_2 \\ c & d \end{vmatrix} = \begin{vmatrix} a_1 & b_1 \\ c & d \end{vmatrix} + \begin{vmatrix} a_2 & b_2 \\ c & d \end{vmatrix}$$

Proof

$$\begin{aligned} \begin{vmatrix} a_1 + a_2 & b_1 + b_2 \\ c & d \end{vmatrix} &= (a_1 + a_2)d - (b_1 + b_2)c \\ &= a_1 d + a_2 d - b_1 c - b_2 c \\ &= (a_1 d - b_1 c) + (a_2 d - b_2 c) \\ &= \begin{vmatrix} a_1 & b_1 \\ c & d \end{vmatrix} + \begin{vmatrix} a_2 & b_2 \\ c & d \end{vmatrix} \end{aligned}$$

THEOREM 7 If to every element of a column (or a row) of a determinant is added the corresponding element of any other column (or row), each multiplied by the same number k, the value of the determinant is not altered.

Proof

$$\begin{aligned} \begin{vmatrix} a + kb & b \\ c + kd & d \end{vmatrix} &= \begin{vmatrix} a & b \\ c & d \end{vmatrix} + \begin{vmatrix} kb & b \\ kd & d \end{vmatrix} \\ &= \begin{vmatrix} a & b \\ c & d \end{vmatrix} + k\begin{vmatrix} b & b \\ d & d \end{vmatrix} = \begin{vmatrix} a & b \\ c & d \end{vmatrix} + k(0) \\ &= \begin{vmatrix} a & b \\ c & d \end{vmatrix} \end{aligned}$$

11.3 Solution of Systems of Two Linear Equations in Two Variables by Determinants

Consider the two linear equations $a_1x + b_1y = c_1$ and $a_2x + b_2y = c_2$:

$$\begin{array}{lcr} a_1x + b_1y = c_1 & \xrightarrow{\times(b_2)} & a_1b_2x + b_1b_2y = b_2c_1 \\ a_2x + b_2y = c_2 & \xrightarrow{\times(-b_1)} & -a_2b_1x - b_1b_2y = -b_1c_2 \\ \hline \text{Adding, we get} & & (a_1b_2 - a_2b_1)x = (b_2c_1 - b_1c_2) \end{array}$$

Or, if $a_1b_2 \neq b_1a_2$,

$$x = \frac{c_1b_2 - b_1c_2}{a_1b_2 - b_1a_2} = \frac{\begin{vmatrix} c_1 & b_1 \\ c_2 & b_2 \end{vmatrix}}{\begin{vmatrix} a_1 & b_1 \\ a_2 & b_2 \end{vmatrix}}$$

Also,

$$\begin{array}{lcr} a_1x + b_1y = c_1 & \xrightarrow{\times(-a_2)} & -a_1a_2x - b_1a_2y = -c_1a_2 \\ a_2x + b_2y = c_2 & \xrightarrow{\times(a_1)} & a_1a_2x + a_1b_2y = a_1c_2 \\ \hline \text{Adding, we get} & & (a_1b_2 - b_1a_2)y = (a_1c_2 - c_1a_2) \end{array}$$

Or, if $a_1b_2 \neq b_1a_2$,

$$y = \frac{a_1c_2 - c_1a_2}{a_1b_2 - b_1a_2} = \frac{\begin{vmatrix} a_1 & c_1 \\ a_2 & c_2 \end{vmatrix}}{\begin{vmatrix} a_1 & b_1 \\ a_2 & b_2 \end{vmatrix}}$$

The determinant $\begin{vmatrix} a_1 & b_1 \\ a_2 & b_2 \end{vmatrix}$ is normally denoted by D.

Thus the solution set of the system

$$\begin{array}{l} a_1x + b_1y = c_1 \\ a_2x + b_2y = c_2 \end{array} \quad \text{is} \quad \left\{ \left(\frac{\begin{vmatrix} c_1 & b_1 \\ c_2 & b_2 \end{vmatrix}}{D}, \frac{\begin{vmatrix} a_1 & c_1 \\ a_2 & c_2 \end{vmatrix}}{D} \right) \right\}$$

This rule for solving linear equations by determinants is called **Cramer's Rule**.

Notes

1. The elements of D are the coefficients of the variables when the equations are in standard form.
2. The numerator for the value of x, denoted by D_x, is the determinant formed by replacing the column of the coefficients of x in D with the column of the constant terms.
3. The numerator for the value of y, denoted by D_y, is the determinant formed by replacing the column of the coefficients of y in D with the column of the constant terms.

EXAMPLE

Solve the following system:

$$2x - y = 7$$
$$3x + 4y = 5$$

Solution

$$D = \begin{vmatrix} 2 & -1 \\ 3 & 4 \end{vmatrix} = 8 - (-3) = 8 + 3 = 11$$

$$D_x = \begin{vmatrix} 7 & -1 \\ 5 & 4 \end{vmatrix} = 28 - (-5) = 28 + 5 = 33$$

$$D_y = \begin{vmatrix} 2 & 7 \\ 3 & 5 \end{vmatrix} = 10 - 21 = -11$$

The system is equivalent to

$$x = \frac{33}{11} = 3 \quad \text{and} \quad y = \frac{-11}{11} = -1$$

The solution set is $\{(3, -1)\}$.

Exercises 11.2–11.3

Evaluate the following determinants:

1. $\begin{vmatrix} 1 & -2 \\ 3 & 5 \end{vmatrix}$ **2.** $\begin{vmatrix} -2 & 3 \\ 1 & 6 \end{vmatrix}$ **3.** $\begin{vmatrix} 8 & 2 \\ 5 & -3 \end{vmatrix}$

4. $\begin{vmatrix} 6 & -11 \\ 4 & 9 \end{vmatrix}$ **5.** $\begin{vmatrix} 2a & 3 \\ -b & 2 \end{vmatrix}$ **6.** $\begin{vmatrix} 5 & -a \\ 6 & b \end{vmatrix}$

7. $\begin{vmatrix} 3a & b \\ 2b & -7a \end{vmatrix}$ **8.** $\begin{vmatrix} -3a & 4b \\ b & 2a \end{vmatrix}$ **9.** $\begin{vmatrix} a & 3b \\ -a & -2b \end{vmatrix}$

Solve the following equations:

10. $\begin{vmatrix} 4x & -1 \\ 8 & 3 \end{vmatrix} = 0$ **11.** $\begin{vmatrix} -3 & 2x \\ 1 & 7 \end{vmatrix} = 0$ **12.** $\begin{vmatrix} 3x & -7 \\ -5 & 4 \end{vmatrix} = 1$

13. $\begin{vmatrix} 3 & 2x \\ 4 & -3 \end{vmatrix} = 7$ **14.** $\begin{vmatrix} 6 & 5 \\ 7 & 6x \end{vmatrix} = -5$ **15.** $\begin{vmatrix} 4 & 6 \\ 3x & 5 \end{vmatrix} = 2$

16. $\begin{vmatrix} 15 & -9 \\ 8 & x \end{vmatrix} = -3$ **17.** $\begin{vmatrix} 8 & -7 \\ 6 & -3x \end{vmatrix} = 6$ **18.** $\begin{vmatrix} 3 & 8x \\ -5 & -4x \end{vmatrix} = -4$

Solve the following systems by determinants:

19. $x - y = 5$
$3x + 2y = 5$

20. $2x + 3y = 8$
$3x - y = 1$

21. $2x + 3y = -1$
$x - 2y = 3$

22. $2x + y = 0$
$3x - 2y = 7$

23. $2x + 5y = 11$
$3x - y = 8$

24. $5x + 2y = 2$
$4x + 3y = -4$

25 $x - 2y = 3$
$3x - 4y = 6$

26. $3x + y = 1$
$x + 2y = 3$

27. $7x - 3y = 10$
$5x + 2y = 3$

28. $2x + y = 4$
$3x - 2y = 27$

29. $7x - 6y = 17$
$3x + y = 18$

30. $2x + 5y = -1$
$3x - 2y = 27$

31. $4x - 9y = -9$
$2x + 6y = 13$

32. $x + y = 1$
$5x - 6y = 27$

33. $4x + 6y = 7$
$3x + 5y = 6$

34. $4x + y = 7$
$2x - y = -1$

35. $3x - y = 14$
$5x - 7y = 2$

36. $5x + 6y = 10$
$4x + 9y = -13$

11.4 Determinants of Order Higher Than Two

A third-order determinant consists of three rows and three columns, for example

$$\begin{vmatrix} a_1 & b_1 & c_1 \\ a_2 & b_2 & c_2 \\ a_3 & b_3 & c_3 \end{vmatrix}$$

The second-order determinant $\begin{vmatrix} b_2 & c_2 \\ b_3 & c_3 \end{vmatrix}$, which is obtained from the original determinant by crossing the elements of the row and the elements of the column where a_1 lies, is called the **minor** of a_1.

The minor of b_3 is $\begin{vmatrix} a_1 & c_1 \\ a_2 & c_2 \end{vmatrix}$.

A **cofactor** of an element is the product of its minor by $(-1)^{i+j}$, where i and j are the numbers of the row and column, respectively, in which the element lies.
For the element a_1, i is 1 and j is 1.
For the element b_3, i is 3 and j is 2.

The value of a determinant is the sum of the products of each element of any row (or any column) by its respective cofactor.

The determinant $\begin{vmatrix} a_1 & b_1 & c_1 \\ a_2 & b_2 & c_2 \\ a_3 & b_3 & c_3 \end{vmatrix}$ stands for the number

$$a_1(-1)^{1+1}\begin{vmatrix} b_2 & c_2 \\ b_3 & c_3 \end{vmatrix} + b_1(-1)^{1+2}\begin{vmatrix} a_2 & c_2 \\ a_3 & c_3 \end{vmatrix} + c_1(-1)^{1+3}\begin{vmatrix} a_2 & b_2 \\ a_3 & b_3 \end{vmatrix}$$

$$= a_1\begin{vmatrix} b_2 & c_2 \\ b_3 & c_3 \end{vmatrix} - b_1\begin{vmatrix} a_2 & c_2 \\ a_3 & c_3 \end{vmatrix} + c_1\begin{vmatrix} a_2 & b_2 \\ a_3 & b_3 \end{vmatrix}$$

$$= a_1(b_2c_3 - c_2b_3) - b_1(a_2c_3 - c_2a_3) + c_1(a_2b_3 - b_2a_3)$$

$$= a_1b_2c_3 - a_1c_2b_3 - b_1a_2c_3 + b_1c_2a_3 + c_1a_2b_3 - c_1b_2a_3$$

EXAMPLE

Find the value of $D = \begin{vmatrix} 2 & -3 & 1 \\ 1 & 2 & 4 \\ 5 & 1 & -6 \end{vmatrix}$.

Solution

Expanding the determinant with respect to the elements in the second column, we have

$$D = (-3)(-1)^{1+2}\begin{vmatrix} 1 & 4 \\ 5 & -6 \end{vmatrix} + 2(-1)^{2+2}\begin{vmatrix} 2 & 1 \\ 5 & -6 \end{vmatrix} + 1(-1)^{3+2}\begin{vmatrix} 2 & 1 \\ 1 & 4 \end{vmatrix}$$

$$= 3\begin{vmatrix} 1 & 4 \\ 5 & -6 \end{vmatrix} + 2\begin{vmatrix} 2 & 1 \\ 5 & -6 \end{vmatrix} - \begin{vmatrix} 2 & 1 \\ 1 & 4 \end{vmatrix}$$

$$= 3(-6 - 20) + 2(-12 - 5) - (8 - 1)$$

$$= 3(-26) + 2(-17) - (7) = -119$$

Note The calculations can be simplified when some of the elements of a row or a column are zeros. We can introduce some zeros by applying Theorem 7.

EXAMPLE

Find the value of $D = \begin{vmatrix} 1 & 2 & -3 \\ 2 & -1 & 2 \\ 3 & 1 & 4 \end{vmatrix}$.

Solution

Adding (-2) times the elements of the first column to the corresponding elements of the second column, we get

$$D = \begin{vmatrix} 1 & 0 & -3 \\ 2 & -5 & 2 \\ 3 & -5 & 4 \end{vmatrix}$$

Adding 3 times the elements of the first column to the corresponding elements of the third column, we get

$$D = \begin{vmatrix} 1 & 0 & 0 \\ 2 & -5 & 8 \\ 3 & -5 & 13 \end{vmatrix}$$

Expanding the determinant with respect to the elements of the first row, we have

$$D = (1)(-1)^{1+1}\begin{vmatrix} -5 & 8 \\ -5 & 13 \end{vmatrix} = 1[(-5)(13) - (8)(-5)] = -25$$

Note The method used in expanding a determinant of order three, by minors, applies for expanding determinants of order n.

EXAMPLE

Find the value of $D = \begin{vmatrix} 2 & -3 & 1 & -1 \\ 1 & 4 & -3 & 2 \\ 3 & -1 & 1 & -3 \\ 1 & 2 & 2 & -4 \end{vmatrix}$.

Solution

Add 2 times the elements of the fourth column to the corresponding elements of the first column:

$$D = \begin{vmatrix} 0 & -3 & 1 & -1 \\ 5 & 4 & -3 & 2 \\ -3 & -1 & 1 & -3 \\ -7 & 2 & 2 & -4 \end{vmatrix}$$

Add 3 times the elements of the third column to the corresponding elements of the second column:

$$D = \begin{vmatrix} 0 & 0 & 1 & -1 \\ 5 & -5 & -3 & 2 \\ -3 & 2 & 1 & -3 \\ -7 & 8 & 2 & -4 \end{vmatrix}$$

Add the elements of the third column to the corresponding elements of the fourth column:

$$D = \begin{vmatrix} 0 & 0 & 1 & 0 \\ 5 & -5 & -3 & -1 \\ -3 & 2 & 1 & -2 \\ -7 & 8 & 2 & -2 \end{vmatrix}$$

Expand with respect to the elements of the first row:

$$D = (1)(-1)^{1+3}\begin{vmatrix} 5 & -5 & -1 \\ -3 & 2 & -2 \\ -7 & 8 & -2 \end{vmatrix}$$

Add (-2) times the elements of the first row to the corresponding elements of the second and third rows:

$$D = \begin{vmatrix} 5 & -5 & -1 \\ -13 & 12 & 0 \\ -17 & 18 & 0 \end{vmatrix}$$

Expand with respect to the elements of the third column:

$$D = (-1)(-1)^{1+3}\begin{vmatrix} -13 & 12 \\ -17 & 18 \end{vmatrix} = (-1)[(-13)(18) - (12)(-17)] = 30$$

Exercise 11.4

Evaluate the following determinants:

1. $\begin{vmatrix} 2 & 1 & -1 \\ 1 & -2 & 1 \\ 3 & -1 & 2 \end{vmatrix}$

2. $\begin{vmatrix} 3 & -1 & 1 \\ 2 & 2 & -1 \\ 1 & -2 & 2 \end{vmatrix}$

3. $\begin{vmatrix} 1 & 4 & -1 \\ 1 & -2 & -3 \\ 2 & 5 & 8 \end{vmatrix}$

4. $\begin{vmatrix} 1 & -1 & 2 \\ 3 & 2 & -1 \\ 2 & -3 & -2 \end{vmatrix}$

5. $\begin{vmatrix} 1 & -2 & 3 \\ 3 & 1 & 2 \\ 4 & -8 & 12 \end{vmatrix}$

6. $\begin{vmatrix} 1 & -2 & -1 \\ 2 & 3 & 1 \\ 1 & -1 & -2 \end{vmatrix}$

7. $\begin{vmatrix} 3 & 2 & -1 \\ 1 & -5 & 2 \\ 4 & 3 & -1 \end{vmatrix}$

8. $\begin{vmatrix} 4 & -3 & -1 \\ 3 & -4 & 2 \\ 1 & 1 & -3 \end{vmatrix}$

9. $\begin{vmatrix} 1 & -1 & -2 \\ 2 & 1 & 3 \\ -1 & 2 & 4 \end{vmatrix}$

10. $\begin{vmatrix} 1 & 1 & -1 \\ 2 & 3 & -2 \\ 2 & -1 & 3 \end{vmatrix}$

11. $\begin{vmatrix} 3 & -2 & 1 \\ -1 & 2 & 5 \\ 1 & 6 & -3 \end{vmatrix}$

12. $\begin{vmatrix} 2 & -2 & 1 \\ -3 & 1 & 4 \\ 1 & -1 & 6 \end{vmatrix}$

13. $\begin{vmatrix} 4 & 2 & -3 \\ 7 & -1 & 1 \\ 3 & 1 & -1 \end{vmatrix}$

14. $\begin{vmatrix} 1 & -1 & 3 \\ 2 & 1 & 4 \\ 1 & -4 & 5 \end{vmatrix}$

15. $\begin{vmatrix} 1 & 3 & -1 & 0 \\ 3 & -2 & 0 & -1 \\ 2 & 0 & 1 & 2 \\ 0 & 1 & -2 & -3 \end{vmatrix}$

16. $\begin{vmatrix} 1 & 1 & -1 & 1 \\ 2 & -1 & 1 & -1 \\ 2 & 1 & 3 & -2 \\ 1 & -2 & -2 & 3 \end{vmatrix}$

17. $\begin{vmatrix} 1 & -1 & 2 & -1 \\ 1 & 2 & -1 & 3 \\ 2 & -1 & -1 & 2 \\ 3 & 1 & 3 & -1 \end{vmatrix}$

18. $\begin{vmatrix} 1 & 1 & 1 & 1 \\ 2 & -1 & -1 & 2 \\ 1 & 2 & -2 & -3 \\ 2 & 5 & 1 & 4 \end{vmatrix}$

19. $\begin{vmatrix} 3 & 1 & -1 & -1 \\ 2 & -3 & -2 & 1 \\ 1 & 3 & 1 & -2 \\ 1 & -2 & 3 & -3 \end{vmatrix}$

20. $\begin{vmatrix} 1 & 2 & 2 & -3 \\ 2 & -1 & -2 & 1 \\ 1 & 1 & 1 & -2 \\ 1 & -3 & -1 & 1 \end{vmatrix}$

21. $\begin{vmatrix} 1 & 2 & -1 & 1 \\ 1 & -1 & 2 & -1 \\ 2 & 1 & 1 & 2 \\ 1 & -3 & 3 & -2 \end{vmatrix}$

11.5 Solution of Systems of *n* Linear Equations in *n* Variables by Determinants

Cramer's Rule for solving a system of two linear equations in two variables may be generalized to find the solution of a system of n linear equations in n variables.

CRAMER'S RULE

Consider the system of n linear equations in n variables:

$$\begin{array}{l} a_1x + b_1y + c_1z + \cdots + h_1w = k_1 \\ a_2x + b_2y + c_2z + \cdots + h_2w = k_2 \\ \vdots \quad\quad \vdots \quad\quad \vdots \quad\quad \vdots \quad\quad \vdots \quad\quad \vdots \\ a_nx + b_ny + c_nz + \cdots + h_nw = k_n \end{array}$$

The determinant of the coefficients D is

$$\begin{vmatrix} a_1 & b_1 & c_1 & \cdots & h_1 \\ a_2 & b_2 & c_2 & \cdots & h_2 \\ \vdots & \vdots & \vdots & \vdots & \vdots \\ a_n & b_n & c_n & \cdots & h_n \end{vmatrix}$$

Let D_x be the determinant obtained from D by replacing the column of the coefficients of x in D by the column of the constant terms, respectively. Let D_y be the determinant obtained from D by replacing the column of the coefficients of y in D by the column of the constant terms respectively, and so on.

If $D \neq 0$, the solution set of the system is

$$\{(x, y, z, \ldots, w)\} = \left\{\left(\frac{D_x}{D}, \frac{D_y}{D}, \frac{D_z}{D}, \ldots, \frac{D_w}{D}\right)\right\}$$

Notes

1. In order to apply Cramer's Rule, it is mandatory that the equations be in standard form; the terms involving the variables must be on one side of the equality sign and the constant terms on the other side.
2. The variables must be in the same order in all equations.
3. Variables missing in an equation must be re-introduced with a zero coefficient.

EXAMPLE

Solve the following system:

$$2x - y + 2z = 8$$
$$x + 2y + 3z = 3$$
$$3x - y - 2z = 1$$

Solution

$$D = \begin{vmatrix} 2 & -1 & 2 \\ 1 & 2 & 3 \\ 3 & -1 & -2 \end{vmatrix} = -27 \qquad D_x = \begin{vmatrix} 8 & -1 & 2 \\ 3 & 2 & 3 \\ 1 & -1 & -2 \end{vmatrix} = -27$$

$$D_y = \begin{vmatrix} 2 & 8 & 2 \\ 1 & 3 & 3 \\ 3 & 1 & -2 \end{vmatrix} = 54 \qquad D_z = \begin{vmatrix} 2 & -1 & 8 \\ 1 & 2 & 3 \\ 3 & -1 & 1 \end{vmatrix} = -54$$

The system is equivalent to

$$x = \frac{-27}{-27} = 1 \qquad y = \frac{54}{-27} = -2 \qquad z = \frac{-54}{-27} = 2$$

The solution set is $\{(x, y, z) = (1, -2, 2)\}$.

EXAMPLE

Solve the following system:

$$x + y - 2z + t = 6$$
$$2x - y - z + 2t = 9$$
$$x + 3y + z - t = -4$$
$$x - 2y + 3z + t = -1$$

Solution

$$D = \begin{vmatrix} 1 & 1 & -2 & 1 \\ 2 & -1 & -1 & 2 \\ 1 & 3 & 1 & -1 \\ 1 & -2 & 3 & 1 \end{vmatrix} = -12$$

$$D_x = \begin{vmatrix} 6 & 1 & -2 & 1 \\ 9 & -1 & -1 & 2 \\ -4 & 3 & 1 & -1 \\ -1 & -2 & 3 & 1 \end{vmatrix} = -24$$

$$D_y = \begin{vmatrix} 1 & 6 & -2 & 1 \\ 2 & 9 & -1 & 2 \\ 1 & -4 & 1 & -1 \\ 1 & -1 & 3 & 1 \end{vmatrix} = 12$$

$$D_z = \begin{vmatrix} 1 & 1 & 6 & 1 \\ 2 & -1 & 9 & 2 \\ 1 & 3 & -4 & -1 \\ 1 & -2 & -1 & 1 \end{vmatrix} = 24$$

$$D_t = \begin{vmatrix} 1 & 1 & -2 & 6 \\ 2 & -1 & -1 & 9 \\ 1 & 3 & 1 & -4 \\ 1 & -2 & 3 & -1 \end{vmatrix} = -12$$

The system is equivalent to

$$x = \frac{-24}{-12} = 2 \qquad y = \frac{12}{-12} = -1 \qquad z = \frac{24}{-12} = -2 \qquad t = \frac{-12}{-12} = 1$$

The solution set is $\{(x, y, z, t) = (2, -1, -2, 1)\}$.

Exercise 11.5

Solve the following systems by determinants:

1. $3x + 2y - z = 0$, $2x - 3y = 8$, $x - y + 2z = 1$

2. $2x - 2y - z = 6$, $x + y + 2z = 0$, $x - 2z = 7$

3. $x + y + 2z = 1$, $x - 2y - 3z = 3$, $2x + 3y - z = 8$

4. $x + y - z = 3$, $2x - y + z = 0$, $x - 2y - 2z = 1$

5. $x + y + z = 3$, $2x - y + 2z = 6$, $x + 2y - 3z = -1$

6. $x + y - 2z = 6$, $3x - y - z = 4$, $4x + 2y + z = 4$

7. $x + 2y - z = 3$, $x - y + 3z = -1$, $2x - y + 2z = 2$

8. $x + 2y - z = 4$, $x - y + 2z = -2$, $x - 3y + 3z = -7$

9. $x - 2y + z = 3$, $2x + y - z = 2$, $x - 3y + 2z = 3$

10. $x + 2y - z = 2$, $x - 4y + 3z = 3$, $3x - 2y - 2z = 3$

11. $2x + y - z = 4$, $x + 2y - 2z = 2$, $x - 3y + z = 4$

12. $3x - y + z = 5$, $x + 4y - 3z = -4$, $2x - 3y + z = 2$

13. $2x - 3y - 2z = 0$, $x + 3y + z = 5$, $x - 2y + 3z = -9$

14. $2x + 3y - z = -5$, $4x - y + 2z = 9$, $2x + y - 3z = -9$

15. $x + 2y - z + t = 3$, $3x - 2z - 2t = 8$, $x - 3y - t = 4$, $2x - y + z - 3t = 0$

16. $x + y - 2z - t = -1$, $y + z - 2t = 7$, $x - y + 3z = 2$, $x + 2y - 3t = 8$

17. $2x + y - z - t = 4$, $x - 2y - 3z - t = 1$, $x + 3y + z + 2t = 9$, $3x - y - z + 2t = 7$

18. $x + 3y - 4z + t = 4$, $2x - y - 8z - t = -2$, $x + y + 2z + 2t = 0$, $x - 2y + 2z + 4t = -1$

19. $x - y + z - t = -1$, $3x + 2y - 3z + t = 10$, $2x + y - 3z - 4t = 16$, $4x - 3y + z - 2t = -2$

20. $x - y + z + t = 4$, $2x - y + 3z - 2t = 11$, $x + 2y - z - t = -2$, $3x + y - z + 2t = 1$

21.
$$\begin{aligned} x + y + z - t &= 5 \\ 2x + y - z - 2t &= 11 \\ x - 2y + z + t &= -7 \\ x - y - 2z + t &= -2 \end{aligned}$$

22.
$$\begin{aligned} x + 2y + z - 2t &= 7 \\ 2x - y - z - 3t &= 0 \\ x + y - z - t &= -1 \\ x - 2y + 2z + t &= -4 \end{aligned}$$

23.
$$\begin{aligned} x - y - z - 2t &= -10 \\ x - y + z + 2t &= 12 \\ 2x + y - z + t &= 15 \\ x + y + 2z - t &= 11 \end{aligned}$$

24.
$$\begin{aligned} 2x + y + z + t &= 4 \\ x - y + 2z + t &= -7 \\ 3x - 2y - z + 2t &= 5 \\ x + 3y + 2z - t &= 7 \end{aligned}$$

Chapter 11 Review

Solve the following systems by determinants:

1.
$$\begin{aligned} x + y &= 4 \\ 2x - y &= -1 \end{aligned}$$

2.
$$\begin{aligned} x + 3y &= 1 \\ 2x + 5y &= 1 \end{aligned}$$

3.
$$\begin{aligned} 3x - 2y &= 1 \\ 2x + y &= 10 \end{aligned}$$

4.
$$\begin{aligned} x - y &= 1 \\ x - 3y &= 5 \end{aligned}$$

5.
$$\begin{aligned} 3x + y &= -1 \\ 2x - y &= 6 \end{aligned}$$

6.
$$\begin{aligned} 3x - y &= 4 \\ x + 3y &= 8 \end{aligned}$$

7.
$$\begin{aligned} 2x + y &= 1 \\ 3x + 2y &= 0 \end{aligned}$$

8.
$$\begin{aligned} 2x - 3y &= 9 \\ x + 2y &= 1 \end{aligned}$$

9.
$$\begin{aligned} 2x - y &= -3 \\ 3x + 4y &= 1 \end{aligned}$$

10.
$$\begin{aligned} x + y &= 2 \\ 3x + y &= -4 \end{aligned}$$

11.
$$\begin{aligned} 2x + y - z &= 6 \\ 3y + 4z &= 1 \\ x - 2y &= 6 \end{aligned}$$

12.
$$\begin{aligned} x - y + z &= 4 \\ x + 2y + 2z &= 2 \\ 2x + y - 2z &= 1 \end{aligned}$$

13.
$$\begin{aligned} x + y - z &= 3 \\ 2x - y - 3z &= 2 \\ x + 2y + z &= 1 \end{aligned}$$

14.
$$\begin{aligned} 3x - y - z &= 1 \\ 2x + y - z &= 2 \\ x + 3y + z &= 5 \end{aligned}$$

15.
$$\begin{aligned} x + 3y - z &= 3 \\ 2x + y - z &= -1 \\ x - 2y - 2z &= 0 \end{aligned}$$

16.
$$\begin{aligned} x + y + 2z &= 2 \\ x + 2y + 3z &= 4 \\ x - y - 4z &= 2 \end{aligned}$$

17.
$$\begin{aligned} 2x + 2y + z &= 4 \\ 3x - y + z &= 1 \\ x - y + 2z &= 3 \end{aligned}$$

18.
$$\begin{aligned} x - y + z &= 2 \\ 2x - y + 3z &= 8 \\ x - 2y + z &= -1 \end{aligned}$$

19.
$$\begin{aligned} x - 2y - z &= 3 \\ 2x - 3y + z &= 1 \\ x + 3y + z &= 4 \end{aligned}$$

20.
$$\begin{aligned} x + y - z &= 3 \\ x - 2y - z &= 6 \\ 2x + y - z &= 4 \end{aligned}$$

21.
$$\begin{aligned} 2x - y - z &= 3 \\ x + 2y + z &= 6 \\ 3x - y - 2z &= 7 \end{aligned}$$

22.
$$\begin{aligned} x - 2y + z &= -3 \\ x + 3y + 3z &= -1 \\ 2x + y - 4z &= 8 \end{aligned}$$

23.
$$\begin{aligned} x + 2y - z &= 1 \\ x - y + 2z &= 0 \\ 3x - 2y - 2z &= 5 \end{aligned}$$

24.
$$\begin{aligned} x + 3y + 3z &= 4 \\ x - 2y + z &= 1 \\ 2x - y + 5z &= 5 \end{aligned}$$

25.
$$\begin{aligned} 2x - y + z &= 4 \\ 2x - 3y - 3z &= 6 \\ 4x + y - 4z &= -1 \end{aligned}$$

26.
$$\begin{aligned} x + y - z + t &= -1 \\ 2x - y + z + t &= 5 \\ x - 2y - z + 2t &= -5 \\ x + y + 2z - 3t &= 12 \end{aligned}$$

27.
$$\begin{aligned} x + 2y + 2z - 3t &= 11 \\ 2x - y - 2z + t &= -10 \\ x + y + z - 2t &= 6 \\ x - 3y - z + t &= -9 \end{aligned}$$

28.
$$\begin{aligned} x - 2y - 2z + t &= 4 \\ 3x - 2y + 4z - t &= 5 \\ x + 6y - 2z + 3t &= 10 \\ x + 4y + 2z - 2t &= 2 \end{aligned}$$

29.
$$\begin{aligned} 2x - 3y + z + t &= 6 \\ x - 2y - 2z - 3t &= -5 \\ x + y + z + 2t &= 3 \\ 3x + y - z - t &= -4 \end{aligned}$$

30.
$$\begin{aligned} x - 3y - z + t &= -8 \\ 3x - y + 2z - t &= 5 \\ x + 2y - 2z + 3t &= -3 \\ 2x + 4y - 3z - 2t &= 11 \end{aligned}$$

31.
$$\begin{aligned} x + y + 2z + t &= 2 \\ x - 2y + z - t &= -7 \\ 2x - y - 3z - 2t &= 5 \\ 2x + 3y - z + 3t &= 18 \end{aligned}$$

Cumulative Review

Chapter 9 Find the distance between the following points:

1. $A(-4, -1)$, $B(2, 1)$
2. $A(3, 5)$, $B(8, -1)$
3. $A(1, -3)$, $B(4, 6)$
4. $A(-2, 3)$, $B(3, -2)$
5. $A\left(-\frac{3}{2}, \frac{5}{3}\right)$, $B\left(\frac{1}{2}, \frac{11}{3}\right)$
6. $A\left(-\frac{7}{3}, -\frac{1}{2}\right)$, $B\left(\frac{2}{3}, \frac{5}{2}\right)$

Show whether the following points are vertices of a right triangle or not:

7. $A(-1, 1)$, $B(2, -1)$, $C(6, 5)$
8. $A(-1, 2)$, $B(3, -1)$, $C(9, 7)$
9. $A(2, 2)$, $B(4, -2)$, $C(5, -4)$
10. $A(-2, 3)$, $B(1, 2)$, $C(10, -1)$
11. $A(-6, 2)$, $B(2, 4)$, $C(5, -8)$
12. $A(-4, 2)$, $B(2, 5)$, $C(4, 1)$

By the use of the distance formula, show whether the following points are collinear or not:

13. $A(-2, -9)$, $B(3, 1)$, $C(4, 3)$
14. $A(-1, 3)$, $B(2, 1)$, $C(5, -1)$
15. $A(-3, 2)$, $B(3, 4)$, $C(6, 5)$
16. $A(-7, 1)$, $B(-1, 3)$, $C(2, -6)$
17. $A(-2, -6)$, $B(4, 3)$, $C(6, 6)$
18. $A(-4, -4)$, $B(2, -1)$, $C(8, 2)$
19. Find n so that the distance between $A(n, 3)$ and $B(1, -1)$ is $n + 1$.
20. Find n so that the distance between $A(n, 5)$ and $B(3, 1)$ is $n - 1$.
21. Find n so that the distance between $A(n, 6)$ and $B(3, 9)$ is $n - 2$.
22. Find n so that the distance between $A(n, 8)$ and $B(4, -4)$ is $n + 4$.

Find the slopes of the lines through the given points:

23. $A(3, 2)$, $B(9, 6)$
24. $A(2, 4)$, $B(10, 4)$
25. $A(-1, 6)$, $B(-1, 2)$
26. $A(-16, 4)$, $B(4, 0)$
27. $A(-2, -7)$, $B(3, 8)$
28. $A(-3, 4)$, $B(-1, -2)$

Find the slopes of the lines represented by the following equations in two ways:

29. $3y - 2x = 0$
30. $2x + 7y = 0$
31. $2x + 7 = 0$
32. $3y - 4 = 0$
33. $x + 4y = 5$
34. $y - 2x = 7$
35. $7x + 8y = 10$
36. $5x + 2y = 3$

By the use of slopes determine whether or not the given points are collinear:

37. $A(-6, 4)$, $B(-1, 1)$, $C(4, -2)$
38. $A(-2, -5)$, $B(1, -1)$, $C(4, 3)$
39. $A(-3, -2)$, $B(1, -1)$, $C(5, 1)$
40. $A(2, 2)$, $B(4, -2)$, $C(7, -3)$
41. $A(-2, 4)$, $B(1, 2)$, $C(7, -2)$
42. $A(1, -2)$, $B(3, 3)$, $C(5, 8)$

Find n for which all three points lie on the same line:

43. $A(n, -4)$, $B(6, 2)$, $C(9, 4)$
44. $A(n, -5)$, $B(5, -1)$, $C(10, 1)$
45. $A(-2, n)$, $B(2, 3)$, $C(6, -3)$
46. $A(6, n)$, $B(-3, -8)$, $C(9, 8)$

Find the equation of the line through the given points:

47. $A(-4, -3)$, $B(2, 6)$
48. $A(-1, 4)$, $B(5, -4)$
49. $A(1, 2)$, $B(2, -4)$
50. $A(-4, -1)$, $B(2, 1)$
51. $A(-2, 3)$, $B(5, 3)$
52. $A(7, -2)$, $B(9, -2)$
53. $A(-1, -2)$, $B(-1, 6)$
54. $A(4, -4)$, $B(4, 7)$

Find the equation of the line through the given point with the prescribed slope:

55. $A(5, 2)$; 0
56. $A(4, 2)$; 5
57. $A(1, 2)$; -3
58. $A(2, 2)$; -4
59. $A(-1, 5)$; $\frac{3}{4}$
60. $A(4, -1)$; $\frac{2}{5}$
61. $A(-2, -1)$; $-\frac{5}{3}$
62. $A(3, -5)$; $-\frac{3}{2}$

Find the equation of the line with the given x- and y-intercepts:

63. 3; 2
64. 1; -2
65. -5; 7
66. -3; -2
67. $\frac{4}{3}$; 3
68. $\frac{7}{2}$; -3
69. $\frac{2}{7}$; $\frac{3}{2}$
70. $-\frac{4}{3}$; $-\frac{7}{3}$

71. Find the equation of the line through the point $A(2, 1)$ and parallel to the line whose equation is $2x - 3y = 6$.
72. Find the equation of the line through the point $A(-1, 3)$ and parallel to the line whose equation is $3x + 2y = 8$.
73. Find the equation of the line through the point $A(5, -3)$ and parallel to the line whose equation is $4x + 3y = -1$.
74. Find the equation of the line through the point $A(4, -2)$ and parallel to the line whose equation is $3x + 5y = 7$.
75. Find the equation of the line through the point $A(-2, -5)$ and parallel to the line through the points $B(-3, -4)$ and $C(6, 2)$.
76. Find the equation of the line through the point $A(-1, -3)$ and parallel to the line through the points $B(1, -1)$ and $C(5, 2)$.
77. Find the equation of the line through the point $A(3, -2)$ and parallel to the line through the points $B(-5, 4)$ and $C(10, -2)$.
78. Find the equation of the line through the point $A(-4, 1)$ and parallel to the line through the points $B(-1, -1)$ and $C(-8, 2)$.

Determine whether the lines represented by each of the following systems of equations intersect at exactly one point, are parallel but do not coincide, or are coincident. Describe each system as consistent and independent, inconsistent, or consistent and dependent:

79. $4x - 2y = 9$
$5x - 3y = 8$

80. $x + 2y = 5$
$3x - 6y = 7$

81. $2x + 3y = 2$
$6x - 9y = 7$

82. $x - 5y = 4$
$7x - 4y = 6$

83. $2x - 3y = 1$
$4x - 6y = 5$

84. $3x + 6y = 2$
$4x + 8y = -1$

85. $7x + 14y = 9$
$2x + 4y = 3$

86. $15x - 10y = 6$
$9x - 6y = 4$

87. $x - 2y = 3$
$5x - 10y = 15$

88. $2x + y = 4$
$6x + 3y = 12$

89. $x + 3y = -1$
$2x + 6y = -2$

90. $3x - 2y = -3$
$9x - 6y = -9$

Solve the following systems of equations graphically:

91. $x + 2y = 7$
$x + y = 4$

92. $2x - y = -5$
$x + 2y = 5$

93. $2x - 3y = 1$
$x - y = 2$

94. $2x + 3y = -1$
$x - 2y = 3$

95. $2x + 5y = 11$
$3x - y = 8$

96. $4x + 3y = -4$
$5x + 2y = 2$

97. $6x - 3y = 4$
$2x - y = 4$

98. $x + 3y = 3$
$2x + 6y = 11$

Solve the following systems of equations by elimination:

99. $2x + y = 4$
$3x - 2y = 27$

100. $4x + y = 24$
$3x - 2y = 7$

101. $2x - 3y = 12$
$4x + 5y = -20$

102. $3x + y = 18$
$7x - 6y = 17$

103. $3x - 2y = 27$
$2x + 5y = -1$

104. $6x - 7y = 10$
$8x - 13y = 6$

105. $6x - 3y = 4$
$2x - y = 3$

106. $2x + y = 3$
$8x + 4y = 9$

107. $3x + y = 1$
$6x + 2y = 5$

108. $2x - 3y = 4$
$4x - 6y = 8$

109. $3x - y = -1$
$6x - 2y = -2$

110. $x - 3y = -2$
$4x - 12y = -8$

Solve the following systems of equations by the substitution method:

111. $4x + 3y = 5$
$3x + y = -5$

112. $3x - 5y = 4$
$4x - y = 11$

113. $3x - 2y = -1$
$2x - 3y = 6$

114. $3x + 2y = 7$
$4x + 3y = 8$

115. $5x + 6y = 10$
$4x + 9y = -13$

116. $3x - 4y = -1$
$4x - 5y = 1$

Solve the following systems of equations:

117. $4(x + 1) - 3(y + 2) = 19$
$5x + 4(y - 3) = -9$

118. $3x - 2(2y + 3) = 4$
$7(x - y) + 2(x + 4y) = 17$

119. $3(2x + y) - 2(x - 2y) = 26$
$2(x - y) - 3(2x + y) = -22$

120. $3(2x + 3y) + 4(3x - y) = -11$
$6(x + y) - (4x + y) = 21$

121. $4(2x + 7y) - (x + y) = 19$
$5(3x + 8y) + 2(x + 2y) = 3$

122. $3(2x - y) - (x - y) = -6$
$5(3x - 2y) + (x - 2y) = -8$

123. $\frac{1}{2}x + \frac{5}{6}y = 3$
$\frac{3}{4}x - \frac{2}{3}y = -7$

124. $\frac{1}{4}x - \frac{1}{3}y = \frac{1}{12}$
$\frac{5}{6}x + \frac{3}{4}y = 4$

125. $\frac{2}{3}x - \frac{3}{4}y = \frac{1}{12}$
$\frac{4}{7}x + \frac{9}{8}y = \frac{37}{56}$

126. $\frac{2}{9}x - \frac{3}{4}y = \frac{43}{6}$
$\frac{5}{6}x + \frac{3}{8}y = \frac{31}{4}$

127. $\frac{3}{8}x - \frac{2}{3}y = \frac{7}{12}$
$\frac{3}{2}x + \frac{8}{9}y = \frac{5}{9}$

128. $\frac{5}{2}x + \frac{2}{3}y = \frac{2}{3}$
$\frac{3}{8}x - \frac{4}{3}y = \frac{9}{4}$

129. $\frac{x + y}{3} + \frac{3x - y}{2} = \frac{7}{2}$
$\frac{x}{3} - \frac{3x - 5y}{6} = \frac{1}{2}$

130. $\frac{2x - y}{3} - \frac{x - y}{2} = \frac{1}{6}$
$\frac{3x - y}{4} - \frac{x - 3y}{3} = \frac{1}{12}$

131. $\frac{x + y}{4} - \frac{3x - y}{9} = \frac{3}{4}$
$\frac{3x - y}{3} - \frac{4y - x}{4} = 1$

132. $\frac{x + 3y}{7} - \frac{x + 2y}{4} = \frac{5}{28}$
$\frac{3x + 2y}{5} - \frac{x - y}{6} = -\frac{1}{6}$

133. $\frac{4}{x} + \frac{3}{y} = -\frac{1}{6}$
$\frac{5}{x} - \frac{2}{y} = \frac{8}{3}$

134. $\frac{7}{x} + \frac{3}{y} = \frac{15}{2}$
$\frac{3}{x} + \frac{4}{y} = \frac{1}{2}$

135. $\frac{3}{4x} - \frac{5}{3y} = \frac{29}{6}$
$\frac{4}{3x} + \frac{1}{4y} = \frac{13}{6}$

136. $\frac{3}{2x} - \frac{7}{3y} = \frac{5}{2}$
$\frac{2}{3x} - \frac{5}{4y} = \frac{19}{24}$

137. $\frac{1}{5x} + \frac{2}{9y} = -1$
$\frac{7}{10x} - \frac{5}{3y} = \frac{23}{6}$

138. $\frac{5}{3x} + \frac{3}{4y} = 1$
$\frac{3}{4x} + \frac{1}{3y} = \frac{1}{2}$

139. $\frac{3}{2x + y} + \frac{2}{x + 2y} = \frac{5}{2}$
$\frac{5}{2x + y} + \frac{3}{x + 2y} = \frac{17}{4}$

140. $\frac{4}{x + 3y} - \frac{3}{2x - y} = 3$
$\frac{6}{x + 3y} + \frac{7}{2x - y} = \frac{2}{3}$

141. $\frac{3}{5x - y} - \frac{2}{2x - 3y} = -\frac{3}{2}$
$\frac{2}{5x - y} + \frac{3}{2x - 3y} = \frac{1}{12}$

142. $\frac{4}{3x + 4y} - \frac{9}{4x + 3y} = \frac{5}{2}$
$\frac{5}{3x + 4y} - \frac{6}{4x + 3y} = 4$

143. Three times a number is 5 more than twice a second number, while 7 times the first number is 2 less than 5 times the second number. Find the two numbers.

144. Five times a number is 2 more than 3 times a second number, while 8 times the first number is 9 less than 5 times the second number. Find the two numbers.

145. A two digit number is 6 more than 4 times the sum of its digits. If the digits are interchanged the new number is 4 more than 6 times the sum of its digits. Find the original number.

146. A two digit number is 3 more than 6 times the sum of its digits. If the digits are interchanged, the new number is 1 more than 8 times the tens digit in the original number. Find the original number.

147. If 1 is added to the numerator and 4 is added to the denominator of a fraction, its value becomes $\frac{3}{4}$. If 3 is subtracted from the numerator and 4 is subtracted from the denominator, its value becomes $\frac{4}{5}$. Find the fractions.

148. If 8 is added to the numerator and 5 is added to the denominator of a fraction, its value becomes $\frac{6}{7}$. If 4 is subtracted from the numerator and 3 is subtracted from the denominator, its value becomes $\frac{4}{5}$. Find the fraction.

149. A woman invested part of her money at 5.5% and the rest at 8%. The income from both investments totaled $2520. If she interchanged her investments, her income would have totaled $2745. How much did she have in each investment?

150. The total return from $11,000 and $21,000 investments was $2710. If the investments were interchanged, the total return would be $2410. How much was the rate of return on each investment?

151. If a 20% acid solution is added to a 50% acid solution, the mixture is a 38% acid solution. If there were 10 more gallons of the 50% acid solution, the new mixture would be a 40% acid solution. How many gallons of each solution do we have?

152. If an 8% silver alloy was combined with a 20% silver alloy, the mixture would contain 10.4% silver. If there were 10 pounds less of the 8% alloy and 10 pounds more of the 20% alloy, the mixture would contain 12.8% silver. How many pounds of each alloy do we have?

153. A man rowed 8 miles up a river in 2 hours and back in 1 hour. Find the rate of the current and the man's rate of rowing in still water.

154. When a man drives from home to work at 60 mph, he arrives 4 minutes earlier than usual, and when he drives at 40 mph, he arrives 6 minutes later than usual. How far is his office from home, and how fast does he usually drive?

155. A fulcrum is placed so that weights of 80 pounds and 120 pounds are in balance. When 100 pounds are added to the 80 pounds, the fulcrum must be moved 1 foot farther from the 120 pounds to preserve the balance. Find the length of the lever.

156. A fulcrum is placed so that weights of 60 pounds and 90 pounds are in balance. When 15 pounds are added to the 60 pounds, the 90 pounds must be moved 2 feet farther from the fulcrum to preserve the balance. Find the original distance between the 60-pound weight and the 90-pound weight.

157. If the length of a rectangle is increased by 2 inches and the width is decreased by 2 inches, the area of the rectangle decreases by 16 square inches. If the length is decreased by 1 inch and the width is increased by 2 inches, the area increases by 20 square inches. Find the area of the original rectangle.

158. If the length of a lot is decreased by 10 feet and the width is increased by 10 feet, the area of the lot increases by 400 square feet. If the length is increased by 10 feet and the width is decreased by 5 feet, the area of the lot stays the same. Find the area of the original lot.

159. A and B working together can do a job in 36 hours. If A works alone for 10 hours and B finishes the job in 75 hours, how many hours would it take each working alone to do the job?

160. A tank can be filled by two pipes running simultaneously for 80 minutes. If the first pipe was turned on for only 1 hour, and the second pipe filled the rest of the tank in 105 minutes, how long would it have taken each pipe separately to fill the tank?

Graph the solution set of each of the following systems of inequalities:

161. $3x + 2y < 6$
$x - y < 3$

162. $3x - y \leq 4$
$x - 2y < 2$

163. $2x + y > 4$
$x + 3y \geq 3$

164. $x - 2y < 3$
$2x - y > -3$

165. $x + 3y > 7$
$x + 2y > -3$

166. $2x + 3y \leq -1$
$x - 2y \geq 4$

167. $x + y < 5$
$3x - 2y > -5$

168. $x - y < 1$
$2x + y \geq 3$

Chapter 10 Find the solution set of each of the following systems:

169. $x + 3y - 2z = 3$
$4x - 2y - z = 5$

170. $2x - 5y + z = 11$
$x + 2y - z = -5$

171. $2x - y - 3z = 6$
$3x + y - 7z = -1$

172. $x - 5y + z = 3$
$x + y - 2z = -6$

173. $x + y - z = -5$
$2x - y + z = -1$

174. $4x + 3y - 2z = 7$
$2x + y - z = 2$

175. $3x - y + 2z = 9$
$6x - 2y + z = 6$

176. $x - 2y - 2z = 10$
$2x + 3y - 4z = -1$

177. $2x + y - z = 6$
$x + 5y - 2z = 18$

178. $x - y - z = -6$
$3x - 2y - 4z = -19$

179.
$$\begin{aligned} x + y - z &= 4 \\ 2x - y + 2z &= -2 \\ x + 3y + z &= 6 \end{aligned}$$

180.
$$\begin{aligned} 2x - y + z &= 0 \\ x + 2y - z &= -2 \\ x + 3y + 2z &= 8 \end{aligned}$$

181.
$$\begin{aligned} 2x + y - z &= 2 \\ x - 2y + 3z &= 7 \\ x - 3y + 2z &= 7 \end{aligned}$$

182.
$$\begin{aligned} 3x + y - 2z &= -1 \\ x - 2y + z &= 6 \\ 2x - y + 3z &= 7 \end{aligned}$$

183.
$$\begin{aligned} x + 3y - z &= 3 \\ 2x - y + z &= 5 \\ x + 2y - 3z &= -2 \end{aligned}$$

184.
$$\begin{aligned} x - 2y - 3z &= 5 \\ 3x + y + 2z &= 5 \\ x - y - z &= 4 \end{aligned}$$

185.
$$\begin{aligned} x - y + z &= 1 \\ 2x - 2y + 2z &= 2 \\ 4x - 4y + 4z &= 4 \end{aligned}$$

186.
$$\begin{aligned} 2x + y - z &= 2 \\ 4x + 2y - 2z &= 4 \\ 6x + 3y - 3z &= 6 \end{aligned}$$

187.
$$\begin{aligned} x + 2y + z &= 4 \\ 2x + 4y + 2z &= 8 \\ 5x + 10y + 5z &= 20 \end{aligned}$$

188.
$$\begin{aligned} 2x - y - z &= 3 \\ 4x - 2y - 2z &= 5 \\ 6x - 3y - 3z &= 7 \end{aligned}$$

189.
$$\begin{aligned} x - 2y - 2z &= 6 \\ 2x - 4y - 4z &= 9 \\ 3x - 6y - 6z &= 11 \end{aligned}$$

190.
$$\begin{aligned} x + 2y - z &= 1 \\ 2x + 4y - 2z &= 3 \\ 3x + 6y - 3z &= 4 \end{aligned}$$

191.
$$\begin{aligned} x - y + z &= 1 \\ 3x - 2y + z &= 1 \\ 5x - 4y + 3z &= 3 \end{aligned}$$

192.
$$\begin{aligned} x + y - 4z &= 8 \\ x - 2y - z &= 2 \\ 4x - 5y - 7z &= 14 \end{aligned}$$

193.
$$\begin{aligned} 2x + y - 5z &= 5 \\ x - y - z &= -2 \\ x + 5y - 4z &= 16 \end{aligned}$$

194.
$$\begin{aligned} x + 2y - 3z &= -8 \\ x - 4y + 3z &= 4 \\ 2x - 5y + 3z &= 2 \end{aligned}$$

Chapter 11 Solve the following systems by determinants:

195.
$$\begin{aligned} 3x - 2y &= 7 \\ 2x - 3y &= 3 \end{aligned}$$

196.
$$\begin{aligned} 3x - y &= -9 \\ 2x + 3y &= 5 \end{aligned}$$

197.
$$\begin{aligned} 4x - 5y &= -3 \\ 5x - 2y &= 9 \end{aligned}$$

198.
$$\begin{aligned} 7x + 2y &= 3 \\ 6x + y &= -1 \end{aligned}$$

199.
$$\begin{aligned} 2x + y + z &= 3 \\ 3x - y + 2z &= -1 \\ x + 2y - z &= 6 \end{aligned}$$

200.
$$\begin{aligned} x - y - 2z &= 5 \\ x + 2y + z &= -1 \\ 2x + y + z &= 2 \end{aligned}$$

201.
$$\begin{aligned} 2x + y - 3z &= -4 \\ x - 2y + z &= 3 \\ 3x - y - z &= -1 \end{aligned}$$

202.
$$\begin{aligned} x - y - 2z &= -1 \\ 3x + 2y + z &= 1 \\ x - 3y - z &= 5 \end{aligned}$$

203.
$$\begin{aligned} x - 3y + 2z &= 14 \\ 2x + y + 3z &= 5 \\ 3x + y - 2z &= -4 \end{aligned}$$

204.
$$\begin{aligned} x - y + 3z &= -10 \\ 3x + y - 2z &= 4 \\ x - 2y + z &= -9 \end{aligned}$$

205.
$$\begin{aligned} x - y + z - t &= -1 \\ 2x - y + 2z + t &= 2 \\ x - 2y - z + 2t &= 3 \\ x + y + z - 2t &= 0 \end{aligned}$$

206.
$$\begin{aligned} x + 2y - 2z + t &= -8 \\ 2x - y - 2z - t &= 0 \\ x + y + z - 2t &= -2 \\ x - 3y + z - 3t &= 9 \end{aligned}$$

207.
$$\begin{aligned} x - 3y + z + t &= -4 \\ x + y + 2z - 3t &= -5 \\ 2x - y - 3z + 2t &= 7 \\ 2x + y - z + t &= 7 \end{aligned}$$

208.
$$\begin{aligned} x - 2y - 2z + t &= -4 \\ x + y - z + 2t &= 6 \\ 3x - y + 3z - t &= -5 \\ x + 3y + 2z - 2t &= 1 \end{aligned}$$

209.
$$\begin{aligned} 3x + y + z - t &= 6 \\ x + 2y - 3z - 2t &= 12 \\ 2x + y + 2z + t &= 0 \\ x - 3y - z + 3t &= -2 \end{aligned}$$

210.
$$\begin{aligned} x + 2y - z - 2t &= -2 \\ 2x - 3y - 2z + t &= 13 \\ x - y + 2z - t &= 0 \\ x + y - 3z + 2t &= 9 \end{aligned}$$

chapter 12

Complex Numbers

12.1 Pure imaginary numbers

12.2 Complex numbers definition and notation

12.3 Operations on complex numbers

12.4 Graphs of complex numbers

12.1 Pure Imaginary Numbers

When the index n in $\sqrt[n]{a}$ is even, a is restricted to the positive real numbers. In the real number system, $\sqrt{-2}$ is not defined. For the square root of a negative number to have meaning, a new unit, called the **imaginary unit**, $\sqrt{-1}$, is introduced and is denoted by i. Since $(\sqrt{a})^2$ was defined to be a, for conformity i is defined so that $i^2 = -1$.

DEFINITION

> If $a \in R$, $a > 0$, we define $\sqrt{-a} = \sqrt{-1}\,\sqrt{a} = i\sqrt{a}$.

EXAMPLES

1. $\sqrt{-2} = i\sqrt{2}$
2. $\sqrt{-9} = i\sqrt{9} = 3i$
3. $\sqrt{-12} = i\sqrt{12} = 2i\sqrt{3}$

DEFINITION

> A number of the form ai, $a \in R$, $i = \sqrt{-1}$, is called a **pure imaginary number**.

Addition and Subtraction of Pure Imaginary Numbers

For real specific and literal numbers, we have $5a + 8a = (5 + 8)a = 13a$. For consistency, the addition of pure imaginary numbers is defined in a similar way. Hence the sum of $5i$ and $8i$ must be $13i$. Accordingly, we define addition as follows.

DEFINITION

> If $a, b \in R$, $i = \sqrt{-1}$, then $ai + bi = (a + b)i$.

Notes

1. Because $0i$ has the same properties as 0, $ai + 0i = (a + 0)i = ai$, it is customary and convenient to represent $0i$ by just 0.
2. Since $ai + (-ai) = [a + (-a)]i = 0i = 0$, $(-ai)$ is the additive inverse of ai.

Subtraction is defined in terms of addition. Thus we have

$$ai - bi = ai + (-b)i = [a + (-b)]i = (a - b)i$$

EXAMPLES

1. $7i + 3i - 4i = (7 + 3 - 4)i = 6i$

2. $\sqrt{-9} + \sqrt{-25} - \sqrt{-72} = i\sqrt{9} + i\sqrt{25} - i\sqrt{72}$
$= 3i + 5i - 6i\sqrt{2}$
$= (3 + 5 - 6\sqrt{2})i = (8 - 6\sqrt{2})i$

3. $\sqrt{1 - 4} - \sqrt{4 - 16} = \sqrt{-3} - \sqrt{-12}$
$= i\sqrt{3} - i\sqrt{12}$
$= i\sqrt{3} - 2i\sqrt{3} = -i\sqrt{3}$

Product of a Real Number and a Pure Imaginary Number

When a is a real number, $2(3a) = 6a$. In order to be consistent, if $a, b \in R$, $i = \sqrt{-1}$, we have

$$a(bi) = (ab)i$$

EXAMPLES

1. $\sqrt{6}\sqrt{-8} = \sqrt{6}(i\sqrt{8}) = i\sqrt{48} = 4i\sqrt{3}$

2. $\sqrt{-10}\sqrt{15} = i\sqrt{10}\sqrt{15} = i\sqrt{150} = 5i\sqrt{6}$

Products of Pure Imaginary Numbers

DEFINITION

The product of two pure imaginary numbers, ai and bi, is defined by
$(ai)(bi) = (a \cdot b)(i \cdot i) = abi^2 = ab(-1) = -ab$

Note If $a > 0$, $b > 0$, then

$$\sqrt{-a}\sqrt{-b} = i\sqrt{a} \cdot i\sqrt{b} = i^2\sqrt{ab} = -\sqrt{ab} \quad \text{not} \quad \sqrt{ab}$$

EXAMPLE

$\sqrt{-6}\sqrt{-48} = i\sqrt{6}(i\sqrt{48}) = i^2\sqrt{288} = -12\sqrt{2}$

From the definition of i and the rules of exponents we have

$i = \sqrt{-1}$
$i^2 = -1$
$i^3 = i \cdot i^2 = i(-1) = -i$
$i^4 = i^2 \cdot i^2 = (-1)(-1) = 1$

$i^5 = i \cdot i^4 = i \cdot 1 = i$
$i^6 = i^2 \cdot i^4 = (-1)(1) = -1$
$i^7 = i^3 \cdot i^4 = (-i)(1) = -i$
$i^8 = i^4 \cdot i^4 = (1)(1) = 1$

Since $i^4 = 1$ for $n \in N$, we have

$$i^{4n} = (i^4)^n = (1)^n = 1$$

For every $m \in N$, $m = 4q + r$, where $0 \leq r < 4$. Hence we may write

$$i^m = i^{4q+r} = i^{4q} \cdot i^r = (i^4)^q \cdot i^r = 1 \cdot i^r = i^r$$

Therefore i^m may always be replaced by i^r, where r is the remainder obtained by dividing m by 4, and thus can be reduced to $i, i^2 = -1, i^3 = -i$, or 1.

We define $i^0 = 1$, so it will be consistent with the earlier laws of exponents and with the above derivations.

EXAMPLES

1. $i^{26} = i^{4 \cdot 6 + 2} = i^2 = -1$
2. $i^{51} = i^{4 \cdot 12 + 3} = i^3 = -i$
3. $i^{125} = i^{4 \cdot 31 + 1} = i^1 = i$

Division of Pure Imaginary Numbers

Since $i^{4q} = 1$ for all $q \in N$, we have $\dfrac{1}{i^{4q}} = \dfrac{1}{1} = 1$.

If $r = 1, 2$, or 3, then

$$\frac{1}{i^r} = \frac{1}{i^r} \cdot \frac{i^{4-r}}{i^{4-r}} = \frac{i^{4-r}}{i^4} = \frac{i^{4-r}}{1} = i^{4-r}$$

If $m \in N$ and $m = 4q + r$, we have

$$\frac{1}{i^m} = \frac{1}{i^{4q+r}} = \frac{1}{i^r} = i^{4-r}$$

where r is the remainder upon dividing m by 4.

To be consistent, we further define $i^{-m} = \dfrac{1}{i^m}$.

EXAMPLES

1. $i^{-26} = \dfrac{1}{i^{26}} = \dfrac{1}{i^2} = i^{4-2} = i^2 = -1$
2. $i^{-37} = \dfrac{1}{i^{37}} = \dfrac{1}{i} = i^{4-1} = i^3 = -i$

Note All the laws for integral exponents apply to i.

From the definitions above, we have

$$ai \div bi = \frac{ai}{bi} = \frac{a}{b}$$

Note that the quotient of two pure imaginary numbers is a real number, just as the product of two pure imaginary numbers also is a real number.

EXAMPLES

1. $\dfrac{\sqrt{-12}}{\sqrt{3}} = \dfrac{i\sqrt{12}}{\sqrt{3}} = i\sqrt{\dfrac{12}{3}} = i\sqrt{4} = 2i$

2. $\dfrac{\sqrt{18}}{\sqrt{-6}} = \dfrac{\sqrt{18}}{i\sqrt{6}} = \dfrac{i^3\sqrt{18}}{i^4\sqrt{6}} = \dfrac{-i\sqrt{18}}{\sqrt{6}} = -i\sqrt{3}$

3. $\dfrac{\sqrt{-35}}{\sqrt{-5}} = \dfrac{i\sqrt{35}}{i\sqrt{5}} = \sqrt{7}$

Notes If $a > 0$ and $b > 0$, we have

1. $\sqrt{-a}\,\sqrt{b} = i\sqrt{a}\,\sqrt{b} = i\sqrt{ab}$
2. $\sqrt{-a}\,\sqrt{-b} = i\sqrt{a}(i\sqrt{b}) = i^2\sqrt{ab} = -\sqrt{ab}$
3. $\dfrac{\sqrt{-a}}{\sqrt{b}} = \dfrac{i\sqrt{a}}{\sqrt{b}} = i\sqrt{\dfrac{a}{b}} = \dfrac{\sqrt{ab}}{b}i$
4. $\dfrac{\sqrt{a}}{\sqrt{-b}} = \dfrac{\sqrt{a}}{i\sqrt{b}} = \dfrac{i^3\sqrt{a}}{i^4\sqrt{b}} = \dfrac{-i\sqrt{a}}{\sqrt{b}} = -\dfrac{\sqrt{ab}}{b}i$
5. $\dfrac{\sqrt{-a}}{\sqrt{-b}} = \dfrac{i\sqrt{a}}{i\sqrt{b}} = \dfrac{\sqrt{a}}{\sqrt{b}} = \dfrac{\sqrt{ab}}{b}$

Exercise 12.1

Express each of the following in the form ai, given that $x > 0$ and $y > 0$:

1. $\sqrt{-16}$ **2.** $\sqrt{-49}$ **3.** $\sqrt{-18}$ **4.** $\sqrt{-108}$

5. $\sqrt{-10}$ **6.** $2\sqrt{-8}$ **7.** $\sqrt{-\dfrac{3}{2}}$ **8.** $\sqrt{-6\dfrac{1}{4}}$

9. $\sqrt{-4-9}$ **10.** $\sqrt{-\dfrac{1}{4}-\dfrac{1}{9}}$ **11.** $\sqrt{-x^2}$ **12.** $\sqrt{-y^3}$

13. $\sqrt{-x^2y^5}$ **14.** $\sqrt{-x^3y^4}$ **15.** $\sqrt{-x^2-y^2}$ **16.** $\sqrt{-(x+y)^2}$

Simplify each of the following to either i, -1, $-i$, or 1:

17. i^{10} 18. i^{29} 19. $-i^{34}$ 20. $-i^{47}$
21. i^{51} 22. i^{61} 23. i^{70} 24. $-i^{1000}$
25. i^{-6} 26. i^{-15} 27. i^{-24} 28. $-i^{-30}$
29. $-i^{-41}$ 30. i^{-87} 31. i^{-70} 32. i^{-52}

Perform the indicated operations and express the answer as a or ai:

33. $\sqrt{-6} + \sqrt{-24} - \sqrt{-54}$
34. $\sqrt{-18} - \sqrt{-128} - \sqrt{-32}$
35. $\sqrt{-12} - \sqrt{-27} + \sqrt{-75}$
36. $\sqrt{-36} - \sqrt{-108} - \sqrt{-96}$
37. $\sqrt{-45} - \sqrt{-80} + \sqrt{-125}$
38. $\sqrt{-294} - \sqrt{-216} + \sqrt{-150}$
39. $\sqrt{4-20} + \sqrt{9-36}$
40. $\sqrt{5-23} - \sqrt{6-56}$
41. $\sqrt{8-28} - \sqrt{5-50}$
42. $\sqrt{9-25} - \sqrt{36-100}$
43. $\sqrt{16-64} + \sqrt{6-81}$
44. $\sqrt{-4-9} - \sqrt{-1-9}$
45. $\sqrt{-9-16} - \sqrt{-25-1}$
46. $\sqrt{-2-7} + \sqrt{-3-13}$

47. $\sqrt{3}\sqrt{-9}$ 48. $\sqrt{3}\sqrt{-27}$ 49. $\sqrt{10}\sqrt{-15}$ 50. $\sqrt{8}\sqrt{-6}$
51. $\sqrt{-21}\sqrt{7}$ 52. $\sqrt{-14}\sqrt{2}$ 53. $\sqrt{-35}\sqrt{7}$ 54. $\sqrt{-12}\sqrt{32}$

55. $\sqrt{-4}\sqrt{-9}$ 56. $\sqrt{-4}\sqrt{-36}$ 57. $\sqrt{-9}\sqrt{-25}$
58. $\sqrt{-36}\sqrt{-16}$ 59. $\sqrt{-9}\sqrt{-12}$ 60. $\sqrt{-16}\sqrt{-8}$
61. $\sqrt{-16}\sqrt{-50}$ 62. $\sqrt{-6}\sqrt{-18}$ 63. $\sqrt{-6}\sqrt{-30}$
64. $\sqrt{-5}\sqrt{-15}$ 65. $\sqrt{-10}\sqrt{-20}$ 66. $\sqrt{-12}\sqrt{-21}$

67. $\dfrac{\sqrt{-9}}{\sqrt{3}}$ 68. $\dfrac{\sqrt{-10}}{\sqrt{2}}$ 69. $\dfrac{\sqrt{-21}}{\sqrt{7}}$ 70. $\dfrac{\sqrt{-42}}{\sqrt{6}}$

71. $\dfrac{\sqrt{30}}{\sqrt{-15}}$ 72. $\dfrac{\sqrt{14}}{\sqrt{-21}}$ 73. $\dfrac{\sqrt{15}}{\sqrt{-6}}$ 74. $\dfrac{\sqrt{45}}{\sqrt{-12}}$

75. $\dfrac{\sqrt{-8}}{\sqrt{-6}}$ 76. $\dfrac{\sqrt{-32}}{\sqrt{-6}}$ 77. $\dfrac{\sqrt{-24}}{\sqrt{-18}}$ 78. $\dfrac{\sqrt{-27}}{\sqrt{-12}}$

12.2 Complex Numbers Definition and Notation

When a, b, and c are real numbers, $a \cdot b + c$ is also a real number. However the expression $(ai)(bi) + ci = -ab + ci$ is neither a real number nor is it a pure imaginary number. Expressions of this type lead us into the realm of complex numbers.

DEFINITION

> A **complex number** is a number of the form $a + bi$ where a and b are real numbers and $i = \sqrt{-1}$. The number a is called the **real part** of the complex number and b is called the **imaginary part**.

The letter z is sometimes used to represent a complex number, that is, $z = a + bi$. It is used for brevity.

The set of complex numbers C is the set

$$C = \{z = a + bi \mid a, b \in R, i = \sqrt{-1}\}$$

When a complex number is written in the form $a + bi$, the complex number is said to be in **simplified form**, or in **standard form**. The form $a + bi$ is sometimes referred to as the **Cartesian** or **rectangular form** of a complex number.

Notes

1. The complex number $a + 0i = a$ is a real number. That is, the set of real numbers, R, is a subset of the set of complex numbers, C.
2. The complex number $0 + bi$, $b \neq 0$, is a pure imaginary number. That is, the set of pure imaginary numbers is a subset of the set of complex numbers.

EXAMPLE Write $i^6 - i^{-10} - i^{15}$ in standard form.

Solution

$$\begin{aligned} i^6 - i^{-10} - i^{15} &= i^2 - \frac{1}{i^{10}} - i^3 \\ &= -1 - \frac{i^2}{i^{12}} - (-i) \\ &= -1 - (-1) + i = 0 + i \end{aligned}$$

EXAMPLE Multiply $\sqrt{-14}(3\sqrt{-6} + 2\sqrt{21})$ and put the answer in standard form.

Solution

$$\begin{aligned} \sqrt{-14}(3\sqrt{-6} + 2\sqrt{21}) &= i\sqrt{14}(3i\sqrt{6} + 2\sqrt{21}) \\ &= 3i^2\sqrt{14 \cdot 6} + 2i\sqrt{14 \cdot 21} \\ &= -6\sqrt{21} + 14i\sqrt{6} \end{aligned}$$

EXAMPLE Multiply $-2i[(-3i)^3 + (-2i)^4]$ and put the answer in standard form.

Solution

$$\begin{aligned} -2i[(-3i)^3 + (-2i)^4] &= -2i[-27i^3 + 16i^4] \\ &= 54i^4 - 32i^5 \\ &= 54 - 32i \end{aligned}$$

Exercise 12.2

Perform the indicated operations and write the answer in standard form:

1. $i^3 + i^4 + i^5$
2. $i^7 + i^9 - i^{11}$
3. $i^8 - i^{10} + i^{-12}$
4. $i^{20} - i^{-23} - i^{26}$
5. $i^{31} + i^{35} + i^{39}$
6. $i^{33} + i^{38} - i^{43}$
7. $i^{-8} + i^{18} - i^{-28}$
8. $i^{15} - i^{-21} - i^{-27}$
9. $i^{-12} - i^{-19} - i^{-26}$

10. $i^{-11} - i^{-23} - i^{-35}$
11. $\sqrt{3}(\sqrt{3} + \sqrt{-2})$
12. $\sqrt{2}(\sqrt{8} + \sqrt{-6})$
13. $\sqrt{3}(\sqrt{12} + \sqrt{-18})$
14. $\sqrt{6}(\sqrt{9} - \sqrt{-12})$
15. $\sqrt{10}(\sqrt{15} - \sqrt{-20})$
16. $\sqrt{8}(\sqrt{10} - \sqrt{-14})$
17. $\sqrt{3}(\sqrt{-6} + \sqrt{24})$
18. $\sqrt{2}(\sqrt{-10} + \sqrt{6})$
19. $\sqrt{3}(\sqrt{-15} + \sqrt{21})$
20. $\sqrt{5}(\sqrt{-30} - \sqrt{5})$
21. $\sqrt{6}(\sqrt{-8} - \sqrt{12})$
22. $\sqrt{15}(\sqrt{-6} - \sqrt{10})$
23. $\sqrt{-2}(\sqrt{2} + \sqrt{-6})$
24. $\sqrt{-3}(\sqrt{3} + \sqrt{-12})$
25. $\sqrt{-10}(\sqrt{2} + \sqrt{-5})$
26. $\sqrt{-14}(\sqrt{7} - \sqrt{-8})$
27. $\sqrt{-6}(\sqrt{2} - \sqrt{-18})$
28. $\sqrt{-8}(\sqrt{16} - \sqrt{-6})$
29. $\sqrt{-6}(\sqrt{-30} + \sqrt{42})$
30. $\sqrt{-5}(\sqrt{-10} + \sqrt{15})$
31. $\sqrt{-10}(\sqrt{-2} + \sqrt{8})$
32. $\sqrt{-12}(\sqrt{-21} - \sqrt{6})$
33. $\sqrt{-18}(\sqrt{-15} - \sqrt{12})$
34. $\sqrt{-24}(\sqrt{-18} - \sqrt{27})$
35. $(-5i)^2 + (-2i)^3$
36. $(-3i)^3 - (-i)^5$
37. $(-4i^2)^3 - (3i^3)^2$
38. $(i^{-2})^3 - (-i^{-3})^4$
39. $2i(-4i)^2 + 6i(2i)^3$
40. $4i(-i)^4 - 3i^3(-i)^7$

12.3 Operations on Complex Numbers

Since the set C is an extension of the set of real numbers as well as an extension of the set of pure imaginary numbers, we desire the arithmetic operations on C to be extensions of the arithmetic operations on the two subsets.

Addition and Subtraction of Complex Numbers

DEFINITION

> The sum of two complex numbers $a + bi$ and $c + di$ is defined by
>
> $(a + bi) + (c + di) = (a + c) + (b + d)i$

EXAMPLE $(3 + 2i) + (7 - 5i) = (3 + 7) + [2 + (-5)]i = 10 - 3i$

For every number $a + bi \in C$, there exists a unique complex number $(-a) + (-bi)$ called the **additive inverse** of $a + bi$

$$(a + bi) + [(-a) + (-bi)] = [a + (-a)] + [b + (-b)]i$$
$$= 0 + 0i = 0$$

Since $-(a + bi)$ is the additive inverse of $(a + bi)$, we have

$$-(a + bi) = (-a) + (-bi)$$

Since subtraction is defined in terms of addition, we have

$$
\begin{aligned}
(a + bi) - (c + di) &= (a + bi) + [-(c + di)] \\
&= (a + bi) + [(-c) + (-di)] \\
&= [a + (-c)] + [b + (-d)]i \\
&= (a - c) + (b - d)i
\end{aligned}
$$

Thus subtraction of two complex numbers is defined by

$$(a + bi) - (c + di) = (a - c) + (b - d)i$$

EXAMPLE Subtract $(-2 + 3i)$ from $(5 - i)$.

Solution

$$
\begin{aligned}
(5 - i) - (-2 + 3i) &= [5 - (-2)] + [-1 - (3)]i \\
&= (5 + 2) + (-1 - 3)i \\
&= 7 - 4i
\end{aligned}
$$

DEFINITION When $a + bi \in C$, $a + bi = 0$, then $a = 0$ and $b = 0$.

THEOREM If $a + bi,\ c + di \in C$, and $a + bi = c + di$, then $a = c$ and $b = d$.

Proof

$$a + bi = c + di$$

$$(a + bi) - (c + di) = 0$$

$$(a - c) + (b - d)i = 0$$

Since if $x + yi = 0$ then $x = 0$ and $y = 0$, we have

$$a - c = 0, \quad \text{that is,} \quad a = c$$

and $$b - d = 0, \quad \text{that is,} \quad b = d$$

Note The set of complex numbers cannot be ordered. A complex number cannot be labeled as positive or negative. Also, we cannot categorize a complex number as greater than or less than another complex number.

EXAMPLE Find the real values of x and y that satisfy $(3 + 2i)x + (2 + i)y = 8 + 5i$.

Solution

$$(3 + 2i)x + (2 + i)y = 8 + 5i$$

$$3x + 2ix + 2y + iy = 8 + 5i$$

$$(3x + 2y) + (2x + y)i = 8 + 5i$$

From the theorem we have

$$3x + 2y = 8 \quad \text{and} \quad 2x + y = 5$$

Solving the above system simultaneously, we get $x = 2$ and $y = 1$.

Exercise 12.3A

Write each of the following complex numbers in standard form:

1. $(2 + \sqrt{-25}) + (6 - \sqrt{-16})$
2. $(-7 + \sqrt{-1}) + (-3 - \sqrt{-49})$
3. $(6 - \sqrt{-4}) + (1 + \sqrt{-9})$
4. $(8 - \sqrt{-18}) + (-5 + \sqrt{-8})$
5. $(-2 - \sqrt{-25}) - (6 + \sqrt{-49})$
6. $(3 + \sqrt{-2}) - (7 + \sqrt{-50})$
7. $(10 - \sqrt{-32}) - (7 - 2\sqrt{-8})$
8. $(-5 + \sqrt{-27}) - (8 + 3\sqrt{-3})$
9. $(-4 + \sqrt{-5}) - (-4 - \sqrt{-20})$
10. $(6 - \sqrt{-36}) - (6 + \sqrt{-72})$
11. $(5 - \sqrt{-24}) + (-5 + 2\sqrt{-6})$
12. $(-1 + \sqrt{-63}) + (1 - 3\sqrt{-7})$

Find the real values of x and y that satisfy the following equations:

13. $2x + 3iy = 9i$
14. $x - 2iy = 6i$
15. $4x - 5iy = -6$
16. $5x - iy = -1$
17. $x + iy = 3 - 2i$
18. $2x - iy = 3i - 4$
19. $(1 + 2i)x + (1 - i)y = 2 + i$
20. $(3 + i)x + (1 + 2i)y = 1 + 3i$
21. $(3 + 2i)x - (1 - i)y = 5i$
22. $(1 + 3i)x + (4 - 4i)y = 5 - 17i$
23. $(7 + 5i)x - (3 - 2i)y = 10 + 3i$
24. $(4 + 3i)x + (3 - 5i)y = 6 + 19i$
25. $2x + y - 4 + (3x - 2y - 27)i = 0$
26. $6x - 7y - 10 - (8x - 13y - 6)i = 0$
27. $7x - 6y - 17 - (3x + y - 18)i = 0$
28. $3x - 2y - 7 + (4x + y - 24)i = 0$
29. $2x - 2iy + 1 = 27i - 5y - 3ix$
30. $19 + 2iy - 6x = y + ix + 12i$
31. $4x + 6iy + 9 = 9y - 2ix + 13i$
32. $3x + 4iy - 9i = 5 - 4y - 9ix$

Multiplication of Complex Numbers

If we treat the two complex numbers $a + bi$ and $c + di$ as binomials, and i as a literal, then

$$\begin{aligned}(a + bi)(c + di) &= a(c + di) + bi(c + di)\\ &= ac + adi + bci + bdi^2\\ &= ac + (ad + bc)i + bdi^2\end{aligned}$$

Since $i^2 = -1$,

$$\begin{aligned}(a + bi)(c + di) &= ac + (ad + bc)i - bd\\ &= (ac - bd) + (ad + bc)i\end{aligned}$$

We are thus motivated to make the following definition.

DEFINITION

> The product of two complex numbers $(a + bi)$ and $(c + di)$ is defined by
>
> $(a + bi)(c + di) = (ac - bd) + (ad + bc)i$

cancel

EXAMPLE Multiply $(2 + 3i)(5 - 2i)$.

Solution

$$\begin{array}{r} 2 + 3i \\ 5 - 2i \\ \hline 10 + 15i \\ - 4i - 6i^2 \\ \hline 10 + 11i - 6i^2 \end{array}$$

Hence $$(2 + 3i)(5 - 2i) = 10 + 11i - 6i^2$$
$$= 10 + 11i + 6 = 16 + 11i$$

Division of Complex Numbers

Consider the two complex numbers $a + bi$ and $a - bi$

$$(a + bi) + (a - bi) = 2a$$
$$(a + bi)(a - bi) = a^2 + b^2$$

DEFINITION

The two complex numbers $a + bi$ and $a - bi$ are called **conjugates** of each other.

When we denote the complex number by z, its conjugate is denoted by $\bar{z}$.

$$(a + bi) \div (c + di) = \frac{a + bi}{c + di} \qquad c + di \neq 0$$

To put the above fraction in standard form we must write an equivalent fraction whose denominator is a real number. This can be accomplished by multiplying the numerator and the denominator of the fraction by the conjugate of the denominator:

$$\frac{a + bi}{c + di} = \frac{(a + bi)(c - di)}{(c + di)(c - di)} = \frac{(ac + bd) + (bc - ad)i}{c^2 + d^2}$$
$$= \frac{ac + bd}{c^2 + d^2} + \frac{bc - ad}{c^2 + d^2}i$$

EXAMPLE Divide $\dfrac{3 - 2i}{2 + i}$.

Solution The conjugate of the denominator is $(2 - i)$:

$$\frac{3 - 2i}{2 + i} = \frac{(3 - 2i)(2 - i)}{(2 + i)(2 - i)} = \frac{6 - 3i - 4i + 2i^2}{4 - i^2}$$
$$= \frac{6 - 7i + 2(-1)}{4 - (-1)} = \frac{4 - 7i}{5} = \frac{4}{5} - \frac{7}{5}i$$

EXAMPLE

Express $\dfrac{6-i}{2-3i}$ in standard form.

Solution

The conjugate of the denominator is $2+3i$:

$$\frac{6-i}{2-3i}=\frac{(6-i)(2+3i)}{(2-3i)(2+3i)}=\frac{12+18i-2i-3i^2}{4-9i^2}$$
$$=\frac{12+16i-3(-1)}{4-9(-1)}$$
$$=\frac{15+16i}{13}=\frac{15}{13}+\frac{16}{13}i$$

Exercise 12.3B

Perform the indicated operations and express the result in standard form:

1. $(1+i)(2-i)$ **2.** $(2+i)(3-i)$ **3.** $(6+i)(4+i)$

4. $(4-i)(1-3i)$ **5.** $(3-2i)(2-3i)$ **6.** $(2+3i)(5+2i)$

7. $(1-4i)(2+4i)$ **8.** $(3-2i)(2-5i)$ **9.** $(1-3i)^2$

10. $(2+3i)^2$ **11.** $(1+\sqrt{-2})(1+\sqrt{-8})$

12. $(2-\sqrt{-3})(3-\sqrt{-12})$ **13.** $(4-\sqrt{-18})(1+\sqrt{-32})$

14. $(5-\sqrt{-6})(2+\sqrt{-24})$ **15.** $(2+i\sqrt{3})^2$

16. $(1-2i\sqrt{2})^2$ **17.** $(3+i\sqrt{6})^2$

18. $(2+i\sqrt{5})^2$ **19.** $(\sqrt{2}-i\sqrt{3})^2$

20. $(\sqrt{3}-2i\sqrt{2})^2$ **21.** $(1+i)(2-i)(3+i)$

22. $(2+i)(1-2i)(3-i)$ **23.** $(1+5i)(2-3i)(4-i)$

24. $(2-3i)(3+4i)(4+i)$ **25.** $(1-i\sqrt{3})^3$

26. $\left(-\dfrac{1}{2}+\dfrac{\sqrt{3}}{2}i\right)^3$ **27.** $(\sqrt{3}-i)^3$

28. $\left(\dfrac{\sqrt{3}}{2}+\dfrac{i}{2}\right)^3$ **29.** $(\sqrt{3}+i)^3$ **30.** $\left(\dfrac{\sqrt{2}}{2}+\dfrac{\sqrt{2}}{2}i\right)^6$

31. $\dfrac{2-i}{4i}$ **32.** $\dfrac{6+5i}{3i}$ **33.** $\dfrac{1}{1+2i}$

34. $\dfrac{1}{3-i}$ **35.** $\dfrac{1}{2-3i}$ **36.** $\dfrac{1}{1+5i}$

37. $\dfrac{1}{5-2i}$ **38.** $\dfrac{1}{1-i\sqrt{2}}$ **39.** $\dfrac{1}{1+i\sqrt{3}}$

40. $\dfrac{1}{2 - i\sqrt{5}}$ **41.** $\dfrac{3i}{2 - i}$ **42.** $\dfrac{4}{3 + i}$

43. $\dfrac{1 + i}{1 - i}$ **44.** $\dfrac{1 - i}{1 + i}$ **45.** $\dfrac{1 - 2i}{3 + 2i}$

46. $\dfrac{2 + 3i}{3 - 2i}$ **47.** $\dfrac{5 - i}{3i - 2}$ **48.** $\dfrac{2 + i}{1 + 3i}$

49. $\dfrac{1 - 4i}{2 + i}$ **50.** $\dfrac{3 + 4i}{1 - 2i}$ **51.** $\dfrac{1 + 2i\sqrt{3}}{3 - i\sqrt{3}}$

52. $\dfrac{4 - i\sqrt{2}}{3 + i\sqrt{2}}$ **53.** $\dfrac{\sqrt{2} + i\sqrt{3}}{\sqrt{2} - i\sqrt{3}}$ **54.** $\dfrac{\sqrt{3} - 2i\sqrt{2}}{\sqrt{3} + 2i\sqrt{2}}$

55. $(1 - i)^{-2}$ **56.** $(3 - 2i)^{-2}$ **57.** $(1 + i\sqrt{2})^{-2}$

58. $(2 - i\sqrt{3})^{-2}$ **59.** $\dfrac{2 - i}{(3 + i)(4 - i)}$ **60.** $\dfrac{1 + 3i}{(1 + 4i)(2 - 4i)}$

61. $\dfrac{3 + i}{(2 + 3i)(4 - 3i)}$ **62.** $\dfrac{2i - 1}{(3i + 5)(2i - 3)}$

Find the complex number z, $z = x + yi$, that satisfies:

63. $3z = 6 + 9i$ **64.** $5z = 10 - 5i$ **65.** $2z + 6 = 14 - 2i$
66. $iz = i + 1$ **67.** $iz = -3i - 1$ **68.** $z + 2iz = 5$
69. $3z - iz = 10$ **70.** $z + iz = 5 - i$ **71.** $2z - iz = 4 + 3i$
72. $4z - 3iz = 17 - 19i$

12.4 Graphs of Complex Numbers

Since a complex number $x + yi$ consists of a real number x and a pure imaginary number yi, we can graph complex numbers in the same manner as we graphed ordered pairs of real numbers.

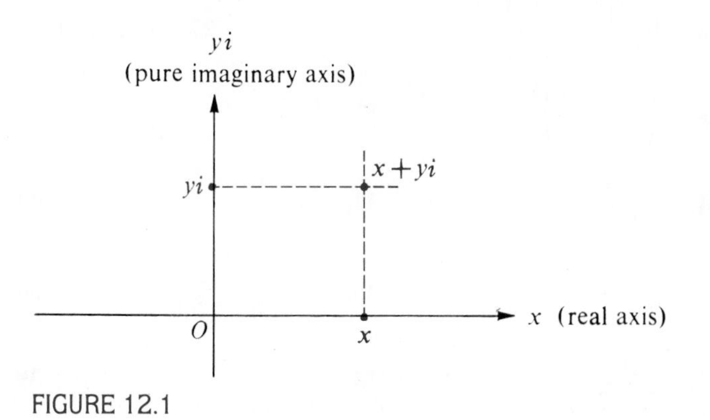

FIGURE 12.1

In a Cartesian coordinate system, let the x-axis represent the real number axis and the y-axis represent the pure imaginary number axis. To graph the complex number $x + yi$, draw a vertical line at the point whose coordinate is x on the real axis, and a horizontal line at the point whose coordinate is yi on the pure imaginary axis, as in Figure 12.1.

The intersection of the two lines is a point representing the number $x + yi$.

Since we can draw the complex number $x + yi$ on a Cartesian coordinate system, the form $x + yi$ is referred to as the Cartesian or rectangular representation of a complex number.

The plane of the set of complex numbers is called the **complex number plane**.

Note The origin represents the complex number $0 + 0i$.

EXAMPLE Plot the numbers $4 + 3i$, $-2 + 5i$, $-3 - i$, and $1 - 2i$.

Solution Refer to Figure 12.2

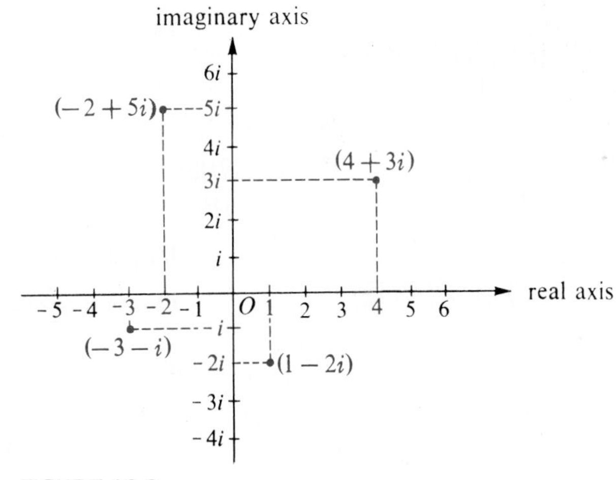

FIGURE 12.2

Exercise 12.4

Plot the following points in a complex plane:

1. 5	**2.** 1	**3.** -2	**4.** -6
5. $-3i$	**6.** $-i$	**7.** i	**8.** $7i$
9. $1 + 2i$	**10.** $3 + i$	**11.** $3 - i$	**12.** $2 - 3i$
13. $-2 - 3i$	**14.** $-4 - 2i$	**15.** $-4 + 2i$	**16.** $-5 + i$

Chapter 12 Review

Find the real values of x and y that satisfy the following equations:

1. $3x + 4i = 4iy - 6$
2. $x - 2iy = 5 + 20i$
3. $7x + 3iy = 21$
4. $2x - iy = 7i$
5. $x(1 + 3i) - y(1 - 2i) = 5i$
6. $x(1 + 2i) - y(1 - i) = 1 + 8i$
7. $x(5 + 3i) - y(1 - i) = 1 + 7i$
8. $x(4 + 2i) + y(1 - i) = 7 - i$
9. $x(12 + 9i) + y(1 + 7i) = 34 + 13i$
10. $x(2 - i) + y(3 - 5i) = 3 - 4i$
11. $3x - y - 14 + (5x - 7y - 2)i = 0$
12. $8x - 7y - 4 + (7x - 4y + 5)i = 0$
13. $2x + 6y - 5 + (7x - y - 1)i = 0$
14. $2x - 3y - 5 + (3x + 4y + 18)i = 0$
15. $5x + 6y - 10 + (4x + 9y + 13)i = 0$
16. $4x + 6y - 7 + (3x + 5y - 6)i = 0$

Find the complex number z that satisfies:

17. $iz = 7 + i$
18. $iz = 2 - 3i$
19. $iz = -3 - i$
20. $iz = 2 + 6i$
21. $z + iz = -1 + 5i$
22. $z - 2iz = -5 - 5i$
23. $z - 3iz = 1 - 13i$
24. $3z + 2iz = -5 + i$
25. $5z + 4iz = 28 + 47i$
26. $3z - 7iz = -27 + 5i$

Perform the indicated operations and express the result in standard form:

27. $i^6 - i^9 + 2i^{13}$
28. $2i^3 + i^7 + 5i^8$
29. $i^{12} + 4i^{17} - i^{21}$
30. $i^{-15} + i^{-18} + i^{-30}$
31. $3i^{-6} - 4i^{-11} - 7i^{-19}$
32. $i^{-26} + 3i^{-31} - 4i^{-37}$
33. $(1 + i)(6 + 2i)$
34. $(3 + 5i)(7 + i)$
35. $(i - 3)(6i + 1)$
36. $(i + 4)(3i - 8)$
37. $(3 - 7i)(6 - i)$
38. $(4 - 5i)(3 - 10i)$
39. $(3 - 2i)^2$
40. $(5 + i)^2$
41. $(1 - i)^3$
42. $(2 + i)^3$
43. $(1 + \sqrt{-6})(2 - \sqrt{-54})$
44. $(3 - \sqrt{-2})(4 - \sqrt{-32})$
45. $(8 + \sqrt{-3})(1 + \sqrt{-12})$
46. $(1 + \sqrt{-5})(2 + \sqrt{-45})$
47. $\dfrac{6 - i}{3i}$
48. $\dfrac{5 + 2i}{-i}$
49. $\dfrac{3}{2 + i}$
50. $\dfrac{-2}{4 - i}$
51. $\dfrac{1 + 2i}{1 - 2i}$
52. $\dfrac{3 - i}{6 + i}$
53. $\dfrac{3i + 1}{5i + 1}$
54. $\dfrac{5 - i}{i - 8}$
55. $\dfrac{4i - 1}{2i - 7}$
56. $\dfrac{i\sqrt{2} - 1}{i\sqrt{2} - 3}$
57. $\dfrac{i^3 - i^5}{3i^2 - 2i^7}$
58. $\dfrac{2i - i^8}{4i^4 - 3i^9}$
59. $(2 + i)^{-2}$
60. $(3 - i)^{-2}$

chapter 13

Quadratic Equations and Inequalities in One Variable

13.1 Introduction

13.2 Solution of quadratic equations by factoring

13.3 Solution of quadratic equations by completing the square

13.4 Solution of quadratic equations by the quadratic formula

13.5 Character of the roots

13.6 Properties of the roots

13.7 Equations that lead to quadratic equations

13.8 Word problems

13.9 Graphs of quadratic equations

13.10 Analytic solution of quadratic inequalities

13.11 Graphical solution of quadratic inequalities

13.1 Introduction

A **polynomial** P in x is an expression of the form

$$a_0x^n + a_1x^{n-1} + a_2x^{n-2} + \cdots + a_n$$

where $a_0, a_1, a_2, \ldots, a_n$ are real numbers and $n \in W$.

If $a_0 \neq 0$, the polynomial is of **degree** n. When $n = 0$, the polynomial is of the form a_0, called a **constant polynomial**, and its degree is zero. When $n = 1$, the polynomial is of the form $a_0x + a_1$, called a **linear polynomial**. When $n = 2$, the polynomial is of the form $a_0x^2 + a_1x + a_2$, a **quadratic polynomial**. The number a_0 is called the **coefficient of the leading term**, and a_n is the **constant term**.

A **polynomial equation** in x is a polynomial in x with $n \in N$ equated to zero. A polynomial equation of the form $ax^2 + bx + c = 0$, where $a \neq 0$, $a, b, c \in R$, and x is the variable, is called a **second-degree equation** or a **quadratic equation** in the variable x.

The form $ax^2 + bx + c = 0$ is called the **standard form of the quadratic equation**. The values of x that satisfy the equation are the **roots** of the equation or the elements of the **solution set** of the equation.

THEOREM If P and Q are polynomials and $P \cdot Q = 0$, then $P = 0$ or $Q = 0$.

Proof If $P \neq 0$, then divide $P \cdot Q = 0$ by P:

$$\frac{P \cdot Q}{P} = \frac{0}{P}, \qquad \text{that is,} \qquad Q = 0$$

Hence if $P \cdot Q = 0$, then $P = 0$ or $Q = 0$

13.2 Solution of Quadratic Equations by Factoring

When the polynomial $ax^2 + bx + c$ can be factored into the product of two linear factors, the quadratic equation $ax^2 + bx + c = 0$ can be solved by equating each factor to zero, separately. Thus the quadratic equation is expressed as two linear equations. The solution set of the quadratic equation is the union of the solution sets of the two linear equations.

EXAMPLE Find the solution set of the equation $x^2 - 3x = 0$.

Solution $x^2 - 3x = x(x - 3) = 0$

Hence $x = 0$

or $x - 3 = 0$, that is, $x = 3$

Therefore, the solution set of the quadratic equation $x^2 - 3x = 0$ is the union of the solution set of the equation $x = 0$ with the solution set of the equation $x = 3$.

Thus the solution set of the quadratic equation is $\{0, 3\}$.

EXAMPLE

Find the solution set of the equation $x^2 - 2x - 24 = 0$.

Solution

$x^2 - 2x - 24 = (x + 4)(x - 6) = 0$

Hence $\quad x + 4 = 0, \quad$ that is, $\quad x = -4$

or $\quad x - 6 = 0, \quad$ that is, $\quad x = 6$

The solution set of the quadratic equation is $\{-4, 6\}$.

EXAMPLE

Find the solution set of $6x^2 - x = 12$.

Solution

First write the equation in standard form:

$6x^2 - x - 12 = 0$

$6x^2 - x - 12 = (2x - 3)(3x + 4) = 0$

Hence $\quad 2x - 3 = 0, \quad$ that is, $\quad x = \frac{3}{2}$

or $\quad 3x + 4 = 0, \quad$ that is, $\quad x = -\frac{4}{3}$

The solution set is $\left\{-\frac{4}{3}, \frac{3}{2}\right\}$.

DEFINITION

When the two roots of the equation are the same, we say that the equation has a **double root**, or a **root of multiplicity two**.

EXAMPLE

Find the solution set of $9x^2 + 6x + 1 = 0$.

Solution

$9x^2 + 6x + 1 = (3x + 1)(3x + 1) = 0$

Hence $\quad 3x + 1 = 0, \quad$ that is, $\quad x = -\frac{1}{3}$

or $\quad 3x + 1 = 0, \quad$ that is, $\quad x = -\frac{1}{3}$

The solution set is $\left\{-\frac{1}{3}, -\frac{1}{3}\right\}$.

Note The solution set is written as $\left\{-\frac{1}{3}, -\frac{1}{3}\right\}$, not $\left\{-\frac{1}{3}\right\}$, to indicate that $-\frac{1}{3}$ is a double root. It also indicates that the original equation is the quadratic equation $9x^2 + 6x + 1 = 0$ and not the linear equation $3x + 1 = 0$.

DEFINITION

> A **pure quadratic equation** is an equation of the form $x^2 - a^2 = 0$.

Solving $x^2 - a^2 = 0$ by factoring, we get

$$x^2 - a^2 = (x - a)(x + a) = 0$$

Hence $x - a = 0$, that is, $x = a$

or $x + a = 0$, that is, $x = -a$

The solution set of the pure quadratic equation $x^2 - a^2 = 0$ is the union of the solution set of the equation $x = +a$ and the solution set of the equation $x = -a$. The two linear equations are frequently written as one equation, $x = \pm a$. The solution set is $\{-a, a\}$.

Hence if $x^2 = a^2$, we write $x = \pm a$

or if $x^2 = a^2$, then $x = \pm\sqrt{a^2}$

EXAMPLE Find the solution set of the equation $2x^2 - 3 = 0$.

Solution

$$2x^2 - 3 = 0$$

$$x^2 = \frac{3}{2}$$

$$x = \pm\sqrt{\frac{3}{2}}$$

$$= \pm\sqrt{\frac{3 \cdot 2}{2 \cdot 2}} = \pm\frac{1}{2}\sqrt{6}$$

The solution set is $\left\{-\frac{1}{2}\sqrt{6}, \frac{1}{2}\sqrt{6}\right\}$.

EXAMPLE Find the solution set of the equation $x^2 + 12 = 0$.

Solution

$$x^2 + 12 = 0$$

$$x^2 = -12$$

$$x = \pm\sqrt{-12} = \pm 2i\sqrt{3}$$

The solution set is $\{-2i\sqrt{3}, 2i\sqrt{3}\}$.

EXAMPLE Solve the equation $(x + 2a)^2 - 9b^2 = 0$ for x.

Solution

$$(x + 2a)^2 - 9b^2 = 0$$
$$(x + 2a)^2 = 9b^2$$
$$x + 2a = \pm 3b$$
$$x = -2a \pm 3b$$

The solution set is $\{-2a + 3b, -2a - 3b\}$.

Note When the roots of a quadratic equation are r_1 and r_2, then $(x - r_1)$ and $(x - r_2)$ are the factors of the quadratic polynomial. Thus a quadratic equation whose roots are r_1 and r_2 is

$$(x - r_1)(x - r_2) = 0 \qquad \text{or} \qquad x^2 - (r_1 + r_2)x + r_1 r_2 = 0$$

EXAMPLE Form a quadratic equation whose roots are $-\frac{2}{3}$ and 5.

Solution The factors of the polynomial are $\left[x - \left(-\frac{2}{3}\right)\right]$ and $(x - 5)$.

A quadratic equation is $\left(x + \frac{2}{3}\right)(x - 5) = 0$.

Expanding we get $\quad x^2 - \frac{13}{3}x - \frac{10}{3} = 0$

or $\quad 3x^2 - 13x - 10 = 0$

EXAMPLE Form a quadratic equation whose roots are $2 + 3i$ and $2 - 3i$.

Solution The factors of the polynomial are

$$[x - (2 + 3i)] \qquad \text{and} \qquad [x - (2 - 3i)]$$

A quadratic equation is

$$[x - (2 + 3i)][x - (2 - 3i)] = 0$$
$$x^2 - 4x + 4 - 9i^2 = 0$$

or $\quad x^2 - 4x + 13 = 0$

Exercise 13.2

Solve the following equations for x:

1. $x^2 + x = 0$ **2.** $x^2 + 2x = 0$ **3.** $x^2 - 4x = 0$

4. $x^2 - 5x = 0$ **5.** $3x^2 + x = 0$ **6.** $2x^2 - 7x = 0$

7. $2x^2 + 6x = 0$
8. $6x^2 + 9x = 0$
9. $4x^2 - 6x = 0$
10. $10x^2 - 15x = 0$
11. $x^2 - 1 = 0$
12. $x^2 - 4 = 0$
13. $x^2 - 25 = 0$
14. $9x^2 - 4 = 0$
15. $4x^2 - 36 = 0$
16. $x^2 - a^2 = 0$
17. $x^2 - 4b^2 = 0$
18. $x^2 - 2 = 0$
19. $x^2 - 3 = 0$
20. $2x^2 - 1 = 0$
21. $3x^2 - 2 = 0$
22. $5x^2 - 3 = 0$
23. $7x^2 - 5 = 0$
24. $2x^2 - 4a^2 = 0$
25. $x^2 - a = 0$
26. $x^2 - a - b = 0$
27. $2x^2 - b = 0$
28. $x^2 - (2a + b)^2 = 0$
29. $x^2 - (3a + 2b)^2 = 0$
30. $x^2 - (a - 2b)^2 = 0$
31. $x^2 - (2a - 3b)^2 = 0$
32. $(x + a)^2 - 4 = 0$
33. $(x - a)^2 - 16 = 0$
34. $(x + a)^2 - b^2 = 0$
35. $(x + 2a)^2 - 4b^2 = 0$
36. $(x - 2a)^2 - 9b^2 = 0$
37. $(2x + a)^2 - b^2 = 0$
38. $(2x - a)^2 - 25b^2 = 0$
39. $x^2 + 2 = 0$
40. $x^2 + 5 = 0$
41. $x^2 + 18 = 0$
42. $x^2 + 32 = 0$
43. $2x^2 + 3 = 0$
44. $3x^2 + 2 = 0$
45. $x^2 + 3x + 2 = 0$
46. $x^2 + 4x + 3 = 0$
47. $x^2 + 5x + 6 = 0$
48. $x^2 + 6x + 8 = 0$
49. $x^2 + x = 2$
50. $x^2 + x = 6$
51. $x^2 + 2x = 3$
52. $x^2 + x = 12$
53. $x^2 + 2x + 1 = 0$
54. $x^2 + 4x + 4 = 0$
55. $x^2 - 4x + 4 = 0$
56. $x^2 - 6x + 9 = 0$
57. $x^2 + 5x - 24 = 0$
58. $x^2 - 8x + 12 = 0$
59. $x^2 + 24 = 10x$
60. $x^2 = 9x - 18$
61. $x^2 = 7x + 18$
62. $x^2 = 5x + 24$
63. $x^2 + 12 = 7x$
64. $x^2 + 3x = 18$
65. $2x^2 - 6x + 4 = 0$
66. $2x^2 = 4x + 6$
67. $3x^2 - 15x + 18 = 0$
68. $4x^2 - 32 = 8x$
69. $2x^2 + 3x + 1 = 0$
70. $3x^2 + 13x + 4 = 0$
71. $2x^2 + 7x + 6 = 0$
72. $3x^2 + 11x + 6 = 0$
73. $2x^2 - 3x = 2$
74. $2x^2 + 11x = 6$
75. $3x^2 + 11x = 4$
76. $6x^2 - 5x = 4$
77. $4x^2 + 5x = 6$
78. $6x^2 = x + 12$
79. $9x^2 - 2 = 3x$
80. $12x^2 = 3 - 5x$
81. $3x^2 = 8 - 10x$
82. $6x^2 = 5x + 6$
83. $6x^2 + 17x = 3$
84. $8x^2 + 6x - 9 = 0$
85. $9x^2 - 18x + 8 = 0$
86. $4x^2 = 19x - 12$
87. $3x^2 = 14x - 8$
88. $6x^2 = 37x - 6$
89. $4x^2 + 12x + 9 = 0$
90. $9x^2 + 12x + 4 = 0$
91. $4x^2 - 4x + 1 = 0$
92. $9x^2 - 24x + 16 = 0$
93. $6x^2 - 13x + 6 = 0$
94. $9x^2 - 15x + 4 = 0$
95. $6x^2 - 19x + 8 = 0$
96. $12x^2 - 20x + 3 = 0$
97. $12x^2 + 25x + 12 = 0$
98. $x^2 - ax - 2a^2 = 0$
99. $x^2 - ax - 6a^2 = 0$
100. $2x^2 + 5ax - 3a^2 = 0$
101. $4x^2 - 4ax - 3a^2 = 0$
102. $6x^2 - 5ax - 4a^2 = 0$
103. $4x^2 + 13ax - 12a^2 = 0$
104. $x^2 + ax + bx + ab = 0$
105. $x^2 - ax + 2bx = 2ab$
106. $x^2 + 2ax - 4bx = 8ab$
107. $4x^2 - ax + 5ab = 20bx$
108. $4x^2 - 6ax + 9ab = 6bx$
109. $6x^2 - 9ax + 8bx = 12ab$
110. $x^2 - a^2 + 2ab - b^2 = 0$
111. $x^2 - 4a^2 - 4ab - b^2 = 0$
112. $4x^2 - 9a^2 - 6ab - b^2 = 0$
113. $9x^2 - a^2 + 4ab - 4b^2 = 0$
114. $x^2 + 4a^2 + 4ax - b^2 = 0$
115. $x^2 + 4a^2 - 4ax - 4b^2 = 0$

Form a quadratic equation with the given numbers as roots:

116. $1;\ 2$ **117.** $2;\ 4$ **118.** $1;\ -2$ **119.** $2;\ -3$

120. $1;\ \frac{1}{2}$ **121.** $2;\ \frac{1}{3}$ **122.** $3;\ -\frac{1}{2}$ **123.** $2;\ -\frac{2}{3}$

124. $\frac{1}{2};\ \frac{1}{3}$ **125.** $\frac{1}{2};\ \frac{1}{4}$ **126.** $\frac{4}{3};\ -\frac{1}{2}$ **127.** $\frac{2}{3};\ -\frac{3}{2}$

128. $a;\ b$ **129.** $2a;\ -b$ **130.** $\sqrt{2};\ \sqrt{3}$ **131.** $\sqrt{2};\ 2\sqrt{2}$

132. $\sqrt{3};\ -2\sqrt{3}$ **133.** $2\sqrt{2};\ -3\sqrt{2}$

134. $\frac{3\sqrt{2}}{2};\ -\frac{\sqrt{2}}{2}$ **135.** $\frac{\sqrt{3}}{2};\ -\frac{3\sqrt{3}}{2}$

136. $1-\sqrt{2};\ 1+\sqrt{2}$ **137.** $4-\sqrt{5};\ 4+\sqrt{5}$

138. $1-2\sqrt{3};\ 1+2\sqrt{3}$ **139.** $1-i;\ 1+i$

140. $2-i;\ 2+i$ **141.** $3-2i;\ 3+2i$

142. $1-i\sqrt{2};\ 1+i\sqrt{2}$ **143.** $2-i\sqrt{3};\ 2+i\sqrt{3}$

13.3 Solution of Quadratic Equations by Completing the Square

The quantity $(x+a)^2$ is a **perfect square**. Since $(x+a)^2 = x^2 + 2ax + a^2$, $x^2 + 2ax + a^2$ is a **perfect square trinomial**.
The expression $x^2 + 2ax$ is not a perfect square, yet if a^2 is added, the result is a perfect square trinomial.

Note that the term a^2 is the square of one-half the coefficient of x.
Likewise, $x^2 - 2ax$ can be made a perfect square trinomial by addition of a^2 since $(x-a)^2 = x^2 - 2ax + a^2$.

Note The term that when added to $x^2 + bx$ will make it a perfect square trinomial is $\left(\frac{b}{2}\right)^2 = \frac{b^2}{4}$.

EXAMPLE Find the term that when added to $x^2 + 6x$ will make it a perfect square trinomial and write it in factored form.

Solution One-half of the coefficient of x is $\frac{6}{2} = 3$.

The desired term is $(3)^2 = 9$:

$x^2 + 6x + 9 = (x+3)^2$

EXAMPLE Find the term that when added to $x^2 - \frac{3}{2}x$ will make it a perfect square trinomial and write it in factored form.

Solution One-half the coefficient of x is $\frac{1}{2}\left(-\frac{3}{2}\right) = -\frac{3}{4}$.

The desired term is $\left(-\frac{3}{4}\right)^2 = \frac{9}{16}$:

$$x^2 - \frac{3}{2}x + \frac{9}{16} = \left(x - \frac{3}{4}\right)^2$$

This method of completing the square enables us to put any quadratic equation in the form of a pure quadratic and thus find the solution set easily.

Consider the equation

$$ax^2 + bx + c = 0 \qquad a \neq 0$$

$$ax^2 + bx = -c$$

Divide both sides of the equation by a:

$$x^2 + \frac{b}{a}x = -\frac{c}{a}$$

The square of one-half the coefficient of x is $\left[\frac{1}{2}\left(\frac{b}{a}\right)\right]^2 = \left[\frac{b}{2a}\right]^2 = \frac{b^2}{4a^2}$.

Thus $\frac{b^2}{4a^2}$ is the term that will make the left side of the equation a perfect square trinomial. Add $\frac{b^2}{4a^2}$ to both sides of the equation:

$$x^2 + \frac{b}{a}x + \frac{b^2}{4a^2} = \frac{b^2}{4a^2} - \frac{c}{a}$$

Factor the left side of the equation:

$$\left(x + \frac{b}{2a}\right)^2 = \frac{b^2 - 4ac}{4a^2}$$

Hence

$$x + \frac{b}{2a} = \pm\sqrt{\frac{b^2 - 4ac}{4a^2}}$$

$$x = -\frac{b}{2a} \pm \sqrt{\frac{b^2 - 4ac}{4a^2}}$$

The solution set is $\left\{-\frac{b}{2a} + \sqrt{\frac{b^2 - 4ac}{4a^2}}, -\frac{b}{2a} - \sqrt{\frac{b^2 - 4ac}{4a^2}}\right\}$.

Note The method of factoring gives the solution set of a quadratic equation only when the quadratic polynomial can be factored.
The method of completing the square gives the solution set of any quadratic equation.

EXAMPLE Solve $2x^2 - 5x - 3 = 0$ by completing the square.

Solution

$$2x^2 - 5x - 3 = 0$$
$$2x^2 - 5x = 3$$
$$x^2 - \frac{5}{2}x = \frac{3}{2}$$

Add $\left[\frac{1}{2}\left(-\frac{5}{2}\right)\right]^2 = \left(-\frac{5}{4}\right)^2 = \frac{25}{16}$ to both sides of the equation:

$$x^2 - \frac{5}{2}x + \frac{25}{16} = \frac{3}{2} + \frac{25}{16}$$
$$\left(x - \frac{5}{4}\right)^2 = \frac{49}{16}$$
$$x - \frac{5}{4} = \pm\frac{7}{4}$$
$$x = \frac{5}{4} \pm \frac{7}{4}$$

The solution set is $\left\{\frac{5}{4} + \frac{7}{4}, \frac{5}{4} - \frac{7}{4}\right\}$ or $\left\{3, -\frac{1}{2}\right\}$.

EXAMPLE Solve $3x^2 - 2x + 1 = 0$ by completing the square.

Solution

$$3x^2 - 2x + 1 = 0$$
$$3x^2 - 2x = -1$$
$$x^2 - \frac{2}{3}x = -\frac{1}{3}$$
$$x^2 - \frac{2}{3}x + \frac{1}{9} = -\frac{1}{3} + \frac{1}{9}$$
$$\left(x - \frac{1}{3}\right)^2 = -\frac{2}{9}$$
$$x - \frac{1}{3} = \pm\sqrt{-\frac{2}{9}}$$
$$x = \frac{1}{3} \pm \frac{\sqrt{2}}{3}i$$

The solution set is $\left\{\frac{1}{3} + \frac{\sqrt{2}}{3}i, \frac{1}{3} - \frac{\sqrt{2}}{3}i\right\}$.

Exercise 13.3

Find the term that when added to each of the following will make it a perfect square trinomial and write it in factored form:

1. $x^2 + 2x$ **2.** $x^2 + 4x$ **3.** $x^2 + 8x$ **4.** $x^2 - 2x$
5. $x^2 - 4x$ **6.** $x^2 - 6x$ **7.** $x^2 + 3x$ **8.** $x^2 + 5x$
9. $x^2 + 9x$ **10.** $x^2 - x$ **11.** $x^2 - 3x$ **12.** $x^2 - 7x$

13. $x^2 + \frac{1}{2}x$ **14.** $x^2 + \frac{4}{3}x$ **15.** $x^2 + \frac{3}{2}x$ **16.** $x^2 - \frac{2}{3}x$

17. $x^2 - \frac{3}{5}x$ **18.** $x^2 - \frac{5}{7}x$ **19.** $x^2 + 2ax$ **20.** $x^2 + 3ax$

21. $x^2 + 5ax$ **22.** $x^2 - ax$ **23.** $x^2 - 7ax$ **24.** $x^2 - 9ax$

25. $x^2 + \frac{2a}{3}x$ **26.** $x^2 + \frac{3a}{5}x$ **27.** $x^2 - \frac{3a}{2}x$ **28.** $x^2 - \frac{5a}{3}x$

29. $x^2 + \frac{b}{2a}x$ **30.** $x^2 - \frac{b}{3a}x$ **31.** $x^2 + \sqrt{3}\,x$

32. $x^2 + \sqrt{2}\,x$ **33.** $x^2 + \sqrt{10}\,x$ **34.** $x^2 - \sqrt{6}\,x$

35. $x^2 - 3\sqrt{2}\,x$ **36.** $x^2 - 2\sqrt{3}\,x$ **37.** $x^2 - \frac{\sqrt{3}}{2}x$

38. $x^2 - \frac{\sqrt{2}}{3}x$ **39.** $x^2 - \frac{\sqrt{5}}{2}x$ **40.** $x^2 - \frac{2\sqrt{3}}{3}x$

Solve the following equations for x by completing the square:

41. $x^2 + 3x + 2 = 0$ **42.** $x^2 + 5x + 4 = 0$ **43.** $x^2 + x - 6 = 0$
44. $x^2 + 2x = 15$ **45.** $x^2 - x = 6$ **46.** $x^2 - 4x + 3 = 0$
47. $x^2 - 5x + 6 = 0$ **48.** $x^2 - 7x + 12 = 0$ **49.** $3x^2 - 7x + 2 = 0$
50. $2x^2 - x - 6 = 0$ **51.** $6x^2 + x - 2 = 0$ **52.** $3x^2 - 5x - 12 = 0$
53. $x^2 + 2x = 0$ **54.** $x^2 + 3x = 0$ **55.** $x^2 - 7x = 0$
56. $x^2 + x - 3 = 0$ **57.** $x^2 - x - 4 = 0$ **58.** $x^2 + 2x - 4 = 0$
59. $x^2 - 2x - 5 = 0$ **60.** $x^2 - 2x - 6 = 0$ **61.** $x^2 + 3x + 1 = 0$
62. $x^2 + 2x - 7 = 0$ **63.** $x^2 + 2x - 9 = 0$ **64.** $x^2 + 4x + 2 = 0$
65. $x^2 - 3x - 2 = 0$ **66.** $x^2 - 3x - 5 = 0$ **67.** $x^2 + 7x + 3 = 0$
68. $2x^2 + x - 2 = 0$ **69.** $2x^2 - x - 4 = 0$ **70.** $2x^2 + 3x - 1 = 0$
71. $2x^2 + 5x + 1 = 0$ **72.** $3x^2 + x - 3 = 0$ **73.** $3x^2 + 7x + 3 = 0$
74. $4x^2 + 9x + 4 = 0$ **75.** $5x^2 - 2x - 1 = 0$ **76.** $x^2 + x + 2 = 0$
77. $x^2 - 2x + 2 = 0$ **78.** $x^2 - 2x + 5 = 0$ **79.** $x^2 + 3x + 7 = 0$
80. $x^2 - 4x + 6 = 0$ **81.** $x^2 + 2x + 4 = 0$ **82.** $x^2 - 3x + 3 = 0$
83. $x^2 - x + 5 = 0$ **84.** $2x^2 - x + 1 = 0$ **85.** $2x^2 + x + 4 = 0$
86. $2x^2 - 5x + 4 = 0$ **87.** $3x^2 + 2x + 2 = 0$ **88.** $3x^2 - 4x + 2 = 0$
89. $5x^2 - 3x + 1 = 0$ **90.** $5x^2 + 4x + 1 = 0$

91. $6x^2 + 5x + 2 = 0$

92. $x^2 + ax - 3a^2 = 0$

93. $x^2 + 2ax - 4a^2 = 0$

94. $x^2 + 3ax - a^2 = 0$

95. $x^2 - 2ax - 5a^2 = 0$

96. $x^2 - 3ax - 2a^2 = 0$

97. $2x^2 - 6ax - 3a^2 = 0$

98. $2x^2 + 3ax - a^2 = 0$

99. $3x^2 + ax - 3a^2 = 0$

100. $x^2 + 3ax + 4a^2 = 0$

101. $x^2 - 3ax + 5a^2 = 0$

102. $2x^2 - 2ax + a^2 = 0$

103. $3x^2 - 4ax + 2a^2 = 0$

104. $3x^2 - 4ax + 4a^2 = 0$

105. $x^2 + \sqrt{3}x - 1 = 0$

106. $x^2 + \sqrt{2}x - 2 = 0$

107. $x^2 - \sqrt{5}x - 5 = 0$

108. $x^2 - 2\sqrt{3}x - 6 = 0$

109. $x^2 - \sqrt{7}x + 4 = 0$

110. $x^2 - 3\sqrt{2}x + 9 = 0$

111. $x^2 + \sqrt{6}x + 6 = 0$

112. $2x^2 - \sqrt{3}x + 6 = 0$

13.4 Solution of Quadratic Equations by the Quadratic Formula

When $ax^2 + bx + c = 0$, $a \neq 0$, is solved by completing the square we get

$$x = -\frac{b}{2a} \pm \sqrt{\frac{b^2 - 4ac}{4a^2}}$$

Simplifying the above equation we obtain

$$x = -\frac{b}{2a} \pm \frac{\sqrt{b^2 - 4ac}}{2a}$$

$$= \frac{-b \pm \sqrt{b^2 - 4ac}}{2a}$$

Hence if $ax^2 + bx + c = 0$, $a \neq 0$

$$x = \frac{-b \pm \sqrt{b^2 - 4ac}}{2a}$$

The expression for x is called the **quadratic formula**.
From the quadratic formula, the solution set of the quadratic equation $ax^2 + bx + c = 0$ is

$$\left\{ \frac{-b + \sqrt{b^2 - 4ac}}{2a}, \frac{-b - \sqrt{b^2 - 4ac}}{2a} \right\}$$

To solve a given quadratic equation by the quadratic formula, compare the given equation with the quadratic equation in standard form, $ax^2 + bx + c = 0$, to find the values of a, b, and c. Then substitute these values in the quadratic formula.

Note that a is the coefficient of x^2, b is the coefficient of x, and c is the constant term when the quadratic equation is written in standard form.

For the equation $2x^2 - 3x + 4 = 0$, $a = 2$, $b = -3$, and $c = 4$
For the equation $4x^2 - 5x = 0$, $a = 4$, $b = -5$, and $c = 0$
For the equation $3x^2 - 7 = 0$, $a = 3$, $b = 0$, and $c = -7$

EXAMPLE

Solve $x^2 - 2x = 15$, by the quadratic formula.

Solution

$$x^2 - 2x = 15$$

$$x^2 - 2x - 15 = 0$$

$$a = 1, \quad b = -2, \quad c = -15$$

Substitute 1 for a, -2 for b, and -15 for c in the quadratic formula:

$$x = \frac{-(-2) \pm \sqrt{(-2)^2 - 4(1)(-15)}}{2(1)}$$

$$= \frac{2 \pm \sqrt{4 + 60}}{2}$$

$$= \frac{2 \pm \sqrt{64}}{2} = \frac{2 \pm 8}{2}$$

$$= \frac{2(1 \pm 4)}{2}$$

$$= 1 \pm 4$$

Thus $r_1 = 1 + 4 = 5$

$r_2 = 1 - 4 = -3$

The solution set is $\{-3, 5\}$.

EXAMPLE

Solve $4x^2 - 12x + 7 = 0$ by the quadratic formula.

Solution

$$4x^2 - 12x + 7 = 0$$

$$a = 4, \quad b = -12, \quad c = 7$$

$$x = \frac{-(-12) \pm \sqrt{(-12)^2 - 4(4)(7)}}{2(4)}$$

$$= \frac{12 \pm \sqrt{144 - 112}}{8}$$

$$= \frac{12 \pm \sqrt{32}}{8}$$

$$= \frac{12 \pm 4\sqrt{2}}{8}$$

$$= \frac{4(3 \pm \sqrt{2})}{8}$$

$$= \frac{3 \pm \sqrt{2}}{2}$$

The solution set is $\left\{\frac{3 + \sqrt{2}}{2}, \frac{3 - \sqrt{2}}{2}\right\}$.

EXAMPLE

Solve $2x^2 - 4x + 5 = 0$ using the quadratic formula.

Solution

$2x^2 - 4x + 5 = 0$

$a = 2, \quad b = -4, \quad c = 5$

$$\begin{aligned} x &= \frac{-(-4) \pm \sqrt{(-4)^2 - 4(2)(5)}}{2(2)} \\ &= \frac{4 \pm \sqrt{16 - 40}}{4} \\ &= \frac{4 \pm \sqrt{-24}}{4} \\ &= \frac{4 \pm 2i\sqrt{6}}{4} \\ &= \frac{2 \pm i\sqrt{6}}{2} \\ &= 1 \pm \frac{\sqrt{6}}{2}i \end{aligned}$$

The solution set is $\left\{1 + \frac{\sqrt{6}}{2}i,\ 1 - \frac{\sqrt{6}}{2}i\right\}$.

EXAMPLE

Solve $2ix^2 - 9x + 35i = 0$, $i = \sqrt{-1}$, by the quadratic formula.

Solution

$2ix^2 - 9x + 35i = 0$

$a = 2i, \quad b = -9, \quad c = 35i$

$$\begin{aligned} x &= \frac{-(-9) \pm \sqrt{(-9)^2 - 4(2i)(35i)}}{2(2i)} \\ &= \frac{9 \pm \sqrt{81 - 280i^2}}{4i} \\ &= \frac{9 \pm \sqrt{81 + 280}}{4i} \\ &= \frac{9 \pm \sqrt{361}}{4i} = \frac{9 \pm 19}{4i} \end{aligned}$$

Thus

$$r_1 = \frac{9 + 19}{4i} = \frac{28}{4i} = \frac{7}{i} = \frac{7i}{i^2} = -7i$$

$$r_2 = \frac{9 - 19}{4i} = \frac{-10}{4i} = -\frac{5i}{2i^2} = \frac{5}{2}i$$

The solution set is $\left\{-7i, \frac{5}{2}i\right\}$.

Exercise 13.4

Solve the following equations for x using the quadratic formula:

1. $x^2 + 5x + 6 = 0$
2. $x^2 + 6x + 8 = 0$
3. $x^2 + 6x + 5 = 0$
4. $x^2 - 3x + 2 = 0$
5. $x^2 - 4x + 3 = 0$
6. $x^2 - 5x + 6 = 0$
7. $x^2 + 2x - 3 = 0$
8. $x^2 + x - 6 = 0$
9. $x^2 - 2x - 3 = 0$
10. $x^2 - 4x + 4 = 0$
11. $x^2 + 6x + 9 = 0$
12. $9x^2 + 12x + 4 = 0$
13. $4x^2 - 12x + 9 = 0$
14. $2x^2 - 5x + 2 = 0$
15. $2x^2 + x - 6 = 0$
16. $4x^2 + 8x + 3 = 0$
17. $x^2 + 5x = 0$
18. $x^2 + 7x = 0$
19. $x^2 - x = 0$
20. $x^2 - 3x = 0$
21. $x^2 - 4 = 0$
22. $x^2 - 36 = 0$
23. $x^2 - 2 = 0$
24. $x^2 - 3 = 0$
25. $x^2 - 12 = 0$
26. $x^2 - 18 = 0$
27. $x^2 + x - 1 = 0$
28. $x^2 + x - 3 = 0$
29. $x^2 + x - 4 = 0$
30. $x^2 + 4x + 1 = 0$
31. $x^2 + x - 5 = 0$
32. $x^2 - 7x + 1 = 0$
33. $x^2 - x - 3 = 0$
34. $x^2 - x - 7 = 0$
35. $x^2 - 2x - 4 = 0$
36. $x^2 - 2x - 7 = 0$
37. $x^2 + 2x - 6 = 0$
38. $2x^2 + x - 2 = 0$
39. $2x^2 - 3x - 1 = 0$
40. $3x^2 + 7x + 1 = 0$
41. $2x^2 + 5x + 1 = 0$
42. $3x^2 - 8x + 2 = 0$
43. $x^2 + 1 = 0$
44. $x^2 + 4 = 0$
45. $x^2 + 9 = 0$
46. $x^2 + 16 = 0$
47. $x^2 + 2 = 0$
48. $x^2 + 3 = 0$
49. $x^2 + 8 = 0$
50. $x^2 + 12 = 0$
51. $x^2 + 18 = 0$
52. $x^2 + 24 = 0$
53. $x^2 - 2x + 5 = 0$
54. $x^2 - 4x + 5 = 0$
55. $x^2 + 2x + 2 = 0$
56. $x^2 - 2x + 10 = 0$
57. $x^2 + 2x + 3 = 0$
58. $x^2 + 3x + 3 = 0$
59. $x^2 + x + 2 = 0$
60. $x^2 - 2x + 6 = 0$
61. $2x^2 - 2x + 1 = 0$
62. $2x^2 + 3x + 2 = 0$
63. $2x^2 + 4x + 3 = 0$
64. $3x^2 - 4x + 3 = 0$
65. $3x^2 - 2x + 7 = 0$
66. $3x^2 + 6x + 8 = 0$
67. $4x^2 + 5x + 2 = 0$
68. $4x^2 - 7x + 4 = 0$
69. $3x^2 + x + 3 = 0$
70. $x^2 + 3ax + 3a^2 = 0$
71. $x^2 - 2ax + 7a^2 = 0$
72. $3x^2 - 2ax + 2a^2 = 0$
73. $3x^2 - 5ax + 4a^2 = 0$
74. $ax^2 + 3x = abx + 3b$
75. $a^2x^2 + 4ax = 2abx + 8b$
76. $abx^2 + 3bx = 4ax + 12$
77. $2ax^2 + 3bx = 2abx + 3b^2$
78. $a^2x^2 - b^2x^2 + ax = 5bx + 6$
79. $4a^2x^2 - b^2x^2 - 2ax + 7bx = 12$
80. $\sqrt{2}x^2 - 3x + \sqrt{2} = 0$
81. $\sqrt{3}x^2 - 4x + \sqrt{3} = 0$
82. $x^2 - \sqrt{3}x + 3 = 0$
83. $x^2 - 2\sqrt{2}x + 4 = 0$
84. $x^2 + \sqrt{6}x + 9 = 0$
85. $x^2 + 2\sqrt{3}x + 7 = 0$
86. $x^2 - 2\sqrt{3}x = 1$
87. $3\sqrt{2}x^2 + x = 2\sqrt{2}$
88. $\sqrt{3}x^2 + \sqrt{13}x = \sqrt{3}$
89. $\sqrt{7}x^2 - 2\sqrt{2}x = \sqrt{7}$
90. $3x^2 + 4 = ix, \quad i = \sqrt{-1}$
91. $2x^2 - 5ix + 7 = 0, \quad i = \sqrt{-1}$
92. $3ix^2 + 2i = 5x, \quad i = \sqrt{-1}$
93. $2ix^2 + 6x + 3i = 0, \quad i = \sqrt{-1}$

13.5 Character of the Roots

The roots r_1 and r_2 of the quadratic equation $ax^2 + bx + c = 0, a \neq 0,$ are

$$r_1 = \frac{-b - \sqrt{b^2 - 4ac}}{2a} \quad \text{and} \quad r_2 = \frac{-b + \sqrt{b^2 - 4ac}}{2a}$$

When $a, b, c \in R$, the character of the roots of the quadratic equation, either real (different or equal) or complex, is determined by the radicand $b^2 - 4ac$. Since the quantity $b^2 - 4ac$ discriminates among the characters of the roots, it is called the **discriminant**.

1. When $b^2 - 4ac > 0$, there are two distinct real roots.
If $a, b, c \in Q$, then the roots are

rational if $b^2 - 4ac$ is a perfect square,
irrational if $b^2 - 4ac$ is not a perfect square.

2. When $b^2 - 4ac = 0$, there is one real root, a double root (or root of multiplicity two).

3. When $b^2 - 4ac < 0$, there are two imaginary roots, conjugate complex roots.

EXAMPLE

By examination of the discriminant, determine the nature of the roots of $x^2 - 8x + 15 = 0$.

Solution

$$x^2 - 8x + 15 = 0$$

$$a = 1, \quad b = -8, \quad c = 15$$

$$\begin{aligned} b^2 - 4ac &= (-8)^2 - 4(1)(15) \\ &= 64 - 60 = 4 > 0 \end{aligned}$$

Since 4 is a perfect square, and $a, b, c \in Q$, there are two distinct rational roots.

EXAMPLE

Without solving, determine the nature of the roots of $x^2 + 4x - 1 = 0$.

Solution

$$x^2 + 4x - 1 = 0$$

$$a = 1, \quad b = 4, \quad c = -1$$

$$\begin{aligned} b^2 - 4ac &= (4)^2 - 4(1)(-1) \\ &= 16 + 4 = 20 > 0 \end{aligned}$$

Since 20 is not a perfect square, there are two distinct irrational roots.

EXAMPLE

Without solving, determine the nature of the roots of $9x^2 - 24x + 16 = 0$.

Solution

$$9x^2 - 24x + 16 = 0$$

$$a = 9, \quad b = -24, \quad c = 16$$

$$\begin{aligned} b^2 - 4ac &= (-24)^2 - 4(9)(16) \\ &= 576 - 576 = 0 \end{aligned}$$

Since $b^2 - 4ac = 0$, there is one real root of multiplicity two.

EXAMPLE Without solving, determine the nature of the roots of $2x^2 + 5x + 4 = 0$.

Solution

$$2x^2 + 5x + 4 = 0$$

$$a = 2, \quad b = 5, \quad c = 4$$

$$\begin{aligned} b^2 - 4ac &= (5)^2 - 4(2)(4) \\ &= 25 - 32 \\ &= -7 < 0 \end{aligned}$$

Since $b^2 - 4ac < 0$, there are two imaginary roots.

EXAMPLE Find all values of k for which $2x^2 + 5x + k = 0$ has two real distinct roots.

Solution

$$\begin{aligned} b^2 - 4ac &= 25 - 4(2)(k) \\ &= 25 - 8k \end{aligned}$$

The equation has two real distinct roots if $b^2 - 4ac > 0$.

Thus $\quad 25 - 8k > 0, \quad$ that is, $\quad k < \dfrac{25}{8}$

EXAMPLE Find all values of k for which $x^2 - 6x + k - 1 = 0$ has one double root.

Solution

$$a = 1, \quad b = -6, \quad c = k - 1$$

$$\begin{aligned} b^2 - 4ac &= 36 - 4(k - 1) \\ &= 36 - 4k + 4 \\ &= 40 - 4k \end{aligned}$$

The equation has one double root if $b^2 - 4ac = 0$.

Thus $\quad 40 - 4k = 0, \quad$ that is, $\quad k = 10.$

EXAMPLE Find all values of k for which $(k + 1)x^2 - 8x + 1 = 0$ has two imaginary roots.

Solution

$$\begin{aligned} b^2 - 4ac &= 64 - 4(k + 1) \\ &= 64 - 4k - 4 \\ &= 60 - 4k \end{aligned}$$

The equation has two imaginary roots if $b^2 - 4ac < 0$.

Thus $\quad 60 - 4k < 0, \quad$ that is, $\quad k > 15$

Exercise 13.5

Determine the nature of the roots of the following equations:

1. $x^2 - 3x + 2 = 0$
2. $x^2 - 5x - 6 = 0$
3. $x^2 + 2x - 8 = 0$
4. $x^2 + 6x - 7 = 0$
5. $x^2 - 4x + 4 = 0$
6. $x^2 - 8x + 16 = 0$
7. $x^2 + 10x + 25 = 0$
8. $x^2 + 12x + 36 = 0$
9. $x^2 + 3x + 1 = 0$
10. $x^2 + 7x - 2 = 0$
11. $x^2 - 6x - 2 = 0$
12. $x^2 - 5x + 3 = 0$
13. $x^2 - 3x + 7 = 0$
14. $x^2 + x + 1 = 0$
15. $x^2 + 4x + 5 = 0$
16. $x^2 - 3x + 6 = 0$
17. $2x^2 - 5x - 9 = 0$
18. $3x^2 - 7x + 3 = 0$
19. $4x^2 - 4x + 1 = 0$
20. $9x^2 + 12x + 4 = 0$
21. $12x^2 + x - 6 = 0$
22. $6x^2 - 5x - 6 = 0$
23. $4x^2 + 2x + 1 = 0$
24. $3x^2 - 4x + 2 = 0$
25. $2x^2 + 3x = 0$
26. $5x^2 - 6x = 0$
27. $4x^2 - 49 = 0$
28. $9x^2 - 25 = 0$
29. $5x^2 - 6 = 0$
30. $3x^2 - 2 = 0$
31. $x^2 + 1 = 0$
32. $x^2 + 4 = 0$
33. $4x^2 + 9 = 0$
34. $7x^2 + 3 = 0$
35. $x^2 + \sqrt{2}\,x - 3 = 0$
36. $x^2 - \sqrt{3}\,x - 5 = 0$
37. $x^2 - \sqrt{6}\,x + 1 = 0$
38. $x^2 - \sqrt{11}\,x + 2 = 0$
39. $\sqrt{2}\,x^2 - 7x + 2\sqrt{2} = 0$
40. $\sqrt{3}\,x^2 + 8x + 2\sqrt{3} = 0$
41. $x^2 - \sqrt{7}\,x + 2 = 0$
42. $3x^2 - \sqrt{5}\,x + 1 = 0$
43. $\sqrt{2}\,x^2 + x + 3\sqrt{2} = 0$
44. $\sqrt{3}\,x^2 - 2x + \sqrt{3} = 0$
45. $3x^2 + 2\sqrt{6}\,x + 2 = 0$
46. $5x^2 + 2\sqrt{10}\,x + 2 = 0$
47. $2x^2 - 2\sqrt{14}\,x + 7 = 0$
48. $6x^2 - 2\sqrt{30}\,x + 5 = 0$

Find all values of k for which the nature of the roots of the given equation is as indicated:

49. $x^2 - 4x + k = 0$
two real distinct roots

50. $x^2 + 5x + k = 0$
two real distinct roots

51. $2x^2 - 3x + k = 0$
two real distinct roots

52. $3x^2 + 2x + k = 0$
two real distinct roots

53. $x^2 - kx - 2 = 0$
two real distinct roots

54. $x^2 + kx - 3 = 0$
two real distinct roots

55. $3x^2 + kx - 5 = 0$
two real distinct roots

56. $2x^2 - kx - 1 = 0$
two real distinct roots

57. $kx^2 + 7x - 1 = 0$
two real distinct roots

58. $kx^2 - 3x - 2 = 0$
two real distinct roots

59. $kx^2 - 2x + 3 = 0$
two real distinct roots

60. $kx^2 - 3x + 8 = 0$
two real distinct roots

61. $x^2 - 2x + k = 0$
one double root

62. $x^2 + 5x + k = 0$
one double root

63. $x^2 + 3x + k = 1$
one double root

64. $x^2 - 6x + k = 2$
one double root

65. $x^2 + kx + 7 = 0$
one double root

66. $x^2 + kx + 8 = 0$
one double root

67. $2x^2 + kx - 3 = 0$
one double root

68. $3x^2 + kx - 10 = 0$
one double root

69. $kx^2 + 3x + 2 = 0$
one double root

70. $kx^2 + 5x + 1 = 0$
one double root

71. $kx^2 + 6x - 2 = 0$
one double root

72. $kx^2 + \sqrt{3}\,x - 1 = 0$
one double root

73. $x^2 + 2x + k = 0$
two imaginary roots

74. $x^2 - 3x + k = 0$
two imaginary roots

75. $x^2 - 4x - 3k = 0$
two imaginary roots

76. $2x^2 + 5x - k = 0$
two imaginary roots

77. $x^2 - kx - 1 = 0$
two imaginary roots

78. $x^2 - kx - 5 = 0$
two imaginary roots

79. $5x^2 + kx - 4 = 0$
two imaginary roots

80. $3x^2 + kx - 8 = 0$
two imaginary roots

81. $kx^2 - 2x + 1 = 0$
two imaginary roots

82. $kx^2 + 6x + 5 = 0$
two imaginary roots

83. $kx^2 + 5\sqrt{2}\,x - 3 = 0$
two imaginary roots

84. $kx^2 + 2\sqrt{3}\,x - 7 = 0$
two imaginary roots

13.6 Properties of the Roots

The roots r_1 and r_2 of the quadratic equation $ax^2 + bx + c = 0$ are

$$r_1 = \frac{-b + \sqrt{b^2 - 4ac}}{2a} \qquad \text{and} \qquad r_2 = \frac{-b - \sqrt{b^2 - 4ac}}{2a}$$

When the roots are added, we get

$$\begin{aligned} r_1 + r_2 &= \frac{-b + \sqrt{b^2 - 4ac}}{2a} + \frac{-b - \sqrt{b^2 - 4ac}}{2a} \\ &= \frac{-2b}{2a} \\ &= -\frac{b}{a} \end{aligned}$$

Hence $\quad r_1 + r_2 = -\dfrac{b}{a}$

That is, the sum of the roots of a quadratic equation is the negative of the quotient obtained by dividing the coefficient of x by the coefficient of x^2.

EXAMPLE Without solving, find the sum of the roots of the equation $2x^2 - 7x + 15 = 0$.

Solution

$$2x^2 - 7x + 15 = 0$$

$$a = 2, \quad b = -7$$

$$\text{The sum of the roots of the equation} = -\frac{b}{a} = -\frac{-7}{2} = \frac{7}{2}.$$

EXAMPLE Determine the value of m so that the sum of the roots of the equation $(m + 1)x^2 - (3m - 1)x - 21 = 0$ is 2.

Solution

$$(m + 1)x^2 - (3m - 1)x - 21 = 0$$

$$a = m + 1, \quad b = -(3m - 1), \quad c = -21$$

$$-\frac{b}{a} = -\frac{-(3m - 1)}{m + 1} = 2$$

or

$$3m - 1 = 2(m + 1)$$

$$3m - 1 = 2m + 2$$

Hence $m = 3$

When the roots are multiplied we get

$$r_1 \cdot r_2 = \frac{-b + \sqrt{b^2 - 4ac}}{2a} \cdot \frac{-b - \sqrt{b^2 - 4ac}}{2a}$$

$$= \frac{(-b + \sqrt{b^2 - 4ac})(-b - \sqrt{b^2 - 4ac})}{4a^2}$$

$$= \frac{b^2 - (b^2 - 4ac)}{4a^2}$$

$$= \frac{4ac}{4a^2} = \frac{c}{a}$$

Hence $r_1 \cdot r_2 = \frac{c}{a}$

That is, the product of the roots of a quadratic equation is equal to the quotient obtained by dividing the constant term by the coefficient of x^2.

Note When we write the equation $ax^2 + bx + c = 0$ in the form

$$x^2 + \frac{b}{a}x + \frac{c}{a} = 0$$

the sum of the roots is the negative (additive inverse) of the coefficient of x and the product of the roots is the constant term.

Remark Given the sum and the product of the roots of a quadratic equation, we can find that equation.

For example, if the sum of the roots of a quadratic equation is $\frac{2}{3}$ and the product of the roots is $\frac{1}{4}$, we have

$$x^2 - \frac{2}{3}x + \frac{1}{4} = 0 \qquad \text{or} \qquad 12x^2 - 8x + 3 = 0$$

EXAMPLE Without solving, find the sum and the product of the roots of the equation $3x^2 - 2x - 7 = 0$.

Solution

$$3x^2 - 2x - 7 = 0$$

$$a = 3, \quad b = -2, \quad c = -7$$

$$\text{The sum of the roots} = -\frac{b}{a} = -\frac{-2}{3} = \frac{2}{3}.$$

$$\text{The product of the roots} = \frac{c}{a} = \frac{-7}{3} = -\frac{7}{3}.$$

EXAMPLE Find the value of m in the equation $4x^2 + 4x + m = 0$ so that one root exceeds the other by 4.

Solution

$$4x^2 + 4x + m = 0$$

$$a = 4, \quad b = 4, \quad c = m$$

$$\text{The sum of the roots} = -\frac{b}{a} = -\frac{4}{4} = -1.$$

$$\text{The product of the roots} = \frac{c}{a} = \frac{m}{4}.$$

Let one root be r; the other root is thus $(r + 4)$:

$$r + (r + 4) = -1$$

$$2r + 4 = -1$$

$$r = -\frac{5}{2}$$

and $$r + 4 = -\frac{5}{2} + 4 = \frac{3}{2}$$

Thus $$\left(-\frac{5}{2}\right)\left(\frac{3}{2}\right) = \frac{m}{4}$$

Hence $$m = -15$$

EXAMPLE

Find the value of m in the equation $2x^2 - 15x - 4m - 1 = 0$ so that one root is $\frac{3}{2}$ of the other.

Solution

$2x^2 - 15x - 4m - 1 = 0$

$a = 2, \quad b = -15, \quad c = -4m - 1$

The sum of the roots $= -\frac{b}{a} = -\frac{-15}{2} = \frac{15}{2}$.

The products of the roots $= \frac{c}{a} = \frac{-4m-1}{2}$.

Let one root be r; the other root is thus $\frac{3}{2}r$

$$r + \frac{3}{2}r = \frac{15}{2}$$

or $$r = 3$$

and $$\frac{3}{2}r = \frac{3}{2}(3) = \frac{9}{2}$$

Thus $$3\left(\frac{9}{2}\right) = \frac{-4m-1}{2}$$

or $$27 = -4m - 1$$

Hence $$m = -7$$

EXAMPLE

Find a quadratic equation whose roots r_1 and r_2 satisfy

$$r_1 + r_2 = \frac{14}{3} \quad \text{and} \quad r_1 - r_2 = \frac{22}{3}$$

Solution

We can find a quadratic equation if we know r_1 and r_2. Solving the two equations

$$r_1 + r_2 = \frac{14}{3} \quad \text{and} \quad r_1 - r_2 = \frac{22}{3}$$

for r_1 and r_2 we get

$$r_1 = 6 \quad \text{and} \quad r_2 = -\frac{4}{3}$$

A quadratic equation is

$$(x - 6)\left(x + \frac{4}{3}\right) = 0, \quad \text{that is,} \quad x^2 - \frac{14}{3}x - 8 = 0$$

or $3x^2 - 14x - 24 = 0$

EXAMPLE

Find a quadratic equation whose roots r_1 and r_2 satisfy

$$r_1 + r_2 = 1 \quad \text{and} \quad \frac{r_1}{r_2} = 2$$

Solution

We can find a quadratic equation if we know the sum of the roots and the product of the roots. Since the sum of the roots is given, we need to calculate the product of the roots.

From $\frac{r_1}{r_2} = 2$ we get $r_1 = 2r_2$.

Substituting $2r_2$ for r_1 in $r_1 + r_2 = 1$ we get

$$2r_2 + r_2 = 1 \quad \text{or} \quad r_2 = \frac{1}{3}$$

Hence $r_1 = \frac{2}{3}$ and $r_1 r_2 = \frac{2}{9}$

A quadratic equation is

$$x^2 - x + \frac{2}{9} = 0 \quad \text{or} \quad 9x^2 - 9x + 2 = 0$$

EXAMPLE

Find a quadratic equation whose roots r_1 and r_2 satisfy

$$r_1 - r_2 = \frac{13}{3} \quad \text{and} \quad r_1 r_2 = -4$$

Solution

We can find a quadratic equation if we know r_1 and r_2.

From $r_1 - r_2 = \frac{13}{3}$ we get $r_1 = r_2 + \frac{13}{3}$.

Substituting $r_2 + \frac{13}{3}$ for r_1 in $r_1 r_2 = -4$ we get

$$r_2\left(r_2 + \frac{13}{3}\right) = -4$$

$$r_2^{\ 2} + \frac{13}{3} r_2 = -4$$

or

$$3r_2^{\ 2} + 13r_2 + 12 = 0$$

$$(3r_2 + 4)(r_2 + 3) = 0$$

Hence $r_2 = -\frac{4}{3}$ or $r_2 = -3$

When $r_2 = -\frac{4}{3}$, then $r_1 = -\frac{4}{3} + \frac{13}{3} = 3$, and we get the equation

$$(x - 3)\left(x + \frac{4}{3}\right) = 0 \quad \text{or} \quad 3x^2 - 5x - 12 = 0$$

When $r_2 = -3$, then $r_1 = -3 + \frac{13}{3} = \frac{4}{3}$, and we get the equation

$$\left(x - \frac{4}{3}\right)(x + 3) = 0 \quad \text{or} \quad 3x^2 + 5x - 12 = 0$$

Exercise 13.6

Without solving, find the sum and product of the roots of each of the following equations:

1. $x^2 + 2x + 3 = 0$
2. $x^2 + 5x + 7 = 0$
3. $x^2 - 4x + 8 = 0$
4. $x^2 - 6x + 10 = 0$
5. $x^2 + 4x - 5 = 0$
6. $x^2 + 3x - 2 = 0$
7. $x^2 - x = 4$
8. $x^2 - 3x = 6$
9. $3x^2 + x + 2 = 0$
10. $2x^2 + x + 5 = 0$
11. $3x^2 + 1 = 2x$
12. $4x^2 + 7 = x$
13. $5x^2 + 6x = 1$
14. $2x^2 + x = 4$
15. $9x^2 = 3x + 5$
16. $4x^2 = 8x + 11$
17. $\sqrt{2}x^2 + x = 3$
18. $x^2 + 2\sqrt{3}x + 4 = 0$
19. $x^2 = \sqrt{3}(x - \sqrt{2})$
20. $x^2 = \sqrt{2}(3 - x)$
21. $2x^2 + x = k$
22. $5x^2 = kx + 4$
23. $kx^2 + x + 5 = 0$
24. $x^2 + k = 2x + 1$
25. $2x^2 + kx = 3 - x$
26. $k(x^2 + x) + 3x^2 + 1 = 0$
27. $x(mx + n) = k$
28. $mx(x - 1) + nx(x - 1) + m + 5 = 0$
29. $m(x^2 - mx + 1) = 2n(x^2 - 2nx - 1)$
30. $m(x^2 - x - 1) + x(x + m^2 + 2) + 1 = 0$

Find the value of m in each of the following equations for the given condition:

31. $x^2 - 4x + m = 0$
one root is $2 + \sqrt{3}$

32. $x^2 + mx - 1 = 0$
one root is $1 - \sqrt{2}$

33. $x^2 - 6x + m = 0$
one root is $3 + i$

34. $x^2 + mx + 5 = 0$
one root is $1 - 2i$

35. $x^2 - 2x + m = 0$
one root is $1 - i\sqrt{2}$

36. $x^2 + mx + 7 = 0$
one root is $2 + i\sqrt{3}$

37. $mx^2 + (6 - m)x = 12$
the sum of the roots is 4

38. $2mx^2 + (m + 5)x = 18$
the sum of the roots is -3

39. $(m+1)x^2-(7m-1)x+21=0$
the sum of the roots is 5

40. $(2m-1)x^2+3(m+1)x+5=0$
the sum of the roots is -2

41. $x^2-8x+m-3=0$
the difference of the roots is 10

42. $x^2-x+m+1=0$
the difference of the roots is 7

43. $16x^2+40x-(2m+1)=0$
the difference of the roots is 3

44. $x^2+6x+m=0$
one root exceeds the other by 2

45. $x^2+3x+m=0$
one root exceeds the other by 7

46. $x^2-8x+m=0$
one root exceeds the other by 6

47. $9x^2+18x+m=0$
one root exceeds the other by $\frac{10}{3}$

48. $2x^2-9x+4m+1=0$
one root is twice the other

49. $2x^2-5x+m+7=0$
one root is $\frac{2}{3}$ of the other

50. $6x^2+7x+m+3=0$
one root is $\frac{3}{4}$ of the other

51. $6x^2-7x+m=1$
one root is $\frac{4}{3}$ of the other

52. $4x^2-9x+m=3$
one root is $\frac{1}{8}$ of the other

53. $2mx^2-19x+m+1=0$
the product of the roots is $\frac{3}{5}$

54. $mx^2-4x-3m+5=0$
the product of the roots is $\left(-\frac{4}{3}\right)$

55. $(m+1)x^2+29x-m-5=0$
the product of the roots is $\left(-\frac{3}{2}\right)$

56. $4x^2-8x-m=0$
the quotient of the roots is $\frac{1}{3}$

57. $16x^2-8x-m+6=0$
the quotient of the roots is (-3)

58. $4x^2+5x+2m+3=0$
the quotient of the roots is 4

Find a quadratic equation whose roots r_1 and r_2 satisfy the given conditions:

59. $r_1+r_2=-\frac{4}{3},\quad r_1r_2=\frac{3}{2}$

60. $r_1+r_2=-1,\quad r_1r_2=-\frac{3}{4}$

61. $r_1+r_2=-\frac{5}{3},\quad r_1r_2=-4$

62. $r_1+r_2=2,\quad r_1r_2=\frac{7}{5}$

63. $r_1+r_2=2,\quad r_1-r_2=2\sqrt{3}$

64. $r_1+r_2=\frac{7}{4},\quad r_1-r_2=\frac{13}{4}$

65. $r_1+r_2=-2,\quad r_1-r_2=-8$

66. $r_1+r_2=-\frac{1}{6},\quad r_1-r_2=\frac{5}{6}$

67. $r_1+r_2=4,\quad \frac{r_1}{r_2}=-2$

68. $r_1+r_2=-10,\quad \frac{r_1}{r_2}=-6$

69. $r_1-r_2=\frac{1}{3},\quad r_1r_2=\frac{10}{3}$

70. $r_1-r_2=9,\quad r_1r_2=-14$

71. $r_1 - r_2 = 5, \quad r_1 r_2 = -6$

72. $r_1 - r_2 = 1, \quad r_1 r_2 = \frac{3}{4}$

73. $r_1 - r_2 = 1, \quad \frac{r_1}{r_2} = 3$

74. $r_1 - r_2 = \frac{5}{2}, \quad \frac{r_1}{r_2} = -4$

75. $r_1 = r_2, \quad r_1 r_2 = \frac{9}{4}$

76. $r_1 = r_2, \quad r_1 r_2 = \frac{1}{4}$

77. $r_1 r_2 = 18, \quad \frac{r_1}{r_2} = \frac{9}{8}$

78. $r_1 r_2 = -1, \quad \frac{r_1}{r_2} = -36$

79. $r_1 r_2 = -1, \quad \frac{r_1}{r_2} = -\frac{9}{16}$

80. $r_1 r_2 = -2, \quad \frac{r_1}{r_2} = -\frac{8}{9}$

13.7 Equations That Lead to Quadratic Equations

We saw that if P and Q are polynomials and $P \cdot Q = 0$, then $P = 0$ or $Q = 0$. This theorem can be extended to the product of any number of polynomials. Given a polynomial equation of degree higher than two, if we can express the polynomial as the product of linear and quadratic polynomials, we can find the roots of the equation.

EXAMPLE Find all roots of $x^6 - 64 = 0$.

Solution

$$x^6 - 64 = 0$$
$$(x^3 + 8)(x^3 - 8) = 0$$
$$(x + 2)(x^2 - 2x + 4)(x - 2)(x^2 + 2x + 4) = 0$$

Equating each factor to zero, we get

$$x + 2 = 0 \qquad x = -2$$

$$x^2 - 2x + 4 = 0 \qquad x = \frac{2 \pm \sqrt{4 - 16}}{2} = 1 \pm i\sqrt{3}$$

$$x - 2 = 0 \qquad x = 2$$

$$x^2 + 2x + 4 = 0 \qquad x = \frac{-2 \pm \sqrt{4 - 16}}{2} = -1 \pm i\sqrt{3}$$

The solution set is $\{-2, 2, 1 + i\sqrt{3}, 1 - i\sqrt{3}, -1 + i\sqrt{3}, -1 - i\sqrt{3}\}$.

Note Any polynomial equation of degree $n \geq 1$ has exactly n roots provided that a root of multiplicity k is counted as k roots.

EXAMPLE Find all roots of $x + x^{\frac{1}{2}} - 6 = 0$.

Solution $x + x^{\frac{1}{2}} - 6$ is of the form $u^2 + u - 6$ whose factors are $(u + 3)(u - 2)$

Hence $x + x^{\frac{1}{2}} - 6 = \left(x^{\frac{1}{2}} + 3\right)\left(x^{\frac{1}{2}} - 2\right) = 0$

$x^{\frac{1}{2}} + 3 = 0,$ that is, $x^{\frac{1}{2}} = -3,$ from which $x = 9$

or $x^{\frac{1}{2}} - 2 = 0,$ that is, $x^{\frac{1}{2}} = 2,$ from which $x = 4$

When $x = 9$, we have

$x + x^{\frac{1}{2}} - 6 = 9 + 3 - 6 = 6 \neq 0$

Thus 9 is not a root.

When $x = 4$ we have

$x + x^{\frac{1}{2}} - 6 = 4 + 2 - 6 = 0$

Thus 4 is a root.

The solution set is $\{4\}$.

EXAMPLE Find all roots of $x^{\frac{2}{3}} - x^{\frac{1}{3}} - 2 = 0$.

Solution $x^{\frac{2}{3}} - x^{\frac{1}{3}} - 2$ is of the form $u^2 - u - 2$ whose factors are $(u - 2)(u + 1)$.

Hence $x^{\frac{2}{3}} - x^{\frac{1}{3}} - 2 = \left(x^{\frac{1}{3}} - 2\right)\left(x^{\frac{1}{3}} + 1\right) = 0$

$x^{\frac{1}{3}} - 2 = 0,$ that is, $x^{\frac{1}{3}} = 2,$ from which $x = 8$

or $x^{\frac{1}{3}} + 1 = 0,$ that is, $x^{\frac{1}{3}} = -1,$ from which $x = -1$

When $x = 8$ we have

$x^{\frac{2}{3}} - x^{\frac{1}{3}} - 2 = 4 - 2 - 2 = 0$

Thus 8 is a root.

When $x = -1$ we have

$x^{\frac{2}{3}} - x^{\frac{1}{3}} - 2 = 1 + 1 - 2 = 0$

Thus -1 is a root.

The solution set is $\{-1, 8\}$.

An equation involving fractions can be put in a simpler form when both sides of the equation are multiplied by the LCD of all the fractions in the equation.

However, when an equation is multiplied by a polynomial in the variable, the resulting equation may not be equivalent to the original equation. This means that the resulting equation may have roots that do not satisfy the original equation. The values obtained for the variable that satisfy the original equation are the roots of the equation.

EXAMPLE

Solve $\dfrac{18}{x^2-4x-5} - \dfrac{14}{x^2-3x-4} = \dfrac{x}{x^2-9x+20}$.

Solution

Begin by factoring the denominators:

$$\frac{18}{(x-5)(x+1)} - \frac{14}{(x-4)(x+1)} = \frac{x}{(x-5)(x-4)}$$

Multiply both sides of the equation by the LCD, $(x-5)(x+1)(x-4)$, and simplify:

$$18(x-4) - 14(x-5) = x(x+1)$$
$$18x - 72 - 14x + 70 = x^2 + x$$
$$x^2 - 3x + 2 = 0$$
$$(x-1)(x-2) = 0$$

Thus $x - 1 = 0$, that is, $x = 1$

or $x - 2 = 0$, that is, $x = 2$

The solution set is $\{1, 2\}$. The check is left as an exercise.

EXAMPLE

Solve $\sqrt{2x+6} = 1 + \sqrt{7-2x}$.

Solution

Square both sides of the equation $(\sqrt{2x+6})^2 = (1+\sqrt{7-2x})^2$.
Note that since we are squaring, the resulting equation may not be equivalent to the original equation.

Since $(1+\sqrt{7-2x})^2 = 1 + 2\sqrt{7-2x} + 7 - 2x$

then $2x + 6 = 1 + 2\sqrt{7-2x} + 7 - 2x$

$$2\sqrt{7-2x} = 4x - 2$$
$$\sqrt{7-2x} = 2x - 1$$
$$(\sqrt{7-2x})^2 = (2x-1)^2$$
$$7 - 2x = 4x^2 - 4x + 1$$
$$4x^2 - 2x - 6 = 0$$
$$2x^2 - x - 3 = 0$$
$$(2x-3)(x+1) = 0$$

Thus $\quad 2x - 3 = 0, \quad$ that is, $\quad x = \frac{3}{2}$

or $\quad x + 1 = 0, \quad$ that is, $\quad x = -1$

We must check these values in the original equation.

For $x = \frac{3}{2}$:

Left Side	*Right Side*
$= \sqrt{2\left(\frac{3}{2}\right) + 6}$	$= 1 + \sqrt{7 - 2\left(\frac{3}{2}\right)}$
$= \sqrt{3 + 6}$	$= 1 + \sqrt{7 - 3}$
$= \sqrt{9}$	$= 1 + \sqrt{4}$
$= 3$	$= 1 + 2$
	$= 3$

Hence $\frac{3}{2}$ is a root of the equation.

For $x = -1$:

Left Side	*Right Side*
$= \sqrt{2(-1) + 6}$	$= 1 + \sqrt{7 - 2(-1)}$
$= \sqrt{-2 + 6}$	$= 1 + \sqrt{7 + 2}$
$= \sqrt{4}$	$= 1 + \sqrt{9}$
$= 2$	$= 1 + 3$
	$= 4$

Hence -1 is not a root of the equation.

Thus the solution set is $\left\{\frac{3}{2}\right\}$.

Exercise 13.7

Find all roots of the following equations:

1. $2x^3 + 5x^2 - 12x = 0$ **2.** $3x^3 - 11x^2 - 4x = 0$

3. $6x^3 - 11x^2 - 10x = 0$ **4.** $6x^3 - x^2 - 2x = 0$

5. $x^3 + 1 = 0$ **6.** $x^3 + 8 = 0$ **7.** $x^3 - 27 = 0$

8. $x^3 - 64 = 0$ **9.** $x^4 - 1 = 0$ **10.** $x^4 - 16 = 0$

11. $x^4 - 4 = 0$ **12.** $x^4 - 9 = 0$ **13.** $x^4 + x^2 = 2$

14. $x^4 + x^2 = 20$ **15.** $x^4 - 6x^2 = 27$

16. $x^4 - 14x^2 - 32 = 0$ **17.** $36x^4 + 5x^2 - 1 = 0$

18. $x^4 - 13x^2 + 4 = 0$ **19.** $x^4 - 21x^2 + 4 = 0$

20. $x^4 - 34x^2 + 1 = 0$ **21.** $x^4 - 3x^2 + 1 = 0$

22. $x^4 - 18x^2 + 1 = 0$ **23.** $x^4 - 3x^2 + 9 = 0$ **24.** $x^4 - 22x^2 + 9 = 0$

25. $x^6 - 1 = 0$ **26.** $x^6 - 729 = 0$ **27.** $216x^6 + 19x^3 = 1$

28. $x - 6x^{\frac{1}{2}} + 5 = 0$ **29.** $x - 2x^{\frac{1}{2}} - 3 = 0$ **30.** $x - 2x^{\frac{1}{2}} - 8 = 0$

31. $2x + 9x^{\frac{1}{2}} + 4 = 0$ **32.** $9x - 9x^{\frac{1}{2}} - 4 = 0$ **33.** $12x - 7x^{\frac{1}{2}} + 1 = 0$

34. $x^{\frac{2}{3}} + x^{\frac{1}{3}} - 2 = 0$ **35.** $x^{\frac{2}{3}} + 4x^{\frac{1}{3}} + 3 = 0$ **36.** $x^{\frac{2}{3}} - 5x^{\frac{1}{3}} + 6 = 0$

37. $2x^{\frac{2}{3}} + 7x^{\frac{1}{3}} - 4 = 0$ **38.** $2x^{\frac{2}{3}} - 7x^{\frac{1}{3}} - 4 = 0$ **39.** $6x^{\frac{2}{3}} - 5x^{\frac{1}{3}} - 6 = 0$

Solve the following equations:

40. $\dfrac{1}{3x + 2} = 4x - 1$ **41.** $\dfrac{5}{2x + 1} = 3x + 2$

42. $\dfrac{9}{4x + 3} = 4 - 2x$ **43.** $\dfrac{x + 16}{4x + 1} = x + 1$

44. $\dfrac{3}{x + 1} + \dfrac{2}{x + 3} = 2$ **45.** $\dfrac{5}{x + 1} + \dfrac{4}{2x - 1} = 3$

46. $\dfrac{5}{x + 1} - \dfrac{2}{3x - 1} = 1$ **47.** $\dfrac{5}{x - 3} - \dfrac{6}{x - 1} = 1$

48. $\dfrac{4}{x - 1} - \dfrac{1}{2x - 5} = 1$ **49.** $\dfrac{7}{x + 2} + \dfrac{8}{2x - 1} = 3$

50. $\dfrac{x}{x^2 + 2x - 3} + \dfrac{2}{x^2 + 5x - 6} = \dfrac{2}{x^2 + 9x + 18}$

51. $\dfrac{x - 6}{x^2 - x - 2} + \dfrac{8}{x^2 - 3x + 2} + \dfrac{4}{x^2 - 1} = 0$

52. $\dfrac{x - 3}{x^2 + 2x - 3} + \dfrac{1}{x^2 + x - 2} = \dfrac{1}{x^2 + 5x + 6}$

53. $\dfrac{x + 8}{x^2 + 2x - 8} + \dfrac{1}{x^2 + 5x + 4} = \dfrac{9}{x^2 - x - 2}$

54. $\dfrac{x - 2}{x^2 + x - 20} - \dfrac{2}{x^2 + 8x + 15} = \dfrac{1}{x^2 - x - 12}$

55. $\dfrac{x + 8}{x^2 - 2x - 8} - \dfrac{1}{x^2 + 5x + 6} = \dfrac{13}{x^2 - x - 12}$

56. $\dfrac{x + 4}{x^2 + 4x - 12} - \dfrac{2}{x^2 - x - 2} = \dfrac{10}{x^2 + 7x + 6}$

57. $\dfrac{x - 2}{x^2 - 8x + 15} + \dfrac{3}{x^2 - 2x - 3} = \dfrac{7}{x^2 - 4x - 5}$

58. $\dfrac{3x}{x^2+x-6}+\dfrac{2}{x^2+4x+3}=\dfrac{x}{x^2-x-2}$

59. $\dfrac{5x}{x^2+x-20}-\dfrac{4x}{x^2+2x-15}=\dfrac{3}{x^2-7x+12}$

60. $\dfrac{x-4}{x^2-4}+\dfrac{10}{2x^2+5x+2}=\dfrac{2}{2x^2-3x-2}$

61. $\dfrac{x+4}{2x^2-x-3}-\dfrac{x+3}{3x^2+2x-1}=\dfrac{7}{6x^2-11x+3}$

62. $\dfrac{3x+1}{3x^2-4x-4}+\dfrac{9}{9x^2-4}=\dfrac{2x-2}{3x^2-8x+4}$

63. $\dfrac{x-11}{2x^2+5x-3}+\dfrac{2}{6x^2-x-1}=\dfrac{x-15}{3x^2+10x+3}$

64. $\dfrac{x+2}{2x^2-5x-3}+\dfrac{1}{8x^2+10x+3}=\dfrac{x+6}{4x^2-9x-9}$

65. $\dfrac{x+1}{2x^2+x-3}+\dfrac{2}{6x^2+7x-3}=\dfrac{x+1}{3x^2-4x+1}$

66. $\dfrac{2x}{2x^2-x-6}+\dfrac{6}{4x^2+4x-3}=\dfrac{x-1}{2x^2-5x+2}$

67. $\dfrac{2x-1}{x^2-x-2}+\dfrac{x+3}{x^2-2x-3}=\dfrac{2x-5}{x^2-5x+6}$

68. $\dfrac{2x+1}{x^2+x-2}+\dfrac{x-5}{x^2+6x+8}=\dfrac{2x+3}{x^2+3x-4}$

69. $\dfrac{2x-10}{x^2-x-12}+\dfrac{2x-6}{x^2-6x+8}=\dfrac{3x-1}{x^2+x-6}$

70. $\sqrt{5x-11}=1+\sqrt{3x-8}$

71. $\sqrt{7x+23}=2+\sqrt{3x+7}$

72. $\sqrt{3x+10}=1+\sqrt{x+7}$

73. $\sqrt{3x+16}=\sqrt{5x+21}-1$

74. $\sqrt{19-12x}=1-\sqrt{16-8x}$

75. $\sqrt{2-3x}=1-\sqrt{3x+3}$

76. $\sqrt{11x+58}=5+\sqrt{x+3}$

77. $\sqrt{5x+31}=4+\sqrt{x+3}$

78. $\sqrt{3x-2}=2+\sqrt{2-x}$

79. $\sqrt{8x+5}=3+\sqrt{2-4x}$

80. $\sqrt{3+3x}=\sqrt{9x-2}-1$

81. $\sqrt{10x+56}=4+\sqrt{2x+8}$

82. $\sqrt{16x-23}=2+\sqrt{8x-15}$

83. $\sqrt{10x+20}=1+\sqrt{6x+13}$

84. $\sqrt{3x+6}=1+\sqrt{x+3}$

85. $\sqrt{x-3}+\sqrt{x+5}=2\sqrt{2x-4}$

86. $\sqrt{x-2}+\sqrt{x+3}=\sqrt{6x-11}$

87. $\sqrt{x-1}+\sqrt{x+2}=\sqrt{8x-7}$

88. $\sqrt{x+6}+\sqrt{x+11}=\sqrt{6x+37}$

89. $\sqrt{x-5}+\sqrt{x-2}=\sqrt{6x-27}$

90. $\sqrt{x-6}+\sqrt{x+2}=\sqrt{6x-26}$

91. $\sqrt{x-4}+2\sqrt{x+1}=\sqrt{13x-40}$

92. $\sqrt{x+5}+\sqrt{1-6x}=\sqrt{24-3x}$

93. $\sqrt{x+4}+\sqrt{3-2x}=\sqrt{x+19}$

94. $\sqrt{x+2}-\sqrt{x+7}=\sqrt{5-2x}$

95. $\sqrt{x-1}-\sqrt{x+8}=\sqrt{11-2x}$

13.8 Word Problems

EXAMPLE

The difference between two natural numbers is 6, and the difference between their reciprocals is $\frac{2}{105}$. Find the two numbers.

Solution

First Number   *Second Number*

x $x + 6$

$$\frac{1}{x} - \frac{1}{x+6} = \frac{2}{105}$$

Note that $\frac{1}{x} > \frac{1}{x+6}$.

$$105(x + 6) - 105x = 2x(x + 6)$$
$$105x + 630 - 105x = 2x^2 + 12x$$
$$2x^2 + 12x - 630 = 0$$
$$x^2 + 6x - 315 = 0$$
$$(x + 21)(x - 15) = 0$$

Thus $x + 21 = 0$, that is, $x = -21$

or $x - 15 = 0$, that is, $x = 15$

The two numbers are 15 and $15 + 6 = 21$.

We reject -21 since it is not a natural number.

EXAMPLE

A man did a job for \$336. It took him 4 hours longer than he expected and so he earned \$2 an hour less than he anticipated. How long did he expect it would take to do the job?

Solution

Let the expected time to do the job be x hours.
The hourly rate he expected to receive minus \$2 is equal to the actual hourly rate he earned:

$$\frac{336}{x} - 2 = \frac{336}{x+4}$$

$$336(x + 4) - 2x(x + 4) = 336x$$
$$336x + 1344 - 2x^2 - 8x = 336x$$
$$x^2 + 4x - 672 = 0$$
$$(x + 28)(x - 24) = 0$$

Thus $x + 28 = 0$, that is, $x = -28$

or $x - 24 = 0$, that is, $x = 24$

The expected time to do the job is 24 hours.

We reject -28 because it does not describe a physical experience.

EXAMPLE

A crew can row 20 miles down a stream and back in $7\frac{1}{2}$ hours. If the rate of the current is 2 mph, find the rate at which the crew can row in still water.

Solution

Let the rate the crew can row in still water be x mph.

The time to row downstream plus the time to row upstream is equal to $7\frac{1}{2}$ hours:

$$\frac{20}{x+2} + \frac{20}{x-2} = 7\frac{1}{2}$$

$$40(x-2) + 40(x+2) = 15(x+2)(x-2)$$

$$40x - 80 + 40x + 80 = 15x^2 - 60$$

$$15x^2 - 80x - 60 = 0$$

$$3x^2 - 16x - 12 = 0$$

$$(3x+2)(x-6) = 0$$

Thus $3x + 2 = 0$, that is, $x = -\frac{2}{3}$

or $x - 6 = 0$, that is, $x = 6$

The rate of rowing in still water is 6 mph.

Exercise 13.8

1. The product of two consecutive natural numbers is 12 more than 6 times the next consecutive number. Find the two numbers.
2. The product of two consecutive even natural numbers is 24 less than 12 times the next even number. Find the two numbers.
3. The sum of two natural numbers is 25, and the sum of their squares is 353. Find the two numbers.
4. The sum of two natural numbers is 41, and the sum of their squares is 953. Find the two numbers.
5. The difference between two natural numbers is 8, and the sum of their squares is 1282. Find the two numbers.
6. The difference between two natural numbers is 11, and the sum of their squares is 821. Find the two numbers.
7. The sum of two natural numbers is 26. The difference between their squares is 55 more than their product. Find the two numbers.
8. The sum of two natural numbers is 48, and the difference between their squares is 36 more than their product. Find the two numbers.
9. The sum of two natural numbers is 32, and the sum of their squares is 4 more than twice their product. Find the two numbers.

10. The sum of two natural numbers is 20, and the sum of their reciprocals is $\frac{5}{24}$. Find the two numbers.

11. The difference between two natural numbers is 6, and the sum of their reciprocals is $\frac{1}{4}$. Find the two numbers.

12. The difference between two natural numbers is 16, and the difference between their reciprocals is $\frac{1}{12}$. Find the two numbers.

13. A ski trip cost \$1600. If there had been 8 fewer members in the club, it would have cost each member \$10 more. How many members are there in the club?

14. A geology trip cost \$288. If there had been 4 more students, it would have cost each student \$1 less. How many students took the trip?

15. It takes A 39 hours longer to do a job than it takes B to do the same job. If A and B working together can do the job in 40 hours, how long does it take each alone to do the job?

16. A man painted a house for \$1200. It took him 10 hours longer than he expected, and so he earned 50¢ an hour less than he anticipated. How long did he expect it would take him to paint the house?

17. A man did a job for \$120. It took him 4 hours less than he expected and so he earned \$1 an hour more than he anticipated. How long did he expect it would take him to finish the job?

18. The length of a rectangle is 6 feet more than its width. The area of the rectangle is 216 square feet. Find the dimensions of the rectangle.

19. The length of a rectangle is 4 feet more than twice its width. The area of the rectangle is 126 square feet. Find the dimensions of the rectangle.

20. If each of two opposite sides of a square is doubled and the other two sides are each decreased by 3 feet, the area of the resulting rectangle is 27 square feet more than the area of the square. Find the sides of the square.

21. A man wishes to construct an open metal box. The box is to have a square base, sides 6 inches high, and a capacity of 384 cubic inches. What size piece of metal should he buy?

22. A man wishes to plow a rectangular field 60 rods by 80 rods in two equal periods of time. How wide a strip must he plow around the field the first period?

23. The base of a triangle is 6 feet less than the height. The area of the triangle is 108 square feet. Find the base and the height of the triangle.

24. The height of a triangle is 4 feet less than twice the base. The area of the triangle is 168 square feet. Find the base and the height of the triangle.

25. A tree was broken over by a storm $\frac{5}{18}$ of the distance from the bottom. If the top touched the ground 60 feet from the base of the tree, what was the height of the tree?

26. A crew can row 30 miles downstream and back in 8 hours. If the rate of the current is 2 mph, find the rate at which the crew can row in still water.

27. A crew can row 12 miles downstream and back in 6 hours. If the rate of the current is $1\frac{1}{2}$ mph, find the rate at which the crew can row in still water.

28. A man rows a boat 20 miles downstream and returns in 11 hours and 20 minutes. If he can row $4\frac{1}{4}$ miles per hour in still water, what is the rate of the current in the river?

29. A plane flies between two cities 400 miles apart. When the tail wind is 40 mph, it reaches its destination $\frac{1}{2}$ hour earlier than without the tail wind. What is the speed of the plane in still air?

30. A plane flies between the two cities 2400 miles apart. When the head wind is 40 mph, it reaches its destination 15 minutes later than if there had been no wind. What is the speed of the plane in still air?

31. By increasing the speed of a train $6\frac{1}{2}$ mph, it was possible to make a run of 468 miles in 48 minutes less time. What was the original speed for the trip?

32. A man lives 15 miles from his office. If he drives 15 mph faster than usual, he will arrive at his office 10 minutes earlier than normal. How fast does he normally drive?

33. The percent markup on the cost of a suit was the same as the cost in dollars. If the suit sold for \$96, what was the cost of the suit?

34. The manager of a theater found that with an admission charge of \$2.50 per person the average daily attendance was 4000, while with every increase of 25¢ the attendance dropped 200. What should the admission price be so that the daily receipts will be a maximum?

35. The manager of a theater found that with an admission charge of \$4 per person the average daily attendance was 2000, while for each 25¢ subtracted from the admission price there was an increase of 200. What should the admission price be so that the daily receipts will be a maximum?

36. A new labor contract provided for an increase in wages of \$1 per hour and a reduction of 5 hours in the work week. A worker who had been receiving \$240 per week would get a \$5 per week raise under the new contract. How long was the work week before the new contract?

37. The distance S in feet traveled by a freely falling body in t seconds with an initial velocity v_0 feet per second is given by the relation $S = v_0t + 16t^2$. An airplane has a device for releasing objects with a downward velocity of 120 feet per second when the plane has an altitude of 8800 feet. How long will it take for an object that is released to reach the ground?

38. When an object is thrown upward with a velocity v_0 feet per second, the height h above the ground in t seconds is given by the relation $h = v_0t - 16t^2$. If a man shoots a gun with a muzzle velocity of 2400 feet per second at a balloon that is 13,824 feet directly above him, how long will it take for the bullet to hit the balloon?

13.9 Graphs of Quadratic Equations

The graph of a quadratic equation $y = ax^2 + bx + c, a \neq 0, a, b, c \in R,$ is the set of points whose coordinates are the ordered pairs (x, y) that satisfy the equation. The graphical representation of the quadratic equation is called a **parabola**.
The ordered pairs can be found by assigning arbitrary values to x and determining the corresponding values for y.

Consider the equation $y = x^2 - 4x - 5$. The ordered pairs $(-2, 7), (-1, 0), (0, -5), (1, -8), (2, -9), (3, -8), (4, -5), (5, 0), (6, 7)$ are solutions of the equation. Construct a table of the ordered pairs.

x	-2	-1	0	1	2	3	4	5	6
y	7	0	-5	-8	-9	-8	-5	0	7

By plotting these ordered pairs of numbers and connecting them by a smooth curve, we obtain the graph of the equation, as shown in Figure 13.1.

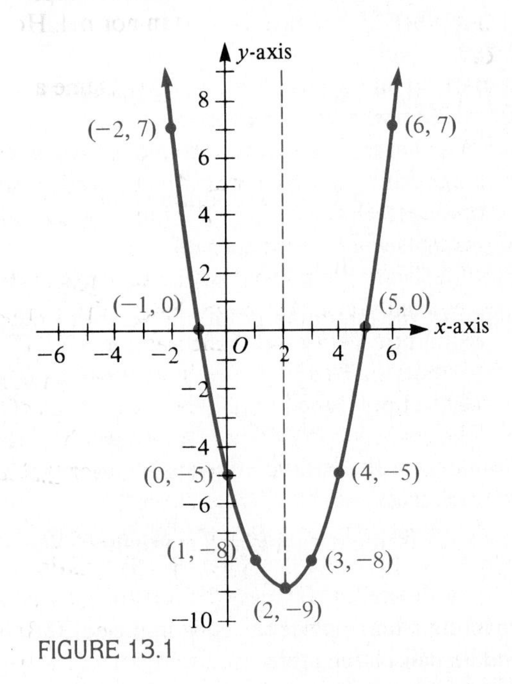

FIGURE 13.1

Notes

1. As x increases the curve descends (that is, y decreases) until at $x = 2$, $y = -9$ the curve stops descending and starts to rise as x increases. The point where the curve stops descending and starts rising is called the **minimum point of the curve**; it is also called the **vertex of the parabola**.

2. The vertical line through the vertex divides the curve into two symmetric branches. The vertical line is called the **line of symmetry**, or the **axis of the parabola.** Any two points on the parabola whose abscissas x_1 and x_2 are symmetric with respect to the axis of the parabola (equidistant from it) have equal ordinates.

Coordinates of the Vertex and Equation of the Line of Symmetry

Consider the equation $y = ax^2 + bx + c, a \neq 0, a, b, c \in R$:

$$\begin{aligned}
y &= a\left[x^2 + \frac{b}{a}x\right] + c \\
&= a\left[x^2 + \frac{b}{a}x + \frac{b^2}{4a^2} - \frac{b^2}{4a^2}\right] + c \\
&= a\left[\left(x + \frac{b}{2a}\right)^2 - \frac{b^2}{4a^2}\right] + c \\
&= a\left(x + \frac{b}{2a}\right)^2 - \frac{b^2}{4a} + c \\
&= a\left(x + \frac{b}{2a}\right)^2 - \frac{b^2 - 4ac}{4a}
\end{aligned}$$

For $a > 0$, since $\left(x + \frac{b}{2a}\right)^2 \geq 0$, the minimum value of y, $y = -\frac{b^2 - 4ac}{4a}$, is attained when $x + \frac{b}{2a} = 0$, that is, $x = -\frac{b}{2a}$.

Thus the coordinates of the minimum point, the vertex of the parabola, are

$$\left(-\frac{b}{2a}, -\frac{b^2 - 4ac}{4a}\right)$$

The equation of the axis of the parabola is $x = -\frac{b}{2a}$.

Remark When $a < 0$, the vertex of the parabola is a **maximum point**, and the parabola opens downward.

Notes

1. In constructing a table, place the coordinates of the vertex of the parabola as the middle pair of the table.

   Since values of x that are symmetric with respect to $-\frac{b}{2a}$ give values of y that are equal, the work is reduced by half.

   Take values of x symmetric with $-\frac{b}{2a}$. The values of y corresponding to these symmetric values of x are equal.

2. The set of ordered pairs $\{(x, y) \mid y = ax^2 + bx + c\}$ is a function. For each value of x there is a unique value of y.

   The domain of the function is $\{x \mid x \in R\}$.
   The range of the function is

$$\left\{y \,\middle|\, y \geq -\frac{b^2 - 4ac}{4a},\ a > 0\right\} \quad \text{or} \quad \left\{y \,\middle|\, y \leq -\frac{b^2 - 4ac}{4a},\ a < 0\right\}$$

EXAMPLE Find the coordinates of the vertex and the equation of the axis of the parabola whose equation is $y = 3x^2 - 4$.

Solution

$$\begin{aligned} y &= 3x^2 - 4 \\ &= 3(x - 0)^2 - 4 \end{aligned}$$

The coordinates of the vertex are $(0, -4)$.
The equation of the axis is $x = 0$.

Remark The set $\{(x, y) \mid y = 3x^2 - 4\}$ is a function.
The domain of the function is $\{x \mid x \in R\}$.
The range of the function is $\{y \mid y \geq -4\}$.

EXAMPLE Find the coordinates of the vertex and the equation of the axis of the parabola whose equation is $y = 3x^2 - x - 2$.

Solution

$$\begin{aligned} y &= 3x^2 - x - 2 \\ &= 3\left(x^2 - \frac{1}{3}x\right) - 2 \\ &= 3\left(x^2 - \frac{1}{3}x + \frac{1}{36} - \frac{1}{36}\right) - 2 \\ &= 3\left[\left(x - \frac{1}{6}\right)^2 - \frac{1}{36}\right] - 2 \\ &= 3\left(x - \frac{1}{6}\right)^2 - \frac{1}{12} - 2 \\ &= 3\left(x - \frac{1}{6}\right)^2 - \frac{25}{12} \end{aligned}$$

The coordinates of the vertex are $\left(\frac{1}{6}, -\frac{25}{12}\right)$.

The equation of the axis is $x = \frac{1}{6}$.

Remark The set $\{(x, y) \mid y = 3x^2 - x - 2\}$ is a function.
The domain of the function is $\{x \mid x \in R\}$.
The range of the function is $\left\{y \,\middle|\, y \geq -\frac{25}{12}\right\}$.

EXAMPLE Graph $y = 4x^2 + 8x - 5$.

Solution

$$\begin{aligned} y &= 4x^2 + 8x - 5 \\ &= 4(x+1)^2 - 9 \end{aligned}$$

The coordinates of the vertex are $(-1, -9)$.

Construct a table with $(-1, -9)$ as the middle pair of the table. Take values for x symmetric with -1, and find the corresponding values for y:

x	-3	-2	-1	0	1
y	7	-5	-9	-5	7

Note that $x = -2$ and $x = 0$ are symmetric with respect to $x = -1$, and thus the y values are equal, $y = -5$.

For $x = -3$ and $x = 1$, $y = 7$.

Plotting the ordered pairs of numbers and connecting them by a smooth curve, we obtain the desired graph, as shown in Figure 13.2.

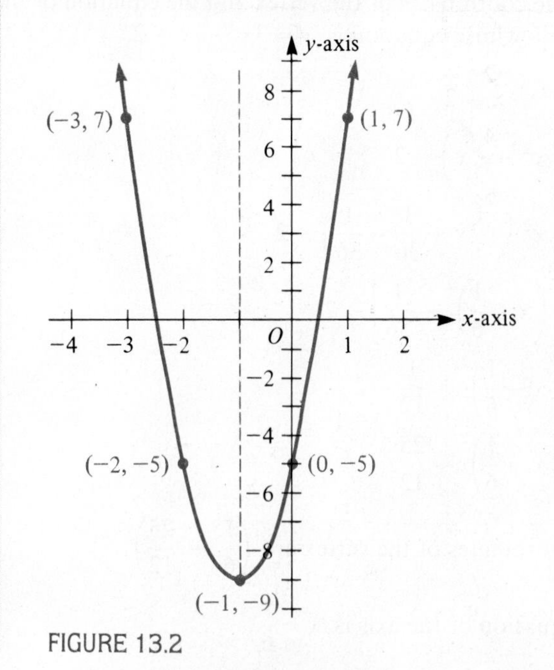

FIGURE 13.2

EXAMPLE Graph $y = -x^2 + 6x - 5$.

Solution

$$\begin{aligned} y &= -x^2 + 6x - 5 \\ &= -(x^2 - 6x) - 5 \\ &= -(x - 3)^2 + 4 \end{aligned}$$

The coordinates of the vertex are $(3, 4)$.

Construct a table with $(3, 4)$ as the middle pair of the table.

x	0	1	2	3	4	5	6
y	-5	0	3	4	3	0	-5

Plotting the ordered pairs of numbers and connecting them by a smooth curve, we obtain the desired graph, as shown in Figure 13.3.

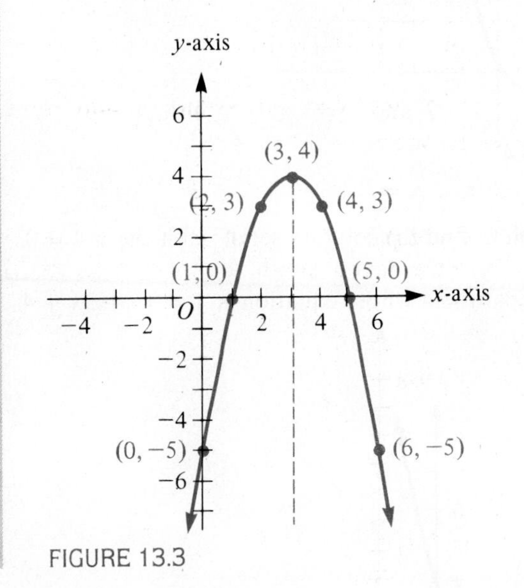

FIGURE 13.3

Remark The set $\{(x, y) \mid y = -x^2 + 6x - 5\}$ is a function.

The domain of the function is $\{x \mid x \in R\}$.

The range of the function is $\{y \mid y \leq 4\}$.

Solution of Quadratic Equations by Graphs

The abscissas of the points of intersection, if they exist, of the graphs of the parabola whose equation is $y = ax^2 + bx + c$ and the straight line whose equation is $y = 0$ (the x-axis) are the real roots of the equation $ax^2 + bx + c = 0$.

EXAMPLE Use a graph to find the solution set of $x^2 + 2x - 3 = 0$.

Solution Graph the parabola whose equation is $y = x^2 + 2x - 3$.

The abscissas of the points of intersection of the parabola with the line $y = 0$, the x-axis, are -3 and 1 (Figure 13.4).

Hence the solution set is $\{-3, 1\}$.

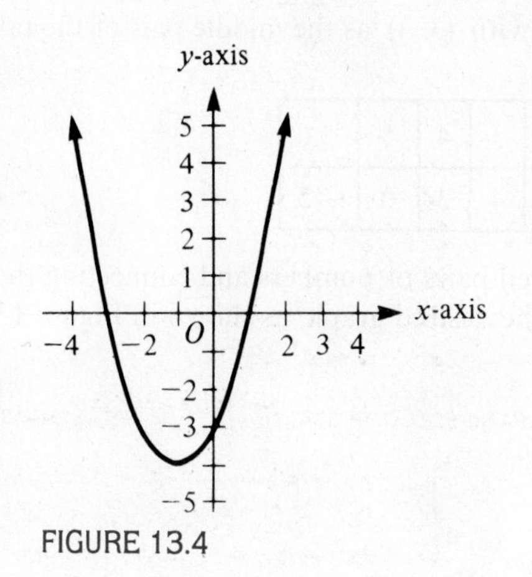

FIGURE 13.4

EXAMPLE Use a graph to find the solution set of $x^2 + 3x + 4 = 0$.

Solution Graph the parabola whose equation is $y = x^2 + 3x + 4$.

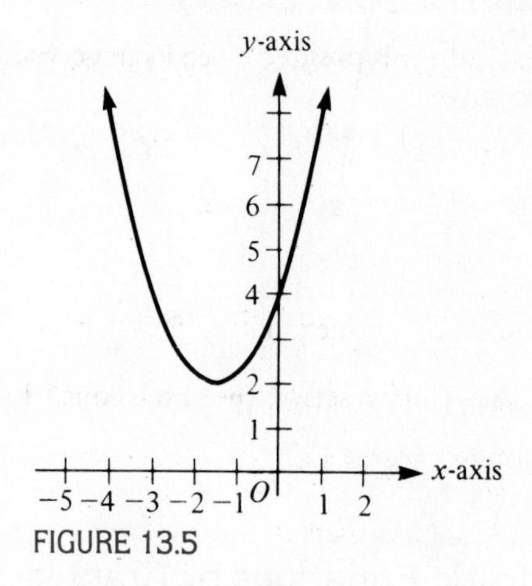

FIGURE 13.5

From Figure 13.5 we find that the parabola does not intercept the x-axis.

Thus the solution set is $\varnothing$.

Exercise 13.9

Find the coordinates of the vertex, the equation of the line of symmetry, and sketch the parabolas. Find the range of the functions whose elements are the solution set of the following equations:

1. $y = x^2$	**2.** $y = x^2 - 1$	**3.** $y = x^2 - 4$
4. $y = x^2 - 9$	**5.** $y = x^2 + 1$	**6.** $y = x^2 + 2$
7. $y = 1 - x^2$	**8.** $y = 4 - x^2$	**9.** $y = 3 - x^2$
10. $y = 6 - x^2$	**11.** $y = x^2 - 2x + 1$	**12.** $y = x^2 - 4x + 4$
13. $y = x^2 + 6x + 9$	**14.** $y = x^2 + 8x + 16$	**15.** $y = x^2 - 3x + 2$
16. $y = x^2 + 2x - 3$	**17.** $y = x^2 + 2x + 3$	**18.** $y = x^2 - 3x + 6$
19. $y = 2x^2 - 10x + 5$	**20.** $y = 2x^2 + 9x + 4$	**21.** $y = 6 + x - 2x^2$
22. $y = 4 + 7x - 2x^2$	**23.** $y = (2x - 3)^2$	**24.** $y = (3x + 4)^2$
25. $y = 2x^2 + 2x + 1$	**26.** $y = 3x^2 - 4x + 2$	**27.** $2y = x^2 + 3x - 5$
28. $3y = x^2 - x + 2$	**29.** $y = x^2 + 2x$	**30.** $y = x^2 + 4x$
31. $y = x^2 + x$	**32.** $y = x^2 + 3x$	**33.** $y = x^2 - 6x$
34. $y = x^2 - 8x$	**35.** $y = x^2 - 5x$	**36.** $y = x^2 - 7x$

By the use of graphs find the solution sets of the following equations:

37. $x^2 - 2x - 1 = 0$	**38.** $x^2 - 4x + 2 = 0$	**39.** $x^2 + 2x + 3 = 0$
40. $2x^2 - 3x - 2 = 0$	**41.** $3x^2 + 22x + 24 = 0$	**42.** $4x^2 - 8x + 3 = 0$
43. $4x^2 + 4x - 15 = 0$	**44.** $9x^2 - 6x + 1 = 0$	**45.** $4x^2 - 4x - 1 = 0$
46. $3x^2 - 5x + 4 = 0$		

13.10 Analytic Solution of Quadratic Inequalities

The product of two polynomials is positive when both polynomials are positive or both polynomials are negative.
That is, if P and Q are two polynomials and $P \cdot Q > 0$, then

either $\quad P > 0 \quad$ and $\quad Q > 0$

or $\quad P < 0 \quad$ and $\quad Q < 0$

Thus, to find the solution set of the inequality $PQ > 0$, we calculate the following.

1. The intersection of the solution sets of the two inequalities

$$P > 0 \quad \text{and} \quad Q > 0$$

2. The intersection of the solution sets of the two inequalities

$$P < 0 \quad \text{and} \quad Q < 0$$

The solution set of the inequality $PQ > 0$ is the union of the solution sets of the two cases.

EXAMPLE Solve the inequality $x^2 + 3x > 10$.

Solution

$$x^2 + 3x > 10$$
$$x^2 + 3x - 10 > 0$$
$$(x + 5)(x - 2) > 0$$

(1) $x + 5 > 0$ and $x - 2 > 0$

$x > -5$ and $x > 2$

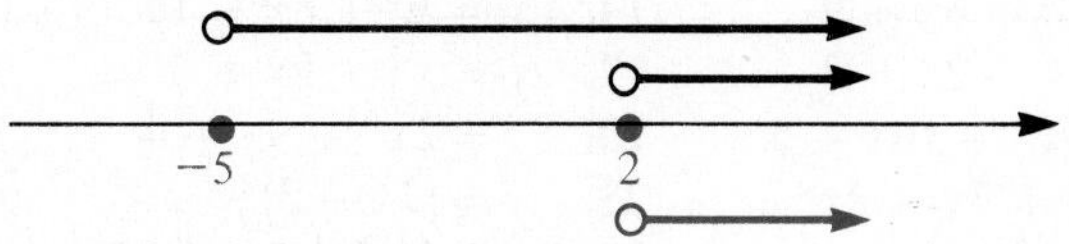

The solution set is $\{x \mid x > -5\} \cap \{x \mid x > 2\} = \{x \mid x > 2\}$.

(2) $x + 5 < 0$ and $x - 2 < 0$

$x < -5$ and $x < 2$

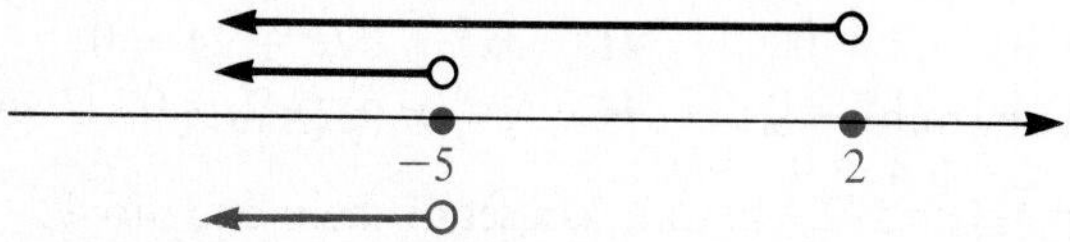

The solution set is $\{x \mid x < -5\} \cap \{x \mid x < 2\} = \{x \mid x < -5\}$.

From (1) and (2) the solution set of the inequality is

$\{x \mid x > 2\} \cup \{x \mid x < -5\} = \{x \mid x > 2 \text{ or } x < -5\}$

Similarly, when $P \cdot Q < 0$, then

either $P > 0$ and $Q < 0$

or $P < 0$ and $Q > 0$

Thus to find the solution set of the inequality $PQ < 0$, we calculate the following.

1. The intersection of the solution sets of the two inequalities

$P > 0$ and $Q < 0$

2. The intersection of the solution sets of the two inequalities

$P < 0$ and $Q > 0$

The solution set of the inequality $PQ < 0$ is the union of the solution sets of the two cases.

EXAMPLE Solve the inequality $x^2 - 2x - 24 \le 0$.

Solution

$$x^2 - 2x - 24 \le 0$$
$$(x + 4)(x - 6) \le 0$$

(1) $x + 4 \ge 0$ and $x - 6 \le 0$

$x \ge -4$ and $x \le 6$

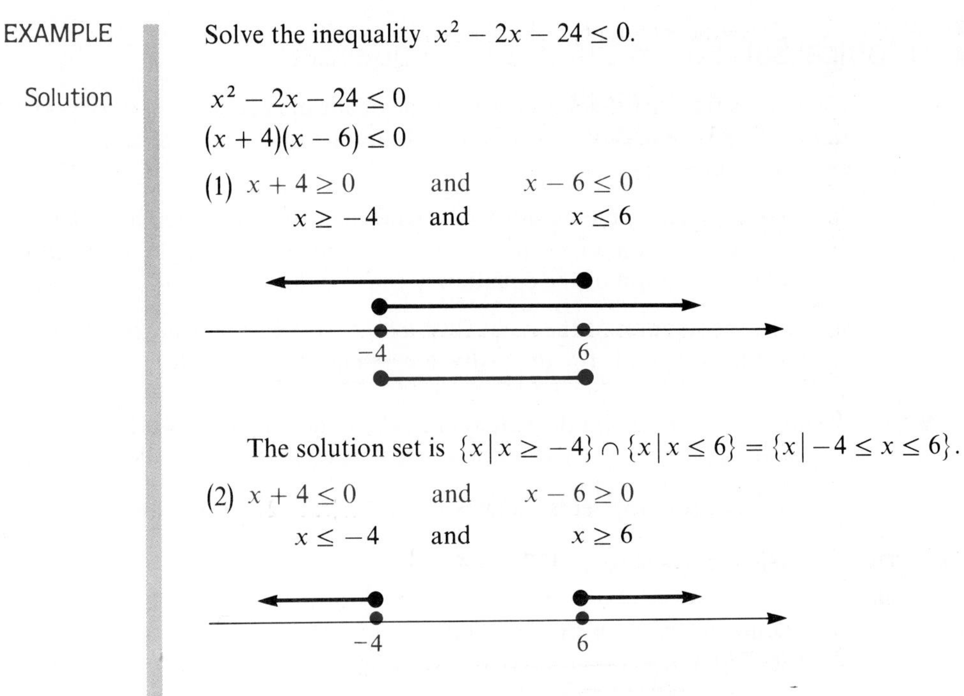

The solution set is $\{x \mid x \ge -4\} \cap \{x \mid x \le 6\} = \{x \mid -4 \le x \le 6\}$.

(2) $x + 4 \le 0$ and $x - 6 \ge 0$

$x \le -4$ and $x \ge 6$

The solution set is $\{x \mid x \le -4\} \cap \{x \mid x \ge 6\} = \varnothing$.

From (1) and (2) the solution set of the inequality is

$$\{x \mid -4 \le x \le 6\} \cup \varnothing = \{x \mid -4 \le x \le 6\}$$

Exercise 13.10

Find the solution set of the following inequalities:

1. $x^2 < 1$
2. $x^2 < 4$
3. $x^2 \le 16$
4. $x^2 \le 2$
5. $x^2 < 6$
6. $x^2 < 12$
7. $x^2 + x < 0$
8. $x^2 + 3x \le 0$
9. $2x^2 - x \le 0$
10. $3x^2 - 2x \le 0$
11. $x^2 + 3x + 2 < 0$
12. $x^2 + 7x + 10 < 0$
13. $x^2 + 3x \le 4$
14. $x^2 - 2x \le 8$
15. $x^2 + 2 \le 3x$
16. $x^2 + 6 < 5x$
17. $2x^2 - 7x < 4$
18. $2x^2 + x < 6$
19. $4x^2 - 3 < 4x$
20. $6x^2 + x \le 2$
21. $9x^2 - 6x \le 8$
22. $6x^2 + 11x < 7$
23. $6x^2 + 35x < 6$
24. $9x^2 - 9x < 10$
25. $x^2 \ge 9$
26. $x^2 \ge 25$
27. $x^2 > 3$
28. $x^2 > 18$
29. $x^2 + 2x > 0$
30. $2x^2 + 7x > 0$
31. $2x^2 - 3x \ge 0$
32. $2x^2 - 5x \ge 0$
33. $x^2 + 4x + 3 > 0$
34. $x^2 + 7x + 12 > 0$
35. $x^2 + x > 6$
36. $x^2 - 3 \ge 2x$
37. $x^2 + 4 \ge 5x$
38. $x^2 + 12 \ge 7x$
39. $2x^2 + 5x > 3$
40. $2x^2 - 3x > 9$
41. $6x^2 - x > 12$
42. $3x^2 + 17x > 6$
43. $9x^2 + 23x - 12 \ge 0$
44. $6x^2 - 13x - 8 \ge 0$
45. $24x^2 - 5x - 1 > 0$
46. $12x^2 + 17x + 6 > 0$

13.11 Graphical Solution of Quadratic Inequalities

The intersection of the solution sets of the quadratic equation $y = ax^2 + bx + c$ and the linear inequality $y > 0$ is the solution set of the inequality $ax^2 + bx + c > 0$.

Since the graphical representation of the inequality $y > 0$ is the half-plane above the x-axis, the solution of the inequality $ax^2 + bx + c > 0$ is the set of values of x for which the graph of the equation $y = ax^2 + bx + c$ lies above the x-axis.

Similarly, the solution of the inequality $ax^2 + bx + c < 0$ is the set of values of x for which the graph of $y = ax^2 + bx + c$ lies below the x-axis.

Note We are only interested in the values of x where the graph crosses the x-axis, where the curve is above the x-axis, and where the curve is below the x-axis, not in the exact shape of the graph.

EXAMPLE Solve graphically $x^2 - 3x - 4 > 0$.

Solution Draw the graph of the equation $y = x^2 - 3x - 4$. See Figure 13.6.

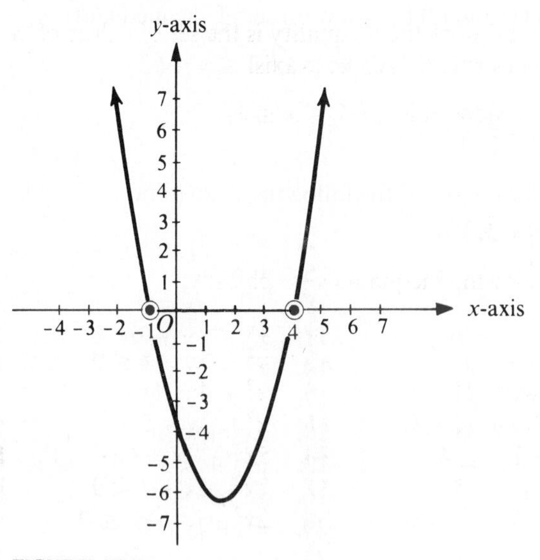

FIGURE 13.6

The solution of the inequality is the set of values of x for which the graph is above the x-axis.

The solution set is $\{x \mid x < -1, \text{ or } x > 4\}$.

EXAMPLE Solve graphically $x^2 - 6x + 8 \le 0$.

Solution Draw the graph of the equation $y = x^2 - 6x + 8$. (See Figure 13.7.)

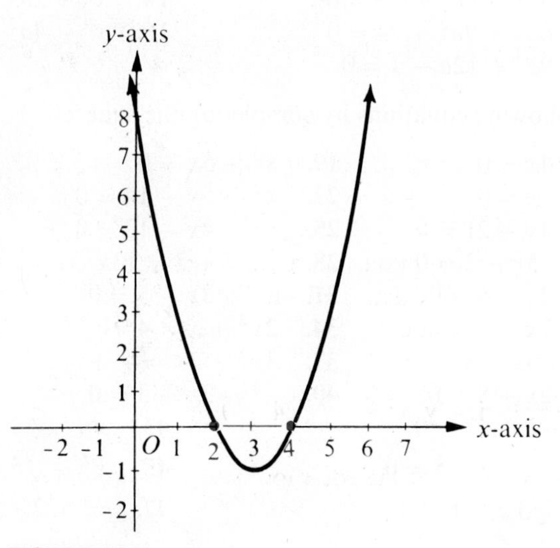

FIGURE 13.7

The solution of the inequality is the set of values of x for which the graph intersects or is below the x-axis.

The solution set is $\{x \mid 2 \le x \le 4\}$.

Exercise 13.11

Solve the following inequalities graphically:

1. $x^2 \ge 2$
2. $2x^2 + 7x < 0$
3. $x^2 - 2x < 3$
4. $x^2 > x + 6$
5. $x^2 - 2x - 8 \le 0$
6. $x^2 + 7x + 12 < 0$
7. $x^2 - 4x \ge 21$
8. $x^2 + 4 > 4x$
9. $x^2 + x + 2 > 0$
10. $x^2 + 8x + 16 \le 0$
11. $x^2 < 3x - 5$
12. $4x^2 - 5x + 1 < 0$
13. $6x^2 - 23x \ge 4$
14. $5x^2 + 7x \ge 6$
15. $4x^2 + 1 > 4x$
16. $12x^2 > 2 - 5x$
17. $2x^2 + x + 1 < 0$
18. $3x^2 + 4x + 5 < 0$
19. $4x^2 + 4x - 1 \le 0$
20. $2x^2 - 2x - 1 \le 0$

Chapter 13 Review

Solve the following equations for x by factoring:

1. $x^2 + 9x + 20 = 0$
2. $x^2 + 4x - 21 = 0$
3. $8x^2 + 14x - 15 = 0$
4. $9x^2 - 6x - 8 = 0$

5. $6x^2 - 23x + 21 = 0$
6. $20x^2 - 33x + 10 = 0$
7. $12x^2 - 35x + 8 = 0$
8. $18x^2 + 3x - 10 = 0$
9. $x^2 + 3ax - 4a^2 = 0$
10. $6x^2 + 11ax + 3a^2 = 0$
11. $x^2 - ax + 2bx = 2ab$
12. $2x^2 + ax = 6bx + 3ab$
13. $x^2 - 4ax + 4a^2 - 25 = 0$
14. $x^2 + 2ax + a^2 - 12 = 0$
15. $x^2 - 6ax + 9a^2 + 24 = 0$
16. $x^2 - 4a^2 - 4a - 1 = 0$
17. $x^2 - 9a^2 + 12a - 4 = 0$

Solve the following equations by completing the square:

18. $x^2 + 4x = 0$
19. $x^2 + 5x = 0$
20. $x^2 - 3x = 0$
21. $x^2 - 7x = 0$
22. $x^2 - 5x - 14 = 0$
23. $x^2 + 3x - 10 = 0$
24. $x^2 + 4x - 21 = 0$
25. $x^2 - 4x - 12 = 0$
26. $3x^2 - 10x - 8 = 0$
27. $2x^2 - 5x - 3 = 0$
28. $12x^2 + 3 = 13x$
29. $6x^2 + 12 = 17x$
30. $x^2 + 2x - 4 = 0$
31. $x^2 + 3x - 5 = 0$
32. $x^2 - 7x - 2 = 0$
33. $x^2 - 2x = 17$
34. $2x^2 + x = 4$
35. $3x^2 + 2x = 6$
36. $2x^2 = 3x + 8$
37. $3x^2 - 3x = 4$
38. $x^2 + 3x + 5 = 0$
39. $x^2 + 2x + 8 = 0$
40. $x^2 - 3x + 3 = 0$
41. $2x^2 + x + 2 = 0$
42. $2x^2 - 3x + 7 = 0$
43. $x^2 - \sqrt{3}x - 6 = 0$
44. $x^2 - 2\sqrt{2}x - 5 = 0$
45. $x^2 + \sqrt{5}x + 1 = 0$
46. $x^2 + \sqrt{3}x + 1 = 0$
47. $x^2 + 2\sqrt{5}x + 7 = 0$

Solve the following equations by the quadratic formula:

48. $x^2 + 6x = 0$
49. $x^2 + 3x = 0$
50. $x^2 - 2x = 0$
51. $2x^2 - 5x = 0$
52. $x^2 + 5 = 0$
53. $2x^2 - 7 = 0$
54. $x^2 + 6x - 16 = 0$
55. $x^2 - 9x + 18 = 0$
56. $x^2 - 2x - 8 = 0$
57. $x^2 - 2x - 15 = 0$
58. $2x^2 + 5x - 12 = 0$
59. $6x^2 + 7x - 3 = 0$
60. $12x^2 - 13x - 4 = 0$
61. $4x^2 + 8x - 5 = 0$
62. $x^2 - 2x - 11 = 0$
63. $x^2 - 5x - 3 = 0$
64. $x^2 + 2x - 7 = 0$
65. $x^2 + 4x - 6 = 0$
66. $2x^2 + 4x - 1 = 0$
67. $3x^2 + 2x - 2 = 0$
68. $x^2 - x + 3 = 0$
69. $x^2 - 2x + 6 = 0$
70. $x^2 + x + 4 = 0$
71. $x^2 + 3x + 3 = 0$
72. $2x^2 - 5x + 4 = 0$
73. $6x^2 - 7x + 4 = 0$
74. $x^2 - \sqrt{3}x + 13 = 0$
75. $3x^2 + 2\sqrt{2}x + 9 = 0$
76. $2\sqrt{3}x^2 - 2x + \sqrt{3} = 0$

Find all values of k for which the nature of the roots of the given equation is as indicated:

77. $x^2 - 3x + k = 0$
two real distinct roots

78. $3x^2 + 5x + k = 0$
two real distinct roots

79. $kx^2 + 7x + 2 = 0$
two real distinct roots

80. $kx^2 - 8x - 3 = 0$
two real distinct roots

81. $x^2 + kx - 8 = 0$
two real distinct roots

82. $5x^2 + kx - 1 = 0$
two real distinct roots

83. $x^2 - 7x + k = 0$
one double root

84. $2x^2 + 5x + k = 0$
one double root

85. $x^2 + kx + 2 = 0$
one double root

86. $x^2 + kx + 9 = 0$
one double root

87. $kx^2 + 3x - 8 = 0$
one double root

88. $kx^2 - 10x + 3 = 0$
one double root

89. $4x^2 + 3x + k = 0$
two imaginary roots

90. $3x^2 - 8x + k = 0$
two imaginary roots

91. $x^2 + kx - 7 = 0$
two imaginary roots

92. $2x^2 + kx - 4 = 0$
two imaginary roots

93. $kx^2 - 5x + 3 = 0$
two imaginary roots

94. $kx^2 + 9x - 2 = 0$
two imaginary roots

Find the value of m in each of the following equations for the given conditions:

95. $x^2 - 2x + m = 0$
one root is $1 - 3i$

96. $x^2 - 8x + m = 0$
one root is $4 + i$

97. $(m + 1)x^2 - (4m - 5)x + 8 = 0$
the sum of the roots is 3.

98. $(m - 1)x^2 + (3m + 1)x = 9$
the sum of the roots is (-4)

99. $4x^2 - 4x + m = 1$
the difference of the roots is 2

100. $18x^2 - 45x + 4m = 3$
the difference of the roots is $\frac{5}{6}$

101. $3x^2 - 26x = m$
one root exceeds the other by $\frac{8}{3}$

102. $x^2 + x = m$
one root exceeds the other by 7

103. $2x^2 + 7x + m + 2 = 0$
one root is $\frac{4}{3}$ the other

104. $6x^2 - 11x + m = 8$
one root is $\frac{3}{8}$ the other

105. $3mx^2 - 35x + 5m = 2$
the product of the roots is $\frac{3}{2}$

106. $(3m - 1)x^2 - 15x = m - 1$
the product of the roots is $\left(-\frac{1}{4}\right)$

107. $4x^2 - 5x + m + 7 = 0$
the quotient of the roots is $\left(-\frac{3}{8}\right)$

108. $9x^2 - 18x + m + 2 = 0$
the quotient of the roots is 2

Find a quadratic equation whose roots r_1 and r_2 satisfy the given conditions:

109. $r_1 + r_2 = 11, \quad r_1 - r_2 = 5$

110. $r_1 + r_2 = -6, \quad r_1 - r_2 = 4$

111. $r_1 + r_2 = 3, \quad \frac{r_1}{r_2} = \frac{2}{7}$

112. $r_1 + r_2 = 2, \quad \frac{r_1}{r_2} = \frac{1}{4}$

113. $r_1 - r_2 = \frac{5}{4}, \quad r_1 r_2 = -\frac{3}{8}$

114. $r_1 - r_2 = -\frac{1}{2}, \quad r_1 r_2 = \frac{3}{16}$

115. $r_1 r_2 = \frac{4}{9}, \quad \frac{r_1}{r_2} = \frac{1}{4}$

116. $r_1 r_2 = -\frac{5}{7}, \quad \frac{r_1}{r_2} = -\frac{35}{9}$

Find all the roots of the following equations:

117. $8x^3 + 2x^2 + 3x = 0$

118. $9x^3 + 3x^2 + 2x = 0$

119. $8x^3 + 1 = 0$

120. $27x^3 + 1 = 0$

121. $8x^3 - 27 = 0$

122. $x^4 - 25 = 0$

123. $x^4 - 81 = 0$

124. $16x^4 - 81 = 0$

125. $2x^4 - 5x^2 - 12 = 0$

126. $3x^4 - 29x^2 + 18 = 0$

127. $x^4 - 14x^2 + 1 = 0$

128. $x^4 - 12x^2 + 16 = 0$

129. $x - 3x^{\frac{1}{2}} + 2 = 0$

130. $x + 5x^{\frac{1}{2}} + 4 = 0$

131. $3x - 10x^{\frac{1}{2}} + 3 = 0$

132. $8x + 2x^{\frac{1}{2}} - 1 = 0$

133. $x^{\frac{2}{3}} - 4x^{\frac{1}{3}} + 3 = 0$

134. $x^{\frac{2}{3}} - 2x^{\frac{1}{3}} - 8 = 0$

135. $6x^{\frac{2}{3}} + 7x^{\frac{1}{3}} - 3 = 0$

136. $12x^{\frac{2}{3}} + 5x^{\frac{1}{3}} - 2 = 0$

Solve the following equations:

137. $\dfrac{7}{x+2} + \dfrac{x}{2x-1} = 2$

138. $\dfrac{x}{x+2} + \dfrac{9}{2x-3} = 2$

139. $\dfrac{4x-1}{6x^2+x-1} + \dfrac{2x+1}{3x^2-13x+4} = \dfrac{3x-3}{2x^2-7x-4}$

140. $\dfrac{4x+33}{4x^2-13x-12} + \dfrac{2x-21}{4x^2-9x-9} = \dfrac{1}{x-3}$

141. $\dfrac{14x-7}{6x^2-11x+3} + \dfrac{7x-15}{2x^2+5x-12} = \dfrac{14x+9}{3x^2+11x-4}$

142. $\dfrac{3x-1}{12x^2+x-20} + \dfrac{2x+4}{3x^2+x-4} = \dfrac{x-4}{4x^2-9x+5}$

143. $\dfrac{3x-5}{24x^2-23x-12} + \dfrac{2x-6}{8x^2+51x+18} = \dfrac{x-2}{3x^2+14x-24}$

144. $\dfrac{2x-5}{2x^2-9x+4} - \dfrac{6x-4}{12x^2-4x-1} = \dfrac{2x-3}{6x^2-23x-4}$

145. $\sqrt{x-1} + \sqrt{x+4} = \sqrt{6x-5}$

146. $\sqrt{x-6} + \sqrt{4x-3} = \sqrt{7x-13}$

147. $\sqrt{x-5} + \sqrt{2x-3} = \sqrt{5x-14}$

148. $\sqrt{x+3} + \sqrt{2x+8} = \sqrt{13x+35}$

149. $\sqrt{x+3} - \sqrt{x+8} = \sqrt{5-4x}$

150. $\sqrt{x+4} - 2\sqrt{x-1} = \sqrt{16-3x}$

Find the coordinates of the vertex, the equation of the line of symmetry, and sketch the parabolas whose equations are the following:

151. $y = x^2 - 4$

152. $y = x^2 - 2x - 8$

153. $y = x^2 - 4x + 3$

154. $y = 2x^2 + 5x - 3$

155. $y = 3 + 2x - x^2$

156. $y = 2 + x - 3x^2$

Solve the following inequalities analytically:

157. $x^2 + x < 2$
158. $x^2 + 2x < 3$
159. $x^2 + 2x > 8$
160. $x^2 - x < 12$
161. $x^2 + 6 < 7x$
162. $2x^2 + x \geq 21$
163. $3x^2 - 8x \geq 16$
164. $2x^2 > 7x + 4$
165. $3x^2 \leq 8x - 4$
166. $2x^2 \leq 30 - 7x$
167. $15x^2 + 11x \geq 12$
168. $6x^2 + 23x \geq 4$
169. $9x^2 < 18x + 7$
170. $3x^2 > 20x + 32$

Solve the following inequalities graphically:

171. $x^2 - 3x > 0$
172. $x^2 - 6x < 0$
173. $x^2 + x \leq 12$
174. $x^2 - 3x > 10$
175. $x^2 - 10x + 25 \leq 0$
176. $2x^2 - 9x \leq 18$
177. $3x^2 - 7x \geq 6$
178. $3x^2 - 17x + 10 < 0$
179. $2x^2 - 11x + 9 > 0$
180. $x^2 + 2x - 4 \geq 0$
181. $x^2 - 4x + 1 \leq 0$
182. $3x^2 - 2x + 3 > 0$
183. $4x^2 - 6x + 5 < 0$
184. $2x^2 - x > 28$
185. $2x^2 + x \geq 15$
186. $3x^2 - 2x \leq 8$

chapter 14

Conic Sections and Solution of Systems of Quadratic Equations in Two Variables

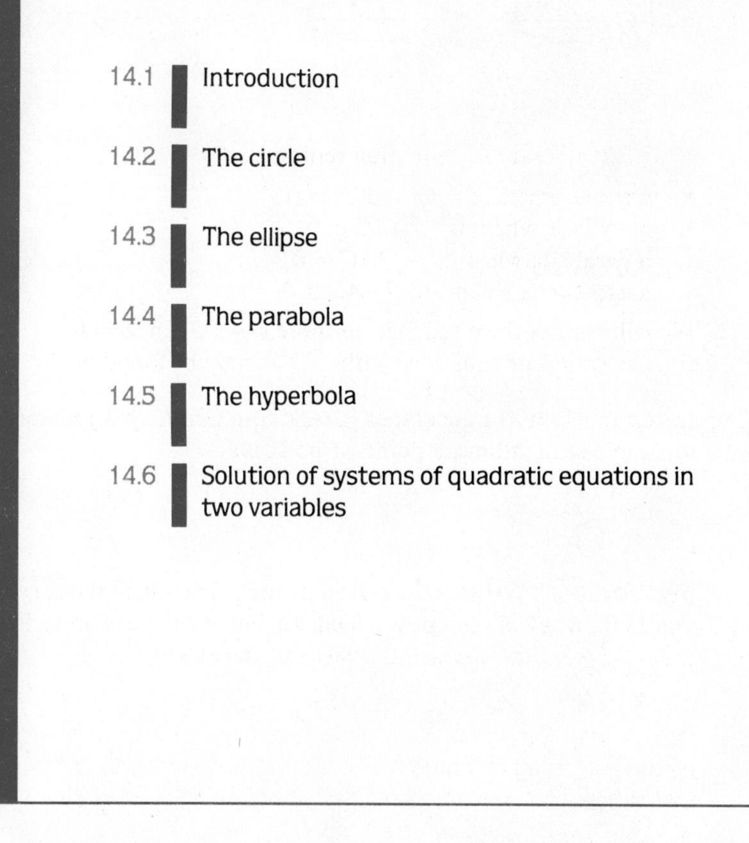

14.1 Introduction

14.2 The circle

14.3 The ellipse

14.4 The parabola

14.5 The hyperbola

14.6 Solution of systems of quadratic equations in two variables

14.1 Introduction

The general form of a **quadratic equation in two variables** is

$$Ax^2 + Bxy + Cy^2 + Dx + Ey + F = 0$$

where $A, B, C, D, E, F \in R$ and A, B, C are not all zero.

The solution set of a quadratic equation in two variables is an infinite set of ordered pairs of numbers, where each pair satisfies the given equation.

A quadratic equation in two variables represents a **conic section**. The term conic section is applied to a **circle**, an **ellipse**, a **parabola**, or a **hyperbola**, since they are the curves obtained when a plane intersects a right circular cone (see Figure 14.1).

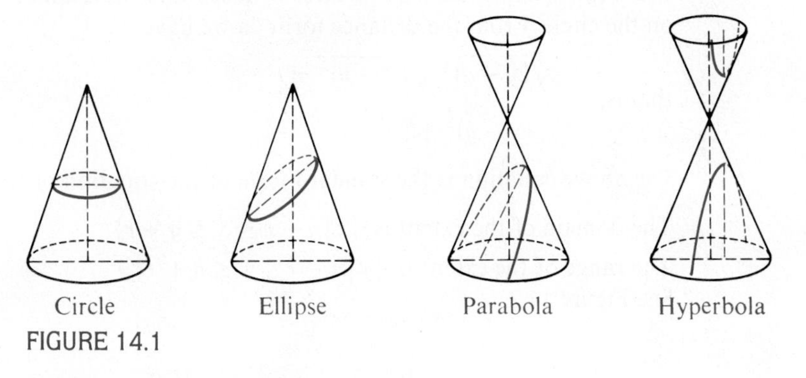

FIGURE 14.1

The general quadratic equation represents

1. a **circle**, when $A = C$ and $B = 0$
2. an **ellipse**, when $B^2 - 4AC < 0$
3. a **parabola**, when $B^2 - 4AC = 0$
4. a **hyperbola**, when $B^2 - 4AC > 0$

We will restrict the discussion to those cases when $B = 0$.
Consideration of equations with $B \neq 0$ may be found in analytic geometry texts.

Note In certain cases, the general quadratic equation may degenerate into two straight lines, one straight line, a point, or no curve.

Extent

The **extent** of a curve is the region of the plane where the curve lies. The set of real values that x can assume is called the **domain of the extent**. The set of real values that y can assume is called the **range of the extent**.

The domain of the graph can be found by expressing y in terms of x and then determining the real values that x can assume that would make y real. The range of the graph can be found by expressing x in terms of y and then determining the real values that y can assume that would make x real.

14.2 The Circle

DEFINITION

> A **circle** is a set of points in a plane equidistant from a fixed point in the plane.
> The fixed point is called the **center** of the circle and the constant distance the **radius** of the circle.

Equation of a Circle

Let $C(g, h)$ be the center of a circle whose radius is r, and let $P(x, y)$ be any point on the circle. From the distance formula we have

$$\sqrt{(x-g)^2 + (y-h)^2} = r$$

that is,

$$(x-g)^2 + (y-h)^2 = r^2$$

The above equation is the **standard form** of the equation of the circle.

The domain of the extent is $\{x \mid g - r \leq x \leq g + r\}$.
The range of the extent is $\{y \mid h - r \leq y \leq h + r\}$.
See Figure 14.2.

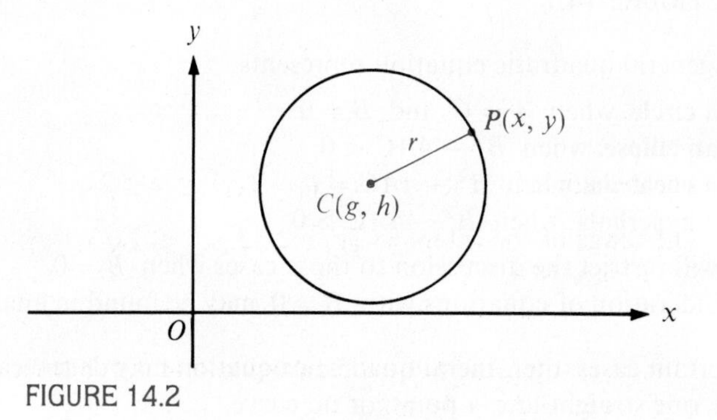

FIGURE 14.2

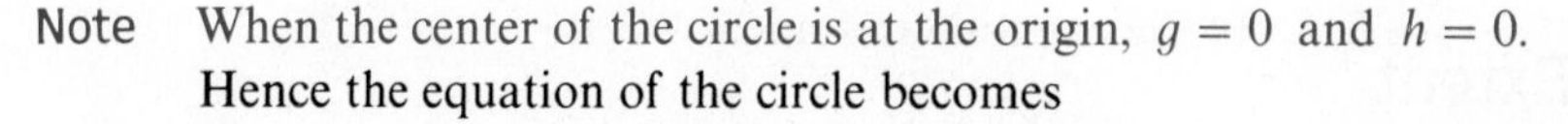

Note When the center of the circle is at the origin, $g = 0$ and $h = 0$. Hence the equation of the circle becomes

$$x^2 + y^2 = r^2$$

Graphing the Circle

We first put the equation in standard form, by completing the square. Locate the center of the circle; that is, find the point $C(g, h)$, and find the radius r. Select values of x in the domain of the extent of the equation. Calculate the corresponding values of y by substituting the values of x in the equation, and plot these points on a Cartesian coordinate system. Draw a smooth curve through the points plotted.

Note Remember to take the same scale on both axes.

EXAMPLE Find the center, the radius, the extents, and graph the circle whose equation is $x^2 + y^2 = 12$.

Solution $C(0, 0)$, $r = \sqrt{12} = 2\sqrt{3}$

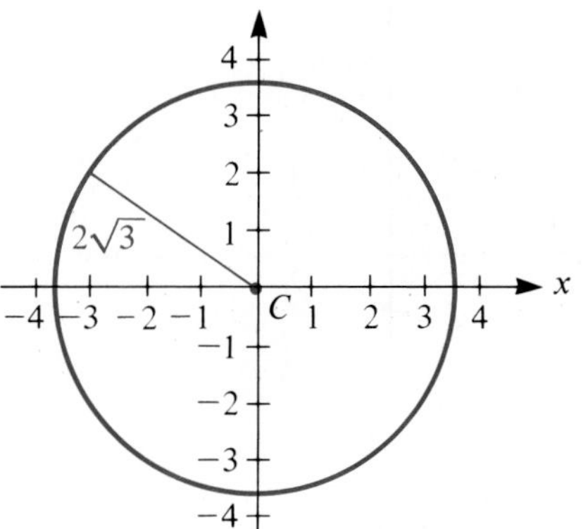

FIGURE 14.3

The domain of the extent is $\{x \mid -2\sqrt{3} \le x \le 2\sqrt{3}\}$.
The range of the extent is $\{y \mid -2\sqrt{3} \le y \le 2\sqrt{3}\}$.
See Figure 14.3.

EXAMPLE Find the center, the radius, the extents, and graph the circle whose equation is $x^2 + y^2 + 3x - 4y = 6$.

Solution We first put the equation in standard form

$$(x^2 + 3x) + (y^2 - 4y) = 6$$

$$\left(x^2 + 3x + \frac{9}{4}\right) - \frac{9}{4} + (y^2 - 4y + 4) - 4 = 6$$

$$\left(x + \frac{3}{2}\right)^2 + (y - 2)^2 = \frac{49}{4}$$

$$C\left(-\frac{3}{2}, 2\right) \qquad r = \sqrt{\frac{49}{4}} = \frac{7}{2}$$

The following is a table of coordinates of points on the graph. Irrational values of y are approximated to their rational equivalents (correct to one decimal place):

x	-5	-3	-3	-1.5	-1.5	0	0	2
y	2	5.2	-1.2	-1.5	5.5	5.2	-1.2	2

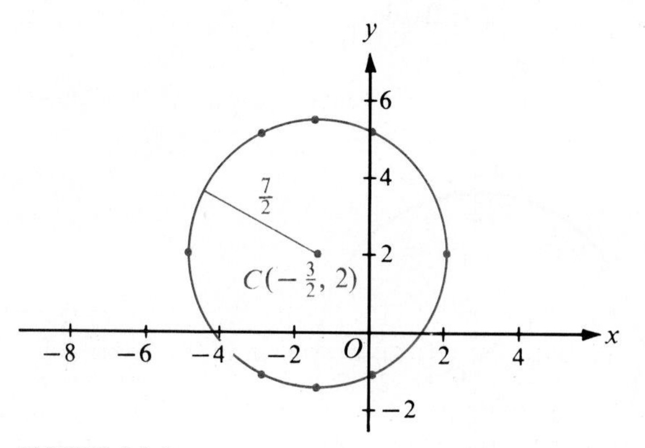

FIGURE 14.4

The domain of the extent is $\{x \mid -5 \leq x \leq 2\}$.

The range of the extent is $\left\{y \,\middle|\, -\dfrac{3}{2} \leq y \leq \dfrac{11}{2}\right\}$.

See Figure 14.4.

Exercise 14.2

Find the center, the radius, the extents, and graph the circles whose equations are as follows:

1. $x^2 + y^2 = 1$ **2.** $x^2 + y^2 = 4$ **3.** $x^2 + y^2 = 9$
4. $x^2 + y^2 = 16$ **5.** $x^2 + y^2 = 25$ **6.** $x^2 + y^2 = 36$
7. $x^2 + y^2 = 49$ **8.** $x^2 + y^2 = 64$ **9.** $x^2 + y^2 = 6$
10. $x^2 + y^2 = 8$ **11.** $x^2 + y^2 = 18$ **12.** $x^2 + y^2 = 24$
13. $x^2 + y^2 + 2x = 15$ **14.** $x^2 + y^2 + 4x = 0$
15. $x^2 + y^2 - 6x + 8 = 0$ **16.** $x^2 + y^2 - 2x = 24$
17. $x^2 + y^2 + 6y = 27$ **18.** $x^2 + y^2 + 4y = 45$
19. $x^2 + y^2 - 8y + 8 = 0$ **20.** $x^2 + y^2 - 6y = 11$
21. $x^2 + y^2 + 2x - 4y = 20$ **22.** $x^2 + y^2 + 4x + 6y = 23$
23. $x^2 + y^2 + 6x - 4y = -4$ **24.** $x^2 + y^2 - 6x + 2y = 39$

25. $x^2 + y^2 - 2x - 2y = 98$

26. $x^2 + y^2 - 4x + 6y = 37$

27. $x^2 + y^2 - 4x - 6y = 14$

28. $2x^2 + 2y^2 - 2x + 14y + 15 = 0$

29. $9x^2 + 9y^2 + 24x + 108y = -232$

30. $16x^2 + 16y^2 + 24x - 128y = 55$

14.3 The Ellipse

DEFINITION

> An **ellipse** is a set of points in a plane each of which has the property that the sum of the distances from any point in the set to two fixed points in the plane is constant.
> The two fixed points are called the **foci** of the ellipse.

Equation of an Ellipse

Let the foci be $F_1(g - k, h)$ and $F_2(g + k, h)$, and the constant distance equal $2a$ where $a > k$. Let $P(x, y)$ be any point on the ellipse. Hence

$$PF_1 + PF_2 = 2a$$

or $$\sqrt{[x - (g - k)]^2 + (y - h)^2} + \sqrt{[x - (g + k)]^2 + (y - h)^2} = 2a$$

See Figure 14.5.

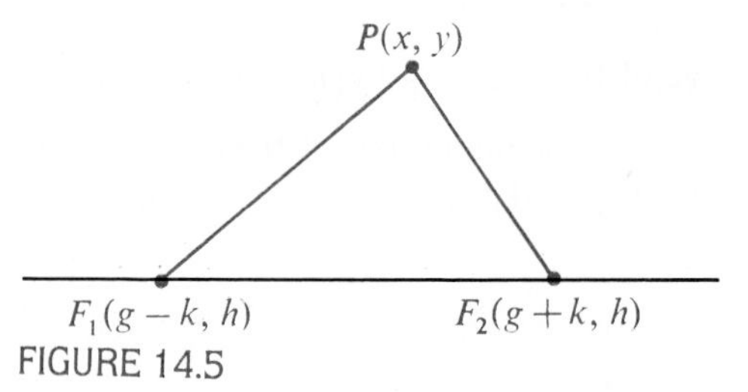

FIGURE 14.5

Simplifying and putting $a^2 - k^2 = b^2$ we get (see Appendix E)

$$\frac{(x - g)^2}{a^2} + \frac{(y - h)^2}{b^2} = 1$$

This equation is the **standard form** of the equation of an ellipse. The point $C(g, h)$ is called the **center** of the ellipse.

Note When the center of the ellipse is at the origin, $g = 0$ and $h = 0$. Hence the equation of the ellipse becomes

$$\frac{x^2}{a^2} + \frac{y^2}{b^2} = 1$$

First: When $a > b$. The foci of the ellipse are at the points

$$F_1(g - \sqrt{a^2 - b^2}, h) \quad \text{and} \quad F_2(g + \sqrt{a^2 - b^2}, h)$$

The points $V_1(g - a, h)$ and $V_2(g + a, h)$ are called the **vertices** of the ellipse. The line segment joining V_1 and V_2 is called the **major axis**, with length $2a$. The major axis is parallel to the x-axis. The foci and the center of the ellipse lie on the major axis. See Figure 14.6.

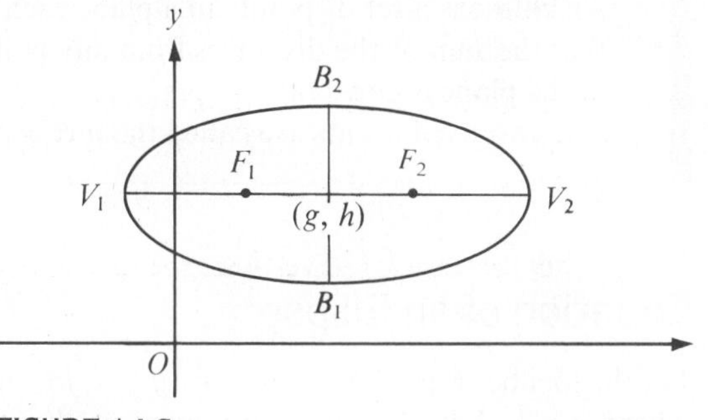

FIGURE 14.6

The line segment joining the points $B_1(g, h - b)$ and $B_2(g, h + b)$ is called the **minor axis** of the ellipse, with length $2b$, and is parallel to the y-axis.

The domain of the extent is $\{x \mid g - a \le x \le g + a\}$.

The range of the extent is $\{y \mid h - b \le y \le h + b\}$.

Second: When $b > a$. The major axis of the ellipse is parallel to the y-axis and the minor axis is parallel to the x-axis. See Figure 14.7.

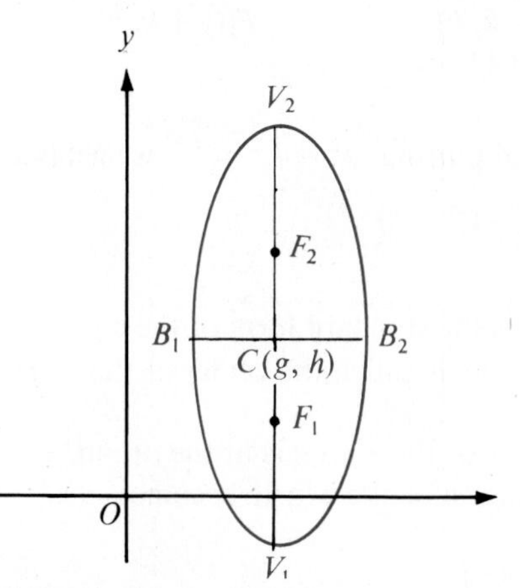

FIGURE 14.7

We also have

$$F_1(g, h - \sqrt{b^2 - a^2}), \quad F_2(g, h + \sqrt{b^2 - a^2})$$

$$V_1(g, h - b), \quad V_2(g, h + b), \quad B_1(g - a, h), \quad B_2(g + a, h)$$

Graphing the Ellipse

First put the equation in standard form. Identify g and h, thus locating the center of the ellipse. Calculate the values of a and b. With the values of $g, h, a,$ and b known, calculate the coordinates of V_1, V_2, B_1, and B_2. Plot these four points.

For a more accurate curve select values of x in the domain of the extent, calculate the values of y associated with them, and plot these points. Draw a smooth curve through the points plotted.

EXAMPLE

Find the center, the vertices, the foci, and graph the ellipse whose equation is $3x^2 + y^2 = 9$.

Solution

$$3x^2 + y^2 = 9$$

$$\frac{3x^2}{9} + \frac{y^2}{9} = 1$$

$$\frac{x^2}{3} + \frac{y^2}{9} = 1$$

$$C(0, 0), \quad a = \sqrt{3}, \quad b = \sqrt{9} = 3$$

$$V_1(0, -3), \quad V_2(0, 3), \quad B_1(-\sqrt{3}, 0), \quad B_2(\sqrt{3}, 0)$$

$$\sqrt{b^2 - a^2} = \sqrt{9 - 3} = \sqrt{6}, \quad F_1(0, -\sqrt{6}), \quad F_2(0, \sqrt{6})$$

See Figure 14.8.

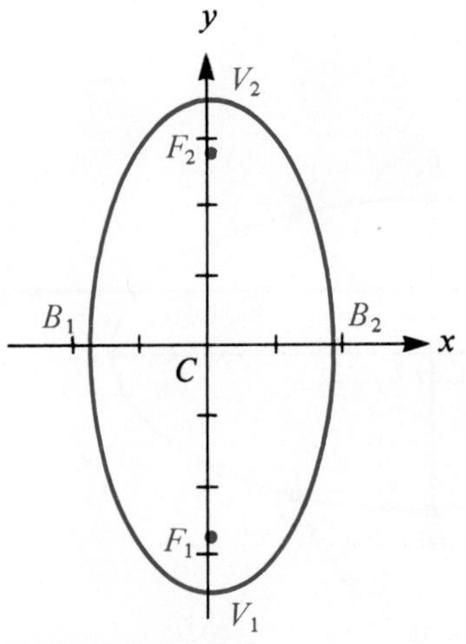

FIGURE 14.8

EXAMPLE Find the center, the vertices, the foci, and graph the ellipse whose equation is $4x^2 - 28x + 16y^2 + 48y = 171$.

Solution We first put the equation in standard form:

$$4(x^2 - 7x) + 16(y^2 + 3y) = 171 \qquad \text{(complete the square)}$$

$$4\left[\left(x - \frac{7}{2}\right)^2 - \frac{49}{4}\right] + 16\left[\left(y + \frac{3}{2}\right)^2 - \frac{9}{4}\right] = 171$$

$$4\left(x - \frac{7}{2}\right)^2 - 49 + 16\left(y + \frac{3}{2}\right)^2 - 36 = 171$$

$$4\left(x - \frac{7}{2}\right)^2 + 16\left(y + \frac{3}{2}\right)^2 = 256$$

$$\frac{\left(x - \frac{7}{2}\right)^2}{\frac{256}{4}} + \frac{\left(y + \frac{3}{2}\right)^2}{\frac{256}{16}} = 1$$

$$\frac{\left(x - \frac{7}{2}\right)^2}{64} + \frac{\left(y + \frac{3}{2}\right)^2}{16} = 1$$

$$C\left(\frac{7}{2}, -\frac{3}{2}\right), \qquad a = \sqrt{64} = 8, \qquad b = \sqrt{16} = 4$$

$$V_1\left(-\frac{9}{2}, -\frac{3}{2}\right), \qquad V_2\left(\frac{23}{2}, -\frac{3}{2}\right), \qquad B_1\left(\frac{7}{2}, -\frac{11}{2}\right), \qquad B_2\left(\frac{7}{2}, \frac{5}{2}\right)$$

$$\sqrt{a^2 - b^2} = 4\sqrt{3}, \qquad F_1\left(\frac{7}{2} - 4\sqrt{3}, \frac{3}{2}\right), \qquad F_2\left(\frac{7}{2} + 4\sqrt{3}, -\frac{3}{2}\right)$$

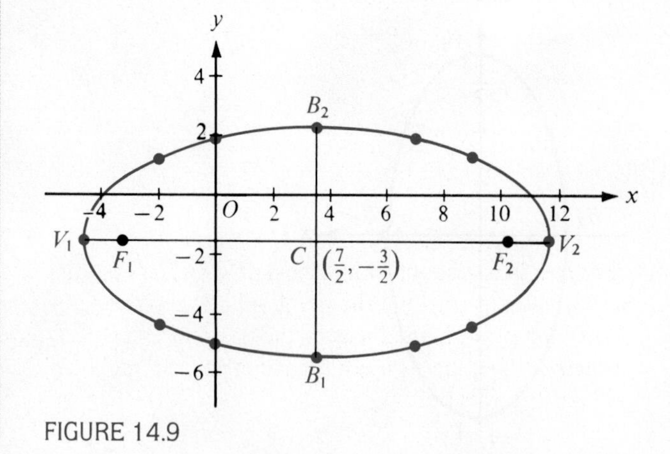

FIGURE 14.9

The following is a table of coordinates of points on the graph (see Figure 14.9). Irrational values of y are approximated to their rational equivalents correct to one decimal place.

x	-2	-2	0	0	3.5	3.5	7	7	9	9
y	1.4	-4.4	-5.1	2.1	-5.5	2.5	-5.1	2.1	1.4	-4.4

Exercise 14.3

Find the center, the vertices, the foci, and sketch the ellipses whose equations are as follows:

1. $x^2 + 4y^2 = 4$
2. $x^2 + 4y^2 = 16$
3. $x^2 + 9y^2 = 9$
4. $x^2 + 9y^2 = 36$
5. $4x^2 + 9y^2 = 36$
6. $x^2 + 16y^2 = 16$
7. $9x^2 + 16y^2 = 144$
8. $4x^2 + y^2 = 4$
9. $9x^2 + y^2 = 9$
10. $4x^2 + y^2 = 16$
11. $9x^2 + y^2 = 36$
12. $16x^2 + y^2 = 16$
13. $9x^2 + 4y^2 = 36$
14. $16x^2 + 9y^2 = 144$
15. $x^2 + 2y^2 = 8$
16. $x^2 + 3y^2 = 12$
17. $28x^2 + 36y^2 = 63$
18. $6x^2 + 10y^2 = 15$
19. $6x^2 + y^2 = 24$
20. $8x^2 + y^2 = 32$
21. $3x^2 + 2y^2 = 4$
22. $9x^2 + 2y^2 = 6$
23. $2x^2 + y^2 + 8x = 0$
24. $3x^2 + y^2 - 6x = 9$
25. $x^2 + 3y^2 - 12y + 9 = 0$
26. $x^2 + 2y^2 + 4y = 6$
27. $9x^2 + 16y^2 - 18x + 32y = 119$
28. $x^2 + 4y^2 + 2x + 16y + 1 = 0$
29. $4x^2 + 9y^2 - 16x - 54y + 61 = 0$
30. $x^2 + 9y^2 + 4x - 18y + 4 = 0$
31. $2x^2 + y^2 + 12x - 8y + 22 = 0$
32. $9x^2 + 2y^2 - 18x + 12y = 9$
33. $4x^2 + 8y^2 + 12x - 40y = 69$

14.4 The Parabola

DEFINITION

A **parabola** is a set of points each of which is equidistant from a fixed point and a line not through the point, all lying in the same plane.
The fixed point is the **focus** of the parabola and the fixed line is the **directrix**. The line through the focus perpendicular to the directrix is the **axis** of the parabola.

Equation of a Parabola

Let $F(g + a, h)$, $a > 0$, be the focus and the line D_1D_2, whose equation is $x = g - a$, be the directrix of the parabola. Let $P(x, y)$ be any point on the parabola. Let A be the foot of the perpendicular from P on D_1D_2. Then A has coordinates $(g - a, y)$.

By definition $AP = FP$. Thus

$$x - (g - a) = \sqrt{[x - (g + a)]^2 + (y - h)^2}$$

See Figure 14.10.

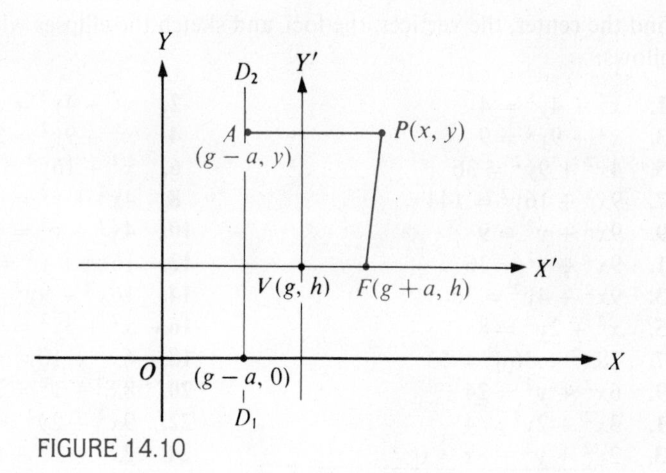

FIGURE 14.10

Simplifying we get (see Appendix E)

$$(y - h)^2 = 4a(x - g)$$

The previous equation is the **standard form** of the equation of a parabola. The point $V(g, h)$ is called the **vertex** of the parabola. The focus lies at $(g + a, h)$ and the equation of the directrix is $x = g - a$. The equation of the axis is $y = h$.

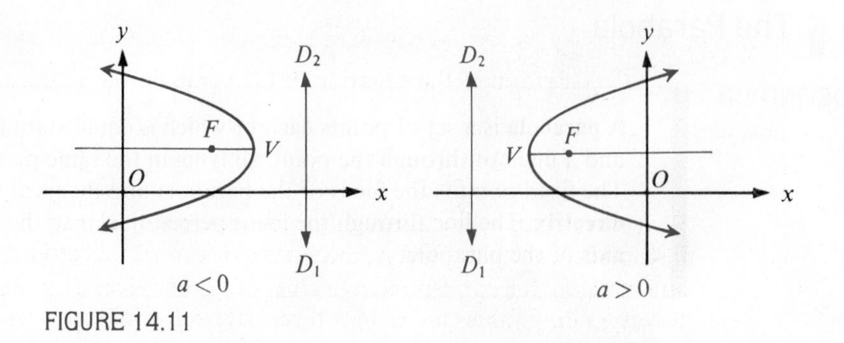

FIGURE 14.11

The domain of the extent is

$$\{x \mid x \geq g\} \text{ if } a > 0, \quad \text{and} \quad \{x \mid x \leq g\} \text{ if } a < 0$$

The range of the extent is $\{y \mid y \in R\}$
See Figure 14.11.

Notes

1. The parabola opens to the right if $a > 0$, and opens to the left when $a < 0$.
2. When the vertex of the parabola is at the origin, that is, $g = 0$ and $h = 0$, the equation of the parabola becomes $y^2 = 4ax$.

Another Form of the Equation of a Parabola

The equation

$$(x - g)^2 = 4a(y - h)$$

represents a parabola with the vertex at (g, h), the focus at $(g, h + a)$, the directrix with equation $y = h - a$, and the axis with equation $x = g$.
See Figure 14.12.

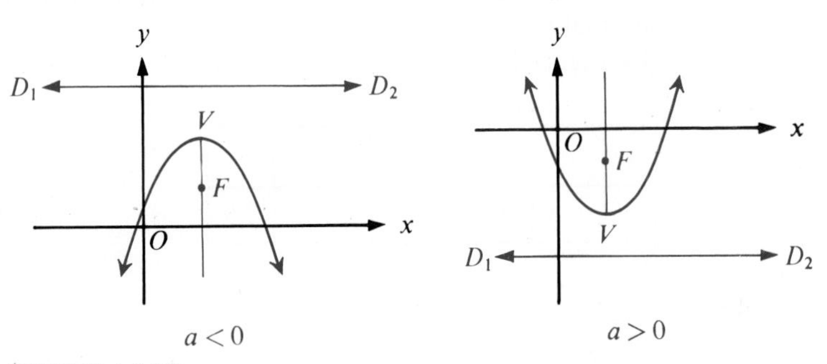

FIGURE 14.12

The domain of the extent is $\{x \mid x \in R\}$.
The range of the extent is

$$\{y \mid y \geq h\} \text{ if } a > 0, \quad \text{and} \quad \{y \mid y \leq h\} \text{ if } a < 0$$

Notes

1. The parabola opens upward if $a > 0$, and opens downward if $a < 0$.
2. This form of the equation of the parabola was discussed in Chapter 13.

Graphing the Parabola

Write the equation in standard form, from which $g, h,$ and a can be found by inspection. Locate the vertex; draw the axis and the directrix. Select values of x and compute the corresponding values of y, or values of y and compute the corresponding values of x. Plot these points and join them by a smooth curve.

EXAMPLE

Sketch the parabola $y^2 - 6x - 6y = 3$.

Solution

We first write the equation in standard form:

$$(y^2 - 6y) - 6x = 3$$
$$(y-3)^2 - 9 - 6x = 3$$
$$(y-3)^2 = 6x + 12$$
$$(y-3)^2 = 6(x+2)$$

Comparing this equation with the equation $(y-h)^2 = 4a(x-g)$, we have

$$g = -2, \qquad h = 3, \qquad a = \frac{3}{2}$$

The vertex is $V(-2, 3)$.

Equation of the axis is $y = 3$.

Equation of the directrix is $x = -\frac{7}{2}$.

The following is a table of coordinates of points on the graph. Irrational values of y are approximated to their rational equivalents correct to one decimal place.
See Figure 14.13.

x	-2	0	0	2	2
y	3	$-.5$	6.5	-1.9	7.9

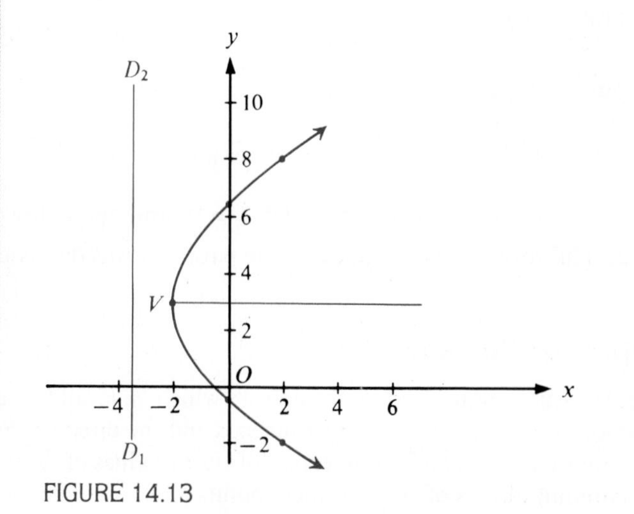

FIGURE 14.13

Exercise 14.4

Find the vertex, the focus, the equation of the directrix, and sketch the parabolas whose equations are as follows:

1. $y^2 = 4x$
2. $y^2 = 8x$
3. $y^2 = 2x$
4. $y^2 = 5x$
5. $y^2 = -12x$
6. $y^2 = -x$
7. $y^2 = -6x$
8. $y^2 = -10x$
9. $x^2 = 4y$
10. $x^2 = 8y$
11. $x^2 = 2y$
12. $x^2 = 3y$
13. $x^2 = -4y$
14. $x^2 = -12y$
15. $x^2 = -2y$
16. $x^2 = -5y$
17. $y^2 = 4x - 4$
18. $y^2 = 8x - 24$
19. $y^2 = x + 1$
20. $y^2 = 6x + 12$
21. $y^2 + 4x = 8$
22. $y^2 + 16x = 16$
23. $y^2 + 12x + 12 = 0$
24. $y^2 + 2x + 8 = 0$
25. $y^2 - 6y + 8x + 9 = 0$
26. $y^2 - 2y + 4x + 1 = 0$
27. $y^2 + 2y + 2x + 1 = 0$
28. $y^2 + 8y + 5x + 16 = 0$
29. $y^2 - 2y - 12x + 1 = 0$
30. $y^2 - 4y - x + 4 = 0$
31. $y^2 + 2y - 4x + 1 = 0$
32. $y^2 + 4y - 3x + 4 = 0$
33. $x^2 + 6x - 2y + 9 = 0$
34. $x^2 + 2x - y + 1 = 0$
35. $x^2 - 4x - 4y + 4 = 0$
36. $x^2 - 2x - 6y + 1 = 0$
37. $x^2 + 6x + 12y + 9 = 0$
38. $x^2 + 2x + 7y + 1 = 0$
39. $x^2 - 4x + 8y + 4 = 0$
40. $x^2 - 8x + 6y + 16 = 0$
41. $y^2 - 4x + 2y + 9 = 0$
42. $y^2 - 6x - 4y - 8 = 0$
43. $y^2 + 4x + 6y + 5 = 0$
44. $y^2 + 6x - 8y + 34 = 0$
45. $x^2 - 2x - 12y + 25 = 0$
46. $x^2 + 4x - y = 0$
47. $x^2 + 12x + y + 30 = 0$
48. $x^2 - 6x + 2y + 2 = 0$

14.5 The Hyperbola

DEFINITION

A **hyperbola** is a set of points in a plane each of which has the property that the difference between the distances from any point in the set to two fixed points in the plane is a constant.

The two fixed points are called the **foci** of the hyperbola.

Equation of the Hyperbola

Let the foci be $F_1(g - k, h)$ and $F_2(g + k, h)$, and the constant distance equal $2a$, $a < k$. Let $P(x, y)$ be any point on the hyperbola. Hence

$$PF_1 - PF_2 = 2a$$

or $$\sqrt{[x - (g - k)]^2 + (y - h)^2} - \sqrt{[x - (g + k)]^2 + (y - h)^2} = 2a$$

See Figure 14.14.

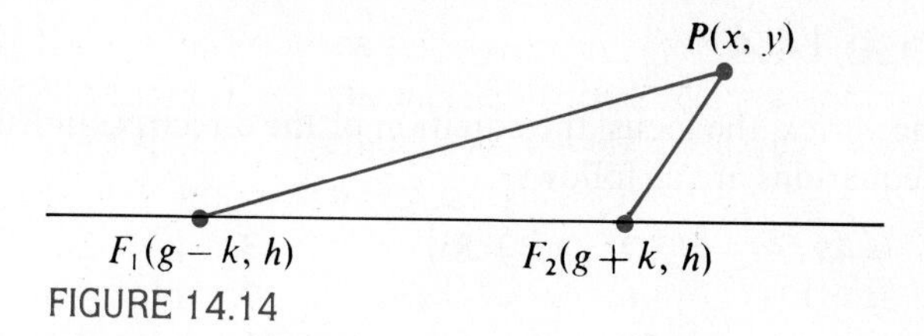

FIGURE 14.14

Simplifying and putting $k^2 - a^2 = b^2$, we get (see Appendix E)

$$\frac{(x-g)^2}{a^2} - \frac{(y-h)^2}{b^2} = 1$$

This equation is the **standard form** of the equation of the hyperbola. The point $C(g, h)$ is the **center** of the hyperbola. The foci are $F_1(g - \sqrt{a^2 + b^2}, h)$ and $F_2(g + \sqrt{a^2 + b^2}, h)$. The lines $x = g$ and $y = h$ are called the **axes of the hyperbola**. The **vertices** are the points $V_1(g - a, h)$ and $V_2(g + a, h)$. The line segment joining the vertices is called the **transverse axis** and is parallel to the x-axis.

The hyperbola has two branches: one opening to the left, through the vertex $V_2(g - a, h)$, and the other opening to the right, through the vertex $V_2(g + a, h)$. See Figure 14.15.

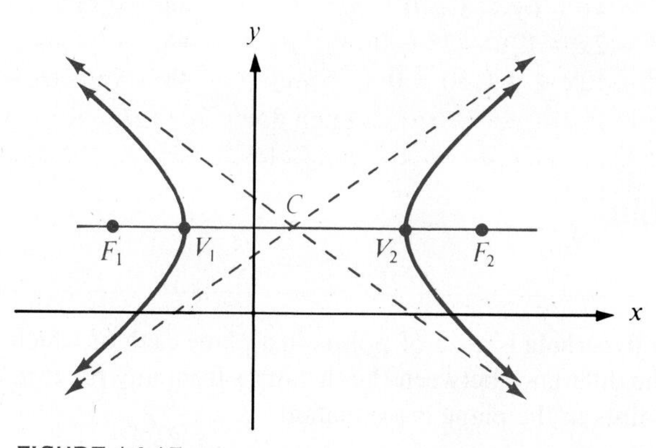

FIGURE 14.15

The two branches of the hyperbola always lie between two lines through its center. The equations of the two lines are obtained by equating the left member of the standard equation of the hyperbola to zero:

$$\frac{(x-g)^2}{a^2} - \frac{(y-h)^2}{b^2} = 0$$

which gives, as equations for the lines,

$$(y - h) = -\frac{b}{a}(x - g) \qquad \text{and} \qquad (y - h) = \frac{b}{a}(x - g)$$

These lines each have the property that as x gets farther and farther from the center of the hyperbola, the distance from the line to the hyperbola gets smaller and smaller. The lines are called the **asymptotes** of the hyperbola.
The domain of the extent is $\{x \mid x \leq g - a \text{ or } x \geq g + a\}$.
The range of the extent is $\{y \mid y \in R\}$.

Note If the center of the hyperbola is at the origin, that is, $g = 0, h = 0$, the equation becomes

$$\frac{x^2}{a^2} - \frac{y^2}{b^2} = 1$$

Another Form of the Equation of a Hyperbola

The equation

$$\frac{(y-h)^2}{b^2} - \frac{(x-g)^2}{a^2} = 1$$

is called the **conjugate hyperbola** to that given above. In this case the transverse axis is parallel to the y-axis.
The vertices are $V_1(g, h-b)$ and $V_2(g, h+b)$.
The foci are $F_1(g, h - \sqrt{a^2+b^2})$ and $F_2(g, h + \sqrt{a^2+b^2})$.
The branches of this hyperbola open down and up instead of to the left and right.
It shares the same center and asymptotes as the hyperbola above.
See Figure 14.16.

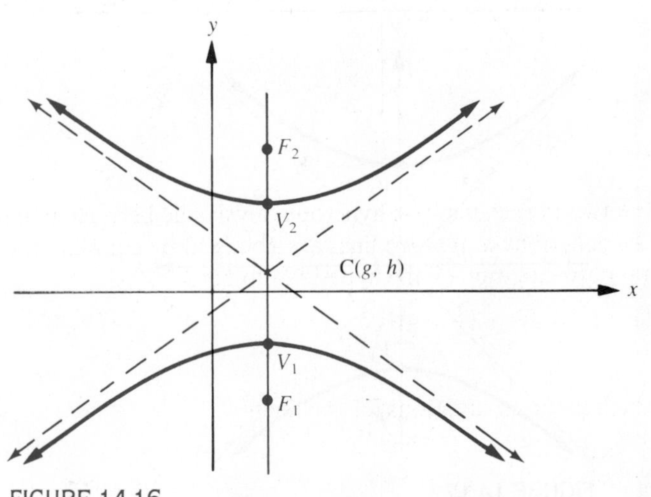

FIGURE 14.16

Graphing the Hyperbola

Write the equation in standard form. Obtain g, h, a, and b by inspection. Plot the center. Determine and plot the vertices. Set the left member of the equation equal to zero, determine the equations of the asymptotes and plot them. Select values of x and compute the corresponding values of y. Plot these points and connect them by a smooth curve.

EXAMPLE

Find the center, the vertices, the foci, the equations of the asymptotes, and sketch the hyperbola whose equation is $9y^2 - 8x^2 = 144$.

Solution

$$9y^2 - 8x^2 = 144$$

$$\frac{9y^2}{144} - \frac{8x^2}{144} = 1$$

$$\frac{y^2}{16} - \frac{x^2}{18} = 1$$

$C(0, 0)$, $\quad b = \sqrt{16} = 4$, $\quad a = \sqrt{18} = 3\sqrt{2}$

$V_1(0, -4)$, $\quad V_2(0, 4)$

$\sqrt{a^2 + b^2} = \sqrt{18 + 16} = \sqrt{34}$ $\quad F_1(0, -\sqrt{34})$,

$F_2(0, \sqrt{34})$

The equations of the asymptotes are $y = \pm\dfrac{2\sqrt{2}}{3}x$.

See Figure 14.17.

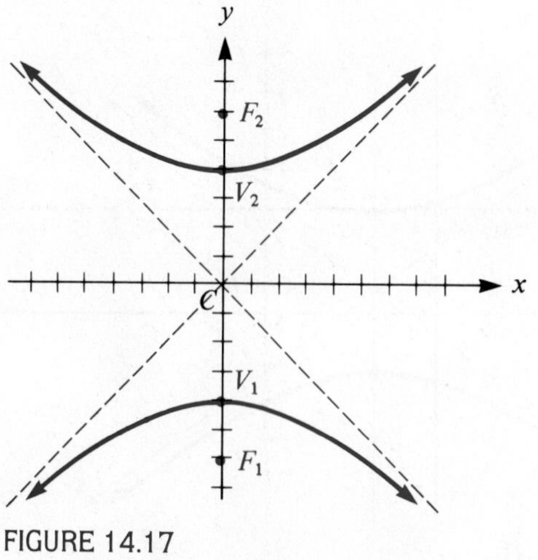

y, x, F_2, V_2, C, V_1, F_1

FIGURE 14.17

EXAMPLE

Find the center, the vertices, the foci, the equations of the asymptotes, and sketch the hyperbola whose equation is

$$9x^2 - 16y^2 + 18x - 64y = 199$$

Solution

Write the equation in standard form:

$$(9x^2 + 18x) - (16y^2 + 64y) = 199$$
$$9(x^2 + 2x) - 16(y^2 + 4y) = 199$$
$$9[(x + 1)^2 - 1] - 16[(y + 2)^2 - 4] = 199$$
$$9(x + 1)^2 - 9 - 16(y + 2)^2 + 64 = 199$$
$$9(x + 1)^2 - 16(y + 2)^2 = 144$$
$$\frac{(x + 1)^2}{16} - \frac{(y + 2)^2}{9} = 1$$

$C(-1, -2)$, $a = \sqrt{16} = 4$, $b = \sqrt{9} = 3$

$V_1(-5, -2)$, $V_2(3, -2)$

$\sqrt{a^2 + b^2} = \sqrt{16 + 9} = 5$ $F_1(-6, -2)$, $F_2(4, -2)$

Equations of the asymptotes are $y = -2 \pm \frac{3}{4}(x + 1)$.

The following is a table of coordinates of points on the graph. Irrational values of y are approximated to their rational equivalents correct to one decimal place. See Figure 14.18.

x	−8	−8	−6	−6	−5	3	4	4	6	6
y	−6.3	2.3	−4.25	.25	−2	−2	−4.25	.25	−6.3	2.3

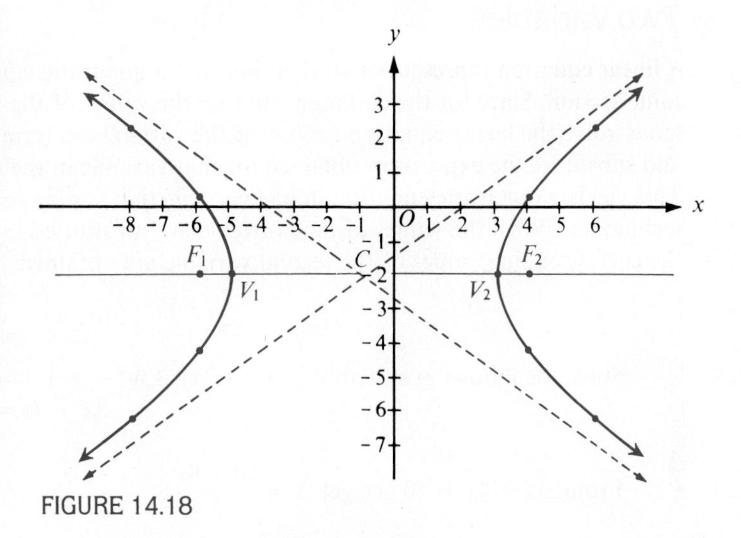

FIGURE 14.18

Exercise 14.5

Find the center, the vertices, the foci, the equations of the asymptotes, and sketch the hyperbolas whose equations are as follows:

1. $x^2 - 4y^2 = 4$ **2.** $x^2 - 4y^2 = 16$ **3.** $9x^2 - 4y^2 = 36$
4. $9x^2 - 8y^2 = 72$ **5.** $16y^2 - 25x^2 = 400$ **6.** $9y^2 - 16x^2 = 144$
7. $3y^2 - 2x^2 = 18$ **8.** $9y^2 - 2x^2 = 36$ **9.** $64x^2 - 81y^2 = 144$
10. $4x^2 - 9y^2 = 1$ **11.** $16y^2 - 25x^2 = 100$ **12.** $25y^2 - 81x^2 = 225$
13. $4x^2 - 9y^2 - 8x = 32$ **14.** $4x^2 - 25y^2 + 16x = 84$
15. $3x^2 - 4y^2 - 8y = 52$ **16.** $2x^2 - y^2 + 6y = 25$
17. $2x^2 - 9y^2 - 4x - 18y = 43$ **18.** $x^2 - 2y^2 + 4x + 4y = 14$
19. $4x^2 - 3y^2 + 8x + 12y = 32$ **20.** $2x^2 - y^2 - 8x - 4y = 14$
21. $4y^2 - 3x^2 - 6x - 16y = 23$ **22.** $2y^2 - x^2 + 4x - 4y = 10$
23. $2y^2 - 3x^2 - 18x + 8y = 43$ **24.** $3y^2 - 4x^2 + 24x + 6y = 81$

14.6 Solution of Systems of Quadratic Equations in Two Variables

The graphical solution of a system of quadratic equations in two variables is the set of coordinates of the points of intersection of the two graphs. The graphical solution gives only the real values of the two variables. Graphing is not always accurate. Even the real values of the solution we obtain are approximate ones. Thus we shall discuss only algebraic solutions since these give the exact values of the variables in the complex number system.

The following is a discussion of special types of quadratic equations.

Solution of a Linear and a Quadratic Equation in Two Variables

A linear equation represents a straight line and a quadratic equation represents a conic section. Since for the common solution the values of the variables are the same, solve the linear equation for one of the variables in terms of the other variable and substitute the expression obtained for that variable in the quadratic equation. This yields a quadratic equation in one variable that can be solved by demonstrated techniques. When the values of that variable are substituted in the linear equation, the corresponding values of the second variable are obtained.

EXAMPLE Solve the following system:

$$x^2 + 2xy + y^2 - x + y = 8$$
$$3x + 4y = 10$$

Solution From $3x + 4y = 10$ we get $x = \dfrac{10 - 4y}{3}$.

Substituting $\dfrac{10-4y}{3}$ for x in $x^2+2xy+y^2-x+y=8$ we get

$$\left(\frac{10-4y}{3}\right)^2+2\left(\frac{10-4y}{3}\right)y+y^2-\left(\frac{10-4y}{3}\right)+y=8$$

$$\frac{100-80y+16y^2}{9}+\frac{20y-8y^2}{3}+y^2-\frac{10-4y}{3}+y=8$$

$$100-80y+16y^2+3(20y-8y^2)+9y^2-3(10-4y)+9y=72$$

Hence $\quad y^2+y-2=0 \quad$ or $\quad (y+2)(y-1)=0$

Thus $\quad y=-2 \quad$ or $\quad y=1$

When $y=2$, $\quad x=\dfrac{10-4(-2)}{3}=6$

when $y=1$ $\quad x=\dfrac{10-4(1)}{3}=2$

Thus the solution set is $\{(2,1),(6,-2)\}$.

Exercise 14.6A

Solve the following systems:

1. $3x^2+y^2=4$, $y-x=0$
2. $x^2+y^2=25$, $x+y=7$
3. $x^2-2y^2=2$, $x-2y=0$
4. $x^2+6xy=-2$, $x+4y=0$
5. $y^2+3xy=-9$, $7y+12x=3$
6. $2x^2-xy=2$, $x-y=-1$
7. $2x^2-xy=6$, $x-y=1$
8. $3x^2-xy=2$, $x-y=-3$
9. $x^2-y^2+x=5$, $2y-x=5$
10. $2x^2-y^2-2y=4$, $x-y=1$
11. $2x^2-y^2+x=2$, $x+y=0$
12. $3x^2-y^2+9y=2$, $x+y=1$
13. $6x^2-5xy+5x=-6$, $x-y+2=0$
14. $3y^2-2xy+8y+3=0$, $x-y=2$
15. $4x^2-2y^2+x+4y=3$, $x-y+1=0$
16. $5x^2-2xy+4x=8$, $x-y=3$
17. $2x^2+y^2+5x-2y=6$, $x+y=2$
18. $x^2+y^2+10x-15y+60=0$, $y-4x=19$
19. $x^2+y^2+x-13y+34=0$, $x+y=5$
20. $x^2+y^2+7x-19y+70=0$, $x+3y=10$
21. $x^2+y^2-4x-8y=5$, $2x-y=5$
22. $x^2+y^2-3x-15y+52=0$, $2x+3y=19$
23. $x^2+y^2-2x-11y+15=0$, $2x+3y=12$
24. $x^2+y^2+3x-4y=10$, $3x+2y+7=0$

25. $x^2 - 3xy - 2y^2 + 4 = 0$
$2y + x = 4$

26. $x^2 + 2xy - y^2 + x - 2y = 3$
$3y - 4x = 1$

27. $x^2 + y^2 + 3x - 6y = 5$
$4y - 7x + 10 = 0$

28. $2x^2 - y^2 - x - y + 3 = 0$
$y + x = 1$

29. $3x^2 + y^2 + 2x + 3y = 26$
$9y - 10x = 47$

30. $x^2 + 3xy + y^2 - 2x = 17$
$y + x = 4$

31. $2x^2 - xy - 4y^2 + y = 16$
$6y + 13x = 33$

32. $x^2 + 4xy - 3y^2 + 2x - 4y + 12 = 0$
$8y - x + 18 = 0$

33. $y^2 - x^2 + 6y = -9$
$x - y = 3$

34. $x^2 - 4y^2 + 4y = 1$
$x - 2y + 1 = 0$

35. $y^2 - x^2 + 2x = 1$
$x - y = 1$

36. $4y^2 - x^2 - 4x = 4$
$2y - x = 2$

37. $x^2 + xy + y^2 - 2x = 7$
$x + y = 6$

38. $x^2 + y^2 - 2x + 2y = 23$
$y - x = 8$

39. $4x^2 + 9y^2 + 16x - 18y = 11$
$x - y = 2$

40. $25x^2 - 16y^2 + 50x + 32y = 391$
$3x - 2y = 1$

Solution of a System of Two Quadratic Equations of the Form $Ax^2 + Ay^2 + Dx + Ey + F = 0$

An equation of the form $Ax^2 + Ay^2 + Dx + Ey + F = 0$, $A \neq 0$, represents a circle.
When the two circles are concentric, that is, when they have the same center but different radii, the solution set is the null set.

If the two circles are not concentric, elimination of the x^2 and y^2 terms (their coefficients are the same) gives a linear equation of the form $ax + by + c = 0$. Solving the linear equation with one of the two quadratics gives the elements of the solution set of the two quadratic equations.

Note When the two circles intersect in two real points, the linear equation is the equation of their common chord. When the two circles are tangent, the linear equation represents their common tangent.

EXAMPLE Solve the following system:

$$x^2 + y^2 + 5x + 3y + 8 = 0$$
$$x^2 + y^2 + 3x + 4y + 4 = 0$$

Solution Eliminate the square terms between the two equations:

$$x^2 + y^2 + 5x + 3y + 8 = 0 \longrightarrow x^2 + y^2 + 5x + 3y + 8 = 0$$
$$x^2 + y^2 + 3x + 4y + 4 = 0 \xrightarrow{\times(-1)} -x^2 - y^2 - 3x - 4y - 4 = 0$$

Adding, we get

$$2x - y + 4 = 0$$

or

$$y = 2x + 4$$

Substituting $(2x + 4)$ for y in the first equation we get

$$x^2 + (2x + 4)^2 + 5x + 3(2x + 4) + 8 = 0$$
$$x^2 + 4x^2 + 16x + 16 + 5x + 6x + 12 + 8 = 0$$
$$5x^2 + 27x + 36 = 0$$
$$(5x + 12)(x + 3) = 0$$

Hence $\quad x = -\frac{12}{5} \quad$ or $\quad x = -3$

Substituting these values of x in $y = 2x + 4$, we get y:

when $x = -\frac{12}{5}$, $\quad y = 2\left(-\frac{12}{5}\right) + 4 = -\frac{4}{5}$

when $x = -3$, $\quad y = 2(-3) + 4 = -2$

The solution set is $\left\{\left(-\frac{12}{5}, -\frac{4}{5}\right), (-3, -2)\right\}$.

Exercise 14.6B

Solve the following systems:

1. $x^2 + y^2 = 5$
   $x^2 + y^2 - x - y = 4$
2. $x^2 + y^2 = 13$
   $x^2 + y^2 - x + y = 12$
3. $x^2 + y^2 = 2$
   $x^2 + y^2 - x - 2y + 1 = 0$
4. $x^2 + y^2 + x = 13$
   $x^2 + y^2 + 2y = 12$
5. $x^2 + y^2 + 2x = 7$
   $x^2 + y^2 + y = 3$
6. $x^2 + y^2 + x = 21$
   $x^2 + y^2 + 2y = 19$
7. $x^2 + y^2 + 3x = 5$
   $x^2 + y^2 - y = 1$
8. $x^2 + y^2 + x = 3$
   $x^2 + y^2 + 3y = 2$
9. $x^2 + y^2 + 3x + 2y = 7$
   $x^2 + y^2 + 3y = 5$
10. $x^2 + y^2 + x - 4y + 4 = 0$
    $x^2 + y^2 - 6y + 7 = 0$
11. $x^2 + y^2 + x - y = 12$
    $x^2 + y^2 - x = 11$
12. $x^2 + y^2 - x + y = 6$
    $x^2 + y^2 - 3x = 1$
13. $x^2 + y^2 + 3x - y = 10$
    $x^2 + y^2 + x - 2y = 5$
14. $x^2 + y^2 - x + 7y = 74$
    $x^2 + y^2 - 2x + 8y = 72$
15. $x^2 + y^2 + 4x + y = 7$
    $x^2 + y^2 + 3x - 2y = 3$
16. $x^2 + y^2 + 3x - 4y = 25$
    $x^2 + y^2 + 2x - 3y = 22$
17. $x^2 + y^2 + 4x - 3y = 25$
    $x^2 + y^2 + 3x - 4y = 23$
18. $x^2 + y^2 - x - 3y + 2 = 0$
    $x^2 + y^2 - 4x - 2y + 3 = 0$
19. $x^2 + y^2 - 2x - 3y = 10$
    $x^2 + y^2 - 5x - 4y = 6$
20. $x^2 + y^2 - x + y = 48$
    $x^2 + y^2 - 2x + 3y = 45$
21. $x^2 + y^2 + 3x - 6y = 55$
    $x^2 + y^2 + 2x - 4y = 56$
22. $x^2 + y^2 + x - 3y - 6 = 0$
    $x^2 + y^2 - 2x + y + 1 = 0$

23. $x^2 + y^2 + 2x - 5y = 4$
$x^2 + y^2 - x - 3y = 0$

24. $x^2 + y^2 + 8x - y = 86$
$x^2 + y^2 + 6x + 2y = 80$

25. $x^2 + y^2 - 2x - 2y = 2$
$x^2 + y^2 + 6x - 2y + 6 = 0$

26. $x^2 + y^2 - 8x - 2y = 3$
$x^2 + y^2 - 2x - 14y + 45 = 0$

27. $x^2 + y^2 + 4x - 10y + 24 = 0$
$x^2 + y^2 - 2x + 2y = 18$

28. $4x^2 + 4y^2 - 32x - 48y = -183$
$4x^2 + 4y^2 + 16x + 16y = 193$

29. $x^2 + y^2 - 2x + 2y + 1 = 0$
$x^2 + y^2 + 2x - 2y + 1 = 0$

30. $x^2 + y^2 - 6x - 4y + 9 = 0$
$x^2 + y^2 + 6x + 4y + 9 = 0$

31. $x^2 + y^2 + 4x + 6y = 7$
$x^2 + y^2 + 4x + 6y = 20$

32. $x^2 + y^2 - 2x + 3y = 5$
$x^2 + y^2 - 2x + 3y = 11$

Solution of a System of Two Quadratic Equations of the Form $Ax^2 + Cy^2 + F = 0$

An equation of the form $Ax^2 + Cy^2 + F = 0$ represents either a circle, an ellipse, or a hyperbola. Elimination of x^2 or y^2 between the two equations yields a quadratic equation in one variable that when solved gives two values for that variable. Each value when substituted in one of the original equations gives two values for the second variable. Note that the two values may coincide.

EXAMPLE

Solve the following system:
$$3x^2 - y^2 = 69$$
$$4x^2 + 3y^2 = 209$$

Solution

Eliminate y^2 between the two equations:

$$3x^2 - y^2 = 69 \xrightarrow{\times(3)} 9x^2 - 3y^2 = 207$$
$$4x^2 + 3y^2 = 209 \longrightarrow 4x^2 + 3y^2 = 209$$

Adding, we get $13x^2 = 416$

or $x^2 = 32$

Hence $x = \pm\sqrt{32} = \pm 4\sqrt{2}$

Substituting $(-4\sqrt{2})$ for x in the first equation, we get

$3(32) - y^2 = 69$ or $y^2 = 27$

Hence $y = \pm 3\sqrt{3}$

Substituting $(4\sqrt{2})$ for x in the first equation, we get

$y = \pm 3\sqrt{3}$

Thus the solution set is

$$\{(-4\sqrt{2}, -3\sqrt{3}), (-4\sqrt{2}, 3\sqrt{3}), (4\sqrt{2}, -3\sqrt{3}), (4\sqrt{2}, 3\sqrt{3})\}$$

Exercise 14.6C

Solve the following systems:

1. $x^2 + y^2 = 9$, $16x^2 - 9y^2 = 144$
2. $x^2 + y^2 = 25$, $3x^2 + 25y^2 = 75$
3. $x^2 + y^2 = 4$, $x^2 + 4y^2 = 16$
4. $y^2 + 8x^2 = 16$, $5y^2 - 16x^2 = 80$
5. $4x^2 - y^2 = -5$, $3x^2 + 2y^2 = 21$
6. $x^2 + 7y^2 = 11$, $2x^2 - y^2 = 7$
7. $x^2 + y^2 = 40$, $3x^2 + y^2 = 112$
8. $5x^2 - 2y^2 = 13$, $x^2 - 3y^2 = -39$
9. $x^2 - 2y^2 = 17$, $3x^2 + y^2 = 79$
10. $x^2 + 8y^2 = 18$, $3x^2 - 4y^2 = 47$
11. $4x^2 - 2y^2 = -9$, $8x^2 - 3y^2 = -9$
12. $8x^2 - 4y^2 = -23$, $12x^2 + 16y^2 = 103$
13. $2x^2 + y^2 = 14$, $3x^2 + 4y^2 = 51$
14. $2x^2 + y^2 = 6$, $3x^2 - 2y^2 = 2$
15. $x^2 - 3y^2 = 3$, $4x^2 - y^2 = 67$
16. $x^2 - 2y^2 = -45$, $2x^2 - y^2 = -18$
17. $4x^2 + 4y^2 = 9$, $3x^2 + 3y^2 = 49$
18. $x^2 + y^2 = 5$, $3x^2 + 4y^2 = 12$
19. $4y^2 + x^2 = 30$, $3y^2 - 2x^2 = 72$
20. $x^2 + 4y^2 = 4$, $4x^2 - 9y^2 = 36$

Solution of a System of Two Quadratic Equations of the Form $Ax^2 + Bxy + Cy^2 + F = 0$

When at least one of the equations $Ax^2 + Bxy + Cy^2 + F = 0$ is such that $B \neq 0$, elimination of F between the two equations gives an equation of the form $ax^2 + bxy + cy^2 = 0$. When $ax^2 + bxy + cy^2 = 0$ is factored we get $(a_1x + b_1y)(a_2x + b_2y) = 0$. Setting each factor to zero we obtain two linear equations. Solving each of the linear equations with one of the quadratics, we obtain the solution set of the given system.

EXAMPLE Solve the following system:

$$x^2 + 4xy + 3y^2 = 10$$
$$x^2 + xy + y^2 = 6$$

Solution Eliminate the constant between the two equations:

$$x^2 + 4xy + 3y^2 = 10 \xrightarrow{\times(-3)} -3x^2 - 12xy - 9y^2 = -30$$
$$x^2 + xy + y^2 = 6 \xrightarrow{\times(5)} 5x^2 + 5xy + 5y^2 = 30$$

Adding, we get $2x^2 - 7xy - 4y^2 = 0$

or $(x - 4y)(2x + y) = 0$

Hence $x - 4y = 0$, $x = 4y$

or $2x + y = 0$, $y = -2x$

First: Solve $x = 4y$ and $x^2 + xy + y^2 = 6$

$$(4y)^2 + (4y)(y) + y^2 = 6$$
$$16y^2 + 4y^2 + y^2 = 6$$
$$21y^2 = 6$$
$$y^2 = \frac{2}{7}$$
$$y = \pm\sqrt{\frac{2}{7}} = \pm\frac{\sqrt{14}}{7}, \qquad x = 4y = \pm\frac{4\sqrt{14}}{7}$$

The solution set is $\left\{\left(\frac{4\sqrt{14}}{7}, \frac{\sqrt{14}}{7}\right), \left(-\frac{4\sqrt{14}}{7}, -\frac{\sqrt{14}}{7}\right)\right\}$.

Second: Solve $y = -2x$ and $x^2 + xy + y^2 = 6$

$$x^2 + x(-2x) + (-2x)^2 = 6$$
$$x^2 - 2x^2 + 4x^2 = 6$$
$$3x^2 = 6$$
$$x^2 = 2$$
$$x = \pm\sqrt{2}, \qquad y = -2x = \mp 2\sqrt{2}$$

The solution set is $\{(\sqrt{2}, -2\sqrt{2}), (-\sqrt{2}, 2\sqrt{2})\}$.

The solution set of the system is the union of the solution sets of first and second cases. Hence the solution set is

$$\left\{\left(\frac{4\sqrt{14}}{7}, \frac{\sqrt{14}}{7}\right), \left(-\frac{4\sqrt{14}}{7}, -\frac{\sqrt{14}}{7}\right), (\sqrt{2}, -2\sqrt{2}), (-\sqrt{2}, 2\sqrt{2})\right\}$$

Exercise 14.6D

Solve the following systems:

1. $x^2 + xy + y^2 = 1$
$2x^2 - y^2 = 1$

2. $4x^2 - 3xy + 3y^2 = 16$
$x^2 + y^2 = 8$

3. $x^2 + xy + 2y^2 = 14$
$x^2 - 2y^2 = 7$

4. $5y^2 - xy - x^2 = 21$
$3y^2 - xy = 14$

5. $3x^2 + xy + 2y^2 = 2$
$4x^2 + 4xy = 3$

6. $3x^2 + xy + 2y^2 = 6$
$3xy + 6y^2 = 2$

7. $x^2 + 4xy + 7y^2 = 9$
$xy + 3y^2 = 3$

8. $2x^2 + xy + y^2 = 2$
$3x^2 + 2xy + 3y^2 = 4$

9. $x^2 + 2xy + 3y^2 = 6$
$2x^2 + 3xy + 5y^2 = 14$

10. $2x^2 - xy + 3y^2 = 12$
$2x^2 + xy + 3y^2 = 16$

11. $x^2 + 2xy + 2y^2 = 1$
$2x^2 + 5xy + 8y^2 = 3$

12. $2x^2 + xy + 3y^2 = 6$
$x^2 + 4xy + y^2 = 6$

13. $4x^2 + 5xy + 3y^2 = 6$
$3x^2 + 4xy + 3y^2 = 5$

14. $2x^2 + xy + 5y^2 = 12$
$x^2 + 2xy + 2y^2 = 3$

15. $4x^2 - 3xy + y^2 = 6$
$3x^2 - 4xy + 2y^2 = 9$

16. $4x^2 + 3xy + 2y^2 = 12$
$3x^2 + 2xy + 6y^2 = 18$

17. $3x^2 + 4xy + y^2 = 4$
$2x^2 - 7xy + 6y^2 = 8$

18. $3x^2 + 6xy - y^2 = 12$
$2x^2 + 5xy = 9$

19. $6x^2 + 5xy - y^2 = -6$
$2x^2 + 5xy + y^2 = -3$

20. $x^2 + 2xy - 2y^2 = 10$
$x^2 + 4xy + y^2 = 15$

Chapter 14 Review

1. Find the center and radius of the circle $x^2 + y^2 + 6x - 8y = 56$.
2. Find the center and radius of the circle $4x^2 + 4y^2 - 16x - 12y = 103$.

Put the equations of the following ellipses in standard form. Find the center, the major axis, the minor axis, the foci, and sketch:

3. $9x^2 + 16y^2 - 18x - 32y = 119$
4. $2x^2 + 3y^2 - 16x = 4$
5. $3x^2 + 2y^2 + 12x + 8y = 28$
6. $14x^2 + 6y^2 - 84x + 12y + 111 = 0$

Sketch the following parabolas. Also find the coordinates of the vertex, the focus, and the equation of the directrix:

7. $y^2 - 4x + 2y = 3$
8. $y^2 - 2x - 6y + 11 = 0$
9. $y^2 + 6x - 8y + 4 = 0$
10. $y^2 + 12x - 2y + 49 = 0$
11. $x^2 + 2x - 3y + 1 = 0$
12. $x^2 - 4x - y + 2 = 0$

Sketch the following hyperbolas. Find the coordinates of the center, the vertices, the foci, and the equations of the asymptotes:

13. $x^2 - 4y^2 - 2x + 8y = 19$
14. $2x^2 - y^2 - 12x - 2y = 1$
15. $3y^2 - 4x^2 + 24y + 16x = 16$
16. $3y^2 - 2x^2 - 6y - 12x = 33$

Solve the following systems:

17. $x^2 + y^2 = 13$
$x - y = 1$

18. $2x^2 + y^2 - 2x - 9y + 14 = 0$
$y - x = 1$

19. $2x^2 + y^2 - 4x - 3y = 34$
$y + 2x = 2$

20. $9x^2 + 6xy - 4y^2 + 12x - 46y = 85$
$3x - 2y = 7$

21. $18x^2 - 12xy - 8y^2 + 3x - 56y = 71$
$3x - 4y = 5$

22. $4x^2 + 6xy - 5y^2 - 2x + 8y + 13 = 0$
$2x + 5y = -1$

23. $x^2 + y^2 + 6x - 2y = 7$
$x^2 + y^2 + 4x - 3y = 3$

24. $x^2 + y^2 + 5x + 3y = 12$
$x^2 + y^2 + 2x + 4y = 4$

25. $x^2 + y^2 + 8x - 2y = 8$
$x^2 + y^2 + 4x - y = 5$

26. $x^2 + y^2 + x - 2y = 20$
$x^2 + y^2 - x - 3y = 10$

27. $2x^2 + 2y^2 - x - 3y = 50$
$3x^2 + 3y^2 - x - 4y = 77$

28. $3x^2 + 3y^2 + 2x + y = 45$
$4x^2 + 4y^2 + 3x + y = 61$

29. $x^2 + 3y^2 = 7$
$4x^2 - 5y^2 = 11$

30. $2x^2 + y^2 = 22$
$4x^2 - 3y^2 = 24$

31. $3x^2 + y^2 = 10$
$x^2 + 2y^2 = 10$

32. $x^2 - 2y^2 = 12$
$2x^2 + y^2 = 84$

33. $3x^2 + 2y^2 = 19$
$x^2 - 5y^2 = 29$

34. $x^2 + 4y^2 = 7$
$2x^2 + 7y^2 = 10$

35. $x^2 + xy + 3y^2 = 5$
$x^2 + 5xy + 4y^2 = 10$

36. $4x^2 + xy + y^2 = 8$
$5x^2 + xy + 2y^2 = 12$

37. $2x^2 + 3xy + 6y^2 = 8$
$x^2 + xy + 4y^2 = 6$

38. $2x^2 + 5xy - 6y^2 + 36 = 0$
$x^2 + 4xy - 3y^2 + 24 = 0$

39. $4x^2 + xy + 2y^2 = -4$
$6x^2 + 2xy + 7y^2 = -12$

40. $x^2 + 6xy + 2y^2 = 6$
$x^2 + 9xy + 4y^2 = 8$

41. The product of two natural numbers is 48. The product of one number and the sum of the two numbers is 84. Find the numbers.

42. The product of two numbers is 72, and the sum of their reciprocals is $\frac{1}{4}$. What are the numbers?

43. Find two numbers such that the sum of their reciprocals is $\frac{5}{6}$ and the product of their reciprocals is $\frac{1}{9}$.

44. Find two natural numbers whose product is 432 and the sum of their squares is 900.

45. Find two natural numbers whose product is 448 and the difference of their squares is 528.

46. Find two positive numbers such that the sum of their squares is 274 and the difference of their squares is 176.

47. Find two positive numbers such that the square of their sum is 432 more than the square of their difference, and the difference of their squares is 63.

48. The product of the digits of a two-digit number is 6 more than 3 times the sum of the digits. If the digits are reversed, the new number is 20 more than the product of the digits. Find the original number.

49. The product of the digits of a two-digit number is 5 less than 3 times the sum of the digits. If the digits are reversed the new number is 3 more than 4 times the sum of the digits. Find the original number.

50. In a certain positive fraction, the numerator is 3 less than the denominator. If the numerator and denominator are interchanged, the fraction is increased by $\frac{39}{40}$. Find the fraction.

51. The area of a rectangle is 972 square inches and the diagonal is 45 inches. Find the length and width of the rectangle.

52. The diagonal of a rectangle is 35 inches. If the length of the rectangle is increased by 12 inches and the width is increased by 9 inches the diagonal is increased by 15 inches. Find the length and width of the original rectangle.

53. A rectangular flower plot has an area of 1120 square feet and is surrounded by a path 6 feet wide. The area of the path is 960 square feet. Find the length and the width of the flower plot.

54. The area of a picture without its border is 3072 square inches. The border is 3 inches wide and its area is 708 square inches. Find the length and width of the picture.

55. A rectangular piece of tin with an area of 448 square inches is made into an open box. By cutting 2-inch squares from each corner of the tin and folding up the sides, the box has a volume of 576 cubic inches. What were the dimensions of the tin?

56. The area of a right triangle is 30 square inches and its perimeter is 30 inches. Find the length of the hypotenuse.

57. A woman planned to spend \$21 to buy some neckties of a certain kind for Christmas presents. She decided, however, to buy some that cost 75¢ apiece less, because she could thereby get one more tie and save \$1.75 What was the price of the cheaper kind of tie?

58. A workman did $\frac{1}{4}$ of a job and a slower workman finished it. Between them they put in 33 hours. If they had worked together they could have done the work in 14 hours 24 minutes. How long would it have taken each of them to do it alone?

Cumulative Review

Chapter 12 Express each of the following in the form ai:

1. $\sqrt{-4}$ **2.** $\sqrt{-64}$ **3.** $\sqrt{-27}$ **4.** $\sqrt{-48}$
5. $\sqrt{-50}$ **6.** $\sqrt{-72}$ **7.** $\sqrt{-75}$ **8.** $\sqrt{-98}$

Simplify each of the following to either i, -1, $-i$ or 1:

9. i^{20} **10.** i^{30} **11.** i^{43} **12.** i^{57}
13. i^{-10} **14.** i^{-37} **15.** i^{-60} **16.** i^{-75}

Perform the indicated operations and express the answer as a or ai:

17. $\sqrt{-9} - \sqrt{-16} + 2\sqrt{25}$ **18.** $\sqrt{-36} - \sqrt{-49} + \sqrt{-121}$
19. $\sqrt{-72} + \sqrt{-8} - \sqrt{-50}$ **20.** $\sqrt{-27} + \sqrt{-12} - \sqrt{-147}$
21. $\sqrt{-54} - \sqrt{-96} - \sqrt{-150}$ **22.** $\sqrt{-45} - \sqrt{-80} - \sqrt{-125}$
23. $\sqrt{-8} - \sqrt{-12} + \sqrt{-48}$ **24.** $\sqrt{-18} + \sqrt{-27} - \sqrt{-108}$
25. $\sqrt{-32}\sqrt{8}$ **26.** $\sqrt{-54}\sqrt{2}$ **27.** $\sqrt{6}\sqrt{-20}$
28. $\sqrt{10}\sqrt{-45}$ **29.** $\sqrt{-3}\sqrt{-12}$ **30.** $\sqrt{-5}\sqrt{-40}$
31. $\sqrt{-8}\sqrt{-50}$ **32.** $\sqrt{-14}\sqrt{-21}$ **33.** $\sqrt{-2} \div \sqrt{3}$
34. $\sqrt{-48} \div \sqrt{8}$ **35.** $\sqrt{-45} \div \sqrt{12}$ **36.** $\sqrt{-147} \div \sqrt{63}$
37. $7 \div \sqrt{-21}$ **38.** $4 \div \sqrt{-32}$ **39.** $\sqrt{18} \div \sqrt{-15}$
40. $\sqrt{15} \div \sqrt{-35}$ **41.** $\sqrt{-2} \div \sqrt{-3}$ **42.** $\sqrt{-3} \div \sqrt{-5}$
43. $\sqrt{-27} \div \sqrt{-32}$ **44.** $\sqrt{-45} \div \sqrt{-28}$

Perform the indicated operations and write the answer in standard form:

45. $(1 + i)(3 + i)$ **46.** $(2 + i)(3 + i)$ **47.** $(1 - i)(4 + i)$
48. $(5 - i)(1 + 2i)$ **49.** $(4 - 3i)(4 + 3i)$ **50.** $(2 - 3i)(1 - 2i)$
51. $(1 - 3i)(4 - 3i)$ **52.** $(1 + i)^2$ **53.** $(2 - i)^2$
54. $(3 + 2i)^2$ **55.** $(\sqrt{2} - i)^2$ **56.** $(1 - 2i\sqrt{3})^2$

57. $\dfrac{5}{2 + i}$ **58.** $\dfrac{2}{1 + 3i}$ **59.** $\dfrac{1}{4 - i}$ **60.** $\dfrac{1}{2 - 3i}$

61. $\dfrac{1 + i}{1 + 2i}$ **62.** $\dfrac{2 - i}{1 + 3i}$ **63.** $\dfrac{2 + i}{2 - 3i}$ **64.** $\dfrac{3 - 2i}{1 - 3i}$

65. $\dfrac{1+2i}{3-2i}$ 66. $\dfrac{2-3i}{3+2i}$ 67. $\dfrac{1+4i}{2-i}$ 68. $\dfrac{3-4i}{1+4i}$

69. $\dfrac{1+i\sqrt{2}}{1-i\sqrt{2}}$ 70. $\dfrac{2-i\sqrt{3}}{2+i\sqrt{3}}$ 71. $(1-i)^{-2}$ 72. $(1+2i)^{-2}$

Find the real values of x and y that satisfy the following equations:

73. $2x - yi = 4 - 3i$
74. $x + 2yi = 3 - 4i$
75. $(1+i)x + (1+3i)y = 1 + 5i$
76. $(1+2i)x + (1+i)y = 3 + 3i$
77. $(2+i)x + (1-2i)y = 4 + 2i$
78. $(1+i)x - (3-2i)y = 4i - 1$
79. $(1+2i)x + (1+3i)y = 2 + 3i$
80. $(2+i)x + (1-2i)y = -1 - 8i$
81. $(2+7i)x + (3+4i)y = 5 - 2i$
82. $(9+5i)x + (7+3i)y = -3 + i$
83. $(3+i)x + 2(2-i)y = 8 + i$
84. $2(2+i)x - 3(1+2i)y = 4 + 5i$

Chapter 13 Solve the following equations for x by factoring:

85. $x^2 + 7x + 10 = 0$
86. $x^2 + 7x + 12 = 0$
87. $x^2 - 4x - 12 = 0$
88. $x^2 - 5x - 24 = 0$
89. $x^2 - 6x + 8 = 0$
90. $x^2 - 13x + 42 = 0$
91. $6x^2 + 17x + 12 = 0$
92. $6x^2 + 31x + 18 = 0$
93. $6x^2 + 19x - 20 = 0$
94. $9x^2 + 25x - 6 = 0$
95. $16x^2 - 26x + 3 = 0$
96. $18x^2 - 45x + 28 = 0$
97. $x^2 - a^2 + 2x + 1 = 0$
98. $x^2 - a^2 - 4x + 4 = 0$
99. $4x^2 - a^2 - 4x + 1 = 0$
100. $x^2 - a^2 + 2a - 1 = 0$
101. $x^2 - a^2 - 4a - 4 = 0$
102. $4x^2 - 4a^2 + 4a - 1 = 0$

Form a quadratic equation with the given numbers as roots:

103. 3; 1 104. 2; 4 105. 2; -1 106. 3; -2

107. -2; -2 108. -1; -4 109. 2; $\dfrac{1}{2}$ 110. $\dfrac{1}{2}$; $\dfrac{2}{3}$

111. $1+2i$; $1-2i$ 112. $1+3i$; $1-3i$ 113. $\sqrt{2}+i$; $\sqrt{2}-i$

Solve the following equations for x by completing the square:

114. $x^2 + 3x + 5 = 0$
115. $x^2 - 2x + 3 = 0$
116. $x^2 - 5x + 7 = 0$
117. $x^2 + 4x + 8 = 0$
118. $x^2 + 2x - 1 = 0$
119. $x^2 + 3x - 2 = 0$
120. $2x^2 - 6x + 3 = 0$
121. $5x^2 - 10x + 3 = 0$
122. $3x^2 - 2x - 2 = 0$
123. $3x^2 - 5x - 1 = 0$
124. $2x^2 - 7x + 4 = 0$
125. $6x^2 - 9x + 2 = 0$
126. $x^2 - 6x + 10 = 0$
127. $x^2 + 3x + 11 = 0$
128. $2x^2 - 3x + 4 = 0$
129. $3x^2 - 5x + 3 = 0$
130. $2x^2 - ax - 4a^2 = 0$
131. $2x^2 + 3ax - a^2 = 0$
132. $3x^2 + 2ax + 2a^2 = 0$
133. $3x^2 - 4ax + 2a^2 = 0$

Solve the following equations by the quadratic formula:

134. $x^2 + 2x - 4 = 0$

135. $x^2 - 4x - 4 = 0$

136. $x^2 + 3x - 2 = 0$

137. $3x^2 + 10x + 6 = 0$

138. $2x^2 - 4x + 3 = 0$

139. $3x^2 - 2x + 6 = 0$

140. $5x^2 + 3x + 9 = 0$

141. $x^2 + \sqrt{2}x - 5 = 0$

142. $x^2 + \sqrt{3}x - 4 = 0$

143. $x^2 - \sqrt{5}x - 2 = 0$

144. $x^2 - 2\sqrt{3}x - 1 = 0$

145. $\sqrt{3}x^2 + 5x - 2\sqrt{3} = 0$

Find all real values of k for which the nature of the roots of the given equation is as indicated:

146. $x^2 + 2x + k = 0$
two real distinct roots

147. $x^2 - 3x + k = 0$
two real distinct roots

148. $2x^2 - kx - 5 = 0$
two real distinct roots

149. $4x^2 + kx - 1 = 0$
two real distinct roots

150. $kx^2 + 3x - 2 = 0$
two real distinct roots

151. $kx^2 + 5x + 3 = 0$
two real distinct roots

152. $x^2 - 4x + k = 0$
one double root

153. $x^2 + 2x + k - 3 = 0$
one double root

154. $x^2 - kx + 2 = 0$
one double root

155. $3x^2 + kx - 5 = 0$
one double root

156. $kx^2 + 4x - 3 = 0$
one double root

157. $kx^2 - x + 1 = 0$
one double root

158. $x^2 + x + k = 0$
two imaginary roots

159. $2x^2 + 3x - k = 0$
two imaginary roots

160. $3x^2 + kx - 1 = 0$
two imaginary roots

161. $5x^2 - kx - 2 = 0$
two imaginary roots

162. $kx^2 + x + 7 = 0$
two imaginary roots

163. $kx^2 - 2x - 3 = 0$
two imaginary roots

Find the value of m in each of the following equations for the given condition:

164. $x^2 - 2x + m = 0$
one root is $1 - \sqrt{3}$

165. $x^2 - 6x + m = 0$
one root is $3 + \sqrt{2}$

166. $x^2 + mx + 5 = 0$
one root is $2 + i$

167. $x^2 + mx + 9 = 0$
one root is $-1 + 2i\sqrt{2}$

168. $mx^2 + (m - 8)x = 7$
sum of the roots is 3

169. $(m - 7)x^2 + (m + 5)x = 1$
sum of the roots is -5

170. $x^2 - 2x + m + 4 = 0$
the difference of the roots is 6

171. $x^2 - 10x + m + 15 = 0$
the difference of the roots is 4

172. $8x^2 - 14x + m = 1$
one root is 6 times the other

173. $9x^2 - 18x + m = 3$
one root is $\frac{1}{2}$ the other

174. $(m+3)x^2 - 16x + m = 12$
the product of the roots is $-\frac{1}{4}$

175. $(m-2)x^2 - 14x + m + 3 = 0$
the product of the roots is $\frac{8}{3}$

176. $8x^2 + 14x = m + 2$
the quotient of the roots is $-\frac{3}{10}$

177. $6x^2 - 5x = m + 1$
the quotient of the roots is $-\frac{3}{8}$

Find a quadratic equation whose roots r_1 and r_2 satisfy the given conditions:

178. $r_1 + r_2 = -7;\ r_1r_2 = 12$

179. $r_1 + r_2 = 4;\ r_1r_2 = -21$

180. $r_1 + r_2 = \frac{5}{3};\ r_1 - r_2 = -\frac{1}{3}$

181. $r_1 + r_2 = \frac{3}{4};\ r_1 - r_2 = -\frac{5}{4}$

182. $r_1 + r_2 = 1;\ \frac{r_1}{r_2} = -3$

183. $r_1 + r_2 = -\frac{8}{3};\ \frac{r_1}{r_2} = -9$

184. $r_1 - r_2 = \frac{5}{6};\ r_1r_2 = -\frac{1}{6}$

185. $r_1 - r_2 = \frac{13}{6};\ r_1r_2 = -1$

186. $r_1 - r_2 = 4;\ \frac{r_1}{r_2} = -\frac{3}{5}$

187. $r_1 - r_2 = \frac{13}{3};\ \frac{r_1}{r_2} = -\frac{4}{9}$

188. $r_1r_2 = -2;\ \frac{r_1}{r_2} = -\frac{8}{9}$

189. $r_1r_2 = -\frac{2}{3};\ \frac{r_1}{r_2} = -\frac{8}{3}$

Find all roots of the following equations:

190. $x^3 - 1 = 0$

191. $x^3 - 8 = 0$

192. $x^3 + 27 = 0$

193. $x^3 + 64 = 0$

194. $x^4 - 144 = 0$

195. $x^4 - 324 = 0$

196. $x^4 + 2x^2 - 24 = 0$

197. $x^4 - 5x^2 - 36 = 0$

198. $x^4 - 15x^2 + 9 = 0$

199. $x^4 - 12x^2 + 4 = 0$

200. $4x^4 - 8x^2 + 1 = 0$

201. $9x^4 + 5x^2 + 1 = 0$

202. $x + 5x^{\frac{1}{2}} + 6 = 0$

203. $3x - 7x^{\frac{1}{2}} + 2 = 0$

204. $4x - 15x^{\frac{1}{2}} + 9 = 0$

205. $12x - 5x^{\frac{1}{2}} - 2 = 0$

206. $x^{\frac{2}{3}} - 3x^{\frac{1}{3}} + 2 = 0$

207, $x^{\frac{2}{3}} - 5x^{\frac{1}{3}} - 36 = 0$

208. $8x^{\frac{2}{3}} - 10x^{\frac{1}{3}} - 3 = 0$

209. $4x^{\frac{2}{3}} + 9x^{\frac{1}{3}} + 2 = 0$

Solve the following equations:

210. $\frac{17x}{2x+3} + \frac{21}{x-4} + 8 = 0$

211. $\frac{18}{x^2 - 3x - 4} + \frac{x}{x^2 - 6x + 8} = \frac{12}{x^2 - x - 2}$

212. $\dfrac{22}{x^2 + x - 12} + \dfrac{3x}{x^2 + 5x + 4} = \dfrac{10}{x^2 - 2x - 3}$

213. $\dfrac{x + 4}{2x^2 - x - 3} - \dfrac{x + 3}{3x^2 + 2x - 1} = \dfrac{7}{6x^2 - 11x + 3}$

214. $\dfrac{2x + 1}{3x^2 - 13x + 4} - \dfrac{4x - 1}{1 - x - 6x^2} = \dfrac{3x - 3}{2x^2 - 7x - 4}$

215. $\dfrac{3x - 1}{12x^2 + x - 20} - \dfrac{2x + 4}{4 - x - 3x^2} = \dfrac{x - 4}{4x^2 - 9x + 5}$

216. $\dfrac{x - 3}{x^2 - 3x - 4} + \dfrac{x + 5}{x^2 - 2x - 3} = \dfrac{x - 5}{x^2 - 7x + 12}$

217. $\dfrac{x - 5}{x^2 - 2x - 3} + \dfrac{x}{x^2 + 3x + 2} = \dfrac{x - 8}{x^2 - x - 6}$

218. $\dfrac{x - 3}{x^2 - 3x + 2} - \dfrac{x - 9}{x^2 + x - 6} = \dfrac{x + 7}{x^2 + 2x - 3}$

219. $\dfrac{x + 4}{x^2 - x - 2} - \dfrac{x + 2}{x^2 - 3x - 4} = \dfrac{x - 6}{x^2 - 6x + 8}$

220. $\sqrt{x - 4} + \sqrt{x - 1} = \sqrt{6x - 21}$

221. $\sqrt{x - 3} + \sqrt{x + 5} = \sqrt{6x - 8}$

222. $\sqrt{x + 2} - \sqrt{x - 3} = \sqrt{15 - 2x}$

223. $\sqrt{x + 1} - \sqrt{x - 4} = \sqrt{17 - 2x}$

224. $\sqrt{x + 3} + \sqrt{x - 4} = \sqrt{6x - 29}$

225. $\sqrt{x + 4} + \sqrt{x + 11} = \sqrt{6x + 19}$

226. The sum of two numbers is 25 and the sum of their squares is 317. Find the two numbers.

227. The difference of two natural number is 4 and the sum of their reciprocals is $\frac{4}{15}$. Find the two numbers.

228. A man painted a house for \$800. It took him 20 hours less than he expected and thus earned \$2 an hour more than he anticipated. How long did he expect it would take him to paint the house?

229. If each of two opposite sides of a square are increased by 5 inches more than twice the side of the square and the other two opposite sides are each decreased by 7 inches, the area of the resulting rectangle is 55 square inches more than the area of the square. Find the side of the square.

230. A crew can row 18 miles down a stream and back in 9 hours. If the rate of the stream is 1.5 miles an hour, find the rate at which the crew can row in still water.

231. A tree was broken over by a storm $\frac{3}{8}$ of the distance from the bottom. If the top touched the ground 40 feet from the foot of the tree, what was the height of the tree?

Find the coordinates of the vertex, the equation of the line of symmetry, and sketch each of the following parabolas whose equations are as follows:

232. $y = 2x^2$ **233.** $y = x^2 + 3$ **234.** $y = 9 - x^2$
235. $y = x^2 + 4x + 4$ **236.** $y = x^2 - x - 2$ **237.** $y = x^2 - x - 6$
238. $y = 6 - x - x^2$ **239.** $y = 4 + 3x - x^2$

Find the solution set of each of the following inequalities:

240. $x^2 < 9$ **241.** $x^2 - 25 < 0$ **242.** $x^2 - 8x \leq -15$
243. $x^2 + 4x \leq -3$ **244.** $x^2 - 2x < 8$ **245.** $x^2 - 2x < 24$
246. $x^2 - 4 > 0$ **247.** $x^2 - 16 > 0$ **248.** $x^2 - 7x > -10$
249. $x^2 + 3x \geq 4$ **250.** $x^2 + 8x \geq -12$ **251.** $x^2 - 11x > -24$

Chapter 14

Find the center, the radius, the extents, and graph the circles whose equations are as follows:

252. $x^2 + y^2 = 10$ **253.** $x^2 + y^2 = 27$
254. $x^2 + y^2 - 2x = 8$ **255.** $x^2 + y^2 - 4x = 12$
256. $x^2 + y^2 - 4y = 0$ **257.** $x^2 + y^2 - 6y = 16$
258. $x^2 + y^2 - 4x + 2y = 44$ **259.** $x^2 + y^2 + 2x - 6y = 26$
260. $x^2 + y^2 + 2x + 4y = 15$ **261.** $x^2 + y^2 - 6x - 2y = 8$

Find the center, the vertices, the foci, and sketch the ellipses whose equations are as follows:

262. $x^2 + 2y^2 = 16$ **263.** $x^2 + 4y^2 = 8$ **264.** $4x^2 + y^2 = 36$
265. $2x^2 + y^2 = 18$ **266.** $2x^2 + 3y^2 = 24$ **267.** $3x^2 + 4y^2 = 36$
268. $4x^2 + 9y^2 - 8x = 32$ **269.** $x^2 + 4y^2 - 4x = 12$
270. $6x^2 + y^2 + 4y = 20$ **271.** $4x^2 + y^2 - 4y = 12$
272. $x^2 + 9y^2 + 2x + 18y = -1$ **273.** $x^2 + 3y^2 - 4x + 6y = 5$

Find the vertex, the focus, the equation of the directrix, and sketch the parabolas whose equations are as follows:

274. $y^2 = 6x$ **275.** $y^2 = 10x$ **276.** $y^2 = -4x$ **277.** $y^2 = -8x$
278. $x^2 = 12y$ **279.** $x^2 = 6y$ **280.** $x^2 = -8y$ **281.** $x^2 = -10y$
282. $y^2 = 4x + 4$ **283.** $y^2 = 8x + 16$
284. $y^2 + 4y + 4x = 8$ **285.** $y^2 - 6y + 6x + 3 = 0$
286. $x^2 = 6y + 12$ **287.** $x^2 = 8y - 16$
288. $x^2 - 2x + 6y + 13 = 0$ **289.** $x^2 + 6x + 4y + 1 = 0$

Find the center, the vertices, the foci, the equation of the asymptotes, and sketch the hyperbolas whose equations are as follows:

290. $x^2 - 2y^2 = 16$ **291.** $x^2 - 4y^2 = 8$ **292.** $4x^2 - y^2 = 36$

293. $y^2 - 2x^2 = 18$ **294.** $4y^2 - 3x^2 = 24$ **295.** $9y^2 - x^2 = 16$

296. $9x^2 - 4y^2 - 18x = 27$ **297.** $4x^2 - y^2 - 16x = 0$

298. $4x^2 - 9y^2 + 16x + 18y = 29$ **299.** $x^2 - 16y^2 + 2x + 96y = 159$

300. $3y^2 - 4x^2 - 12y + 8x = 4$ **301.** $9y^2 - 2x^2 + 36y - 8x + 10 = 0$

Solve the following systems:

302. $x^2 + xy - y^2 + 2x - 4y = 5$
$x - y = 1$

303. $x^2 - 2xy + 4y^2 - 2x + 14y + 9 = 0$
$x - 2y = 3$

304. $x^2 - xy + y^2 - x + 5y + 4 = 0$
$x - y = 3$

305. $x^2 - 2xy - 4y^2 + 3x + 28y = 42$
$x + 2y = 6$

306. $x^2 - 3xy + 9y^2 - 3x + 27y + 14 = 0$
$x - 3y = 5$

307. $x^2 + 4xy - 16y^2 - x - 28y = 8$
$x - 4y = 3$

308. $x^2 + y^2 + 2x + y = 3$
$x^2 + y^2 + x + 2y = 1$

309. $x^2 + y^2 - x + 2y = 8$
$x^2 + y^2 - 3x + y = 4$

310. $x^2 + y^2 - x + 4y = 7$
$x^2 + y^2 - 2x + y = 2$

311. $x^2 + y^2 + 3x - 5y = -2$
$x^2 + y^2 - x - 4y = -4$

312. $x^2 + y^2 + 3x + 5y = 22$
$x^2 + y^2 + x + 2y = 17$

313. $x^2 + y^2 + 4x - 2y = 15$
$x^2 + y^2 + 2x + y = 8$

314. $9x^2 + 2y^2 = 13$
$4x^2 + y^2 = 6$

315. $3x^2 + y^2 = 2$
$2x^2 + 2y^2 = 3$

316. $5x^2 + 2y^2 = 5$
$7x^2 + 3y^2 = 3$

317. $4x^2 + y^2 = 4$
$6x^2 + 3y^2 = 7$

318. $2x^2 + 3xy + y^2 = 3$
$x^2 + 2xy + 3y^2 = 3$

319. $3x^2 + 2xy + y^2 = 8$
$2x^2 - xy - y^2 = 8$

320. $2x^2 - xy + 2y^2 = 2$
$2x^2 + xy + 2y^2 = 3$

321. $3x^2 - 2xy + y^2 = 3$
$4x^2 + xy + 5y^2 = 6$

322. $2x^2 + 5xy - y^2 = 6$
$2x^2 + xy - 2y^2 = 4$

323. $2x^2 - xy + 3y^2 = 1$
$3x^2 - 5xy + 10y^2 = 3$

chapter 15

Functions

15.1 Properties of functions

15.2 Function notation

15.3 Algebra of functions

15.4 Functions defined by equations

15.5 Exponential functions

15.6 Inverse functions

15.1 Properties of Functions

The concept of **function** is one of the most important concepts in the study of mathematics. An understanding of this concept is necessary for those with an interest in mathematics, physics, engineering, statistics, or any discipline dealing with quantitative relationships.

Any set of ordered pairs of numbers (x, y) is called a **relation**. The set whose elements are the first coordinates of the pairs is called the **domain of the relation**. The set whose elements are the second coordinates of the pairs is called the **range of the relation**.

A **function** is a set of ordered pairs of numbers (x, y) such that for every x there is a unique y.

A relation assigns to each member x of its domain one or more members of its range. A function is a relation that assigns to each member of its domain a single member of its range.

The set $\{(-2, 1), (-1, 4), (0, 3), (2, 8), (5, 10)\}$ is a function.
The set $\{(0, 4), (1, 4), (2, 7), (3, 7), (6, 8)\}$ is a function.
The set $\{(2, -1), (2, -3), (4, 0), (5, 6), (7, 9)\}$ is a relation but not a function.

The function $\{(1, 4), (2, 7), (3, 10), (4, 13), (5, 16)\}$ has for its domain the set $\{1, 2, 3, 4, 5\}$ and for its range the set $\{4, 7, 10, 13, 16\}$.

Consider the function $\{(1, 5), (2, 10), (3, 15), (4, 20), \ldots\}$. The elements of the range are five times the elements of the domain. This function can be written as $\{(x, y) \mid y = 5x, x \in N\}$.

The set $\{(x, y) \mid y = 3x + 2, x \in W\}$ in expanded form is $\{(0, 2), (1, 5), (2, 8), (3, 11), \ldots\}$.
This set is a function whose domain is $\{0, 1, 2, 3, \ldots\}$ and whose range is $\{2, 5, 8, 11, \ldots\}$. The above function assigns to the number 0 the number 2, to the number 1 the number 5, to the number 2 the number 8, and so on.

Note: An element of the range of a function may be paired with more than one element of the domain; however, an element of the domain has exactly one element of the range paired with it.

The assignment of the elements of the range to the elements of the domain may be an algebraic expression or just an arbitrary pairing.

Exercise 15.1

Determine which of the following relations are functions and which are not. For each function, find its domain and range:

1. $\{(-3, 0), (-2, 2), (0, 4), (3, 6), (5, 7)\}$
2. $\{(3, 1), (4, 1), (5, 2), (6, 2), (7, 3)\}$

3. $\{(1, 3), (2, 5), (2, 7), (3, 8), (4, 9)\}$
4. $\{(0, 2), (1, 3), (1, 4), (2, 5), (3, 10)\}$
5. $\{(-2, 0), (-1, 0), (0, 2), (1, 2), (2, 4)\}$
6. $\{(0, 1), (1, 2), (2, 3), (3, 4), \ldots\}$
7. $\{(x, y\} \mid y = 2x + 1, x \in W\}$
8. $\{(x, y) \mid y = x + 3, x \in N\}$
9. $\{(x, y) \mid y = x^2, x \in N\}$
10. $\{(x, y) \mid y = x^2 + 1, x \in W\}$

List the values that x cannot take on so that the following relations will be functions:

11. $\{(x, 2), (2, 3), (3, 4), (4, 5)\}$
12. $\{(1, 2), (x, 4), (3, 6), (4, 8)\}$
13. $\{(-10, 7), (0, 1), (x, 20), (-7, -30)\}$
14. $\{(3, 2), (x, 3), (4, 4), (5, 4)\}$
15. $\{(-1, -7), (x, -7), (0, 0), (1, 7)\}$

15.2 Function Notation

Functions are denoted by letters, such as $f, g, h,$ or sometimes $f_1, f_2, f_3, \ldots$. When x is any member of the domain of a function f, then $f(x)$, read "f of x," or "f at x," is the element of the range assigned to x by the function f.

Let the function f have for its domain the set of integers, and for its rule the assignment to each integer of its square. Some of the elements of the function are $(-3, 9), (-2, 4), (0, 0), (4, 16)$.

If x is any element of the domain, then the corresponding element of the range is x^2. That is, f at x equals x^2, or simply $f(x) = x^2$. Thus the function f can be written as

$$\{(x, x^2) \mid x \in I\} \quad \text{or} \quad \{(x, f(x)) \mid f(x) = x^2, x \in I\}$$

From this notation we see that a function can sometimes be defined by its domain and by an equation that defines the rule of association of the elements of the domain with the elements of the range.

EXAMPLE

Find the elements of the range of the function f defined by the equation $f(x) = 3x^2 - 5$ that correspond to the elements $1, 3, a, a + b$ of the domain.

Solution

To find the elements of the range that correspond to the elements $1, 3, a, a + b$ of the domain, calculate $f(1), f(3), f(a), f(a + b)$:

$$f(x) = 3x^2 - 5$$
$$f(1) = 3(1)^2 - 5 = 3 - 5 = -2$$
$$f(3) = 3(3)^2 - 5 = 27 - 5 = 22$$
$$f(a) = 3(a)^2 - 5 = 3a^2 - 5$$
$$f(a + b) = 3(a + b)^2 - 5 = 3a^2 + 6ab + 3b^2 - 5$$

The elements of the range are $-2, 22, 3a^2 - 5$, and $3a^2 + 6ab + 3b^2 - 5$, respectively.

DEFINITION

> The function $\{(x, a) \mid x \in R\}$ assigns the number a to every real number x. A function whose range is exactly one number is called a **constant function**.

DEFINITION

> The function $\{(x, x) \mid x \in R\}$ assigns each real number to itself. Such a function is called an **identity function.**

Consider the function f defined by the equation $f(x) = 2x^3$, whose domain is the set of integers. The function can be expressed as

$$\{(x, f(x)) \mid f(x) = 2x^3, x \in I\}$$

Note that x represents the elements in the domain, and $f(x)$ represents the elements in the range that are assigned to each element x in the domain by the rule $f(x) = 2x^3$.

To find some elements of the function, assign to x numbers from the set I and calculate the corresponding elements of the range, that is, $f(x)$.

For $x = -2$, $\quad f(-2) = 2(-2)^3 = -16$

For $x = -1$, $\quad f(-1) = 2(-1)^3 = -2$

For $x = 0$, $\quad f(0) = 2(0)^3 = 0$

For $x = 1$, $\quad f(1) = 2(1)^3 = 2$

For $x = 2$, $\quad f(2) = 2(2)^3 = 16$

For $x = 3$, $\quad f(3) = 2(3)^3 = 54$

The elements of the function are the pairs $(x, 2x^3)$ or $(x, f(x))$ where $x \in I$. Thus some of the elements of the function are
$(-2, -16), (-1, -2), (0, 0), (1, 2), (2, 16), (3, 54)$.

The domain of the function is the set $\{x \mid x \in I\}$, and the range of the function is the set $\{f(x) \mid f(x) = 2x^3, x \in I\}$.

15.3 Algebra of Functions

From this notation we see that a function can sometimes be defined by its domain and by an equation that defines the rule of association of the elements of the domain with the elements of the range.

Sum of Two Functions The sum of two functions f_1 and f_2, denoted by $f_1 + f_2$, is defined by

$$(f_1 + f_2)(x) = f_1(x) + f_2(x)$$

Product of Two Functions The product of two functions f_1 and f_2, denoted by $f_1 \cdot f_2$, is defined by

$$(f_1 \cdot f_2)(x) = f_1(x) \cdot f_2(x)$$

EXAMPLE

Given $f_1(x) = x - 2$, $f_2(x) = 4x + 8$, find $(f_1 + f_2)(x)$ and $(f_1 \cdot f_2)(x)$.

Solution

$$(f_1 + f_2)(x) = f_1(x) + f_2(x)$$
$$= (x - 2) + (4x + 8) = 5x + 6$$
$$(f_1 \cdot f_2)(x) = f_1(x) \cdot f_2(x)$$
$$= (x - 2) \cdot (4x + 8) = 4x^2 - 16$$

Composite Functions Sometimes it is necessary to apply one function after another. As an example, suppose that $f(x) = 3x + 5$ and $g(x) = x^2 + 3$. What is the result of applying first g, then f to 3, or in other words, what is $f(g(3))$?

We first calculate $g(3)$: $\quad g(3) = (3)^2 + 3 = 12$

Now, we calculate $f(12)$: $\quad f(12) = 3(12) + 5 = 41$

Hence $\quad f(g(3)) = 41$

Sometimes $f(g(x))$ is written as $(f \circ g)(x)$ and read as "f circle g." It is called a **function of a function** or a **composite function**.

For the composite function $f(g(x))$, x must be in the domain of g and the range of g must be contained in the domain of f. Otherwise $f(g(x))$ has no meaning.

EXAMPLE

Given $f(x) = x^2 + 5x$, $g(x) = 2x + 1$, find

$(f \cdot g)(x)$, $\quad f(g(x))$, $\quad$ and $\quad g(f(x))$

Solution

$$(f \cdot g)(x) = f(x) \cdot g(x)$$
$$= (x^2 + 5x) \cdot (2x + 1) = 2x^3 + 11x^2 + 5x$$

To find $f(g(x))$, substitute $g(x)$ for x in $f(x)$:

$$f(g(x)) = [g(x)]^2 + 5[g(x)]$$
$$= (2x + 1)^2 + 5(2x + 1)$$
$$= (4x^2 + 4x + 1) + (10x + 5)$$
$$= 4x^2 + 14x + 6$$

To find $g(f(x))$, substitute $f(x)$ for x in $g(x)$:

$$g(f(x)) = 2[f(x)] + 1$$
$$= 2(x^2 + 5x) + 1$$
$$= 2x^2 + 10x + 1$$

Exercises 15.2–15.3

Write the rule for each of the functions in Problems 1–8:

1. $\{(1, 2), (2, 4), (3, 6), (4, 8), \ldots\}$
2. $\{(1, 4), (2, 8), (3, 12), (4, 16), \ldots\}$
3. $\{(1, 0), (2, 1), (3, 2), (4, 3), \ldots\}$
4. $\{(0, 6), (1, 7), (2, 8), (3, 9), \ldots\}$
5. $\left\{(0, 0), \left(1, \frac{2}{3}\right), \left(2, \frac{4}{3}\right), (3, 2), \left(4, \frac{8}{3}\right), \ldots\right\}$
6. $\{(0, -1), (1, 1), (2, 3), (3, 5), \ldots\}$
7. $\{\ldots, (-2, -10), (-1, -5), (0, 0), (1, 5), (2, 10), \ldots\}$
8. $\{\ldots, (-2, -3), (-1, -1), (0, 1), (1, 3), (2, 5), \ldots\}$
9. If $f(x) = 3x^2 + x - 8$, find $f(-4)$, $f(0)$, and $f(3)$.
10. If $f(x) = 6x^2 + 11x - 10$, find $f\left(-\frac{5}{2}\right)$, $f\left(\frac{2}{3}\right)$, and $f(2)$.
11. If $f(x) = ax^2 + bx + c$, find $f\left(-\frac{b}{2a}\right)$, $f(y)$, and, $f(x + 2)$.
12. If $g(x) = 2x^2 - 4x$, find $g(2)$, $g(x - 1)$, and, $g(2x + 1)$.
13. If $F(x) = 3(x - a)^2 + 2(x - a)$, find $F(-a)$, $F(a)$, $F(2a)$, and $F(a - 5)$.
14. If $g(x) = \frac{2x - 3}{8}$, find $g(-3)$, $g(0)$, and $g(x - 3)$.
15. If $g(x) = \frac{x}{2x + 1}$, find $g\left(-\frac{1}{2}\right)$, $g\left(\frac{3}{5}\right)$, and, $g(2x^2)$.
16. If $h(x) = \frac{3x - 1}{5x - 2}$, find $h(0)$, $h\left(\frac{2}{5}\right)$, and, $h(x + 1)$.
17. If $F(x) = 3x^2 - 2x$, check the truth of $F(3) - F(2) = F(1)$.
18. If $F(x) = x^2 + 2$, check the truth of $F(a + b) = F(a) + F(b)$.
19. If $f(x) = 4x - 1$, $g(x) = x^2 - 2$, find $(f + g)(x)$, $(f \cdot g)(x)$, $f(g(x))$, and $g(f(x))$.
20. If $f(x) = 3x^2 + x - 1$, $g(x) = x - 3$, find $(f + g)(x)$, $(f \cdot g)(x)$, $(f \circ g)(x)$, and $(g \circ f)(x)$.

15.4 Functions Defined by Equations

The definition of a function does not have any restriction on the domain, range, or the rule of association of the elements of the range to the elements of the domain. In many functions, the domain and range are subsets of the set of real numbers, and the rule of association is expressed by an equation.

When a function in this book is defined by an equation without mention of its domain or range, we will consider the domain and range to be subsets of the real numbers.

Thus the equation $f(x) = x^2 + 1$, or $y = x^2 + 1$, describes the function

$$\{(x, f(x)) \mid f(x) = x^2 + 1, x \in R\}$$

or

$$\{(x, y) \mid y = x^2 + 1, x \in R\}$$

When a function is defined by an equation, analysis enables us to determine the domain and the range.

The following examples illustrate how to determine the domain and range of functions defined by equations.

EXAMPLE

Find the domain and range of the function f defined by $f(x) = x^2 + 6$.

Solution

The set of numbers that can be assigned to x for which $f(x)$ exists determines the domain of the function.

Since $f(x)$, that is, $x^2 + 6$, exists for all real values of x, the domain of the function is the set of real numbers.

The set of numbers that $x^2 + 6$ assumes when x is a real number defines the range.

Since $x^2 \geq 0$, for all real values of x, $x^2 + 6$ takes all real values greater than or equal to 6.

The domain is the set $\{x \mid x \in R\}$.

The range is the set $\{f(x) \mid f(x) \geq 6, f(x) \in R\}$.

EXAMPLE

Find the domain and range of the function defined by $y = \sqrt{x^2 - 4}$.

Solution

The quantity $\sqrt{x^2 - 4}$ is a real number when $x^2 - 4 \geq 0$.

Solving the inequality $x^2 - 4 \geq 0$ we get $x \leq -2$ or $x \geq 2$.

Hence the domain is $\{x \mid x \leq -2 \text{ or } x \geq 2, x \in \mathrm{R}\}$.

For all the values of x in the domain, $y = \sqrt{x^2 - 4} \geq 0$.

Hence, the range is $\{y \mid y \geq 0, y \in R\}$.

EXAMPLE

Find the domain and range of the function defined by

$y = \sqrt{16 - x^2}$

Solution

The quantity $\sqrt{16 - x^2}$ is a real number when $16 - x^2 \geq 0$.

Solving the inequality $16 - x^2 \geq 0$ we get $-4 \leq x \leq 4$.

Hence the domain is $\{x \mid -4 \leq x \leq 4\}$.

The set of values that y assumes for the values of x in the domain is $0 \leq y \leq 4$.

Hence the range is $\{y \mid 0 \leq y \leq 4\}$.

Note The range of some functions can be found by expressing x in terms of y and then determining those values of y for which x exists.

EXAMPLE Find the domain and range of the function defined by $y = \dfrac{4x - 1}{2x + 1}$.

Solution The quantity $\dfrac{4x-1}{2x+1}$ exists for all real values of x except $x = -\dfrac{1}{2}$

$\left(\text{since when } x = -\dfrac{1}{2},\ 2x + 1 = 0, \text{ and } \dfrac{4x-1}{2x+1} \text{ is not defined}\right)$.

The domain of the function is $\left\{x \,\middle|\, x \neq -\dfrac{1}{2}, x \in R\right\}$.

When $x \neq -\dfrac{1}{2}$ and $x \in R$, we have

$$y = \frac{4x-1}{2x+1}$$

or

$$y(2x+1) = 4x - 1$$
$$2xy + y = 4x - 1$$
$$x(2y - 4) = -y - 1$$

or

$$x = \frac{-y-1}{2y-4}$$

When $y = 2$, x is not defined.

Hence the range of the function is $\{y \mid y \neq 2, y \in R\}$.

DEFINITION

A function defined by an equation of the form

$y = ax + b, \qquad a, b \in R, a \neq 0$

is called a **linear function.**

When $a = 0$, we have the **constant function**.

DEFINITION

A function defined by an equation of the form

$y = ax^2 + bx + c, \qquad a, b, c \in R, a \neq 0$

is called a **quadratic function.**

An equation defines a function if for every value of x there corresponds only one value for y.

Note Not every equation defines a function.
The equation $y^2 = 4x$ defines a relation but not a function. Corresponding to $x = 1$ there are two values for y, -2, and $+2$. That is, the equation assigns two distinct numbers of the range to the same number of the domain.

Exercise 15.4

Determine which of the following equations define functions. For each function find its domain and range:

1. $y = 3x - 2$ **2.** $y = 5x + 1$ **3.** $y = x^2 + 4$
4. $y = 2x^2 - 5$ **5.** $y = 9 - 4x^2$ **6.** $y = 8 - 3x^2$
7. $y = \sqrt{x + 15}$ **8.** $y = \sqrt{x - 6}$ **9.** $y = \sqrt{7 - 2x}$
10. $y = \sqrt{3 - 4x}$ **11.** $y = \sqrt{x^2 + 4}$ **12.** $y = \sqrt{4x^2 + 1}$
13. $y = \sqrt{x^2 - 1}$ **14.** $y = \sqrt{x^2 - 9}$ **15.** $y = \sqrt{x^2 - 16}$
16. $y = \sqrt{2x^2 - 7}$ **17.** $y = \sqrt{1 - x^2}$ **18.** $y = \sqrt{4 - x^2}$
19. $y = \sqrt{9 - x^2}$ **20.** $y = \sqrt{1 - 4x^2}$ **21.** $y = \sqrt{12 - x^2}$
22. $y = \sqrt{10 - 3x^2}$ **23.** $y = \dfrac{3x + 1}{x}$ **24.** $y = \dfrac{2x - 3}{5x - 8}$
25. $y = \dfrac{4 - x}{3 + x}$ **26.** $y = \dfrac{3 - 5x}{4 - 7x}$ **27.** $y = \pm\sqrt{x^2 + 1}$
28. $y = \pm\sqrt{x^2 - 4}$ **29.** $y^2 = 5x$ **30.** $y^2 = 2x + 7$
31. $y^2 = -4x$ **32.** $y^2 = 3x^2$
33. $y^3 = x - 4$ **34.** $y^3 = -3x + 1$

15.5 Exponential Functions

The equation $y = a^x, a > 0, a \neq 1$, defines an **exponential function**.
When $a = 1$, then $y = 1^x = 1$, and the equation defines a **constant function**, which is not considered an exponential function.

Note If $a > 0, a \neq 1$, the quantity a^x is a definite number. That is, when $a > 0, b > 0, a \neq 1, b \neq 1$, we have the following.

1. The equation $a^x = a^z$ is equivalent to the equation $x = z$.
2. If $x \neq 0$, the equation $a^x = b^x$ is equivalent to the equation $a = b$.

EXAMPLE Find the solution set of $2^x = 8$.

Solution $2^x = 8 = 2^3$

Since the base is the same, the exponential equation $2^x = 2^3$ is equivalent to the equation $x = 3$.

The solution set is $\{3\}$.

EXAMPLE Find the solution set of $2^{4x} \cdot 4^{x-3} = 8^{x+3}$.

Solution $4^{x-3} = (2^2)^{x-3} = 2^{2x-6}$ and $8^{x+3} = (2^3)^{x+3} = 2^{3x+9}$

Hence $2^{4x} \cdot 2^{2x-6} = 2^{3x+9}$

or $2^{4x+2x-6} = 2^{3x+9}$

$$2^{6x-6} = 2^{3x+9}$$

Thus the given exponential equation is equivalent to the equation

$6x - 6 = 3x + 9$, that is, $x = 5$

The solution set is $\{5\}$.

Consider the exponential function defined by $y = 2^x$. Below is a table of ordered pairs (x, y) that satisfy the equation, that is, some elements of the function.

x	-3	-2	-1	0	1	2	3
y	$\frac{1}{8}$	$\frac{1}{4}$	$\frac{1}{2}$	1	2	4	8

Plot the ordered pairs on a Cartesian coordinate system and draw a smooth curve through the points, as shown in Figure 15.1.

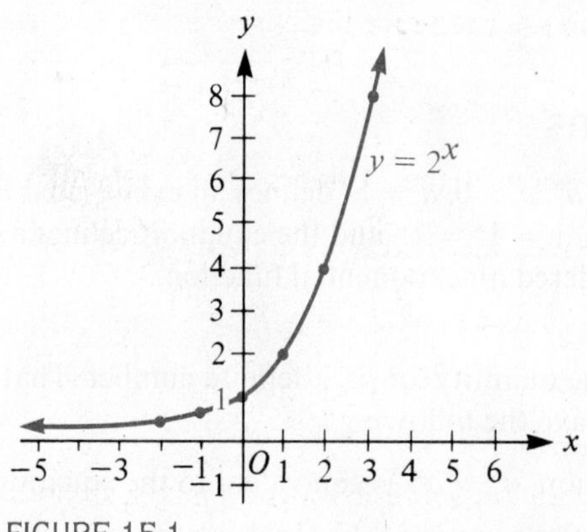

FIGURE 15.1

DEFINITION

> When a point moves on a curve such that the distance between the point and a given line becomes smaller and remains smaller than any preassigned positive number, the line is called an **asymptote** to the curve.

Figure 15.2 shows graphs of some exponential functions.

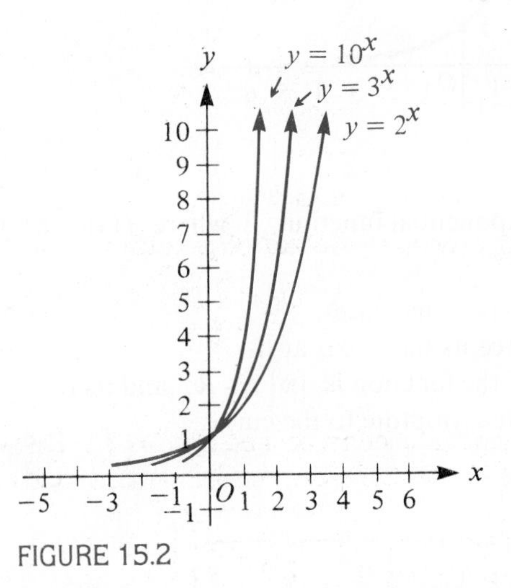

FIGURE 15.2

The graph of an exponential function f where $f(x) = a^x$, $a > 1$, has the following properties:

1. $f(x)$ increases as x increases.
2. The curve intercepts the y-axis at 1.
3. The domain of the function is $\{x \mid x \in R\}$ and its range is $\{f(x) \mid f(x) > 0\}$.
4. The x-axis is an asymptote to the curve.

Consider the exponential function defined by $y = \left(\frac{1}{2}\right)^x$.

Below is a table of some of the elements of the function:

x	-3	-2	-1	0	1	2	3
y	8	4	2	1	$\frac{1}{2}$	$\frac{1}{4}$	$\frac{1}{8}$

Plotting these ordered pairs and joining them by a smooth curve, we get the graph in Figure 15.3.

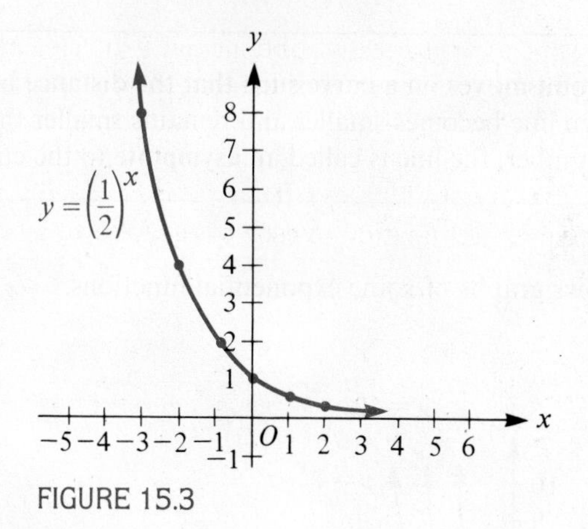

FIGURE 15.3

The graph of an exponential function f, where $f(x) = a^x, 0 < a < 1$, has the following properties:

1. $f(x)$ decreases as x increases.
2. The curve intercepts the y-axis at 1.
3. The domain of the function is $\{x \mid x \in R\}$ and its range is $\{f(x) \mid f(x) > 0\}$.
4. The x-axis is an asymptote to the curve.

Exercise 15.5

Find the solution set of the following exponential equations:

1. $3^x = 81$ **2.** $2^x = 32$ **3.** $5^x = 625$ **4.** $7^x = 343$

5. $27^x = 9$ **6.** $64^x = 2$ **7.** $125^x = 5$ **8.** $49^x = 7$

9. $32^x = \frac{1}{4}$ **10.** $243^x = \frac{1}{3}$ **11.** $9^x = \frac{1}{27}$ **12.** $625^x = \frac{1}{125}$

13. $10^x = .01$ **14.** $2^{x+3} = \frac{1}{8}$ **15.** $2^{2x-3} = 32 \cdot 2^x$

16. $125 \cdot 5^{2x} = 625$ **17.** $2^x \cdot 4^{x+1} = 8$ **18.** $3^{3x+1} \cdot 9^{2-x} = 27$

19. $3^{2x} \cdot 27 = 81^x$ **20.** $5^{x-2} \cdot \frac{1}{125} = \frac{1}{25^x}$

21. Sketch the graph of the function $\{(x, y) \mid y = 2^x, x \in R\}$, and from the graph find a rational approximation to each of the following:

(a) $\sqrt{2}$ **(b)** $\sqrt[3]{2}$ **(c)** $2^{\sqrt{2}}$ **(d)** $2^{\sqrt[3]{2}}$

22. Sketch the graph of the function $\{(x, y) \mid y = 3^x, x \in R\}$, and from the graph find a rational approximation to each of the following:

(a) $\sqrt{3}$ **(b)** $\sqrt[3]{3}$ **(c)** $3^{\sqrt{3}}$ **(d)** $3^{\sqrt[3]{3}}$

23. From the graphs and the results obtained in Problems 21 and 22, find a rational approximation to each of the following:

(a) $2^{\sqrt{3}}$ (b) $2^{\sqrt[3]{3}}$ (c) $3^{\sqrt{2}}$ (d) $3^{\sqrt[3]{2}}$

24. Sketch the graph of the function $\{(x, y) \mid y = 5^x, x \in R\}$, and from the graph find a rational approximation to each of the following:

(a) $\sqrt{5}$ (b) $\sqrt[3]{5}$

15.6 Inverse Functions

Consider the functions

$$f: \{(1, 3), (2, 6), (3, 9), (4, 12), (5, 15)\}$$

and $$g: \{(3, 1), (6, 2), (9, 3), (12, 4), (15, 5)\}$$

The function g is obtained from the function f by interchanging the first and second coordinates of the pairs. That is, if the function f assigns to the number x the number y, the function g assigns to the number y the number x. The functions f and g are called **inverses**.

Note that the domain of f is the range of g, and the range of f is the domain of g.
When the coordinates of pairs in the elements of a function are interchanged and the new relation is a function, each function is called the inverse of the other.

The inverse of a function f is denoted by f^{-1}, read "f inverse." The -1 is not an exponent.

Note If $(x, y) \in f$ that is, $f(x) = y$
then $(y, x) \in f^{-1}$ that is, $f^{-1}(y) = x$
Also, $f^{-1}(f(x)) = f^{-1}(y) = x$
and $f(f^{-1}(y)) = f(x) = y$

A function has an inverse if each element in the domain is paired with one element of the range, and every element of the range is paired with one element of the domain.

To find the inverse of a function defined by a set of ordered pairs, interchange the first and second coordinates of the pairs.

EXAMPLE The inverse of the function $\{(1, 3), (2, 5), (3, 7), (4, 9)\}$ is the function $\{(3, 1), (5, 2), (7, 3), (9, 4)\}$.

An equation defines a function if for every value of x there corresponds only one value for y. In addition, the function possesses an inverse if for every value of y there corresponds only one value of x.

When a function is represented by an equation and the function has an inverse, the inverse function can be found by interchanging x and y in the equation and then expressing y in terms of x. The new equation defines the inverse function. The domain of the inverse function is the range of the function and the range of the inverse function is the domain of the function.

EXAMPLE

Consider the function f defined by the equation $y = 2x + 1$. Interchanging x and y, we get $x = 2y + 1$.

Expressing y in terms of x, we get $y = \dfrac{x-1}{2}$.

The function defined by $y = \dfrac{x-1}{2}$ is f^{-1}.

Some elements of the function f are $(3, 7), (10, 21), (4, 9), (-3, -5)$, and the pairs $(7, 3), (21, 10), (9, 4), (-5, -3)$ are elements of f^{-1}.

Not every function defined by an equation possesses an inverse. The equation $y = x^2$ defines a function. Every element of the domain of the function is associated with one element of its range. The equation $y^2 = x$, or $y = \pm\sqrt{x}$, which is obtained from $y = x^2$ by interchanging x and y, does not define a function.
The equation $y = \pm\sqrt{x}$ assigns two values to y for every positive value of x. Thus the function given by $y = x^2$ defined on its maximal domain does not possess an inverse. On the other hand, if we take the domain of the function to be $\{x \mid x \geq 0\}$, the function has an inverse: $y = \sqrt{x}$. The range of the inverse function is $\{y \mid y \geq 0\}$.

EXAMPLE

The function defined by $y = (x - 2)^2 - 3,\ x \geq 2$ has an inverse.

The inverse function is defined by $y = 2 + \sqrt{x + 3}$.

The domain of the inverse function is $\{x \mid x \geq -3\}$.

The range of the inverse function is $\{y \mid y \geq 2\}$.

Geometrically, a curve is a function if each line $x = a$, where a is an element of the domain of the function, intercepts the curve in only one point. A function has an inverse if each line $y = b$, where b is an element of the range of the function, intercepts the curve in only one point.

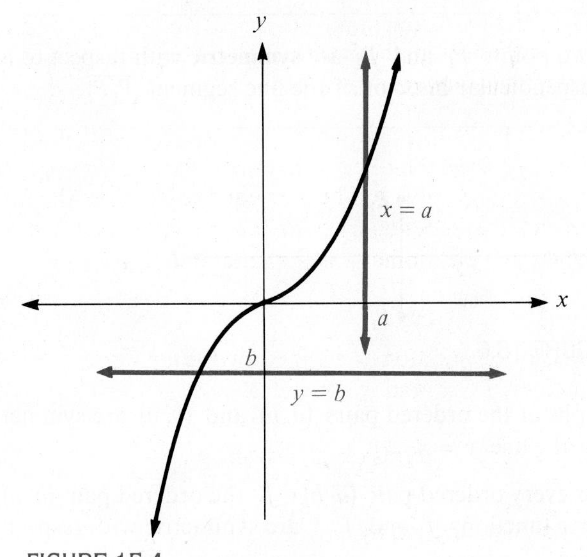

FIGURE 15.4

The curve shown in Figure 15.4 is a function since each line $x = a$, where a is an element of the domain, intercepts the curve in only one point. The function possesses an inverse since each line $y = b$, where b is an element of the range, intercepts the curve in one point.

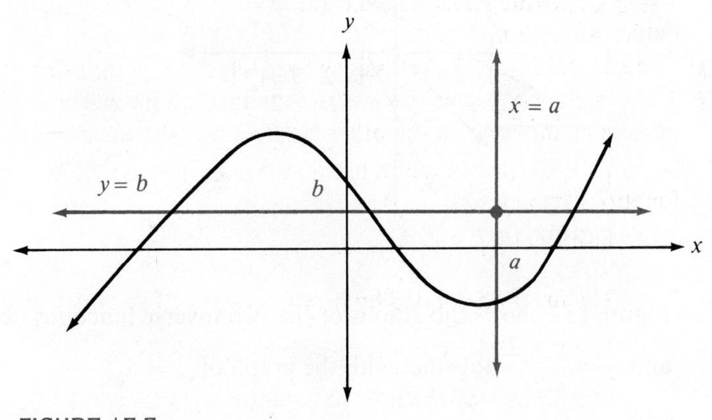

FIGURE 15.5

The curve shown in Figure 15.5 is a function since each line $x = a$, where a is an element of the domain of the function, intercepts the curve in exactly one point. The function does not possess an inverse since for some values of b, where b is in the range of the function, the line $y = b$ intercepts the curve in more than one point.

DEFINITION

> Two points P_1 and P_2 are **symmetric** with respect to a line L, if L is the perpendicular bisector of the line segment $\overline{P_1P_2}$.

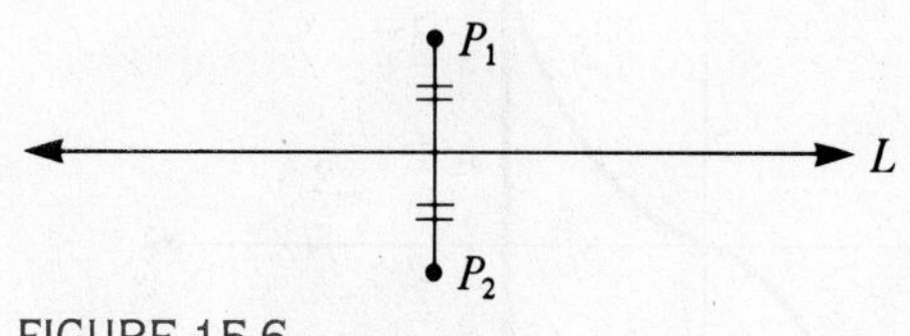

FIGURE 15.6

The graphs of the ordered pairs (a, b) and (b, a) are symmetric with respect to the graph of the line $y = x$.

Since for every ordered pair $(a, b) \in f$, the ordered pair $(b, a) \in f^{-1}$. The graphs of the inverse functions f and f^{-1} are symmetric with respect to the graph of the line $y = x$.

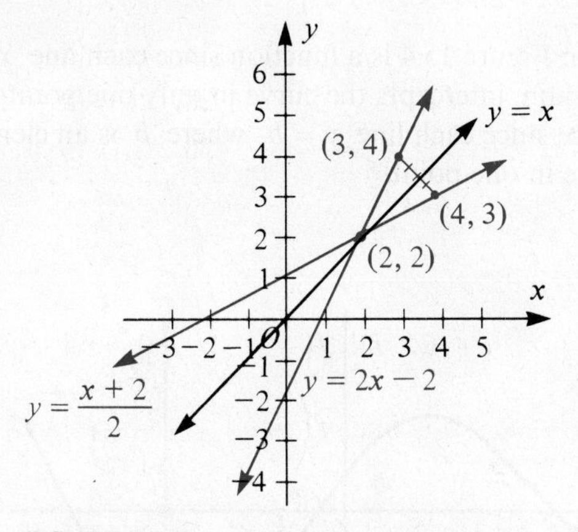

FIGURE 15.7

Figure 15.7 shows the graphs of the two inverse functions defined by $y = 2x - 2$ and $y = \dfrac{x+2}{2}$ together with the graph of $y = x$.

Exercise 15.6

Find an expression for the inverse of the function expressed by:

1. $\{(0, 0), (1, 1), (2, 4), (3, 9), (4, 16)\}$ **2.** $\{(0, 0), (1, 1), (2, 8), (3, 27), (4, 64)\}$

3. $\{(-3, 0), (-1, 2), (3, 6), (7, 10), (12, 15)\}$

4. $\{(-5, 23), (-2, 5), (0, -7), (2, -19), (4, -31)\}$
5. $\{(1, 2), (3, 5), (5, 8), (7, 11), (11, 17)\}$
6. $\{(1, 0), (2, 1), (3, 8), (4, 27), (5, 64)\}$
7. $\{(-1, 0), (0, 2), (1, 16), (2, 54), (3, 128)\}$
8. $\{(-2, 0), (-1, 1), (0, 4), (1, 9), (2, 16)\}$
9. $y = 7 - 3x$
10. $y = 3x - 8$
11. $y = -6x - 7$
12. $y = \dfrac{x - 3}{2}$
13. $y = \dfrac{2x - 1}{3}$
14. $y = \dfrac{2 - 5x}{4}$
15. $y = \dfrac{5 - 2x}{3}$
16. $y = \dfrac{2}{3}(2x - 1)$
17. $y = -\dfrac{4}{3}(x + 1)$
18. $y = \dfrac{7}{6}(x - 2)$
19. $y = -\dfrac{5}{7}(x + 3)$
20. $y = -\dfrac{2}{3}(x - 4)$

The following equations define functions. Find equations defining the inverse functions. Find the domain and range of the inverse functions:

21. $y = 2x^3$
22. $y = x^3 - 2$
23. $y = x^3 + 3$
24. $y = (x - 1)^3$
25. $y = (x + 1)^2 + 2,\ x \geq -1$
26. $y = (x - 1)^2 - 1,\ x \geq 1$
27. $y = (x + 2)^2 - 4,\ x \geq -2$
28. $y = (x - 2)^2 + 1,\ x \leq 2$
29. $y = (x - 3)^2 - 1,\ x \leq 3$
30. $y = (x + 3)^2 - 3,\ x \leq -3$

Chapter 15 Review

1. If $f(x) = 2x^3 - 3x + 6$, find $f(-3)$, $f(-1)$, and $f(2)$.
2. If $f(x) = 6x^2 + x - 15$, find $f\left(-\dfrac{5}{3}\right)$, $f\left(-\dfrac{1}{2}\right)$, and $f\left(\dfrac{3}{2}\right)$.
3. If $f(x) = x^2 - bx + c$, find $f\left(\dfrac{b}{2}\right)$, $f(-y)$, and $f(x + 1)$.
4. If $f(x) = (x + a)^2 + 3(x + a)$, find $f(x - 2a)$, $f(x - a)$, $f(b - a)$, and $f(-2a)$.
5. If $f(x) = (x - a)^2 - 4(x - a)$, find $f(a)$, $f(-a)$, $f(2a)$, $f(x + a)$, and $f(x + 2a)$.
6. If $f(x) = \dfrac{7x + 1}{2x - 1}$, find $f\left(-\dfrac{1}{2}\right)$, $f\left(\dfrac{1}{2}\right)$, $f(2x)$, and $f(3x^2)$.
7. If $f(x) = x^2 - 2x - 1$, check the truth of $f(5) - f(3) = f(2)$.
8. If $f(x) = 3x^2 + x$, check the truth of $f(a + b) - f(a) = f(b)$.
9. If $f(x) = 2x - 7$, $g(x) = x - 1$, find $(f + g)(x)$, $(f \cdot g)(x)$, $f(g(x))$, and $(g \circ f)(x)$.

10. If $f(x) = x^2 + x - 4$, $g(x) = 3x + 1$, find $(f + g)(x)$, $(f \cdot g)(x)$, $f(g(x))$, and $g(f(x))$.

Determine which of the following equations define functions. For each function, find its domain and range:

11. $y = 6x + 11$
12. $y = 3 - 2x$
13. $y = x^2 - 32$
14. $y = 5 - 4x^2$
15. $y = \sqrt{x + 9}$
16. $y = \sqrt{2x + 5}$
17. $y = \sqrt{4 - x}$
18. $y = \sqrt{6 - 2x}$
19. $y = \sqrt{x^2 + 6}$
20. $y = \sqrt{2x^2 + 3}$
21. $y = \sqrt{x^2 - 25}$
22. $y = \sqrt{4x^2 - 1}$
23. $y = \sqrt{49 - 4x^2}$
24. $y = \sqrt{36 - 8x^2}$
25. $y = \dfrac{5x - 7}{x}$
26. $y = \dfrac{3 - x}{2x - 9}$
27. $y^2 = 4 - 3x$
28. $y^2 = x - 7$

Find the solution set of the following equations:

29. $2^{x+3} \cdot 8 = 16^{x+3}$
30. $3^{2x} \cdot 9^{x-1} = 27^x$
31. $2^x\left(\dfrac{1}{4}\right)^{x-2} = \dfrac{1}{64}$
32. $5^{2-x} \cdot 25^{x+2} = \left(\dfrac{1}{125}\right)^{x-1}$

33. Sketch the graph of the function $\left\{(x, y) \middle| y = \left(\frac{1}{2}\right)^x, x \in R\right\}$ and from the graph find a rational approximation to each of the following:

(a) $\sqrt{\dfrac{1}{2}}$
(b) $\sqrt[3]{\dfrac{1}{2}}$

Find $f^{-1}(x)$ given the following:

34. $f(x) = \dfrac{x + 1}{6}$
35. $f(x) = \dfrac{3x + 1}{2}$
36. $f(x) = \dfrac{2x - 3}{4}$
37. $f(x) = \dfrac{2}{3}(8 - 5x)$

chapter 16

Logarithms

16.1 Logarithmic functions

16.2 Graphs of logarithmic functions

16.3 Properties of logarithms

16.4 Common logarithms

16.5 Natural logarithms

16.6 Computations with logarithms

16.1 Logarithmic Functions

An equation defines a function, if for every value of x there corresponds only one value for y. The function possesses an inverse, if for every value of y there corresponds only one value for x.

The exponential function defined by $y = a^x$ with $a \in R$, $a > 0, a \neq 1$, has the set of real numbers for its domain and the set of positive real numbers as its range. This exponential function has an inverse, since for every $y > 0$ there corresponds only one value of x. That is if $y = a^{x_1} = a^{x_2}$, then $x_1 = x_2$.

Geometrically, we can see that the exponential function defined by $y = a^x$ has an inverse. Every line $y = b, b \in R, b > 0$, intersects the graph of $y = a^x$ in one and only one point (refer back to Figures 15.2 and 15.3).

The inverse of the function defined by $y = a^x$ can be obtained by interchanging x and y and solving the resulting equation for y. By interchanging x and y in $y = a^x$ we obtain $x = a^y$. We have no means at our disposal to symbolically or algebraically solve the equation $x = a^y$ for y in terms of x. Therefore, we require a special notation.

To express y explicitly in terms of x we write $y = \log_a x$, which is read "y is equal to the logarithm of the number x to the base a." A function defined by an equation of the form $y = \log_a x$ is called a **logarithmic function**.
Since the equation $y = \log_a x$ is another form of the equation $x = a^y$, the logarithmic function defined by $y = \log_a x$ is the inverse of the exponential function defined by $y = a^x$.

Because the exponential function is not defined when $a \leq 0$ or $a = 1$, we must observe the same restrictions for the logarithmic function.

DEFINITION

If $a, b, c \in R$, $a > 0, b > 0, b \neq 1$,

$\log_b a = c$ is equivalent to $a = b^c$

EXAMPLES

1. Since $3^4 = 81$, $\log_3 81 = 4$

2. Since $(16)^{\frac{3}{4}} = 8$, $\log_{16} 8 = \frac{3}{4}$

3. $\log_2 64 = 6$ is equivalent to $64 = 2^6$

4. $\log_2 \frac{1}{8} = -3$ is equivalent to $\frac{1}{8} = 2^{-3}$

EXAMPLE Solve for x: $\log_{81} 27 = x$

Solution $\log_{81} 27 = x$ is equivalent to $81^x = 27$

Hence $3^{4x} = 3^3$

Thus $4x = 3$, that is, $x = \frac{3}{4}$

EXAMPLE Solve for x: $\log_{16} .125 = x$

Solution $\log_{16} .125 = x$ is equivalent to $16^x = .125$

Hence $16^x = \frac{125}{1000}$

or $16^x = \frac{1}{8}$ or $2^{4x} = 2^{-3}$

Thus $4x = -3$, that is, $x = -\frac{3}{4}$

EXAMPLE Solve for x: $\log_x \frac{1}{3} = -\frac{1}{5}$

Solution $\log_x \frac{1}{3} = -\frac{1}{5}$ is equivalent to $x^{-\frac{1}{5}} = \frac{1}{3}$

Hence $\left[x^{-\frac{1}{5}}\right]^{-5} = \left(\frac{1}{3}\right)^{-5}$

or

$$\begin{aligned} x &= (3^{-1})^{-5} \\ &= 3^5 \\ &= 243 \end{aligned}$$

EXAMPLE Solve for x: $\log_x 128 = -7$

Solution $\log_x 128 = -7$ is equivalent to $x^{-7} = 128$

Hence $x^{-7} = 2^7$

or $(x^{-7})^{-\frac{1}{7}} = (2^7)^{-\frac{1}{7}}$

Thus

$$\begin{aligned} x &= 2^{-1} \\ &= \frac{1}{2} \end{aligned}$$

EXAMPLE Solve for x: $\log_{216} x = \frac{1}{3}$

Solution $\log_{216} x = \frac{1}{3}$ is equivalent to

$$x = 216^{\frac{1}{3}}$$

or $$x = (2^3 \cdot 3^3)^{\frac{1}{3}} = 2 \cdot 3 = 6$$

EXAMPLE Solve for x: $\log_{64} x = -\frac{5}{6}$

Solution $\log_{64} x = -\frac{5}{6}$ is equivalent to

$$x = (64)^{-\frac{5}{6}}$$

or $$x = (2^6)^{-\frac{5}{6}} = 2^{-5} = \frac{1}{32}$$

Exercise 16.1

Write the following equations in logarithmic form:

1. $2^3 = 8$ **2.** $5^4 = 625$ **3.** $81^{\frac{1}{4}} = 3$

4. $3^{-4} = \frac{1}{81}$ **5.** $6^{-2} = \frac{1}{36}$ **6.** $7^{-3} = \frac{1}{343}$

7. $10^{-4} = .0001$ **8.** $10^{.301} = 2$ **9.** $10^{.699} = 5$

10. $x^4 = y$ **11.** $a^{-6} = b$ **12.** $x^{-n} = z$

Write the following equations in exponential form:

13. $\log_3 9 = 2$ **14.** $\log_4 8 = \frac{3}{2}$ **15.** $\log_{125} 5 = \frac{1}{3}$

16. $\log_{100} 10 = \frac{1}{2}$ **17.** $\log_3 \frac{1}{27} = -3$ **18.** $\log_2 \frac{1}{64} = -6$

19. $\log_{10} .01 = -2$ **20.** $\log_{10} 3 = .477$ **21.** $\log_{10} 6 = .778$

22. $\log_{10} 452 = 2.655$ **23.** $\log_y x = 3$ **24.** $\log_y x = -5$

Find the value of x in each of the following:

25. $\log_2 8 = x$ **26.** $\log_3 243 = x$ **27.** $\log_{49} 343 = x$

28. $\log_{64} 16 = x$ **29.** $\log_{27} 3 = x$ **30.** $\log_{36} 6 = x$

31. $\log_{49} 7 = x$

32. $\log_{32} 4 = x$

33. $\log_{256} 8 = x$

34. $\log_2 \frac{1}{2} = x$

35. $\log_3 \frac{1}{9} = x$

36. $\log_9 \frac{1}{3} = x$

37. $\log_{128} \frac{1}{64} = x$

38. $\log_{81} \frac{1}{27} = x$

39. $\log_{1000} .1 = x$

40. $\log_{64} .5 = x$

41. $\log_8 .125 = x$

42. $\log_{125} .04 = x$

43. $\log_{\frac{1}{2}} 4 = x$

44. $\log_{\frac{1}{3}} 81 = x$

45. $\log_{\frac{1}{5}} 125 = x$

46. $\log_{\frac{1}{16}} 64 = x$

47. $\log_x 49 = 2$

48. $\log_x 1000 = 3$

49. $\log_x 16 = \frac{4}{3}$

50. $\log_x 81 = \frac{4}{5}$

51. $\log_x 125 = \frac{3}{2}$

52. $\log_x 32 = \frac{5}{7}$

53. $\log_x 3 = \frac{1}{2}$

54. $\log_x 5 = \frac{1}{3}$

55. $\log_x 64 = \frac{3}{2}$

56. $\log_x 9 = \frac{2}{3}$

57. $\log_x \frac{1}{8} = -3$

58. $\log_x \frac{1}{81} = -4$

59. $\log_x \frac{1}{3} = -\frac{1}{4}$

60. $\log_x \frac{1}{2} = -\frac{1}{3}$

61. $\log_x .001 = -3$

62. $\log_x .25 = -\frac{2}{5}$

63. $\log_x 8 = -\frac{3}{2}$

64. $\log_x 49 = -2$

65. $\log_x 9 = -\frac{2}{3}$

66. $\log_x 32 = -\frac{5}{6}$

67. $\log_4 x = 3$

68. $\log_{11} x = 2$

69. $\log_5 x = 4$

70. $\log_{81} x = \frac{1}{2}$

71. $\log_{16} x = \frac{3}{2}$

72. $\log_8 x = \frac{2}{3}$

73. $\log_8 x = \frac{7}{3}$

74. $\log_9 x = \frac{5}{2}$

75. $\log_{27} x = \frac{4}{3}$

76. $\log_{128} x = \frac{1}{7}$

77. $\log_{243} x = \frac{1}{5}$

78. $\log_{343} x = \frac{2}{3}$

79. $\log_{64} x = \frac{1}{6}$

80. $\log_5 x = -4$

81. $\log_{100} x = -\frac{1}{2}$

82. $\log_{16} x = -\frac{3}{4}$

83. $\log_9 x = -\frac{3}{2}$

84. $\log_{243} x = -\frac{2}{5}$

85. $\log_{\frac{1}{2}} x = -3$

86. $\log_{\frac{1}{3}} x = -2$

87. $\log_{\frac{1}{25}} x = -1$

88. $\log_{\frac{1}{27}} x = \frac{4}{3}$

89. $\log_{\frac{1}{8}} x = \frac{2}{3}$

90. $\log_{\frac{1}{32}} x = \frac{4}{5}$

16.2 Graphs of Logarithmic Functions

The graph of the logarithmic function defined by the equation $y = \log_a x$ can be plotted by considering the equivalent equation $x = a^y$. Select values of y, $y \in R$, and compute the corresponding values of x. See Figure 16.1.

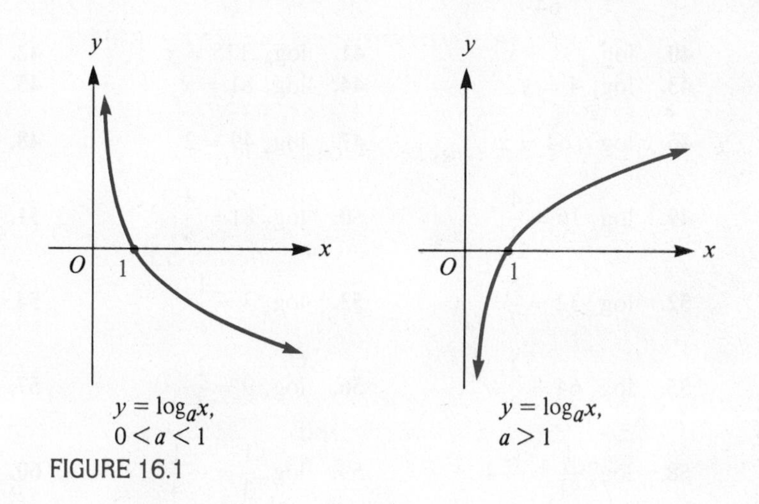

FIGURE 16.1

16.3 Properties of Logarithms

The following are some of the properties of logarithms that can be derived from the properties of exponents.

1. $\log_a 1 = 0$ since $a^0 = 1$

That is, the logarithm of the number 1 to any base, a, is equal to zero.

EXAMPLES $\log_{10} 1 = 0$, $\log_{213} 1 = 0$, $\log_{\sqrt{6}} 1 = 0$

2. $\log_a a = 1$ since $a^1 = a$

That is, the logarithm of a number to the same number as its base is equal to one.

EXAMPLES $\log_{10} 10 = 1$, $\log_{728} 728 = 1$, $\log_{\sqrt{5}} \sqrt{5} = 1$

THEOREM 1 $\log_c ab = \log_c a + \log_c b$

Proof

Let $\log_c a = x$ and $\log_c b = y$

Hence $a = c^x$ and $b = c^y$

$$a \cdot b = c^x \cdot c^y$$

$$ab = c^{x+y}$$

Thus $\log_c ab = x + y$

or $\log_c ab = \log_c a + \log_c b$

The logarithm of the product of two numbers is equal to the sum of the logarithms of the two numbers.

EXAMPLES

1. $\log_7 6 = \log_7 2 \cdot 3 = \log_7 2 + \log_7 3$

2. $\log_5 \frac{8}{9} + \log_5 \frac{27}{32} = \log_5 \frac{8}{9} \times \frac{27}{32} = \log_5 \frac{3}{4}$

THEOREM 2 $\log_c \frac{a}{b} = \log_c a - \log_c b$

Proof

Let $\log_c a = x$ and $\log_c b = y$

Hence $a = c^x$ and $b = c^y$

$$\frac{a}{b} = \frac{c^x}{c^y}$$

$$\frac{a}{b} = c^{x-y}$$

Thus $\log_c \frac{a}{b} = x - y$

or $\log_c \frac{a}{b} = \log_c a - \log_c b$

The logarithm of the quotient of two numbers is equal to the logarithm of the dividend minus the logarithm of the divisor.

EXAMPLES

1. $\log_3 \frac{5}{7} = \log_3 5 - \log_3 7$

2. $\log_a \frac{14}{25} - \log_a \frac{49}{45} = \log_a \frac{14}{25} \div \frac{49}{45}$

$$= \log_a \frac{14}{25} \times \frac{45}{49}$$

$$= \log_a \frac{18}{35}$$

THEOREM 3 $\log_c(a)^b = b \log_c a$

Proof

Let $\log_c a = x$

Hence $a = c^x$

$(a)^b = (c^x)^b$

$(a^b) = c^{bx}$

Thus $\log_c a^b = bx$

or $\log_c a^b = b \log_c a$

The logarithm of a number raised to a power is equal to the power multiplied by the logarithm of the number.

EXAMPLES

1. $\log_{13} 8 = \log_{13} 2^3 = 3 \log_{13} 2$

2.
$$
\begin{aligned}
-3 \log_a 5 &= \log_a 5^{-3} \\
&= \log_a \frac{1}{5^3} \\
&= \log_a \frac{1}{125}
\end{aligned}
$$

From Theorems 1, 2, and 3 we have

$$\log_k \frac{a^m b^n}{c^p d^q} = \log_k a^m b^n - \log_k c^p d^q \qquad \text{(Theorem 2)}$$

$$= (\log_k a^m + \log_k b^n) - (\log_k c^p + \log_k d^q) \qquad \text{(Theorem 1)}$$

$$= \log_k a^m + \log_k b^n - \log_k c^p - \log_k d^q$$

$$\log_k \frac{a^m b^n}{c^p d^q} = m \log_k a + n \log_k b - p \log_k c - q \log_k d \qquad \text{(Theorem 3)}$$

EXAMPLES

1. $$\log_{10} \frac{23^2 \cdot 17^{\frac{1}{2}}}{58^{\frac{2}{3}}} = 2 \log_{10} 23 + \frac{1}{2} \log_{10} 17 - \frac{2}{3} \log_{10} 58$$

2.
$$
\begin{aligned}
&3 \log_k a - 2 \log_k b + 4 \log_k c - 5 \log_k d \\
&= \log_k a^3 - \log_k b^2 + \log_k c^4 - \log_k d^5 \\
&= \log_k \frac{a^3 c^4}{b^2 d^5}
\end{aligned}
$$

THEOREM 4 $\log_b a \times \log_c b = \log_c a$

Proof

Let	$\log_b a = x$	and	$\log_c b = y$
Hence	$a = b^x$	and	$b = c^y$

Substituting c^y for b in $a = b^x$ we get

$$a = (c^y)^x = c^{xy}$$

Thus $\log_c a = xy$

or $\log_c a = \log_b a \times \log_c b$

EXAMPLES

1. $\log_7 5 = \log_{10} 5 \times \log_7 10$

2. $\log_e 25 = \log_{10} 25 \times \log_e 10$

THEOREM 5 $\log_b a = \dfrac{1}{\log_a b}$

Proof

From Theorem 4 we have

$$\log_b a \times \log_a b = \log_a a$$

or $\log_b a \times \log_a b = 1$

Dividing both sides by $\log_a b$ we get

$$\log_b a = \frac{1}{\log_a b}$$

EXAMPLES

1. $\log_3 10 = \dfrac{1}{\log_{10} 3}$

2. $\log_6 15 = \log_{10} 15 \times \log_6 10$

$$= \log_{10} 15 \times \frac{1}{\log_{10} 6}$$

$$= \frac{\log_{10} 15}{\log_{10} 6}$$

Note Given the logarithms of numbers to a certain base, we can find the logarithm of a number to a different base by applying Theorems 4 and 5.

For example, given the logarithms of numbers to the base 10, we can approximate the logarithms of numbers to any base:

$$\log_2 29 = \frac{\log_{10} 29}{\log_{10} 2}, \qquad \log_7 51 = \frac{\log_{10} 51}{\log_{10} 7}, \qquad \log_{2.718} 23 = \frac{\log_{10} 23}{\log_{10} 2.718}$$

THEOREM 6 $(a)^{\log_a b} = b$

Proof

Let	$\log_a b = x$	(1)
Hence	$b = a^x$	(2)
	$(a)^{\log_a b} = a^x$	from (1)
	$(a)^{\log_a b} = b$	from (2)

EXAMPLES **1.** $2^{\log_2 7} = 7$ **2.** $e^{\log_e x} = x$

Exercises 16.2–16.3

Given $\log_{10} 2 = .301$, $\log_{10} 3 = .477$, and $\log_{10} 5 = .699$, find the following:

1. $\log_{10} 6$ **2.** $\log_{10} 15$ **3.** $\log_{10} \frac{3}{2}$ **4.** $\log_{10} \frac{5}{2}$

5. $\log_{10} \frac{2}{3}$ **6.** $\log_{10} \frac{5}{3}$ **7.** $\log_{10} \frac{2}{5}$ **8.** $\log_{10} \frac{3}{5}$

9. $\log_{10} 8$ **10.** $\log_{10} 243$ **11.** $\log_{10} 25$ **12.** $\log_{10} \sqrt{2}$

13. $\log_{10} \sqrt{3}$ **14.** $\log_{10} \sqrt{5}$ **15.** $\log_{10} 24$ **16.** $\log_{10} 72$

17. $\log_{10} 40$ **18.** $\log_{10} 75$ **19.** $\log_{10} 375$ **20.** $\log_{10} \sqrt{6}$

21. $\log_{10} \sqrt[3]{4}$ **22.** $\log_{10} \sqrt[3]{9}$ **23.** $\log_{10} \sqrt[3]{25}$ **24.** $\log_{10} \sqrt[4]{8}$

25. $\log_{10} \frac{1}{4}$ **26.** $\log_{10} \frac{1}{27}$ **27.** $\log_{10} \frac{16}{81}$ **28.** $\log_{10} \frac{25}{8}$

29. $\log_2 10$ **30.** $\log_3 10$ **31.** $\log_5 10$ **32.** $\log_2 3$

33. $\log_2 5$ **34.** $\log_3 2$ **35.** $\log_3 5$ **36.** $\log_5 2$

37. $\log_5 3$ **38.** $\log_2 100$ **39.** $\log_3 .01$ **40.** $\log_5 12$

41. $\log_9 16$ **42.** $\log_4 125$ **43.** $\log_6 4$ **44.** $\log_{15} 45$

45. $\log_{12} 72$ **46.** $\log_{18} 16$ **47.** $\log_{24} 25$ **48.** $\log_{54} 36$

49. $\log_{\frac{2}{3}} 6$ **50.** $\log_{\frac{2}{5}} 15$ **51.** $\log_{\frac{3}{2}} 24$ **52.** $\log_{\frac{3}{5}} 18$

53. $\log_{\sqrt{2}} 3$ **54.** $\log_{\sqrt{2}} 5$ **55.** $\log_{\sqrt{3}} 4$ **56.** $\log_{\sqrt{3}} 6$

Use the properties of logarithms to transform the left member into the right member in each of the following. Assume for each problem that the base is a number greater than zero and different from one:

57. $\log \frac{32}{39} + \log \frac{51}{64} + \log \frac{52}{17} = \log 2$ **58.** $\log \frac{33}{38} + \log \frac{57}{26} + \log \frac{13}{22} = \log \frac{9}{8}$

59. $\log \frac{15}{16} - \log \frac{18}{7} + \log \frac{9}{35} = \log \frac{3}{32}$

60. $\log \frac{21}{26} - \log \frac{14}{39} + \log \frac{8}{9} = \log 2$

61. $\log \frac{44}{15} + \log \frac{25}{11} - \log \frac{5}{3} = 2 \log 2$

62. $\log \frac{36}{85} + \log \frac{34}{27} - \log \frac{8}{15} = 0$

63. $\log \frac{16}{27} - \log \frac{8}{9} - \log \frac{2}{3} = 0$

64. $\log \frac{3}{4} + \log \frac{5}{9} - \log \frac{25}{6} = -\log 10$

65. $\log \frac{33}{52} - \log \frac{11}{13} - \log \frac{3}{2} = -\log 2$

66. $\log \frac{7}{24} - \log \frac{16}{9} - \log \frac{21}{32} = -2 \log 2$

67. $\log a^4 + \log 27a - \log 3a^3 = 2 \log 3a$

68. $\log a^3 + \log b^2 - \log a^4 - \log b^3 = -\log ab$

69. $\log a^2 + \log 3b^2 - \log a - \log \frac{b^3}{6} = \log \frac{18a}{b}$

70. $\log 2a^3 - \log 2\sqrt{a^3} + \log \sqrt{a} = 2 \log a$

71. $\log \sqrt{a^5} + \log \sqrt[3]{b} - \log \sqrt{a} - \log \sqrt[3]{b^7} = 2 \log \frac{a}{b}$

72. $\log \sqrt[4]{a^3} - \log b + \log \sqrt{b} - \log \sqrt[4]{a} = \frac{1}{2} \log \frac{a}{b}$

73. $\log a^3 - \log \sqrt{a^3} - \log \frac{4}{a^5} - \log \frac{\sqrt{a^5}}{4} = 4 \log a$

74. $\log \frac{\sqrt[3]{a^5}}{3} - \log \frac{\sqrt[3]{a^7}}{6} - \log \frac{2}{a^4} = \frac{10}{3} \log a$

75. $3 \log a + \frac{5}{6} \log b - \frac{7}{3} \log a - \frac{2}{3} \log b = \frac{1}{6} \log a^4 b$

76. $\frac{7}{4} \log a + \frac{3}{2} \log b - \frac{5}{2} \log a - 3 \log b = -\frac{3}{4} \log ab^2$

77. $\frac{2}{5} \log a + \frac{9}{5} \log b - 2 \log a - \log b = -\frac{4}{5} \log \frac{a^2}{b}$

78. $\frac{2}{3}\log a + 3\log b - \frac{5}{3}\log b - \frac{4}{3}\log a = \frac{2}{3}\log\frac{b^2}{a}$

79. $\log\left(a - \frac{1}{a}\right) = \log(a+1) + \log(a-1) - \log a$

80. $\log\frac{a^3+1}{a^3-a^2+a} = \log(a+1) - \log a$

81. $\log\frac{a^{\frac{3}{2}} - a^{\frac{1}{2}}}{a^2-2a+1} = \frac{1}{2}\log a - \log(a-1)$

82. $\log\frac{(a+1)^{\frac{3}{2}}}{a^4-a^2} = \frac{1}{2}\log(a+1) - 2\log a - \log(a-1)$

83. $\log\frac{x+y}{x^{-1}+y^{-1}} = \log x + \log y$

84. $\log\frac{x^2+y}{x^{-2}+y^{-1}} = 2\log x + \log y$

85. $\log\frac{x+y}{x^{-2}-y^{-2}} = 2\log xy - \log(y-x)$

86. $\log\frac{x^{-3}-1}{x^{-1}-1} = \log(1+x+x^2) - 2\log x$

87. $\log\frac{\sqrt{a^2+b^2}-a}{\sqrt{a^2+b^2}+a} = 2\log(\sqrt{a^2+b^2}-a) - 2\log b$

88. $\log\frac{a+\sqrt{a^2-b^2}}{a-\sqrt{a^2-b^2}} = 2\log(a+\sqrt{a^2-b^2}) - 2\log b$

89. $\log\frac{\sqrt{a}+\sqrt{b}}{\sqrt{a}-\sqrt{b}} = 2\log(\sqrt{a}+\sqrt{b}) - \log(a-b)$

90. $\log\frac{\sqrt{a+b}-\sqrt{a-b}}{\sqrt{a+b}+\sqrt{a-b}} = \log(a-\sqrt{a^2-b^2}) - \log b$

16.4 Common Logarithms

The most convenient logarithms for computations use the number 10 as base. These logarithms are called **common logarithms**. Thus we omit writing the base and write simply $\log a$ to mean $\log_{10} a$.

Because of the wide use and the availability of electronic calculators, explanation of how to use the logarithmic tables will be given in Appendix D.

To find the common logarithm of a number by using a calculator, enter the number first, then press the key that reads "log."

EXAMPLES

1. $\log 562 = 2.7497363$
2. $\log 78.4 = 1.8943161$
3. $\log 6 = .7781513$
4. $\log .34 = -.4685211$
5. $\log .00237 = -2.6252517$

Remember that $\log 4.87 = .687529$ means $4.87 = 10^{.687529}$.

When the common logarithm of a number is given, the number corresponding to that logarithm can be found by using a calculator as follows:
Enter the logarithm on the calculator; first press the key that reads inverse "INV," then press the key that reads "log."

EXAMPLES

1. $\log x = 3.87$ $\quad x = 7413.1024$
2. $\log x = 1.642$ $\quad x = 43.85307$
3. $\log x = .7385$ $\quad x = 5.476461$
4. $\log x = -1.2016$ $\quad x = .0628637$
5. $\log x = -2.05634$ $\quad x = .0087833$

Exercise 16.4

Evaluate the following logarithms:

1. log 2.703 **2.** log 3.798 **3.** log 58.61 **4.** log 79.36
5. log 100.6 **6.** log 524.5 **7.** log 3842 **8.** log 7901
9. log .4902 **10.** log .6179 **11.** log .01634 **12.** log .06237
13. log .004213 **14.** log .007961 **15.** log .0002169

Solve the following for x:

16. $\log x = 1.7943$ **17.** $\log x = 1.9866$ **18.** $\log x = 2.1640$
19. $\log x = 3.71634$ **20.** $\log x = 4.28945$ **21.** $\log x = 2.59368$
22. $\log x = .4879$ **23.** $\log x = .39176$ **24.** $\log x = .06934$
25. $\log x = -.6235$ **26.** $\log x = -.7961$ **27.** $\log x = -2.3805$
28. $\log x = -1.7186$ **29.** $\log x = -3.5960$ **30.** $\log x = -2.0749$

16.5 Natural Logarithms

In mathematics as well as other fields of study a number between 2 and 3 appears very often.

Consider the quantity $\left(1+\frac{1}{n}\right)^n$ and let n grow very large.

When $n = 1000$, $\left(1+\frac{1}{n}\right)^n = 2.7169239$

When $n = 10^6$, $\left(1+\frac{1}{n}\right)^n = 2.7182805$

As n increases, the quantity $\left(1+\frac{1}{n}\right)^n$ approaches an irrational number denoted by the letter e.

A ten decimal approximation to e is 2.7182818285.
The number e is used as a base of the logarithmic function, called **natural logarithms**, and is denoted by "ln." Hence by $\ln x$ is meant $\log_e x$.

Note Even though the natural logarithms can be used in computations, it is easier to use common logarithms.

To find the natural logarithm of a number with a calculator, you enter the number first, then press the key that reads "ln."

EXAMPLES

1. $\ln 564.3 = 6.335586$
2. $\ln 78.941 = 4.3687007$
3. $\ln 6.3825 = 1.8535599$
4. $\ln .4672 = -.7609979$
5. $\ln .003627 = -5.6193494$

The inverse of the function defined by $y = \ln x$ is the exponential function defined by $y = e^x$.

Again, when the natural logarithm of a number is given (that is, if the value of $\ln x$ is given), we can find x by entering the value of $\ln x$, then pressing the key that reads "INV," and then the key that reads "ln."

EXAMPLES

1. $\ln x = 8.792$ $\quad x = 6581.3818$
2. $\ln x = 5.4317$ $\quad x = 228.53743$

3. $\ln x = .8124$ $\qquad x = 2.2533094$

4. $\ln x = -.03479$ $\qquad x = .9658082$

5. $\ln x = -2.3684$ $\qquad x = .0939681$

Exercise 16.5

Evaluate the following logarithms:

1. ln 865.2	**2.** ln 257.34	**3.** ln 458.62
4. ln 94.687	**5.** ln 39.731	**6.** ln 24.683
7. ln 7.6942	**8.** ln 5.986	**9.** ln 1.7472
10. ln .8104	**11.** ln .4078	**12.** ln .09806
13. ln .01635	**14.** ln .005954	**15.** ln .008183

Solve the following for x:

16. $\ln x = 6.621$	**17.** $\ln x = 4.719$	**18.** $\ln x = 5.94$
19. $\ln x = 1.8753$	**20.** $\ln x = 2.3928$	**21.** $\ln x = 1.763$
22. $\ln x = .6451$	**23.** $\ln x = .0727$	**24.** $\ln x = .0058$
25. $\ln x = -.369$	**26.** $\ln x = -.863$	**27.** $\ln x = -1.416$
28. $\ln x = -2.8537$	**29.** $\ln x = -2.045$	**30.** $\ln x = -3.7801$

16.6 Computations with Logarithms

By means of the theorems on logarithms, we can reduce problems of multiplication to addition, division to subtraction, and powers to multiplication, thus simplifying computations with very large numbers.

EXAMPLE Find the value of $\sqrt[3]{(.9674)^2}$.

Solution Change the radical to an exponent:

Let $y = \sqrt[3]{(.9674)^2} = (.9674)^{\frac{2}{3}}$.

Take the logarithm of both sides:

$$\log y = \log(.9674)^{\frac{2}{3}} = \frac{2}{3}\log .9674$$

$$= \frac{2}{3}(-.0143939) = -.0095959$$

Hence $\qquad y = .9781468$

EXAMPLE

Find the value of $\dfrac{\sqrt[4]{(.3964)^3}(7.948)^2}{\sqrt{(923.1)^3}}$.

Solution

Let
$$y = \frac{\sqrt[4]{(.3964)^3}(7.948)^2}{\sqrt{(923.1)^3}} = \frac{(.3964)^{\frac{3}{4}}(7.948)^2}{(923.1)^{\frac{3}{2}}}.$$

$$\begin{aligned}
\log y &= \log \frac{(.3964)^{\frac{3}{4}}(7.948)^2}{(923.1)^{\frac{3}{2}}} \\
&= \log(.3964)^{\frac{3}{4}} + \log(7.948)^2 - \log(923.1)^{\frac{3}{2}} \\
&= \frac{3}{4}\log .3964 + 2\log 7.948 - \frac{3}{2}\log 923.1 \\
&= \frac{3}{4}(-.4018664) + 2(.9002579) - \frac{3}{2}(2.9652488) \\
&= -.3013998 + 1.8005157 - 4.4478731 \\
&= -2.9487572
\end{aligned}$$

Hence $\qquad y = .0011252$

EXAMPLE

Find the value of $\dfrac{\sqrt{28.3} + \sqrt[3]{7.82}}{(.568)^2}$.

Solution

Note that $\log(a+b) \neq \log a + \log b$.

Let $y_1 = \sqrt{28.3}$ and $y_2 = \sqrt[3]{7.82}$

$$\begin{aligned}
y_1 &= (28.3)^{\frac{1}{2}} \\
\log y_1 &= \log(28.3)^{\frac{1}{2}} \\
&= \frac{1}{2}\log 28.3 \\
&= \frac{1}{2}(1.4517864) \\
&= .7258932 \\
y_1 &= 5.3197744
\end{aligned}$$

$$\begin{aligned}
y_2 &= (7.82)^{\frac{1}{3}} \\
\log y_2 &= \log(7.82)^{\frac{1}{3}} \\
&= \frac{1}{3}\log 7.82 \\
&= \frac{1}{3}(.8932068) \\
&= .2977356 \\
y_2 &= 1.9848861
\end{aligned}$$

$$\begin{aligned}
y &= \frac{\sqrt{28.3} + \sqrt[3]{7.82}}{(.568)^2} = \frac{5.3197744 + 1.9848861}{.322624} \\
&= 22.641405
\end{aligned}$$

EXAMPLE Find $\log_6 53$.

Solution
$$\log_6 53 = \frac{\log_{10} 53}{\log_{10} 6} = \frac{1.7242759}{.7781513} = 2.2158621$$

EXAMPLE Solve for x: $2^x = 9$.

Solution Taking the logarithm of both sides we get

$$\log 2^x = \log 9$$
$$x \log 2 = \log 9$$
$$x = \frac{\log 9}{\log 2} = \frac{.9542425}{.30103} = 3.169925$$

The solution set is $\{3.169925\}$.

EXAMPLE Solve for x: $2^{x+4} \cdot 3^{2x+1} = 12^{x+2}$.

Solution Take the logarithm of both sides:

$$\log 2^{x+4} + \log 3^{2x+1} = \log 12^{x+2}$$
$$(x+4)\log 2 + (2x+1)\log 3 = (x+2)\log 12$$
$$(x+4)(.30103) + (2x+1)(.4771213) = (x+2)(1.0791812)$$
$$.30103x + 1.20412 + .9542426x + .4771213 = 1.0791812x + 2.1583624$$
$$.1760914x = .4771211$$
$$x = \frac{.4771211}{.1760914} = 2.7095082$$

The solution set is $\{2.7095082\}$.

EXAMPLE Solve for x: $\log(x+1) = 1.4$

Solution

$$x + 1 = 25.118864$$
$$x = 24.118864$$

The solution set is $\{24.118864\}$.

Exercise 16.6

Use logarithms to perform the following computations:

1. $\sqrt[3]{79.43}$ **2.** $\sqrt[3]{.8571}$ **3.** $\sqrt[4]{4814}$ **4.** $\sqrt[4]{51.76}$

5. $\sqrt[3]{(66.15)^2}$ **6.** $\sqrt[3]{(406.2)^2}$ **7.** $\sqrt[4]{(2.356)^3}$ **8.** $\sqrt[4]{(.946)^3}$

9. $(2.361)^2\sqrt[3]{67.83}$ **10.** $(.7926)^3\sqrt{92.72}$ **11.** $\dfrac{(34.78)^2}{\sqrt{80.31}}$

12. $\dfrac{(6.134)^2}{\sqrt[3]{1.237}}$ **13.** $\dfrac{\sqrt{98.49}}{(2.164)^2}$ **14.** $\dfrac{\sqrt[3]{7.318}}{(.8645)^2}$

15. $\dfrac{(2.718)^2\sqrt{45.79}}{\sqrt[3]{.5142}}$ **16.** $\dfrac{(67.12)\sqrt[3]{.03211}}{\sqrt{91.38}}$ **17.** $\dfrac{\sqrt[3]{(42.96)^2}}{(.7033)^2\sqrt{294.5}}$

18. $\dfrac{\sqrt[3]{(3.541)^2}}{(6.304)^3\sqrt{.04866}}$ **19.** $\dfrac{\sqrt[4]{(.4758)^3}}{(5.119)^2\sqrt[3]{91.06}}$ **20.** $\dfrac{\sqrt{8.739}}{(.4418)^4\sqrt[3]{(55.27)^2}}$

21. $\dfrac{\sqrt{.3032}}{\sqrt[3]{6.803}\ \sqrt[4]{.4698}}$ **22.** $\dfrac{\sqrt[3]{94.79}}{\sqrt{2405}\ \sqrt[5]{(8.176)^3}}$ **23.** $\dfrac{\sqrt{89.21} + \sqrt{9.042}}{\sqrt{4.713}}$

24. $\dfrac{\sqrt{5.105} - \sqrt[3]{334.3}}{\sqrt{.1138}}$ **25.** $\dfrac{\sqrt{57.63}}{\sqrt{4.232} + \sqrt{8.608}}$ **26.** $\dfrac{\sqrt{.2201}}{\sqrt{6.799} - \sqrt[3]{707.4}}$

Evaluate the following logarithms:

27. $\log_7 29$ **28.** $\log_4 38$ **29.** $\log_5 6.14$ **30.** $\log_{30} 96.7$

31. $\log_{38.7} 21.6$ **32.** $\log_{.721} 63.9$

33. $\log_{1.85} .429$ **34.** $\log_{.822} .0971$

Solve each of the following equations for x:

35. $2^x = 6$ **36.** $3^x = 17$ **37.** $7^x = 352$

38. $28^x = 773$ **39.** $3^{x+1} = 2^{x+2}$ **40.** $3^{2x-1} = 6^x$

41. $5^{3x-1} = 7^{2x}$ **42.** $12^{x-1} \cdot 5^{2x} = 15^{x+2}$

43. $6^{x+4} \cdot 7^x = 12^{2x+1}$ **44.** $3^{2x-3} \cdot 7^{x+2} = 21^{x+1}$

45. $7^{x-2} \cdot 2^{3x-1} = 11^{x-1}$ **46.** $\log(x + 2) = 1.5119$

47. $\log(x - 3) = 1.6656$ **48.** $\log(2x + 5) = 1.4669$

49. $\log(3x + 1) = 1.7007$ **50.** $\log x - \log(x + 1) = -.899$

51. $\log(x - 3) - \log(x + 5) = -.27$ **52.** $\log(x + 9) - \log(x + 2) = .1523$

53. $\log(x + 16) - \log(x - 7) = .6064$

Chapter 16 Review

Find $f^{-1}(x)$ given the following:

1. $f(x) = 2^x$ **2.** $f(x) = 3^x$ **3.** $f(x) = 2^{3x}$

4. $f(x) = 5^{2x}$ **5.** $f(x) = 3^{-2x}$ **6.** $f(x) = 7^{-4x}$

Without the use of a calculator find the value of x in each of the following:

7. $\log_{10} x = -2$ **8.** $\log_{32} x = -\dfrac{1}{5}$ **9.** $\log_{64} x = \dfrac{2}{3}$

10. $\log_{216} x = \frac{1}{3}$ **11.** $\log_{\frac{1}{2}} x = 5$ **12.** $\log_{\frac{1}{9}} x = -2$

13. $\log_x 8 = \frac{3}{7}$ **14.** $\log_x 9 = 4$ **15.** $\log_x 7 = -\frac{1}{3}$

16. $\log_x 25 = -2$ **17.** $\log_x \frac{1}{3} = -1$ **18.** $\log_x \frac{1}{4} = \frac{2}{5}$

19. $\log_{64} 4 = x$ **20.** $\log_{27} 243 = x$ **21.** $\log_4 \frac{1}{8} = x$

22. $\log_{625} \frac{1}{25} = x$ **23.** $\log_2 .25 = x$ **24.** $\log_{\frac{1}{25}} 25 = x$

25. $\log_{\frac{1}{27}} 9 = x$ **26.** $\log_{\frac{1}{4}} \frac{1}{2} = x$

Use the properties of logarithms to transform the left member into the right member in each of the following. Assume for each problem the base is a number greater than zero and different from 1:

27. $\log \frac{45}{16} + \log \frac{7}{12} - \log \frac{105}{8} = -3 \log 2$

28. $\log \frac{28}{27} - \log \frac{64}{15} - \log \frac{7}{16} = \log \frac{5}{9}$

29. $\log \frac{18}{11} - \log \frac{36}{44} + \log \frac{9}{8} = -2 \log \frac{2}{3}$

30. $\log \frac{49}{40} + \log \frac{96}{343} - \log \frac{18}{35} = -\log \frac{3}{2}$

31. $\log 12a^5 + \log 21b^4 - \log 9a^4 - \log 16b^6 = \log \frac{7a}{4b^2}$

32. $\log \sqrt[3]{a^4} - \log \sqrt{b} - \log \sqrt[3]{a^{10}} + \log \sqrt{b^5} = 2 \log \frac{b}{a}$

33. $\frac{2}{3} \log a + \frac{1}{6} \log b - \frac{2}{3} \log b - \frac{1}{6} \log a = -\frac{1}{2} \log \frac{b}{a}$

34. $\frac{5}{2} \log a + \frac{7}{2} \log \frac{1}{a} + \frac{1}{4} \log b - \frac{3}{4} \log \frac{1}{b} = -\log \frac{a}{b}$

35. $\log \frac{a^{-2} - a^2}{a^{-1} + a} = \log(a + 1) + \log(1 - a) - \log a$

36. $\log \frac{\sqrt{a + 2} + \sqrt{a - 2}}{\sqrt{a + 2} - \sqrt{a - 2}} = 2 \log(\sqrt{a + 2} + \sqrt{a - 2}) - 2 \log a$

Use logarithms to perform the following computations:

37. $\sqrt[4]{356.2}$ **38.** $\sqrt[3]{(.01651)^2}$ **39.** $\sqrt[4]{(46.73)^3}$

40. $(7.393)^2\sqrt[5]{6.865}$ **41.** $\dfrac{\sqrt{43.64}}{(2.196)^2}$ **42.** $\dfrac{(.9718)^2}{\sqrt[3]{68.12}}$

43. $\dfrac{\sqrt{29.74}}{\sqrt[3]{146.9}}$ **44.** $\dfrac{(.8376)^2\sqrt{44.44}}{\sqrt[3]{455.6}}$ **45.** $\dfrac{(19.87)^2\sqrt[3]{(1.363)^2}}{\sqrt{795.9}}$

46. $\dfrac{\sqrt{95.71}}{(24.05)^2\sqrt[3]{(6.064)^2}}$ **47.** $\dfrac{\sqrt[3]{29.72}}{(7.427)^3\sqrt[4]{(.6107)^3}}$ **48.** $\dfrac{\sqrt{451.8}\,\sqrt[3]{(.7364)^2}}{\sqrt[4]{(68.84)^3}(2.005)^2}$

49. $\dfrac{\sqrt{.5145}+\sqrt{40.82}}{\sqrt{8.998}}$ **50.** $\dfrac{\sqrt{8.802}-\sqrt{60.16}}{\sqrt{5763}}$ **51.** $\dfrac{\sqrt{732.4}}{\sqrt{278.8}+\sqrt[3]{.8235}}$

52. $\dfrac{\sqrt[3]{84.32}}{\sqrt{99.78}-\sqrt[3]{.2304}}$ **53.** $\dfrac{\sqrt{12.39}+\sqrt{6.474}}{\sqrt{8.621}+\sqrt{.5999}}$ **54.** $\dfrac{\sqrt{49.28}-\sqrt{14.61}}{\sqrt[3]{7.356}+\sqrt[3]{50.13}}$

Solve each of the following equations for x:

55. $12^x = 64$ **56.** $6.12^x = 79.8$

57. $11^{2x+1} = 15^{x+10}$ **58.** $7^{3x} = 6^{4x-5}$

59. $2^{x+3} \cdot 3^{x-1} = 6^{2x-1}$ **60.** $5^{2x+1} \cdot 7^{x+2} = 35^{3x}$

61. $\log(2x-1) = 1.9253$ **62.** $\log(3x+4) = 1.4382$

63. $\log 2x - \log(x+6) = -.7447$ **64.** $\log(x+4) - \log(x-3) = 1.8212$

65. An adiabatic expansion is one that is not accompanied by any gain or loss of heat. The law for adiabatic expansion of a gas is $pv^{1.4} = c$, where p is the pressure, v is the volume, and c is a constant. Find the pressure when $v = 8.9$ cubic feet and $c = 400$.

66. The acidity or alkalinity, called the pH, of a solution is determined by $\text{pH} = -\log H$ where H is the hydrogen ion concentration of the solution. When the pH is 7, the solution is neutral; less than 7 the solution is acid; and greater than 7 the solution is alkaline. Find the pH of a solution when $H = 2.4 \times 10^{-8}$.

67. A radioactive substance decays according to the formula $m = m_0(2.3)^{-.4t}$, where m_0 is the number of grams of the substance present originally and m is the number of grams left after t years. How much of 35.6 grams of the substance will remain after 8 years?

68. Bacteria in a certain culture grow according to the relation $N = 200e^{.54t}$ where N is the number of bacteria present t hours after the culture was started. How many bacteria are present 15 hours after the culture was started?

chapter 17

Progressions

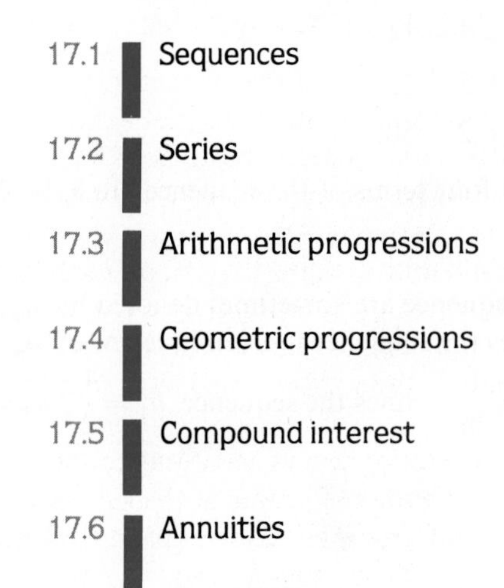

17.1 Sequences

17.2 Series

17.3 Arithmetic progressions

17.4 Geometric progressions

17.5 Compound interest

17.6 Annuities

17.1 Sequences

A **sequence** is the set of the elements of the range of a function, f, whose domain is the set of natural numbers. The elements of the range, $f(1), f(2), f(3), \ldots, f(n), \ldots$ are called the **terms of the sequence.** $f(1)$ is the **first term,** $f(2)$ the **second term,** and in general $f(n)$ is the ***n*th term.**

EXAMPLE

Find the first four terms of the sequence defined by $f(n) = 2n - 1$.

Solution

$f(1) = 2(1) - 1 = 1$
$f(2) = 2(2) - 1 = 3$
$f(3) = 2(3) - 1 = 5$
$f(4) = 2(4) - 1 = 7$

Thus the first four terms of the sequence are 1, 3, 5, 7.

EXAMPLE

Find the first four terms of the sequence defined by $f(n) = 3n^2 + 2$.

Solution

$f(1) = 3(1)^2 + 2 = 5$
$f(2) = 3(2)^2 + 2 = 14$
$f(3) = 3(3)^2 + 2 = 29$
$f(4) = 3(4)^2 + 2 = 50$

Thus the first four terms of the sequence are 5, 14, 29, 50.

The terms of the sequence are sometimes denoted by $a_1, a_2, a_3, \ldots, a_n, \ldots,$ and the rule that defines the sequence is given in terms of a_n.

For example, $a_n = \frac{1}{2n}$ defines the sequence $a_1 = \frac{1}{2}$, $a_2 = \frac{1}{4}$, $a_3 = \frac{1}{6}, \ldots$.

17.2 Series

The indicated sum, if it exists, of the terms of a sequence is called a **series.** Thus for the sequence $a_1, a_2, a_3, \ldots, a_n, \ldots,$ we have the series $a_1 + a_2 + a_3 + \cdots + a_n + \cdots$.

The sum of the first n terms of a series is denoted by S_n, $S_n = a_1 + a_2 + \cdots + a_n$. S_n is called the ***n*th partial sum.**
Thus $S_1 = a_1$, $S_2 = a_1 + a_2$, $S_3 = a_1 + a_2 + a_3, \ldots,$ etc.
Note that the partial sums $S_1, S_2, S_3, \ldots, S_n, \ldots,$ by themselves form a sequence that is called a **sequence of partial sums.**

EXAMPLE Find the first four terms of the sequence S_n where $a_n = 3n + 1$.

Solution

$$a_1 = 4, \quad a_2 = 7, \quad a_3 = 10, \quad a_4 = 13, \ldots.$$

$$S_1 = a_1 = 4$$

$$S_2 = a_1 + a_2 = 4 + 7 = 11$$

$$S_3 = a_1 + a_2 + a_3 = 4 + 7 + 10 = 21$$

$$S_4 = a_1 + a_2 + a_3 + a_4 = 4 + 7 + 10 + 13 = 34$$

The first four terms of the sequence are 4, 11, 21, 34.

Summation Notation

Partial sums are defined in terms of a sequence, which in turn is defined by a rule. The partial sums can be expressed directly by a rule using what is called the **summation notation**.

For the summation notation we use the Greek letter $\sum$, sigma. Instead of writing $a_1 + a_2 + a_3 + a_4 + a_5$, we write

$$\sum_{i=1}^{5} a_i$$

which is read "the sum of a_i for i running from 1 to 5, $i \in N$." The **summation symbol** is $\sum$, i is called the **index of summation**, and the set of values that i takes is the **range of the summation**.

EXAMPLE Write $\sum_{i=1}^{7} i$ in expanded form.

Solution Substitute $1, 2, 3, \ldots, 7$ for i and add the resulting terms:

$$\sum_{i=1}^{7} i = 1 + 2 + 3 + 4 + 5 + 6 + 7$$

EXAMPLE Write $\sum_{k=1}^{4} k(k+1)$ in expanded form.

Solution

$$\sum_{k=1}^{4} k(k+1) = 1 \cdot 2 + 2 \cdot 3 + 3 \cdot 4 + 4 \cdot 5$$

EXAMPLE Write $\sum_{i=1}^{n} i^2$ in expanded form.

Solution

$$\sum_{i=1}^{n} i^2 = 1^2 + 2^2 + 3^3 + \cdots + n^2$$

EXAMPLE Write $\sum_{k=1}^{5} (-1)^{k+1} \cdot \frac{1}{k}$ in expanded form.

Solution

$$\sum_{k=1}^{5} (-1)^{k+1} \cdot \frac{1}{k}$$

$$= (-1)^2 \cdot \frac{1}{1} + (-1)^3 \cdot \frac{1}{2} + (-1)^4 \cdot \frac{1}{3} + (-1)^5 \cdot \frac{1}{4} + (-1)^6 \cdot \frac{1}{5}$$

$$= 1 - \frac{1}{2} + \frac{1}{3} - \frac{1}{4} + \frac{1}{5}$$

Exercises 17.1–17.2

Find the first five terms of the sequence defined by the following:

1. $f(n) = 3n$ **2.** $f(n) = 5n$ **3.** $f(n) = 4n + 1$

4. $f(n) = 1 - 3n$ **5.** $f(n) = 1 - 6n$ **6.** $f(n) = 2 - 7n$

7. $f(n) = 2n^2$ **8.** $f(n) = n^2 + 1$ **9.** $f(n) = 3n^2 - 1$

10. $f(n) = n^3 + 1$ **11.** $f(n) = \frac{2}{n}$ **12.** $f(n) = \frac{1}{n^2}$

13. $a_n = \frac{n}{n+1}$ **14.** $a_n = \frac{n}{2n-1}$ **15.** $a_n = 3^{n-1}$

16. $a_n = 3 \cdot 2^n$ **17.** $a_n = \frac{1}{2^n}$ **18.** $a_n = \frac{27}{3^n}$

19. $a_n = (-1)^n \frac{2}{n}$ **20.** $a_n = (-1)^{n+1} \frac{3}{2n}$ **21.** $a_n = (-1)^n \frac{1}{4n-1}$

22. $a_n = (-1)^{n-1} \frac{2}{n^2}$ **23.** $a_n = (-1)^{2n} \frac{1}{n^3}$ **24.** $a_n = (-1)^{2n+1} \frac{1}{2^n}$

Find the first four terms of the sequence S_n in the following:

25. $a_n = n + 2$ **26.** $a_n = 7n$ **27.** $a_n = 2n + 3$ **28.** $a_n = 4n - 1$

29. $a_n = 2 - 3n$ **30.** $a_n = 3 - 5n$ **31.** $a_n = n^2 + 1$ **32.** $a_n = 2n^2 - 1$

33. $a_n = 2^n$ **34.** $a_n = 3^n - 15$ **35.** $a_n = \frac{n}{2}$ **36.** $a_n = \frac{2n}{3}$

37. $a_n = (-1)^n n$ **38.** $a_n = (-1)^{n-1} 3n$ **39.** $a_n = (-1)^{n+1} n - 2$

40. $a_n = 10 - (-1)^n n^2$

Write the following in expanded form:

41. $\sum_{i=1}^{6} i$ **42.** $\sum_{i=1}^{5} 4i$ **43.** $\sum_{k=1}^{5} (2k - 1)$ **44.** $\sum_{k=1}^{5} (4k - 3)$

45. $\sum_{i=1}^{n} (i+3)$ **46.** $\sum_{i=1}^{n} (1-3i)$ **47.** $\sum_{i=1}^{5} \frac{i-2}{8}$ **48.** $\sum_{i=2}^{6} \frac{3i-1}{5}$

49. $\sum_{k=4}^{8} \frac{4k-1}{3}$ **50.** $\sum_{k=1}^{4} k(k+2)$ **51.** $\sum_{k=1}^{4} 3k(k+1)$

52. $\sum_{k=1}^{4} (2k-1)(2k+1)$ **53.** $\sum_{k=1}^{3} k(k+1)(k+2)$ **54.** $\sum_{k=1}^{4} k^2$

55. $\sum_{k=1}^{n} k^3$ **56.** $\sum_{k=1}^{n} 3^k$ **57.** $\sum_{i=1}^{4} \frac{1}{i(i+1)}$

58. $\sum_{i=1}^{4} \frac{1}{(3i-1)(3i+1)}$ **59.** $\sum_{k=1}^{n} \frac{1}{2^k}$ **60.** $\sum_{k=1}^{7} (-1)^{k+1}k$

61. $\sum_{k=1}^{5} (-1)^{k-1}\frac{1}{k}$ **62.** $\sum_{k=1}^{5} (-1)^{2k-1}\frac{1}{k^2}$

17.3 Arithmetic Progressions

An **arithmetic progression** is a sequence of numbers in which the difference between any term and the preceding term is always the same. The difference between the terms is called the **common difference**.

For example, the sequence 1, 5, 9, 13, . . . , is an arithmetic progression; the common difference is 4.

When the first term is denoted by a_1 and the common difference by d, the terms of the arithmetic progression are

$$a_1 = a_1, \quad a_2 = a_1 + d, \quad a_3 = a_2 + d = a_1 + 2d$$

$$a_4 = a_3 + d = a_1 + 3d = a_1 + (4-1)d$$

$$a_5 = a_1 + 4d = a_1 + (5-1)d, \ldots$$

$$a_n = a_1 + (n-1)d, \ldots$$

EXAMPLE

If $a_1 = 2$ and $d = 3$, the arithmetic progression is defined by

$$a_n = 2 + (n-1)(3) \quad \text{or} \quad a_n = 3n - 1$$

The terms of the arithmetic progression are obtained by substituting 1, 2, 3, . . . , for n.

Thus the arithmetic progression is 2, 5, 8, 11, . . . , $3n - 1$, ...

Note Given some terms of an arithmetic progression, we find the common difference d from the relation

$$d = a_n - a_{n-1}$$

EXAMPLES

1. The common difference for the arithmetic progression

$3, 9, 15, 21, \ldots,$ is 6

2. The common difference for the arithmetic progression

$3a - b, 4a + 2b, 5a + 5b, \ldots,$ is $a + 3b$

Note Given the first term a_1 and the common difference d of an arithmetic progression, we find the nth term from the relation

$$a_n = a_1 + (n - 1)d$$

EXAMPLE Find the fiftieth term of the arithmetic progression whose first term is 4 and whose common difference is 7.

Solution $a_{50} = 4 + (50 - 1)(7) = 347$

Sum of an Arithmetic Progression

The sum S_n of the first n terms of an arithmetic progression is

$$S_n = a_1 + (a_1 + d) + (a_1 + 2d) + \cdots + (a_n - 2d) + (a_n - d) + a_n \qquad (1)$$

Note that $a_{n-1} = a_n - d$ and $a_{n-2} = a_{n-1} - d = a_n - 2d$.

If we rewrite the terms of the sum in reverse order, that is, the nth term first and first term last, we have

$$S_n = a_n + (a_n - d) + (a_n - 2d) + \cdots + (a_1 + 2d) + (a_1 + d) + a_1 \qquad (2)$$

Adding (1) and (2) column by column, we get

$$2S_n = (a_1 + a_n) + (a_1 + a_n) + \cdots + (a_1 + a_n) \qquad (3)$$

Since there are n terms in the right member of (3), each of the form $(a_1 + a_n)$, we have

$$2S_n = n(a_1 + a_n) \qquad \text{or} \qquad S_n = \frac{n}{2}(a_1 + a_n)$$

EXAMPLE Find S_{100} if $a_1 = 1$ and $a_{100} = 298$.

Solution $S_{100} = \frac{100}{2}(1 + 298) = 14{,}950$

Notes

1. Given the first term a_1, the nth term a_n, and the number of terms n of an arithmetic progression, we find the sum of the first n terms from the relation

$$S_n = \frac{n}{2}(a_1 + a_n)$$

2. Given the first term a_1, the common difference d, and the number of terms n of an arithmetic progression, we find the sum of the first n terms by substituting $a_1 + (n-1)d$ for a_n in the above relation:

$$S_n = \frac{n}{2}[a_1 + a_1 + (n-1)d]$$

or

$$S_n = \frac{n}{2}[2a_1 + (n-1)d]$$

EXAMPLE

Find $\sum_{i=1}^{k} 6i$.

Solution

The terms in the expansion of $\sum_{i=1}^{k} 6i$ form an arithmetic progression for which $a_1 = 6$, $a_n = 6k$, $n = k$.

Thus $\quad \sum_{i=1}^{k} 6i = S_k = \frac{k}{2}(6 + 6k) = 3k(k+1)$

EXAMPLE

Given $a_1 = 80$ and $d = -11$, find a_{14} and S_{17}.

Solution

$$\begin{aligned} a_{14} &= 80 + (14-1)(-11) \\ &= -63 \end{aligned}$$

$$\begin{aligned} S_{17} &= \frac{17}{2}[2(80) + (17-1)(-11)] \\ &= \frac{17}{2}(160 - 176) = -136 \end{aligned}$$

EXAMPLE

The twenty-second and thirty-eighth terms of an arithmetic progression are 29 and 77, respectively. Find the tenth term.

Solution

$a_{22} = 29, \quad a_{38} = 77$

$$a_{22} = a_1 + 21d = 29 \tag{1}$$

$$a_{38} = a_1 + 37d = 77 \tag{2}$$

Solving Eqs. (1) and (2) for a_1 and d, we get

$$a_1 = -34, \qquad d = 3$$

Thus $\quad a_{10} = -34 + (10-1)(3) = -7$

EXAMPLE

Given $a_1 = 14, d = \frac{3}{2}$, and $S_n = 180$, find n and a_n.

Solution

$$S_n = \frac{n}{2}\left[28 + (n-1)\left(\frac{3}{2}\right)\right] = 180$$

Simplifying the above equation we get

$$3n^2 + 53n - 720 = 0$$
$$(3n + 80)(n - 9) = 0$$

Hence $n = 9$

We reject $-\frac{80}{3}$, since n has to be a natural number.

Thus $a_9 = 14 + 8\left(\frac{3}{2}\right) = 26$

EXAMPLE

Given $d = -3, a_n = -20$, and $S_n = -42$, find a_1 and n.

Solution

$$a_n = a_1 + (n-1)(-3) = -20$$

or

$$a_1 = 3n - 23 \tag{1}$$

$$S_n = \frac{n}{2}(a_1 - 20) = -42$$

or

$$na_1 - 20n = -84 \tag{2}$$

Substituting $3n - 23$ for a_1 in Eq. (2) we get

$$n(3n - 23) - 20n = -84$$

or

$$3n^2 - 43n + 84 = 0$$
$$(3n - 7)(n - 12) = 0$$

Hence $n = 12$

We reject $\frac{7}{3}$, since n has to be a natural number.

Substituting 12 for n in Eq. (1) we get

$a_1 = 3(12) - 23 = 13$

Notes

1. When an even number of terms forms an arithmetic progression, the computations can be simplified if we let the middle terms be $a - d$ and $a + d$, respectively. The common difference is $2d$ and the terms of the progression are

$$\ldots, a - 3d, a - d, a + d, a + 3d, \ldots.$$

2. When an odd number of terms forms an arithmetic progression, the computations can be simplified if we let the middle term be a, and take d as the common difference. The terms of the progression are

   $\ldots, a - 2d, a - d, a, a + d, a + 2d, \ldots.$

EXAMPLE

The sum of four numbers in an arithmetic progression is 68. Find the numbers if the sum of their squares is 1336.

Solution

Let the four numbers be $a - 3d, a - d, a + d, a + 3d$:

$(a - 3d) + (a - d) + (a + d) + (a + 3d) = 68$

Hence $a = 17$

Thus the numbers are $17 - 3d, 17 - d, 17 + d, 17 + 3d$:

$(17 - 3d)^2 + (17 - d)^2 + (17 + d)^2 + (17 + 3d)^2 = 1336$

Simplifying the above equation we get

$d^2 = 9$ or $d = \pm 3$

Taking $d = 3$, we get

$17 - 3(3),\ 17 - (3),\ 17 + (3),\ 17 + 3(3)$

The numbers are 8, 14, 20, 26.

When $d = -3$ we get the same numbers, but in the reverse order.

The terms a_1 and a_n of the arithmetic progression $a_1, a_2, \ldots, a_n$ are called the **extremes**, and the intermediate terms, $a_2, \ldots, a_{n-1}$, are called the **arithmetic means**.

To insert k arithmetic means between two numbers b and c is to find k numbers that together with b and c form an arithmetic progression, whose first term is b and whose $(k + 2)$ term is c.

EXAMPLE

Insert five arithmetic means between -7 and 11.

Solution

Since there are to be five means, we will have seven terms, where $a_1 = -7$ and $a_7 = 11$.

Since $a_7 = a_1 + 6d$ we have

$11 = -7 + 6d, \qquad d = 3$

Hence the five arithmetic means are $-4, -1, 2, 5, 8$.

Exercise 17.3

Determine which of the following sequences (Problems 1–6) are arithmetic progressions and find the common difference of each such progression:

1. 2, 6, 10, 14
2. 36, 28, 20, 12
3. 1.4, 2.8, 5.6, 11.2
4. $1 - \sqrt{2}, 3 - 2\sqrt{2}, 5 - 3\sqrt{2}$
5. $a + 3b, 3a + 2b, 5a + b$
6. $a + 2b, a^2 + 2ab, a^3 + 2a^2b$
7. Find the tenth and fourteenth terms of the arithmetic progression $-7, -4, -1, \ldots$.
8. Find the fifteenth and twenty-sixth terms of the arithmetic progression 18, 11, 4, ...
9. Find the seventh and thirty-second terms of the arithmetic progression $10, 11\frac{1}{2}, 13, 14\frac{1}{2}, \ldots$.
10. Find the twentieth and fifty-first terms of the arithmetic progression 20, 19.5, 19, 18.5, ...
11. Find $\sum_{i=1}^{20} (i + 2)$
12. Find $\sum_{i=1}^{32} (2i - 3)$
13. Find $\sum_{i=1}^{16} (30 - 7i)$
14. Find $\sum_{i=1}^{12} (60 - 13i)$
15. Find $\sum_{i=1}^{k} 2i$
16. Find $\sum_{i=1}^{k} (2i - 1)$
17. Find $\sum_{i=1}^{k} 3i$
18. Find $\sum_{i=1}^{k} 5i$

In each of the following problems certain elements a_1, d, a_n, n, S_n of an arithmetic progression are given. Find the indicated elements whose values are not given:

19. $a_1 = 4, d = 6$; a_9; S_{13}
20. $a_1 = 3, d = 8$; a_{11}; S_{15}
21. $a_1 = 40, d = -3$; a_{20}; S_{20}
22. $a_1 = 60, d = -6$; a_{30}; S_{30}
23. $a_1 = 1, d = 3, a_n = 37$; n; S_n
24. $a_1 = -13, d = 7, a_n = 106$; n; S_n
25. $a_1 = 82, d = -13, a_n = -9$; n; S_n
26. $a_1 = 46, d = -9, a_n = -125$; n; S_n
27. $a_1 = 12, S_{28} = 1092$; d; a_{28}
28. $a_1 = -20, S_{27} = 337\frac{1}{2}$; d; a_{27}
29. $a_1 = 103, S_{15} = 285$; d; a_{15}
30. $a_1 = 71, S_{20} = -100$; d; a_{20}
31. $d = 2, S_{50} = 2500$; a_1; a_{50}
32. $d = 5, S_{32} = 2576$; a_1; a_{32}
33. $d = \frac{7}{2}, S_{41} = 1230$; a_1; a_{41}
34. $d = -\frac{3}{2}, S_{35} = \frac{35}{2}$; a_1; a_{35}
35. $a_1 = -1, a_n = -53, S_n = -729$; n; d
36. $a_1 = -43, a_n = 62, S_n = 209$; n; d
37. $a_1 = 37, a_n = 66, S_n = 1545$; n; d
38. $a_1 = 12, a_n = 38, S_n = 1325$; n; d
39. $a_{15} = 45, S_{15} = 255$; a_1; d
40. $a_{24} = 163, S_{24} = 1980$; a_1; d

41. $a_{35} = 61, S_{35} = 647\frac{1}{2};\quad a_1; d$

42. $a_{43} = -84, S_{43} = -451\frac{1}{2};\quad a_1; d$

43. $a_{15} = 59, a_{26} = 103;\quad a_{60}$

44. $a_{30} = 63, a_{82} = 167;\quad a_7$

45. $a_{11} = 73, a_{48} = 332;\quad a_{27}$

46. $a_{17} = -39, a_{54} = -150;\quad a_4$

47. $a_9 = -34, a_{21} = -130;\quad a_{39}$

48. $a_{12} = 161, a_{41} = -274;\quad a_{33}$

49. $a_1 = 4, d = .4, S_n = 30;\quad n; a_n$

50. $a_1 = 6, d = -\frac{5}{3}, S_n = -15;\quad n; a_n$

51. $a_1 = 12, d = \frac{3}{4}, S_n = 67\frac{1}{2};\quad n; a_n$

52. $a_1 = 15, d = -\frac{9}{2}, S_n = 10\frac{1}{2};\quad n; a_n$

53. $d = -2, a_n = -7, S_n = 48;\quad a_1; n$

54. $d = 4, a_n = 34, S_n = 162;\quad a_1; n$

55. $d = 3, a_n = 1, S_n = -76;\quad a_1; n$

56. $d = -4, a_n = -56, S_n = 60;\quad a_1; n$

57. Insert five arithmetic means between 13 and 49.

58. Insert six arithmetic means between -30 and 61.

59. Insert four arithmetic means between $10 + 3\sqrt{2}$ and $8\sqrt{2}$.

60. Insert four arithmetic means between $10x - 6y$ and $4y - 5x$.

61. How many numbers between 20 and 125 are exactly divisible by 3? Find the sum of these numbers.

62. How many numbers between 30 and 150 are exactly divisible by 4? Find the sum of these numbers.

63. How many numbers between 60 and 230 are exactly divisible by 7? Find the sum of these numbers.

64. How many numbers between 70 and 400 are exactly divisible by 11? Find the sum of these numbers.

65. The sum of three numbers in an arithmetic progression is 27. Find the numbers if the sum of their squares is 293.

66. The sum of three numbers in an arithmetic progression is 45. Find the numbers if the sum of their squares is 707.

67. The sum of four numbers in an arithmetic progression is 48. Find the numbers if the sum of their squares is 656.

68. The sum of four numbers in an arithmetic progression is 54. Find the numbers if the sum of their squares is 744.

69. The sum of four numbers in an arithmetic progression is 40. Find the numbers if the product of the first and fourth plus the product of the second and third is 160.

70. The sum of four numbers in an arithmetic progression is 34. Find the numbers if the product of the first and fourth plus the product of the second and third is 122.

71. The sum of five numbers in an arithmetic progression is 40. Find the numbers if the sum of their squares is 410.

72. The sum of five numbers in an arithmetic progression is 35. Find the numbers if the sum of their squares is 285.

17.4 Geometric Progressions

A **geometric progression** is a sequence of numbers (none of which is zero) in which the ratio of any term to the preceding term is always the same. The ratio of the terms is called the **common ratio**.

The sequence 1, 3, 9, 27, . . . , is a geometric progression with common ratio 3.

When the first term is denoted by a and the common ratio by r, the terms of the geometric progression can be written as

$$a_1 = a_1, \quad a_2 = a_1 r, \quad a_3 = a_2 r = a_1 r^2, \quad a_4 = a_3 r = a_1 r^3,$$
$$a_5 = a_1 r^4 = a_1 r^{5-1}, \ldots, \quad a_n = a_1 r^{n-1}, \ldots.$$

EXAMPLE

If $a_1 = 2$ and $r = 3$, the nth term of the geometric progression is given by $a_n = 2 \cdot 3^{n-1}$.
Thus the geometric progression is

$$2, 6, 18, 54, \ldots, 2 \cdot 3^{n-1}, \ldots.$$

Note Given some terms of a geometric progression, we find the common ratio r from the relation

$$r = \frac{a_n}{a_{n-1}}$$

EXAMPLES

1. The common ratio for the geometric progression

$$2, \frac{4}{3}, \frac{8}{9}, \frac{16}{27}, \ldots, \quad \text{is} \quad \frac{2}{3}$$

2. The common ratio for the geometric progression

$$a, -a^{\frac{3}{2}}, a^2, -a^{\frac{5}{2}}, \ldots, \quad \text{is} \quad -a^{\frac{1}{2}}$$

Note Given the first term a_1 and the common ratio r of a geometric progression, we find the nth term from the relation

$$a_n = a_1 r^{n-1}$$

EXAMPLE

Find the tenth term of the geometric progression whose first term is $\frac{1}{8}$ and whose common ratio is 2.

Solution

$$a_{10} = \frac{1}{8}(2)^{10-1} = 64$$

Sum of a Geometric Progression

The sum of the first n terms of a geometric progression is

$$S_n = a_1 + a_1 r + a_1 r^2 + \cdots + a_1 r^{n-2} + a_1 r^{n-1} \quad (1)$$

Multiply both members of (1) by r:

$$rS_n = a_1 r + a_1 r^2 + a_1 r^3 + \cdots + a_1 r^{n-1} + a_1 r^n \quad (2)$$

Rewrite (1) and (2) so that like terms in r occur in columns, and subtract:

$$\begin{array}{rcl} S_n &=& a_1 + a_1 r + a_1 r^2 + \cdots + a_1 r^{n-2} + a_1 r^{n-1} + 0 \\ rS_n &=& 0 + a_1 r + a_1 r^2 + \cdots + a_1 r^{n-2} + a_1 r^{n-1} + a_1 r^n \\ \hline S_n - rS_n &=& a_1 + 0 + 0 + \cdots + 0 + 0 - a_1 r^n \end{array}$$

Hence $\quad S_n - rS_n = a_1 - a_1 r^n$

$$S_n(1 - r) = a_1(1 - r^n)$$

or

$$S_n = \frac{a_1(1 - r^n)}{1 - r} \quad \text{if } r \neq 1$$

EXAMPLE

Find S_{10} if $a_1 = 243$ and $r = \frac{1}{3}$.

Solution

$$S_{10} = \frac{243\left[1 - \left(\frac{1}{3}\right)^{10}\right]}{1 - \frac{1}{3}} = \frac{243 - 243\left(\frac{1}{3}\right)^{10}}{\frac{2}{3}}$$

$$= \frac{243 - \frac{1}{243}}{\frac{2}{3}} = 364\frac{40}{81}$$

Notes

1. Given the first term a_1, the common ratio r, where $r \neq 1$, and the number of terms n of a geometric progression, we find the sum of the first n terms from the relation

$$S_n = \frac{a_1(1 - r^n)}{1 - r} \qquad r \neq 1$$

2. When $r = 1$, the terms of the geometric progression are all the same as the first term a_1, and the sum of n terms is

$$S_n = na_1 \qquad r = 1$$

3. Since $a_1 r^n = a_n r$, the sum of the first n terms of a geometric progression whose first term is a_1, whose nth term is a_n, and whose common ratio is r can be found from the relation

$$S_n = \frac{a_1 - a_n r}{1 - r} \qquad r \neq 1$$

EXAMPLE

Find S_n if $a_1 = \frac{1}{128}$, $r = 2$, and $a_n = 256$.

Solution

$$S_n = \frac{\frac{1}{128} - 256(2)}{1 - 2} = \frac{256(2) - \frac{1}{128}}{2 - 1} = 511\frac{127}{128}$$

EXAMPLE

The fifth and eighth terms of a geometric progression are $\frac{1}{8}$ and 1, respectively. Find the thirteenth term.

Solution

$$a_5 = \frac{1}{8} \quad \text{and} \quad a_8 = 1.$$

Hence $\quad a_1 r^4 = \frac{1}{8} \quad \text{and} \quad a_1 r^7 = 1$

$$\frac{a_1 r^7}{a_1 r^4} = \frac{1}{\frac{1}{8}}$$

or $\quad r^3 = 8$

$$r^3 - 8 = 0 \quad \text{or} \quad (r - 2)(r^2 + 2r + 4) = 0$$

Thus $\quad r = 2$

We consider only the real values of r:

$$a_1(2)^4 = \frac{1}{8} \quad \text{or} \quad a_1 = \frac{1}{128}$$

Hence $\quad a_{13} = \frac{1}{128}(2)^{12} = 32$

EXAMPLE

For a geometric progression where $a_1 = 8$, $a_n = \frac{8}{243}$, and $S_n = \frac{2912}{243}$, find r and n.

Solution

Using the relation $S_n = \frac{a_1 - a_n r}{1 - r}$, we get

$$\frac{2912}{243} = \frac{8 - \frac{8}{243}r}{1 - r}$$

or $\quad 2912(1 - r) = 243\left(8 - \frac{8}{243}r\right)$

or $\quad r = \frac{1}{3}$

Substituting $\frac{1}{3}$ for r in $a_n = a_1 r^{n-1}$, we get

$$\frac{8}{243} = 8\left(\frac{1}{3}\right)^{n-1}$$

$$\frac{1}{243} = \left(\frac{1}{3}\right)^{n-1}$$

$$\left(\frac{1}{3}\right)^5 = \left(\frac{1}{3}\right)^{n-1}$$

or $\quad n = 6$

Hence $\quad r = \frac{1}{3} \quad$ and $\quad n = 6$

EXAMPLE For a geometric progression where $a_1 = 1701$, $r = \frac{1}{3}$, and $S_n = \frac{7651}{3}$, find n and a_n.

Solution

$$S_n = \frac{7651}{3} = \frac{1701\left[1 - \left(\frac{1}{3}\right)^n\right]}{1 - \frac{1}{3}}$$

or
$$\frac{7651}{3 \times 1701} = \frac{1 - \left(\frac{1}{3}\right)^n}{\frac{2}{3}}$$

or
$$\frac{1093}{3 \times 243} \cdot \frac{2}{3} = 1 - \left(\frac{1}{3}\right)^n$$

or
$$\frac{2186}{2187} = 1 - \left(\frac{1}{3}\right)^n$$

or
$$\left(\frac{1}{3}\right)^n = 1 - \frac{2186}{2187}$$

$$= \frac{1}{2187}$$

$$= \left(\frac{1}{3}\right)^7$$

Hence $\quad n = 7$

Thus
$$a_7 = 1701\left(\frac{1}{3}\right)^6$$

$$= \frac{7}{3}$$

EXAMPLE

For a geometric progression where $a_1 = \frac{256}{243}$ and $S_3 = \frac{208}{243}$, find r and a_3.

Solution

$$S_3 = \frac{\frac{256}{243}(1 - r^3)}{1 - r} = \frac{208}{243}$$

or
$$256(1 - r^3) = 208(1 - r)$$

or
$$16(1 - r)(1 + r + r^2) = 13(1 - r)$$

$$16(1 - r)(1 + r + r^2) - 13(1 - r) = 0$$
$$(1 - r)[16(1 + r + r^2) - 13] = 0$$
$$(1 - r)(16r^2 + 16r + 3) = 0$$
$$(1 - r)(4r + 3)(4r + 1) = 0$$

Thus $\quad r = 1 \quad$ or $\quad r = -\frac{3}{4} \quad$ or $\quad r = -\frac{1}{4}$

We reject $r = 1$ since the sum $S_3 = 3a_1 = 3\left(\frac{256}{243}\right) = \frac{256}{81} \neq \frac{208}{243}$.

For $\quad r = -\frac{3}{4}, \quad a_3 = \frac{16}{27}$

For $\quad r = -\frac{1}{4}, \quad a_3 = \frac{16}{243}$

EXAMPLE

If three numbers in arithmetic progression are increased by 1, 3, and 41, respectively, the resulting numbers are in geometric progression. Find the original numbers if their sum is 72.

Solution

Let the three numbers be $a - d, a, a + d$:

$(a - d) + a + (a + d) = 72$

Hence $\quad a = 24$

Thus the numbers are $24 - d, 24, 24 + d$.

Adding 1, 3, and 41 to these numbers, respectively, we get $25 - d, 27, 65 + d$.

In order that $25 - d$, 27, and $65 + d$ form a geometric progression, we must have

$$\frac{27}{25 - d} = \frac{65 + d}{27}$$

or $\quad 1625 - 40d - d^2 = 729$

or $\quad d^2 + 40d - 896 = 0$

or $\quad (d - 16)(d + 56) = 0$

Hence $\quad d = 16 \quad$ or $\quad d = -56$

For $d = 16$, the numbers are
$24 - 16, 24, 24 + 16$ or $8, 24, 40$.
For $d = -56$, the numbers are
$24 + 56, 24, 24 - 56$ or $80, 24, -32$.

Thus the numbers are $8, 24, 40$ or $80, 24, -32$.

The terms a_1 and a_n of the geometric progression $a_1, a_2, \ldots, a_n$ are called the **extremes**, while the intermediate terms $a_2, \ldots, a_{n-1}$ are called the **geometric means**.

To insert k geometric means between two numbers b and c means to find k numbers that together with b and c form a geometric progression whose first term is b and whose $(k+2)$ term is c.

Note We will consider only real values of r here.

EXAMPLE Insert three geometric means between $\frac{2}{3}$ and $\frac{54}{625}$.

Solution $a_1 = \frac{2}{3}, \; a_5 = \frac{54}{625}$

Since $a_5 = a_1 r^4$,

$$\frac{54}{625} = \frac{2}{3} r^4 \qquad r^4 = \frac{81}{625} \qquad \text{or} \qquad r^4 - \left(\frac{3}{5}\right)^4 = 0$$

The real values of r are $\pm\frac{3}{5}$. Hence the geometric means are

$$\frac{2}{5}, \frac{6}{25}, \frac{18}{125} \qquad \text{or} \qquad -\frac{2}{5}, \frac{6}{25}, -\frac{18}{125}$$

Exercise 17.4

Determine which of the following sequences are geometric progressions, and give the value of the common ratio for each such progression:

1. $1, 2, 4, 8$

2. $1, 4, 16, 64$

3. $2, 6, 18, 54$

4. $1, \frac{1}{4}, \frac{1}{9}, \frac{1}{16}$

5. $4, 2, 1, \frac{1}{2}$

6. $1, -\frac{1}{2}, \frac{1}{4}, -\frac{1}{8}$

7. $\frac{3}{2}, \frac{1}{2}, \frac{3}{10}, \frac{3}{14}$

8. $-2, \frac{2}{3}, -\frac{2}{9}, \frac{2}{27}$

9. $\frac{4}{9}, -\frac{2}{3}, 1, -\frac{3}{2}$

10. $\frac{9}{16}, -\frac{3}{8}, \frac{1}{4}, -\frac{1}{6}$

Find the indicated terms of the following geometric progressions:

11. $8, 24, 72, \ldots;\quad a_5; a_8$

12. $27, 18, 12, \ldots;\quad a_7; a_{11}$

13. $-16, 8, -4, \ldots;\quad a_9; a_{14}$

14. $1, \sqrt{3}, 3, \ldots;\quad a_8; a_{15}$

15. $6, -6\sqrt{2}, 12, \ldots;\quad a_{11}; a_{16}$

16. $32, 16\sqrt{2}, 16, \ldots;\quad a_{14}; a_{23}$

In each of the following certain elements of a geometric progression are given. Find the indicated elements whose values are not given:

17. $a_1 = 8, r = \frac{3}{2};\quad a_6; S_6$

18. $a_1 = 81, r = \frac{1}{3};\quad a_5; S_5$

19. $a_1 = \frac{27}{16}, r = -\frac{2}{3};\quad a_7; S_7$

20. $a_1 = 64, r = -\frac{1}{2};\quad a_9; S_9$

21. $r = 2, S_5 = 372;\quad a_1; a_5$

22. $r = -3, S_6 = 364;\quad a_1; a_6$

23. $r = -\frac{1}{2}, S_8 = 340;\quad a_1; a_8$

24. $r = -\frac{3}{2}, S_7 = 926;\quad a_1; a_7$

25. $a_7 = 384, r = -2;\quad a_1; S_7$

26. $a_8 = 324, r = 3;\quad a_1; S_8$

27. $a_6 = \frac{9}{2}, r = -\frac{3}{2};\quad a_1; S_6$

28. $a_7 = \frac{1}{18}, r = -\frac{2}{3};\quad a_1; S_7$

29. $a_1 = \frac{1}{128}, r = 4, a_n = 2;\quad n; S_n$

30. $a_1 = 486, r = \frac{1}{3}, a_n = 2;\quad n; S_n$

31. $a_1 = \frac{9}{64}, r = \frac{2}{3}, a_n = \frac{1}{36};\quad n; S_n$

32. $a_1 = 144, r = -\frac{1}{2} a_n = \frac{9}{4};\quad n; S_n$

33. $a_5 = 48, a_{12} = 6144;\quad a_8$

34. $a_3 = 50, a_6 = \frac{2}{5};\quad a_9$

35. $a_7 = 3, a_{10} = -\frac{3}{8};\quad a_4$

36. $a_3 = \frac{9}{16}, a_6 = \frac{1}{6};\quad a_{11}$

37. $r = 2, a_n = 112, S_n = 217;\quad n; a_1$

38. $r = 3, a_n = 486, S_n = 728;\quad n; a_1$

39. $r = -\frac{1}{2}, a_n = -20, S_n = 420;\quad n; a_1$

40. $r = \frac{3}{2}, a_n = \frac{3}{8}, S_n = \frac{2059}{1944};\quad n; a_1$

41. $a_1 = 405, a_n = 5, S_n = 605;\quad r; n$

42. $a_1 = \frac{9}{256}, a_n = \frac{9}{4}, S_n = \frac{1143}{256};\quad r; n$

43. $a_1 = \frac{27}{64}, a_n = \frac{16}{9}, S_n = \frac{3367}{576};\quad r; n$

44. $a_1 = \frac{16}{243}, a_n = 27, S_n = \frac{5371}{243};\quad r; n$

45. $a_1 = 729, r = \frac{1}{3}, S_n = 1089;\quad n; a_n$

46. $a_1 = 1024, r = \frac{1}{2}, S_n = 2040; n; a_n$

47. $a_1 = 3, r = -2, S_n = 129;\quad n; a_n$

48. $a_1 = 5, r = -3, S_n = -910;\quad n; a_n$

49. $a_1 = 40, S_4 = 75;\quad a_4; r$

50. $a_1 = \frac{1}{64}, S_6 = \frac{63}{64};\quad a_6; r$

51. $a_1 = 162, S_3 = 234;\quad a_3; r$

52. $a_1 = \frac{4}{9}, S_3 = \frac{7}{9};\quad a_3; r$

53. Insert three geometric means between $\frac{3}{2}$ and $\frac{8}{27}$.

54. Insert three geometric means between 1 and 4.

55. Insert four geometric means between 48 and $-\frac{3}{2}$.

56. Insert five geometric means between 8 and 1.

57. If three numbers in an arithmetic progression are increased by 5, 4, and 12, respectively, the resulting numbers are in a geometric progression. Find the original numbers if their sum is 42.

58. If three numbers in an arithmetic progression are increased by 3, 5, and 71, respectively, the resulting numbers are in a geometric progression. Find the original numbers if their sum is 129.

59. If three numbers in an arithmetic progression are increased by 4, 5, and 12, respectively, the resulting numbers are in a geometric progression. Find the original numbers if their sum is 93.

60. If three numbers in an arithmetic progression are increased by 66, 33, and 8, respectively, the resulting numbers are in a geometric progression. Find the original numbers if their sum is 45.

17.5 Compound Interest

For **simple interest**, if P is the principal, i is the rate of interest per year, and t is the number of years, the amount accumulated A (that is, the principal plus interest) at the end of t years is given by

$$A = P + Pit = P(1 + it)$$

For **compound interest**, at the end of each unit of time (called a **period**) the interest is computed and added to the principal to form the principal for the next period. If P is the **principal** (sometimes called the **present value**), i is the **rate of interest per period**, and P_n is the **amount accumulated** at the end of n interest periods, we have

$$P_1 = P + Pi = P(1 + i)$$

$$P_2 = P_1 + P_1 i = P_1(1 + i) = P(1 + i)^2$$

$$P_3 = P_2 + P_2 i = P_2(1 + i) = P(1 + i)^3$$

In general,

$$P_n = P(1 + i)^n$$

Notes

1. The i used in the above formulas is the interest rate per period, not the yearly rate. To obtain i from the yearly rate, divide the yearly rate by the number of periods per year.

   For example, if the yearly interest rate is 6% compounded quarterly, then $i = \frac{6}{4}\% = 1\frac{1}{2}\%$ since there are 4 quarters in each year.

2. The n in the formula for P_n is the number of periods involved, not the number of years. If money is invested at 6% compounded semiannually for 10 years, then $n = 10 \times 2 = 20$.

3. The amounts P_n are terms of a geometric progression with a common ratio $(1 + i)$.

There are tables for computing $(1 + i)^n$ to the nearest cent. For our purposes we can find an approximate answer either by the use of logarithms or by the binomial formula.

EXAMPLE

Find the amount of $1000 after 10 years at 6% compounded monthly.

Solution

$$P = \$1000, \qquad i = \frac{6}{12}\% = \frac{1}{2}\%, \qquad n = 10 \times 12 = 120 \text{ periods}$$

$$\begin{aligned} P_{10} &= 1000\left(1 + \frac{1}{2}\%\right)^{120} \\ &= 1000(1 + .005)^{120} \\ &= 1000(1.819397) \\ &= \$1819.40 \end{aligned}$$

Investment tables are used in calculating $(1 + .005)^{120}$.

Note

If interest is compounded quarterly, the amount $P_{10} = \$1814.02$.
If interest is compounded yearly, $P_{10} = \$1790.85$.
Actually there is little gained by increasing the frequency of compounding beyond the monthly limit.

The present value of a future sum can be calculated from the relation

$$P = \frac{P_n}{(1 + i)^n} = P_n(1 + i)^{-n}$$

EXAMPLE How much money should be invested now at 5% compounded quarterly so that there will be \$12,000 fifteen years from now?

Solution

$$P_n = \$12{,}000, \qquad i = 1\frac{1}{4}\%, \qquad n = 60$$

$$P = 12{,}000\left(1 + 1\frac{1}{4}\%\right)^{-60}$$

$$= 12{,}000(.4745676)$$

$$= \$5694.81$$

Investment tables are used to arrive at the answer.

17.6 Annuities

Equal payments of a certain amount of money at regular intervals of time are called an **annuity**. The amount of each payment is called the **periodic rent**, denoted by R. The time between payments is called the **payment interval**. The length of time for which the payments continue is called the **term** of the annuity. The value to which a series of payments will accumulate at a given time is called the **accumulated amount** of the annuity. The present value of a series of future payments is called the **present value of the annuity**.

Consider deposits of the amount R for 10 periods at $i\%$ per period at compound interest. To find the amount accumulated after the tenth deposit we have:
The tenth deposit will be drawn right after it has been deposited; thus it earns no interest.
The ninth deposit earns interest for one period.
The eighth deposit accumulates interest for two periods.
The seventh deposit accumulates interest for three periods, and so on.
Thus the accumulated amount of the annuity is given by

$$S_{10} = R + R(1 + i) + R(1 + i)^2 + \cdots + R(1 + i)^9$$

The accumulated amount is the sum of a geometric progression with $a_1 = R$, $r = 1 + i$, and $n = 10$.

Hence $$S_{10} = \frac{R[1 - (1 + i)^{10}]}{1 - (1 + i)} = \frac{R[(1 + i)^{10} - 1]}{i}$$

In general, if the periodic deposit is R, the rate of interest per period is $i\%$ and the number of periods is n, then the accumulated amount S_n right after the nth payment is

$$S_n = \frac{R[(1 + i)^n - 1]}{i}$$

Note There are tables for evaluating $\frac{(1 + i)^n - 1}{i}$ to the nearest cent.

EXAMPLE A man makes a deposit of \$300 every three months to a bank that pays 5% compounded quarterly. What is the amount to his credit immediately after the fortieth deposit?

Solution Here $R = \$300$, $i = \frac{5}{4}\% = 1.25\%$, $n = 40$:

$$S_n = \frac{300[(1 + 1.25\%)^{40} - 1]}{1.25\%}$$
$$= 300(51.489557)$$
$$= \$15{,}446.87$$

Investment tables are used to arrive at the answer.

The **present value** of an annuity, A_n, is that sum of money which, if invested now at compound interest at $i\%$ per period for n periods, will be equal at the end of n periods to the accumulated value of an annuity with rent R at $i\%$ per period for n periods.
That is,

$$A_n(1 + i)^n = S_n$$

Since $S_n = \frac{R[(1 + i)^n - 1]}{i}$, we have

$$A_n = \frac{R[(1 + i)^n - 1]}{i(1 + i)^n}$$
$$= \frac{R[(1 + i)^n - 1]}{i}(1 + i)^{-n}$$

or $$A_n = \frac{R[1 - (1 + i)^{-n}]}{i}$$

EXAMPLE Find the present value of an annuity of \$600 invested semiannually for 10 years at 6% compounded semiannually.

Solution $R = 600$, $i = 3\%$, $n = 20$

$$A_n = \frac{600[1 - (1 + 3)^{-20}]}{3\%}$$
$$= 600(14.877475)$$
$$= \$8926.48$$

Note The value of $\frac{1 - (1 + i)^{-n}}{i}$ was arrived at from investment tables.

Exercises 17.5–17.6

The answers for the following problems were arrived at by the use of investment tables:

1. Find the accumulated amount of \$12,000 at the end of 30 years at 6% compounded quarterly.
2. Find the accumulated amount of \$4000 at the end of 30 years at 8% compounded semiannually.
3. What is the present value of \$10,000 due at the end of 15 years if the money is worth 10% compounded semiannually?
4. What is the present value of \$100,000 due at the end of 20 years if the money is worth 6% compounded quarterly?
5. What is the present value of \$8000 due at the end of 5 years if the money is worth 7% compounded semiannually?
6. What is the discount on \$5000 due at the end of 2 years if the money is worth 6% compounded monthly?
7. In how many years will \$4000 amount to \$10,000 with the money earning 8% compounded annually?
8. In how many years will \$6000 amount to \$24,000 with the money earning 7% compounded annually?
9. At what rate of interest will \$1000 be \$2000 at the end of 14 years if the interest is compounded annually?
10. At what rate of interest will \$2000 be \$6000 at the end of 20 years if interest is compounded annually?
11. A man deposits \$500 at the end of each year for 15 years in a bank that pays 6% interest compounded annually. Find the amount in his account just after the last deposit.
12. A man deposits \$100 at the end of each month for 10 years in a bank that pays 6% interest compounded monthly. Find the amount in his account just after the last deposit.
13. What is the amount accumulated by an investment of \$1000 at the end of each half year for 8 years, if the money is worth 5% compounded semiannually?
14. A house is sold for a down payment of \$10,000 with \$200 to be paid at the end of each month for 30 years. What is the present value of the house if the money is worth 6% compounded monthly?
15. A car is bought for a down payment of \$1500 with \$150 to be paid at the end of each month for 3 years. What is the present value of the car if the money is worth 9% compounded monthly?
16. A man buys a lot of land and agrees to pay \$1200 down and \$250 at the end of each month for 6 years. What is the cash value of the lot if the money is worth 6% compounded monthly?
17. How much should a man pay for an annuity that will pay \$600 at the end of each month for 10 years if the money is worth 6% compounded monthly?
18. How much should a man pay for an annuity that will pay \$800 at the end of each month for 5 years if the money is worth 9% compounded monthly?

19. If a monthly pension for the next 5 years is purchased today for \$20,000, with the money worth 6% compounded monthly, what amount will be received each month from the pension.

20. If a monthly pension for the next 10 years is purchased today for \$30,000, with the money worth 6% compounded monthly, what amount will be received each month from the pension?

Chapter 17 Review

Find the first five terms of the sequence defined by the following:

1. $f(n) = 4n - 1$

2. $f(n) = \dfrac{n^2 + 1}{2}$

3. $a_n = (-1)^{n+1}\dfrac{1}{2n - 1}$

4. $a_n = (-1)^n \dfrac{n}{3n - 1}$

Find the first four terms of the sequence S_n where

5. $a_n = n^3 - 5$

6. $a_n = \dfrac{3n^2 - 1}{2}$

7. $a_n = (-1)^n \cdot 2^n$

8. $a_n = (-1)^{n+1}n^2 + 1$

Write the following in expanded form:

9. $\sum_{i=1}^{5} (4i + 1)$

10. $\sum_{i=1}^{4} i(2i + 3)$

11. $\sum_{i=1}^{5} \dfrac{i}{2^i}$

12. $\sum_{k=1}^{5} k^{k-2}$

In each of the following certain elements of an arithmetic progression are given. Find the indicated elements whose values are not given:

13. $a_1 = 4, d = 9; \quad a_{15}; S_{12}$

14. $a_1 = -15, d = \dfrac{7}{2}, a_n = 69; \quad n; S_n$

15. $a_1 = 39, S_{20} = 210; \quad d; a_{20}$

16. $d = \dfrac{2}{3}, S_{43} = 645; \quad a_1; a_{43}$

17. $a_1 = 50, a_n = -13, S_n = 851; \quad n; d$

18. $a_{64} = 61\dfrac{1}{2}, S_{64} = 1920; \quad a_1; d$

19. $a_{20} = 10, a_{85} = -120; \quad a_{60}$

20. $a_1 = -9, d = \dfrac{1}{3}, S_n = 546; \quad n; a_n$

21. $d = -3, a_n = 20, S_n = 970; \quad a_1; n$

22. Insert six arithmetic means between 1 and 50.

23. How many numbers between 15 and 200 are exactly divisible by 9? Find the sum of these numbers.

24. The sum of four numbers in an arithmetic progression is 70. Find the numbers if the sum of their squares is 1830.

In each of the following certain elements of a geometric progression are given. Find the indicated elements whose values are not given:

25. $a_1 = 6, r = \frac{1}{2}$; $a_7; S_7$

26. $r = -2, S_8 = -\frac{85}{256}$; $a_1; a_8$

27. $a_6 = \frac{8}{3}, r = \frac{2}{3}$; $a_1; S_6$

28. $a_1 = \frac{64}{81}, r = -\frac{3}{2}, a_n = -6$; $n; S_n$

29. $a_7 = 16, a_{10} = 1024$; a_3

30. $r = \frac{1}{6}, a_n = 3, S_n = 777$; $n; a_1$

31. $a_1 = -\frac{64}{81}, a_n = \frac{3}{16}, S_n = -\frac{481}{1296}$; $n; r$

32. $a_1 = -384, r = -\frac{1}{2}, S_n = -255$; $n; a_n$

33. $a_1 = 686, S_3 = 798$; $a_3; r$

34. Insert four geometric means between 9 and $\frac{1024}{27}$

35. If three numbers in arithmetic progression are increased by 1, 7, and 25, respectively, the resulting numbers are in geometric progression. Find the original numbers if their sum is 51.

Cumulative Review

Chapter 15

1. If $f(x) = 2x^2 - x + 5$, find $f(-2)$, $f(0)$, and $f(4)$.
2. If $f(x) = 3x^2 + x - 7$, find $f\left(-\frac{2}{3}\right)$, $f\left(\frac{1}{2}\right)$, and $f\left(\frac{4}{3}\right)$.
3. If $f(x) = x^2 - 3x$, find $f(3)$, $f(x + 1)$, and $f(2x - 1)$.
4. If $f(x) = 2(x + a)^2 + 3(x + a)$, find $f(-a)$, $f(2a)$, and $f(a - 1)$.
5. If $f(x) = x^2 + 4x$, check the truth of $f(4) - f(1) = f(3)$.
6. If $f(x) = 3x^2 - x$, check the truth of $f(a + b) = f(a) + f(b)$.
7. If $f(x) = 2x + 1$, $g(x) = x^2 - 1$, find $(f + g)(x)$, $(f \cdot g)(x)$, $f(g(x))$, and $g(f(x))$.
8. If $f(x) = x + 2$, $g(x) = x^2 - x$, find $(f + g)(x)$, $(f \cdot g)(x)$, $(f \circ g)(x)$, and $(g \circ f)(x)$.

Determine which of the following equations define functions. For each function find its domain and range:

9. $y = 4x + 3$
10. $y = 2x - 1$
11. $y = x^2 + 8$
12. $y = 2x^2 + 6$
13. $y = x^2 - 1$
14. $y = 4x^2 - 3$
15. $y = 8 - x^2$
16. $y = 1 - 4x^2$
17. $y = \sqrt{x + 4}$
18. $y = \sqrt{x - 2}$
19. $y = \sqrt{9 - x}$
20. $y = \sqrt{16 - x}$
21. $y = \sqrt{x^2 + 1}$
22. $y = \sqrt{9x^2 + 1}$
23. $y = \sqrt{x^2 - 8}$
24. $y = \sqrt{x^2 - 12}$
25. $y = \sqrt{4x^2 - 1}$
26. $y = \sqrt{25x^2 - 1}$
27. $y = \sqrt{36 - x^2}$
28. $y = \sqrt{9 - 4x^2}$
29. $y = \sqrt{16 - 9x^2}$
30. $y = \sqrt{4 - 25x^2}$
31. $y = \dfrac{x - 2}{3x + 1}$
32. $y = \dfrac{x + 4}{2x - 1}$

Find the solution set of the following exponential equations:

33. $2^x = 16$
34. $3^x = 27$
35. $2^x = \dfrac{1}{4}$
36. $3^x = \dfrac{1}{81}$
37. $4^x = 32$
38. $9^x = 27$
39. $81^x = 3$
40. $64^x = \dfrac{1}{16}$

41. $3^{2x+1} = 81^x$ **42.** $2^{3x-2} = 64 \cdot 2^x$

43. $3^x \cdot 27^{x-1} = 27$ **44.** $2^{3x+1} \cdot 4^{3-x} = 8$

The following equations define functions. Find equations defining the inverse functions:

45. $y = 4x - 1$ **46.** $y = 2x + 3$ **47.** $y = 5 - 2x$

48. $y = 8 - 3x$ **49.** $y = \frac{1}{3}(2x - 5)$ **50.** $y = \frac{1}{3}(x - 6)$

51. $y = \frac{1}{4}(x - 3)$ **52.** $y = \frac{1}{6}(4x - 5)$ **53.** $y = \frac{1}{2}(4 - 3x)$

54. $y = \frac{1}{7}(3 - 9x)$ **55.** $y = \frac{1}{5}(6 - x)$ **56.** $y = \frac{1}{4}(2 - 7x)$

57. $y = x^3 + 1$ **58.** $y = 2x^3 + 3$ **59.** $y = x^3 - 8$

60. $y = 4x^3 - 5$ **61.** $y = (x - 2)^3$ **62.** $y = (x + 3)^3$

Chapter 16

Find the value of x in each of the following:

63. $\log_8 2 = x$ **64.** $\log_{243} 3 = x$ **65.** $\log_{16} 64 = x$

66. $\log_{32} 128 = x$ **67.** $\log_9 243 = x$ **68.** $\log_{125} 625 = x$

69. $\log_{32} \frac{1}{64} = x$ **70.** $\log_{27} \frac{1}{81} = x$ **71.** $\log_x 16 = \frac{2}{3}$

72. $\log_x 27 = \frac{3}{2}$ **73.** $\log_x 125 = \frac{3}{4}$ **74.** $\log_x 4 = \frac{1}{2}$

75. $\log_x 25 = -2$ **76.** $\log_x 9 = -\frac{2}{3}$ **77.** $\log_x 32 = -\frac{5}{7}$

78. $\log_x 81 = -\frac{4}{5}$ **79.** $\log_{49} x = \frac{3}{2}$ **80.** $\log_{81} x = \frac{3}{4}$

81. $\log_{125} x = \frac{2}{3}$ **82.** $\log_{128} x = \frac{3}{7}$ **83.** $\log_8 x = -\frac{4}{3}$

84. $\log_{81} x = -\frac{1}{2}$ **85.** $\log_{\frac{1}{2}} x = -4$ **86.** $\log_{\frac{4}{9}} x = -\frac{1}{2}$

Given $\log_n 2 = a$, $\log_n 3 = b$, find each of the following in terms of a and b:

87. $\log_4 24$ **88.** $\log_8 36$ **89.** $\log_9 48$ **90.** $\log_6 18$

91. $\log_6 24$ **92.** $\log_{12} 72$ **93.** $\log_{12} 96$ **94.** $\log_{36} 162$

95. $\log_{18} \frac{3}{4}$ **96.** $\log_{24} \frac{8}{9}$ **97.** $\log_{24} \frac{27}{16}$ **98.** $\log_{54} \frac{32}{81}$

Use the properties of logarithms to transform the left member into the right member in each of the following. Assume for each problem the base is a number greater than zero and different from one:

99. $\log \frac{52}{15} - \log \frac{44}{35} + \log \frac{33}{91} = 0$

100. $\log \frac{81}{85} - \log \frac{63}{68} + \log \frac{35}{36} = 0$

101. $\log \frac{46}{27} + \log \frac{76}{69} - \log \frac{38}{81} = 2 \log 2$

102. $\log \frac{49}{72} + \log \frac{10}{21} - \log \frac{35}{48} = -2 \log \frac{3}{2}$

103. $\log 12a^2 + \log 6a^3 - \log 48a^4 = \log \frac{3a}{2}$

104. $\log 8a + \log 27a^2 - \log 24a^5 = -2 \log \frac{a}{3}$

105. $\log 3a^2 - \log 2ab^4 + \log 6a^3b^2 = 2 \log \frac{3a^2}{b}$

106. $\log 18a^2 - \log 16\sqrt{a^3} + \log 4\sqrt{a} = \log \frac{9a}{2}$

By the use of a calculator evaluate the following logarithms:

107. $\log 4.32$ **108.** $\log 1.64$ **109.** $\log 23.7$ **110.** $\log 63.2$
111. $\log 231.4$ **112.** $\log 723.5$ **113.** $\log 3819$ **114.** $\log 1568$
115. $\log .793$ **116.** $\log .286$ **117.** $\log .053$ **118.** $\log .094$
119. $\ln 342$ **120.** $\ln 794$ **121.** $\ln 84.7$ **122.** $\ln 23.5$
123. $\ln 9.36$ **124.** $\ln 1.74$ **125.** $\ln .94$ **126.** $\ln .32$
127. $\ln .089$ **128.** $\ln .061$ **129.** $\ln .0048$ **130.** $\ln .0073$

Solve the following for x using a calculator:

131. $\log x = 1.34$ **132.** $\log x = 1.79$ **133.** $\log x = 2.63$
134. $\log x = 3.75$ **135.** $\log x = .468$ **136.** $\log x = .071$
137. $\log x = -1.31$ **138.** $\log x = -2.49$ **139.** $\log x = -.86$
140. $\ln x = 7.61$ **141.** $\ln x = 5.28$ **142.** $\ln x = 9.42$
143. $\ln x = .49$ **144.** $\ln x = .307$ **145.** $\ln x = .062$
146. $\ln x = -2.08$ **147.** $\ln x = -4.91$ **148.** $\ln x = -.734$

Use logarithms to perform the following computations:

149. $\sqrt[3]{479.3}$ **150.** $\sqrt[3]{(.691)^2}$ **151.** $\sqrt[4]{(26.3)^3}$

152. $(3.79)^2\sqrt[5]{68.5}$ **153.** $\frac{(1.92)^3}{\sqrt{28.3}}$ **154.** $\frac{\sqrt{7.29}}{\sqrt[3]{46.8}}$

155. $\dfrac{(87.1)^2 \sqrt[3]{(.39)^2}}{\sqrt{579.4}}$

156. $\dfrac{\sqrt[3]{72.8}}{(2.74)^3 \sqrt[4]{(1.63)^3}}$

157. $\dfrac{\sqrt{.415} + \sqrt{63.4}}{\sqrt{7.81}}$

158. $\dfrac{\sqrt{12.6} - \sqrt{67.2}}{\sqrt{531}}$

159. $\dfrac{\sqrt{327}}{\sqrt{87.8} + \sqrt[3]{.283}}$

160. $\dfrac{\sqrt{53.7} - \sqrt{18.6}}{\sqrt[3]{47.9} + \sqrt[3]{6.48}}$

Solve each of the following equations for x:

161. $3.29^x = 61.7$

162. $12^{2x+1} = 17^{x+8}$

163. $5^{x+2} \cdot 3^{x-1} = 15^{2x-3}$

164. $7^{x+3} \cdot 3^{2x+1} = 21^{3x}$

165. $\log(3x - 1) = 1.146$

166. $\log(2x + 3) = 1.322$

167. $\ln(x + 2) = 1.61$

168. $\ln(2x - 3) = 3.22$

169. $\log 3x - \log(x + 2) = .903$

170. $\log(x - 4) - \log(x + 1) = .477$

171. $\log_2 x + \log_2(x - 2) = 3$

172. $\log_2 x + \log_2(x - 6) = 4$

173. $\log_3 x + \log_3(x - 8) = 2$

174. $\log_3 x + \log_3(x + 6) = 3$

Chapter 17

Write the following in expanded form:

175. $\sum_{i=1}^{5} (4i + 1)$

176. $\sum_{i=1}^{6} (1 - 5i)$

177. $\sum_{k=1}^{4} 2k(k + 3)$

178. $\sum_{k=1}^{4} (k + 1)(2k - 1)$

179. $\sum_{i=1}^{4} \dfrac{1}{i(2i + 1)}$

180. $\sum_{i=1}^{5} (-1)^i \cdot 2^{i+1}$

In each of the following certain elements of an arithmetic progression are given. Find the indicated elements whose values are not given:

181. $a_1 = 4, d = 2;\quad a_{15}; S_{20}$

182. $a_1 = 7, d = 3;\quad a_{20}; S_{16}$

183. $a_1 = 43, d = -3, a_n = -26;\quad n; S_n$

184. $a_1 = 8, d = 5, a_n = 93;\quad n; S_n$

185. $a_1 = 32, S_{35} = -70;\quad d; a_{35}$

186. $a_1 = 98, S_{14} = 644;\quad d; a_{14}$

187. $d = 6, S_{24} = 1368;\quad a_1; a_{24}$

188. $d = -2, S_{40} = -120;\quad a_1; a_{40}$

189. $a_1 = -20, a_n = 48, S_n = 252;\quad n; d$

190. $a_1 = 12, a_n = -14, S_n = -27;\quad n; d$

191. $a_{12} = 54, S_{12} = 384;\quad a_1; d$

192. $a_{20} = -26, S_{20} = 240;\quad a_1; d$

193. $a_{10} = 58, a_{25} = 148;\quad a_{36}$

194. $a_8 = 19, a_{40} = -77;\quad a_{28}$

195. $a_1 = 70, d = -2, S_n = 1240;\quad n; a_n$

196. $a_1 = -30, d = 5, S_n = 75;\quad n; a_n$

197. $d = 8, a_n = 124, S_n = 768;\quad a_1; n$

198. $d = -6, a_n = -74, S_n = 390;\quad a_1; n$

In each of the following certain elements of a geometric progression are given. Find the indicated elements whose values are not given:

199. $a_1 = 27, r = \frac{2}{3}$; a_6; S_6

200. $a_1 = 32, r = -\frac{3}{2}$; a_8; S_8

201. $r = 2, S_7 = 508$; a_1; a_7

202. $r = -\frac{1}{3}, S_8 = \frac{1640}{27}$; a_1; a_8

203. $a_6 = \frac{1}{2}, r = \frac{1}{2}$; a_1; S_6

204. $a_8 = 2187, r = \frac{3}{2}$; a_1; S_8

205. $a_1 = \frac{16}{27}, r = -\frac{3}{2}, a_n = 3$; n; S_n

206. $a_1 = 6480, r = \frac{1}{6}, a_n = 5$; n; S_n

207. $a_4 = 16, a_{10} = 128$; a_8

208. $a_5 = 45, a_9 = 405$; a_{12}

209. $r = \frac{3}{2}, a_n = \frac{243}{4}, S_n = \frac{665}{4}$; n; a_1

210. $r = \frac{1}{3}, a_n = 1, S_n = 121$; n; a_1

211. $a_1 = -2, a_n = 486, S_n = 364$; r; n

212. $a_1 = 64, a_n = \frac{1}{4}, S_n = \frac{513}{12}$; r; n

213. $a_1 = 6, r = -2, S_n = 258$; n; a_n

214. $a_1 = \frac{4}{27}, r = 3, S_n = \frac{13,120}{27}$; n; a_n

215. $a_1 = \frac{1}{128}, S_3 = \frac{21}{128}$; a_3; r

216. $a_1 = 144, S_3 = 172$; a_3; r

appendix A

Interval Notation

Bounded Intervals

The statement $3 < x < 10$ means that x takes all real values between the number 3 and the number 10, but not the number 3 or the number 10.

The set of real numbers between 3 and 10 is called an **interval**. When the interval does not include the end points, it is called an **open interval**. In set notation, we write $\{x \mid 3 < x < 10\}$. The interval representation is $(3, 10)$; the parentheses indicate an open interval.

The statement $-2 \leq x \leq 6$ represents an interval, but this time the end points are included. When the interval contains its end points, it is called a **closed interval**. In set notation, we write $\{x \mid -2 \leq x \leq 6\}$. The interval representation is $[-2, 6]$; the brackets indicate a closed interval.
In $-1 \leq x < 4$, the variable x takes the values in the open interval $(-1, 4)$, and also the value -1, but not the value 4. The interval is called **half open** or **half closed**. The set notation is $\{x \mid -1 \leq x < 4\}$. The interval notation is $[-1, 4)$; the bracket indicates that the number -1 is included while the parenthesis indicates that the number 4 is not included.

Infinite Intervals

For $x > 2$, the variable x takes on all real values greater than 2. It is an open interval bounded on the left by the number 2 but not bounded on the right by any real number. To represent the right end of the interval we use the symbol $+\infty$, read **plus infinity**, to indicate that the value of x is not bounded on the right by any real number.
Note that $+\infty$ is not a number, it is a symbol indicating the lack of a specific upper bound. The interval notation is $(2, \infty)$. An interval is always open at plus infinity.
Similarly, $x \leq 5$ represents an interval which is bounded on the right by the number 5, but not bounded on the left. To indicate that the interval is not bounded on the left, we use the symbol $-\infty$, read **minus infinity**. The interval notation is $(-\infty, 5]$. An interval is always open at minus infinity.

appendix B

Binomial Expansion

Consider the following expansions:

$$(a+b)^1 = a + b$$

$$(a+b)^2 = a^2 + 2ab + b^2$$
$$= a^2b^0 + \frac{2}{1}a^{2-1}b^1 + \frac{2 \cdot 1}{1 \cdot 2}a^0b^2$$

$$(a+b)^3 = a^3 + 3a^2b + 3ab^2 + b^3$$
$$= a^3b^0 + \frac{3}{1}a^{3-1}b^1 + \frac{3 \cdot 2}{1 \cdot 2}a^{3-2}b^2 + \frac{3 \cdot 2 \cdot 1}{1 \cdot 2 \cdot 3}a^0b^3$$

$$(a+b)^4 = a^4 + 4a^3b + 6a^2b^2 + 4ab^3 + b^4$$
$$= a^4b^0 + \frac{4}{1}a^{4-1}b^1 + \frac{4 \cdot 3}{1 \cdot 2}a^{4-2}b^2 + \frac{4 \cdot 3 \cdot 2}{1 \cdot 2 \cdot 3}a^{4-3}b^3 + \frac{4 \cdot 3 \cdot 2 \cdot 1}{1 \cdot 2 \cdot 3 \cdot 4}a^0b^4$$

$$(a+b)^5 = a^5 + 5a^4b + 10a^3b^2 + 10a^2b^3 + 5ab^4 + b^5$$
$$= a^5b^0 + \frac{5}{1}a^{5-1}b^1 + \frac{5 \cdot 4}{1 \cdot 2}a^{5-2}b^2 + \frac{5 \cdot 4 \cdot 3}{1 \cdot 2 \cdot 3}a^{5-3}b^3 + \frac{5 \cdot 4 \cdot 3 \cdot 2}{1 \cdot 2 \cdot 3 \cdot 4}a^{5-4}b^4 + \frac{5 \cdot 4 \cdot 3 \cdot 2 \cdot 1}{1 \cdot 2 \cdot 3 \cdot 4 \cdot 5}a^0b^5$$

From the above we notice the following general information about the expansion of $(a+b)^n$, $n = 1, 2, 3, 4, 5$.

1. The first term is a^nb^0 and the last term is a^0b^n.
2. The exponents of a decrease by 1 from one term to the next.
   The exponents of b increase by 1 from one term to the next.
   The sum of the exponents of a and b in every term is n.
3. The coefficient of the second term in the expansion is $\frac{n}{1}$.

   The coefficient of the third term is $\frac{n(n-1)}{1 \cdot 2}$.

   The coefficient of the fourth term is $\frac{n(n-1)(n-2)}{1 \cdot 2 \cdot 3}$.

The information we obtained from these expansions can be proved to be true for any value of $n \in N$.
The general formula for the expansion of $(a+b)^n$, $n \in N$, is known as the **binomial theorem**.

THE BINOMIAL THEOREM

If $n \in N$, then

$$(a+b)^n = a^n + \frac{n}{1}a^{n-1}b + \frac{n(n-1)}{1 \cdot 2}a^{n-2}b^2 + \frac{n(n-1)(n-2)}{1 \cdot 2 \cdot 3}a^{n-3}b^3 + \cdots + b^n$$

EXAMPLE Expand $(x - 2y)^5$.

Solution Take $a = x$, $b = -2y$, and $n = 5$ in the binomial theorem:

$$(x - 2y)^5 = x^5 + \frac{5}{1}x^4(-2y) + \frac{5 \cdot 4}{1 \cdot 2}x^3(-2y)^2 + \frac{5 \cdot 4 \cdot 3}{1 \cdot 2 \cdot 3}x^2(-2y)^3 + \frac{5 \cdot 4 \cdot 3 \cdot 2}{1 \cdot 2 \cdot 3 \cdot 4}x(-2y)^4 + (-2y)^5$$
$$= x^5 - 10x^4y + 40x^3y^2 - 80x^2y^3 + 80xy^4 - 32y^5$$

EXAMPLE Find the first four terms in the expansion of $\left[\frac{x^2}{2} - \frac{y}{x}\right]^{10}$.

Solution Take $a = \frac{x^2}{2}$, $b = -\frac{y}{x}$, and $n = 10$ in the binomial theorem:

$$\left[\frac{x^2}{2} - \frac{y}{x}\right]^{10} = \left[\frac{x^2}{2}\right]^{10} + \frac{10}{1}\left[\frac{x^2}{2}\right]^9\left(-\frac{y}{x}\right) + \frac{10 \cdot 9}{1 \cdot 2}\left[\frac{x^2}{2}\right]^8\left(-\frac{y}{x}\right)^2 + \frac{10 \cdot 9 \cdot 8}{1 \cdot 2 \cdot 3}\left[\frac{x^2}{2}\right]^7\left(-\frac{y}{x}\right)^3 + \cdots$$
$$= \frac{1}{1024}x^{20} - \frac{5}{256}x^{17}y + \frac{45}{256}x^{14}y^2 - \frac{15}{16}x^{11}y^3 + \cdots$$

Notes To **round off a number**, the following rules are obeyed:

1. When the first digit of the part to be discarded is less than 5, delete the digits of the discarded part.

   EXAMPLE 7.16448 = 7.164 to three decimal places.

2. When the first digit in the part to be discarded is greater than 5, or when that digit is 5 and the remaining digits of the part to be discarded are not all zeros, increase the last digit by one.

   EXAMPLES 378.362 = 378.4 to one decimal place.
   2.4652 = 2.47 to two decimal places.

3. When the digit to be discarded is 5, add one to the last retained digit if that digit is an odd number, otherwise leave the retained part alone.

   EXAMPLES 76.35 = 76.4 to one decimal place.
   8.965 = 8.96 to two decimal places.

EXAMPLE Find the value of $(1.02)^6$ correct to four decimal places.

Solution Write $(1.02)^6$ as $(1+.02)^6$ and take $a=1$, $b=.02$, and $n=6$ in the binomial theorem:

$$\begin{aligned}(1.02)^6 &= (1+.02)^6 \\ &= 1+\frac{6}{1}(.02)+\frac{6\cdot 5}{1\cdot 2}(.02)^2+\frac{6\cdot 5\cdot 4}{1\cdot 2\cdot 3}(.02)^3 \\ &\qquad +\frac{6\cdot 5\cdot 4\cdot 3}{1\cdot 2\cdot 3\cdot 4}(.02)^4+\cdots \\ &= 1+.12+.006+.00016+.0000024+\cdots \\ &= 1.1262\end{aligned}$$

Remark The remaining terms in the expansion are less than .00004.

Notes
1. Write the expansion first, then simplify.
2. To calculate $(.99)^n$, write it as $(1-.01)^n$.

Exercise

Expand the following by the binomial theorem and simplify:

1. $(x+1)^3$ **2.** $(x+y)^3$ **3.** $(x-y)^4$ **4.** $(x-1)^4$
5. $(x+y)^5$ **6.** $(x-y)^5$ **7.** $(x+1)^6$ **8.** $(x-1)^6$
9. $(x-2)^4$ **10.** $(x+2)^4$ **11.** $(x+3)^4$ **12.** $(x-3)^4$
13. $(2x-1)^4$ **14.** $(2x+1)^4$ **15.** $(2x+3)^4$ **16.** $(2x-3)^4$
17. $(x-2y)^4$ **18.** $(2-x^2)^4$ **19.** $(3x+y^{-1})^4$
20. $(2x^{-2}+y^{-1})^4$ **21.** $(x^{-1}-y^{-2})^4$ **22.** $(x^2-x^{-1}y)^5$
23. $(x^2y^{-1}+x^{-1}y^2)^5$ **24.** $(3x^{-1}+2y^{-2})^5$ **25.** $(x^{-2}+3y^{-1})^5$
26. $(xy^2-2y^{-1})^6$ **27.** $(2x-x^{-1}y)^7$ **28.** $(x-1)^8$
29. $(x-2)^8$ **30.** $(x^{-1}+y^{-2})^8$ **31.** $\left(x+2^{\frac{1}{2}}\right)^4$
32. $\left(x^{\frac{1}{2}}-1\right)^4$ **33.** $\left(x^{\frac{1}{2}}-2\right)^4$ **34.** $\left(x^{\frac{1}{2}}+y^{\frac{1}{2}}\right)^6$

Write and simplify the first five terms in the expansion of the following:

35. $(x+2)^{10}$ **36.** $(x-2x^{-1})^{10}$ **37.** $(x-x^{-1}y)^{10}$
38. $(x^2+x^{-1}y)^{10}$ **39.** $(x+1)^{12}$ **40.** $(x-1)^{12}$

Using the binomial theorem, calculate the following to the nearest thousandth:

41. $(1.01)^{10}$ **42.** $(1.02)^8$ **43.** $(1.03)^6$ **44.** $(.99)^{10}$
45. $(.98)^8$ **46.** $(.97)^6$

appendix C

Synthetic Division

Division of a polynomial $P(x)$ by a linear polynomial of the form $(x - r)$ is simplified by a special procedure called **synthetic division**. Such a procedure is essential because of the relation between the roots of a polynomial equation and factors of the form $(x - r)$.

Analysis of a division problem illustrates the principle behind synthetic division.

EXAMPLE Divide $(2x^4 - 13x^3 + 26x^2 - 31x + 40)$ by $(x - 4)$.

Solution

$$\begin{array}{r|l} & 2x^3 - 5x^2 + 6x - 7 \\ \hline x - 4 & (2x^4) - 13x^3 + 26x^2 - 31x + 40 \\ & 2x^4 - 8x^3 \\ \hline & (-5x^3) \\ & -5x^3 + 20x^2 \\ \hline & (+6x^2) \\ & +6x^2 - 24x \\ \hline & (-7x) \\ & -7x + 28 \\ \hline & (+12) \end{array}$$

The division operation can still be written in the form

$$\begin{array}{r|ll} & 2x^3 - 5x^2 + 6x - 7 & \text{(quotient)} \\ \hline x - 4 & 2x^4 - 13x^3 + 26x^2 - 31x + 40 & \text{(dividend)} \\ & \quad - 8x^3 + 20x^2 - 24x + 28 & \\ \hline & 2x^4 - 5x^3 + 6x^2 - 7x + \underbrace{12}_{\text{(remainder)}} & \end{array}$$

Notes

1. The first term in the bottom line is the same as the first term in the polynomial.
2. When we divide the terms in the bottom line, except the remainder, by x, we get the terms in the top line, which is the quotient.

Eliminating the top line and writing only the coefficients of x and the constant in an array, we have the following arrangement:

$$\begin{array}{r|rrrrrl} -4 & 2 & -13 & +26 & -31 & +40 & \\ & & -8 & +20 & -24 & +28 & \text{(subtract)} \\ \hline \times & 2 & -5 & 6 & -7 & 12 & \end{array}$$

Remarks

1. The top line consists of the coefficients of the polynomial arranged in descending powers of the variable and the constant.
2. The -4 is from the divisor $x - 4$.

3. Each number in the third line is the result of subtracting the number in the second line from the one above it in the first line.
4. Each number in the second line is the product of the number in the preceding column of the third line and the number (-4).

Since multiplying by (-4) and then subtracting is the same as multiplying by $(+4)$ and adding, we have

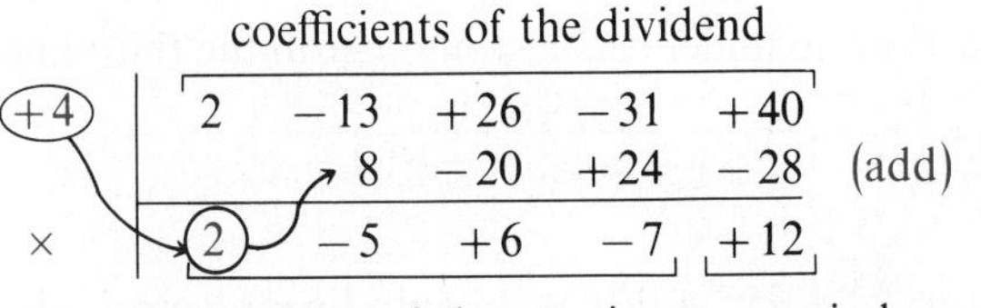

To divide a polynomial $P(x)$ by a divisor $(x - r)$ using synthetic division:

1. In the first line, write the coefficients of the polynomial $P(x)$ arranged in descending powers of x, and the constant, putting zeros for the coefficients of missing powers.
2. Write the first coefficient in the third line, below its position in the first line.
3. Write the product of the multiplier r and this coefficient in the second line beneath the second coefficient of the first line. Add these to get the second number in the third line. Proceed in the same manner to get subsequent numbers in rows two and three.

If the remainder is zero, then $(x - r)$ is a factor of the polynomial $P(x)$, and the quotient is called a **depressed polynomial**.
In the case of an equation, if the remainder is zero, r is a root of the equation $P(x) = 0$. When the quotient is equated to zero, it is called a **depressed equation**.

Note Any root of a depressed equation is a root of the original equation.

REMAINDER THEOREM When a polynomial $P(x)$, of degree greater than or equal to one, is divided by a divisor of the form $(x - r)$ until the remainder does not involve x, the remainder is equal to $P(r)$.

EXAMPLE Divide $(x^5 - 4x^3 + x^2 + 4x - 6)$ by $(x + 2)$.

Solution The coefficients of the terms of the polynomial are $1, 0, -4, 1, 4, -6$ and $r = -2$.

$$\begin{array}{r|rrrrrr} -2 & 1 & 0 & -4 & 1 & 4 & -6 \\ & & -2 & 4 & 0 & -2 & -4 \\ \hline & 1 & -2 & 0 & 1 & 2 & -10 \end{array}$$

Hence the quotient is $x^4 - 2x^3 + x + 2$ and the remainder is -10.

EXAMPLE

If $f(x) = 4x^4 - 7x^2 - 5x - 15$, find $f\left(-\frac{3}{2}\right)$ by means of synthetic division.

Solution

$$\begin{array}{r|rrrrr} -\frac{3}{2} & 4 & 0 & -7 & -5 & -15 \\ & & -6 & 9 & -3 & 12 \\ \hline & 4 & -6 & 2 & -8 & -3 \end{array}$$

Since the remainder (the last number in the third line) is -3, we have $f\left(-\frac{3}{2}\right) = -3$.

EXAMPLE

Factor $x^3 + 7x^2 + 17x + 15$ into linear factors, given that $(x + 3)$ is one of the factors.

Solution

$$\begin{array}{r|rrrr} -3 & 1 & 7 & 17 & 15 \\ & & -3 & -12 & -15 \\ \hline & 1 & 4 & 5 & 0 \end{array}$$

$$(x^3 + 7x^2 + 17x + 15) = (x + 3)(x^2 + 4x + 5)$$

For $\quad x^2 + 4x + 5 = 0, \quad x = \dfrac{-4 \pm \sqrt{16 - 20}}{2} = -2 \pm i$

Hence $\quad (x^3 + 7x^2 + 17x + 15) = (x + 3)[x - (-2 + i)][x - (-2 - i)]$

$$= (x + 3)(x + 2 - i)(x + 2 + i)$$

EXAMPLE

By use of synthetic division find m such that $\frac{1}{4}$ is a root of $4x^4 + 23x^3 + mx^2 + 16x - 3 = 0$.

Solution

$$\begin{array}{r|ccccc} \frac{1}{4} & 4 & 23 & m & 16 & -3 \\ & & 1 & 6 & \frac{1}{4}(m+6) & \frac{1}{4}\left[\frac{1}{4}(m+6)+16\right] \\ \hline & 4 & 24 & (m+6) & \frac{1}{4}(m+6)+16 & \frac{1}{4}\left[\frac{1}{4}(m+6)+16\right]-3 \end{array}$$

For $\frac{1}{4}$ to be a root we must have

$$\frac{1}{4}\left[\frac{1}{4}(m + 6) + 16\right] - 3 = 0$$

or

$$\frac{1}{16}(m + 6) + 4 - 3 = 0$$

or

$$m = -22$$

Exercise

Find the quotient and remainder by synthetic division:

1. $(6x^4 - 16x^3 + 9x^2 - 10x + 8) \div (x - 2)$

2. $(4x^4 + 16x^3 - x^2 - 2x + 8) \div (x + 4)$

3. $(3x^4 - 21x^3 + 31x^2 - 25) \div (x - 5)$

4. $(3x^5 + 4x^4 + 3x^2 - 4x + 16) \div (x + 2)$

5. $(x^5 - 8x^3 - 10x^2 + 18x + 6) \div (x - 3)$

6. $(x^5 - 2x^4 - 44x^3 + 24x^2 + 3x - 18) \div (x + 6)$

By means of synthetic division:

7. If $f(x) = x^4 + x^3 - 3x^2 - x - 2$, find $f(-2)$.

8. If $f(x) = 2x^4 - 5x^3 - 17x^2 + 22x - 6$, find $f(4)$.

9. If $f(x) = 2x^4 - 3x^3 - 5x^2 - 13x + 11$, find $f\left(\frac{1}{2}\right)$.

10. If $f(x) = 12x^4 + 19x^3 - x^2 + 19x + 7$, find $f\left(-\frac{1}{3}\right)$.

11. If $f(x) = 4x^4 - 6x^3 + 2x^2 + x + 5$, find $f\left(-\frac{1}{2}\right)$.

12. If $f(x) = 4x^4 + 7x^3 - 34x^2 - 28x + 9$, find $f\left(\frac{1}{4}\right)$.

By means of synthetic division decide whether each statement is true or false; if true factor the polynomial into linear factors:

13. $(x + 2)$ is a factor of $x^3 - 7x - 8$

14. $(x - 2)$ is a factor of $x^3 - 3x^2 - 10x + 20$

15. $(x - 1)$ is a factor of $x^3 - x^2 - 3x + 3$

16. $(x + 2)$ is a factor of $x^3 + 2x^2 + 9x + 18$

17. $(x - 4)$ is a factor of $x^3 - 2x^2 - 6x - 8$

18. $(x - 3)$ is a factor of $x^3 - 5x^2 + 5x + 3$

By the use of synthetic division find m such that the following is true:

19. 2 is a root of $x^4 - x^3 + mx^2 + 16x - 12 = 0$.

20. -3 is a root of $2x^4 + 5x^3 + mx^2 + 5x - 12 = 0$.

21. $-\frac{1}{2}$ is a root of $6x^4 + 7x^3 + mx^2 - 11x - 5 = 0$.

22. $\frac{1}{3}$ is a root of $6x^4 - 14x^3 + mx^2 - 27x + 7 = 0$.

23. $\frac{1}{6}$ is a root of $18x^4 - 15x^3 + 26x^2 + mx + 8 = 0$.

24. $\frac{3}{2}$ is a root of $4x^4 - 8x^3 + x^2 + mx + 6 = 0$.

appendix D

Use of a Common Logarithms Table

Any positive number in decimal notation can be written as the product of a number between 1 and 10 and a power of 10. For example,

$$628.4 = 6.284 \times 10^2 \qquad .00472 = 4.72 \times 10^{-3}$$

The decimal point is always placed after the leftmost digit. This notation is called the **scientific notation for a number**.

Using scientific notation and the laws for logarithms, we can easily reduce finding the logarithm of a number to a table look-up scheme.
Consider the logarithm of the number 2564:

$$\begin{aligned} \log 2564 &= \log(2.564 \times 10^3) \\ &= \log 2.564 + \log 10^3 \\ &= \log 2.564 + 3 \end{aligned}$$

Note also that

$$\begin{aligned} \log .0002564 &= \log(2.564 \times 10^{-4}) \\ &= \log 2.564 + \log 10^{-4} \\ &= \log 2.564 - 4 \end{aligned}$$

So we see that a table of logarithms need only contain values for numbers between 1 and 10. A number outside this range can be obtained by adding an integer to the logarithm corresponding to the power of 10 that appears when the number is expressed in scientific notation.

Since $\log 1 = 0$ and $\log 10 = 1$, we conclude that the common logarithm of a number x, where $1 < x < 10$, lies between 0 and 1.

The common logarithm of a number consists of two parts, a positive decimal fraction and an integer. The decimal fraction is called the **mantissa**, and depends only on the sequence of the digits in the number; that is, it is independent of the position of the decimal point. The integer part of the common logarithm is called the **characteristic**. It is the power of 10 when the number is written in the scientific form, and it depends only on the position of the decimal point in the number.

The mantissa of the common logarithm of 3 is the same as the mantissa of 3000, and is the same as the mantissa of .0003 However, the characteristics of their common logarithms differ and are, respectively, 0, 3, and -4.

The characteristics of the logarithms of the numbers 432, 56.8, 2.94, .387, and .00791 are, respectively, 2, 1, 0, -1, and -3.

The mantissa of the logarithm of the number 432 is .6355 and the characteristic is 2; thus

$$\begin{aligned} \log 432 &= .6355 + 2 \\ &= 2.6355 \end{aligned}$$

The characteristic is combined with the mantissa since both numbers are positive.

The mantissa of the logarithm of the number .00791 is .8982 and the characteristic is -3; thus

$$\log .00791 = .8982 - 3$$

If we perform the indicated subtraction, we obtain a negative number. However, the table of logarithms contains only positive mantissas. Hence, we customarily maintain a positive mantissa in the following way by writing

$$\begin{aligned} \log .00791 &= .8982 - \overbrace{3} \\ &= \underbrace{.8982 + 7} - 10 \\ &= 7.8982 - 10 \end{aligned}$$

The last form, a positive number minus 10 or an integral multiple of 10, is preferred since it makes computations simpler.

Thus we say that the characteristics of the common logarithms of the numbers .01, .37, and .0006 are $8 - 10, 9 - 10,$ and $6 - 10,$ respectively.

Figure A.1 is a section of the table found in the front endpapers of this book. The table gives the approximate values of the mantissa of the common logarithms of the numbers from 100 to 999.

N	0	1	2	3	4	5	6	7	8	9
55										
56										
57										
58				.7657						
59										
60										

FIGURE A.1

In order to find the mantissa of the logarithm of a number, we seek the first two digits of the number in the first column under *N*. This locates the row in which the mantissa is to be found. The third digit in the number is found in the top row and determines the column in which the mantissa is located.

To find the mantissa of the logarithm of the number 583, we find 58 in the leftmost column, then the 3 in the first row. The mantissa is found at the intersection of row 58 and column 3, which is .7657

To find the mantissa for the logarithm of the number 376, read the number at the intersection of row 37 and column 6, giving .5752
The number has to be rounded off to three digits.

EXAMPLE Find log 78.4.

Solution The characteristic of log 78.4 is 1 and the mantissa is .8943.

Thus $\log 78.4 = 1.8943$

EXAMPLE Find log 6.

Solution The characteristic of log 6 is 0 and the mantissa is the same as the mantissa of the logarithm of the number 600, which is .7782.

Thus $\log 6 = .7782$

EXAMPLE Find log .00237.

Solution The characteristic of log .00237 is $7 - 10$ and the mantissa is .3747.

Thus $\log .00237 = 7.3747 - 10$

The number corresponding to a given logarithm is called the inverse logarithm.

Since $\log 516 = 2.7126$,

$$\text{inverse } \log 2.7126 = 516 = 5.16 \times 10^2$$

Since $\log .0143 = 8.1553 - 10$,

$$\text{inverse } \log(8.1553 - 10) = .0143 = 1.43 \times 10^{-2}$$

To find the inverse logarithm of a number, we first determine the characteristic and the mantissa. Locate the mantissa in the body of the table, or take the closest table entry to the mantissa.
The first two digits of the inverse logarithm appear in the same row under N, and the third digit appears in the same column in the top row.
Place the decimal point after the first digit from the left (as in scientific form) and multiply the number by 10 to a power equaling that of the characteristic.

EXAMPLE Given $\log x = 1.6454$ find x.

Solution The characteristic is 1 and the mantissa is .6454.

By examination of the table we find that .6454 is in line with $N = 44$ and in the column under 2.

Thus $x = 4.42 \times 10^1 = 44.2$

EXAMPLE Given $\log x = 3.8287$ find x.

Solution The characteristic is 3 and the mantissa is .8287.

By examination of the table we find that .8287 is in line with $N = 67$ and in the column under 4.

Thus $x = 6.74 \times 10^3 = 6740$

EXAMPLE Given $\log x = 8.4409 - 10$ find x.

Solution The characteristic is $8 - 10 = -2$ and the mantissa is .4409.

By examination of the table we find that .4409 is in line with $N = 27$ and in the column under 6.

Thus $x = 2.76 \times 10^{-2} = .0276$

Note The logarithm of a number consists of a characteristic, which is an integer, and a mantissa, which is a positive decimal. The table gives positive mantissas.

When the logarithm is given as a negative number, addition of $10k - 10k$, where $k \in N$, yields a positive number minus 10 or integral multiple of 10; thus we can find the inverse logarithm of a negative number.

EXAMPLE Given $\log x = -2.0526$ find x.

Solution

$$\log x = \underbrace{-2.0526 + 10}_{} - 10$$
$$= 7.9474 - 10$$

Hence $x = 8.86 \times 10^{-3} = .00886$

Note that $-2.0526 = -2 - .0526$.

appendix E

Theorems and Proofs

Page 106

THEOREM 1 Let $a, b, c \in R$; if $a > b$ and $b > c$, then $a > c$.

Proof

$a > b$ means $a = b + k_1$ where $k_1 \in R, k_1 > 0$

Also, $b > c$ means $b = c + k_2$ where $k_2 \in R, k_2 > 0$

Thus $a = b + k_1 = c + k_2 + k_1 = c + (k_2 + k_1)$

But $k_1 > 0$ and $k_2 > 0$. Hence $k_1 + k_2 > 0$

Therefore, $a > c$.

THEOREM 2 Let $a, b, c, d \in R$; if $a > b$ and $c > d$, then $a + c > b + d$.

Proof

$a > b$ means $a = b + k_1$ where $k_1 \in R, k_1 > 0$

Also, $c > d$ means $c = d + k_2$ where $k_2 \in R, k_2 > 0$

Thus $(a + c) = b + k_1 + d + k_2 = (b + d) + k_1 + k_2$

But $k_1 > 0$ and $k_2 > 0$. Hence $k_1 + k_2 > 0$

Therefore, $a + c > b + d$.

Page 107

THEOREM 4 Let $a, b, c, d \in R$, $a, b, c, d > 0$; if $a > b$ and $c > d$, then $ac > bd$.

Proof

$a > b$ means $a = b + k_1$ where $k_1 \in R, k_1 > 0$

Also, $c > d$ means $c = d + k_2$ where $k_2 \in R, k_2 > 0$

Thus $ac = (b + k_1)(d + k_2) = bd + (k_1 d + k_2 b + k_1 k_2)$

But $k_1, k_2, b, d > 0$. Hence $k_1 d + k_2 b + k_1 k_2 > 0$

Therefore, $ac > bd$.

COROLLARY Let $a, b \in R$; if $a > b$ where $ab > 0$, then $\frac{1}{a} < \frac{1}{b}$.

Proof

If $a > b$ where $ab > 0$, then from Theorem 5

$$a\left(\frac{1}{ab}\right) > b\left(\frac{1}{ab}\right) \quad \text{or} \quad \frac{1}{b} > \frac{1}{a}$$

Hence $\frac{1}{a} < \frac{1}{b}$

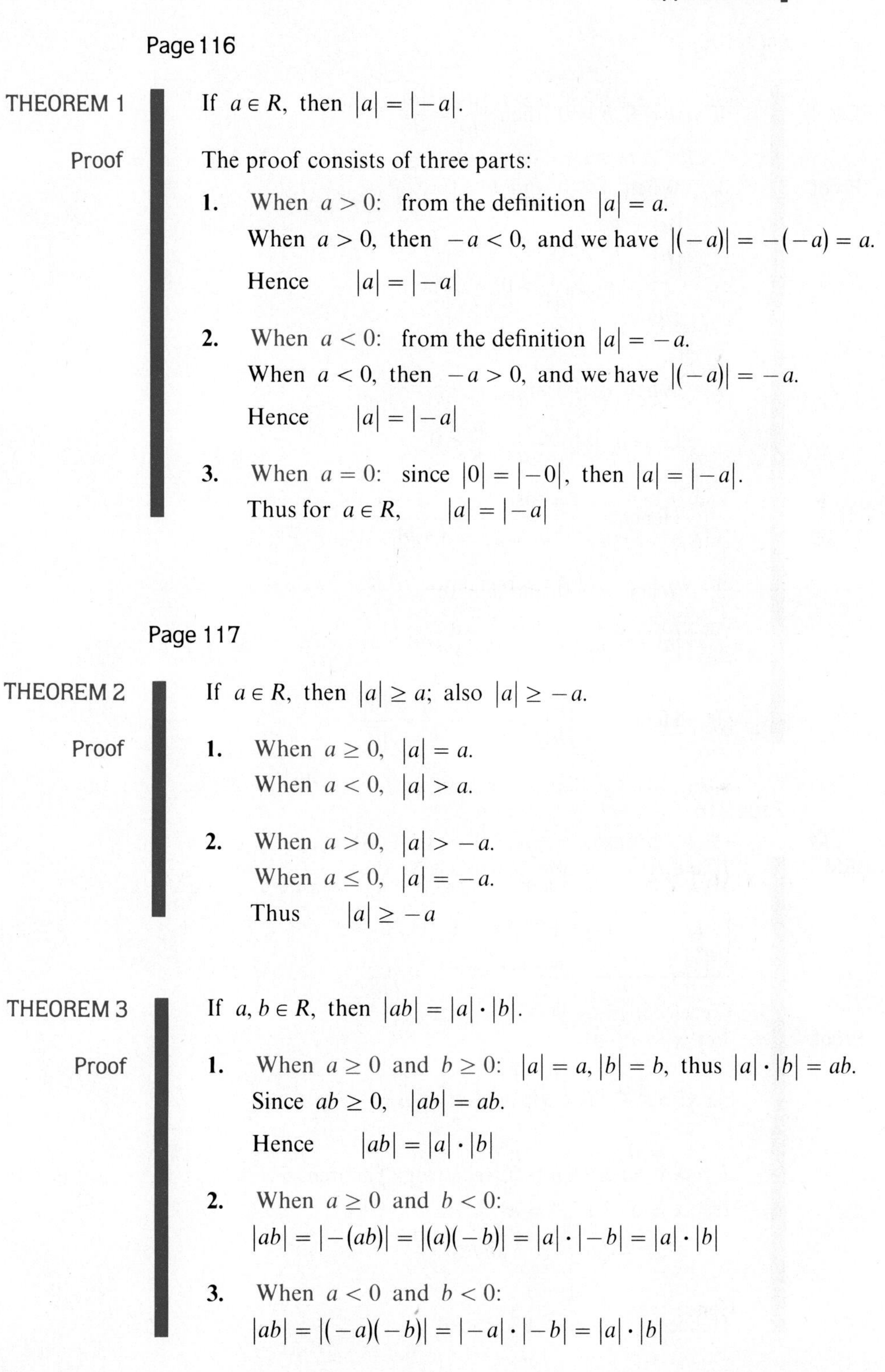

Page 116

THEOREM 1 If $a \in R$, then $|a| = |-a|$.

Proof The proof consists of three parts:

1. When $a > 0$: from the definition $|a| = a$.
When $a > 0$, then $-a < 0$, and we have $|(-a)| = -(-a) = a$.
Hence $|a| = |-a|$

2. When $a < 0$: from the definition $|a| = -a$.
When $a < 0$, then $-a > 0$, and we have $|(-a)| = -a$.
Hence $|a| = |-a|$

3. When $a = 0$: since $|0| = |-0|$, then $|a| = |-a|$.
Thus for $a \in R$, $|a| = |-a|$

Page 117

THEOREM 2 If $a \in R$, then $|a| \geq a$; also $|a| \geq -a$.

Proof

1. When $a \geq 0$, $|a| = a$.
When $a < 0$, $|a| > a$.

2. When $a > 0$, $|a| > -a$.
When $a \leq 0$, $|a| = -a$.
Thus $|a| \geq -a$

THEOREM 3 If $a, b \in R$, then $|ab| = |a| \cdot |b|$.

Proof

1. When $a \geq 0$ and $b \geq 0$: $|a| = a, |b| = b$, thus $|a| \cdot |b| = ab$.
Since $ab \geq 0$, $|ab| = ab$.
Hence $|ab| = |a| \cdot |b|$

2. When $a \geq 0$ and $b < 0$:
$|ab| = |-(ab)| = |(a)(-b)| = |a| \cdot |-b| = |a| \cdot |b|$

3. When $a < 0$ and $b < 0$:
$|ab| = |(-a)(-b)| = |-a| \cdot |-b| = |a| \cdot |b|$

THEOREM 4 If $a, b \in R, b \neq 0$, then $\left|\frac{a}{b}\right| = \frac{|a|}{|b|}$.

Proof

1. When $a > 0$ and $b > 0$:

$$\left|\frac{a}{b}\right| = \frac{a}{b}, \quad |a| = a, \quad |b| = b$$

Hence $\left|\frac{a}{b}\right| = \frac{|a|}{|b|}$

2. When $a > 0$ and $b < 0$:

$$|a| = a, \ |b| = -b, \quad \frac{a}{b} < 0$$

Hence $\left|\frac{a}{b}\right| = -\frac{a}{b} = \frac{a}{(-b)} = \frac{|a|}{|b|}$

3. When $a < 0$ and $b < 0$:

$$|a| = -a, \ |b| = -b, \quad \frac{a}{b} > 0$$

Hence $\left|\frac{a}{b}\right| = \frac{a}{b} = \frac{(-a)}{(-b)} = \frac{|a|}{|b|}$

Page 216

THEOREM 1 If $a \in R, a > 0$ and $p, q, r, s \in N$, then

$$\boxed{a^{\frac{p}{q}} \cdot a^{\frac{r}{s}} = a^{\frac{p}{q}+\frac{r}{s}}}$$

Proof

Let $x = a^{\frac{p}{q}} \cdot a^{\frac{r}{s}}$

$$x^{qs} = \left(a^{\frac{p}{q}} \cdot a^{\frac{r}{s}}\right)^{qs} = \left(a^{\frac{p}{q}}\right)^{qs} \cdot \left(a^{\frac{r}{s}}\right)^{qs}$$

$$= a^{\frac{pqs}{q}} \cdot a^{\frac{rqs}{s}} = a^{ps} \cdot a^{rq} = a^{ps+rq}$$

or $x = a^{(ps+rq)\cdot\frac{1}{qs}} = a^{\frac{ps+rq}{qs}} = a^{\frac{ps}{qs}+\frac{rq}{qs}}$

$$x = a^{\frac{p}{q}+\frac{r}{s}}$$

Hence $a^{\frac{p}{q}} \cdot a^{\frac{r}{s}} = a^{\frac{p}{q}+\frac{r}{s}}$

THEOREM 2 If $a \in R$, $a > 0$, and $p, q, r, s \in N$, then $\left(a^{\frac{p}{q}}\right)^{\frac{r}{s}} = a^{\frac{pr}{qs}}$.

Proof Let $x = \left(a^{\frac{p}{q}}\right)^{\frac{r}{s}}$.

$$x^{qs} = \left[\left(a^{\frac{p}{q}}\right)^{\frac{r}{s}}\right]^{qs} = \left(a^{\frac{p}{q}}\right)^{\frac{r}{s}\cdot qs} = \left(a^{\frac{p}{q}}\right)^{rq} = a^{\frac{prq}{q}} = a^{pr}$$

or $x = a^{\frac{pr}{qs}}$.

Hence $\left(a^{\frac{p}{q}}\right)^{\frac{r}{s}} = a^{\frac{pr}{qs}}$

THEOREM 3 If $a, b \in R$, $a, b > 0$, and $p, q \in N$, then $(ab)^{\frac{p}{q}} = a^{\frac{p}{q}} \cdot b^{\frac{p}{q}}$.

Let $x = a^{\frac{p}{q}} \cdot b^{\frac{p}{q}}$.

$$x^q = \left[a^{\frac{p}{q}} \cdot b^{\frac{p}{q}}\right]^q = \left(a^{\frac{p}{q}}\right)^q \cdot \left(b^{\frac{p}{q}}\right)^q = a^p \cdot b^p = (ab)^p$$

or $x = (ab)^{\frac{p}{q}}$

Hence $(ab)^{\frac{p}{q}} = a^{\frac{p}{q}} \cdot b^{\frac{p}{q}}$

Page 220

THEOREM 4 If $a \in R$, $a > 0$, and $p, q, r, s \in N$, then

$$\frac{a^{\frac{p}{q}}}{a^{\frac{r}{s}}} = \begin{cases} a^{\frac{p}{q}-\frac{r}{s}} & \text{when } \frac{p}{q} > \frac{r}{s} \\ 1 & \text{when } \frac{p}{q} = \frac{r}{s} \\ \frac{1}{a^{\frac{r}{s}-\frac{p}{q}}} & \text{when } \frac{p}{q} < \frac{r}{s} \end{cases}$$

Proof Let $x = \frac{a^{\frac{p}{q}}}{a^{\frac{r}{s}}}$.

$$x^{qs} = \left(\frac{a^{\frac{p}{q}}}{a^{\frac{r}{s}}}\right)^{qs} = \frac{\left(a^{\frac{p}{q}}\right)^{qs}}{\left(a^{\frac{r}{s}}\right)^{qs}} = \frac{a^{ps}}{a^{qr}}$$

If $q, s \in N$ and $\frac{p}{q} > \frac{r}{s}$, then by multiplying both sides of the inequality by $qs > 0$ we get $ps > rq$.

Thus $\quad x^{qs} = a^{ps - qr}$

or $\quad x = a^{\frac{(ps-qr)}{qs}} = a^{\frac{p}{q} - \frac{r}{s}}$

Hence $\quad \dfrac{a^{\frac{p}{q}}}{a^{\frac{r}{s}}} = a^{\frac{p}{q} - \frac{r}{s}} \quad$ when $\frac{p}{q} > \frac{r}{s}$

If $\frac{p}{q} = \frac{r}{s}$, obviously $\dfrac{a^{\frac{p}{q}}}{a^{\frac{r}{s}}} = \dfrac{a^{\frac{p}{q}}}{a^{\frac{p}{q}}} = 1$.

Similarly, we can show that if $\frac{p}{q} < \frac{r}{s}$, then $\dfrac{a^{\frac{p}{q}}}{a^{\frac{r}{s}}} = \dfrac{1}{a^{\frac{r}{s} - \frac{p}{q}}}$.

THEOREM 5 If $a, b \in R, a > 0, b > 0$, and $p, q \in N$, then $\left(\dfrac{a}{b}\right)^{\frac{p}{q}} = \dfrac{a^{\frac{p}{q}}}{b^{\frac{p}{q}}}$.

Proof Let $x = \dfrac{a^{\frac{p}{q}}}{b^{\frac{p}{q}}}$.

Then $\quad x^q = \left(\dfrac{a^{\frac{p}{q}}}{b^{\frac{p}{q}}}\right)^q = \dfrac{\left(a^{\frac{p}{q}}\right)^q}{\left(b^{\frac{p}{q}}\right)^q} = \dfrac{a^p}{b^p} = \left(\dfrac{a}{b}\right)^p$

or $\quad x = \left(\dfrac{a}{b}\right)^{\frac{p}{q}}$

Hence $\quad \left(\dfrac{a}{b}\right)^{\frac{p}{q}} = \dfrac{a^{\frac{p}{q}}}{b^{\frac{p}{q}}}$.

Page 225

The rules for exponents are valid when a zero exponent occurs.

Proofs

1. $a^0 \cdot a^m = 1 \cdot a^m = a^m$; also $a^{0+m} = a^m$.

Hence $\quad a^0 \cdot a^m = a^{0+m}$

2. $(a^m)^0 = 1$, considering a^m as one number; also $a^{m \cdot 0} = a^0 = 1$.

Hence $(a^m)^0 = a^{m \cdot 0}$

$(a^0)^m = (1)^m = 1$; also $a^{0 \cdot m} = a^0 = 1$

Hence $(a^0)^m = a^{0 \cdot m}$

3. $(ab)^0 = 1$, considering ab as one number; also $a^0 b^0 = 1 \cdot 1 = 1$.

Hence $(ab)^0 = a^0 b^0$

4. $\dfrac{a^m}{a^0} = \dfrac{a^m}{1} = a^m$; also $a^{m-0} = a^m$.

Hence $\dfrac{a^m}{a^0} = a^{m-0}$

$\dfrac{a^0}{a^0} = \dfrac{1}{1} = 1 \qquad \dfrac{a^0}{a^m} = \dfrac{1}{a^m}$ Also $\dfrac{1}{a^{m-0}} = \dfrac{1}{a^m}$

Hence $\dfrac{a^0}{a^m} = \dfrac{1}{a^{m-0}}$

5. $\left(\dfrac{a}{b}\right)^0 = 1$ considering $\dfrac{a}{b}$ as one number; also $\dfrac{a^0}{b^0} = \dfrac{1}{1} = 1$.

Hence $\left(\dfrac{a}{b}\right)^0 = \dfrac{a^0}{b^0}$

Page 225

According to the definition of negative exponents, $a \neq 0, a^{-n} = \dfrac{1}{a^n}$, the rules for exponents are still valid.

Proofs

1. $a^m \cdot a^{-n} = a^{m-n}$ since $a^m \cdot a^{-n} = a^m \cdot \dfrac{1}{a^n} = \dfrac{a^m}{a^n} = a^{m-n}$

$a^{-m} \cdot a^{-n} = a^{-m-n}$ since $a^{-m} \cdot a^{-n} = \dfrac{1}{a^m} \cdot \dfrac{1}{a^n} = \dfrac{1}{a^{m+n}} = a^{-(m+n)} = a^{-m-n}$

2. $(a^m)^{-n} = a^{-mn}$ since $(a^m)^{-n} = \dfrac{1}{(a^m)^n} = \dfrac{1}{a^{mn}} = a^{-mn}$

$(a^{-m})^n = a^{-mn}$ since $(a^{-m})^n = \left(\dfrac{1}{a^m}\right)^n = \dfrac{1}{a^{mn}} = a^{-mn}$

$(a^{-m})^{-n} = a^{mn}$ since $(a^{-m})^{-n} = \dfrac{1}{(a^{-m})^n} = \dfrac{1}{a^{-mn}} = a^{mn}$

3. $(ab)^{-m} = a^{-m}b^{-m}$ since $(ab)^{-m} = \dfrac{1}{(ab)^m} = \dfrac{1}{a^m b^m} = a^{-m}b^{-m}$

4. $\dfrac{a^{-m}}{a^n} = a^{-m-n}$ since $\dfrac{a^{-m}}{a^n} = \dfrac{\frac{1}{a^m}}{a^n} = \dfrac{1}{a^m \cdot a^n} = \dfrac{1}{a^{m+n}} = a^{-m-n}$

$\dfrac{a^m}{a^{-n}} = a^{m+n}$ since $\dfrac{a^m}{a^{-n}} = \dfrac{a^m}{\frac{1}{a^n}} = a^m \cdot a^n = a^{m+n}$

$\dfrac{a^{-m}}{a^{-n}} = a^{-m+n} = \dfrac{1}{a^{-n+m}}$ since $\dfrac{a^{-m}}{a^{-n}} = \dfrac{a^{-m}}{\frac{1}{a^n}} = a^{-m} \cdot a^n = a^{-m+n}$

and $\dfrac{a^{-m}}{a^{-n}} = \dfrac{\frac{1}{a^m}}{a^{-n}} = \dfrac{1}{a^{-n} \cdot a^m} = \dfrac{1}{a^{-n+m}}$

5. $\left(\dfrac{a}{b}\right)^{-n} = \dfrac{a^{-n}}{b^{-n}} = \dfrac{b^n}{a^n}$ since $\left(\dfrac{a}{b}\right)^{-n} = \dfrac{1}{\left(\frac{a}{b}\right)^n} = \dfrac{1}{\frac{a^n}{b^n}} = \dfrac{b^n}{a^n}$

and $\dfrac{a^{-n}}{b^{-n}} = \dfrac{\frac{1}{a^n}}{\frac{1}{b^n}} = \dfrac{b^n}{a^n}$

Page 443

EQUATION OF A ELLIPSE

$$\sqrt{[x-(g-k)]^2+(y-h)^2}+\sqrt{[x-(g+k)]^2+(y-h)^2}=2a$$

When this equation is simplified, we get

$$x^2(a^2-k^2)-2gx(a^2-k^2)+g^2(a^2-k^2)+a^2(y-h)^2=a^2(a^2-k^2)$$

Putting $a^2-k^2=b^2$, we get

$$b^2x^2-2gb^2x+b^2g^2+a^2(y-h)^2=a^2b^2$$

$$b^2(x^2-2gx+g^2)+a^2(y-h)^2=a^2b^2$$

$$b^2(x-g)^2+a^2(y-h)^2=a^2b^2$$

or

$$\frac{(x-g)^2}{a^2}+\frac{(y-h)^2}{b^2}=1$$

Page 448

EQUATION OF A PARABOLA

$$x - (g - a) = \sqrt{[x - (g + a)]^2 + (y - h)^2}$$

$$[x - (g - a)]^2 = [x - (g + a)]^2 + (y - h)^2$$

When the above equation is simplified we get

$$(y - h)^2 = 4a(x - g)$$

Page 451

EQUATION OF A HYPERBOLA

$$\sqrt{[x - (g - k)]^2 + (y - h)^2} - \sqrt{[x - (g + k)]^2 + (y - h)^2} = 2a$$

When the above equation is simplified, we get

$$(k^2 - a^2)x^2 - 2xg(k^2 - a^2) + g^2(k^2 - a^2) - a^2(y - h)^2 = a^2(k^2 - a^2)$$

Putting $k^2 - a^2 = b^2$, we get

$$b^2x^2 - 2xgb^2 + g^2b^2 - a^2(y - h)^2 = a^2b^2$$

$$b^2(x^2 - 2xg + g^2) - a^2(y - h)^2 = a^2b^2$$

$$b^2(x - g)^2 - a^2(y - h)^2 = a^2b^2$$

or

$$\frac{(x - g)^2}{a^2} - \frac{(y - h)^2}{b^2} = 1$$

Answers to Odd Numbered Exercises

Exercise 1.3, page 7

1. {Mon., Tues., Wed., Thurs., Fri., Sat., Sun.}
3. {Africa, Antarctica, Asia, Australia, Europe, N. America, S. America}
5. {George Washington, John Adams, Thomas Jefferson, James Madison, James Monroe}
7. $\{M, i, s, p\}$
9. {2, 3, 4, . . .}
11. {0, 4, 8, 12, . . .}
13. {1, 5, 9, 13, . . .}
15. {4, 9, 14, 19, . . .}
17. {8, 12, 16, 20, 24, 28}
19. {7, 14, 21, 28}
21. Yes
23. No
25. No
27. {1}, {2}, {1, 2}, $\varnothing$
29. $\{a\} \subset A$
31. $\{a, b\} \subset A$
33. $\{c\} \not\subset A$
35. No
37. No
39. Yes
41. Yes
43. No
45. {15}
47. {6, 12, 18, 24, . . .}
49. {0, 15, 30, 45, . . .}
51. {1, 2, 3, 4, 5, 6, 8, 10}
53. {1, 2, 3, 4, 5, 6, 7, 8, 9, 10}
55. {1, 3, 5}
57. {1, 2, 3, 4, 5, 6, 7, 8, 9, 10}
59. {1, 3, 5}
61. {1, 2, 3, 4, 5, 7, 9}

Exercise 1.6, page 16

1. -4
3. -11
5. 12
7. 11
9. -16
11. -9
13. 12
15. 11
17. -7
19. 18
21. 8
23. -27
25. -24
27. 40
29. 45
31. -23
33. -17
35. -11
37. -35
39. -16
41. 14
43. 12
45. -23
47. -5
49. 2
51. -12
53. 6
55. -6
57. 27
59. 4
61. 7
63. 23
65. 18
67. -16
69. -11
71. $2 \cdot 2 \cdot 3$
73. $2 \cdot 13$
75. $2 \cdot 2 \cdot 3 \cdot 3$
77. $2 \cdot 2 \cdot 2 \cdot 2 \cdot 3$
79. $2 \cdot 2 \cdot 2 \cdot 2 \cdot 2 \cdot 2$
81. $2 \cdot 2 \cdot 3 \cdot 7$
83. $2 \cdot 2 \cdot 3 \cdot 3 \cdot 3$
85. 137
87. $2 \cdot 2 \cdot 2 \cdot 3 \cdot 7$
89. $2 \cdot 2 \cdot 3 \cdot 3 \cdot 7$

Exercise 1.7, page 27

1. $\frac{55}{24}$
3. $\frac{23}{15}$
5. $\frac{11}{42}$
7. $-\frac{83}{72}$
9. $-\frac{103}{72}$
11. $\frac{3}{8}$
13. $\frac{3}{4}$
15. 1
17. $\frac{11}{15}$
19. $-\frac{1}{36}$
21. $-\frac{3}{8}$
23. $-\frac{7}{8}$

25. $\frac{1}{3}$ **27.** $\frac{5}{12}$ **29.** 1 **31.** $\frac{8}{3}$

33. $\frac{1}{3}$ **35.** $\frac{3}{4}$ **37.** $\frac{2}{5}$ **39.** $\frac{1}{6}$

41. $\frac{3}{2}$ **43.** $\frac{5}{24}$ **45.** $\frac{7}{18}$ **47.** $\frac{35}{48}$

49. $\frac{2}{3}$ **51.** $\frac{19}{24}$ **53.** $\frac{1}{2}$ **55.** 0

57. $\frac{6}{25}$ **59.** $\frac{18}{5}$ **61.** $\frac{51}{25}$ **63.** $\frac{173}{125}$

65. .28 **67.** .088 **69.** $.428571\overline{428571}$

71. $.63\overline{63}$ **73.** $3\frac{4}{5}$ **75.** $4\frac{2}{9}$

77. $32\frac{1}{6}$ **79.** $25\frac{5}{7}$ **81.** $7\frac{4}{15}$ **83.** $12\frac{17}{27}$

85. $24\frac{6}{37}$ **87.** $32\frac{17}{48}$ **89.** $\frac{23}{4}$ **91.** $\frac{39}{8}$

93. $\frac{133}{5}$ **95.** $\frac{131}{8}$ **97.** $\frac{165}{13}$ **99.** $\frac{654}{19}$

101. $\frac{1016}{37}$ **103.** $\frac{1339}{47}$

Chapter 1 Review, page 30

1. {1, 2, 3, 4, 5, 6} **3.** {1, 2, 3, 4, 6} **5.** {4, 6}
7. {2, 4, 6} **9.** $\varnothing$ **11.** {2, 4, 6}
13. Yes **15.** Yes **17.** No **19.** 24
21. 26 **23.** -46 **25.** 33 **27.** 75

29. -17 **31.** 0 **33.** $\frac{7}{12}$ **35.** -1

37. $-\frac{3}{8}$ **39.** $\frac{3}{10}$ **41.** $\frac{1}{4}$ **43.** $\frac{17}{16}$

45. $-\frac{2}{9}$ **47.** $\frac{15}{16}$ **49.** $\frac{14}{3}$ **51.** $\frac{9}{8}$

53. $\frac{3}{2}$ **55.** $-\frac{31}{36}$ **57.** $-\frac{35}{24}$ **59.** $\frac{53}{48}$

61. $\frac{91}{36}$ 63. $\frac{5}{8}$ 65. $\frac{37}{25}$ 67. $\frac{7}{125}$

69. $\frac{53}{125}$ 71. .5625 73. .064

75. $.571428\overline{571428}$ 77. $1.81\overline{818}$ 79. $14\frac{5}{9}$

81. $19\frac{13}{15}$ 83. $16\frac{4}{29}$ 85. $37\frac{2}{47}$ 87. $\frac{99}{8}$

89. $\frac{247}{14}$ 91. $\frac{415}{21}$ 93. $\frac{1499}{35}$

Exercise 2.2, page 34

1. 4 **3.** 6 **5.** 1 **7.** 1
9. -4 **11.** -4 **13.** 5 **15.** -5
17. 7 **19.** 11 **21.** -5 **23.** -16
25. -7 **27.** -12 **29.** 3 **31.** -11
33. 88 **35.** -27 **37.** -128 **39.** 0
41. $\frac{1}{2}$ **43.** 0 **45.** -1 **47.** $-\frac{1}{12}$

Exercises 2.3–2.4, page 36

1. $4a$ **3.** $-3x + 3$ **5.** $-ab - 7b$ **7.** $3a - 2b$
9. $-3y$ **11.** $5y - 2x - 4$ **13.** $3a + 2b$ **15.** $x + 3y - 1$
17. $2xy + 3yz + z + 6$ **19.** $2(a + b) + (c - d)$ **21.** $4(x - y) + 2(z + t)$
23. $-4a$ **25.** $-12a$ **27.** $7a$ **29.** $-3a$
31. $xy - x$ **33.** $2ax - 2a$ **35.** $-x + 7y$ **37.** $a + 13$
39. $6x - 4y - 11$ **41.** $5x + y - z - 1$ **43.** $5ab - 5$
45. 6 **47.** $3y - 2x$ **49.** $x + 5y - 7$
51. $4abc - 3a - 2b + c$ **53.** $2(x + 2y) - 11(a - b)$
55. $2x - 4y + 8$ **57.** $5x - y - 6$

Exercise 2.5, page 38

1. $2x - 2$ **3.** $7 - x$ **5.** $5x + 12$ **7.** $7x - 2y$
9. $2 - 2x$ **11.** $12 - 2x$ **13.** $5x - 4$ **15.** $-2x - 22$
17. $4x + 4$ **19.** $3x - 2y$ **21.** $4y$ **23.** $-2y$

25. $2x + 2y$ **27.** $7x - 10y$ **29.** $16 - x$ **31.** $-8x - 6$
33. $8x - 23$ **35.** $5x - 15$ **37.** $11 - 3x$ **39.** $2x - 7$
41. $2x - 5$ **43.** $2x + 3$ **45.** $14x - 32$
47. $3x + (5y + 6z + 7); 3x - (-5y - 6z - 7)$
49. $6x + (-y + z - 4); 6x - (y - z + 4)$

Exercise 2.6A, page 40

1. 2^4 **3.** a^6 **5.** $(xy)^4$ **7.** $(-3)^4$
9. $3(-a)^3$ **11.** $-(-7)^5$ **13.** $-2(-x)^3y^2$ **15.** $5^2 + 2^4$
17. $(-x)^3 - y^3$ **19.** $(2x - 1)^4$ **21.** $-2 \cdot 2 \cdot 2 \cdot 2 \cdot 2$
23. $-2 \cdot 2 \cdot a \cdot a \cdot a$ **25.** $x(-y)(-y)(-y)(-y)$
27. $-x \cdot x \cdot (-y)(-y)(-y)$ **29.** $a(-b^3)(-b^3)$
31. $(x - 1)(x - 1)(x - 1)(x - 1)$ **33.** $x \cdot x \cdot x + x \cdot x$
35. $x \cdot x \cdot x - 4$ **37.** 17 **39.** 16 **41.** 49
43. 27 **45.** -125 **47.** -16 **49.** -108
51. -225 **53.** -144 **55.** 2^5 **57.** $2^4 \cdot 3$
59. $2^3 \cdot 3^2$ **61.** $2^5 \cdot 3$ **63.** $3^3 \cdot 5$ **65.** $2^3 \cdot 7^2$

Exercise 2.6B, page 43

1. $2^4 = 16$ **3.** $-2^6 = -64$ **5.** $-2^7 = -128$
7. a^6 **9.** $-a^5$ **11.** $-a^7$ **13.** $3a^7$
15. $-3a^6$ **17.** $-a^4b^2$ **19.** $-a^4b^3$ **21.** a^6
23. $-a^8$ **25.** $a^4 + 2a^3$ **27.** $9a^3$ **29.** $8a^5$
31. 2^{n+4} **33.** a^{n+4} **35.** a^{n+3} **37.** 3^{5n}
39. a^{4n} **41.** a^{2n+4} **43.** $(a - 1)^6$
45. $-3(2x - y)^5$ **47.** $5(x - 2y)^8$ **49.** a^4b
51. $2a^4b^2$ **53.** $-3a^3b^2$ **55.** $-a^5b^3$ **57.** $-12a^3b^4$
59. $-24x^4y^4$ **61.** $6a^5b^6$ **63.** $-48a^6b^6$ **65.** $60a^5b^7c^2$
67. $36a^2b^{12}c^5$ **69.** $2^6 = 64$ **71.** a^{12} **73.** a^{3n}
75. a^{3n+6} **77.** a^{3n+3} **79.** a^{3n^2} **81.** a^{n^3}
83. 2^8 **85.** 2^6 **87.** $-a^6$ **89.** a^{12}
91. a^{4n} **93.** $8a^6$ **95.** a^3b^6 **97.** a^nb^{2n}
99. $a^{2n+2}b^{n+1}$ **101.** a^4b^6 **103.** $-64a^3$ **105.** $-64a^3b^9$
107. $16a^6b^2$ **109.** $-a^8b^{12}c^4$ **111.** $-4a^7b^3$ **113.** $48a^8$
115. $4a^8b^9$ **117.** $a^{16}b^6c^6$ **119.** a^8b^9 **121.** $-64a^{13}b^6$
123. $-12a^8b^{10}c^4$ **125.** $8a^{16}b^{11}c^8$ **127.** $-81a^{10}b^{20}c^{20}$
129. $-41{,}472a^8b^5c^{18}$ **131.** $x^5y^7(x^2 + y^2)^9$
133. $-16a^9b^{21}(a - 3b)^{17}$ **135.** $-8x^5$
137. $3x^6$ **139.** $-7x^7y$ **141.** $-60a^7b^6$ **143.** $-65a^6b^6$

Exercise 2.6C, page 46

1. $3a + 12$ **3.** $8a - 24$ **5.** $-15a - 5$ **7.** $-4a + 6$
9. $xy + 2x$ **11.** $3xy - 4x$ **13.** $-3xy - 9x$ **15.** $-2xy + x$
17. $3a^3 + 3a$ **19.** $a^3 - 2a$ **21.** $-7a^3 - 21a$
23. $-2a^3 + 6a$ **25.** $x^4 - x^3 + x^2$ **27.** $2x^5 - 5x^4 - 3x^3$
29. $x^4 - 2x^3y - 3x^2y^2$ **31.** $-2x^3y + 2xy^3 - 10xy$
33. $-5x^5y^2 + 5x^4y^3 + 5x^3y$ **35.** $x^{n+2} + x^{n+1} - 2x^n$
37. $x^{n+3} - 2x^{n+2} - x^{n+1}$ **39.** $x^{3n} + x^{2n} - x^n$
41. $x^{3n+1} - x^{2n+1} - 2x^{n+1}$ **43.** $3(x + 2)^3 - 10(x + 2)^2$
45. $2(x + 1)^4 - 8(x + 1)^3$ **47.** $(x - y)^4 - 2(x - y)^3 - 3(x - y)^2$
49. $12a^3 - 4a^2b$ **51.** $81a^7b^4 - 81a^6b^5 - 81a^4b^7$
53. $x^4y^5 - 7x^3y^6 + 8x^2y^4$ **55.** $5x + 1$
57. $5x - 2$ **59.** $7x + 3$ **61.** $4a^5 + a^3$ **63.** $6a^3 - 32a^2$

Exercise 2.6D, page 48

1. $x^2 + 5x + 4$ **3.** $x^2 - x - 6$ **5.** $x^2 - 3x - 28$
7. $x^2 - 13x + 36$ **9.** $2x^2 + 11x + 12$ **11.** $5x^2 - 23x - 10$
13. $10x^2 + 13x + 4$ **15.** $9x^2 + 12x - 32$ **17.** $15x^2 - 32x + 16$
19. $30 + 7x - 2x^2$ **21.** $24 - 2x - 15x^2$ **23.** $9 - 6x - 8x^2$
25. $5 + 18x - 8x^2$ **27.** $12 + 13x - 35x^2$ **29.** $16x^2 + 24x + 9$
31. $16x^2 - 56x + 49$ **33.** $9x^2 + 6xy + y^2$
35. $25x^2 - 40xy + 16y^2$ **37.** $3x^2 - 2xy - y^2$
39. $3x^2y^2 - 16xy + 16$ **41.** $x^3 - 6x^2 + 3x - 18$
43. $x^4 - x^2 - 20$ **45.** $6x^5 + 3x^3 + 4x^2 + 2$
47. $6x^5 + 8x^3 - 3x^2 - 4$ **49.** $2x^3 + 5x^2 - 7x - 12$
51. $2x^3 - 11x^2 + 18x - 9$ **53.** $27x^3 + 8$
55. $8x^3 - 27y^3$ **57.** $x^4 - 9x^2 - 12x - 4$
59. $6x^4 - 13x^3 - x - 12$ **61.** $4x^4 - 4x^3 + 5x^2 - 2x + 1$
63. $3x^2 + 6x - 72$ **65.** $-12x^2 - 76x + 56$
67. $27x^2 - 27x - 30$ **69.** $-56x^2 + 132x - 72$
71. $4ax^2 - 38ax + 48a$ **73.** $x^3 - 3x^2 - 10x + 24$
75. $4x^3 - 16x^2 - 9x + 36$ **77.** $12x^3 + 13x^2 - 20x + 4$
79. $x^3 - 6x^2 + 12x - 8$ **81.** $8x^3 - 36x^2 + 54x - 27$
83. $x^{2n} + 2x^n - 24$ **85.** $2x^{2n} + 7x^n - 4$
87. $12x^{2n} - 23x^n + 10$ **89.** $x^{n+2} - 6x^{n+1} + 8x^n + 3x - 12$
91. $x^{n+3} + 6x^{n+2} + 9x^{n+1} - x - 3$ **93.** $x^{n+3} + 64x^n - 3x^2 + 12x - 48$
95. $4x^{4n} - 4x^{2n} + 9$ **97.** $11x + 6$ **99.** $2x^2 - 23$
101. $5x - 24$ **103.** $-5x - 17$ **105.** $-16x$

Exercise 2.7A, page 52

1. $2^4 = 16$ **3.** $\frac{1}{2^8} = \frac{1}{256}$ **5.** 1

7. $-\frac{1}{3^5} = -\frac{1}{243}$ 9. $-\frac{1}{3}$ 11. $5^2 = 25$

13. a^3 15. $\frac{1}{a^6}$ 17. $-a^3$ 19. $\frac{1}{a}$

21. $-\frac{1}{a}$ 23. $-\frac{1}{a^4}$ 25. a^n 27. a^2

29. $\frac{1}{a^3}$ 31. $(x + y)^2$ 33. $\frac{1}{x + 2y}$ 35. $\frac{1}{(x + y)^3}$

37. -1 39. $y - x$ 41. $\frac{3}{2}$ 43. $\frac{25}{3}$

45. ab^3 47. b^3 49. $-\frac{a^6}{3b^5}$ 51. $\frac{2ac^2}{3b^6}$

53. $-\frac{2a^2b^5}{3c^2}$ 55. $\frac{64}{9}$ 57. $-\frac{27}{8}$ 59. $\frac{27a^9}{8}$

61. $\frac{4}{9a^2}$ 63. $-\frac{27a^3}{64}$ 65. $\frac{a^4c^4}{b^8}$ 67. $\frac{16b^8}{81a^8c^8}$

69. $\frac{(2b + c)^4}{a^4}$ 71. $\frac{-(b + c)^6}{64a^9}$ 73. $\frac{b^3}{a^3(y - x)^3}$ 75. $\frac{a^4}{b^2(y - x)^4}$

77. $2^2 = 4$ 79. $\frac{9}{2}$ 81. $\frac{81}{160}$ 83. $\frac{512}{729}$

85. $\frac{125}{32}$ 87. $\frac{88}{9}$ 89. $\frac{1}{16}$ 91. $\frac{b^2}{a^4}$

93. $\frac{24a^5c^4}{b^3}$ 95. $-\frac{27a}{800b^3}$ 97. $-\frac{64c^3}{a^2b^2}$ 99. $\frac{16c^2}{b^4}$

101. $5a^4$ 103. $10a^4$ 105. $-23a^4$ 107. $7a^4$
109. $4a^2$

Exercise 2.7B, page 55

1. $2x + 1$ 3. $1 - 2x$ 5. $x - 3$ 7. $3 + 2x$

9. $2x^2 - 1$ 11. $2 - x$ 13. $x - 2a$ 15. $-\frac{1}{a} + \frac{2}{b}$

17. $-2x^2 - x + 1$ 19. $-2x + 1 - \frac{4}{x}$ 21. $-\frac{x^2}{y^2} + 1 - \frac{y^2}{x^2}$

23. $\frac{2a}{b} - 4b + \frac{b^3}{a}$ 25. $3(a + b) - 1$ 27. $-a(x - 5y) + b$

29. $-(x-y)^2-(x-y)$ **31.** $\dfrac{x-2y}{b}-\dfrac{(x-2y)^2}{a}$

33. $-a^2-a+6$ **35.** a^2-a-2 **37.** x^2-3x+1
39. $x+2x^2-x^3$ **41.** $2x^3+x-4$ **43.** $-2x^2+6$
45. x^2-5x+6 **47.** 6 **49.** $2x^{n+1}-12$

Exercise 2.7C, page 59

1. $x+8$ **3.** $x+4$ **5.** $x+2$ **7.** $x+6$
9. $x-4$ **11.** $2x+4$ **13.** $4x+3$ **15.** $3x-2$
17. $3x+4$ **19.** $2x+3$ **21.** $4x+3$

23. $6x+1+\dfrac{2}{3x+5}$ **25.** $2x+3+\dfrac{8}{3x-4}$ **27.** $3x+1-\dfrac{2}{3x+4}$

29. $2x-3-\dfrac{10}{2x-5}$ **31.** $3x+4-\dfrac{4}{2x-9}$

33. $5x^2-6x+2-\dfrac{6}{2x^2+x-3}$ **35.** $3x-2$

37. $6x+7$ **39.** $9x^2-3x+1$ **41.** $16x^2+12x+9$
43. x^2-3x-2 **45.** $2x^2-x+4$ **47.** $2x^2-x-2$

49. $3x^2-2x+5+\dfrac{4x-5}{3x^2+2x-5}$ **51.** $4x^2+2x-5+\dfrac{2x-3}{6x^2-3x+4}$

53. $x^3-3x^2+x-1+\dfrac{-3x+2}{2x^2+6x-1}$ or $x^3-3x^2+x-1-\dfrac{3x-2}{2x^2+6x-1}$

55. $3x^3-2x^2+2x-1-\dfrac{2x+4}{2x^2-x-2}$

57. $3x^4-4x^3+x+5$ **59.** $7x^3-2x^2y+xy^2-y^3$
61. $2x^3-7x^2y+8xy^2-y^3$ **63.** $x^3-2x^2y+5xy^2-4y^3$
65. $4x^3-2x^2y-3xy^2+2y^3$ **67.** $4x^2-2xy+y^2$
69. $x^4+4x^2y^2+16y^4$ **71.** $(x+y)-4$
73. $2(2x+y)+3$ **75.** $9(x-3y)+4$ **77.** $x-3y+2$
79. $3x-2y+4$ **81.** $3x^2-3$ **83.** 3

Chapter 2 Review, page 61

1. $2x+5y-1$ **3.** $2xy+5yz+3z+8$
5. $3x^3+2x^2$ **7.** $-x$ **9.** $5x^2$
11. $-4x+2y-15$ **13.** $-2x^3+3x^2-8x-1$
15. x^2-x **17.** $10x+10$ **19.** $xy-y$ **21.** $6x-15$

23. $2x^2 + 23x + 15$ **25.** -1 **27.** 2
29. -5 **31.** -84 **33.** 23 **35.** 36
37. 2 **39.** $\frac{5}{2}$ **41.** $-\frac{5}{2}$ **43.** $-2x^5y^3$
45. $6x^5yz^4$ **47.** $4x^9y^4$ **49.** $125x^9y^6z^3$
51. $81x^2y^6$ **53.** $-64x^{12}y^3$ **55.** $-x^{10}y^{10}$
57. $-98x^{10}y^5z^{14}$ **59.** $5x^4y^6$ **61.** $-65x^{12}y^{12}z^6$
63. $x^{n+4} - 2x^{n+3} - x^{n+2}$ **65.** $3x^5 - x^2$
67. $6x^2 - 19x + 10$ **69.** $x^{2n} + 2x^n - 8$
71. $8x^{2n} - 14x^n + 3$ **73.** $x^4 + x^3 - 4x - 16$
75. $x^{n+3} - 9x^{n+1} - 4x + 12$ **77.** $x^4 - 4x^2 + 4x - 1$
79. $x^4 - 11x^2 + 25$ **81.** $-3x^3 - 7x^2 + 6x$
83. $3x^3 - 25x^2 + 56x - 16$ **85.** $2x^2 + 3x + 1$
87. -2 **89.** $-16x$ **91.** $\frac{1}{a^2}$ **93.** $\frac{y}{x^2z^2}$
95. $\frac{16y^{16}}{81x^8z^{12}}$ **97.** $-\frac{27x^3z^{12}}{8y^{12}}$ **99.** $\frac{16x^4z^3}{9y^3}$ **101.** $\frac{4x^6z^4}{243y^3}$
103. $x - 8$ **105.** $3x + 2$ **107.** $2x - 3$ **109.** $3x - 4$
111. $4x^2 + 2x + 1$ **113.** $x^2 + 2x - 3$ **115.** $3x^2 + x + 4$
117. $x^3 - x^2 - 2x + 1 + \frac{2x + 5}{2x^2 - 3x + 1}$
119. $x^3 - 3x^2 - 2x + 4 - \frac{2x + 3}{x^2 - x + 2}$

Exercise 3.3A, page 72

1. $\{3\}$ **3.** $\{-2\}$ **5.** $\{0\}$ **7.** $\left\{-\frac{7}{4}\right\}$
9. $\{-9\}$ **11.** $\{-5\}$ **13.** $\{4\}$ **15.** $\left\{\frac{1}{6}\right\}$
17. $\{-6\}$ **19.** $\left\{\frac{2}{3}\right\}$ **21.** $\{2\}$ **23.** $\{-3\}$
25. $\left\{\frac{5}{6}\right\}$ **27.** $\left\{\frac{4}{3}\right\}$ **29.** $\{3\}$ **31.** $\{2\}$
33. $\{3\}$ **35.** $\{-3\}$ **37.** $\{1\}$ **39.** $\{3\}$
41. $\left\{\frac{1}{3}\right\}$ **43.** $\{5\}$ **45.** $\left\{-\frac{1}{4}\right\}$ **47.** $\left\{-\frac{1}{4}\right\}$

49. $\{6\}$ **51.** $\{12\}$ **53.** $\{0\}$ **55.** $\{2\}$

57. $\{-4\}$ **59.** $\{2\}$ **61.** $\{6\}$ **63.** $\{-2\}$

65. $\{3\}$ **67.** $\left\{\frac{7}{3}\right\}$ **69.** $\left\{\frac{3}{2}\right\}$ **71.** $\{3\}$

73. $\left\{\frac{1}{3}\right\}$ **75.** $\left\{\frac{1}{2}\right\}$ **77.** $\{-8\}$ **79.** $\left\{\frac{4}{3}\right\}$

81. $\left\{-\frac{4}{9}\right\}$ **83.** $\left\{-\frac{5}{3}\right\}$ **85.** $\{0\}$ **87.** $\{0\}$

89. $\{x \mid x \in R\}$ **91.** $\{x \mid x \in R\}$ **93.** $\varnothing$ **95.** $\varnothing$

Exercise 3.3B, page 75

1. $\{1\}$ **3.** $\{-1\}$ **5.** $\left\{-\frac{3}{2}\right\}$ **7.** $\left\{\frac{3}{2}\right\}$

9. $\{1\}$ **11.** $\left\{\frac{5}{4}\right\}$ **13.** $\left\{-\frac{5}{2}\right\}$ **15.** $\left\{-\frac{1}{2}\right\}$

17. $\left\{\frac{3}{2}\right\}$ **19.** $\{2\}$ **21.** $\left\{\frac{3}{5}\right\}$ **23.** $\{-1\}$

25. $\left\{\frac{4}{3}\right\}$ **27.** $\{6\}$ **29.** $\{-13\}$ **31.** $\{3\}$

33. $\left\{-\frac{3}{2}\right\}$ **35.** $\{-4\}$ **37.** $\{6\}$ **39.** $\left\{-\frac{7}{2}\right\}$

41. $\{-1\}$ **43.** $\left\{\frac{3}{2}\right\}$ **45.** $\left\{-\frac{2}{3}\right\}$ **47.** $\left\{\frac{3}{2}\right\}$

49. $\{-1\}$ **51.** $\left\{\frac{1}{2}\right\}$ **53.** $\{-1\}$ **55.** $\left\{-\frac{1}{3}\right\}$

57. $\left\{\frac{3}{2}\right\}$ **59.** $\{-2\}$ **61.** $\left\{\frac{6}{5}\right\}$ **63.** $\{1\}$

65. $\{-1\}$ **67.** $\{1\}$ **69.** $\{-12\}$ **71.** $\{-6\}$

73. $\{8\}$ **75.** $\{7\}$ **77.** $\{-8\}$ **79.** $\{5\}$

81. $\{-10\}$ **83.** $\{3\}$ **85.** $\{8\}$ **87.** $\{500\}$

89. $\{600\}$ **91.** $\{3000\}$ **93.** $\{4500\}$ **95.** $\varnothing$

97. $\{x \mid x \in R\}$ **99.** $\{0\}$ **101.** $\{x \mid x \in R\}$ **103.** $\varnothing$

Exercise 3.4A, page 83

1. 12 **3.** 13 **5.** 224; 32 **7.** 12; 18
9. 28; 46 **11.** 37; 59 **13.** 16; 19 **15.** 30; 23
17. 19; 24; 36 **19.** 8; 9; 10 **21.** 21; 23; 25 **23.** 31; 32
25. 56; 58 **27.** 21; 27 **29.** 34; 39 **31.** 9; 11; 13
33. 86 **35.** 37 **37.** 719 **39.** 365
41. 724

Exercise 3.4B, page 88

1. 60% **3.** $700 **5.** $15.84 **7.** $1646
9. $1850 **11.** $1050 **13.** $320 **15.** $6500
17. $7500 at 8%; $12,500 at 11.5% **19.** $12,000; $26,000
21. $42,000 at 9%; $18,000 at 12% **23.** $9000
25. $45,000 at 25%; $7500 at 10% **27.** 9%; 8.5%
29. 9%; 9.75% **31.** 6 liters
33. 80% **35.** 15% **37.** 80 lbs
39. 4000 liters at 24%; 1600 liters at 45%

Exercise 3.4C, page 91

1. 23 nickels; 17 dimes **3.** 44 nickels; 32 quarters
5. 24 at 17¢; 36 at 22¢ **7.** 7 at 5¢; 14 at 17¢; 24 at 22¢
9. 16 at 5¢; 28 at 17¢; 32 at 22¢ **11.** 6 nickels; 22 dimes; 14 quarters
13. 16 nickels; 6 dimes; 12 quarters
15. 80 pounds at $2.40; 40 pounds at $3.60
17. 16 pounds
19. 9000 at $10.50; 27,000 at $7.50; 12,000 at $4.25

Exercise 3.4D, page 94

1. 48 mph; 52 mph **3.** 2.2 mph; 2.3 mph **5.** $17\frac{1}{2}$ hours
7. 43.2 mph **9.** 16 mph **11.** 20 miles

Exercise 3.4E, page 96

1. 20 C; 293 K **3.** 65 C; 338 K **5.** −30 C; 243 K
7. 86 F **9.** −58 F **11.** 257 F
13. 59 F **15.** 800.6 F

Exercise 3.4F, page 98

1. 54 pounds **3.** 120 pounds; 96 pounds
5. 6 feet **7.** 80 pounds; 100 pounds

Exercise 3.4G, page 99

1. 11 inches; 7 inches **3.** 1408 square feet **5.** 360 square feet
7. 11 inches **9.** 19 inches; 12 inches
11. 90°; 40°; 50° **13.** 56°; 34°

Chapter 3 Review, page 100

1. $\left\{\frac{9}{2}\right\}$ **3.** $\left\{\frac{10}{3}\right\}$ **5.** $\{3\}$ **7.** $\{0\}$
9. $\left\{\frac{1}{4}\right\}$ **11.** $\left\{-\frac{9}{4}\right\}$ **13.** $\{2\}$ **15.** $\{-8\}$
17. $\{2\}$ **19.** $\{1\}$ **21.** $\{5\}$ **23.** $\{-5\}$
25. $\{2\}$ **27.** $\left\{\frac{5}{2}\right\}$ **29.** $\{0\}$ **31.** $\{-2\}$
33. $\{-1\}$ **35.** $\{-1\}$ **37.** $\left\{\frac{9}{8}\right\}$ **39.** $\left\{\frac{8}{3}\right\}$
41. $\{-8\}$ **43.** $\left\{\frac{9}{4}\right\}$ **45.** $\{1\}$ **47.** $\{-1\}$
49. $\{3\}$ **51.** $\{-1\}$ **53.** $\{1\}$ **55.** $\{4\}$
57. $\{-1\}$ **59.** 34; 52 **61.** 19; 23; 32 **63.** 26; 32
65. 149 **67.** $250 **69.** 40% **71.** $21,000
73. 12 nickels; 18 dimes; 14 quarters **75.** 240 pounds; 360 pounds
77. 12 miles **79.** 140 feet; 200 feet **81.** 400
83. 9200 pounds **85.** $1.25

Exercise 4.3, page 112

1. $\{x \mid x < 1\}$ **3.** $\{x \mid x \geq -5\}$ **5.** $\{x \mid x \leq -2\}$ **7.** $\{x \mid x > 5\}$
9. $\left\{x \,\middle|\, x < -\frac{1}{2}\right\}$ **11.** $\left\{x \,\middle|\, x \geq -\frac{1}{2}\right\}$ **13.** $\left\{x \,\middle|\, x < \frac{1}{4}\right\}$
15. $\{x \mid x > 3\}$ **17.** $\{x \mid x \leq 4\}$ **19.** $\{x \mid x \leq 0\}$ **21.** $\{x \mid x > 1\}$

23. $\{x \mid x < 2\}$ **25.** $\{x \mid x < -3\}$ **27.** $\left\{x \,\middle|\, x \le -\frac{4}{3}\right\}$

29. $\{x \mid x \le 4\}$ **31.** $\left\{x \,\middle|\, x < -\frac{2}{5}\right\}$ **33.** $\{x \mid x < -1\}$

35. $\{x \mid x < -2\}$ **37.** $\{x \mid x > 2\}$ **39.** $\left\{x \,\middle|\, x < \frac{1}{2}\right\}$ **41.** $\{x \mid x \ge 3\}$

43. $\{x \mid x \le -2\}$ **45.** $\{x \mid x > 4\}$ **47.** $\{x \mid x < -2\}$ **49.** $\{x \mid x \in R\}$

51. $\{x \mid x \in R\}$ **53.** $\varnothing$ **55.** $\varnothing$ **57.** $\{x \mid x \in R\}$

Exercise 4.4, page 115

1. $\{x \mid -3 < x < 2\}$ **3.** $\{x \mid -1 \le x < 1\}$ **5.** $\{x \mid x > 2\}$

7. $\{x \mid x \le -3\}$ **9.** $\{3\}$ **11.** $\{1\}$ **13.** $\varnothing$

15. $\varnothing$ **17.** $\left\{x \,\middle|\, \frac{21}{11} < x \le 4\right\}$ **19.** $\left\{x \,\middle|\, \frac{3}{8} < x < 3\right\}$

21. $\{x \mid 2 < x < 10\}$ **23.** $\{x \mid 1 < x < 9\}$ **25.** $\{x \mid 6 < x < 9\}$

27. $\{x \mid -6 < x \le 8\}$ **29.** $\{x \mid -7 < x \le -5\}$ **31.** $\{x \mid -1 \le x < 5\}$

33. $\{x \mid -5 \le x \le 2\}$ **35.** $\{x \mid -1 \le x \le 3\}$ **37.** $\{x \mid 0 < x < 2\}$

39. $\{x \mid -2 < x < 3\}$ **41.** $\{x \mid -8 < x < 1\}$ **43.** $\{x \mid -4 < x < 2\}$

45. $\{x \mid x > 6\}$ **47.** $\{x \mid x < -2\}$

Exercise 4.6, page 123

1. $\{-1, 1\}$ **3.** $\{-6, 6\}$ **5.** $\varnothing$ **7.** $\varnothing$

9. $\left\{-\frac{3}{2}, \frac{3}{2}\right\}$ **11.** $\{-2, 2\}$ **13.** $\{-2, 2\}$ **15.** $\{7\}$

17. $\{-3\}$ **19.** $\left\{-\frac{3}{4}\right\}$ **21.** $\{2, 4\}$ **23.** $\{0, 4\}$

25. $\{-8, 0\}$ **27.** $\{-1, 2\}$ **29.** $\left\{-\frac{1}{3}, 1\right\}$ **31.** $\left\{-\frac{5}{3}, 3\right\}$

33. $\{-3, 2\}$ **35.** $\{0, 4\}$ **37.** $\{-4, 12\}$ **39.** $\left\{-\frac{7}{2}, \frac{13}{2}\right\}$

41. $\left\{\frac{3}{2}, 2\right\}$ **43.** $\left\{\frac{4}{7}, 2\right\}$ **45.** $\left\{\frac{3}{2}, 3\right\}$ **47.** $\left\{\frac{1}{7}, 1\right\}$

49. $\{-1, 13\}$ **51.** $\left\{-\frac{1}{2}, 3\right\}$ **53.** $\left\{-\frac{2}{5}, 1\right\}$ **55.** $\left\{\frac{2}{9}, \frac{4}{7}\right\}$

57. $\left\{1, \frac{5}{2}\right\}$ **59.** $\left\{-\frac{8}{3}\right\}$ **61.** $\{-1\}$ **63.** $\left\{-\frac{4}{5}\right\}$

65. $\left\{\frac{14}{5}\right\}$ **67.** $\left\{\frac{7}{5}\right\}$ **69.** $\{0\}$ **71.** $\varnothing$

73. $\varnothing$ **75.** $\varnothing$ **77.** $\left\{x \middle| x \geq \frac{1}{2}\right\}$ **79.** $\{x \mid x \leq 2\}$

81. $\left\{x \middle| x \leq \frac{4}{3}\right\}$ **83.** $\left\{x \middle| x \leq \frac{1}{3}\right\}$ **85.** $\left\{x \middle| x \geq \frac{2}{5}\right\}$ **87.** $\left\{x \middle| x \geq \frac{3}{2}\right\}$

89. $\{-2, 0\}$ **91.** $\{0, 10\}$ **93.** $\left\{-3, -\frac{1}{2}\right\}$

Exercise 4.7, page 125

1. $\{x \mid -1 < x < 1\}$ **3.** $\left\{x \middle| -\frac{3}{2} \leq x \leq \frac{3}{2}\right\}$ **5.** $\{x \mid 0 \leq x \leq 2\}$

7. $\{x \mid -1 < x < 9\}$ **9.** $\left\{x \middle| -\frac{7}{3} < x < 3\right\}$ **11.** $\{x \mid -9 \leq x \leq 5\}$

13. $\{x \mid -6 \leq x \leq 3\}$ **15.** $\{x \mid -5 < x < -2\}$ **17.** $\{x \mid 7 < x < 9\}$

19. $\{x \mid -2 < x < 3\}$ **21.** $\{x \mid -3 \leq x \leq 8\}$

23. $\{x \mid x < -1 \text{ or } x > 1\}$ **25.** $\left\{x \middle| x \leq -\frac{5}{3} \text{ or } x \geq \frac{5}{3}\right\}$

27. $\{x \mid x \leq 2 \text{ or } x \geq 4\}$ **29.** $\{x \mid x < -1 \text{ or } x > 2\}$

31. $\left\{x \middle| x < -1 \text{ or } x > \frac{11}{3}\right\}$ **33.** $\left\{x \middle| x \in R, x \neq \frac{5}{2}\right\}$

35. $\{x \mid x \in R\}$ **37.** $\{x \mid x \leq -6 \text{ or } x \geq -2\}$

39. $\{x \mid x < -4 \text{ or } x > -3\}$ **41.** $\{x \mid x < -2 \text{ or } x > 4\}$

43. $\{x \mid x < -1 \text{ or } x > 2\}$ **45.** $\left\{x \middle| x \leq -\frac{4}{5} \text{ or } x \geq 2\right\}$

47. $\{x \mid x > 9\}$ **49.** $\{x \mid x > 1\}$ **51.** $\{x \mid x > 3\}$

53. $\{x \mid -1 < x < 3\}$ **55.** $\{x \mid 1 < x < 5\}$ **57.** $\left\{x \middle| 2 < x < \frac{12}{5}\right\}$

59. $\varnothing$ **61.** $\left\{x \middle| x < \frac{3}{2}\right\}$ **63.** $\left\{x \middle| x < \frac{3}{4}\right\}$

65. $\left\{x \middle| x < -\frac{5}{3}\right\}$ **67.** $\left\{x \middle| x < -\frac{3}{5} \text{ or } x > \frac{5}{2}\right\}$

69. $\left\{x \middle| x < -\frac{4}{5} \text{ or } x > 2\right\}$ **71.** $\{x \mid x < 1 \text{ or } x > 5\}$

73. $\{x \mid x \in R\}$ **75.** $\{x \mid x \in R\}$

Chapter 4 Review, page 126

1. $\{x \mid x < -2\}$ **3.** $\{x \mid x \leq -3\}$ **5.** $\{x \mid x > 3\}$

7. $\{x \mid x \geq 2\}$ **9.** $\{x \mid x > 1\}$ **11.** $\{x \mid x < -1\}$

13. $\{x \mid x < 3\}$ **15.** $\left\{x \middle| x \leq \frac{1}{2}\right\}$ **17.** $\{x \mid x < -1\}$

19. $\left\{x \middle| x > \frac{1}{3}\right\}$ **21.** $\{x \mid x > -2\}$ **23.** $\left\{x \middle| x \geq \frac{1}{2}\right\}$

25. $\left\{x \middle| x \leq \frac{2}{3}\right\}$ **27.** $\{x \mid x < -3\}$ **29.** $\{x \mid x \leq 2\}$

31. $\{x \mid x \leq -2\}$ **33.** $\{x \mid x > 3\}$ **35.** $\{x \mid 2 < x < 5\}$

37. $\{x \mid -5 \leq x < 1\}$ **39.** $\{x \mid -1 < x \leq 3\}$ **41.** $\{x \mid 0 < x < 2\}$

43. $\{x \mid -2 \leq x \leq 2\}$ **45.** $\{x \mid -1 < x < 0\}$ **47.** $\{x \mid -13 < x < 1\}$

49. $\{x \mid -6 < x < 1\}$ **51.** $\left\{x \middle| x > \frac{13}{2}\right\}$ **53.** $\left\{x \middle| x < -\frac{1}{2}\right\}$

55. $\{x \mid 2 < x < 6\}$ **57.** $\{x \mid -3 < x \leq 1\}$ **59.** $\{x \mid x \geq 1\}$

61. $\{x \mid x \leq -6\}$ **63.** $\{4\}$ **65.** $\{-1\}$

67. $\{-4, 4\}$ **69.** $\varnothing$ **71.** $\left\{-\frac{3}{2}, \frac{3}{2}\right\}$ **73.** $\{-2, 8\}$

75. $\{4, 6\}$ **77.** $\{-6, 0\}$ **79.** $\{-2, 3\}$ **81.** $\left\{\frac{7}{3}, 5\right\}$

83. $\{0, 7\}$ **85.** $\left\{-\frac{13}{5}, 5\right\}$ **87.** $\{-1, 4\}$ **89.** $\left\{\frac{2}{3}, 3\right\}$

91. $\left\{\frac{2}{3}\right\}$ **93.** $\{1\}$ **95.** $\{3\}$ **97.** $\varnothing$

99. $\varnothing$ **101.** $\left\{x \middle| x \geq \frac{3}{2}\right\}$ **103.** $\left\{x \middle| x \geq \frac{5}{3}\right\}$

105. $\left\{x \middle| x \geq -\frac{1}{2}\right\}$ **107.** $\{x \mid x \geq 6\}$

109. $\{x \mid x < -2 \text{ or } x > 2\}$ **111.** $\{x \mid -3 \leq x \leq 3\}$

113. $\{x \mid -2 < x < 0\}$ **115.** $\{x \mid x < -1 \text{ or } x > 5\}$

117. $\{x \mid -4 \le x \le -1\}$

119. $\{x \mid 0 \le x \le 1\}$

121. $\{x \mid x \le -3 \text{ or } x \ge 2\}$

123. $\{x \mid x < 0 \text{ or } x > 7\}$

125. $\left\{x \,\middle|\, -\frac{2}{7} \le x \le 4\right\}$

127. $\{x \mid x < -4 \text{ or } x > 5\}$

129. $\left\{x \,\middle|\, x \le -\frac{4}{7} \text{ or } x \ge 2\right\}$

131. $\{x \mid x > 2\}$

133. $\{x \mid x > 7\}$

135. $\left\{x \,\middle|\, \frac{1}{4} < x < \frac{15}{2}\right\}$

137. $\left\{x \,\middle|\, -3 < x < \frac{4}{3}\right\}$

139. $\varnothing$

141. $\varnothing$

143. $\left\{x \,\middle|\, x < \frac{3}{8}\right\}$

145. $\left\{x \,\middle|\, x < -4 \text{ or } x > \frac{10}{3}\right\}$

147. $\{x \mid x < -1 \text{ or } x > 2\}$

149. $\{x \mid x \in R\}$

Cumulative Review, page 129

Chapter 1

1. -20 **3.** 32 **5.** 0 **7.** 12

9. 11 **11.** 7 **13.** $-\frac{5}{12}$ **15.** $-\frac{7}{24}$

17. $-\frac{7}{12}$ **19.** $\frac{4}{9}$ **21.** $\frac{3}{10}$ **23.** $\frac{24}{35}$

25. $\frac{4}{3}$ **27.** $\frac{4}{9}$ **29.** $\frac{2}{3}$ **31.** $\frac{17}{12}$

33. $\frac{5}{36}$ **35.** $-\frac{23}{36}$ **37.** $\frac{11}{12}$ **39.** $\frac{37}{18}$

41. $1.1\overline{1}$ **43.** $1.083\overline{3}$ **45.** $.42\overline{42}$ **47.** $.243\overline{243}$

Chapter 2

49. $7 - 4x$ **51.** $10x - 3$ **53.** $3x$

55. $3y - 2x - 9$ **57.** 14 **59.** 36

61. 0 **63.** $-12x^4y^3z^5$ **65.** $-8x^3y^6$

67. $81x^8y^4$ **69.** $432x^8y^9$ **71.** $16x^8y^5z^5$

73. $72a^8(x-1)^8$ **75.** $-17x^6$ **77.** $2x^3 - 5x + 3$

79. $x^4 + 2x^3 - x^2 - 2x + 1$ **81.** 3

83. $-5x$ **85.** $-\frac{8x^2}{3y^3}$ **87.** $\frac{z^6}{27x^3y^6}$ **89.** $\frac{16}{9}$

91. $\frac{b^5}{a^2}$ **93.** $\frac{27b^5c^4}{80a^2}$ **95.** $-5x^5$ **97.** $\frac{17}{3}x^3y^2$

99. $2x^2 + x - 1$ **101.** $x^2 + x - 2 - \frac{x - 1}{x^2 + 2x - 3}$

103. $3x^2 + 2x + 3$ **105.** $x^3 - 2x^2 + 3x - 1$

107. $2x^2 + xy + y^2$ **109.** $2x^2 - 6xy + y^2$

Chapter 3

111. $\{-1\}$ **113.** $\{-5\}$ **115.** $\left\{\frac{1}{2}\right\}$ **117.** $\{0\}$

119. $\left\{\frac{5}{4}\right\}$ **121.** $\{2\}$ **123.** $\{-2\}$ **125.** $\{6\}$

127. $\left\{\frac{2}{3}\right\}$ **129.** $\left\{\frac{6}{7}\right\}$ **131.** $\{4\}$ **133.** $\{-1\}$

135. $\{22{,}000\}$ **137.** $\{4000\}$ **139.** $\{4500\}$ **141.** $\varnothing$

143. $\{x \mid x \in R\}$ **145.** 55; 37 **147.** 17; 23 **149.** 746

151. \$215 **153.** \$140 **155.** \$22,000 **157.** 15 gallons

159. 81% **161.** 8 at 5¢; 12 at 17¢; 28 at 22¢

163. 12 miles **165.** 77 F **167.** 7 feet

169. 120 feet; 200 feet

Chapter 4

171. $\{x \mid x > 6\}$ **173.** $\{x \mid x < 3\}$ **175.** $\{x \mid x \leq 1\}$ **177.** $\{x \mid x \geq 3\}$

179. $\{x \mid x > 4\}$ **181.** $\{x \mid x < -1\}$ **183.** $\{x \mid x < 2\}$

185. $\{x \mid x \leq -1\}$ **187.** $\left\{x \,\middle|\, x < -\frac{1}{3}\right\}$ **189.** $\{x \mid x > -5\}$

191. $\{x \mid x < -2\}$ **193.** $\{x \mid x \geq -1\}$ **195.** $\{x \mid x \in R\}$

197. $\{x \mid x \in R\}$ **199.** $\varnothing$ **201.** $\varnothing$

203. $\{x \mid 2 < x < 5\}$ **205.** $\{x \mid -3 \leq x < 1\}$ **207.** $\{x \mid 3 < x \leq 6\}$

209. $\{x \mid x > 3\}$ **211.** $\{x \mid x \geq 1\}$ **213.** $\{x \mid x < -1\}$

215. $\{x \mid x \leq -2\}$ **217.** $\{3\}$ **219.** $\varnothing$

221. $\{x \mid -2 < x < 5\}$ **223.** $\{x \mid 2 < x < 6\}$ **225.** $\{x \mid 2 < x < 7\}$

227. $\{x \mid -5 < x < -1\}$ **229.** $\{x \mid -3 < x < 2\}$ **231.** $\{x \mid x > 6\}$

233. $\{x \mid x \leq -2\}$ **235.** $\{x \mid 2 < x < 4\}$ **237.** $\{-5, 7\}$

239. $\{-11, 5\}$ **241.** $\{-7\}$ **243.** $\{-2, 3\}$ **245.** $\left\{-4, \frac{3}{2}\right\}$

247. $\left\{-4, \frac{20}{3}\right\}$ **249.** $\{-1, 5\}$ **251.** $\left\{-\frac{3}{10}, \frac{5}{4}\right\}$ **253.** $\left\{-\frac{3}{5}\right\}$

255. $\{-1\}$ **257.** $\varnothing$ **259.** $\left\{x \middle| x \geq \frac{1}{3}\right\}$ **261.** $\left\{x \middle| x \geq \frac{4}{3}\right\}$

263. $\left\{x \middle| -\frac{7}{3} < x < 2\right\}$ **265.** $\left\{x \middle| -4 < x < -\frac{8}{3}\right\}$

267. $\{x \mid 2 < x < 3\}$ **269.** $\left\{x \middle| 1 < x < \frac{5}{2}\right\}$

271. $\left\{x \middle| -\frac{10}{3} < x < 4\right\}$ **273.** $\left\{x \middle| x < -1 \text{ or } x > \frac{3}{5}\right\}$

275. $\left\{x \middle| x < \frac{2}{5} \text{ or } x > 2\right\}$ **277.** $\left\{x \middle| x < \frac{10}{3} \text{ or } x > 4\right\}$

279. $\left\{x \middle| x \in R, x \neq -\frac{3}{4}\right\}$ **281.** $\left\{x \middle| -\frac{2}{3} < x < 3\right\}$

283. $\{x \mid -1 < x < 5\}$ **285.** $\{x \mid x > 6\}$ **287.** $\varnothing$

289. $\{x \mid x < 0 \text{ or } x > 2\}$ **291.** $\left\{x \middle| x < -\frac{4}{9} \text{ or } x > 2\right\}$

293. $\left\{x \middle| x < -\frac{5}{4}\right\}$ **295.** $\{x \mid x \in R\}$

Exercise 5.1, page 139

1. 3 **3.** 6 **5.** 5 **7.** 12
9. 12 **11.** 9 **13.** x^2y **15.** $5xy^2$
17. $3(x + 1)$ **19.** $x(x + 3)$ **21.** $19(x - 1)$
23. $x(x - 2)$ or $x(2 - x)$ **25.** $(x + 4)$
27. $(x - 2)$ **29.** $(x - 3)$ or $(3 - x)$ **31.** $7(x + 1)$
33. $9(2x + 3)$ **35.** $5(3x - 1)$ **37.** $3(3 - x)$ **39.** $3x(2x + 1)$
41. $3y(x - 2)$ **43.** $4x(2y + 3)$ **45.** $xy(x + y)$ **47.** $7y(3y - 2x)$
49. $6xy(3 - 4xy)$ **51.** $9x^2(3x - 2)$ **53.** $xy^2(2x + y)$
55. $6(x^2 - 2x + 3)$ **57.** $x(x^2 - x + 1)$ **59.** $3x(2x^2 + 3x - 1)$
61. $(2x + 1)(3 + x)$ **63.** $(x + 6)(x + 7)$ **65.** $4(x + 1)(x + 3)$
67. $(2x + 1)^2(2x + 3)$ **69.** $2(x - 4)(x - 1)$ **71.** $4(x - 3)(2x - 3y)$
73. $-3x(3x + 1)(5x + 2)$ **75.** $-(x - 2)(x + 7)$ **77.** $2(x - 4)(5 - x)$
79. $-(x + 1)(x + 3)$ **81.** $(x + 2)(x + 5)$ **83.** $(x - 2)(x - 5)$
85. $(x - 4)(x - y)$ **87.** $2(x - 7)(2x + y)$ **89.** $x(2x - 1)(x + 1)$

Exercise 5.2A, page 142

1. $(x + 2)(x - 2)$
3. $(x + 5)(x - 5)$
5. $(x + 9)(x - 9)$
7. $(x + 15)(x - 15)$
9. $x^2 + 36$
11. $(4 + x)(4 - x)$
13. $(13 + x)(13 - x)$
15. $(6x + 1)(6x - 1)$
17. $(3x + 5)(3x - 5)$
19. $(4x + 3)(4x - 3)$
21. $(2 + xy)(2 - xy)$
23. $(x + yz)(x - yz)$
25. $(2 + x^2)(2 - x^2)$
27. $3(x + 2y)(x - 2y)$
29. $9(x + 2y)(x - 2y)$
31. $x(x + 1)(x - 1)$
33. $x^2(x + 1)(x - 1)$
35. $5x(y + 2x)(y - 2x)$
37. $(x^2 + y^2)(x + y)(x - y)$
39. $(4x^2 + 25y^2)(2x + 5y)(2x - 5y)$
41. $(x^4 + 16y^2)(x^2 + 4y)(x^2 - 4y)$
43. $3(x^2 + 4y^2z^2)(x + 2yz)(x - 2yz)$
45. $9(4x^2 + 9)(2x + 3)(2x - 3)$
47. $3x^2(9x^2 + 1)(3x + 1)(3x - 1)$
49. $(x + y + 2)(x + y - 2)$
51. $(x - 2y + 6)(x - 2y - 6)$
53. $(x + 2y + z)(x + 2y - z)$
55. $5(x - y + 3)(x - y - 3)$
57. $3[x(x - y) + 2z][x(x - y) - 2z]$
59. $(x + y + 1)(x - y - 1)$
61. $(3x + y + 3)(3x - y - 3)$
63. $(x + y - 1)(x - y + 1)$
65. $(x^2 + 2y - 1)(x^2 - 2y + 1)$
67. $(x + 2y - 6)(x - 2y + 6)$
69. $(3x + 4y - 24)(3x - 4y + 24)$
71. $-(2x - y + 2)(y + 2)$
73. $(3x + y + 4)(x + y - 4)$
75. $(x - 2y - 1)(x - 4y + 1)$
77. $(x + 7y - 6)(x + y + 6)$
79. $(x - 2y)(x - 2y + 2)(x - 2y - 2)$
81. $(4x - y)(4x - y + 4)(4x - y - 4)$
83. $2(x - 2y)(2x - 4y + 3)(2x - 4y - 3)$

Exercise 5.2B, page 145

1. $(x + 1)(x^2 - x + 1)$
3. $(x + 3)(x^2 - 3x + 9)$
5. $(x + 6)(x^2 - 6x + 36)$
7. $(x - 2)(x^2 + 2x + 4)$
9. $(1 - 2x)(1 + 2x + 4x^2)$
11. $(3 - x)(9 + 3x + x^2)$
13. $(4x - y)(16x^2 + 4xy + y^2)$
15. $3(x + 2)(x^2 - 2x + 4)$
17. $4(x + 2y)(x^2 - 2xy + 4y^2)$
19. $2(2 - x)(4 + 2x + x^2)$
21. $2(5x - 1)(25x^2 + 5x + 1)$
23. $2(2x + 3y)(4x^2 - 6xy + 9y^2)$
25. $3(3x + 2y)(9x^2 - 6xy + 4y^2)$
27. $x(x + 2)(x^2 - 2x + 4)$
29. $xy^2(x - y)(x^2 + xy + y^2)$
31. $2x(3x + y)(9x^2 - 3xy + y^2)$
33. $(x^2 + y)(x^4 - x^2y + y^2)$
35. $x^3(1 - x)(1 + x + x^2)$
37. $(x^2 + 1)(x^4 - x^2 + 1)$
39. $(x^2 + 2y^2)(x^4 - 2x^2y^2 + 4y^4)$
41. $x^2(x^2 + y^2)(x^4 - x^2y^2 + y^4)$
43. $(2 + x)(4 - 2x + x^2)(2 - x)(4 + 2x + x^2)$
45. $(x + 3)(x^2 - 3x + 9)(x - 3)(x^2 + 3x + 9)$
47. $x^2(y + x)(y^2 - yx + x^2)(y - x)(y^2 + yx + x^2)$

49. $[(x + 1) + y][(x + 1)^2 - y(x + 1) + y^2]$
51. $[(x - 2) + 2y][(x - 2)^2 - 2y(x - 2) + 4y^2]$
53. $[(x - 4) - 4y][(x - 4)^2 + 4y(x - 4) + 16y^2]$
55. $[3(x - 2) - y][9(x - 2)^2 + 3y(x - 2) + y^2]$
57. $[x + (y - 3)][x^2 - x(y - 3) + (y - 3)^2]$
59. $[x + 3(y - 2)][x^2 - 3x(y - 2) + 9(y - 2)^2]$
61. $[3x - (2y + 1)][9x^2 + 3x(2y + 1) + (2y + 1)^2]$
63. $[x - 4(y - 2)][x^2 + 4x(y - 2) + 16(y - 2)^2]$
65. $x[x - (y - 1)][x^2 + x(y - 1) + (y - 1)^2]$
67. $(x + 2y - 1)(x^2 + xy + y^2 + x - y + 1)$
69. $(x - 4y + 4)(x^2 - 2xy + 4y^2 - 4x - 8y + 16)$

Exercise 5.3A, page 148

1. $(x + 2)(x + 3)$ **3.** $(x + 2)^2$ **5.** $(x + 3)(x + 8)$
7. $(x + 1)(x + 12)$ **9.** $(x + 2)(x + 9)$ **11.** $(x - 2)(x - 1)$
13. $(x - 3)(x - 5)$ **15.** $(x - 4)(x - 5)$ **17.** $(x - 3)(x - 15)$
19. $(x - 5)(x - 9)$ **21.** $(x + 2)(x - 1)$ **23.** $(x + 4)(x - 1)$
25. $(x + 12)(x - 3)$ **27.** $(x + 7)(x - 5)$ **29.** $(x + 9)(x - 6)$
31. $(x - 3)(x + 1)$ **33.** $(x - 6)(x + 2)$ **35.** $(x - 5)(x + 1)$
37. $(x - 10)(x + 3)$ **39.** $(x - 10)(x + 1)$ **41.** $x^2 + 2x + 24$
43. $x^2 - 3x + 10$ **45.** $x^2 + 6x - 8$ **47.** $x^2 - 4x - 3$
49. $(x + 3)(x + 4)$ **51.** $(x + 3)(x + 5)$ **53.** $(x - 5)(x - 1)$
55. $(x + 4)(x - 2)$ **57.** $(x + 7)(x - 2)$ **59.** $(x - 5)(x + 2)$
61. $(x + 4)(x + 6)$ **63.** $(x + 2)(x + 10)$ **65.** $(x - 5)(x - 6)$
67. $(x - 2)(x - 18)$ **69.** $(x + 9)(x - 1)$ **71.** $(x + 7)(x - 6)$
73. $(x + 3)(x - 9)$ **75.** $(x + 2)(x - 10)$ **77.** $x^2 + 4x + 12$
79. $x^2 + 8x - 12$ **81.** $x^2 - 10x - 16$ **83.** $(x + 3)(x + 12)$
85. $(x - 7)(x - 6)$ **87.** $(x - 24)(x - 3)$ **89.** $(x + 12)(x - 6)$
91. $(x - 10)(x + 8)$ **93.** $(x - 21)(x + 3)$ **95.** $(x + 6y)(x + y)$
97. $(x + 5y)^2$ **99.** $(x - 7y)(x - 3y)$ **101.** $(x - 4y)(x - 7y)$
103. $(x + 7y)(x - y)$ **105.** $(x + 8y)(x - 5y)$ **107.** $(x - 6y)(x + 3y)$
109. $(x - 15y)(x + 2y)$ **111.** $2(x + 4)^2$ **113.** $b^2(x + 12)(x + 4)$
115. $3(x - 3)(x - 8)$ **117.** $(xy - 8)(xy - 9)$
119. $b^2(bx - 5)(bx - 14)$ **121.** $4(x + 5)(x - 4)$
123. $x^2(x + 8)(x - 4)$ **125.** $(x^2 + 16)(x^2 - 5)$
127. $ax(x - 15)(x + 3)$ **129.** $y^2(x - 24y)(x + 3y)$

131. $(x^2 - 2)(x + 1)(x - 1)$ **133.** $(x^2 + 3)(x + 2)(x - 2)$
135. $(x^2 + 1)(x + 3)(x - 3)$ **137.** $(x^2 - 3)(x + 5)(x - 5)$
139. $(x + 1)^2(x - 1)^2$ **141.** $(x + 5)(x - 5)(x + 1)(x - 1)$
143. $(x + 5)(x - 5)(x + 2)(x - 2)$ **145.** $(x^3 - 3)(x + 1)(x^2 - x + 1)$
147. $(x + 3)(x^2 - 3x + 9)(x + 1)(x^2 - x + 1)$
149. $(x^3 + 3)(x - 1)(x^2 + x + 1)$ **151.** $(x - y - 2)(x - y - 1)$
153. $(x + y - 5)(x + y - 1)$ **155.** $(x - y + 2)(x - y + 3)$
157. $(x + 3y + 1)(x + 3y - 7)$ **159.** $(2x + y + 5)(2x + y - 2)$

Exercise 5.3B, page 155

1. $(2x + 1)(x + 1)$ **3.** $(2x + 1)^2$ **5.** $(2x + 7)(x + 3)$
7. $(2x + 1)(3x + 2)$ **9.** $(4x + 1)(x + 2)$ **11.** $(2x - 1)(x - 2)$
13. $(2x - 3)(2x - 1)$ **15.** $(3x - 1)(x - 1)$ **17.** $(2x - 1)(x - 6)$
19. $(4x - 1)(x - 3)$ **21.** $(2x - 1)(x + 1)$ **23.** $(2x + 3)(x - 1)$
25. $(3x - 1)(x + 3)$ **27.** $(2x - 1)(x + 6)$ **29.** $(4x - 1)(x + 3)$
31. $(2x + 1)(x - 2)$ **33.** $(2x + 1)(x - 6)$ **35.** $(3x + 1)(x - 2)$
37. $(3x + 2)(x - 2)$ **39.** $(3x - 1)(4x + 1)$ **41.** $(2x + 3)(x + 3)$
43. $(4x + 3)(x + 2)$ **45.** $(4x - 3)(x - 3)$ **47.** $(4x - 3)(x + 2)$
49. $(2x + 1)(5x - 1)$ **51.** $(2x - 1)(4x + 1)$ **53.** $(2x + 1)(5x + 4)$
55. $(6x + 7)(2x + 1)$ **57.** $(2x - 5)(2x - 1)$ **59.** $(6x + 7)(2x - 1)$
61. $(3x + 7)(2x - 1)$ **63.** $(4x + 3)(x - 3)$ **65.** $4x^2 + 17x - 18$
67. $6x^2 - x + 12$ **69.** $8x^2 - 22x - 15$ **71.** $(3x + 10)(2x + 1)$
73. $(3x + 2)(x + 3)$ **75.** $(2x - 1)(4x - 7)$ **77.** $(3x - 2)(x - 4)$
79. $(2x - 1)(4x + 9)$ **81.** $(2x - 3)(2x + 5)$ **83.** $(2x + 1)(5x - 3)$
85. $(6x + 1)(2x - 1)$ **87.** $(2xy + 3)(2xy + 5)$ **89.** $(9x + 4y)(x + 3y)$
91. $(2x - 7y)(2x - 3y)$ **93.** $(2xy^2 - 3)(3xy^2 - 4)$
95. $(3x - 2y)(4x - 3y)$ **97.** $(3x + 5y^2)(3x - y^2)$
99. $(2xy^2 + 3)(3xy^2 - 4)$ **101.** $(2xy - 9)(2xy + 1)$
103. $(3x - 4y)(3x + 2y)$ **105.** $(2x + 3y^2)(2x - 7y^2)$
107. $3(3x + 4)(3x + 1)$ **109.** $4(2x + 3)(3x + 2)$
111. $x(6x - 1)(x - 3)$ **113.** $2(3x - 2)(3x - 1)$
115. $x^2(4x - 1)(2x - 3)$ **117.** $x^2(3x + 5)(3x - 1)$
119. $4(2x - 5)(2x + 1)$ **121.** $x^2(3x + 1)(x - 5)$ **123.** $(4 - x)(1 + 6x)$
125. $(3 - 2x)(1 + 6x)$ **127.** $(8 + 9x)(3 - 2x)$ **129.** $(3 + 8x)(1 - 3x)$
131. $(2 - 3x)(3 + 4x)$ **133.** $(2x^2 - 1)(x + 1)(x - 1)$
135. $(3x^2 + 1)(x + 3)(x - 3)$ **137.** $(x^2 + 1)(4x + 1)(4x - 1)$

139. $(x^2 + 3)(3x + 2)(3x - 2)$

141. $(2x^2 - 1)(x + 6)(x - 6)$

143. $(x + 2)(x - 2)(3x + 1)(3x - 1)$

145. $(2x^3 + 1)(x + 1)(x^2 - x + 1)$

147. $(x^3 - 4)(2x + 1)(4x^2 - 2x + 1)$

149. $(2x^3 - 3)(x + 3)(x^2 - 3x + 9)$

151. $(x^3 + 2)(3x - 1)(9x^2 + 3x + 1)$

153. $(x^3 + 2)(4x - 1)(16x^2 + 4x + 1)$

155. $(x - 1)(x^2 + x + 1)(2x + 1)(4x^2 - 2x + 1)$

157. $(3x + 3y + 5)(x + y - 1)$

159. $(4x + 2y + 1)(2x + y - 3)$

161. $(2x - 4y - 3)(3x - 6y - 2)$

163. $(12x + 4y - 3)(6x + 2y + 3)$

165. $[3(x + y) + 4][2(x + y) - 3]$

167. $[3(x - y) + 4][3(x - y) + 2]$

169. $[4(x + 2y) + 9][(x + 2y) + 2]$

171. $\frac{1}{4}(2x - 1)(x - 4)$

173. $\frac{1}{9}(3x - 1)(3x + 5)$

175. $\frac{1}{12}(2x + 9)(2x - 3)$

177. $\frac{1}{6}(2x + 3)(x - 1)$

179. $\frac{1}{12}(3x + 2)(x - 2)$

Exercise 5.4, page 159

1. $(2x^2 + 2x + 1)(2x^2 - 2x + 1)$

3. $(x^4 + 2x^2 + 2)(x^4 - 2x^2 + 2)$

5. $(x^2 + 4x + 8)(x^2 - 4x + 8)$

7. $(x^2 + x + 1)(x^2 - x + 1)$

9. $(x^2 + 2x + 4)(x^2 - 2x + 4)$

11. $(x^2 + x + 4)(x^2 - x + 4)$

13. $(x^2 + 4x + 2)(x^2 - 4x + 2)$

15. $(x^2 + 2x - 1)(x^2 - 2x - 1)$

17. $(x^2 + x - 4)(x^2 - x - 4)$

19. $(x^2 + 4x + 5)(x^2 - 4x + 5)$

21. $(x^2 + 3x + 4)(x^2 - 3x + 4)$

23. $(x^2 + 5x - 4)(x^2 - 5x - 4)$

25. $(x + 3)(x - 3)(x + 1)(x - 1)$

27. $(3x^2 + 2x + 1)(3x^2 - 2x + 1)$

29. $(4x^2 + 3x + 2)(4x^2 - 3x + 2)$

31. $(3x^2 + 4x - 1)(3x^2 - 4x - 1)$

33. $(2x^2 + 4x + 1)(2x^2 - 4x + 1)$

35. $(2x^2 + 4x + 3)(2x^2 - 4x + 3)$

37. $(4x^2 + 5x - 2)(4x^2 - 5x - 2)$

39. $(5x^2 + 4x + 1)(5x^2 - 4x + 1)$

41. $(5x^2 + 6x + 2)(5x^2 - 6x + 2)$

43. $9x^4 - 36x^2 + 25$

45. $(x^4 - x^2 + 1)(x^2 + x + 1)(x^2 - x + 1)$

Exercise 5.5, page 162

1. $(x + y + z)(x + y - z)$

3. $(2x - y + 2z)(2x - y - 2z)$

5. $(x + 2 + y)(x + 2 - y)$

7. $(y + 3 + 3x)(y + 3 - 3x)$

9. $(2x + y + 5)(2x + y - 5)$

11. $(2x + 2y + 5)(2x + 2y - 5)$

13. $2(x - y + 3)(x - y - 3)$

15. $x(x + y + 4)(x + y - 4)$

17. $(x + y + z)(x - y - z)$

19. $(3x + y + 3)(3x - y - 3)$

21. $(5x + 3y + 3)(5x - 3y - 3)$ **23.** $(2 + x + 2y)(2 - x - 2y)$
25. $(3 + 2x + 2y)(3 - 2x - 2y)$ **27.** $2(x + y + 4)(x - y - 4)$
29. $x(x + y + 1)(x - y - 1)$ **31.** $(y + x - 2z)(y - x + 2z)$
33. $(3x^2 + 3y - 2)(3x^2 - 3y + 2)$ **35.** $(4x + y^2 - 4)(4x - y^2 + 4)$
37. $(1 + x^2 - 2y)(1 - x^2 + 2y)$ **39.** $(3 + 6x - 2y)(3 - 6x + 2y)$
41. $(5 + x - 3y)(5 - x + 3y)$ **43.** $3(y + x - z)(y - x + z)$
45. $4(1 + 2x - 4y)(1 - 2x + 4y)$ **47.** $x(2 + x - y)(2 - x + y)$
49. $(3x - y)(x - 2)$ **51.** $(x - 3)(x + y)$ **53.** $(2x - z)(3y - 7)$
55. $(9x + 4)(3x - 2y)$ **57.** $7(2a - b)(x - y)$ **59.** $4(3x - y)(x^2 + y^2)$
61. $(x - y)(x^2 + xy + y^2 - 1)$ **63.** $(x - 5y)(x^2 + 5xy + 25y^2 - 1)$
65. $(4x + y)(16x^2 - 4xy + y^2 - 1)$ **67.** $(2x + y)(4x^2 - 2xy + y^2 - 3)$
69. $(x - 2y)(x^2 + 2xy + 4y^2 + 3)$
71. $(2x - 3y)(4x^2 + 6xy + 9y^2 - 2x - 3y)$
73. $(x + 5y)(x - 5y + x^2 - 5xy + 25y^2)$
75. $(x - 3)(x + 3)^2$ **77.** $(x - 2)(2x + 1)(2x - 1)$
79. $(x + 1)^2(x^2 - x + 1)$ **81.** $(x - 2)(x + 3)(x^2 - 3x + 9)$

Chapter 5 Review, page 163

1. $(x + 1)(x - 1)$ **3.** $(x + 6)(x - 6)$ **5.** $(2x + 3)(2x - 3)$
7. $2(4 + xy)(4 - xy)$ **9.** $3(5x + 2y^2)(5x - 2y^2)$
11. $3x^2(2 + x)(2 - x)$ **13.** $(4x^2 + y^2)(2x + y)(2x - y)$
15. $(x + 1 + y)(x + 1 - y)$ **17.** $(x - 2 + 2y)(x - 2 - 2y)$
19. $(x + 2y + 1)(x - 2y - 1)$ **21.** $(x + 5)(x^2 - 5x + 25)$
23. $(3x + y)(9x^2 - 3xy + y^2)$ **25.** $(2 - x)(4 + 2x + x^2)$
27. $4(2x + 1)(4x^2 - 2x + 1)$ **29.** $x^3(x + y)(x^2 - xy + y^2)$
31. $9x(x - 2y^2)(x^2 + 2xy^2 + 4y^4)$
33. $[x + (y - 1)][x^2 - x(y - 1) + (y - 1)^2]$
35. $[(2x - 1) + 2y][(2x - 1)^2 - 2y(2x - 1) + 4y^2]$
37. $[x - 3(y + 1)][x^2 + 3x(y + 1) + 9(y + 1)^2]$
39. $6[(x - 1) - y][(x - 1)^2 + y(x - 1) + y^2]$
41. $(x + 2)(x + 7)$ **43.** $(x + 3)(x + 15)$ **45.** $(x - 4)(x - 1)$
47. $(x - 3)(x - 9)$ **49.** $(x + 5)(x - 2)$ **51.** $(x + 9)(x - 2)$
53. $(x - 5)(x + 4)$ **55.** $(x - 9)(x + 8)$ **57.** $(x + 3)(x + 20)$
59. $(x - 3)(x - 18)$ **61.** $(x - 2)(x - 21)$ **63.** $(x + 13)(x - 2)$
65. $(x + 5)(x - 10)$ **67.** $(x - 10)(x + 4)$ **69.** $(x + 4)(x + 15)$
71. $(x - 4)(x - 10)$ **73.** $(x + 24)(x - 2)$ **75.** $(x - 27)(x + 2)$

77. $(3x+1)(4x+3)$ **79.** $(2x+1)(3x+7)$ **81.** $(3x-1)(4x-5)$
83. $(2x-1)(3x-8)$ **85.** $(2x-1)(x+4)$ **87.** $(3x-1)(x+2)$
89. $(4x-1)(x+2)$ **91.** $(3x-1)(4x+3)$ **93.** $(2x+1)(x-3)$
95. $(3x+1)(x-3)$ **97.** $(2x+1)(x-6)$ **99.** $(4x+1)(x-3)$
101. $(2x+1)(4x-5)$ **103.** $(4x-3)(x+3)$ **105.** $(2x+5)(2x-1)$
107. $(2x-1)(5x+1)$ **109.** $(2x+1)(3x-7)$ **111.** $(2x-y)(3x+8y)$
113. $(3x^2-1)(x^2+5)$ **115.** $(2x^2+1)(2x+3)(2x-3)$
117. $(2x+3y)(2x-5y)$ **119.** $x^2(2x-3)(2x+7)$
121. $xy(3x+4)(3x-1)$ **123.** $y(6x+1)(x-6)$
125. $3(3x-5)(3x+1)$ **127.** $(3+2x)(1-6x)$ **129.** $(2+3x)(1-4x)$
131. $(3-x)(4+9x)$ **133.** $(4-3x)(1+6x)$
135. $(4x^2+1)(x+4)(x-4)$ **137.** $(x^2+3)(6x+1)(6x-1)$
139. $(2x+1)(2x-1)(3x+2)(3x-2)$ **141.** $(3x^3+1)(2x-1)(4x^2+2x+1)$
143. $(2x^3+1)(x+4)(x^2-4x+16)$
145. $(x-1)(x^2+x+1)(2x-1)(4x^2+2x+1)$
147. $(x-1)(x^2+x+1)(3x+1)(9x^2-3x+1)$
149. $(x-y-3)(x-y-1)$ **151.** $(3x-y-12)(3x-y-1)$
153. $(3x+3y-8)(4x+4y-3)$ **155.** $(12x+12y-1)(x+y+4)$
157. $[3(2x-y)-4][(2x-y)+6]$ **159.** $[3(x-y)-5][7(x-y)+9]$
161. $[8(x+y)-9][(x+y)+1]$ **163.** $(x^2+2x+6)(x^2-2x+6)$
165. $(x^2+3x-2)(x^2-3x-2)$ **167.** $(x^2+5x-7)(x^2-5x-7)$
169. $(x^2+5x+7)(x^2-5x+7)$ **171.** $(2x^2+4x-3)(2x^2-4x-3)$
173. $(2x^2+3x-6)(2x^2-3x-6)$ **175.** $(a-2b)(7x-8)$
177. $(2a+3b)(9y-11z)$ **179.** $6(4a-3b)(5x+2y)$
181. $(2x+y)(2x-y-1)$
183. $(5x-3y)(25x^2+15xy+9y^2-5x-3y)$
185. $(x^2+2y^2)(x^4-2x^2y^2+4y^4+1)$
187. $(x+2+3y)(x+2-3y)$ **189.** $(x-3+2y)(x-3-2y)$
191. $2(8+6x-3y)(8-6x+3y)$ **193.** $(x+2)^2(x^2-2x+4)$
195. $(x-1)^2(x^2+x+1)$

Exercise 6.1, page 170

1. $\frac{1}{x^2}$ **3.** x^5 **5.** $\frac{3x^6}{2}$ **7.** $\frac{2x^3}{3y}$

9. $-\frac{5z}{3xy}$ **11.** $\frac{3x}{4y}$ **13.** $\frac{16}{81x^4y^8}$ **15.** $-\frac{z^6}{x^6y^3}$

17. $-\frac{1}{x^2}$ **19.** $\frac{8}{405x^2y^9}$ **21.** $\frac{x^2}{3(x-y)}$ **23.** $\frac{3x-2}{3-x}$

25. $-\frac{x-4}{x+5}$ **27.** $\frac{1-2x}{1+x}$ **29.** $2x+1$ **31.** $\frac{x-2}{x}$

33. $\frac{3(x-1)}{4x}$ **35.** $\frac{y}{y+1}$ **37.** $\frac{2}{x+1}$ **39.** $\frac{3(x-3)}{x(x+3)}$

41. $\frac{2}{3x}$ **43.** $\frac{x^2+4}{x^2-4}$ **45.** $\frac{2x-y}{3}$ **47.** $\frac{2x-3}{2x+3}$

49. $-(3x+1)$ **51.** x^2-2x+4 **53.** $\frac{2x+1}{4x^2+2x+1}$

55. $\frac{x+1}{x+3}$ **57.** $\frac{x-2}{x-1}$ **59.** $\frac{x+6}{x-4}$

61. $\frac{(x-8)(x+4)}{2(x+8)(x-3)}$ **63.** $\frac{x-12}{x(x+7)}$ **65.** $\frac{x-4}{x+3}$

67. $\frac{x+8y}{x+12y}$ **69.** $\frac{x+1}{x-1}$ **71.** $\frac{x+3}{x+4}$

73. $\frac{x-2}{x-3}$ **75.** $\frac{2x+1}{3x+2}$ **77.** $\frac{2(2x+1)}{4x-3}$ **79.** $\frac{x+3}{x(x+2)}$

81. $\frac{3x-1}{1-2x}$ **83.** $\frac{4+x}{2-x}$ **85.** $-\frac{6x+5}{x+8}$ **87.** $\frac{x^2-x-3}{x^2+3}$

89. $\frac{3x+7}{4x+9y}$ **91.** $\frac{x+3y}{x+3y-2}$ **93.** $\frac{x+y-2}{x-y+2}$ **95.** $\frac{1}{x-4}$

Exercise 6.2A, page 174

1. $\frac{3}{x}$ **3.** $\frac{5}{x^2}$ **5.** $\frac{1}{x}$ **7.** $\frac{x+2}{x^2}$

9. 1 **11.** 2 **13.** 3 **15.** 1

17. 2 **19.** 1 **21.** $\frac{3}{2}$ **23.** $\frac{3}{x}$

25. $\frac{x+1}{2(x+2)}$ **27.** $\frac{1}{9x^2+3x+1}$ **29.** $\frac{3}{x+3}$

31. $\frac{x+2}{2x+1}$ **33.** $\frac{x+4}{x+2}$ **35.** $\frac{x-3}{x-4}$ **37.** $-\frac{x}{x+4}$

39. $\frac{x+2}{x+1}$ 41. $\frac{3x+1}{x-4}$ 43. $\frac{x}{x+y+3}$ 45. $\frac{x}{x+y+1}$

Exercise 6.2B, page 177

1. 120 3. 360 5. 210 7. $6x^2$
9. $60xy^2$ 11. x^4y^3 13. x^4y^3z 15. $2x^2(x+1)$
17. $x^2(x+1)(x-2)$ 19. $(x+3)^2(x-1)^2$
21. $(x-3)(x+1)(x-4)$ 23. $(x+2)(x-6)(x+3)$
25. $(x+2)(x-1)(x+3)$ or $(x+2)(1-x)(x+3)$
27. $24(x+1)$ 29. $12x(x-3)$ 31. $(x^2+4)(x+2)^2$
33. $x^2(x+3)(x-2)$ 35. $12x(3x-1)(2x-1)$
37. $(x+3)(x-2)(x+6)$ 39. $(3x+2)(x-1)(x+4)$
41. $(x-2)(x-3)(x+3)$ 43. $(x^2+x+1)(x-1)^2$
45. $(x+3)(x^2-3x+9)(x-3)^2$

Exercise 6.2C, page 181

1. $\frac{1}{12}$ 3. $\frac{35}{24}$ 5. $\frac{31}{6x}$ 7. $-\frac{1}{10x}$

9. $\frac{x+8}{4x^2}$ 11. $\frac{40x+3}{60x^2}$ 13. $\frac{5y-9x}{2xy}$ 15. $\frac{15y-8x}{12xy}$

17. $\frac{10x+7}{12x}$ 19. $\frac{15x-5}{36x}$ 21. $\frac{x-16}{6x}$ 23. $\frac{5}{14}$

25. $\frac{x^2-2}{(x-1)(x-2)}$ 27. $\frac{x(3x-2)}{(x+2)(x-2)}$ 29. $\frac{2x^2+1}{(2x-1)(x+1)}$

31. $\frac{-3(x-4)}{(x+4)(x-2)}$ 33. $\frac{6}{x-4}$ 35. $\frac{2}{2x-1}$

37. $\frac{3}{x+3}$ 39. $\frac{2x}{(x+1)(x-1)}$ 41. $\frac{4x-13}{(x-3)(x-4)}$

43. $\frac{3x-5}{(x-1)(x-2)}$ 45. $\frac{3x+1}{(2x+3)(x-2)}$ 47. $\frac{(x+7)}{(x-3)(x+2)}$

49. $\frac{x+11}{(x-3)(x+4)}$ 51. $\frac{4}{x-4}$ 53. $\frac{6}{x-4}$

55. $\frac{6}{2x+1}$ 57. $\frac{2}{x-1}$ 59. $\frac{2}{x+3}$ 61. $\frac{5}{2x-1}$

Exercise 6.3, page 184

1. $\frac{1}{9}$ **3.** 1 **5.** $\frac{x^2}{8}$ **7.** $\frac{3}{7x}$

9. $\frac{12b^2x^3}{a^3y^2}$ **11.** $\frac{a^2x}{b^2y^2z}$ **13.** $-x^5y$ **15.** $\frac{5x}{27y}$

17. $\frac{a^2x^5}{36y}$ **19.** $-\frac{x}{108y}$ **21.** $-\frac{189y^6}{2x^2}$ **23.** $\frac{x-4}{2(x+3)}$

25. $\frac{4x-1}{7(2x-1)}$ **27.** $\frac{y(x+1)}{x(x+2)}$ **29.** $x+1$ **31.** $x+3$

33. $\frac{x+6}{x+3}$ **35.** $\frac{x+4}{x+1}$ **37.** $\frac{x-4}{x-5}$ **39.** $\frac{x-3}{x+2}$

41. $\frac{x+3}{x-3}$ **43.** $\frac{4x-1}{x-3}$ **45.** $\frac{3+x}{1-2x}$ **47.** $\frac{3+x}{4-x}$

49. $\frac{2x-1}{2x+3}$ **51.** $\frac{x+y-2}{x-y+1}$ **53.** $\frac{x-3}{x+4}$

Exercise 6.4, page 188

1. $\frac{14}{15}$ **3.** $\frac{1}{3}$ **5.** 12 **7.** $\frac{8x^3y}{5}$

9. $\frac{x}{6ab}$ **11.** $\frac{8}{xy^6}$ **13.** $\frac{bxy^5}{6a^4}$ **15.** $\frac{y^4}{a^2}$

17. $\frac{2}{x^2}$ **19.** abx^2 **21.** $\frac{ax^2y^3}{b^4}$ **23.** $\frac{y}{a}$

25. $\frac{1}{2}$ **27.** $-\frac{3}{4x}$ **29.** 1 **31.** 1

33. 1 **35.** 1 **37.** 1 **39.** $\frac{x+6y}{x-y}$

41. $\frac{x+6}{x+2}$ **43.** $\frac{2x+3}{2x-5}$ **45.** $\frac{2x+3}{3x+2}$ **47.** $-\frac{x+5}{6x+1}$

49. $x-2y$ **51.** $\frac{3x-2}{x-2}$ **53.** $\frac{3x-2}{x-2}$ **55.** 1

Exercise 6.5, page 194

1. $\frac{5x-1}{(x-2)(x+1)}$ **3.** $\frac{x^2+5x}{(4x-1)(3x+1)}$ **5.** $\frac{5x}{(2x-3)(3x-2)}$

7. $\frac{6}{x-1}$ 9. $\frac{x}{2x+3}$ 11. $2(x-3)$

13. $\frac{2}{x(x-2)}$ 15. $\frac{7}{3(x+4)}$ 17. $(x+3)(x-3)$

19. $(x+7)(x-7)$ 21. $(2x-9)(x-6)$ 23. $\frac{x}{x-1}$

25. $\frac{x}{3x-1}$ 27. $\frac{x^2+2x+4}{x^2}$ 29. $\frac{x+2}{x+1}$ 31. $\frac{x+5}{x+4}$

33. $\frac{(x-2)(x+1)}{(x+3)(x-7)}$ 35. 2 37. 25

39. $\frac{2}{3}$ 41. $\frac{2-x}{2x}$ 43. $\frac{x}{x+2}$ 45. $\frac{1-x+x^2}{x^2}$

47. $\frac{x-1}{x+2}$ 49. $-\frac{x+6}{x+3}$ 51. $\frac{x+3}{x+4}$ 53. $\frac{x-2}{x+5}$

55. $\frac{3x-1}{3x+1}$ 57. $\frac{x+2}{x+3}$ 59. $\frac{x+1}{x-2}$ 61. $\frac{x+2}{x-1}$

63. $\frac{x-3}{x+2}$ 65. $\frac{x+2}{x+1}$ 67. $\frac{x+2}{x+1}$ 69. $\frac{x-1}{x+1}$

71. $\frac{(x+4)(x+3)}{(x-5)(x+5)}$ 73. $\frac{x^2+4}{4x}$ 75. $x+2$

77. $\frac{(x-4)(x-2)}{x-6}$

Exercise 6.6, page 198

1. $\left\{\frac{3-y}{2}\right\}$ 3. $\left\{\frac{y+4}{2}\right\}$ 5. $\left\{\frac{4y-8}{3}\right\}$

7. $\left\{\frac{-3y-10}{4}\right\}$ 9. $\left\{\frac{y-6}{3}\right\}$ 11. $\left\{\frac{2y+3}{5}\right\}$

13. $\left\{\frac{a+2}{a} \middle| a \neq 0\right\}$ 15. $\left\{\frac{2a-3}{2a} \middle| a \neq 0\right\}$ 17. $\left\{\frac{3a+b}{a} \middle| a \neq 0\right\}$

19. $\left\{\frac{y+3a}{b} \middle| b \neq 0\right\}$ 21. $\left\{\frac{1}{a} \middle| a \neq 0\right\}$ 23. $\left\{\frac{2a}{a+3} \middle| a \neq -3\right\}$

25. $\left\{\frac{3a}{2-3a} \middle| a \neq \frac{2}{3}\right\}$ 27. $\left\{\frac{4a}{4a+1} \middle| a \neq -\frac{1}{4}\right\}$ 29. $\left\{\frac{3}{2a+b} \middle| b \neq -2a\right\}$

31. $\left\{\frac{a}{a+2b} \middle| a \neq -2b\right\}$ **33.** $\left\{\frac{3b}{a+3b} \middle| a \neq -3b\right\}$ **35.** $\left\{b \middle| a \neq \frac{3}{2}\right\}$

37. $\left\{-2 \middle| a \neq \frac{1}{4}\right\}$ **39.** $\left\{2a+1 \middle| a \neq \frac{1}{2}\right\}$

41. $\{a^2 - a + 1 \mid a \neq -1\}$ **43.** $\left\{\frac{b}{2a+6b} \middle| a \neq -3b\right\}$

45. $\{5 \mid a \neq -3\}$ **47.** $\left\{2 \middle| a \neq \frac{1}{2}\right\}$ **49.** $\left\{\frac{1}{3} \middle| a \neq -2\right\}$

51. $\left\{\frac{a-2}{2} \middle| a \neq -2\right\}$ **53.** $\left\{1 + 2a + 4a^2 \middle| a \neq \frac{1}{2}\right\}$

55. $\{a^2 - 2ab + 4b^2 \mid a \neq -2b\}$ **57.** $\left\{\frac{2a-3b}{2} \middle| a \neq -\frac{3}{2}b\right\}$

59. $\{a - 1 \mid a \neq -4\}$ **61.** $\{3a + 2 \mid a \neq 3\}$ **63.** $\left\{2a - 1 \middle| a \neq -\frac{1}{3}\right\}$

65. $\frac{2A}{b}$ **67.** $\frac{Fd^2}{km_2}$ **69.** $\frac{9}{5}C + 32$

71. $\frac{A-P}{Pt}$; $\frac{A}{1+rt}$ **73.** $\frac{2S_n}{a_1 + a_n}$; $\frac{2S_n}{n} - a_n$

75. $\frac{a_n - a_1}{n-1}$; $\frac{a_n - a_1 + d}{d}$ **77.** $a_n r - S_n(r-1)$; $\frac{S_n - a_1}{S_n - a_n}$

Exercise 6.7, page 201

1. $\{1\}$ **3.** $\left\{\frac{2}{9}\right\}$ **5.** $\{-8\}$ **7.** $\{2\}$

9. $\left\{-\frac{1}{3}\right\}$ **11.** $\left\{\frac{5}{4}\right\}$ **13.** $\left\{\frac{7a}{4}\right\}$ **15.** $\left\{\frac{a}{9}\right\}$

17. $\left\{\frac{2}{7}\right\}$ **19.** $\left\{\frac{6}{11}\right\}$ **21.** $\{11\}$ **23.** $\{9\}$

25. $\left\{\frac{1}{2}\right\}$ **27.** $\left\{\frac{3}{2}\right\}$ **29.** $\{-8\}$

31. $\{7\}$ **33.** $\{-3\}$ **35.** $\{-1\}$

37. $\{-5\}$ **39.** $\varnothing$ **41.** $\varnothing$

43. $\left\{x \middle| x \in R, x \neq 3, x \neq \frac{3}{2}\right\}$ **45.** $\left\{x \middle| x \in R, x \neq -2, x \neq \frac{2}{3}\right\}$

47. $\left\{\frac{5}{2}\right\}$ **49.** $\left\{-\frac{5}{4}\right\}$ **51.** $\left\{-\frac{1}{4}\right\}$

53. $\left\{\frac{7}{2}\right\}$ **55.** $\left\{-\frac{51}{2}\right\}$ **57.** $\left\{\frac{9}{2}\right\}$

59. $\{x \mid x \in R, x \neq -3, x \neq -1, x \neq 2\}$ **61.** $\varnothing$

Exercise 6.8, page 207

1. 19 **3.** 12 **5.** 13 **7.** $\frac{7}{10}$

9. $\frac{3}{8}$ **11.** $\frac{18}{59}$ **13.** 59; 13 **15.** 131; 43

17. 59 **19.** 74 **21.** 249 **23.** 21 hours

25. 144 hours **27.** 20 hours; 30 hours **29.** 12 minutes

31. 12 minutes **33.** 4 minutes; 12 minutes

35. 54 mph

Chapter 6 Review, page 209

1. $\frac{9x^2}{4y^3}$ **3.** $\frac{z^7}{y}$ **5.** $-\frac{x+1}{x(1+x+x^2)}$

7. $-\frac{2x+3}{3x+4}$ **9.** $\frac{25}{12x}$ **11.** $\frac{17}{12}$

13. $\frac{5x}{(x-2)(x+3)}$ **15.** $\frac{3}{x-2}$ **17.** $\frac{4}{x-4}$

19. $\frac{3x+5}{(x+3)(x+1)}$ **21.** $\frac{2x-7}{(x-4)(x-3)}$ **23.** $\frac{3}{x+1}$

25. $\frac{4}{x-2}$ **27.** $\frac{5}{2x+1}$ **29.** $\frac{2}{3x+2}$ **31.** $\frac{8z^2}{9ac^3x}$

33. $\frac{c^3xy^3}{a^2b^3}$ **35.** $\frac{8cx^4}{3a^4yz^3}$ **37.** $\frac{2(2x-1)}{x(x+3)}$ **39.** $x-1$

41. $\frac{x+2}{x+1}$ **43.** $\frac{x-3}{2x+1}$ **45.** $\frac{5y^5}{8ab^2c^3}$ **47.** $\frac{x-3}{x+3}$

49. 1 **51.** $\frac{x-3}{x+3}$ **53.** $-\frac{3x+1}{x+6}$

55. $\dfrac{(x-3)(x+2)}{(2x+1)(x+4)}$ **57.** $\dfrac{(x-5)(2x+1)}{(x-2)(x+3)}$ **59.** $\dfrac{x-1}{(x-4)(x-3)}$

61. $\dfrac{9x-20}{(3x-2)(4x-5)}$ **63.** $(x+1)(x+3)$ **65.** $(x-1)(x+3)$

67. $\dfrac{(x+4)(2x+1)}{(x+6)(x+3)}$ **69.** $\dfrac{(x+2)(x+3)}{(2x+3)(x+1)}$ **71.** $\dfrac{x-3}{x-2}$

73. $\dfrac{x-2}{x+2}$ **75.** $\dfrac{2x-3}{2x+5}$ **77.** $\dfrac{(x-1)(x+3)}{(3x-2)(x+1)}$

79. $x+3$ **81.** $\left\{\dfrac{5a}{a+3}\,\middle|\, a \neq -3\right\}$ **83.** $\{4 \mid a \neq 2\}$

85. $\left\{2a-1 \,\middle|\, a \neq -\dfrac{1}{2}\right\}$ **87.** $\left\{a+b \,\middle|\, a \neq \dfrac{2}{3}b\right\}$

89. $\left\{\dfrac{a-2}{2}\,\middle|\, a \neq -3\right\}$ **91.** $\{7\}$

93. $\{-1\}$ **95.** $\varnothing$

97. $\left\{x \,\middle|\, x \in R, x \neq -3, x \neq \dfrac{5}{2}\right\}$ **99.** $\left\{\dfrac{4}{3}\right\}$

101. $\{1\}$ **103.** $\{-2\}$ **105.** 12

107. $\dfrac{11}{15}$ **109.** 46; 187 **111.** 10 hours

113. 10 minutes; 15 minutes

Exercise 7.1A, page 218

1. 4 **3.** 25 **5.** $2^{\frac{5}{6}}$ **7.** $2^{\frac{5}{2}}$

9. $2^{\frac{7}{2}}$ **11.** $x^{\frac{7}{5}}$ **13.** $x^{\frac{2}{3}}$ **15.** $x^{\frac{5}{2}}$

17. $x^{\frac{3}{2}}$ **19.** $3a^{\frac{5}{2}}b$ **21.** $a^2b^{\frac{1}{2}}$ **23.** $a^{\frac{3}{2}}b^{\frac{4}{3}}$

25. $6ab$ **27.** $a^4b^{\frac{2}{3}}$ **29.** $5x^{\frac{5}{6}}y^{\frac{7}{4}}$ **31.** $6a^{\frac{5}{2}}b^{\frac{7}{6}}$

33. $4a^{\frac{5}{2}}b^{\frac{7}{12}}$ **35.** 2 **37.** 625 **39.** x^2

41. x^6 **43.** x^6 **45.** x^6 **47.** x^{2n}

49. $x^{\frac{3}{2}}$ **51.** $x^{\frac{1}{3}}$ **53.** x^3 **55.** 2

57. 6 **59.** 3 **61.** 5 **63.** 2

65. 4 **67.** 32 **69.** 32

71. Not a real number **73.** -2 **75.** 8

77. 108 **79.** $4a^3b^{\frac{7}{2}}$ **81.** $81a^{16}b^9$ **83.** a^4b^7

85. $175a^3b^4$ **87.** $288a^7b^{14}$ **89.** $a^{\frac{1}{2}}b^{\frac{3}{4}}$ **91.** ab^3

93. $4ab$ **95.** $a^{3n+4m}b^{2m+6n}$ **97.** $a^{\frac{3}{2}} - a$

99. $a^{\frac{7}{3}} + 2a^2$ **101.** $2a^{\frac{5}{4}} - 3a^{\frac{1}{4}}$ **103.** $a + a^{\frac{1}{2}}b^{\frac{1}{2}}$ **105.** $ab^{\frac{2}{3}} - a^{\frac{2}{3}}b$

107. $6a^{\frac{1}{2}}b^{\frac{1}{2}} - 12a^{\frac{5}{6}}b^{\frac{5}{6}}$ **109.** $a^{\frac{11}{4}} - 2a^{\frac{5}{2}} + a^{\frac{9}{4}}$

111. $2x^{\frac{3}{2}} + 4x + x^{\frac{1}{2}} + 2$ **113.** $x - y$

115. $20x^{\frac{4}{3}} - 14x^{\frac{2}{3}}y^{\frac{1}{4}} + 2y^{\frac{1}{2}}$ **117.** $a - 4a^{\frac{1}{2}}b^{\frac{1}{2}} + 4b$

119. $x + y$ **121.** $8x^2 - 27y^2$ **123.** $x + 5x^{\frac{1}{2}} + 9$

125. $x^{\frac{4}{3}} + 6x + 9x^{\frac{2}{3}} - 1$ **127.** $3x^{\frac{4}{3}} + 2x - 6x^{\frac{2}{3}} - x^{\frac{1}{3}} + 2$

129. $x^{\frac{1}{2}} + 3x^{\frac{1}{3}} + 3x^{\frac{1}{6}} + 1$

Exercise 7.1B, page 222

1. $3^{\frac{3}{4}}$ **3.** $7^{\frac{5}{3}}$ **5.** $2^{\frac{1}{6}}$ **7.** $\frac{1}{2^{\frac{1}{2}}}$

9. $\frac{1}{3^{\frac{1}{8}}}$ **11.** $\frac{1}{3^{\frac{3}{2}}}$ **13.** $x^{\frac{1}{3}}$ **15.** $x^{\frac{4}{3}}$

17. $\frac{1}{x^{\frac{1}{2}}}$ **19.** $\frac{3a}{4b}$ **21.** $\frac{3a^{\frac{1}{2}}}{2b^{\frac{1}{2}}}$ **23.** $\frac{b^{\frac{1}{3}}}{a^{\frac{3}{2}}c^{\frac{4}{3}}}$

25. $\frac{a^{\frac{1}{3}}}{b^{\frac{1}{2}}}$ **27.** $\frac{a^{\frac{1}{2}}}{b^2}$ **29.** $a^{\frac{3}{4}}$ **31.** $\frac{1}{a^{\frac{3}{2}}}$

33. $\frac{1}{a^{\frac{1}{12}}}$ **35.** ab **37.** $\frac{b^{\frac{7}{4}}}{a^4}$ **39.** $\frac{a^{\frac{1}{3}}}{b^{\frac{1}{6}}}$

41. 1 **43.** $2^{\frac{1}{2}}$ **45.** 81 **47.** $\frac{64}{a^2b}$

49. $\frac{a}{b^2}$ 51. $\frac{1}{4a^{\frac{5}{2}}b^5}$ 53. $\frac{125a}{32b^7c^2}$ 55. $\frac{1}{x^{\frac{1}{8}}y^{\frac{1}{5}}}$

57. $\frac{1}{x^{\frac{4}{3}}}$ 59. $x^{\frac{1}{2}} - x^{\frac{3}{2}}$ 61. $2x^{\frac{2}{3}} - x^{\frac{1}{3}}$ 63. $-2b^{\frac{1}{2}} + 3c^{\frac{2}{3}}$

65. $6b^{\frac{1}{3}} - 7b^{\frac{1}{6}} + 4$ 67. $x^{\frac{2}{9}} - 5x^{\frac{1}{9}} + 6$ 69. $2x^{\frac{1}{2}} + 5$

71. $4x^{\frac{1}{2}} - 5y^{\frac{1}{2}}$ 73. $x^{\frac{1}{4}} + y^{\frac{1}{4}}$ 75. $2x^{\frac{3}{4}} - 3y^{\frac{3}{4}}$ 77. $x + x^{\frac{1}{2}} + 3$

79. $x + 4x^{\frac{1}{2}} - 2$ 81. $3x + 3x^{\frac{1}{2}} - 2$ 83. $4x + 2x^{\frac{1}{2}} - 3$

85. $x^{\frac{3}{4}} + x^{\frac{1}{2}} + 2$ 87. $3x^{\frac{4}{3}} + x^{\frac{1}{3}} - 3$ 89. $x^{\frac{3}{2}} - xy^{\frac{1}{2}} + x^{\frac{1}{2}}y - y^{\frac{3}{2}}$

Exercise 7.2, page 230

1. 1 3. 1 5. 1
7. Indeterminate 9. 3 11. 9
13. 1 15. 1 17. 5 19. a^5
21. $6 - 3a$ 23. 27 25. 1

27. $4a^2 + 4a + 1$ 29. $9a^2 - 12a + 4$ 31. $\frac{3}{2}$

33. $\frac{3}{a}$ 35. $\frac{a^2}{b}$ 37. $\frac{1}{4a^3}$ 39. $\frac{1}{x^5y^2}$

41. $\frac{x}{3y}$ 43. $\frac{1}{3} - \frac{1}{25}$ 45. $\frac{1}{x^2} - \frac{1}{y}$ 47. $\frac{2}{x} - \frac{3}{y^2}$

49. $\frac{49}{4}$ 51. $\frac{1}{64}$ 53. 1 55. $\frac{1}{9}$

57. $\frac{1}{243}$ 59. x^3 61. x^n 63. $\frac{1}{x^{3n}}$

65. $\frac{3y}{2x}$ 67. $-\frac{2y}{9x^3}$ 69. $-\frac{3}{8x^2}$ 71. $-\frac{25x^2y^3}{4}$

73. $\frac{2b}{a^3c}$ 75. $-\frac{3a^4}{2b^3c^3}$ 77. $\frac{1}{81}$ 79. 9

81. $\frac{1}{x^9}$ 83. x^{12} 85. $\frac{1}{x^{3n}}$ 87. x^{2n^2}

89. $\frac{y^2}{9x^4}$ 91. $-\frac{y^3}{4x^2}$ 93. $\frac{x^{\frac{1}{2}}y}{2}$ 95. $\frac{y^{\frac{3}{2}}}{2x}$

97. $\dfrac{xy^4}{729z^4}$ **99.** $\dfrac{1}{3x^2y}$ **101.** $x^{\frac{1}{2}}y^{\frac{1}{2}} + 1 - \dfrac{1}{x^{\frac{1}{2}}y^{\frac{1}{2}}}$

103. $\dfrac{1}{x^2} - \dfrac{1}{y^2}$ **105.** $\dfrac{2}{x^3} - \dfrac{7}{x^2} + \dfrac{2}{x} + 3$ **107.** $\dfrac{3}{x^3} - \dfrac{11}{x^2} - \dfrac{6}{x} + 8$

109. $\dfrac{4}{x^2} + \dfrac{4}{xy} + \dfrac{1}{y^2}$ **111.** $\dfrac{1}{x} + \dfrac{8}{x^{\frac{1}{2}}y^{\frac{1}{2}}} + \dfrac{16}{y}$

113. $\dfrac{1}{x^2} - \dfrac{3}{x^{\frac{4}{3}}y^{\frac{2}{3}}} + \dfrac{3}{x^{\frac{2}{3}}y^{\frac{4}{3}}} - \dfrac{1}{y^2}$ **115.** 2

117. $\dfrac{27}{2}$ **119.** 32 **121.** $\dfrac{1}{8}$ **123.** $3 + 9 = 12$

125. $27 - 16 = 11$ **127.** $\dfrac{a^2b^2}{3}$ **129.** $\dfrac{2b}{3a^2}$

131. $\dfrac{1}{x^6}$ **133.** $\dfrac{1}{x^4}$ **135.** x^3y^2 **137.** $\dfrac{y^4}{x^6}$

139. $\dfrac{12x^5}{y^4}$ **141.** $-\dfrac{y^9}{2x^4}$ **143.** $\dfrac{4x^3}{9y^7}$ **145.** $\dfrac{x^6}{y^{16}}$

147. $\dfrac{3x^3}{2y^5}$ **149.** $\dfrac{8y^9}{27x^6}$ **151.** $\dfrac{9x^2}{4y^2}$ **153.** $\dfrac{z^3}{xy^5}$

155. $\dfrac{x^3}{6y^2}$ **157.** $\dfrac{x^{10}y^5}{32z^{13}}$ **159.** $\dfrac{224z^7}{81x^2y^9}$ **161.** $\dfrac{16x^{13}a^6}{9y^8}$

163. $\dfrac{9x^2}{a^4y^4}$ **165.** $\dfrac{xy^2}{3y^2 - x}$ **167.** $y + x$ **169.** x^2y

171. $\dfrac{1}{3x}$ **173.** $2x - 1$ **175.** $\dfrac{x-2}{2x-5}$ **177.** $\dfrac{x-2}{x-3}$

179. $\dfrac{1 - x + x^2}{x^2}$ **181.** $\dfrac{16y^4 - 4x^2y^2 + x^4}{x^4y^4}$

183. $\dfrac{x}{x+1}$ **185.** x **187.** $\dfrac{x-2}{2(x-1)^{\frac{3}{2}}}$

189. $\dfrac{2x+6}{3(x+2)^{\frac{4}{3}}}$ **191.** $\dfrac{2x+9}{3(x+4)^{\frac{4}{3}}}$ **193.** $-\dfrac{3}{x^2(x^2+3)^{\frac{1}{2}}}$

195. $\dfrac{3x^3 - 2x}{(3x^2-1)^{\frac{3}{2}}}$ **197.** $\dfrac{6 - 5x^2}{3x^4(x-1)^{\frac{2}{3}}}$ **199.** $\dfrac{2}{x^{\frac{1}{2}}(x+4)^{\frac{3}{2}}}$

201. $\dfrac{2}{3x^{\frac{1}{3}}(x^2+1)^{\frac{4}{3}}}$

203. $\dfrac{3}{(x+4)^{\frac{3}{2}}(x-2)^{\frac{1}{2}}}$

205. $-\dfrac{4x}{(x^2-3)^{\frac{3}{2}}(x^2+1)^{\frac{1}{2}}}$

207. 2.7×10^1

209. 4.5×10^3 **211.** 1.3×10^0 **213.** 9.21×10^{-1}

215. 7.4×10^{-2} **217.** 4.9×10^{-3}

Chapter 7 Review, page 234

1. 5 **3.** 9 **5.** 64 **7.** 32

9. $a^{\frac{7}{3}}$ **11.** $a^{\frac{3}{4}}$ **13.** a^3 **15.** a^2

17. $-a^2$ **19.** $a^{\frac{1}{3}}$ **21.** $a^{\frac{7}{2}}b^{\frac{7}{2}}$ **23.** ab

25. $a^2b^3c^4$ **27.** $6a^{\frac{5}{6}}b^{\frac{13}{20}}c^{\frac{7}{6}}$ **29.** $4a^{\frac{13}{3}}b^{\frac{7}{2}}$ **31.** $256a^{\frac{7}{2}}b^{\frac{11}{2}}$

33. $5^{\frac{1}{2}}x^{\frac{3}{2}}y^{\frac{3}{2}}z^{\frac{3}{2}}$ **35.** $250a^{13}b^5$ **37.** $-2a^{\frac{13}{4}}$ **39.** 0

41. $5x^{\frac{5}{2}}$ **43.** $15x - 7x^{\frac{1}{2}} - 2$

45. $24x^{\frac{1}{3}} + 2x^{\frac{1}{6}}y^{\frac{1}{3}} - y^{\frac{2}{3}}$ **47.** $x^{\frac{2}{3}} - 12x^{\frac{1}{3}} + 16$

49. $\dfrac{1}{a^{\frac{1}{6}}}$ **51.** $\dfrac{1}{a^{\frac{2}{9}}}$ **53.** $\dfrac{a^{15}}{b^4}$ **55.** $\dfrac{a^6}{b^4}$

57. $\dfrac{a^{\frac{1}{4}}}{b^{\frac{1}{2}}}$ **59.** $a^{\frac{1}{7}}b^{\frac{1}{5}}$ **61.** $\dfrac{2}{5a^{\frac{1}{6}}b^{\frac{1}{4}}c^2}$ **63.** $\dfrac{27a^2b}{2}$

65. $\dfrac{a^2b}{c}$ **67.** $\dfrac{729b}{64a}$ **69.** $3x^{\frac{1}{2}} + 4y^{\frac{1}{2}}$

71. $x^{\frac{3}{2}} + 2x - 3x^{\frac{1}{2}}$ **73.** $x^{\frac{1}{2}} + 2x^{\frac{1}{4}} + 4$ **75.** $\dfrac{b}{3a^3c}$

77. $\dfrac{a^7}{32b^{11}}$ **79.** $\dfrac{a^8c}{432b^3}$ **81.** $\dfrac{b^5}{2a^2c^4}$ **83.** $\dfrac{9a}{4b^2c^3}$

85. $\dfrac{3}{x^2} + \dfrac{5}{xy} - \dfrac{2}{y^2}$ **87.** $\dfrac{1}{9x^2} - \dfrac{4}{3xy} + \dfrac{4}{y^2}$

89. $\dfrac{1}{a^4} - \dfrac{3}{a^{\frac{8}{3}}b^{\frac{4}{3}}} + \dfrac{3}{a^{\frac{4}{3}}b^{\frac{8}{3}}} - \dfrac{1}{b^4}$ **91.** $\dfrac{64b^{32}}{9a^{12}}$

93. $\frac{4}{3a^2}$ **95.** $\frac{2b^3}{5a^2}$ **97.** $\frac{1}{9}$ **99.** $\frac{1}{3^{14}}$

101. $2 + 3x$ **103.** $\frac{5 + x}{3 + 4x}$ **105.** $\frac{3 - x}{1 + 4x}$ **107.** 1

109. $\frac{x - 4}{2(x - 2)^{\frac{3}{2}}}$ **111.** $\frac{8x - 9}{3(4x - 3)^{\frac{4}{3}}}$ **113.** $\frac{3}{(x^2 + 3)^{\frac{3}{2}}}$

Exercises 8.1–8.2, page 242

1. 3 **3.** Not a real number **5.** 2

7. -5 **9.** 2 **11.** 2 **13.** 2

15. x **17.** x^4y **19.** $(x + 2)$ **21.** xy^2

23. x^4y^3 **25.** x^4 **27.** $-x^5$ **29.** $\sqrt{5}$

31. $\sqrt{11}$ **33.** $\sqrt[3]{7}$ **35.** $\sqrt{5}$ **37.** $2\sqrt{3}$

39. $2\sqrt{6}$ **41.** $12\sqrt{2}$ **43.** $4\sqrt{3}$ **45.** $4\sqrt{15}$

47. $-6\sqrt{2}$ **49.** $-12\sqrt{5}$ **51.** $6\sqrt{3}$ **53.** 5

55. $2\sqrt{10}$ **57.** $2\sqrt[3]{4}$ **59.** $-3\sqrt[3]{2}$ **61.** $3\sqrt[3]{3}$

63. $5\sqrt[3]{2}$ **65.** $\sqrt[3]{35}$ **67.** $\sqrt[3]{19}$ **69.** $2\sqrt[4]{2}$

71. $2\sqrt[4]{8}$ **73.** $3\sqrt[4]{2}$ **75.** $-2\sqrt[5]{3}$ **77.** $2\sqrt[3]{2}$

79. $2\sqrt[3]{3}$ **81.** $x^2\sqrt{x}$ **83.** $x^4\sqrt{2x}$ **85.** $2xy\sqrt{x}$

87. $y^4\sqrt{6x}$ **89.** $3xy^2\sqrt{xy}$ **91.** $xy^3z\sqrt{xy}$ **93.** $3x^2y^3\sqrt{3x}$

95. $x\sqrt{x + 3}$ **97.** $x^2(x + 1)^2\sqrt{x + 1}$ **99.** $\sqrt{x^2 + 4}$

101. $-x^3\sqrt[3]{2}$ **103.** $2xy^2\sqrt[3]{x}$ **105.** $-x^2y\sqrt[3]{z}$

107. $-x^2y^2z\sqrt[3]{x^2}$ **109.** $\sqrt[3]{8x^3 + 27}$ **111.** $x\sqrt[4]{x^2y}$

113. $2xy\sqrt[4]{2y}$ **115.** $\sqrt[4]{x^4 + y^4}$ **117.** $-x^2y\sqrt[5]{y}$ **119.** $xy\sqrt[3]{y}$

121. $125\sqrt[n]{5}$ **123.** $4\sqrt[n]{2}$ **125.** $2\sqrt[5]{8}$ **127.** $xy^2\sqrt[n]{x^2y}$

129. $x^2y\sqrt[3n]{y^{2n+1}}$ **131.** $x^2y^2\sqrt{y}$ **133.** $x^2y^3\sqrt[n]{xy^2}$

135. $x\sqrt{1 - y^2}$ **137.** $x\sqrt[3]{y + 1}$ **139.** $3\sqrt[4]{x^2 - y^2}$

Exercise 8.3, page 244

1. $2\sqrt{2}$ **3.** $-5\sqrt[3]{5}$ **5.** $7\sqrt{x}$ **7.** 0

9. $5 - \sqrt{2}$ **11.** $-7\sqrt{3}$ **13.** $4\sqrt{2} + 2\sqrt{3}$ **15.** $\sqrt[3]{3}$

17. $5\sqrt[3]{4}$ **19.** $-2\sqrt{2} - 8\sqrt[3]{3}$ **21.** $2\sqrt{6}$

23. $5\sqrt{5} - 4\sqrt[3]{5}$ **25.** $7x\sqrt{x}$ **27.** $6\sqrt{3x} + 9\sqrt{2x}$

29. 0 **31.** $4x\sqrt[3]{x^2y}$

33. $(6x - 3y)\sqrt{x} + (2x - 6y)\sqrt[3]{2x}$ **35.** $8x^2y\sqrt{xy} + xy\sqrt[3]{xy}$

37. $4xy\sqrt{3x}$ **39.** $8x\sqrt{3x} - 2x\sqrt[3]{3x}$ **41.** 0

43. 0
45. $4\sqrt{x^2 - y^2}$
47. $(5xy + 8x - 7y)\sqrt{7x}$
49. $(2xy + 5y^2 - 6x)\sqrt[n]{x}$
51. $-x\sqrt{2x}$
53. $10y^2\sqrt[n]{2x}$

Exercise 8.4, page 248

1. $\sqrt{6}$
3. $\sqrt{14}$
5. $\sqrt[3]{4}$
7. $\sqrt[3]{12}$
9. 2
11. 10
13. 2
15. 5
17. $10\sqrt{3}$
19. $-3\sqrt{7}$
21. $7\sqrt{2}$
23. $3\sqrt{10}$
25. $5\sqrt{6}$
27. $4\sqrt[3]{3}$
29. $-5\sqrt[3]{3}$
31. $-11\sqrt[3]{2}$
33. $3\sqrt[4]{2}$
35. $-2\sqrt[5]{3}$
37. $2\sqrt[n]{8}$
39. $2\sqrt[4]{2}$
41. $2\sqrt[6]{108}$
43. $5\sqrt[6]{540}$
45. $3\sqrt[2n]{243}$
47. $\sqrt{6xy}$
49. $2x$
51. $x + 1$
53. $x\sqrt{6y}$
55. $x\sqrt[3]{2}$
57. $2x\sqrt[3]{3y^2}$
59. $x\sqrt[5]{x}$
61. $\sqrt{x^2 + x}$
63. $3\sqrt{2x + 3}$
65. $3\sqrt{x(x - 1)}$
67. $\sqrt{(x + 1)(x + 2)}$
69. $\sqrt{(x + 3)(x - 3)}$
71. $(x - 2)\sqrt{x + 2}$
73. $(x - 4)\sqrt{x^2 + 4x + 16}$
75. $x^2y\sqrt[n]{x^3y^2}$
77. $xy\sqrt[6]{xy^2}$
79. $xy\sqrt[12]{16x^5y}$
81. $2xy^2\sqrt[2n]{8x^3}$
83. $3 - \sqrt{3}$
85. $4\sqrt{3} + 6$
87. $3 - \sqrt{6}$
89. $2\sqrt{3} + 2\sqrt{5}$
91. $3\sqrt{5} - 3\sqrt{7}$
93. $2\sqrt{35} - 5\sqrt{6}$
95. $3\sqrt[3]{7} - 3\sqrt[3]{4}$
97. $3\sqrt[3]{5} + 5\sqrt[3]{9}$
99. $3\sqrt{2x} - 6\sqrt{y}$
101. $x\sqrt{2} + \sqrt{x}$
103. $2x\sqrt{5} + 2\sqrt{xy}$
105. $x\sqrt{y} - y\sqrt{x}$
107. $6x\sqrt{15y} - 5y\sqrt{2x}$
109. $2\sqrt{3x(x - 1)} + 3\sqrt{2x(x + 1)}$
111. $2x - 4 + 6\sqrt{(x - 2)(x + 1)}$
113. $3a\sqrt[3]{11b^2} - 3b\sqrt[3]{5ab}$
115. $3b\sqrt[3]{2a^2} + 3a\sqrt[3]{b^2}$
117. 1
119. -3
121. 11
123. -1
125. 10
127. $13 + 7\sqrt{6}$
129. $43 + 6\sqrt{15}$
131. $10 - 7\sqrt{6}$
133. $8 + 2\sqrt{15}$
135. $39 - 8\sqrt{14}$
137. 11
139. $x - 1$
141. $x - 9$
143. $x - 3$
145. $x - y$
147. $2x + 3y + 5\sqrt{xy}$
149. $x + y + 2\sqrt{xy}$
151. $x + y - 2\sqrt{xy}$
153. x
155. $x - 7$
157. $x + 2\sqrt{x - 1}$
159. $2x + 3 - 4\sqrt{2x - 1}$
161. $x + 15 - 8\sqrt{x - 1}$
163. $16x - 39 - 24\sqrt{x - 3}$
165. 1
167. $2x + 2$
169. $x + 2 - 5\sqrt{x(x - 1)}$
171. $2x - 1 + 2\sqrt{x(x - 1)}$
173. $6x + 12 + 4\sqrt{2x(x + 3)}$
175. $10x - 27 - 6\sqrt{x(x - 3)}$
177. $2x + 2\sqrt{(x + 1)(x - 1)}$
179. $5x + 11 + 4\sqrt{(x + 3)(x - 1)}$
181. $10x - 3 - 4\sqrt{(2x + 1)(2x - 1)}$
183. $a + b$
185. $6 - 4\sqrt{5}$
187. $7 + 2\sqrt{15}$

Exercise 8.5, page 256

1. $\sqrt{3}$ **3.** $2\sqrt{5}$ **5.** 2 **7.** 4

9. $\sqrt[3]{4}$ **11.** 2 **13.** $\frac{2\sqrt{3}}{3}$ **15.** $\frac{3\sqrt{5}}{5}$

17. $\frac{\sqrt{2}}{2}$ **19.** $\frac{\sqrt{5}}{2}$ **21.** $\frac{2\sqrt{3}}{3}$ **23.** $\frac{2\sqrt[3]{9}}{3}$

25. $\frac{3\sqrt[3]{2}}{2}$ **27.** $\sqrt[3]{4}$ **29.** $\sqrt[4]{8}$ **31.** $\frac{\sqrt[4]{2}}{2}$

33. $2\sqrt[5]{16}$ **35.** $\frac{\sqrt[5]{27}}{3}$ **37.** $-3\sqrt[5]{4}$ **39.** $4\sqrt[5]{2}$

41. $\frac{\sqrt{10}}{2}$ **43.** $\frac{\sqrt{10}}{5}$ **45.** $\frac{\sqrt{3}}{3}$ **47.** $\frac{\sqrt{5}}{5}$

49. $\frac{\sqrt{10}}{5}$ **51.** $\frac{2\sqrt{3}}{3}$ **53.** $\frac{2\sqrt{15}}{15}$ **55.** $\frac{\sqrt[3]{4}}{2}$

57. $\frac{\sqrt[3]{2}}{2}$ **59.** $\frac{\sqrt[3]{175}}{5}$ **61.** $\frac{\sqrt{6}}{6}$ **63.** $\frac{\sqrt[6]{648}}{3}$

65. $\frac{\sqrt[6]{72}}{2}$ **67.** $\frac{\sqrt[4]{3}}{3}$ **69.** $\frac{\sqrt[6]{32}}{2}$ **71.** $\frac{\sqrt{3x}}{x}$

73. $\frac{2\sqrt{x}}{x}$ **75.** $\frac{\sqrt{2x}}{2}$ **77.** $\frac{\sqrt{3x}}{x}$ **79.** $\frac{\sqrt{6}}{3}$

81. $\frac{\sqrt{6xy}}{3y}$ **83.** $\frac{2\sqrt{y}}{3xy^2}$ **85.** $\frac{2x\sqrt{3ax}}{9a^2b}$ **87.** $\frac{2x\sqrt{5ay}}{5ab^2}$

89. $\frac{2\sqrt{x-2}}{x-2}$ **91.** $\frac{\sqrt{2(x-1)}}{x-1}$ **93.** $\frac{\sqrt{2x+1}}{2x+1}$

95. $\frac{\sqrt{2x(2x-1)}}{2x-1}$ **97.** $\frac{\sqrt{2(x+3)}}{x+3}$ **99.** $\frac{\sqrt{x-3}}{x-3}$

101. $\frac{\sqrt{(x+2)(x+3)}}{x+3}$ **103.** $\frac{\sqrt[3]{3x^2y^2}}{xy}$ **105.** $\frac{\sqrt[4]{8x^2y}}{2x}$

107. $\frac{\sqrt[3]{9x^2}}{3x}$ **109.** $\frac{\sqrt[3]{18x^2}}{3x}$ **111.** $\frac{ab^2\sqrt[4]{axy^3}}{xy^3}$

113. $\frac{y^2\sqrt[5]{3x^2b^3}}{3ab^3}$ **115.** $\frac{a\sqrt[5]{ac^2(a^2-b^2)}}{c^2(a+b)^2}$

117. $\frac{bc^2\sqrt[3]{ac(b^2+c^2)^2}}{a^2(b^2+c^2)}$ **119.** $\frac{\sqrt[6]{2xy^5}}{y}$

121. $\dfrac{\sqrt[4]{4x^3}}{4x}$ **123.** 5 **125.** $3\sqrt{2}-\sqrt{7}$ **127.** $\sqrt{5}-\sqrt{6}$

129. $\sqrt{2}-\sqrt{3}$ **131.** $\dfrac{\sqrt{35}}{7}-\dfrac{\sqrt{15}}{3}$ **133.** $\dfrac{2\sqrt{x}}{x}+1$

135. $\dfrac{\sqrt{2y}}{2y}+\dfrac{\sqrt{2x}}{2x}$ **137.** $\dfrac{\sqrt{10y}}{5y}+\dfrac{\sqrt{6x}}{3x}$ **139.** $\dfrac{\sqrt{21y}}{7y}+\dfrac{\sqrt{6x}}{2x}$

141. $\dfrac{\sqrt[3]{18y^2}}{3y}+\dfrac{\sqrt[3]{2x^2}}{x}$ **143.** $\dfrac{4-\sqrt{5}}{11}$ **145.** $\dfrac{4+\sqrt{7}}{9}$

147. $\dfrac{\sqrt{3}+3}{-2}$ **149.** $5-\sqrt{10}$ **151.** $\dfrac{5\sqrt{3}+3\sqrt{5}}{2}$

153. $2\sqrt{3}+2\sqrt{2}-3-\sqrt{6}$ **155.** $\dfrac{7+2\sqrt{10}}{3}$

157. $\dfrac{19-3\sqrt{42}}{17}$ **159.** $\dfrac{2\sqrt{x}+2\sqrt{y}+\sqrt{xy}+x}{x-y}$

161. $\dfrac{3x+2y+2\sqrt{6xy}}{3x-2y}$ **163.** $\dfrac{2x-9+2\sqrt{x(x-9)}}{9}$

165. $\dfrac{2x+y+2\sqrt{x(x+y)}}{y}$ **167.** $2\sqrt[3]{2}+2+\sqrt[3]{4}$

169. $\dfrac{\sqrt[3]{12}-\sqrt[3]{6}+\sqrt[3]{3}}{3}$ **171.** $\dfrac{5\sqrt[3]{4}-2\sqrt[3]{25}}{3}$

173. $\dfrac{\sqrt{6}+\sqrt{3}-3}{2}$ **175.** $\dfrac{7\sqrt{3}-3\sqrt{14}+\sqrt{21}}{6}$

177. $\dfrac{2\sqrt{5}+\sqrt{14}-\sqrt{6}}{2}$ **179.** $\dfrac{40\sqrt{7}+42\sqrt{5}-35\sqrt{14}-7\sqrt{10}}{68}$

181. $\dfrac{\sqrt{210}+\sqrt{110}-10}{20}$ **183.** $\dfrac{7}{3\sqrt{3}-6\sqrt{2}+\sqrt{6}-4}$

185. $\dfrac{-15}{9+4\sqrt{6}}$ **187.** $\dfrac{3}{8+\sqrt{10}}$ **189.** $\dfrac{x-y}{x+y+2\sqrt{xy}}$

191. $\dfrac{2x-3y}{x\sqrt{6}+y\sqrt{6}-5\sqrt{xy}}$ **193.** $\dfrac{3}{x-3-\sqrt{x^2-6x}}$

Exercise 8.6, page 261

1. $\{3\}$ **3.** $\left\{\dfrac{15}{2}\right\}$ **5.** $\{13\}$ **7.** $\{-5\}$

9. $\varnothing$ **11.** $\{3\}$ **13.** $\{15\}$ **15.** $\{-3\}$
17. $\{6\}$ **19.** $\{7\}$ **21.** $\{3\}$ **23.** $\{6\}$
25. $\{5\}$ **27.** $\{5\}$ **29.** $\{-5\}$ **31.** $\{-4\}$
33. $\{9\}$ **35.** $\varnothing$ **37.** $\varnothing$ **39.** $\varnothing$
41. $\{5\}$ **43.** $\{2\}$ **45.** $\{-6\}$

Chapter 8 Review, page 262

1. $4\sqrt{3}$ **3.** $\sqrt{2}$ **5.** $\sqrt[3]{5}$
7. $4\sqrt{6} - \sqrt[3]{14}$ **9.** $6\sqrt{6}$
11. $(2x^2 - 3y^2 + 4x^2y^2)\sqrt{xy}$ **13.** 0
15. 0 **17.** $6\sqrt[3]{2x^2y}$
19. $3x^2y^3\sqrt[n]{x} - 5xy^2\sqrt[n]{y^2} + 4xy^4\sqrt[n]{x^2y}$
21. $21\sqrt{2}$ **23.** $3\sqrt{7}$ **25.** $-5\sqrt[3]{4}$ **27.** $6\sqrt[12]{96}$
29. $3x\sqrt{2}$ **31.** $7xy^3\sqrt{6x}$ **33.** $7\sqrt{x+1}$ **35.** $2\sqrt{x(x-2)}$
37. $(2x-1)\sqrt{2x+1}$ **39.** $-3x\sqrt[3]{2}$ **41.** $7x^3y^2\sqrt[3]{3x^2y}$
43. $2xy\sqrt[12]{128x^6y^{10}}$ **45.** $2x^3y^2\sqrt[n]{4}$ **47.** $5\sqrt{14} - 2\sqrt{15}$
49. $3x\sqrt{2} - 2\sqrt{3x}$ **51.** $1 + \sqrt{7}$ **53.** -2
55. $7 - 2\sqrt{10}$ **57.** $3x - 2$ **59.** $7 - x + 4\sqrt{3-x}$
61. $4x + 13 - 12\sqrt{x+1}$ **63.** $2x + 2 + 2\sqrt{(x+3)(x-1)}$
65. $2x + 2 - 2\sqrt{(x+4)(x-2)}$ **67.** $10x - 40 + 6\sqrt{x^2 - 25}$
69. $6x + 9 - 4\sqrt{(2x-3)(x+3)}$ **71.** 17
73. $x - 8y$ **75.** $1 + 2\sqrt{6}$ **77.** $\dfrac{4xy^2\sqrt[3]{a^2by}}{3ab^3}$
79. $\dfrac{\sqrt[6]{288x^5y^2}}{2y}$ **81.** $\sqrt[2n]{2^{n-4}x^{n-5}y^7}$ **83.** $\dfrac{1}{27}$
85. 3 **87.** 3 **89.** $\dfrac{\sqrt[n]{125}}{25}$ **91.** $\dfrac{3\sqrt{21} - 13}{5}$
93. $\dfrac{2x + 3 - 2\sqrt{x^2 + 3x}}{3}$ **95.** $\dfrac{x^2 + \sqrt{x^4 - y^4}}{y^2}$
97. $\dfrac{\sqrt[3]{18} - \sqrt[3]{12} + 2}{5}$ **99.** $\dfrac{4\sqrt[3]{6} - 4\sqrt[3]{4} - \sqrt[3]{9}}{13}$
101. $\dfrac{12\sqrt{6} + 14\sqrt{3} - 15\sqrt{2} - 29}{23}$ **103.** $\sqrt{50}$
105. $\sqrt{45}$ **107.** $\sqrt[3]{\dfrac{1}{4}}$ **109.** $\sqrt[3]{\dfrac{28}{9}}$ **111.** $\sqrt[4]{16x}$

113. $\sqrt{\dfrac{3x^3}{2y}}$ **115.** $\sqrt{\dfrac{2x^3y}{3}}$ **117.** $\sqrt{\dfrac{2x}{5y^3}}$ **119.** $\sqrt{\dfrac{3(x+1)}{x-1}}$

121. $\sqrt[3]{\dfrac{5}{24xy}}$ **123.** $\{7\}$ **125.** $\{6\}$ **127.** $\varnothing$

129. $\{2\}$ **131.** $\{7\}$

Cumulative Review, page 265

Chapter 5

1. $(x+y)(x+y+1)$ **3.** $(x-y)(x-y-4)$ **5.** $(x-y)(x-y-2)$

7. $(x-3)^2(x-1)$ **9.** $(x-y)^2(x-y-4)$

11. $(2x+1)(2x-1)$ **13.** $x^2(4x+1)(4x-1)$

15. $(x^2+1)(x+1)(x-1)$ **17.** $(x-1+y)(x-1-y)$

19. $(2x+y+1)(2x-y-1)$ **21.** $(2x+1)(4x^2-2x+1)$

23. $3x(3x+1)(9x^2-3x+1)$ **25.** $(3x-1)(9x^2+3x+1)$

27. $(x+1)(x^2-x+1)(x-1)(x^2+x+1)$

29. $(x+2)(x+4)$ **31.** $(x+7)(x+9)$ **33.** $(x-3)(x-5)$

35. $(x+10)(x-8)$ **37.** $(x+7)(x-5)$ **39.** $(x-4)(x-9)$

41. $(x+3)(x-6)$ **43.** $(x+6)(x-12)$

45. $(x^2+4)(x+1)(x-1)$ **47.** $(x+2)(x-2)(x+1)(x-1)$

49. $(2x+1)(3x+2)$ **51.** $(3x+2)^2$ **53.** $(2x-1)(3x-4)$

55. $x^2(3x-2)(x-2)$ **57.** $(3x-2)(x+3)$ **59.** $(3x+2)(4x-1)$

61. $(x-5)(3x+2)$ **63.** $(3x+2)(x-3)$ **65.** $(2+x)(3-4x)$

67. $(3+4x)(1-x)$ **69.** $(2x^2+3)(x+2)(x-2)$

71. $(2x^3+3)(x-2)(x^2+2x+4)$ **73.** $[6(x-y)-1][(x-y)-4]$

75. $(9x^2+6x+2)(9x^2-6x+2)$ **77.** $(x^2+3x-1)(x^2-3x-1)$

79. $(2x^2+5x-1)(2x^2-5x-1)$ **81.** $(3x^2+3x+2)(3x^2-3x+2)$

83. $(x-2y+3)(x-2y-3)$ **85.** $(x+3y+z)(x+3y-z)$

87. $(2x-5+3y)(2x-5-3y)$ **89.** $(x+y+4)(x-y-4)$

91. $(3x+2y-1)(3x-2y+1)$ **93.** $(3x+4)(x-y^2)$

95. $(7x+4)(3x-2y)$ **97.** $(x-2y)(x+2y-3)$

99. $(x-y)(x^2+xy+y^2+2x+2y)$

Chapter 6

101. $\dfrac{5x^4}{4y^3z}$ **103.** $\dfrac{y^{12}}{x^3}$ **105.** $\dfrac{16}{189x^2y^{15}}$ **107.** $-(x-2)^2$

109. $\frac{2x+3}{x-4}$ **111.** $\frac{2x+3}{2x-5}$ **113.** $-\frac{2x+3}{3x+2}$ **115.** $\frac{x+y+4}{x+y+2}$

117. $\frac{3x-8}{(x-4)(x-2)}$ **119.** $\frac{2x+3}{(x+5)(x-2)}$ **121.** $\frac{3}{x-2}$

123. $\frac{3}{x-1}$ **125.** $\frac{2(x+4)}{(x-2)(x-3)}$ **127.** $\frac{2}{3x-2}$

129. $\frac{4x+1}{3x+1}$ **131.** $\frac{2(3x+8)}{3(3x+2)}$ **133.** $-\frac{x+5}{x+6}$

135. $\frac{(x^2+2)(x+3)}{(x+2)(x^2+3)}$ **137.** 1 **139.** $\frac{4}{y}$

141. $\frac{x+1}{x-1}$ **143.** $\frac{3x-2}{2x-1}$ **145.** $-\frac{x-2}{x+2}$

147. $\frac{2x+2y-3}{x+y-6}$ **149.** $\frac{(4x-3)(3x-2)}{(2x-9)(3x+2)}$ **151.** $\frac{2x+11}{(x-2)(2x+1)}$

153. $(x+1)(x-1)$ **155.** $(2x-1)(x+2)$ **157.** $\frac{4x-1}{2(7x-1)}$

159. $\frac{x-7}{x+1}$ **161.** $\frac{x+2}{x-1}$ **163.** $\frac{2x+5}{2x+7}$

165. $\frac{(x+1)(x-2)}{(x-3)(2x-3)}$ **167.** $\left\{\frac{2}{a+5} \middle| a \neq -5\right\}$ **169.** $\{2 \mid a \neq 4\}$

171. $\{a-1 \mid a \neq -2\}$ **173.** $\left\{a-2 \middle| a \neq -\frac{3}{2}\right\}$ **175.** $\{-14\}$

177. $\{-13\}$ **179.** $\{-4\}$ **181.** $\{3\}$

185. $\{x \mid x \in R, x \neq -2, x \neq -1, x \neq 3\}$

185. $\varnothing$ **187.** 7 **189.** 165; 47 **191.** 647

193. 132 hours **195.** 10 minutes; 15 minutes

Chapter 7

197. $x^2y^{\frac{2}{3}}$ **199.** xy **201.** $x^{\frac{9}{2}}y^6$ **203.** $xy^{\frac{7}{5}}$

205. $3x^{\frac{3}{2}} + 6x - x^{\frac{1}{2}} - 2$ **207.** $2x - 5x^{\frac{1}{2}} - 12$

209. $2x^{\frac{4}{3}} - 3x - 3x^{\frac{2}{3}} - 3x^{\frac{1}{3}} - 1$ **211.** $\frac{y}{x^{\frac{3}{5}}}$

213. $x^{\frac{1}{2}}y^3$ **215.** $\frac{x}{y}$ **217.** $\frac{x^{\frac{1}{6}}}{y^{\frac{1}{2}}}$ **219.** $2x^{\frac{1}{2}} - 1$

221. $x^{\frac{1}{2}} - 2x^{\frac{1}{4}} + 1$ **223.** $x + 2x^{\frac{1}{2}} - 3$ **225.** $3x - 2x^{\frac{1}{2}} + 1$

227. $2x - x^{\frac{1}{2}} - 3$ **229.** $\frac{2xz}{y^2}$ **231.** $2x^2$

233. $\frac{x^2}{y^{10}}$ **235.** $\frac{2x}{y}$ **237.** $\frac{3}{x^2} - \frac{11}{x} - 4$

239. $\frac{27}{x^3} + \frac{1}{y^3}$ **241.** $\frac{y^2}{x^5}$ **243.** xy^4 **245.** $\frac{x^3y^4}{-z^{15}}$

247. $\frac{2y^2 + x}{3y^2 - x}$ **249.** $\frac{x}{3 - x}$ **251.** $\frac{x - 1}{(2x - 1)^{\frac{3}{2}}}$ **253.** $\frac{5x^2 - 12x}{3(x - 2)^{\frac{4}{3}}}$

Chapter 8

255. $6\sqrt{3}$ **257.** $-2\sqrt[3]{5}$ **259.** $4xy\sqrt{3yz}$

261. $2x^3z^2\sqrt{3yz}$ **263.** $xy^2z^2\sqrt[3]{y^2}$ **265.** $-2xy^2\sqrt[3]{2x}$

267. $2x\sqrt{xy}$ **269.** $2x\sqrt[3]{y^2}$ **271.** $2xy^3\sqrt[n]{x}$

273. $x^2y^3\sqrt[2n]{xy^3}$ **275.** $6\sqrt{6} + 6$ **277.** $15\sqrt{2} - 11$

279. $\sqrt{3} - \sqrt[3]{4}$ **281.** $3xy\sqrt{x}$ **283.** $(2x + 2y)\sqrt{x + y}$

285. $-\sqrt{x}$ **287.** $-4xy\sqrt[3]{xy}$ **289.** $7\sqrt{6}$

291. -4 **293.** $3\sqrt[3]{10}$ **295.** $3\sqrt[4]{5}$ **297.** $5\sqrt[6]{40}$

299. $(x + 2)\sqrt{x - 2}$ **301.** $2x\sqrt[4]{5x}$ **303.** $2x\sqrt[4]{x}$

305. $2\sqrt{21} + 7\sqrt{6}$ **307.** $42 - 13\sqrt{6}$ **309.** $4x^2 - y$

311. $x + 1 - 4\sqrt{x - 3}$ **313.** $2x - 2\sqrt{(x + 2)(x - 2)}$

315. $5x + 11 + 4\sqrt{(x + 3)(x + 2)}$ **317.** $\sqrt{3}$

319. $\frac{2\sqrt[3]{25}}{5}$ **321.** $2\sqrt[4]{27}$ **323.** $\frac{\sqrt[5]{4}}{2}$ **325.** $\frac{\sqrt{6}}{2}$

327. $\frac{\sqrt{30}}{5}$ **329.** $\frac{\sqrt[3]{4}}{2}$ **331.** $\frac{\sqrt[6]{243}}{3}$ **333.** $\frac{3a^2\sqrt{10y}}{10xy^3}$

335. $\frac{xy\sqrt[3]{by}}{2a^2b^2}$ **337.** $\frac{a\sqrt[4]{bxy^3}}{2xy^2}$ **339.** $2\sqrt{2} - 2$ **341.** $2\sqrt{3} + 3$

343. $\sqrt{6} - 2$ **345.** $-3 - 2\sqrt{2}$ **347.** $\frac{9 + 5\sqrt{6}}{23}$ **349.** $\{1\}$

351. $\{13\}$ **353.** $\{9\}$ **355.** $\{7\}$ **357.** $\{7\}$

359. $\{12\}$

Exercises 9.1–9.2, page 278

1. No **3.** $\{1, 2, 3, 4\}, \{3, 4, 5\}$ **5.** $\{3, 4, 5, 6\}, \{1\}$

7. $\{1, 2, 3, 4, \ldots\}, \{5, 9, 13, 17, \ldots\}$ **9.** $\{0, 1, 2, 3, \ldots\}, \{0, 6, 12, 18, \ldots\}$

11. I **13.** IV **15.** III **17.** II

19. 33

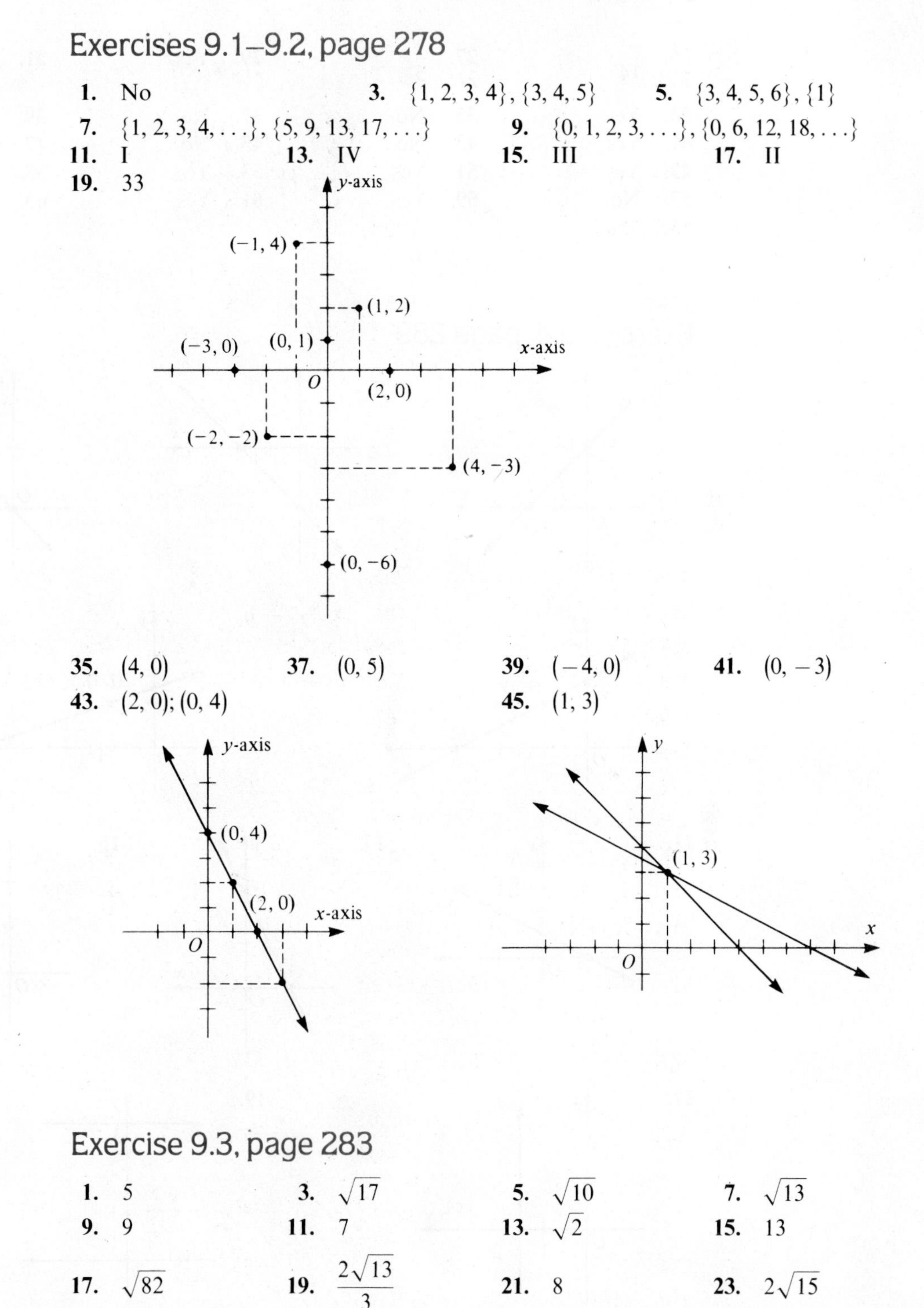

35. $(4, 0)$ **37.** $(0, 5)$ **39.** $(-4, 0)$ **41.** $(0, -3)$

43. $(2, 0); (0, 4)$ **45.** $(1, 3)$

Exercise 9.3, page 283

1. 5 **3.** $\sqrt{17}$ **5.** $\sqrt{10}$ **7.** $\sqrt{13}$

9. 9 **11.** 7 **13.** $\sqrt{2}$ **15.** 13

17. $\sqrt{82}$ **19.** $\dfrac{2\sqrt{13}}{3}$ **21.** 8 **23.** $2\sqrt{15}$

25. $\frac{23}{14}$ **27.** $\frac{3}{5}$ **29.** Yes **31.** Yes

33. Yes **35.** No **37.** No **39.** Yes

41. Yes **43.** No **45.** No **47.** Yes

49. Yes **51.** Yes **53.** Yes **55.** No

57. No **59.** Yes **61.** Yes **63.** No

65. Yes

Exercise 9.4, page 289

1.

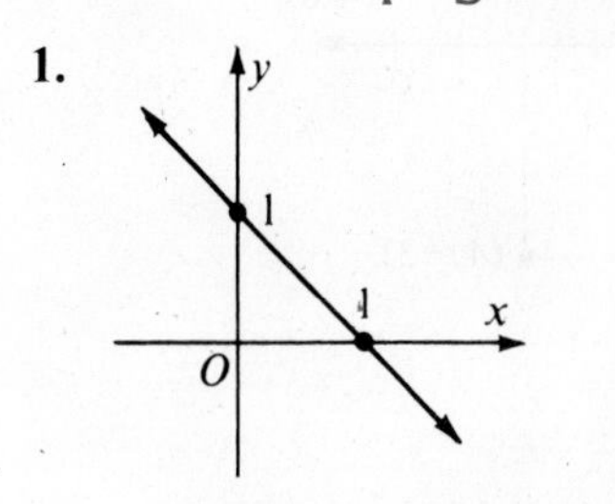

3. **5.**

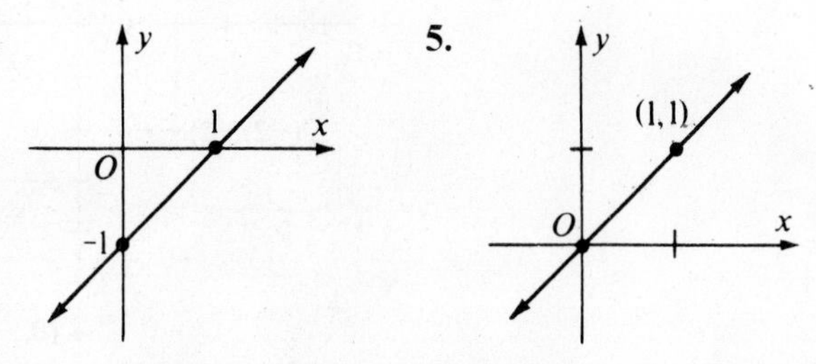

7.

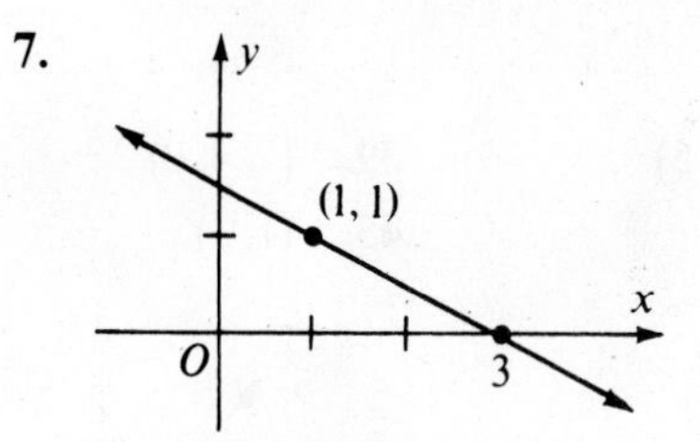

9.

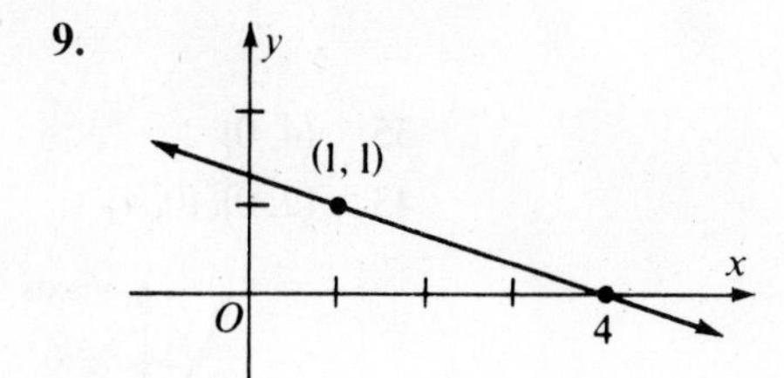

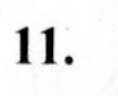

11.

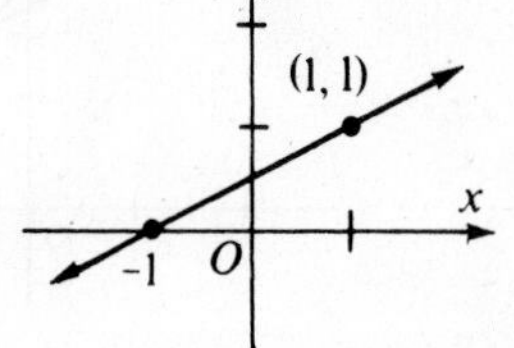

13. **15.**

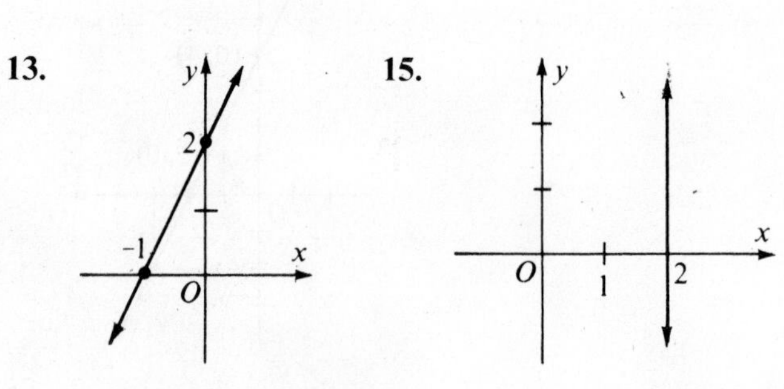

17.

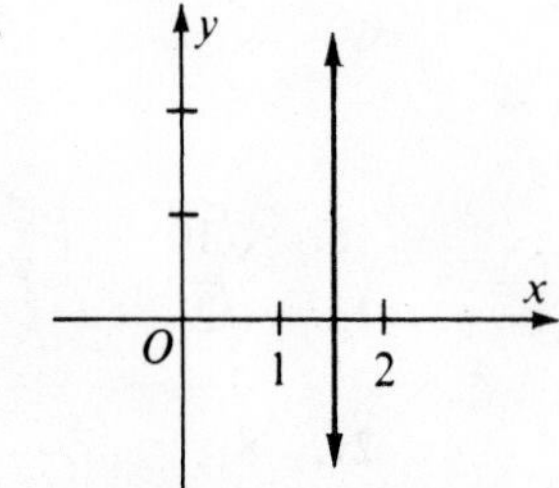

19.

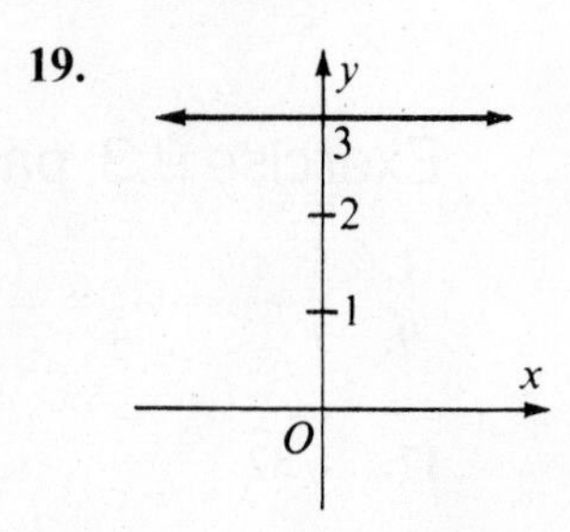

21. **23.**

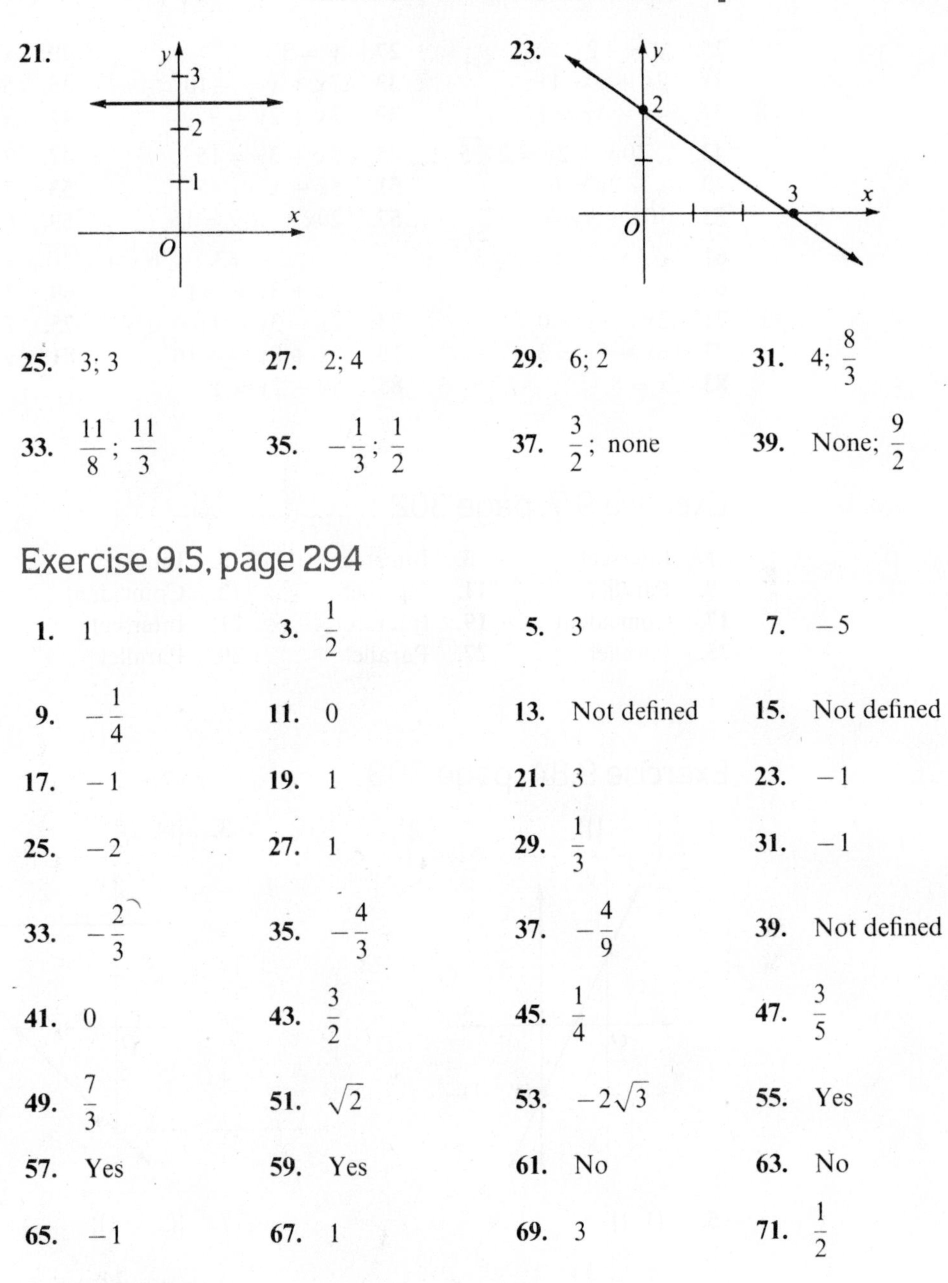

25. 3; 3 **27.** 2; 4 **29.** 6; 2 **31.** 4; $\frac{8}{3}$

33. $\frac{11}{8}$; $\frac{11}{3}$ **35.** $-\frac{1}{3}$; $\frac{1}{2}$ **37.** $\frac{3}{2}$; none **39.** None; $\frac{9}{2}$

Exercise 9.5, page 294

1. 1 **3.** $\frac{1}{2}$ **5.** 3 **7.** -5

9. $-\frac{1}{4}$ **11.** 0 **13.** Not defined **15.** Not defined

17. -1 **19.** 1 **21.** 3 **23.** -1

25. -2 **27.** 1 **29.** $\frac{1}{3}$ **31.** -1

33. $-\frac{2}{3}$ **35.** $-\frac{4}{3}$ **37.** $-\frac{4}{9}$ **39.** Not defined

41. 0 **43.** $\frac{3}{2}$ **45.** $\frac{1}{4}$ **47.** $\frac{3}{5}$

49. $\frac{7}{3}$ **51.** $\sqrt{2}$ **53.** $-2\sqrt{3}$ **55.** Yes

57. Yes **59.** Yes **61.** No **63.** No

65. -1 **67.** 1 **69.** 3 **71.** $\frac{1}{2}$

Exercise 9.6, page 297

1. $x - y = 0$ **3.** $3x - y = 0$ **5.** $2x + 5y = 10$

7. $x - 2y = 3$ **9.** $y - 2x = 3$ **11.** $2x - y = -1$

13. $y = -2$ **15.** $y = 3$ **17.** $x = -1$

19. $\sqrt{3}x + y = 2$ **21.** 1 **23.** $y = 2$

25. $y = -2$ **27.** $y = 5$ **29.** $x - y = 2$
31. $2x + y = 11$ **33.** $7x + y = -10$ **35.** $5x + y = 2$
37. $4x - 3y = 11$ **39.** $3x + 2y = 9$ **41.** $x + 7y = -20$
43. $\sqrt{10}\,x - 2y = 2\sqrt{5}$ **45.** $5x + 3y = 15$ **47.** $7x - 4y = -28$
49. $x - 2y = 6$ **51.** $5x + y = -5$ **53.** $21x + 10y = 6$
55. $10x - 9y = -6$ **57.** $20x - 18y = 15$ **59.** $6x + 40y = -15$
61. $\sqrt{3}\,x + 2y = 2\sqrt{3}$ **63.** $\sqrt{6}\,x + \sqrt{10}\,y = 2\sqrt{15}$
65. $x - 3y = 1$ **67.** $2x + 3y = -1$ **69.** $3x + 7y = 2$
71. $2x - 3y = 0$ **73.** $2x - 3y = 11$ **75.** $7x - 4y = 16$
77. $6x - 5y = 3$ **79.** $7x + 8y = -16$ **81.** $y = -3$
83. $x = 8$ **85.** $5x - 2y = 1$

Exercise 9.7, page 302

1. Intersect **3.** Intersect **5.** Intersect **7.** Parallel
9. Parallel **11.** Parallel **13.** Coincident **15.** Coincident
17. Coincident **19.** Intersect **21.** Intersect **23.** Intersect
25. Parallel **27.** Parallel **29.** Parallel **31.** Intersect

Exercise 9.8A, page 303

1. $\{(1, -1)\}$

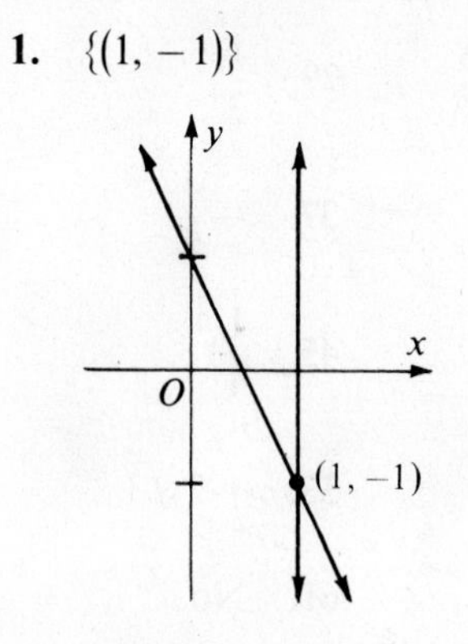

3. $\{(4, 2)\}$

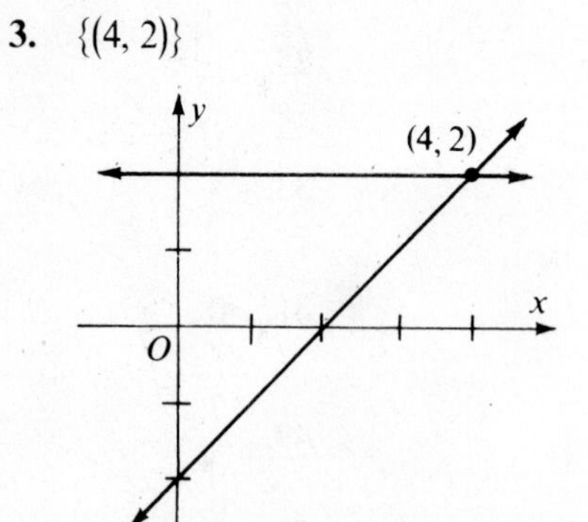

5. $\{(1, 1)\}$

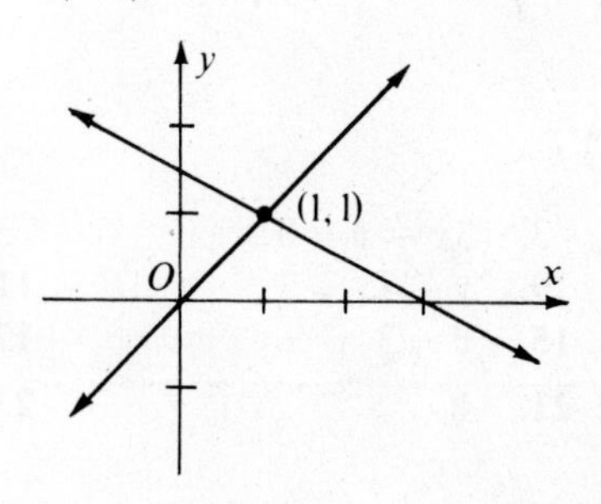

7. $\{(2, -1)\}$

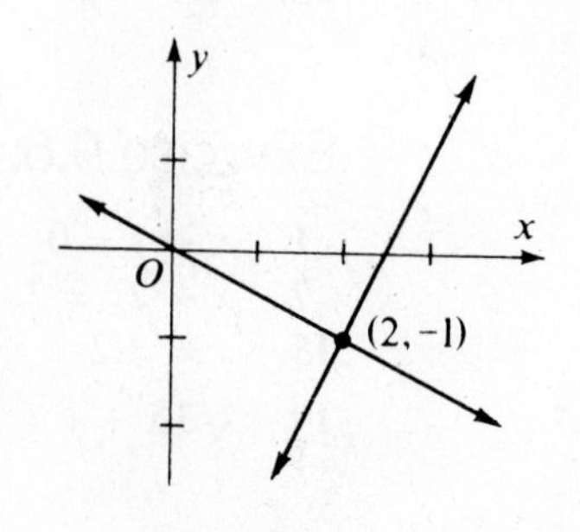

9. $\{(2, 0)\}$ **11.** $\{(2, 1)\}$

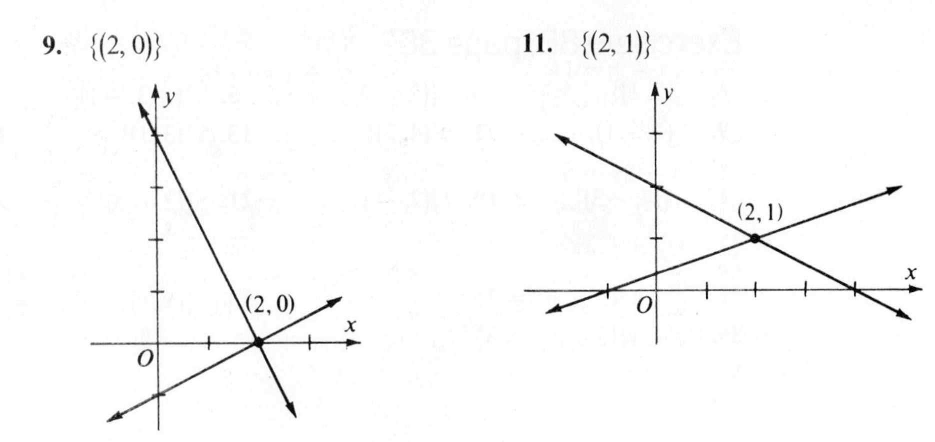

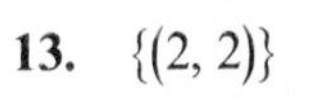

13. $\{(2, 2)\}$

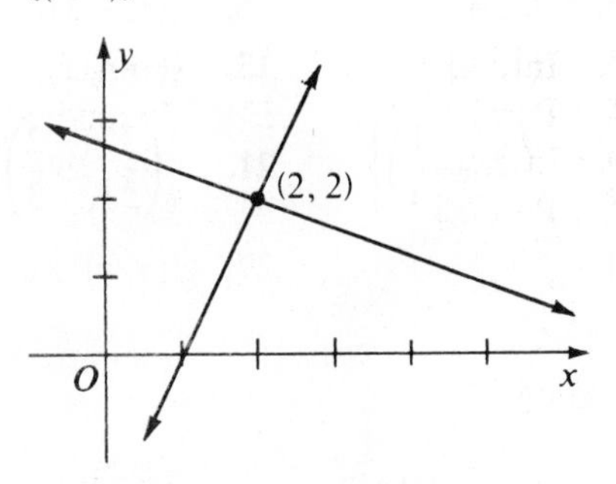

15. $\{(3, -1)\}$

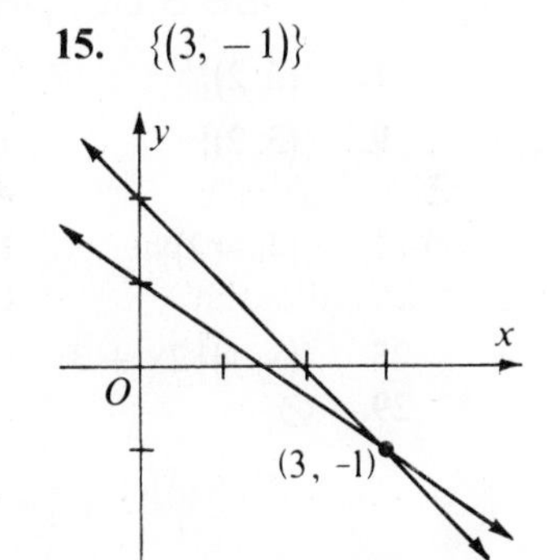

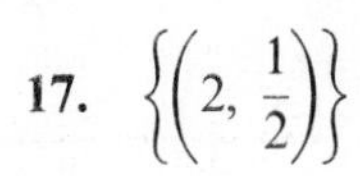

17. $\left\{\left(2, \frac{1}{2}\right)\right\}$

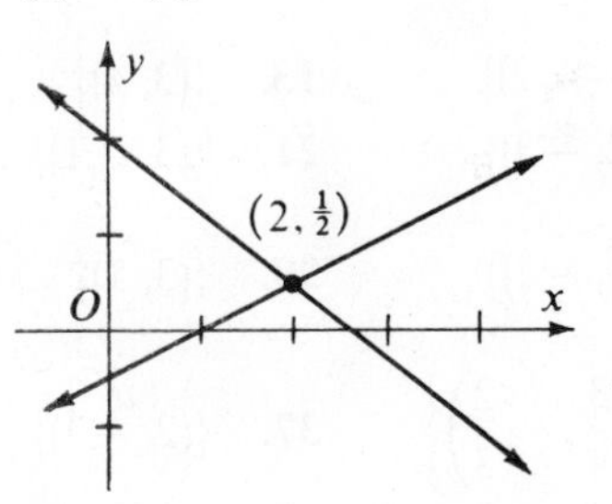

19. $\varnothing$

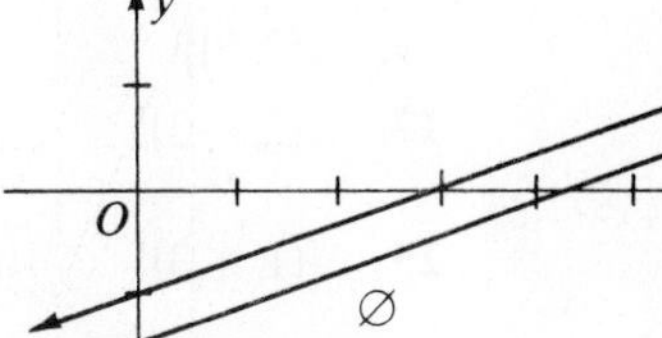

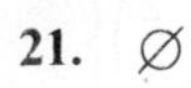

21. $\varnothing$

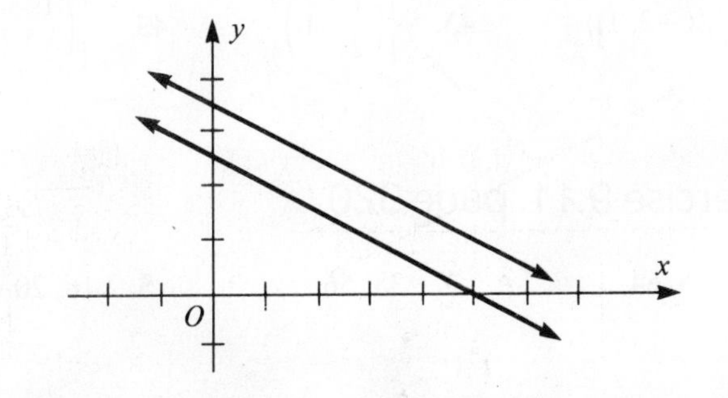

Exercise 9.8B, page 307

1. $\{(2, 1)\}$
3. $\{(2, -2)\}$
5. $\{(-3, -1)\}$
7. $\{(3, -2)\}$
9. $\{(1, -1)\}$
11. $\{(4, 2)\}$
13. $\{(3, 1)\}$
15. $\{(-3, 3)\}$
17. $\{(4, -3)\}$
19. $\{(2, -1)\}$
21. $\{(2, -3)\}$
23. $\left\{\left(\frac{2}{3}, \frac{1}{5}\right)\right\}$
25. $\varnothing$
27. $\varnothing$
29. $\varnothing$
31. $\{(x, y) \mid x + 4y = 1\}$
33. $\{(x, y) \mid x + 3y = -2\}$
35. $\{(x, y) \mid 2x + y = 3\}$

Exercise 9.8C, page 309

1. $\{(2, 2)\}$
3. $\{(2, 4)\}$
5. $\{(3, -1)\}$
7. $\{(1, 1)\}$
9. $\{(3, 2)\}$
11. $\{(3, 3)\}$
13. $\{(-4, 1)\}$
15. $\{(4, 3)\}$
17. $\{(4, -2)\}$
19. $\left\{\left(2, -\frac{1}{3}\right)\right\}$
21. $\left\{\left(\frac{3}{2}, -\frac{2}{3}\right)\right\}$
23. $\left\{\left(-\frac{5}{3}, \frac{1}{2}\right)\right\}$
25. $\{(x, y) \mid 2x + 3y = 1\}$
27. $\{(x, y) \mid 3x + y = -5\}$
29. $\varnothing$

Exercises 9.9–9.10, page 313

1. $\{(1, -2)\}$
3. $\{(1, -3)\}$
5. $\left\{\left(2, \frac{1}{3}\right)\right\}$
7. $\{(2, 1)\}$
9. $\{(2, 3)\}$
11. $\{(-4, 2)\}$
13. $\{(3, 4)\}$
15. $\{(6, 3)\}$
17. $\{(2, -3)\}$
19. $\{(2, -3)\}$
21. $\{(3, -1)\}$
23. $\{(2, -1)\}$
25. $\{(1, -3)\}$
27. $\{(1, -2)\}$
29. $\{(3, 5)\}$
31. $\left\{\left(\frac{1}{5}, -\frac{1}{2}\right)\right\}$
33. $\left\{\left(\frac{2}{3}, \frac{2}{5}\right)\right\}$
35. $\left\{\left(\frac{3}{4}, -\frac{2}{3}\right)\right\}$
37. $\{(2, -3)\}$
39. $\{(2, -1)\}$
41. $\{(-3, 1)\}$
43. $\left\{\left(\frac{1}{2}, 1\right)\right\}$
45. $\left\{\left(\frac{1}{2}, -\frac{2}{3}\right)\right\}$
47. $\left\{\left(\frac{5}{6}, -\frac{1}{6}\right)\right\}$

Exercise 9.11, page 320

1. 25; 64
3. 38; 56
5. 16; 20
7. $\frac{7}{3}; \frac{9}{4}$

9. $\frac{2}{9}$; $\frac{3}{11}$ **11.** 58 **13.** 529 **15.** $\frac{3}{14}$

17. $\frac{36}{59}$ **19.** $\frac{5}{7}$ **21.** $\frac{17}{20}$ **23.** 20; 52

25. 42; 60 **27.** \$12,000; \$18,000 **29.** 5%; $8\frac{1}{2}$%

31. 25¢; 39% **33.** 15¢; 31¢

35. 20 pounds; 25 pounds **37.** 16 nickels; 3 quarters

39. $1\frac{1}{2}$ mph; $4\frac{1}{2}$ mph **41.** 30 mph; 240 mph

43. 32 years old; 46 years old **45.** 12 feet

47. 18 feet **49.** 5500 square feet **51.** 36 hours; 45 hours

53. 70 minutes; 126 minutes **55.** \$64,000; \$36,000; \$50,000

Exercises 9.12–9.13, page 330

1. **3.** **5.**

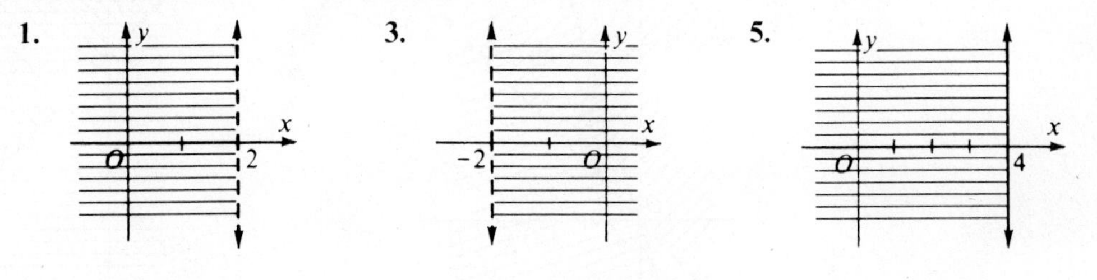

7.

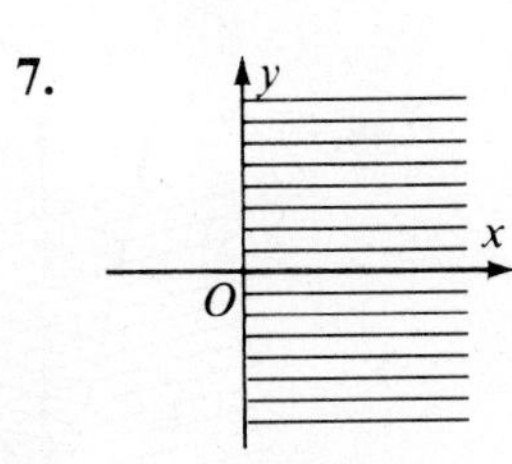

9.

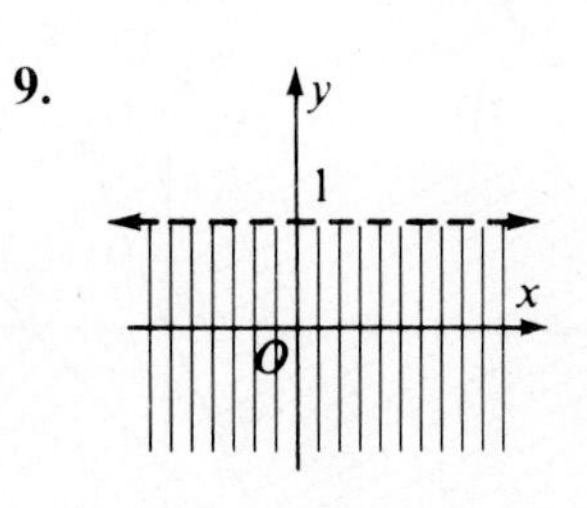

11.

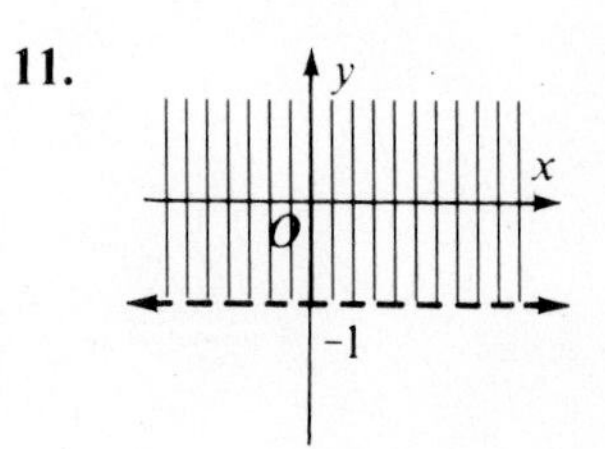

13.

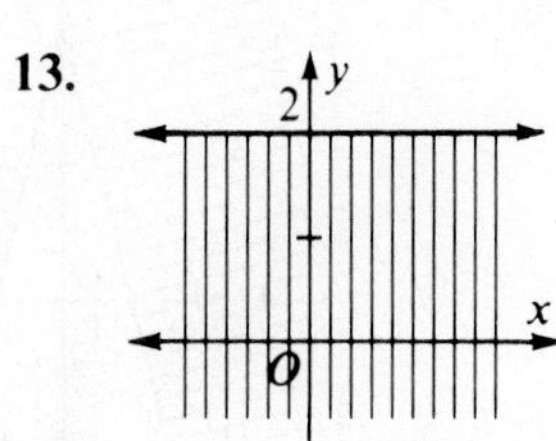

15.

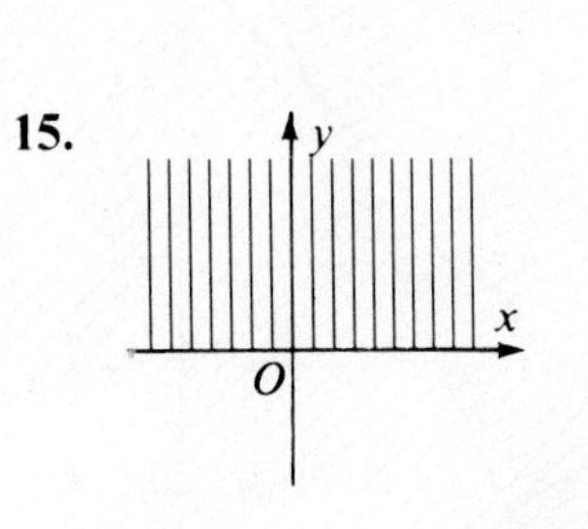

17.

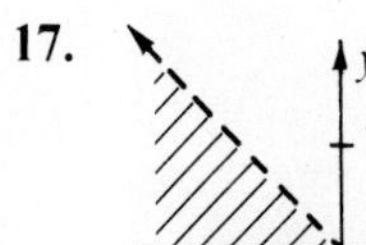

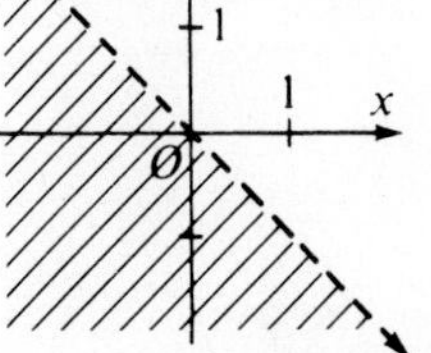

19.

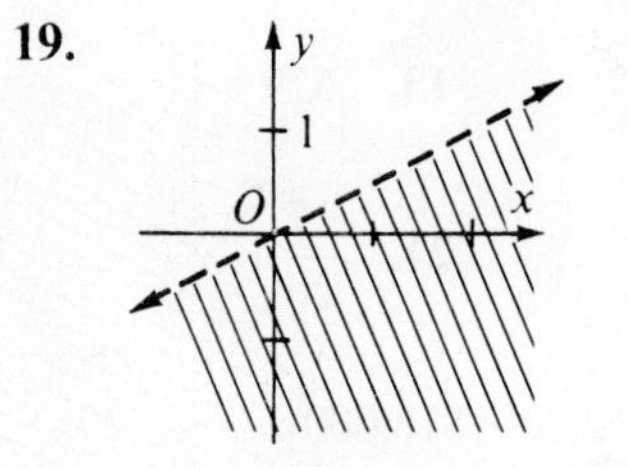

21.

23.

25.

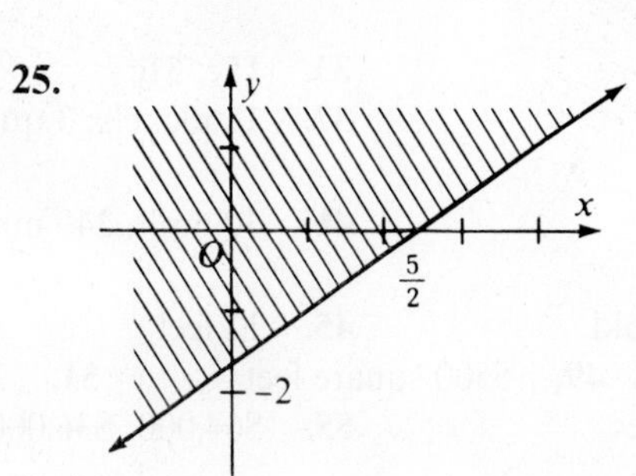

27.

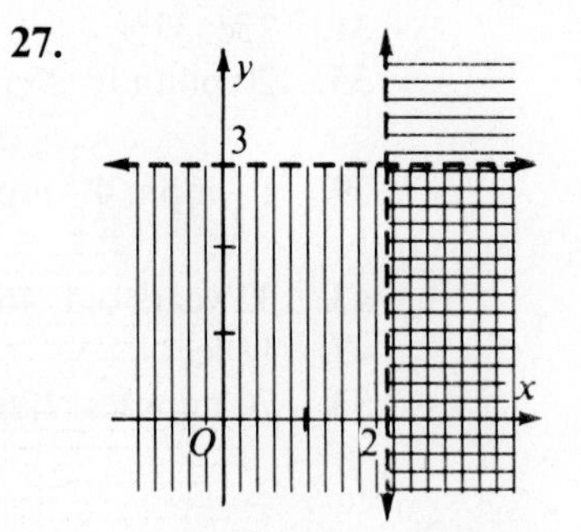

29.

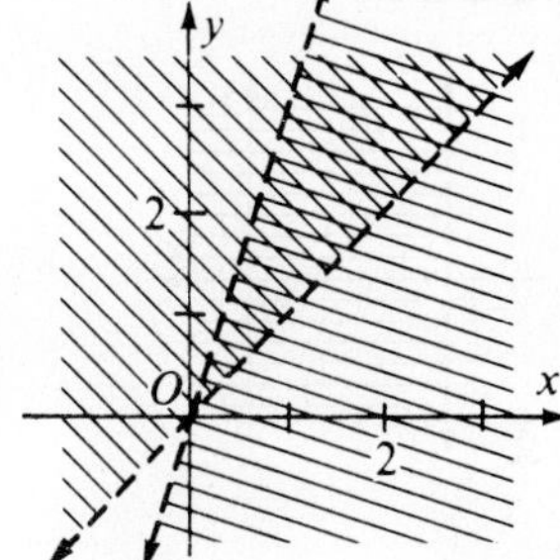

31.

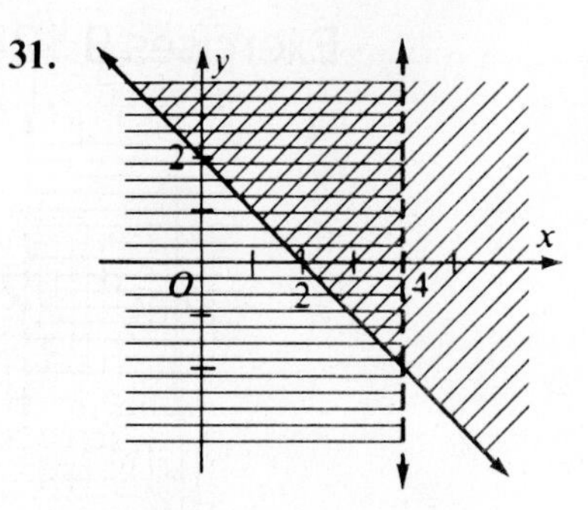

33.

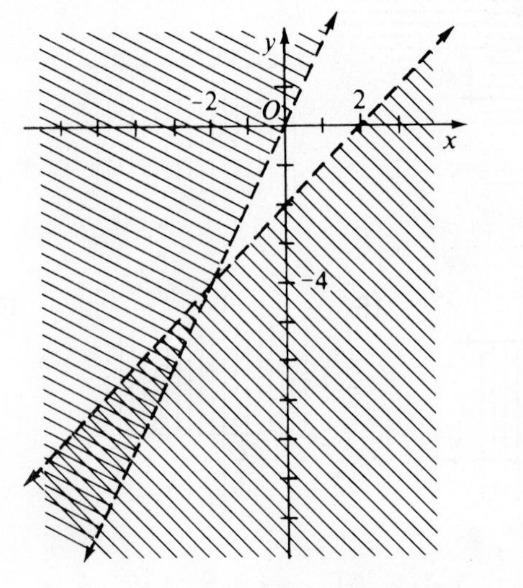

35.

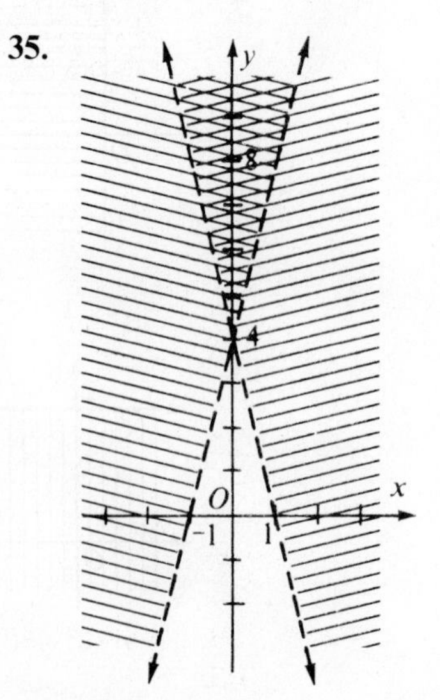

37. 39. 41. 43.

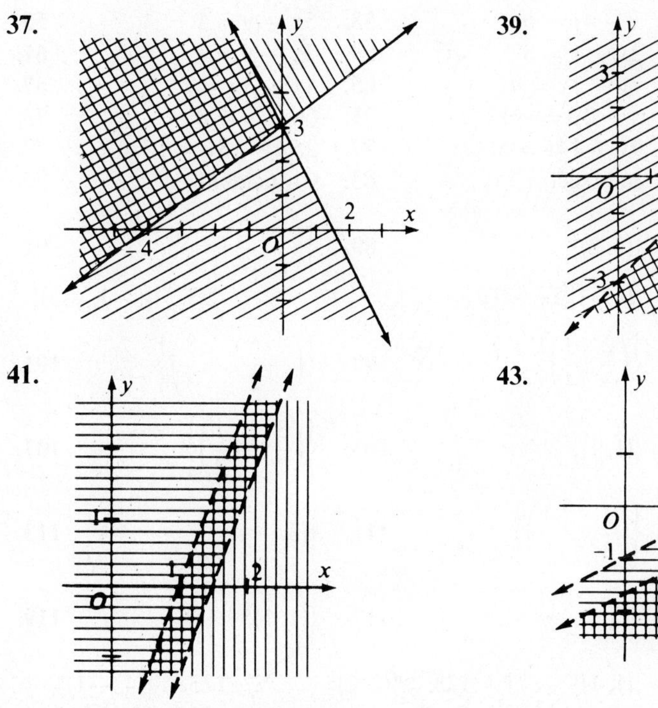

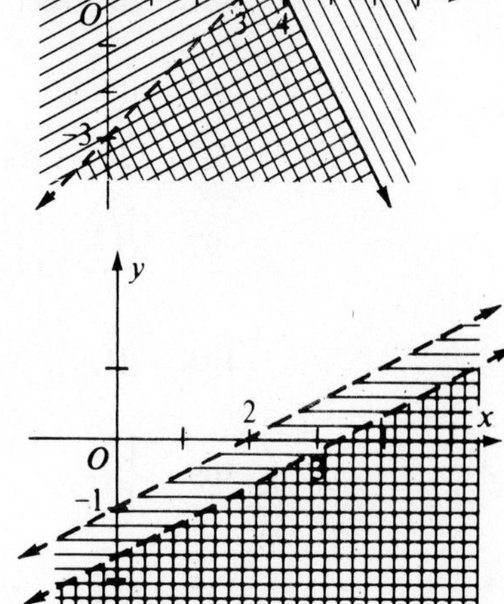

45.

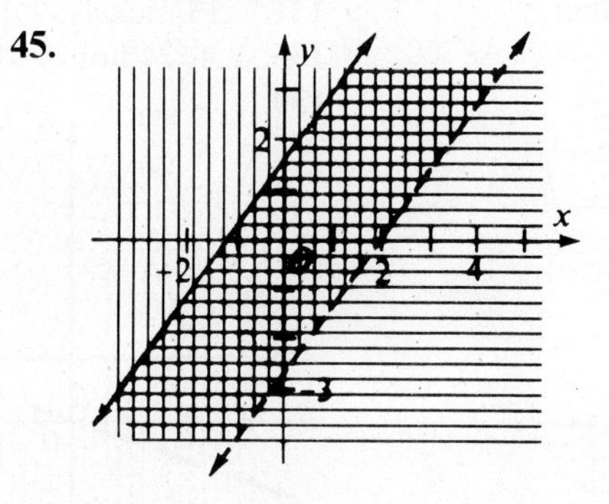

Chapter 9 Review, page 330

1. Yes **3.** Yes **5.** Yes **7.** No
9. No **11.** 4 **13.** 6 **15.** Yes
17. Yes **19.** Yes **21.** No **23.** No
25. $-\frac{1}{2}$ **27.** $-\frac{1}{3}$ **29.** $\frac{1}{2}$ **31.** 2
33. $\frac{4}{7}$ **35.** $-\frac{2\sqrt{7}}{7}$ **37.** Not defined
39. Not defined **41.** 0 **43.** -3
45. 5 **47.** -1 **49.** $2x + y = 0$

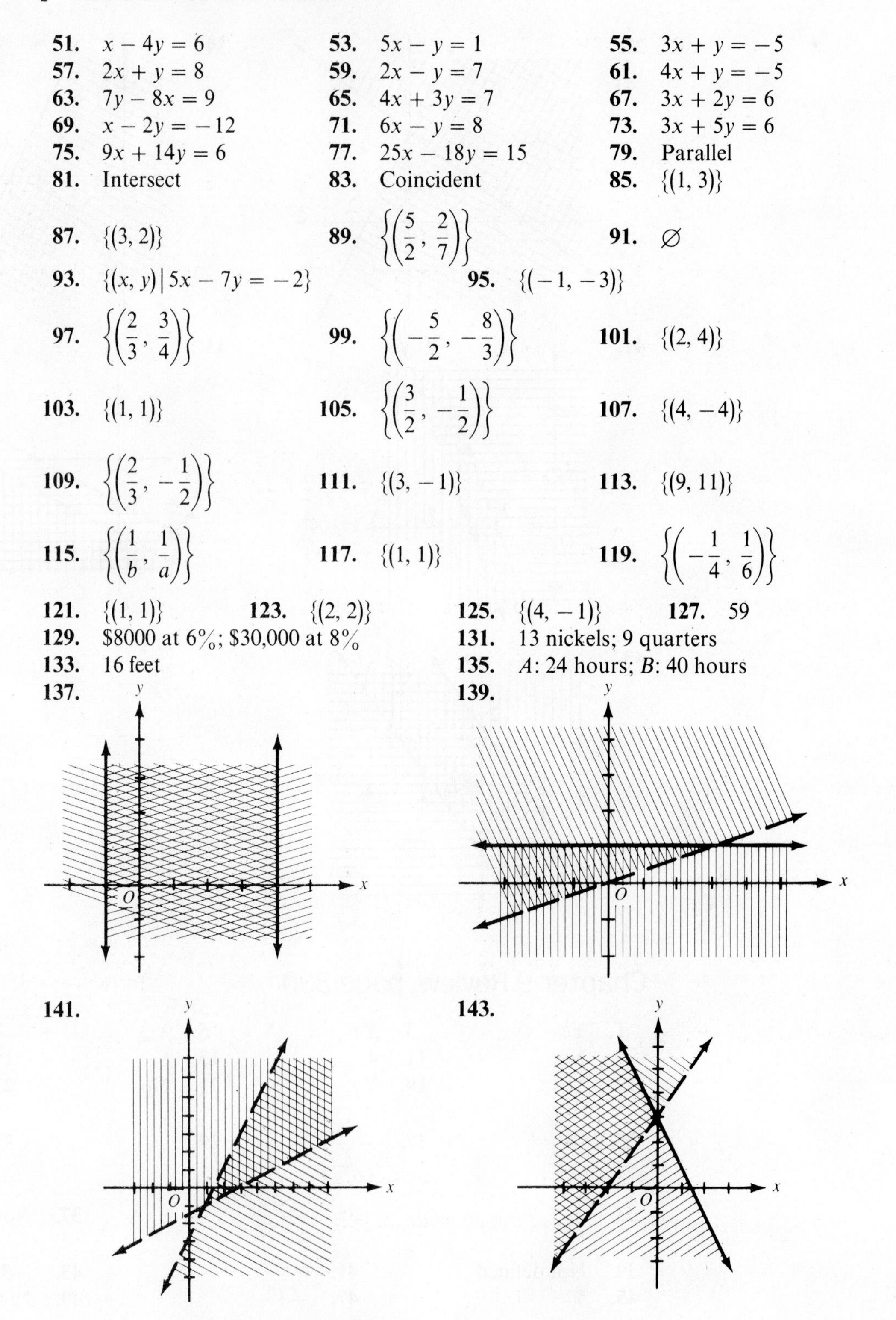

51. $x - 4y = 6$ **53.** $5x - y = 1$ **55.** $3x + y = -5$
57. $2x + y = 8$ **59.** $2x - y = 7$ **61.** $4x + y = -5$
63. $7y - 8x = 9$ **65.** $4x + 3y = 7$ **67.** $3x + 2y = 6$
69. $x - 2y = -12$ **71.** $6x - y = 8$ **73.** $3x + 5y = 6$
75. $9x + 14y = 6$ **77.** $25x - 18y = 15$ **79.** Parallel
81. Intersect **83.** Coincident **85.** $\{(1, 3)\}$
87. $\{(3, 2)\}$ **89.** $\left\{\left(\frac{5}{2}, \frac{2}{7}\right)\right\}$ **91.** $\varnothing$
93. $\{(x, y) \mid 5x - 7y = -2\}$ **95.** $\{(-1, -3)\}$
97. $\left\{\left(\frac{2}{3}, \frac{3}{4}\right)\right\}$ **99.** $\left\{\left(-\frac{5}{2}, -\frac{8}{3}\right)\right\}$ **101.** $\{(2, 4)\}$
103. $\{(1, 1)\}$ **105.** $\left\{\left(\frac{3}{2}, -\frac{1}{2}\right)\right\}$ **107.** $\{(4, -4)\}$
109. $\left\{\left(\frac{2}{3}, -\frac{1}{2}\right)\right\}$ **111.** $\{(3, -1)\}$ **113.** $\{(9, 11)\}$
115. $\left\{\left(\frac{1}{b}, \frac{1}{a}\right)\right\}$ **117.** $\{(1, 1)\}$ **119.** $\left\{\left(-\frac{1}{4}, \frac{1}{6}\right)\right\}$
121. $\{(1, 1)\}$ **123.** $\{(2, 2)\}$ **125.** $\{(4, -1)\}$ **127.** 59
129. \$8000 at 6%; \$30,000 at 8% **131.** 13 nickels; 9 quarters
133. 16 feet **135.** A: 24 hours; B: 40 hours
137. **139.**
141. **143.**

145.

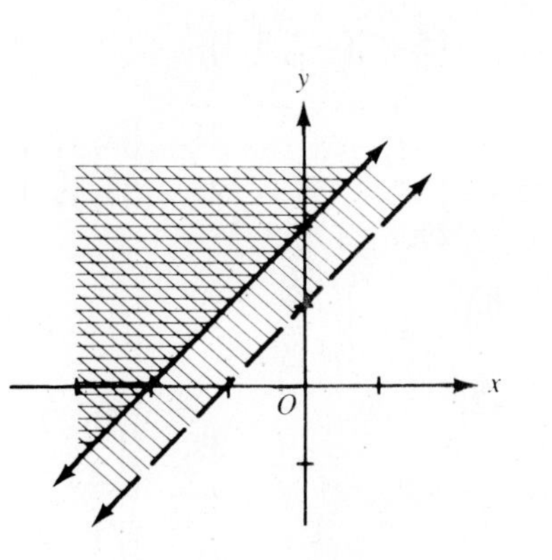

147.

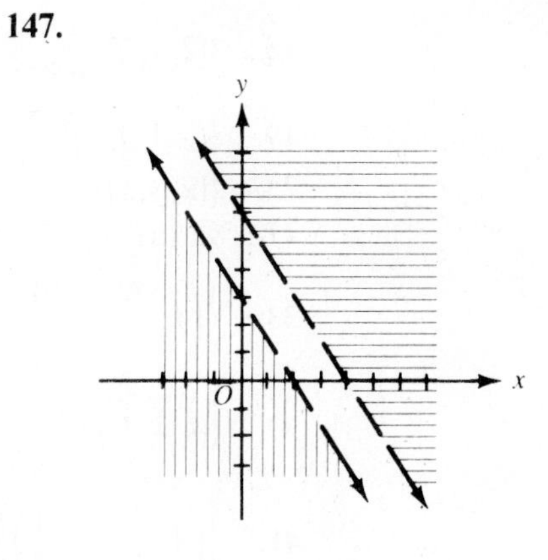

Exercises 10.1–10.2, page 339

1. Parallel **3.** Intersect **5.** Coincide **7.** Parallel
9. Intersect **11.** Coincide

The ordered numbers are (x, y, z):

13. $\{(k, -k + 1, 3k - 2) \mid k \in R\}$

15. $\left\{\left(\dfrac{3k + 2}{2}, k; \dfrac{k - 1}{2}\right) \middle| k \in R\right\}$

17. $\{(2 - k, 4 - 3k, k) \mid k \in R\}$

19. $\left\{\left(k, \dfrac{k - 6}{5}, \dfrac{9 - k}{3}\right) \middle| k \in R\right\}$

21. $\{(1, 2k, k) \mid k \in R\}$

23. $\{(k, 4, 3k) \mid k \in R\}$

25. $\{(2k - 1, k, 2) \mid k \in R\}$

27. $\left\{\left(\dfrac{k + 4}{2}, k, \dfrac{2k - 5}{3}\right) \middle| k \in R\right\}$

29. $\{(x, y, z) \mid x + 3y - 7z = 4\}$

31. $\varnothing$

33. $\{(4 - k, k, 4k + 19) \mid k \in R\}$

35. $\left\{\left(k, \dfrac{-3k + 11}{5}, \dfrac{8k + 15}{12}\right) \middle| k \in R\right\}$

37. $\varnothing$

39. $\left\{\left(-5k - 1, k, \dfrac{1 - 7k}{2}\right) \middle| k \in R\right\}$

Exercise 10.3, page 345

The ordered numbers are (x, y, z):

1. $\{(1, 6, 9)\}$ **3.** $\{(2, 1, -3)\}$ **5.** $\{(3, -2, 3)\}$
7. $\{(0, 3, 2)\}$ **9.** $\{(1, 2, -3)\}$ **11.** $\{(1, 1, 1)\}$

13. $\{(2, 1, 2)\}$ **15.** $\{(-2, 3, 1)\}$ **17.** $\left\{\left(2, 1, -\frac{1}{3}\right)\right\}$

19. $\{(-1, 1, 2)\}$ **21.** $\{(x, y, z) \mid 2x - y + 3z = 1\}$

23. $\{(x, y, z) \mid 2x - y + z = 5\}$ **25.** $\{(x, y, z) \mid x - 2y + z = 3\}$

27. $\varnothing$ **29.** $\varnothing$ **31.** $\varnothing$

33. $\left\{\left(k, \frac{8k - 10}{3}, \frac{7k - 8}{3}\right) \middle| k \in R\right\}$ **35.** $\{(1 - k, 2 - 2k, k) \mid k \in R\}$

37. $\{(k + 3, k, -k - 3) \mid k \in R\}$ **39.** $\left\{\left(\frac{1}{2}, \frac{1}{3}, -\frac{1}{2}\right)\right\}$

41. $\left\{\left(1, \frac{1}{2}, -\frac{1}{3}\right)\right\}$ **43.** $\left\{\left(\frac{1}{2}, \frac{2}{3}, 1\right)\right\}$

Exercise 10.4, page 348

1. 6; 8; 11 **3.** 27; 39; 62 **5.** 263 **7.** 476

9. 20¢; 25¢; 29¢ **11.** 15¢; 19¢; 27¢ **13.** 10%; 12%; 16%

15. 4 hours; 5 hours; 10 hours **17.** 9 hours; 12 hours; 18 hours

19. 2:1:1 **21.** 2:1:2

Chapter 10 Review, page 350

The ordered numbers are (x, y, z):

1. $\left\{\left(k, \frac{3 - 2k}{4}, \frac{1 - k}{5}\right) \middle| k \in R\right\}$ **3.** $\left\{\left(\frac{k + 6}{3}, k, 5 - 2k\right) \middle| k \in R\right\}$

5. $\left\{\left(\frac{3k - 2}{4}, 1 - 6k, k\right) \middle| k \in R\right\}$ **7.** $\{(-1, k, 2k - 1) \mid k \in R\}$

9. $\{(k, -2, 3k + 4) \mid k \in R\}$ **11.** $\{(k, 2k, 1 - k) \mid k \in R\}$

13. $\{(k, 6 - 7k, 7 - 5k) \mid k \in R\}$ **15.** $\{(3, 1, -2)\}$

17. $\varnothing$ **19.** $\left\{\left(k, \frac{5 - 4k}{4}, \frac{2k + 1}{2}\right) \middle| k \in R\right\}$

21. $\{(x, y, z) \mid x - 2y + z = 2\}$ **23.** $\{(4, 1, 1)\}$

25. $\varnothing$ **27.** $\left\{\left(k, \frac{4k + 1}{3}, \frac{5k - 4}{3}\right) \middle| k \in R\right\}$

29. $\{(2, 3, 3)\}$ **31.** $\{(6, 2, 1)\}$

33. $\left\{\left(1, 2, -\frac{1}{2}\right)\right\}$ **35.** $\left\{\left(-1, \frac{1}{4}, \frac{2}{3}\right)\right\}$

Exercises 11.2–11.3, page 356

1. 11 **3.** -34 **5.** $4a + 3b$

7. $-21a^2 - 2b^2$ **9.** ab **11.** $\left\{-\frac{21}{2}\right\}$

13. $\{-2\}$ **15.** $\{1\}$ **17.** $\left\{\frac{3}{2}\right\}$ **19.** $\{(3, -2)\}$

21. $\{(1, -1)\}$ **23.** $\{(3, 1)\}$ **25.** $\left\{\left(0, -\frac{3}{2}\right)\right\}$ **27.** $\{(1, -1)\}$

29. $\{(5, 3)\}$ **31.** $\left\{\left(\frac{3}{2}, \frac{5}{3}\right)\right\}$ **33.** $\left\{\left(-\frac{1}{2}, \frac{3}{2}\right)\right\}$ **35.** $\{(6, 4)\}$

Exercise 11.4, page 360

1. -10 **3.** -66 **5.** 0 **7.** -8
9. -1 **11.** -120 **13.** -10 **15.** -2
17. 32 **19.** -6 **21.** -6

Exercise 11.5, page 363

The ordered numbers are (x, y, z, t):

1. $\{(1, -2, -1)\}$ **3.** $\{(2, 1, -1)\}$ **5.** $\{(2, 0, 1)\}$
7. $\{(2, 0, -1)\}$ **9.** $\{(1, -2, -2)\}$ **11.** $\{(2, -1, -1)\}$
13. $\{(1, 2, -2)\}$ **15.** $\{(2, -1, -2, 1)\}$ **17.** $\{(1, 2, -2, 2)\}$
19. $\{(1, 3, -1, -2)\}$ **21.** $\{(1, 2, -1, -3)\}$ **23.** $\{(5, 4, 3, 4)\}$

Chapter 11 Review, page 364

The ordered numbers are (x, y, z, t):

1. $\{(1, 3)\}$ **3.** $\{(3, 4)\}$ **5.** $\{(1, -4)\}$ **7.** $\{(2, -3)\}$
9. $\{(-1, 1)\}$ **11.** $\{(4, -1, 1)\}$ **13.** $\{(-1, 2, -2)\}$
15. $\{(-2, 1, -2)\}$ **17.** $\{(0, 1, 2)\}$ **19.** $\{(3, 1, -2)\}$

21. $\{(2, 3, -2)\}$

23. $\left\{\left(1, -\frac{1}{3}, -\frac{2}{3}\right)\right\}$

25. $\left\{\left(\frac{3}{4}, -2, \frac{1}{2}\right)\right\}$

27. $\{(-2, 1, 1, -3)\}$

29. $\{(0, -1, 2, 1)\}$

31. $\{(2, 3, -2, 1)\}$

Cumulative Review, page 366

Chapter 9

1. $2\sqrt{10}$ **3.** $3\sqrt{10}$ **5.** $2\sqrt{2}$ **7.** Yes
9. No **11.** Yes **13.** Yes **15.** No
17. Yes **19.** 4 **21.** 7 **23.** $\frac{2}{3}$
25. Not defined **27.** 3 **29.** $\frac{2}{3}$
31. Not defined **33.** $-\frac{1}{4}$ **35.** $-\frac{7}{8}$
37. Yes **39.** No **41.** Yes **43.** -3
45. 9 **47.** $3x - 2y = -6$ **49.** $6x + y = 8$
51. $y = 3$ **53.** $x = -1$ **55.** $y = 2$
57. $3x + y = 5$ **59.** $3x - 4y = -23$ **61.** $5x + 3y = -13$
63. $2x + 3y = 6$ **65.** $7x - 5y = -35$ **67.** $9x + 4y = 12$
69. $21x + 4y = 6$ **71.** $2x - 3y = 1$ **73.** $4x + 3y = 11$
75. $2x - 3y = 11$ **77.** $2x + 5y = -4$ **79.** Intersect
81. Intersect **83.** Parallel **85.** Parallel
87. Coincident **89.** Coincident
91. $\{(1, 3)\}$ **93.** $\{(5, 3)\}$

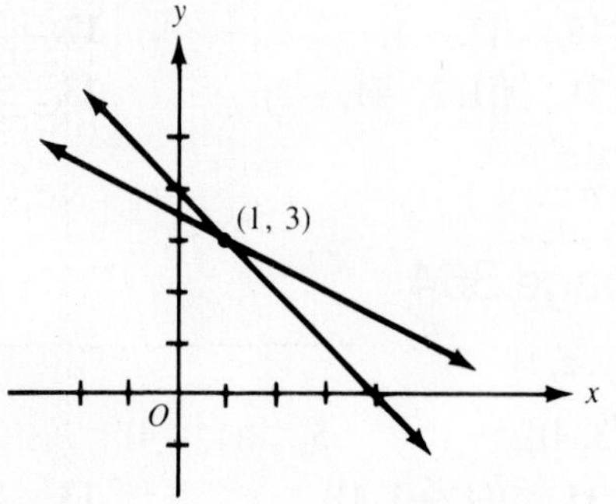

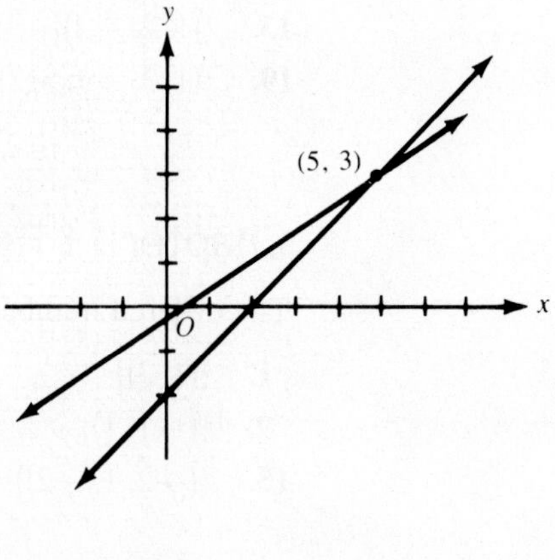

y (1, 3) O x

y (5, 3) O x

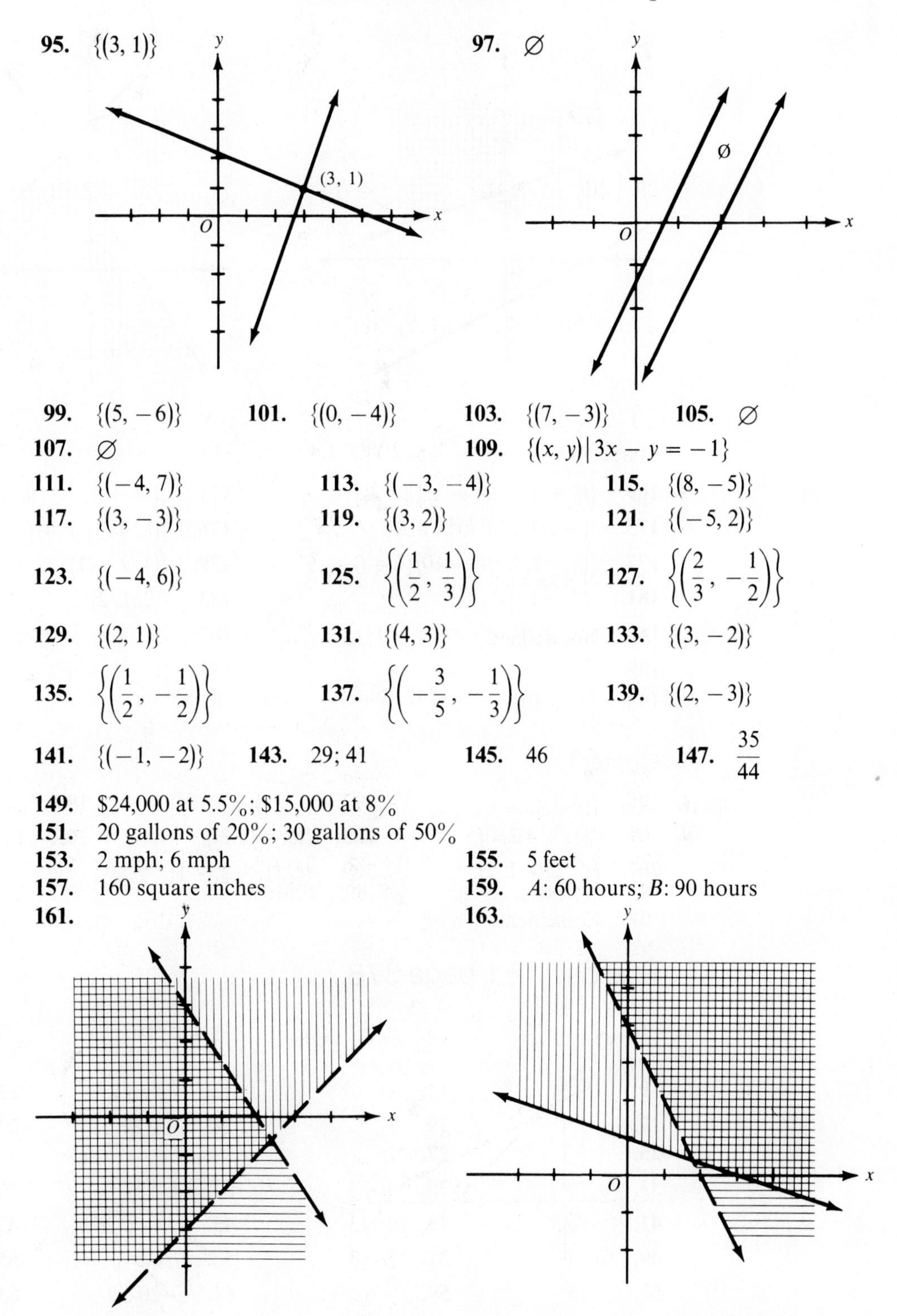

95. $\{(3, 1)\}$

97. $\varnothing$

99. $\{(5, -6)\}$ **101.** $\{(0, -4)\}$ **103.** $\{(7, -3)\}$ **105.** $\varnothing$

107. $\varnothing$ **109.** $\{(x, y) \mid 3x - y = -1\}$

111. $\{(-4, 7)\}$ **113.** $\{(-3, -4)\}$ **115.** $\{(8, -5)\}$

117. $\{(3, -3)\}$ **119.** $\{(3, 2)\}$ **121.** $\{(-5, 2)\}$

123. $\{(-4, 6)\}$ **125.** $\left\{\left(\frac{1}{2}, \frac{1}{3}\right)\right\}$ **127.** $\left\{\left(\frac{2}{3}, -\frac{1}{2}\right)\right\}$

129. $\{(2, 1)\}$ **131.** $\{(4, 3)\}$ **133.** $\{(3, -2)\}$

135. $\left\{\left(\frac{1}{2}, -\frac{1}{2}\right)\right\}$ **137.** $\left\{\left(-\frac{3}{5}, -\frac{1}{3}\right)\right\}$ **139.** $\{(2, -3)\}$

141. $\{(-1, -2)\}$ **143.** 29; 41 **145.** 46 **147.** $\frac{35}{44}$

149. \$24,000 at 5.5%; \$15,000 at 8%

151. 20 gallons of 20%; 30 gallons of 50%

153. 2 mph; 6 mph **155.** 5 feet

157. 160 square inches **159.** *A*: 60 hours; *B*: 90 hours

161.

163.

165. **167.**

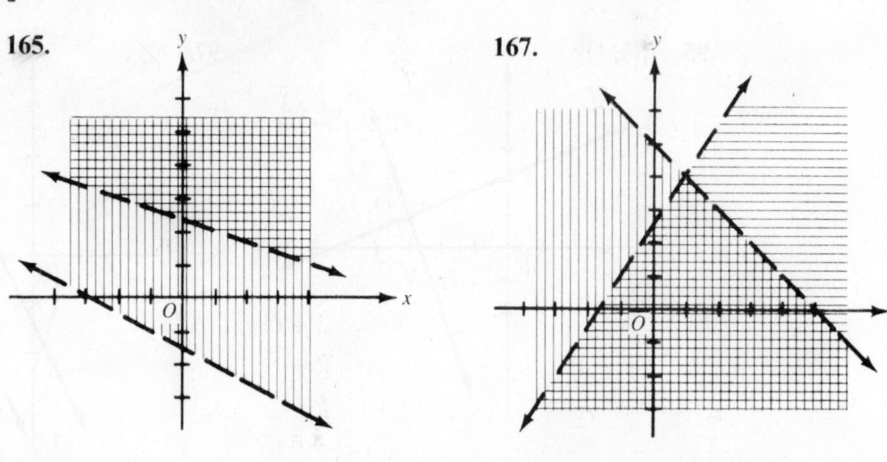

Chapter 10

169. $\{(k+1, k, 2k-1) \mid k \in R\}$
171. $\{(2k+1, k-4, k) \mid k \in R\}$
173. $\{(-2, k-3, k) \mid k \in R\}$
175. $\{(k, 3k-1, 4) \mid k \in R\}$
177. $\{(k-2, k, 3k-10) \mid k \in R\}$
179. $\{(1, 2, -1)\}$
181. $\{(2, -1, 1)\}$
183. $\{(2, 1, 2)\}$
185. $\{(x, y, z) \mid x - y + z = 1\}$
187. $\{(x, y, z) \mid x + 2y + z = 4\}$
189. $\varnothing$
191. $\{(k, 2k, k+1) \mid k \in R\}$
193. $\{(1, 3, 0)\}$

Chapter 11

195. $\{(3, 1)\}$
197. $\{(3, 3)\}$
199. $\{(1, 2, -1)\}$
201. $\{(-1, -2, 0)\}$
203. $\{(1, -3, 2)\}$
205. $\{(2, 1, -1, 1)\}$
207. $\{(1, 2, -1, 2)\}$
209. $\{(2, 1, -2, -1)\}$

Exercise 12.1, page 378

1. $4i$
3. $3i\sqrt{2}$
5. $i\sqrt{10}$
7. $\frac{i\sqrt{6}}{2}$
9. $i\sqrt{13}$
11. xi
13. $xy^2i\sqrt{y}$
15. $i\sqrt{x^2+y^2}$
17. -1
19. 1
21. $-i$
23. -1
25. -1
27. 1
29. i
31. -1
33. 0
35. $4i\sqrt{3}$
37. $4i\sqrt{5}$
39. $(4+3\sqrt{3})i$
41. $-i\sqrt{5}$
43. $9i\sqrt{3}$
45. $(5-\sqrt{26})i$
47. $3i\sqrt{3}$
49. $5i\sqrt{6}$
51. $7i\sqrt{3}$
53. $7i\sqrt{5}$
55. -6
57. -15
59. $-6\sqrt{3}$
61. $-20\sqrt{2}$
63. $-6\sqrt{5}$

65. $-10\sqrt{2}$ **67.** $i\sqrt{3}$ **69.** $i\sqrt{3}$ **71.** $-i\sqrt{2}$

73. $-\frac{\sqrt{10}}{2}i$ **75.** $\frac{2\sqrt{3}}{3}$ **77.** $\frac{2\sqrt{3}}{3}$

Exercise 12.2, page 380

1. $1 + 0i$ **3.** $3 + 0i$ **5.** $0 - 3i$
7. $-1 + 0i$ **9.** $2 - i$ **11.** $3 + i\sqrt{6}$
13. $6 + 3i\sqrt{6}$ **15.** $5\sqrt{6} - 10i\sqrt{2}$ **17.** $6\sqrt{2} + 3i\sqrt{2}$
19. $3\sqrt{7} + 3i\sqrt{5}$ **21.** $-6\sqrt{2} + 4i\sqrt{3}$ **23.** $-2\sqrt{3} + 2i$
25. $-5\sqrt{2} + 2i\sqrt{5}$ **27.** $6\sqrt{3} + 2i\sqrt{3}$ **29.** $-6\sqrt{5} + 6i\sqrt{7}$
31. $-2\sqrt{5} + 4i\sqrt{5}$ **33.** $-3\sqrt{30} - 6i\sqrt{6}$ **35.** $-25 + 8i$
37. $73 + 0i$ **39.** $48 - 32i$

Exercise 12.3A, page 383

1. $8 + i$ **3.** $7 + i$ **5.** $-8 - 12i$ **7.** $3 + 0i$

9. $0 + 3i\sqrt{5}$ **11.** $0 + 0i$ **13.** 0, 3 **15.** $-\frac{3}{2}, 0$

17. 3, −2 **19.** 1, 1 **21.** 1, 3 **23.** 1, −1

25. 5, −6 **27.** 5, 3 **29.** 7, −3 **31.** $\frac{3}{2}, \frac{5}{3}$

Exercise 12.3B, page 385

1. $3 + i$ **3.** $23 + 10i$ **5.** $0 - 13i$ **7.** $18 - 4i$
9. $-8 - 6i$ **11.** $-3 + 3i\sqrt{2}$ **13.** $28 + 13i\sqrt{2}$ **15.** $1 + 4i\sqrt{3}$
17. $3 + 6i\sqrt{6}$ **19.** $-1 - 2i\sqrt{6}$ **21.** $8 + 6i$ **23.** $75 + 11i$

25. $-8 + 0i$ **27.** $0 - 8i$ **29.** $0 + 8i$ **31.** $-\frac{1}{4} - \frac{i}{2}$

33. $\frac{1}{5} - \frac{2}{5}i$ **35.** $\frac{2}{13} + \frac{3}{13}i$ **37.** $\frac{5}{29} + \frac{2}{29}i$ **39.** $\frac{1}{4} - \frac{\sqrt{3}}{4}i$

41. $-\frac{3}{5} + \frac{6}{5}i$ **43.** $0 + i$ **45.** $-\frac{1}{13} - \frac{8}{13}i$ **47.** $-1 - i$

49. $-\frac{2}{5}-\frac{9}{5}i$ **51.** $-\frac{1}{4}+\frac{7\sqrt{3}}{12}i$ **53.** $-\frac{1}{5}+\frac{2\sqrt{6}}{5}i$ **55.** $0+\frac{i}{2}$

57. $-\frac{1}{9}-\frac{2\sqrt{2}}{9}i$ **59.** $\frac{5}{34}-\frac{3}{34}i$ **61.** $\frac{57}{325}-\frac{i}{325}$ **63.** $2+3i$

65. $4-i$ **67.** $-3+i$ **69.** $3+i$ **71.** $1+2i$

Exercise 12.4, page 387

1 – 15

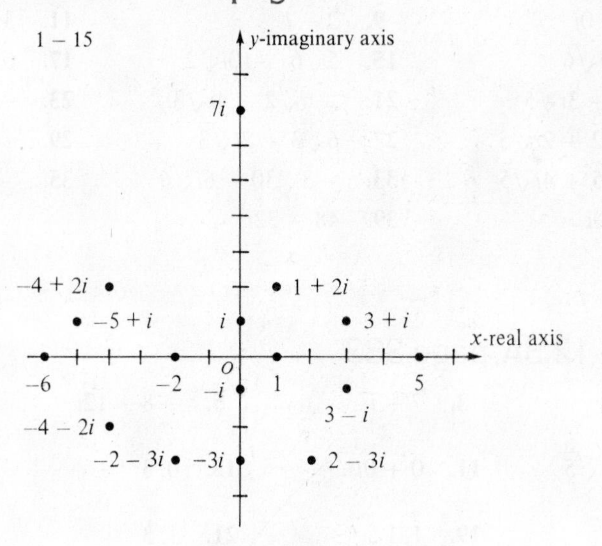

Chapter 12 Review, page 388

1. $-2, 1$ **3.** $3, 0$ **5.** $1, 1$ **7.** $1, 4$

9. $3, -2$ **11.** $6, 4$ **13.** $\frac{1}{4}, \frac{3}{4}$ **15.** $8, -5$

17. $1-7i$ **19.** $-1+3i$ **21.** $2+3i$ **23.** $4-i$
25. $8+3i$ **27.** $-1+i$ **29.** $1+3i$ **31.** $-3-11i$
33. $4+8i$ **35.** $-9-17i$ **37.** $11-45i$ **39.** $5-12i$

41. $-2-2i$ **43.** $20-i\sqrt{6}$ **45.** $2+17i\sqrt{3}$ **47.** $-\frac{1}{3}-2i$

49. $\frac{6}{5}-\frac{3}{5}i$ **51.** $-\frac{3}{5}+\frac{4}{5}i$ **53.** $\frac{8}{13}-\frac{1}{13}i$ **55.** $\frac{15}{53}-\frac{26}{53}i$

57. $-\frac{4}{13}+\frac{6}{13}i$ **59.** $\frac{3}{25}-\frac{4}{25}i$

Exercise 13.2, page 393

1. $\{-1, 0\}$ **3.** $\{0, 4\}$ **5.** $\left\{-\frac{1}{3}, 0\right\}$ **7.** $\{-3, 0\}$

9. $\left\{0, \frac{3}{2}\right\}$ **11.** $\{-1, 1\}$ **13.** $\{-5, 5\}$ **15.** $\{-3, 3\}$

17. $\{-2b, 2b\}$ **19.** $\{-\sqrt{3}, \sqrt{3}\}$ **21.** $\left\{-\frac{\sqrt{6}}{3}, \frac{\sqrt{6}}{3}\right\}$

23. $\left\{-\frac{\sqrt{35}}{7}, \frac{\sqrt{35}}{7}\right\}$ **25.** $\{-\sqrt{a}, \sqrt{a}\}$ **27.** $\left\{-\frac{\sqrt{2b}}{2}, \frac{\sqrt{2b}}{2}\right\}$

29. $\{-3a - 2b, 3a + 2b\}$ **31.** $\{-2a + 3b, 2a - 3b\}$

33. $\{a - 4, a + 4\}$ **35.** $\{-2a - 2b, -2a + 2b\}$

37. $\left\{\frac{-a-b}{2}, \frac{-a+b}{2}\right\}$ **39.** $\{-i\sqrt{2}, i\sqrt{2}\}$

41. $\{-3i\sqrt{2}, 3i\sqrt{2}\}$ **43.** $\left\{-\frac{i}{2}\sqrt{6}, \frac{i}{2}\sqrt{6}\right\}$ **45.** $\{-2, -1\}$

47. $\{-3, -2\}$ **49.** $\{-2, 1\}$ **51.** $\{-3, 1\}$ **53.** $\{-1, -1\}$

55. $\{2, 2\}$ **57.** $\{-8, 3\}$ **59.** $\{4, 6\}$ **61.** $\{-2, 9\}$

63. $\{3, 4\}$ **65.** $\{1, 2\}$ **67.** $\{2, 3\}$ **69.** $\left\{-1, -\frac{1}{2}\right\}$

71. $\left\{-2, -\frac{3}{2}\right\}$ **73.** $\left\{-\frac{1}{2}, 2\right\}$ **75.** $\left\{-4, \frac{1}{3}\right\}$ **77.** $\left\{-2, \frac{3}{4}\right\}$

79. $\left\{-\frac{1}{3}, \frac{2}{3}\right\}$ **81.** $\left\{-4, \frac{2}{3}\right\}$ **83.** $\left\{-3, \frac{1}{6}\right\}$ **85.** $\left\{\frac{2}{3}, \frac{4}{3}\right\}$

87. $\left\{\frac{2}{3}, 4\right\}$ **89.** $\left\{-\frac{3}{2}, -\frac{3}{2}\right\}$ **91.** $\left\{\frac{1}{2}, \frac{1}{2}\right\}$

93. $\left\{\frac{2}{3}, \frac{3}{2}\right\}$ **95.** $\left\{\frac{1}{2}, \frac{8}{3}\right\}$ **97.** $\left\{-\frac{4}{3}, -\frac{3}{4}\right\}$

99. $\{-2a, 3a\}$ **101.** $\left\{-\frac{a}{2}, \frac{3a}{2}\right\}$ **103.** $\left\{-4a, \frac{3a}{4}\right\}$

105. $\{-2b, a\}$ **107.** $\left\{\frac{a}{4}, 5b\right\}$ **109.** $\left\{-\frac{4b}{3}, \frac{3a}{2}\right\}$

111. $\{-2a - b, 2a + b\}$ **113.** $\left\{-\frac{a-2b}{3}, \frac{a-2b}{3}\right\}$

115. $\{2a - 2b, 2a + 2b\}$ **117.** $x^2 - 6x + 8 = 0$

119. $x^2 + x - 6 = 0$ **121.** $3x^2 - 7x + 2 = 0$ **123.** $3x^2 - 4x - 4 = 0$
125. $8x^2 - 6x + 1 = 0$ **127.** $6x^2 + 5x - 6 = 0$
129. $x^2 - 2ax + bx - 2ab = 0$ **131.** $x^2 - 3\sqrt{2}x + 4 = 0$
133. $x^2 + \sqrt{2}x - 12 = 0$ **135.** $4x^2 + 4\sqrt{3}x - 9 = 0$ **137.** $x^2 - 8x + 11 = 0$
139. $x^2 - 2x + 2 = 0$ **141.** $x^2 - 6x + 13 = 0$ **143.** $x^2 - 4x + 7 = 0$

Exercise 13.3, page 398

1. $1; (x+1)^2$ **3.** $16; (x+4)^2$ **5.** $4; (x-2)^2$

7. $\frac{9}{4}; \left(x + \frac{3}{2}\right)^2$ **9.** $\frac{81}{4}; \left(x + \frac{9}{2}\right)^2$ **11.** $\frac{9}{4}; \left(x - \frac{3}{2}\right)^2$

13. $\frac{1}{16}; \left(x + \frac{1}{4}\right)^2$ **15.** $\frac{9}{16}; \left(x + \frac{3}{4}\right)^2$ **17.** $\frac{9}{100}; \left(x - \frac{3}{10}\right)^2$

19. $a^2; (x+a)^2$ **21.** $\frac{25a^2}{4}; \left(x + \frac{5a}{2}\right)^2$ **23.** $\frac{49a^2}{4}; \left(x - \frac{7a}{2}\right)^2$

25. $\frac{a^2}{9}; \left(x + \frac{a}{3}\right)^2$ **27.** $\frac{9a^2}{16}; \left(x - \frac{3a}{4}\right)^2$ **29.** $\frac{b^2}{16a^2}; \left(x + \frac{b}{4a}\right)^2$

31. $\frac{3}{4}; \left(x + \frac{\sqrt{3}}{2}\right)^2$ **33.** $\frac{5}{2}; \left(x + \frac{\sqrt{10}}{2}\right)^2$ **35.** $\frac{9}{2}; \left(x - \frac{3\sqrt{2}}{2}\right)^2$

37. $\frac{3}{16}; \left(x - \frac{\sqrt{3}}{4}\right)^2$ **39.** $\frac{5}{16}; \left(x - \frac{\sqrt{5}}{4}\right)^2$ **41.** $\{-2, -1\}$

43. $\{-3, 2\}$ **45.** $\{-2, 3\}$ **47.** $\{2, 3\}$

49. $\left\{\frac{1}{3}, 2\right\}$ **51.** $\left\{-\frac{2}{3}, \frac{1}{2}\right\}$ **53.** $\{-2, 0\}$ **55.** $\{0, 7\}$

57. $\left\{\frac{1}{2} - \frac{\sqrt{17}}{2}, \frac{1}{2} + \frac{\sqrt{17}}{2}\right\}$ **59.** $\{1 - \sqrt{6}, 1 + \sqrt{6}\}$

61. $\left\{-\frac{3}{2} - \frac{\sqrt{5}}{2}, -\frac{3}{2} + \frac{\sqrt{5}}{2}\right\}$ **63.** $\{-1 - \sqrt{10}, -1 + \sqrt{10}\}$

65. $\left\{\frac{3}{2} - \frac{\sqrt{17}}{2}, \frac{3}{2} + \frac{\sqrt{17}}{2}\right\}$ **67.** $\left\{-\frac{7}{2} - \frac{\sqrt{37}}{2}, -\frac{7}{2} + \frac{\sqrt{37}}{2}\right\}$

69. $\left\{\frac{1}{4} - \frac{\sqrt{33}}{4}, \frac{1}{4} + \frac{\sqrt{33}}{4}\right\}$ **71.** $\left\{-\frac{5}{4} - \frac{\sqrt{17}}{4}, -\frac{5}{4} + \frac{\sqrt{17}}{4}\right\}$

73. $\left\{-\frac{7}{6} - \frac{\sqrt{13}}{6}, -\frac{7}{6} + \frac{\sqrt{13}}{6}\right\}$ **75.** $\left\{\frac{1}{5} - \frac{\sqrt{6}}{5}, \frac{1}{5} + \frac{\sqrt{6}}{5}\right\}$

77. $\{1 - i, 1 + i\}$

79. $\left\{-\frac{3}{2} - \frac{i\sqrt{19}}{2}, -\frac{3}{2} + \frac{i\sqrt{19}}{2}\right\}$

81. $\{-1 - i\sqrt{3}, -1 + i\sqrt{3}\}$

83. $\left\{\frac{1}{2} - \frac{i\sqrt{19}}{2}, \frac{1}{2} + \frac{i\sqrt{19}}{2}\right\}$

85. $\left\{-\frac{1}{4} - \frac{i\sqrt{31}}{4}, -\frac{1}{4} + \frac{i\sqrt{31}}{4}\right\}$

87. $\left\{-\frac{1}{3} - \frac{i\sqrt{5}}{3}, -\frac{1}{3} + \frac{i\sqrt{5}}{3}\right\}$

89. $\left\{\frac{3}{10} - \frac{i\sqrt{11}}{10}, \frac{3}{10} + \frac{i\sqrt{11}}{10}\right\}$

91. $\left\{-\frac{5}{12} - \frac{i\sqrt{23}}{12}, -\frac{5}{12} + \frac{i\sqrt{23}}{12}\right\}$

93. $\{-a - \sqrt{5}\,a, -a + \sqrt{5}\,a\}$

95. $\{a - \sqrt{6}\,a, a + \sqrt{6}\,a\}$

97. $\left\{\frac{3a}{2} - \frac{\sqrt{15}\,a}{2}, \frac{3a}{2} + \frac{\sqrt{15}\,a}{2}\right\}$

99. $\left\{-\frac{a}{6} - \frac{\sqrt{37}\,a}{6}, -\frac{a}{6} + \frac{\sqrt{37}\,a}{6}\right\}$

101. $\left\{\frac{3a}{2} + \frac{\sqrt{11}\,a}{2}i, \frac{3a}{2} - \frac{\sqrt{11}\,a}{2}i\right\}$

103. $\left\{\frac{2a}{3} + \frac{\sqrt{2}\,a}{3}i, \frac{2a}{3} - \frac{\sqrt{2}\,a}{3}i\right\}$

105. $\left\{-\frac{\sqrt{3}}{2} + \frac{\sqrt{7}}{2}, -\frac{\sqrt{3}}{2} - \frac{\sqrt{7}}{2}\right\}$

107. $\left\{\frac{\sqrt{5}}{2} - \frac{5}{2}, \frac{\sqrt{5}}{2} + \frac{5}{2}\right\}$

109. $\left\{\frac{\sqrt{7}}{2} + \frac{3}{2}i, \frac{\sqrt{7}}{2} - \frac{3}{2}i\right\}$

111. $\left\{-\frac{\sqrt{6}}{2} + \frac{3\sqrt{2}}{2}i, -\frac{\sqrt{6}}{2} - \frac{3\sqrt{2}}{2}i\right\}$

Exercise 13.4, page 402

1. $\{-3, -2\}$ **3.** $\{-5, -1\}$ **5.** $\{1, 3\}$ **7.** $\{-3, 1\}$

9. $\{-1, 3\}$ **11.** $\{-3, -3\}$ **13.** $\left\{\frac{3}{2}, \frac{3}{2}\right\}$ **15.** $\left\{-2, \frac{3}{2}\right\}$

17. $\{-5, 0\}$ **19.** $\{0, 1\}$ **21.** $\{-2, 2\}$ **23.** $\{-\sqrt{2}, \sqrt{2}\}$

25. $\{-2\sqrt{3}, 2\sqrt{3}\}$

27. $\left\{-\frac{1}{2} + \frac{\sqrt{5}}{2}, -\frac{1}{2} - \frac{\sqrt{5}}{2}\right\}$

29. $\left\{-\frac{1}{2} + \frac{\sqrt{17}}{2}, -\frac{1}{2} - \frac{\sqrt{17}}{2}\right\}$

31. $\left\{-\frac{1}{2} + \frac{\sqrt{21}}{2}, -\frac{1}{2} - \frac{\sqrt{21}}{2}\right\}$

33. $\left\{\frac{1}{2} + \frac{\sqrt{13}}{2}, \frac{1}{2} - \frac{\sqrt{13}}{2}\right\}$

35. $\{1 + \sqrt{5}, 1 - \sqrt{5}\}$

37. $\{-1 + \sqrt{7}, -1 - \sqrt{7}\}$

39. $\left\{\frac{3}{4} + \frac{\sqrt{17}}{4}, \frac{3}{4} - \frac{\sqrt{17}}{4}\right\}$

41. $\left\{-\frac{5}{4} + \frac{\sqrt{17}}{4}, -\frac{5}{4} - \frac{\sqrt{17}}{4}\right\}$

43. $\{-i, i\}$

45. $\{-3i, 3i\}$

47. $\{-i\sqrt{2}, i\sqrt{2}\}$

49. $\{-2i\sqrt{2}, 2i\sqrt{2}\}$

51. $\{-3i\sqrt{2}, 3i\sqrt{2}\}$

53. $\{1 + 2i, 1 - 2i\}$

55. $\{-1 + i, -1 - i\}$

57. $\{-1 + i\sqrt{2}, -1 - i\sqrt{2}\}$

59. $\left\{-\frac{1}{2} + \frac{i\sqrt{7}}{2}, -\frac{1}{2} - \frac{i\sqrt{7}}{2}\right\}$

61. $\left\{\frac{1}{2} + \frac{i}{2}, \frac{1}{2} - \frac{i}{2}\right\}$

63. $\left\{-1 + \frac{i\sqrt{2}}{2}, -1 - \frac{i\sqrt{2}}{2}\right\}$

65. $\left\{\frac{1}{3} + \frac{2i\sqrt{5}}{3}, \frac{1}{3} - \frac{2i\sqrt{5}}{3}\right\}$

67. $\left\{-\frac{5}{8} + \frac{i\sqrt{7}}{8}, -\frac{5}{8} - \frac{i\sqrt{7}}{8}\right\}$

69. $\left\{\frac{-1 + i\sqrt{35}}{6}, \frac{-1 - i\sqrt{35}}{6}\right\}$

71. $\{a + \sqrt{6}\,ai, a - \sqrt{6}\,ai\}$

73. $\left\{\frac{5a + \sqrt{23}\,ai}{6}, \frac{5a - \sqrt{23}\,ai}{6}\right\}$

75. $\left\{-\frac{4}{a}, \frac{2b}{a}\right\}$

77. $\left\{-\frac{3b}{2a}, b\right\}$

79. $\left\{-\frac{3}{2a-b}, \frac{4}{2a+b}\right\}$

81. $\left\{\frac{\sqrt{3}}{3}, \sqrt{3}\right\}$

83. $\{\sqrt{2} + i\sqrt{2}, \sqrt{2} - i\sqrt{2}\}$

85. $\{-\sqrt{3} + 2i, -\sqrt{3} - 2i\}$

87. $\left\{-\frac{2\sqrt{2}}{3}, \frac{\sqrt{2}}{2}\right\}$

89. $\left\{\frac{\sqrt{14} - 3\sqrt{7}}{7}, \frac{\sqrt{14} + 3\sqrt{7}}{7}\right\}$

91. $\left\{-i, \frac{7i}{2}\right\}$

93. $\left\{\frac{3 - \sqrt{15}}{2}i, \frac{3 + \sqrt{15}}{2}i\right\}$

Exercise 13.5, page 405

1. Two rational
3. Two rational
5. One double
7. One double
9. Two irrational
11. Two irrational
13. Two imaginary
15. Two imaginary
17. Two irrational
19. One double
21. Two rational
23. Two imaginary
25. Two rational
27. Two rational
29. Two irrational
31. Two imaginary
33. Two imaginary
35. Two real
37. Two real
39. Two real
41. Two imaginary
43. Two imaginary
45. One double
47. One double

49. $k < 4$

51. $k < \frac{9}{8}$

53. $k \in R$

55. $k \in R$

57. $k > -\frac{49}{4}$

59. $k < \frac{1}{3}$

61. $k = 1$

63. $k = \frac{13}{4}$

65. $k = \pm 2\sqrt{7}$ **67.** $\varnothing$ **69.** $k = \frac{9}{8}$ **71.** $k = -\frac{9}{2}$

73. $k > 1$ **75.** $k < -\frac{4}{3}$ **77.** $\varnothing$ **79.** $\varnothing$

81. $k > 1$ **83.** $k < -\frac{25}{6}$

Exercise 13.6, page 411

1. $-2; 3$ **3.** $4; 8$ **5.** $-4; -5$ **7.** $1; -4$

9. $-\frac{1}{3}; \frac{2}{3}$ **11.** $\frac{2}{3}; \frac{1}{3}$ **13.** $-\frac{6}{5}; -\frac{1}{5}$ **15.** $\frac{1}{3}; -\frac{5}{9}$

17. $-\frac{\sqrt{2}}{2}; -\frac{3\sqrt{2}}{2}$ **19.** $\sqrt{3}; \sqrt{6}$ **21.** $-\frac{1}{2}; -\frac{k}{2}$

23. $-\frac{1}{k}; \frac{5}{k}$ **25.** $-\frac{k+1}{2}; -\frac{3}{2}$ **27.** $-\frac{n}{m}; -\frac{k}{m}$

29. $m + 2n; \frac{m+2n}{m-2n}$ **31.** 1 **33.** 10

35. 3 **37.** -2 **39.** 3 **41.** -6
43. 5 **45.** -10 **47.** -16 **49.** -4
51. 3 **53.** 5 **55.** 7 **57.** 9
59. $6x^2 + 8x + 9 = 0$ **61.** $3x^2 + 5x - 12 = 0$ **63.** $x^2 - 2x - 2 = 0$
65. $x^2 + 2x - 15 = 0$ **67.** $x^2 - 4x - 32 = 0$
69. $3x^2 + 11x + 10 = 0; 3x^2 - 11x + 10 = 0$
71. $x^2 - x - 6 = 0; x^2 + x - 6 = 0$ **73.** $4x^2 - 8x + 3 = 0$
75. $4x^2 + 12x + 9 = 0; 4x^2 - 12x + 9 = 0$
77. $2x^2 + 17x + 36 = 0; 2x^2 - 17x + 36 = 0$
79. $12x^2 + 7x - 12 = 0; 12x^2 - 7x - 12 = 0$

Exercise 13.7, page 416

1. $\left\{-4, 0, \frac{3}{2}\right\}$ **3.** $\left\{-\frac{2}{3}, 0, \frac{5}{2}\right\}$

5. $\left\{-1, \frac{1+i\sqrt{3}}{2}, \frac{1-i\sqrt{3}}{2}\right\}$ **7.** $\left\{3, \frac{-3+3i\sqrt{3}}{2}, \frac{-3-3i\sqrt{3}}{2}\right\}$

9. $\{-1, 1, -i, i\}$ **11.** $\{-\sqrt{2}, \sqrt{2}, -i\sqrt{2}, i\sqrt{2}\}$

13. $\{-1, 1, -i\sqrt{2}, i\sqrt{2}\}$ **15.** $\{-3, 3, -i\sqrt{3}, i\sqrt{3}\}$

17. $\left\{-\frac{1}{3}, \frac{1}{3}, -\frac{i}{2}, \frac{i}{2}\right\}$

19. $\left\{\frac{-5+\sqrt{17}}{2}, \frac{-5-\sqrt{17}}{2}, \frac{5+\sqrt{17}}{2}, \frac{5-\sqrt{17}}{2}\right\}$

21. $\left\{\frac{-1+\sqrt{5}}{2}, \frac{-1-\sqrt{5}}{2}, \frac{1+\sqrt{5}}{2}, \frac{1-\sqrt{5}}{2}\right\}$

23. $\left\{\frac{-3+i\sqrt{3}}{2}, \frac{-3-i\sqrt{3}}{2}, \frac{3+i\sqrt{3}}{2}, \frac{3-i\sqrt{3}}{2}\right\}$

25. $\left\{-1, 1, \frac{1+i\sqrt{3}}{2}, \frac{1-i\sqrt{3}}{2}, \frac{-1+i\sqrt{3}}{2}, \frac{-1-i\sqrt{3}}{2}\right\}$

27. $\left\{-\frac{1}{2}, \frac{1}{3}, \frac{1}{4}+\frac{i\sqrt{3}}{4}, \frac{1}{4}-\frac{i\sqrt{3}}{4}, -\frac{1}{6}+\frac{i\sqrt{3}}{6}, -\frac{1}{6}-\frac{i\sqrt{3}}{6}\right\}$

29. $\{9\}$ **31.** $\varnothing$ **33.** $\left\{\frac{1}{16}, \frac{1}{9}\right\}$ **35.** $\{-27, -1\}$

37. $\left\{-64, \frac{1}{8}\right\}$ **39.** $\left\{-\frac{8}{27}, \frac{27}{8}\right\}$ **41.** $\left\{-\frac{3}{2}, \frac{1}{3}\right\}$ **43.** $\left\{-\frac{5}{2}, \frac{3}{2}\right\}$

45. $\left\{-\frac{1}{6}, 2\right\}$ **47.** $\{-2, 5\}$ **49.** $\left\{-\frac{5}{6}, 3\right\}$ **51.** $\{-3, -2\}$

53. $\{-6, 5\}$ **55.** $\{1, 2\}$ **57.** $\{1, 4\}$ **59.** $\{-3, 5\}$

61. $\{-2, 1\}$ **63.** $\{-5, 4\}$ **65.** $\{-2, 3\}$ **67.** $\{1\}$

69. $\{-1\}$ **71.** $\{-2, -1\}$ **73.** $\{3\}$ **75.** $\varnothing$

77. $\{-3, 1\}$ **79.** $\left\{\frac{1}{2}\right\}$ **81.** $\{-4, -2\}$ **83.** $\left\{\frac{1}{2}\right\}$

85. $\{3, 4\}$ **87.** $\{2\}$ **89.** $\{5, 6\}$ **91.** $\{8\}$

93. $\left\{-3, -\frac{8}{3}\right\}$ **95.** $\varnothing$

Exercise 13.8, page 420

1. 8, 9 **3.** 8, 17 **5.** 21, 29 **7.** 9, 17
9. 15, 17 **11.** 6, 12 **13.** 40 members
15. 104 hours, 65 hours **17.** 24 hours **19.** 18 feet, 7 feet
21. 20 square inches **23.** 12 feet, 18 feet **25.** 90 feet

27. $4\frac{1}{2}$ mph **29.** 160 mph **31.** $58\frac{1}{2}$ mph **33.** \$60

35. \$3.25 **37.** 20 seconds

Exercise 13.9, page 429

1. $V(0, 0)$; $x = 0$; $\{y \mid y \geq 0\}$

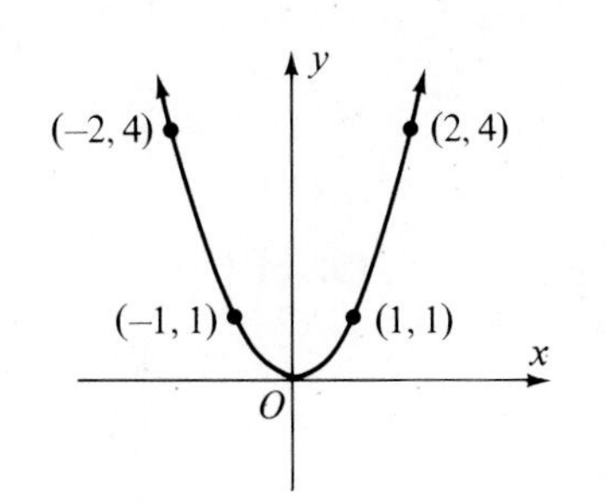

3. $V(0, -4)$; $x = 0$; $\{y \mid y \geq -4\}$

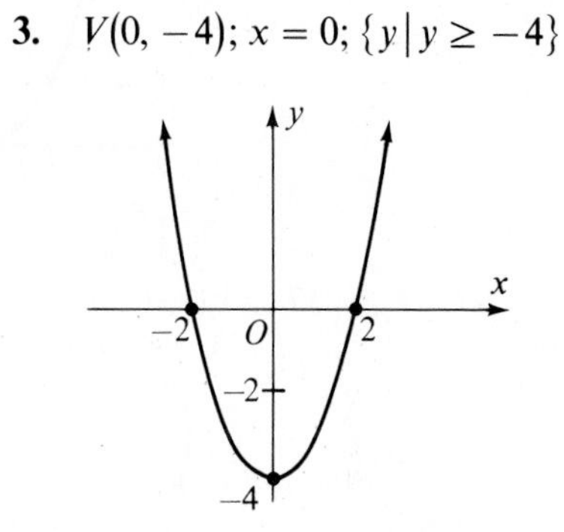

5. $V(0, 1)$; $x = 0$; $\{y \mid y \geq 1\}$

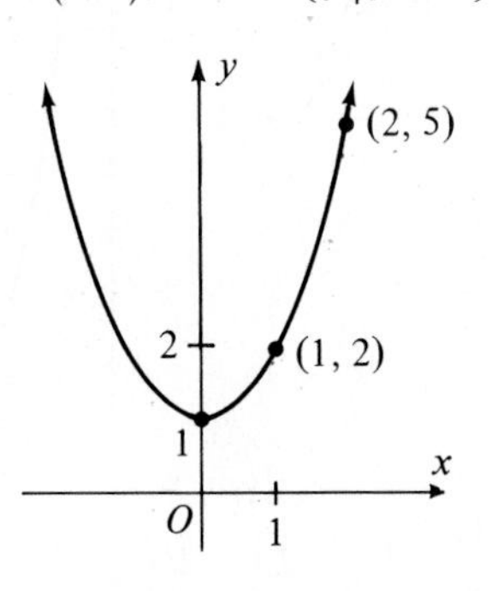

7. $V(0, 1)$; $x = 0$; $\{y \mid y \leq 1\}$

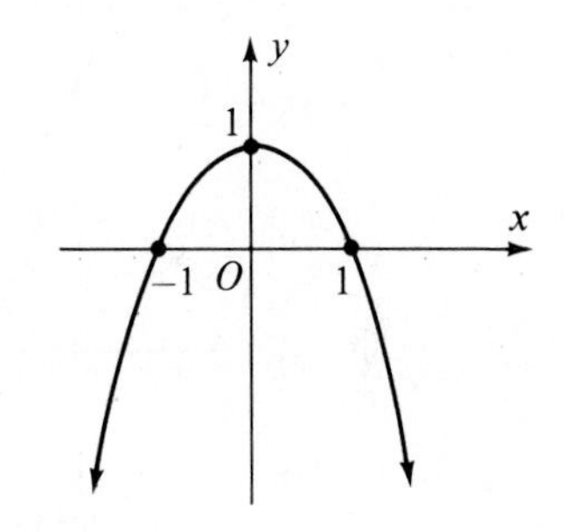

9. $V(0, 3)$; $x = 0$; $\{y \mid y \leq 3\}$

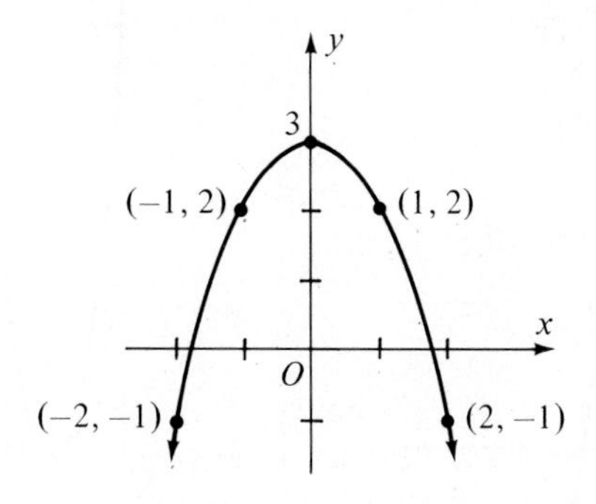

11. $V(1, 0)$; $x = 1$; $\{y \mid y \geq 0\}$

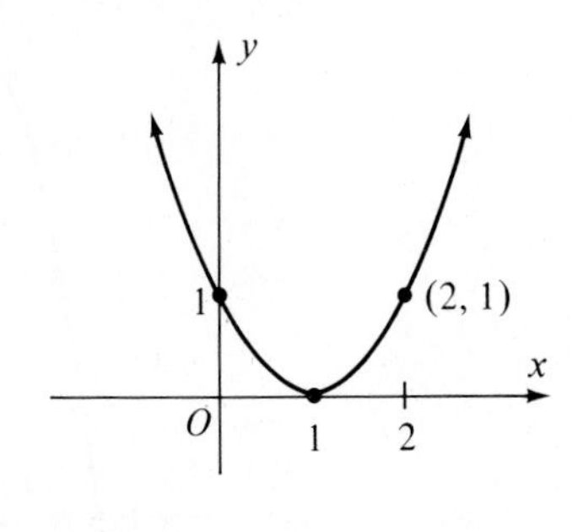

13. $V(-3, 0);\ x = -3;\ \{y \mid y \geq 0\}$

15. $V\left(\frac{3}{2}, -\frac{1}{4}\right);\ x = \frac{3}{2};\ \left\{y \,\middle|\, y \geq -\frac{1}{4}\right\}$

17. $V(-1, 2);\ x = -1;\ \{y \mid y \geq 2\}$

19. $V\left(\frac{5}{2}, -\frac{15}{2}\right);\ x = \frac{5}{2};\ \left\{y \,\middle|\, y \geq -\frac{15}{2}\right\}$

21. $V\left(\frac{1}{4}, \frac{49}{8}\right);\ x = \frac{1}{4};\ \left\{y \,\middle|\, y \leq \frac{49}{8}\right\}$

23. $V\left(\frac{3}{2}, 0\right);\ x = \frac{3}{2};\ \{y \mid y \geq 0\}$

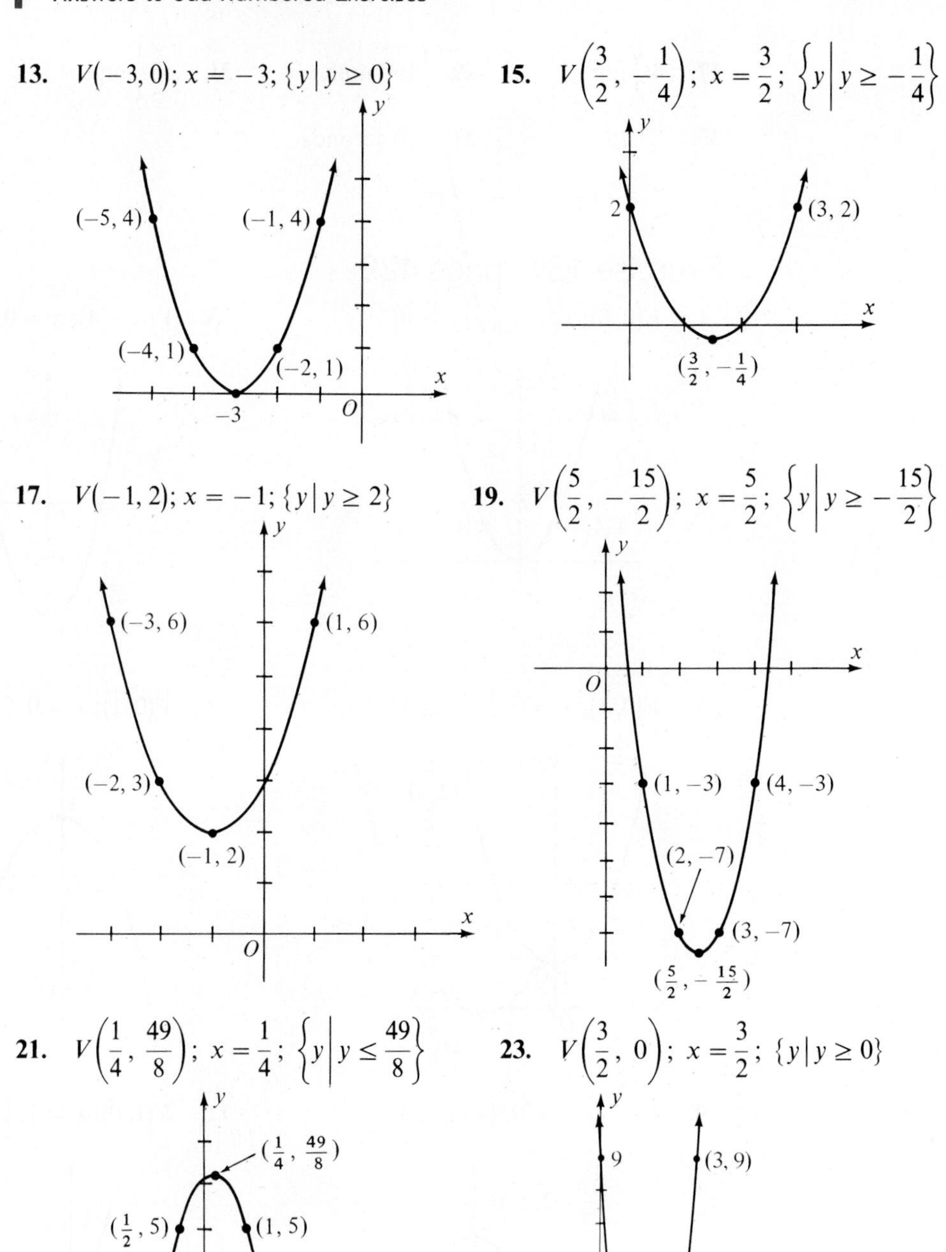

25. $V\left(-\frac{1}{2}, \frac{1}{2}\right)$; $x = -\frac{1}{2}$; $\left\{y \,\middle|\, y \geq \frac{1}{2}\right\}$

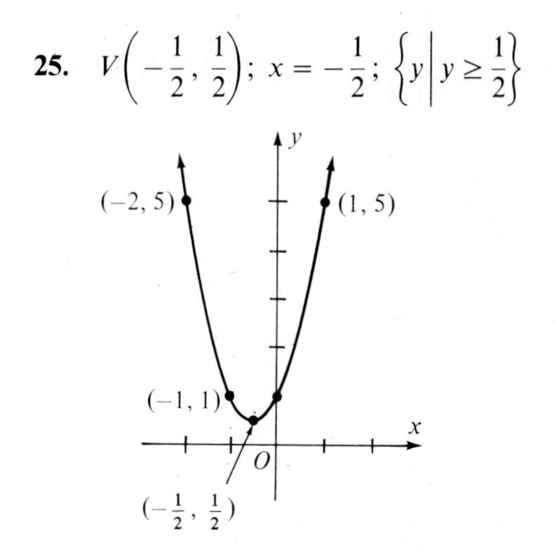

27. $V\left(-\frac{3}{2}, -\frac{29}{8}\right)$; $x = -\frac{3}{2}$; $\left\{y \,\middle|\, y \geq -\frac{29}{8}\right\}$

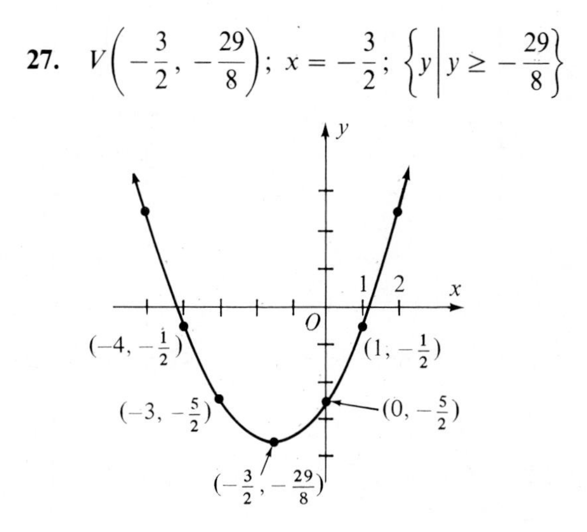

29. $V(-1, -1)$; $x = -1$; $\{y \mid y \geq -1\}$

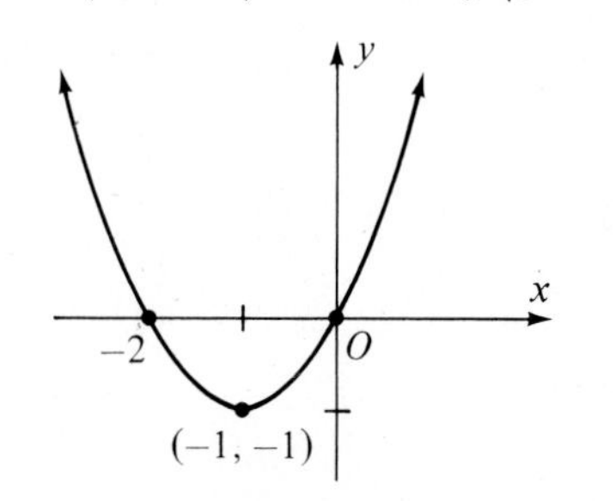

31. $V\left(-\frac{1}{2}, -\frac{1}{4}\right)$; $x = -\frac{1}{2}$; $\left\{y \middle| y \geq -\frac{1}{4}\right\}$

33. $V(3, -9)$; $x = 3$; $\{y \mid y \geq -9\}$

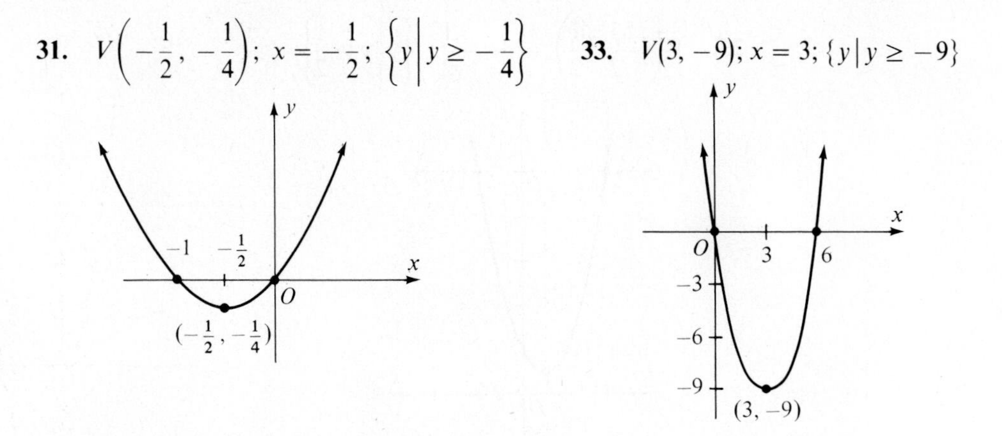

35. $V\left(\frac{5}{2}, -\frac{25}{4}\right)$; $x = \frac{5}{2}$; $\left\{y \middle| y \geq -\frac{25}{4}\right\}$

37. $\{-.4, 2.4\}$

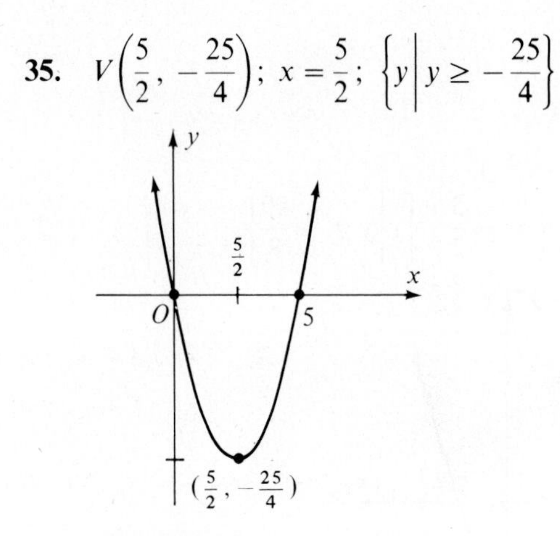

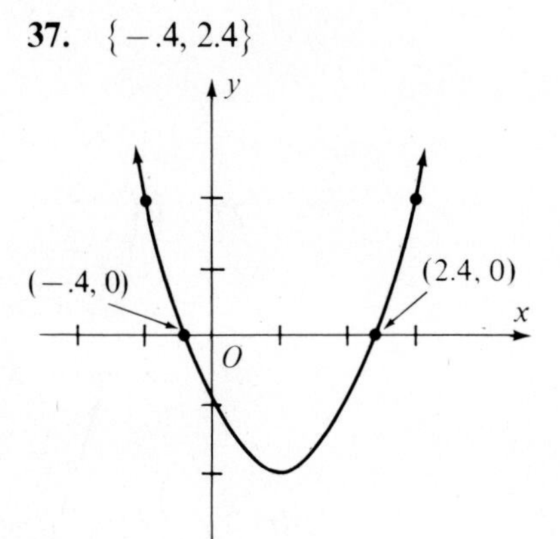

39. $\varnothing$

41. $\{-6, -1.3\}$

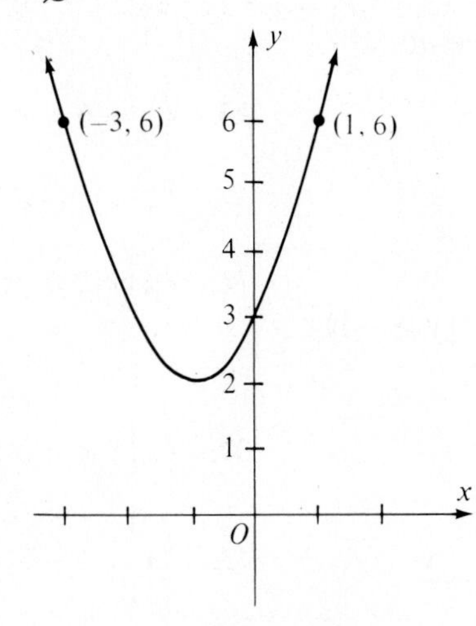

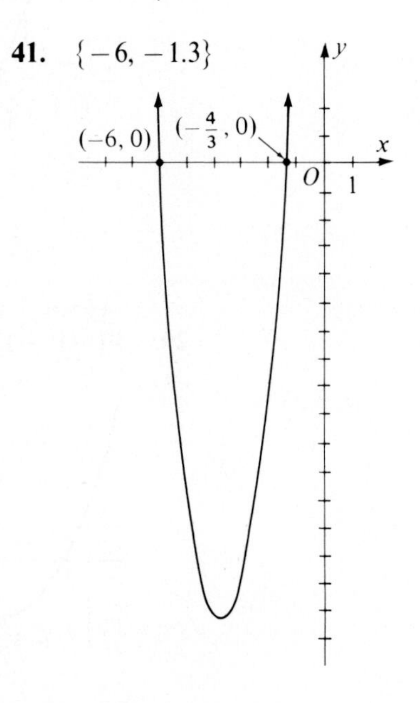

43. $\{-2.5, 1.5\}$

45. $\{-.2, 1.2\}$

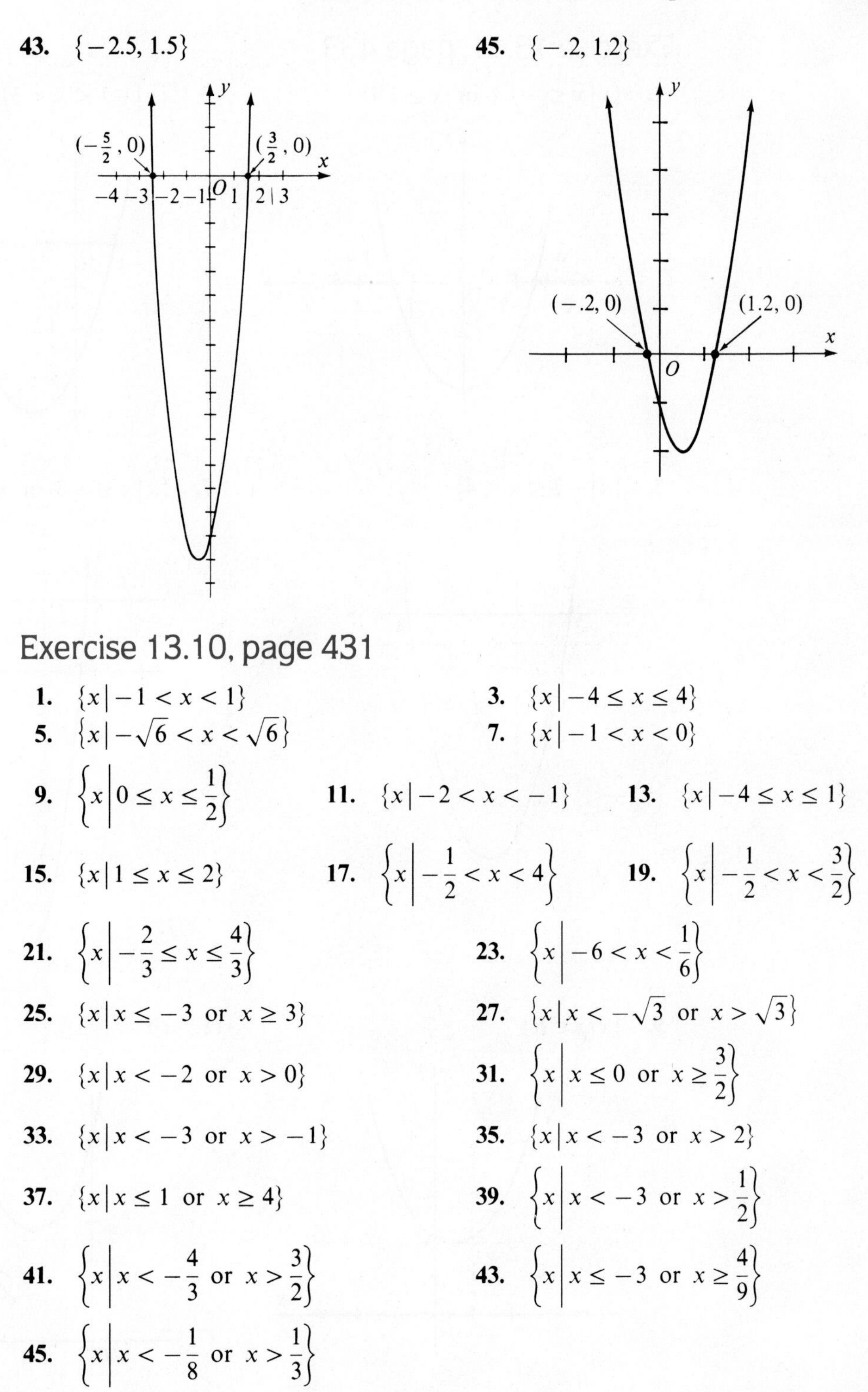

Exercise 13.10, page 431

1. $\{x \mid -1 < x < 1\}$

3. $\{x \mid -4 \le x \le 4\}$

5. $\{x \mid -\sqrt{6} < x < \sqrt{6}\}$

7. $\{x \mid -1 < x < 0\}$

9. $\left\{x \,\middle|\, 0 \le x \le \frac{1}{2}\right\}$

11. $\{x \mid -2 < x < -1\}$

13. $\{x \mid -4 \le x \le 1\}$

15. $\{x \mid 1 \le x \le 2\}$

17. $\left\{x \,\middle|\, -\frac{1}{2} < x < 4\right\}$

19. $\left\{x \,\middle|\, -\frac{1}{2} < x < \frac{3}{2}\right\}$

21. $\left\{x \,\middle|\, -\frac{2}{3} \le x \le \frac{4}{3}\right\}$

23. $\left\{x \,\middle|\, -6 < x < \frac{1}{6}\right\}$

25. $\{x \mid x \le -3 \text{ or } x \ge 3\}$

27. $\{x \mid x < -\sqrt{3} \text{ or } x > \sqrt{3}\}$

29. $\{x \mid x < -2 \text{ or } x > 0\}$

31. $\left\{x \,\middle|\, x \le 0 \text{ or } x \ge \frac{3}{2}\right\}$

33. $\{x \mid x < -3 \text{ or } x > -1\}$

35. $\{x \mid x < -3 \text{ or } x > 2\}$

37. $\{x \mid x \le 1 \text{ or } x \ge 4\}$

39. $\left\{x \,\middle|\, x < -3 \text{ or } x > \frac{1}{2}\right\}$

41. $\left\{x \,\middle|\, x < -\frac{4}{3} \text{ or } x > \frac{3}{2}\right\}$

43. $\left\{x \,\middle|\, x \le -3 \text{ or } x \ge \frac{4}{9}\right\}$

45. $\left\{x \,\middle|\, x < -\frac{1}{8} \text{ or } x > \frac{1}{3}\right\}$

Exercise 13.11, page 433

1. $\{x \mid x \leq -1.4 \text{ or } x \geq 1.4\}$

3. $\{x \mid -1 < x < 3\}$

5. $\{x \mid -2 \leq x \leq 4\}$

7. $\{x \mid x \leq -3 \text{ or } x \geq 7\}$

9. $\{x \mid x \in R\}$

11. $\varnothing$

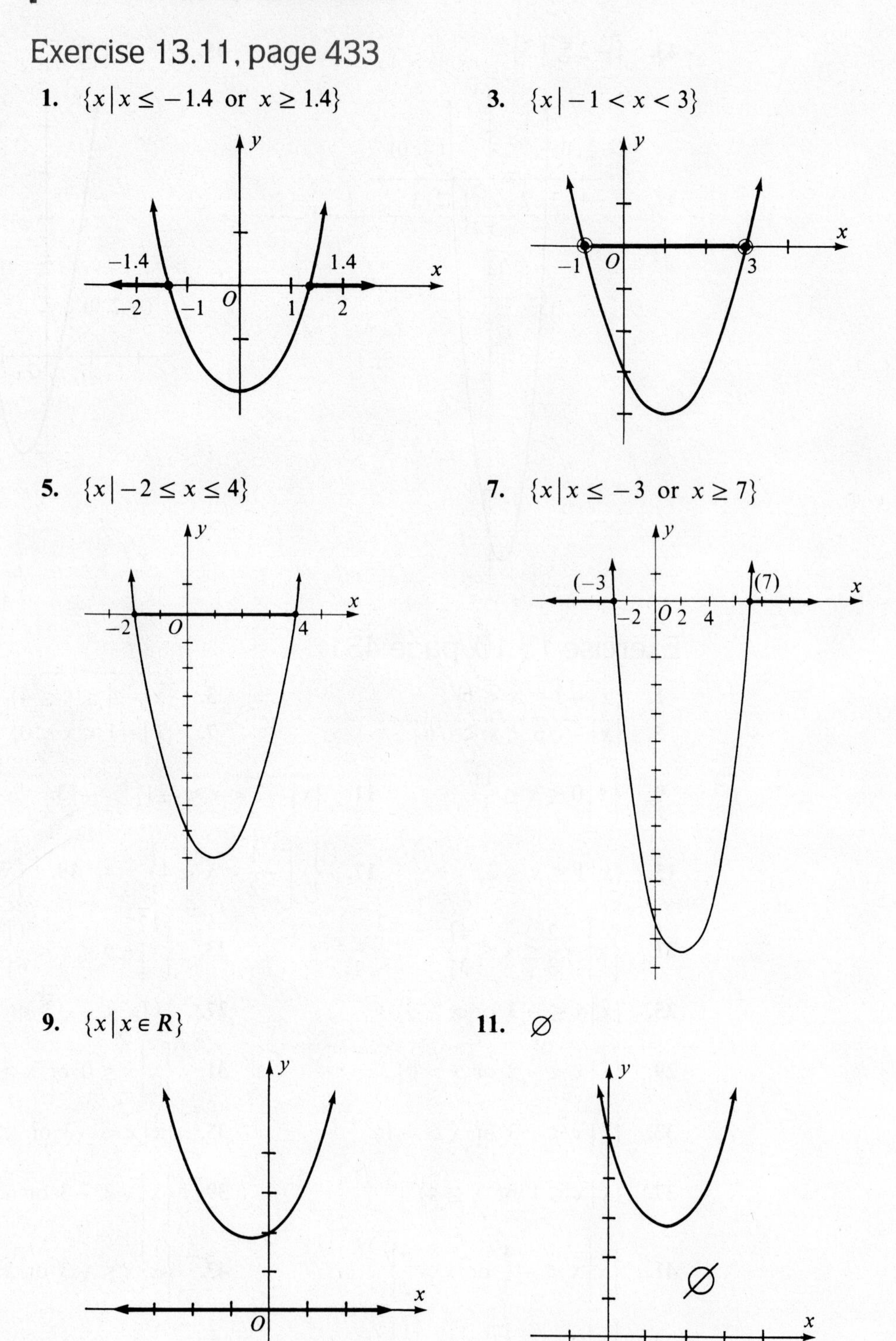

13. $\left\{x \middle| x \leq -\frac{1}{6} \text{ or } x \geq 4\right\}$

15. $\left\{x \middle| x \neq \frac{1}{2},\ x \in R\right\}$

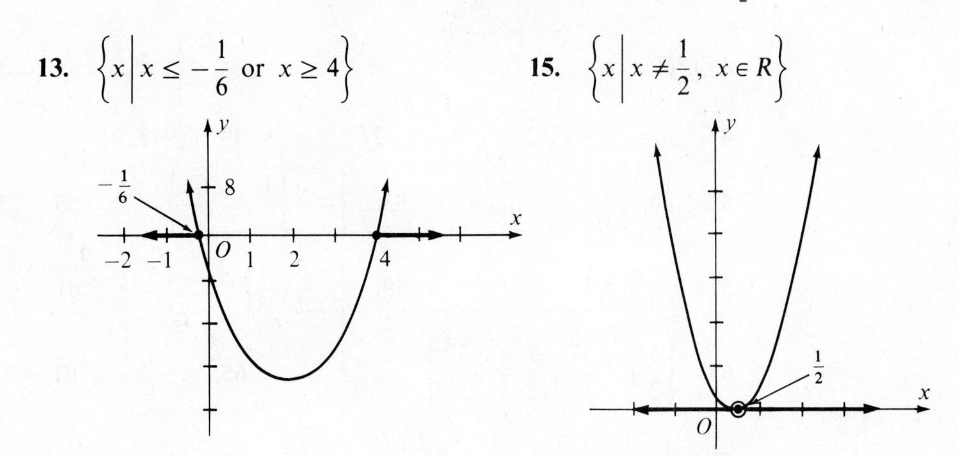

17. $\varnothing$

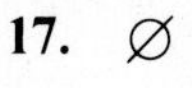

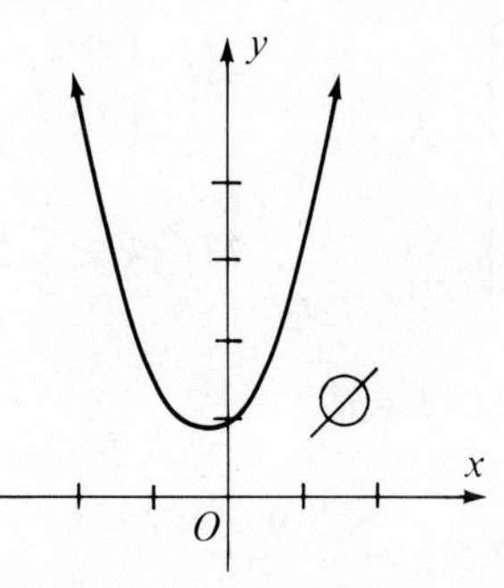

19. $\{x \mid -1.2 \leq x \leq .2\}$

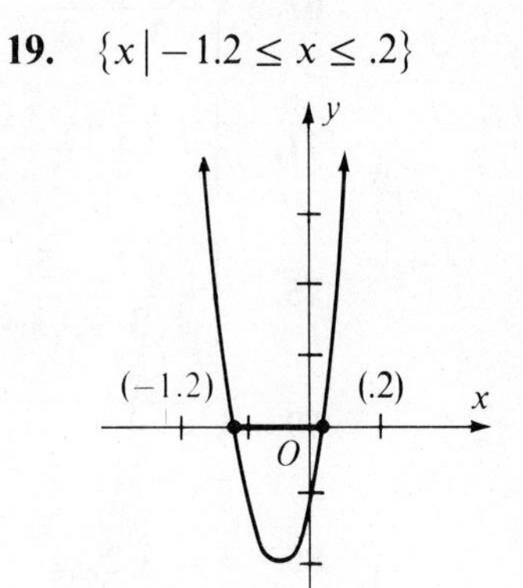

Chapter 13 Review, page 433

1. $\{-5, -4\}$ **3.** $\left\{-\frac{5}{2}, \frac{3}{4}\right\}$ **5.** $\left\{\frac{3}{2}, \frac{7}{3}\right\}$ **7.** $\left\{\frac{1}{4}, \frac{8}{3}\right\}$

9. $\{-4a, a\}$ **11.** $\{-2b, a\}$ **13.** $\{2a - 5, 2a + 5\}$

15. $\{3a + 2i\sqrt{6}, 3a - 2i\sqrt{6}\}$ **17.** $\{-3a + 2, 3a - 2\}$

19. $\{-5, 0\}$ **21.** $\{0, 7\}$ **23.** $\{-5, 2\}$ **25.** $\{-2, 6\}$

27. $\left\{-\frac{1}{2}, 3\right\}$ **29.** $\left\{\frac{4}{3}, \frac{3}{2}\right\}$

31. $\left\{-\frac{3}{2} + \frac{\sqrt{29}}{2}, -\frac{3}{2} - \frac{\sqrt{29}}{2}\right\}$ **33.** $\{1 + 3\sqrt{2}, 1 - 3\sqrt{2}\}$

35. $\left\{-\frac{1}{3} + \frac{\sqrt{19}}{3}, -\frac{1}{3} - \frac{\sqrt{19}}{3}\right\}$ **37.** $\left\{\frac{1}{2} + \frac{\sqrt{57}}{6}, \frac{1}{2} - \frac{\sqrt{57}}{6}\right\}$

39. $\{-1 + i\sqrt{7}, -1 - i\sqrt{7}\}$ **41.** $\left\{-\frac{1}{4} + \frac{i\sqrt{15}}{4}, -\frac{1}{4} - \frac{i\sqrt{15}}{4}\right\}$

43. $\{-\sqrt{3}, 2\sqrt{3}\}$

45. $\left\{-\frac{\sqrt{5}}{2}-\frac{1}{2}, -\frac{\sqrt{5}}{2}+\frac{1}{2}\right\}$

47. $\{-\sqrt{5}+i\sqrt{2}, -\sqrt{5}-i\sqrt{2}\}$

49. $\{-3, 0\}$

51. $\left\{0, \frac{5}{2}\right\}$

53. $\left\{-\frac{\sqrt{14}}{2}, \frac{\sqrt{14}}{2}\right\}$

55. $\{3, 6\}$

57. $\{-3, 5\}$

59. $\left\{-\frac{3}{2}, \frac{1}{3}\right\}$

61. $\left\{-\frac{5}{2}, \frac{1}{2}\right\}$

63. $\left\{\frac{5}{2}+\frac{\sqrt{37}}{2}, \frac{5}{2}-\frac{\sqrt{37}}{2}\right\}$

65. $\{-2+\sqrt{10}, -2-\sqrt{10}\}$

67. $\left\{-\frac{1}{3}+\frac{\sqrt{7}}{3}, -\frac{1}{3}-\frac{\sqrt{7}}{3}\right\}$

69. $\{1+i\sqrt{5}, 1-i\sqrt{5}\}$

71. $\left\{-\frac{3}{2}+\frac{i\sqrt{3}}{2}, -\frac{3}{2}-\frac{i\sqrt{3}}{2}\right\}$

73. $\left\{\frac{7}{12}+\frac{i\sqrt{47}}{12}, \frac{7}{12}-\frac{i\sqrt{47}}{12}\right\}$

75. $\left\{-\frac{\sqrt{2}}{3}+\frac{5i}{3}, -\frac{\sqrt{2}}{3}-\frac{5i}{3}\right\}$

77. $k<\frac{9}{4}$

79. $k<\frac{49}{8}$

81. $k \in R$

83. $k=\frac{49}{4}$

85. $k=\pm 2\sqrt{2}$

87. $k=-\frac{9}{32}$

89. $k>\frac{9}{16}$

91. $\varnothing$

93. $k>\frac{25}{12}$

95. 10

97. 8

99. -2

101. -51

103. 4

105. 4

107. -13

109. $x^2-11x+24=0$

111. $9x^2-27x+14=0$

113. $8x^2-2x-3=0; 8x^2+2x-3=0$

115. $9x^2-15x+4=0; 9x^2+15x+4=0$

117. $\left\{0, \frac{-1+i\sqrt{23}}{8}, \frac{-1-i\sqrt{23}}{8}\right\}$

119. $\left\{-\frac{1}{2}, \frac{1+i\sqrt{3}}{4}, \frac{1-i\sqrt{3}}{4}\right\}$

121. $\left\{\frac{3}{2}, \frac{-3+3i\sqrt{3}}{4}, \frac{-3-3i\sqrt{3}}{4}\right\}$

123. $\{-3, 3, -3i, 3i\}$

125. $\left\{-2, 2, -\frac{\sqrt{6}}{2}i, \frac{\sqrt{6}}{2}i\right\}$

127. $\{-2-\sqrt{3}, -2+\sqrt{3}, 2-\sqrt{3}, 2+\sqrt{3}\}$

129. $\{1, 4\}$

131. $\left\{\frac{1}{9}, 9\right\}$

133. $\{1, 27\}$

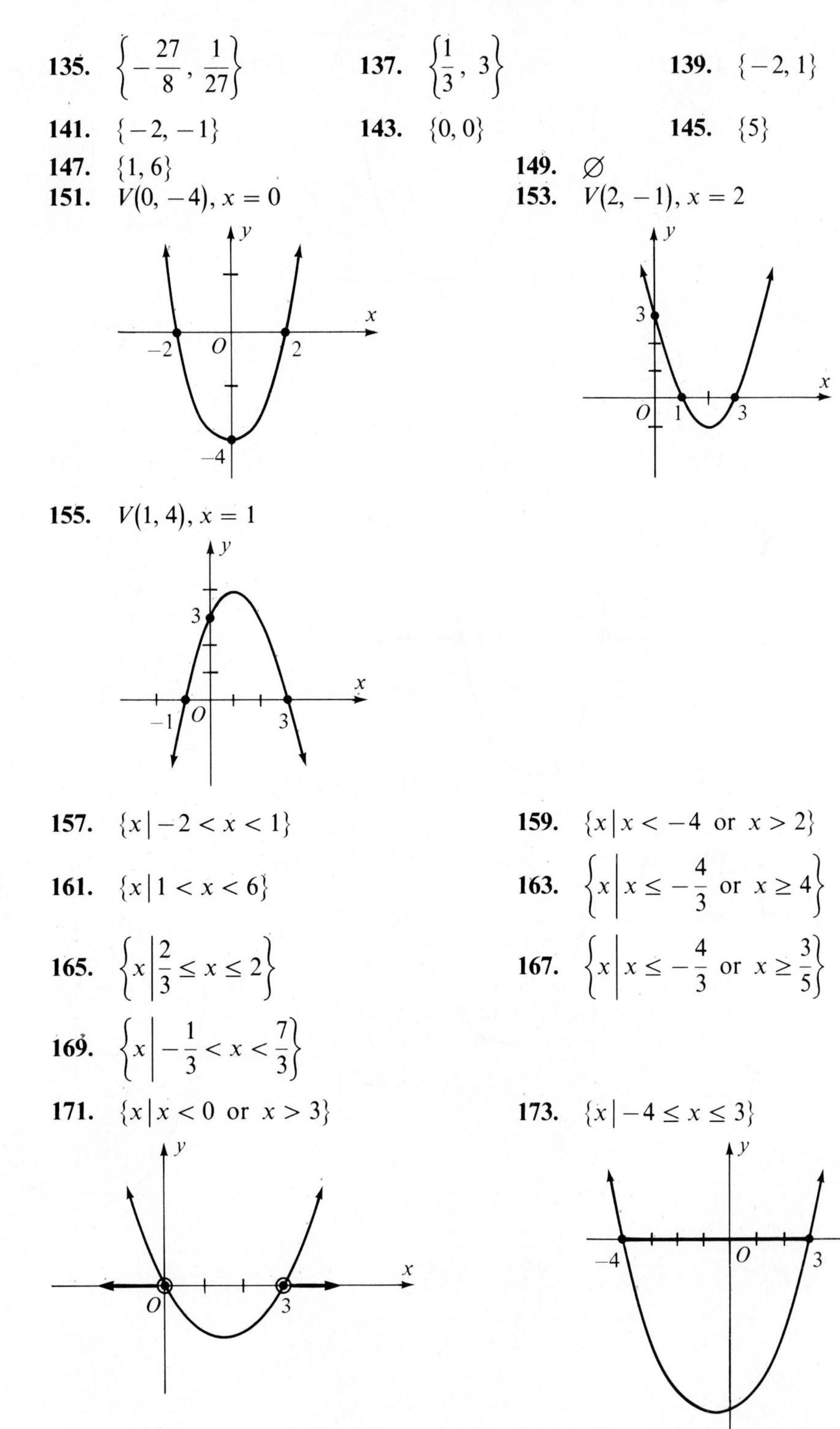

135. $\left\{-\frac{27}{8}, \frac{1}{27}\right\}$

137. $\left\{\frac{1}{3}, 3\right\}$

139. $\{-2, 1\}$

141. $\{-2, -1\}$

143. $\{0, 0\}$

145. $\{5\}$

147. $\{1, 6\}$

149. $\varnothing$

151. $V(0, -4), x = 0$

153. $V(2, -1), x = 2$

155. $V(1, 4), x = 1$

157. $\{x \mid -2 < x < 1\}$

159. $\{x \mid x < -4 \text{ or } x > 2\}$

161. $\{x \mid 1 < x < 6\}$

163. $\left\{x \,\middle|\, x \leq -\frac{4}{3} \text{ or } x \geq 4\right\}$

165. $\left\{x \,\middle|\, \frac{2}{3} \leq x \leq 2\right\}$

167. $\left\{x \,\middle|\, x \leq -\frac{4}{3} \text{ or } x \geq \frac{3}{5}\right\}$

169. $\left\{x \,\middle|\, -\frac{1}{3} < x < \frac{7}{3}\right\}$

171. $\{x \mid x < 0 \text{ or } x > 3\}$

173. $\{x \mid -4 \leq x \leq 3\}$

175. $\{5\}$

177. $\left\{x \,\middle|\, x \leq -\frac{2}{3} \text{ or } x \geq 3\right\}$

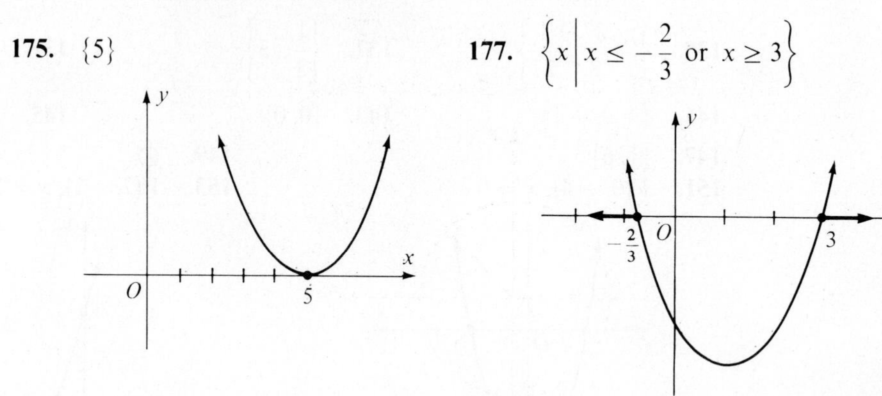

179. $\left\{x \,\middle|\, x < 1 \text{ or } x > \frac{9}{2}\right\}$

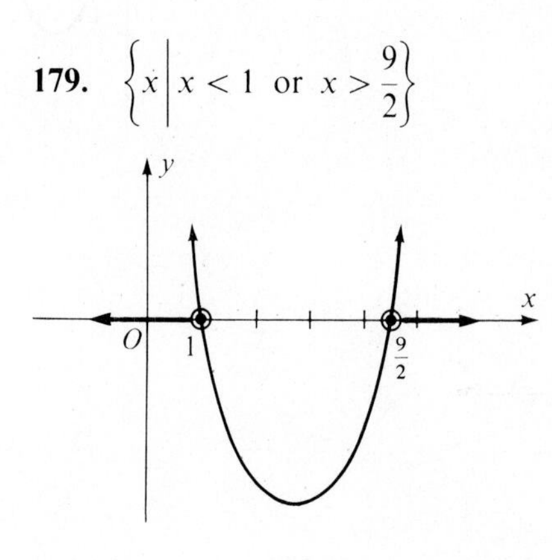

181. $\{x \mid .3 \leq x \leq 3.7\}$

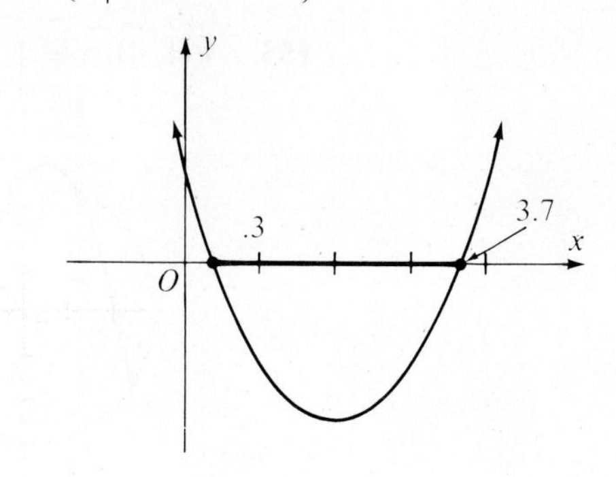

183. $\varnothing$

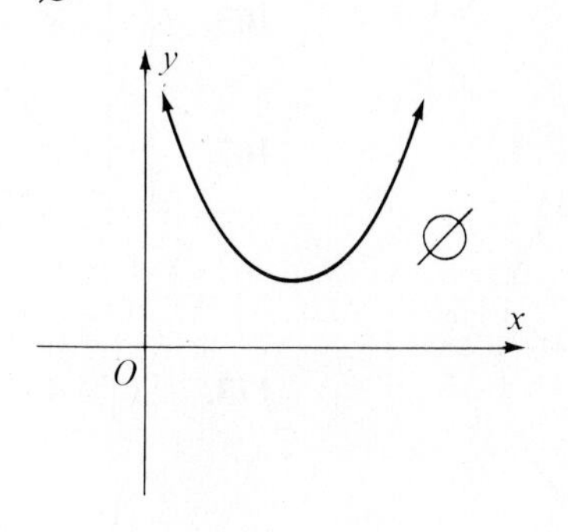

185. $\left\{x \,\middle|\, x \leq -3 \text{ or } x \geq \frac{5}{2}\right\}$

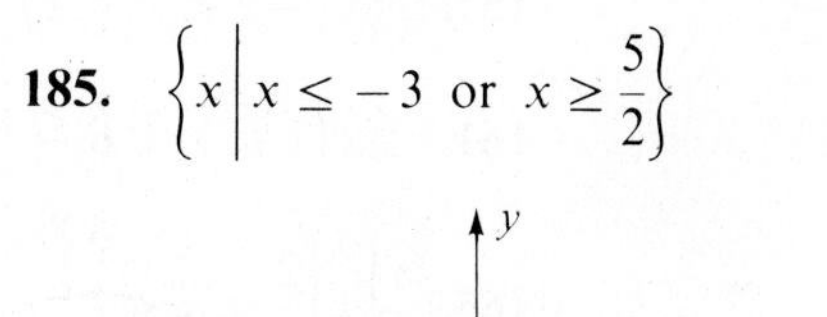

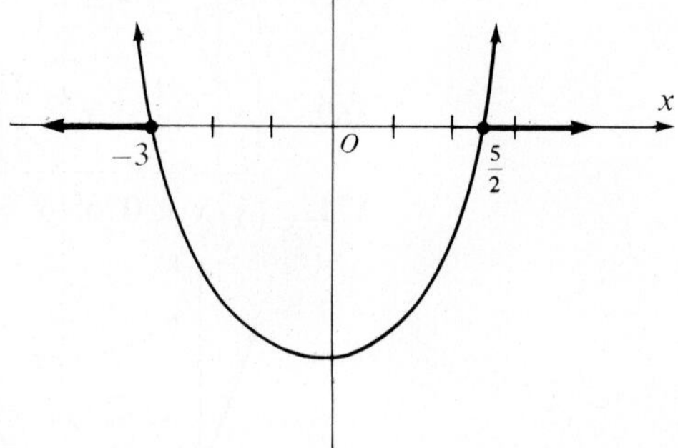

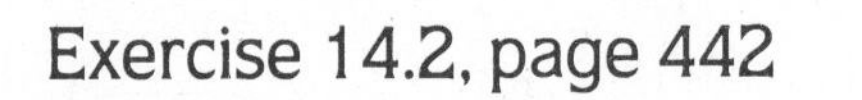

Exercise 14.2, page 442

1. $C(0, 0)$; 1; $-1 \le x \le 1$; $-1 \le y \le 1$

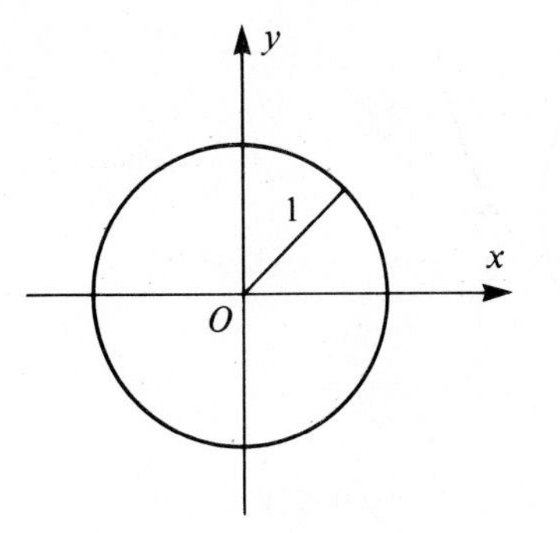

3. $C(0, 0)$; 3; $-3 \le x \le 3$; $-3 \le y \le 3$

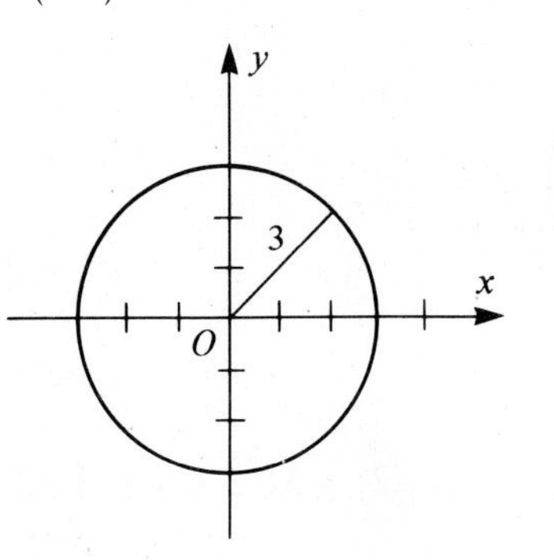

5. $C(0, 0)$; 5; $-5 \le x \le 5$; $-5 \le y \le 5$

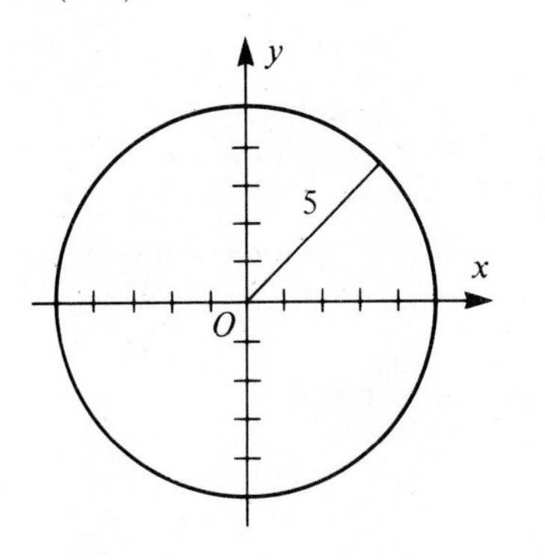

7. $C(0, 0)$; 7; $-7 \leq x \leq 7$; $-7 \leq y \leq 7$

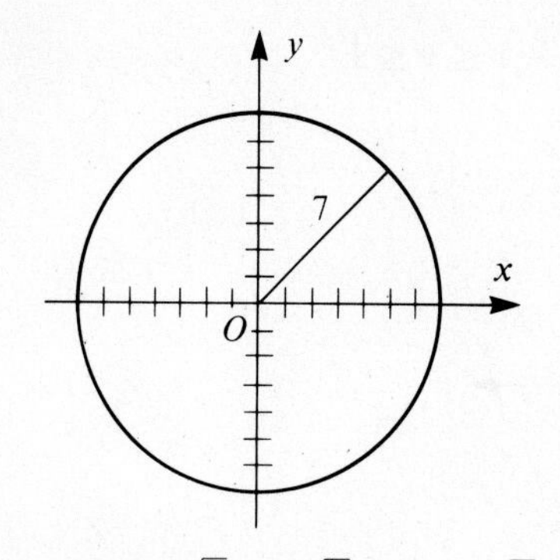

9. $C(0, 0)$; $\sqrt{6}$; $-\sqrt{6} \leq x \leq \sqrt{6}$; $-\sqrt{6} \leq y \leq \sqrt{6}$

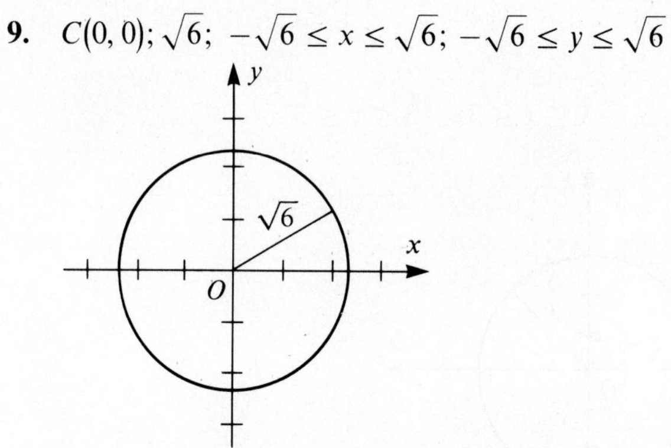

11. $C(0, 0)$; $3\sqrt{2}$; $-3\sqrt{2} \leq x \leq 3\sqrt{2}$; $-3\sqrt{2} \leq y \leq 3\sqrt{2}$

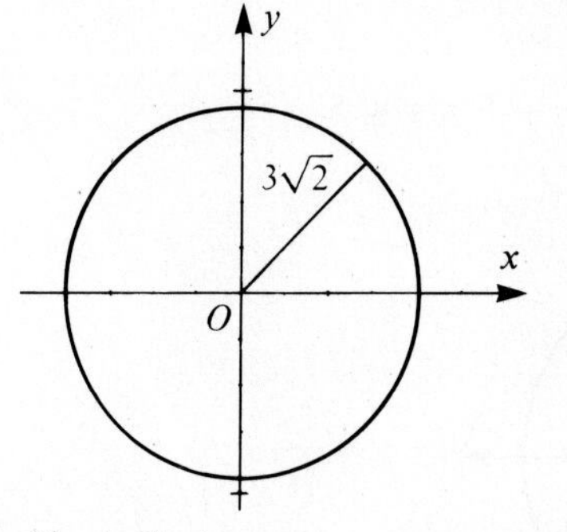

13. $C(-1, 0)$; 4; $-5 \leq x \leq 3$; $-4 \leq y \leq 4$

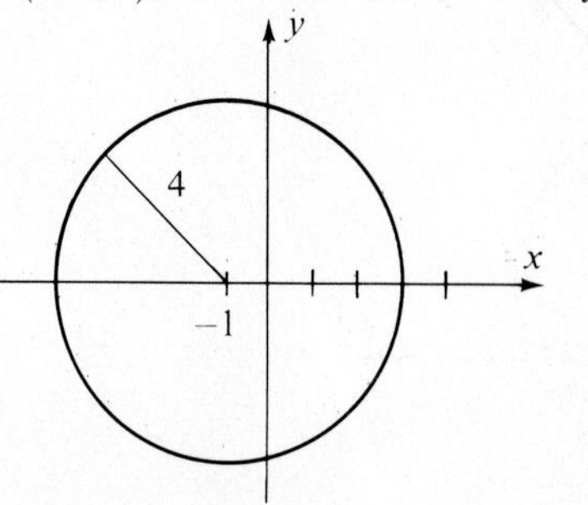

15. $C(3, 0)$; 1; $2 \le x \le 4$; $-1 \le y \le 1$

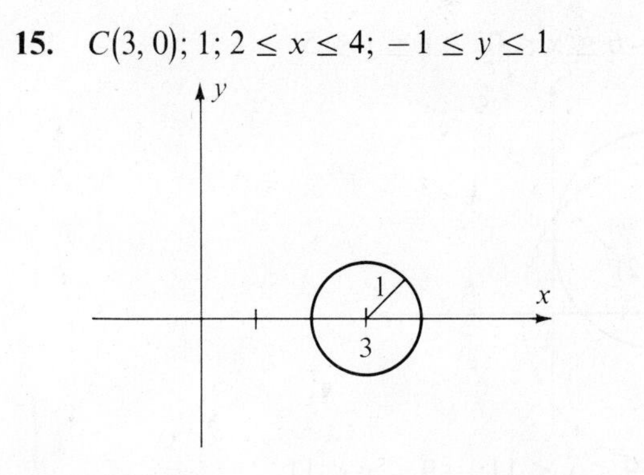

17. $C(0, -3)$; 6; $-6 \le x \le 6$; $-9 \le y \le 3$

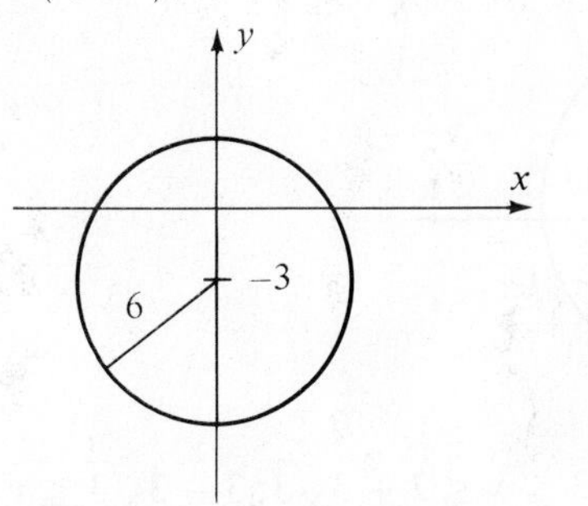

19. $C(0, 4)$; $2\sqrt{2}$; $-2\sqrt{2} \le x \le 2\sqrt{2}$; $4 - 2\sqrt{2} \le y \le 4 + 2\sqrt{2}$

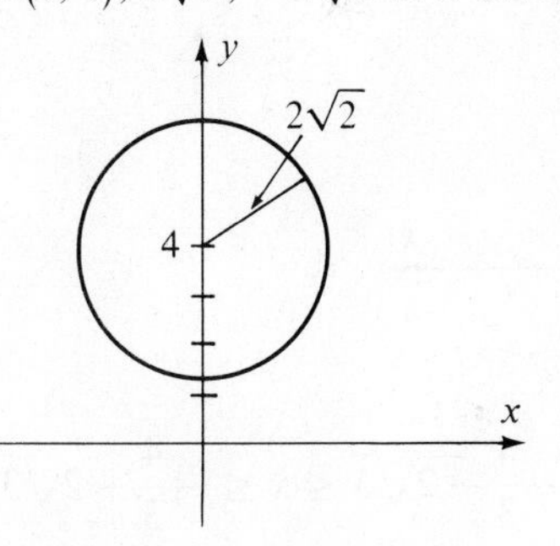

21. $C(-1, 2)$; 5; $-6 \le x \le 4$; $-3 \le y \le 7$

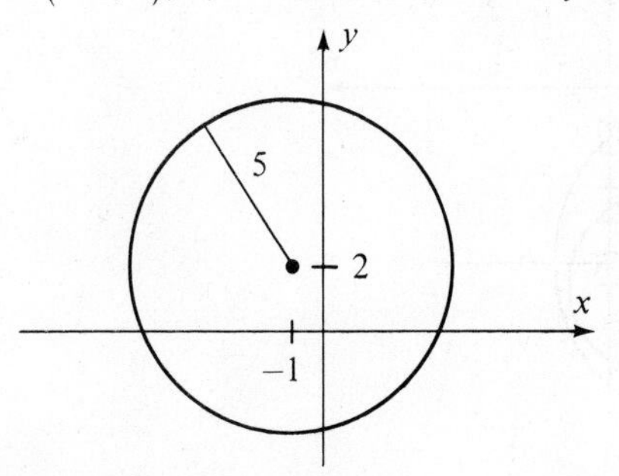

23. $C(-3, 2)$; 3; $-6 \le x \le 0$; $-1 \le y \le 5$

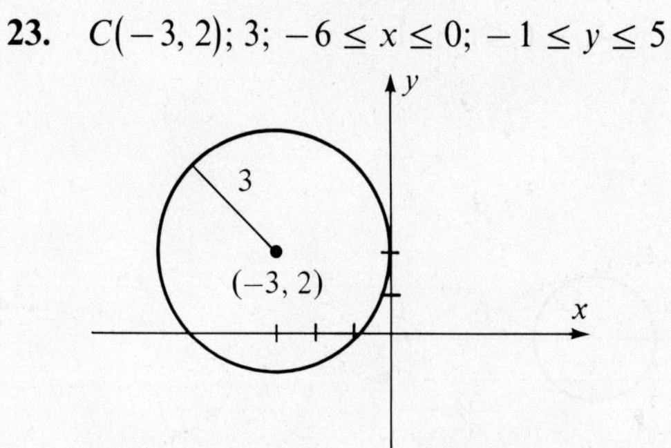

25. $C(1, 1)$; 10; $-9 \le x \le 11$; $-9 \le y \le 11$

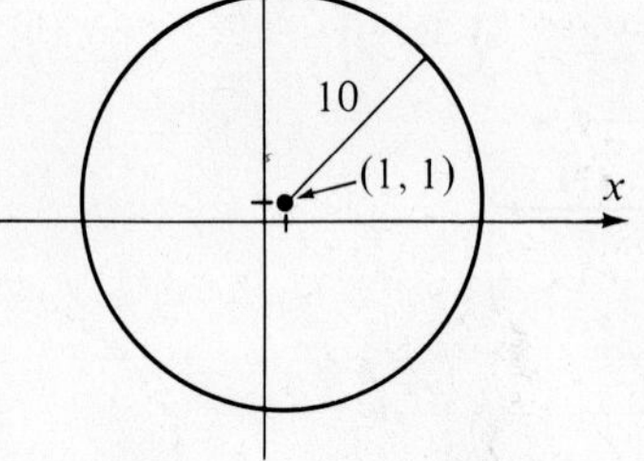

27. $C(2, 3)$; $3\sqrt{3}$; $2 - 3\sqrt{3} \le x \le 2 + 3\sqrt{3}$; $3 - 3\sqrt{3} \le y \le 3 + 3\sqrt{3}$

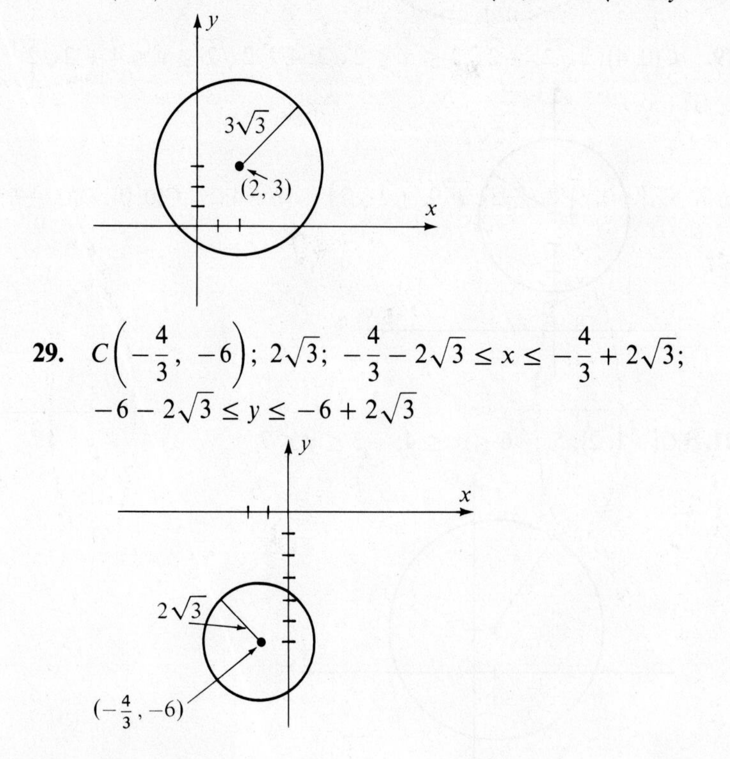

29. $C\left(-\frac{4}{3}, -6\right)$; $2\sqrt{3}$; $-\frac{4}{3} - 2\sqrt{3} \le x \le -\frac{4}{3} + 2\sqrt{3}$;
$-6 - 2\sqrt{3} \le y \le -6 + 2\sqrt{3}$

Exercise 14.3, page 447

1. $C(0, 0)$; $V(\pm 2, 0)$; $F(\pm\sqrt{3}, 0)$

3. $C(0, 0)$; $V(\pm 3, 0)$; $F(\pm 2\sqrt{2}, 0)$

5. $C(0, 0)$; $V(\pm 3, 0)$; $F(\pm\sqrt{5}, 0)$

7. $C(0, 0)$; $V(\pm 4, 0)$; $F(\pm\sqrt{7}, 0)$

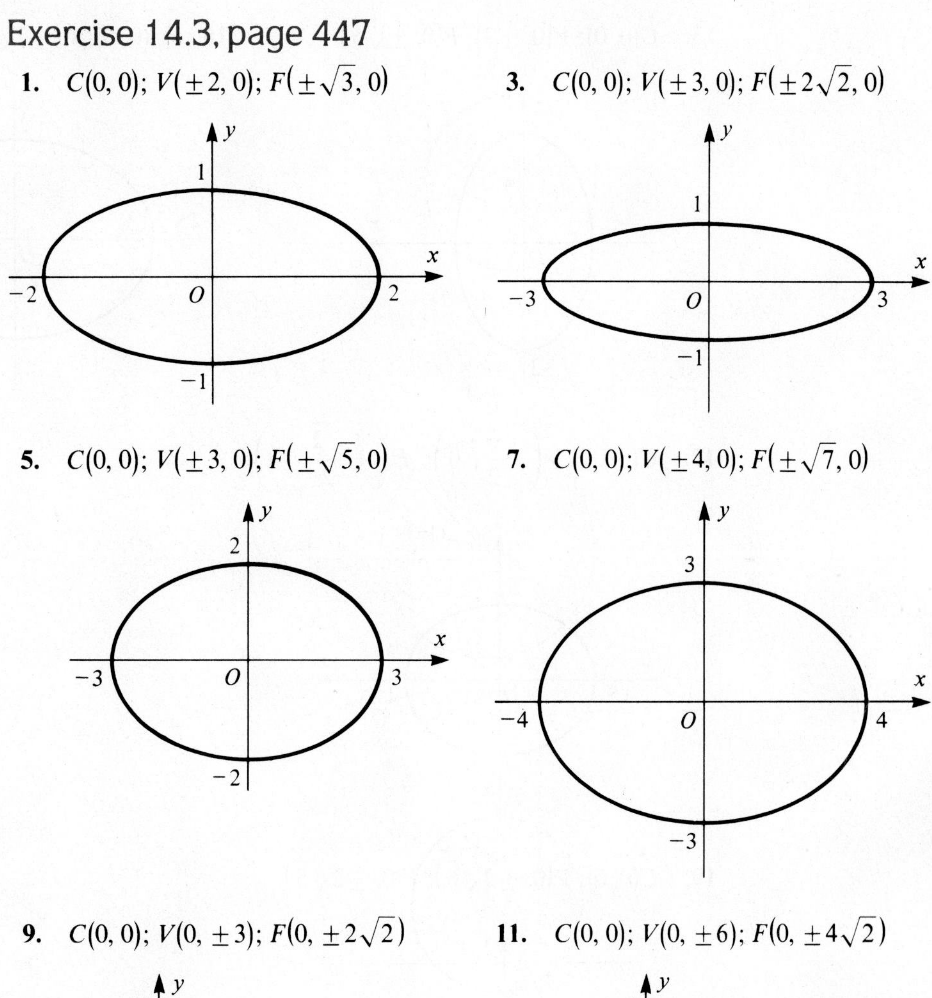

9. $C(0, 0)$; $V(0, \pm 3)$; $F(0, \pm 2\sqrt{2})$

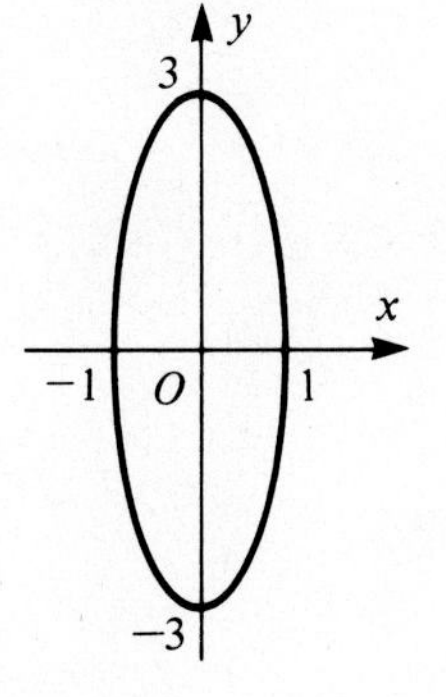

11. $C(0, 0)$; $V(0, \pm 6)$; $F(0, \pm 4\sqrt{2})$

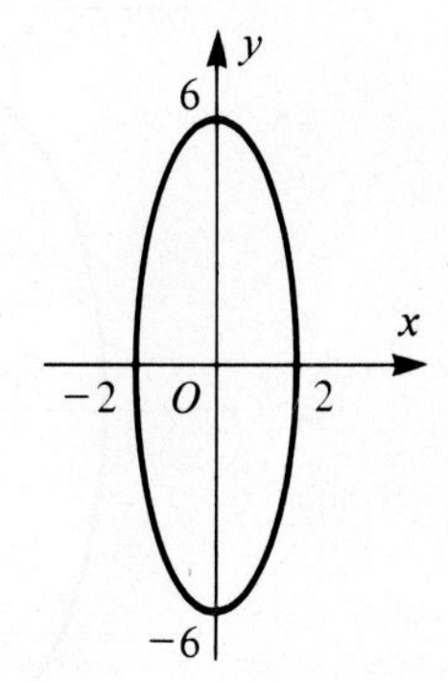

13. $C(0, 0)$; $V(0, \pm 3)$; $F(0, \pm\sqrt{5})$

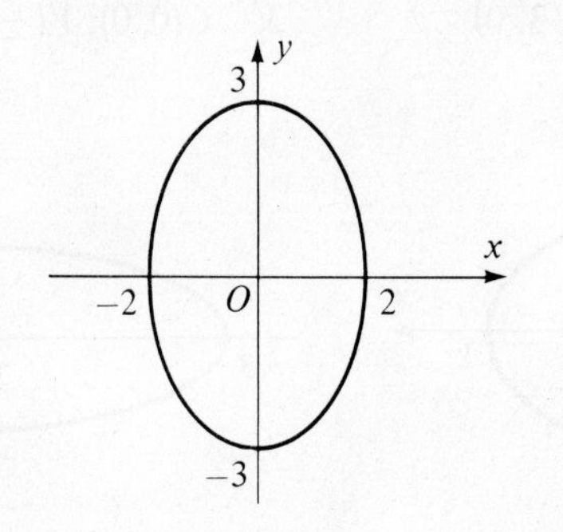

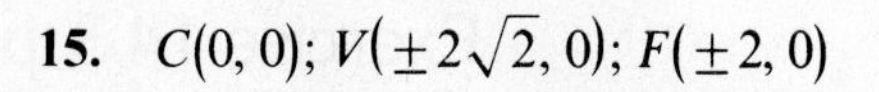

15. $C(0, 0)$; $V(\pm 2\sqrt{2}, 0)$; $F(\pm 2, 0)$

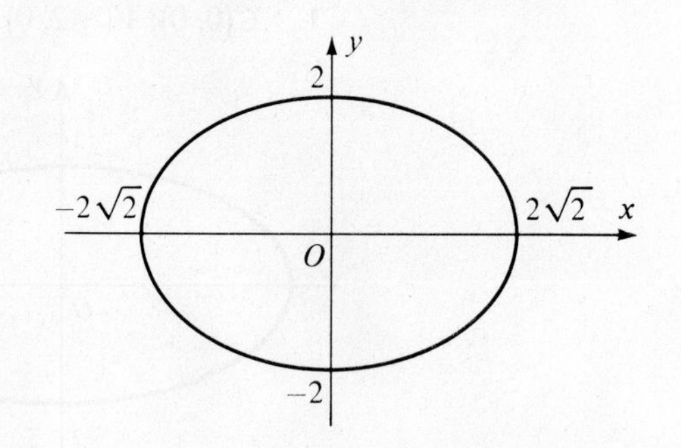

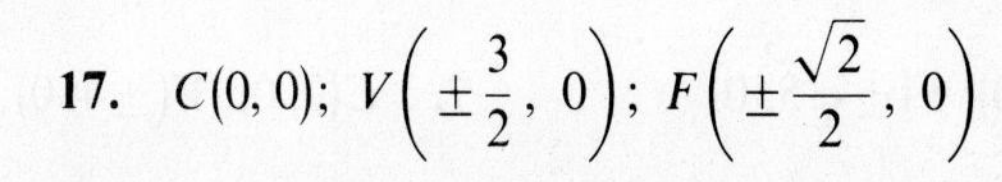

17. $C(0, 0)$; $V\left(\pm\frac{3}{2}, 0\right)$; $F\left(\pm\frac{\sqrt{2}}{2}, 0\right)$

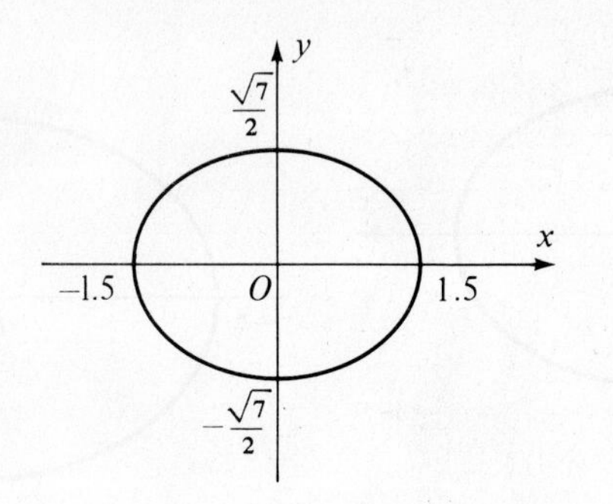

19. $C(0, 0)$; $V(0, \pm 2\sqrt{6})$; $F(0, \pm 2\sqrt{5})$

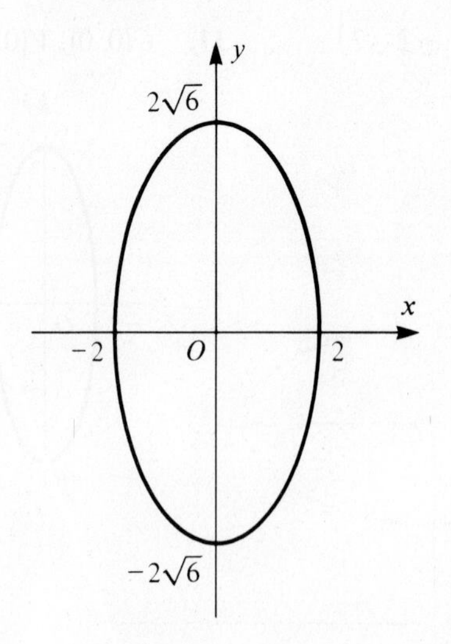

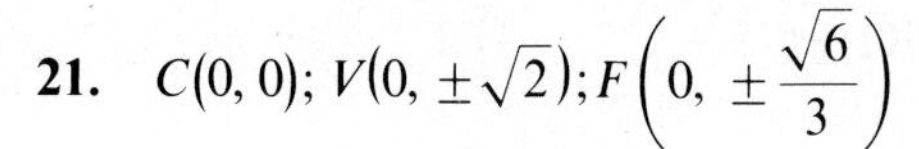

21. $C(0, 0)$; $V(0, \pm\sqrt{2})$; $F\left(0, \pm\dfrac{\sqrt{6}}{3}\right)$

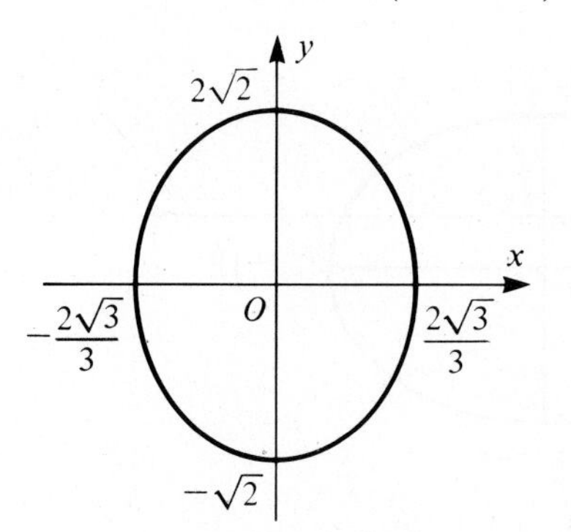

23. $C(-2, 0)$; $V(-2, \pm 2\sqrt{2})$; $F(-2, \pm 2)$

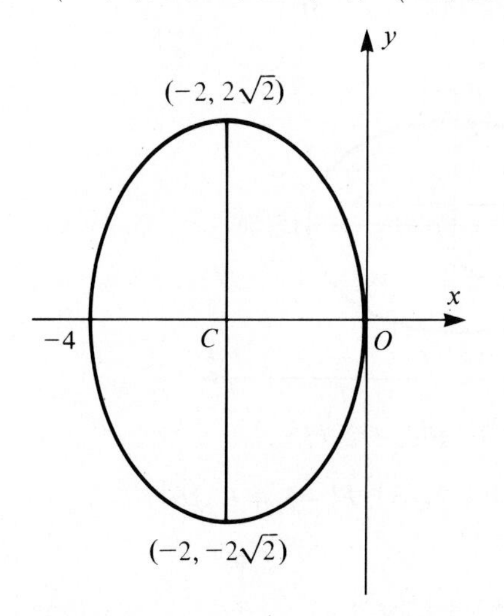

25. $C(0, 2)$; $V(\pm\sqrt{3}, 2)$; $F(\pm\sqrt{2}, 2)$

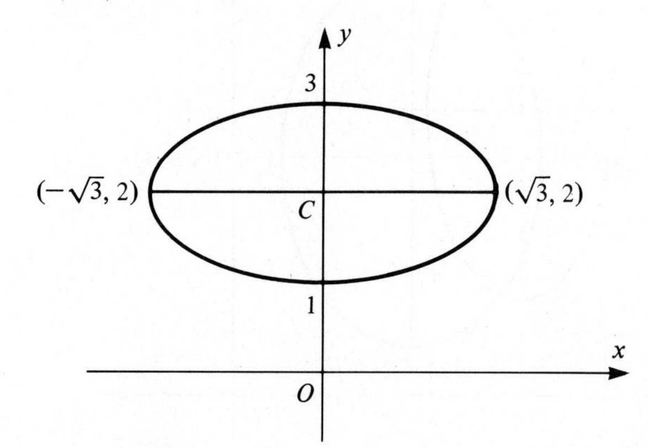

27. $C(1, -1)$; $V(1 \pm 4, -1)$; $F(1 \pm \sqrt{7}, -1)$

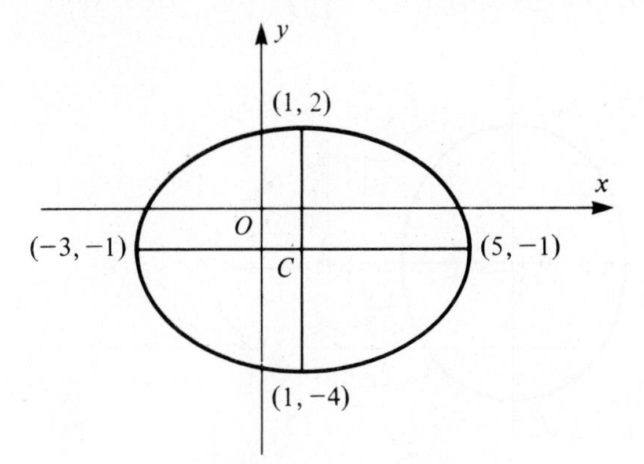

29. $C(2, 3)$; $V(2 \pm 3, 3)$; $F(2 \pm \sqrt{5}, 3)$

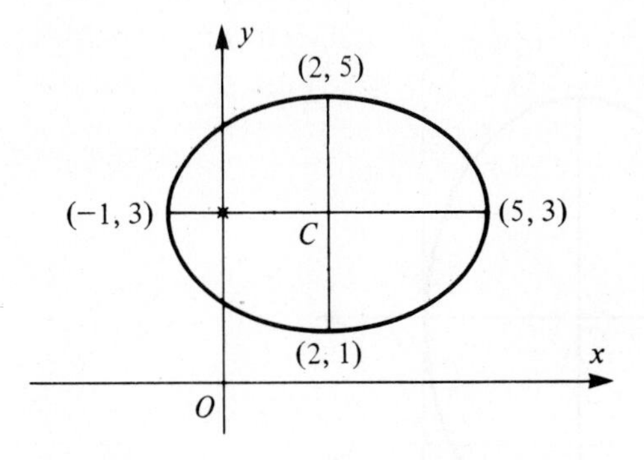

31. $C(-3, 4)$; $V(-3, 4 \pm 2\sqrt{3})$; $F(-3, 4 \pm \sqrt{6})$

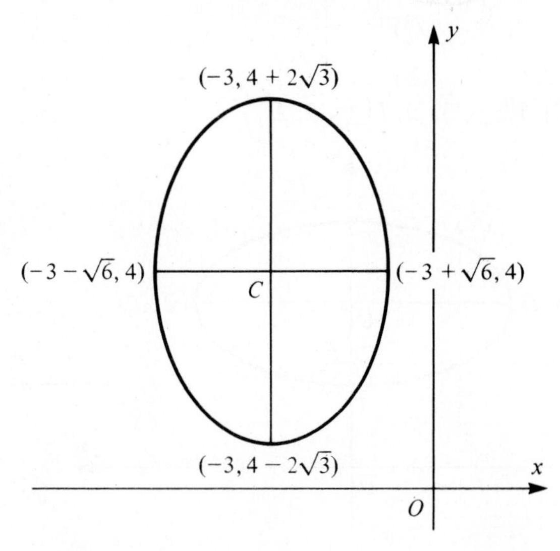

33. $C\left(-\frac{3}{2}, \frac{5}{2}\right)$; $V\left(-\frac{3}{2} \pm 4\sqrt{2}, \frac{5}{2}\right)$; $F\left(-\frac{3}{2} \pm 4, \frac{5}{2}\right)$

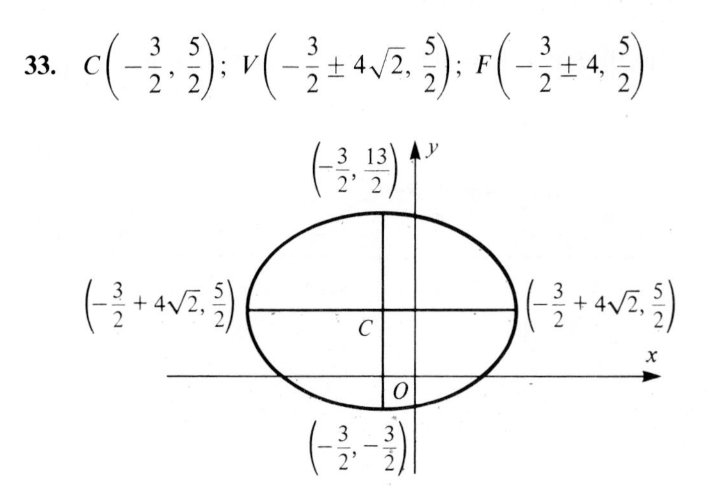

Exercise 14.4, page 451

1. $V(0, 0)$, $F(1, 0)$, $x = -1$

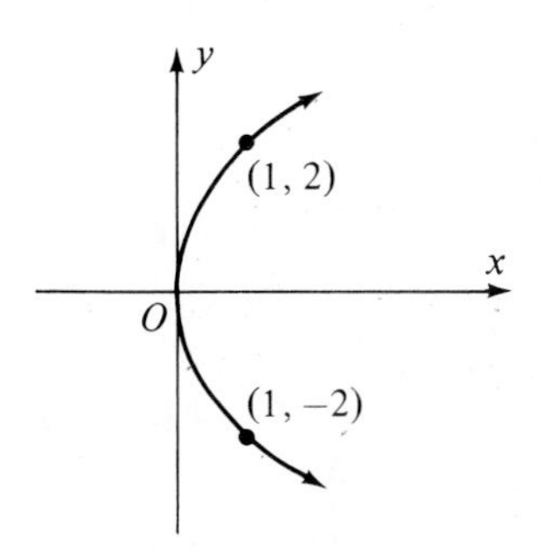

3. $V(0, 0)$, $F\left(\frac{1}{2}, 0\right)$, $x = -\frac{1}{2}$

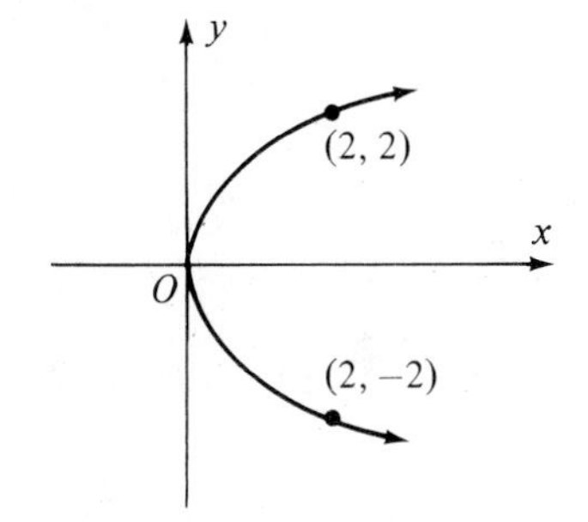

5. $V(0, 0)$, $F(-3, 0)$, $x = 3$

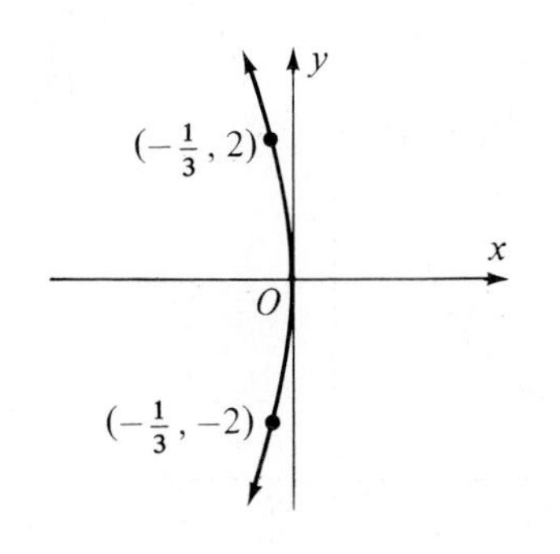

7. $V(0, 0)$, $F\left(-\frac{3}{2}, 0\right)$, $x = \frac{3}{2}$

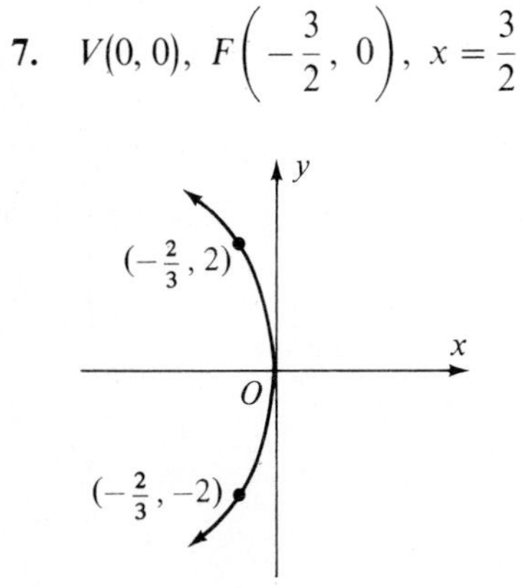

9. $V(0, 0)$, $F(0, 1)$, $y = -1$

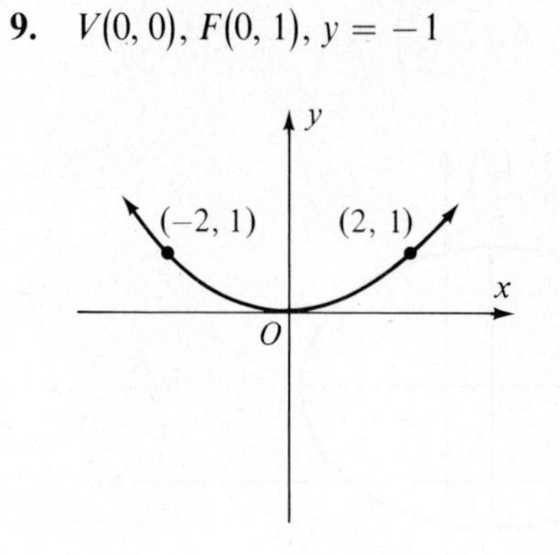

11. $V(0, 0)$, $F\left(0, \frac{1}{2}\right)$, $y = -\frac{1}{2}$

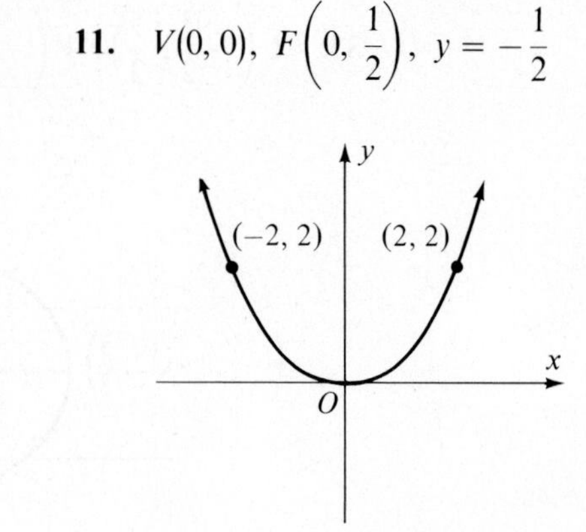

13. $V(0, 0)$, $F(0, -1)$, $y = 1$

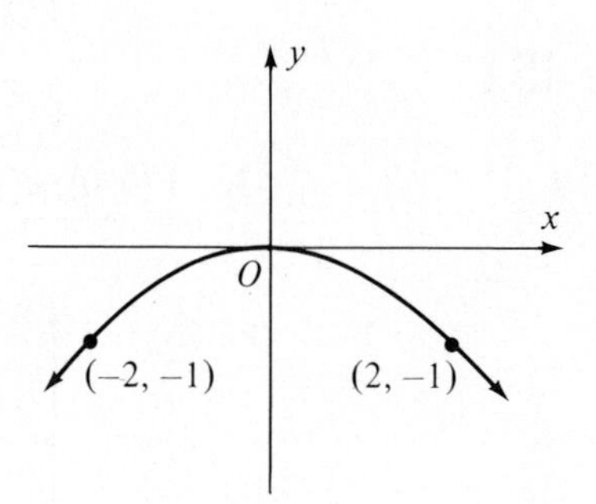

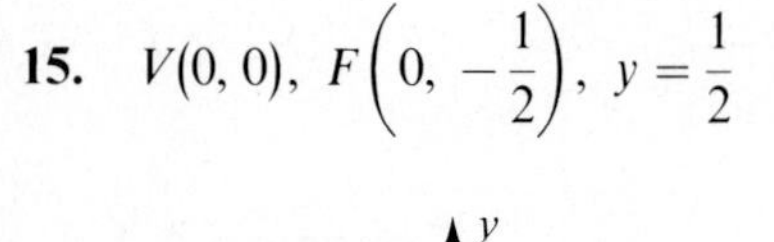

15. $V(0, 0)$, $F\left(0, -\frac{1}{2}\right)$, $y = \frac{1}{2}$

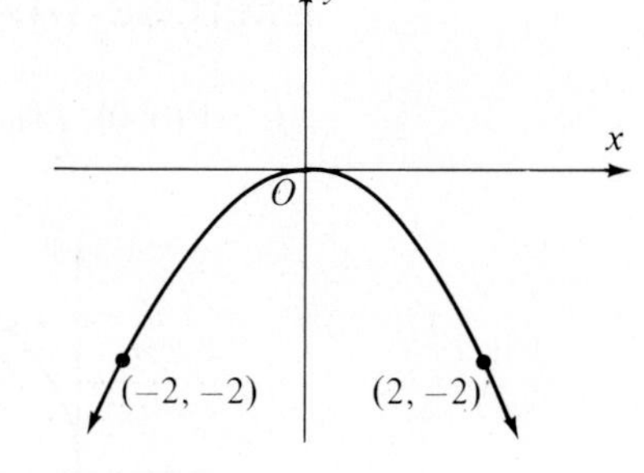

17. $V(1, 0)$, $F(2, 0)$, $x = 0$

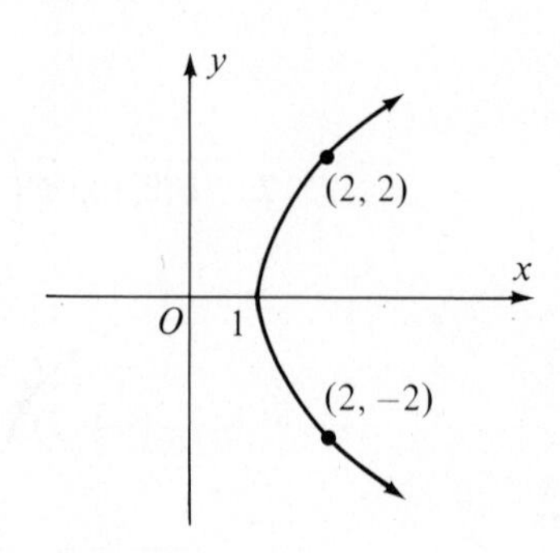

19. $V(-1, 0)$, $F\left(-\frac{3}{4}, 0\right)$, $x = -\frac{5}{4}$

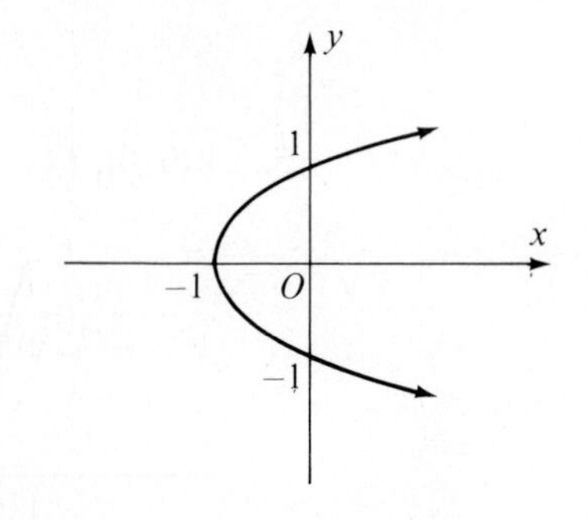

21. $V(2, 0)$, $F(1, 0)$, $x = 3$

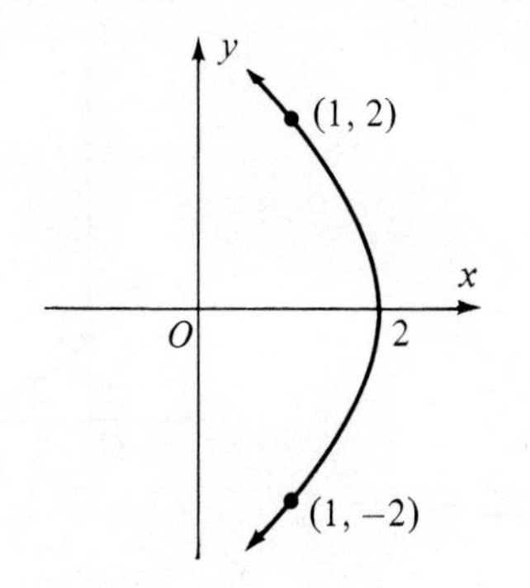

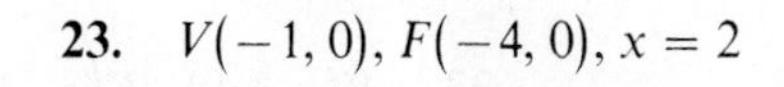

23. $V(-1, 0)$, $F(-4, 0)$, $x = 2$

$(-\frac{4}{3}, 2)$, -1, O, $(-\frac{4}{3}, -2)$

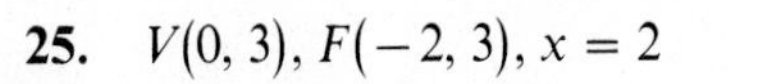

25. $V(0, 3)$, $F(-2, 3)$, $x = 2$

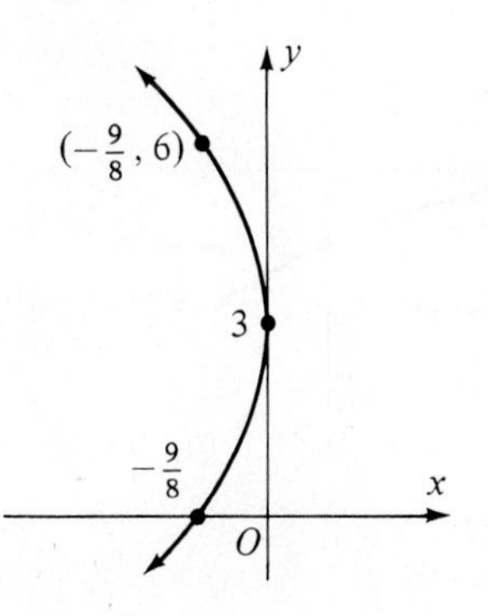

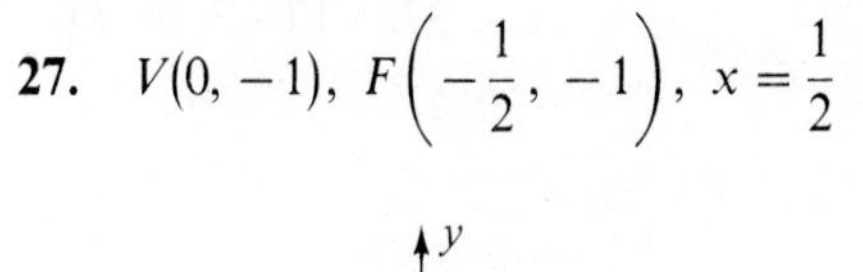

27. $V(0, -1)$, $F\left(-\dfrac{1}{2}, -1\right)$, $x = \dfrac{1}{2}$

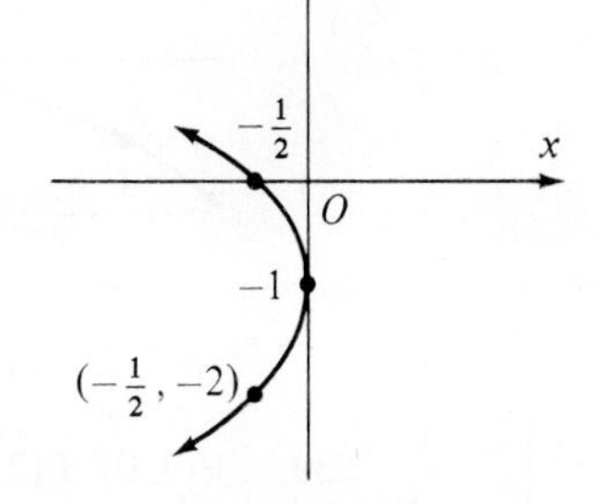

29. $V(0, 1)$, $F(3, 1)$, $x = -3$

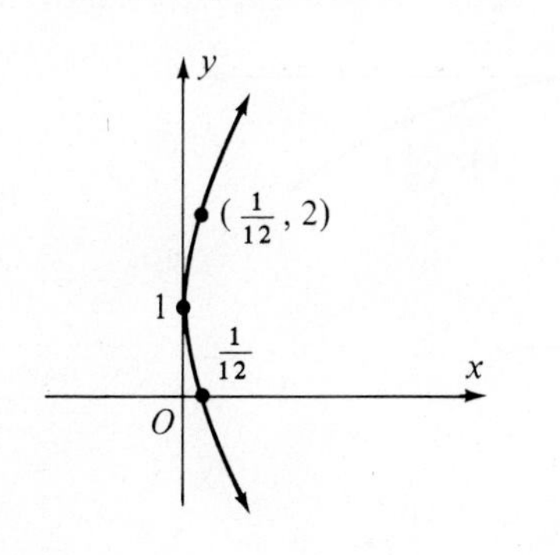

31. $V(0, -1)$, $F(1, -1)$, $x = -1$

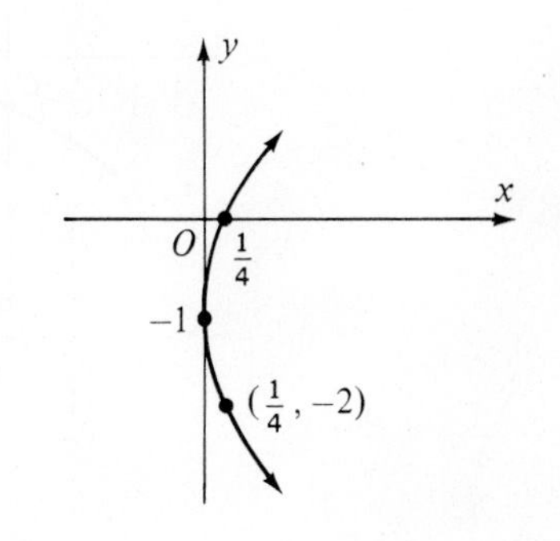

33. $V(-3, 0)$, $F\left(-3, \dfrac{1}{2}\right)$, $y = -\dfrac{1}{2}$

35. $V(2, 0)$, $F(2, 1)$, $y = -1$

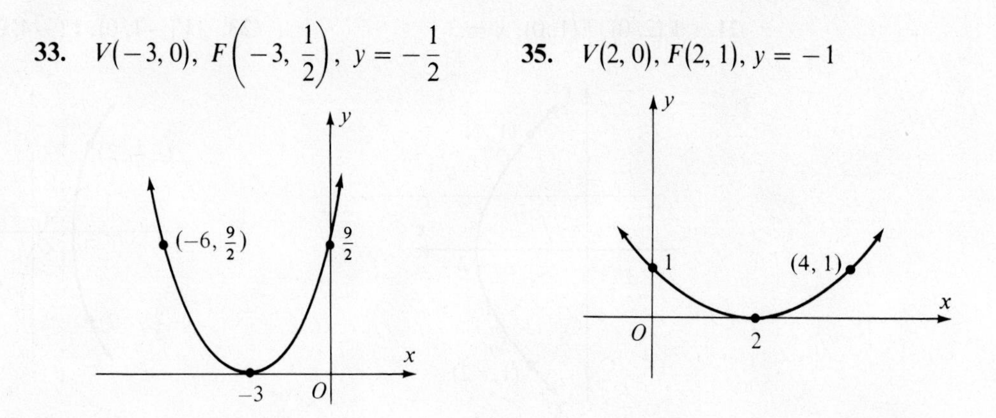

37. $V(-3, 0)$, $F(-3, -3)$, $y = 3$

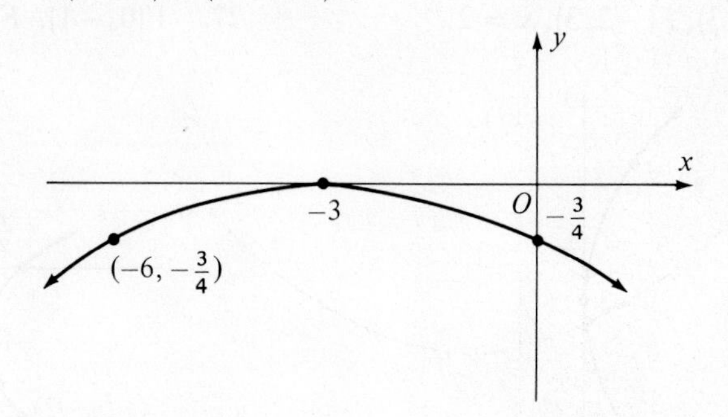

39. $V(2, 0)$, $F(2, -2)$, $y = 2$

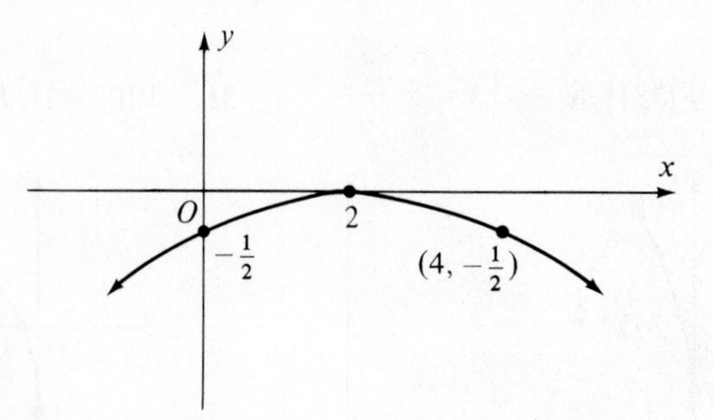

41. $V(2, -1), F(3, -1), x = 1$

43. $V(1, -3), F(0, -3), x = 2$

45. $V(1, 2), F(1, 5), y = -1$

47. $V(-6, 6),\ F\left(-6, \frac{23}{4}\right),\ y = \frac{25}{4}$

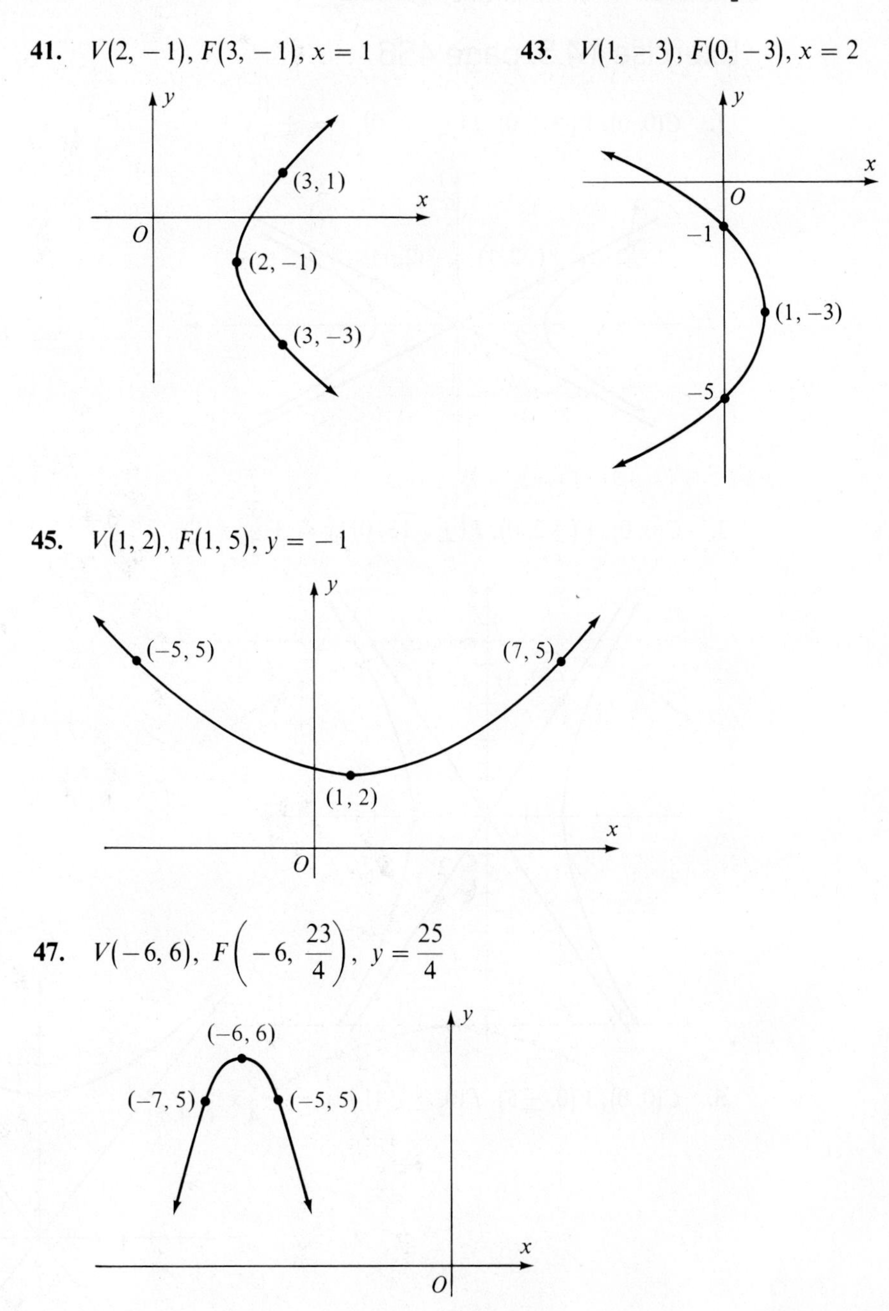

Exercise 14.5, page 456

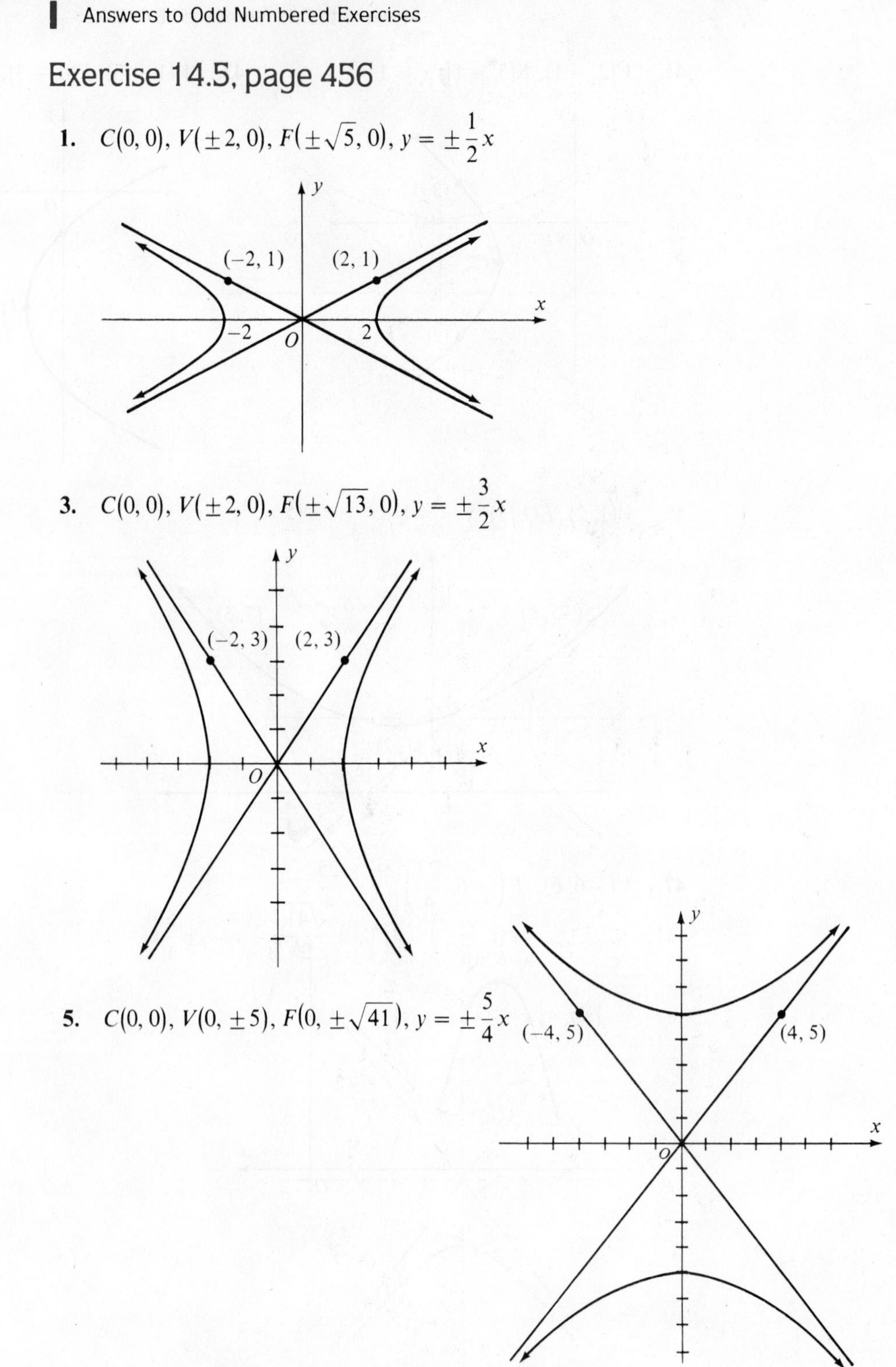

1. $C(0, 0)$, $V(\pm 2, 0)$, $F(\pm\sqrt{5}, 0)$, $y = \pm\frac{1}{2}x$

3. $C(0, 0)$, $V(\pm 2, 0)$, $F(\pm\sqrt{13}, 0)$, $y = \pm\frac{3}{2}x$

5. $C(0, 0)$, $V(0, \pm 5)$, $F(0, \pm\sqrt{41})$, $y = \pm\frac{5}{4}x$

7. $C(0, 0)$, $V(0, \pm\sqrt{6})$, $F(0, \pm\sqrt{15})$, $y = \pm\dfrac{\sqrt{6}}{3}x$

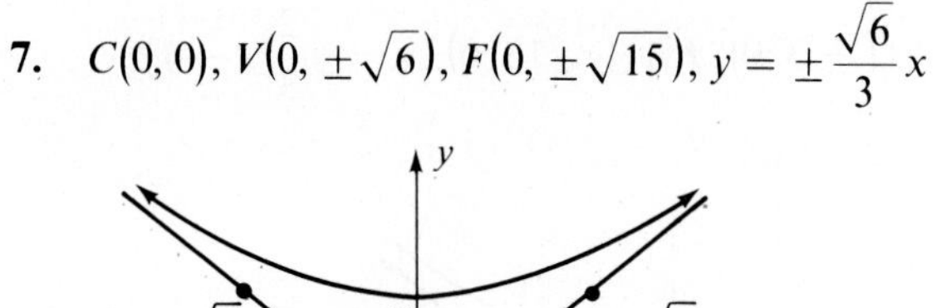
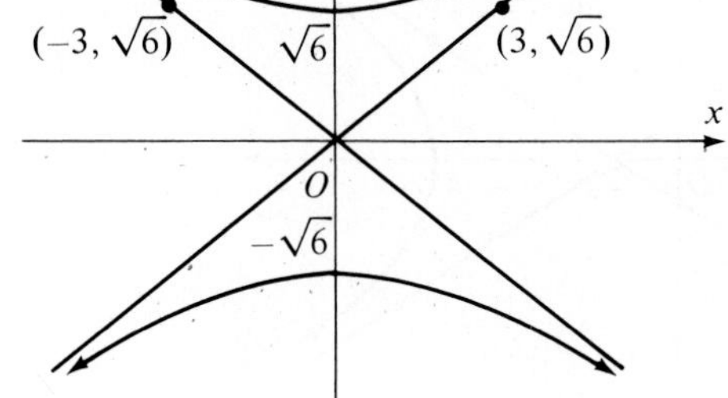

9. $C(0, 0)$, $V\left(\pm\dfrac{3}{2}, 0\right)$, $F\left(\pm\dfrac{\sqrt{145}}{6}, 0\right)$, $y = \pm\dfrac{8}{9}x$

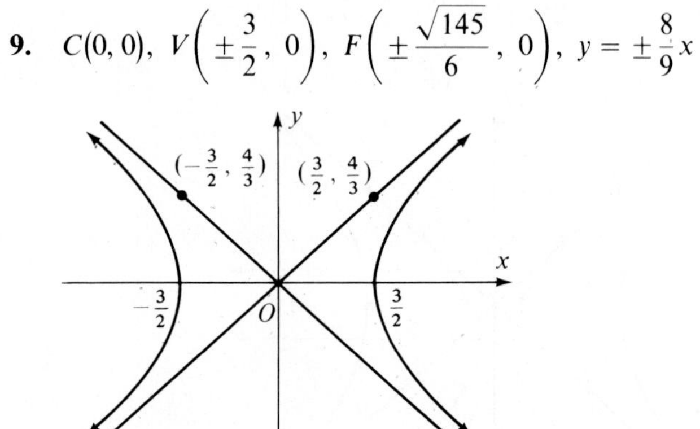

11. $C(0, 0)$, $V\left(0, \pm\dfrac{5}{2}\right)$, $F\left(0, \pm\dfrac{\sqrt{41}}{2}\right)$, $y = \pm\dfrac{5}{4}x$

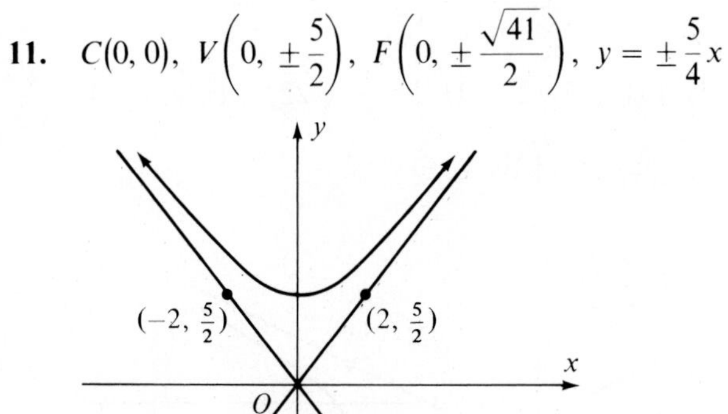

13. $C(1, 0)$, $V(1 \pm 3, 0)$, $F(1 \pm \sqrt{13}, 0)$, $y = \pm \frac{2}{3}(x - 1)$

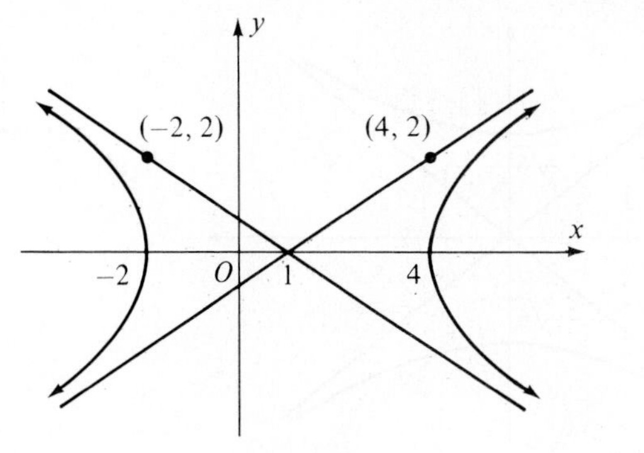

15. $C(0, -1)$, $V(\pm 4, -1)$, $F(\pm 2\sqrt{7}, -1)$, $y = -1 \pm \frac{\sqrt{3}}{2}x$

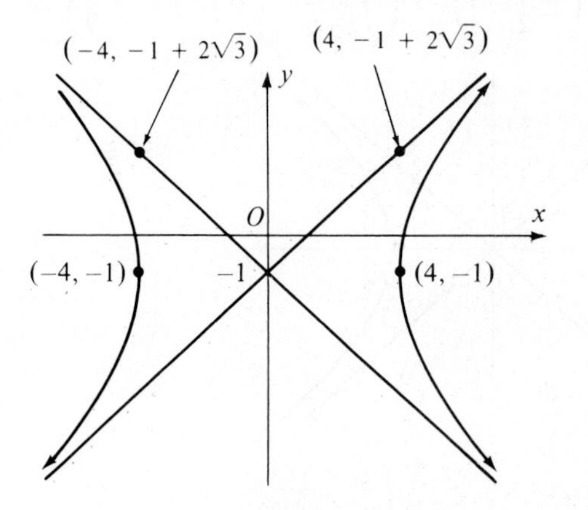

17. $C(1, -1)$, $V(1 \pm 3\sqrt{2}, -1)$, $F(1 \pm \sqrt{22}, -1)$, $y = -1 \pm \frac{\sqrt{2}}{3}(x - 1)$

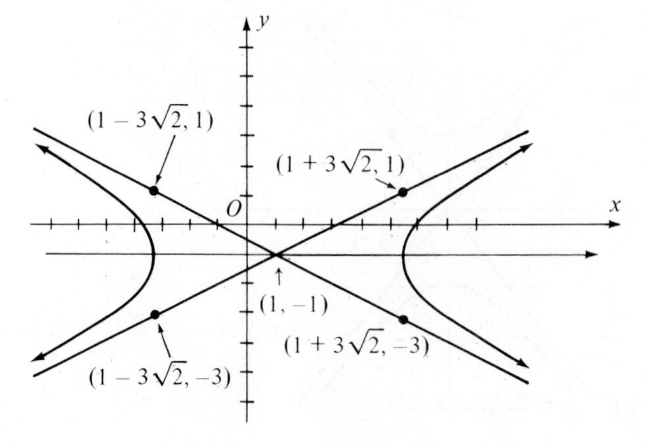

19. $C(-1, 2), V(-1 \pm \sqrt{6}, 2), F(-1 \pm \sqrt{14}, 2), y = 2 \pm \frac{2\sqrt{3}}{3}(x + 1)$

21. $C(-1, 2), V(-1, 2 \pm 3), F(-1, 2 \pm \sqrt{21}), y = 2 \pm \frac{\sqrt{3}}{2}(x + 1)$

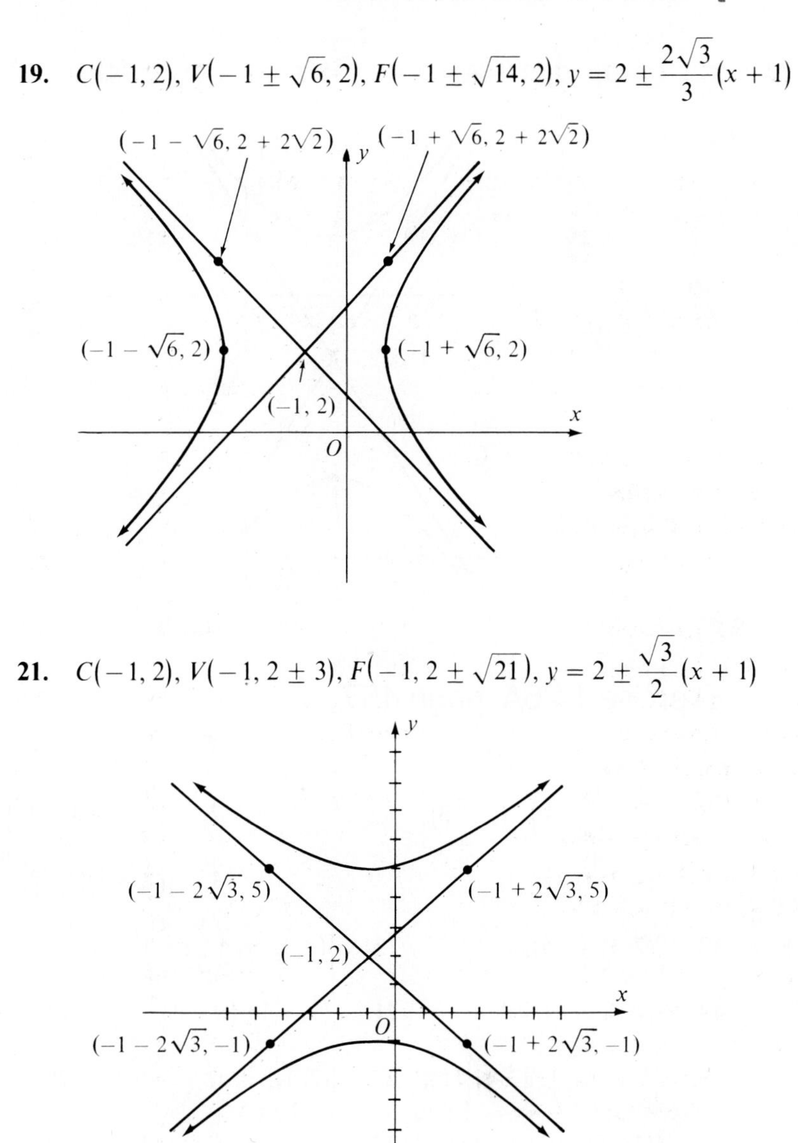

23. $C(-3, -2)$, $V(-3, -2 \pm 2\sqrt{3})$, $F(-3, -2 \pm 2\sqrt{5})$, $y = -2 \pm \frac{\sqrt{6}}{2}(x + 3)$

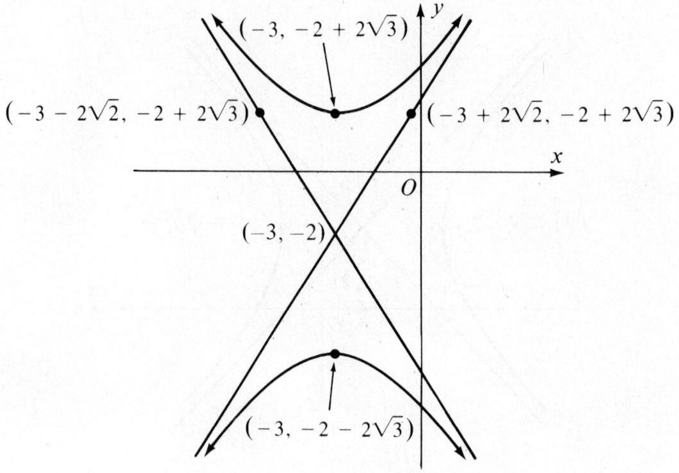

Exercise 14.6A, page 457

1. $\{(-1, -1), (1, 1)\}$

3. $\{(2, 1), (-2, -1)\}$

5. $\left\{\left(-\frac{25}{12}, 4\right), (2, -3)\right\}$

7. $\{(-3, -4), (2, 1)\}$

9. $\{(-3, 1), (5, 5)\}$

11. $\{(1, -1), (-2, 2)\}$

13. $\{(2, 4), (3, 5)\}$

15. $\left\{(-1, 0), \left(\frac{1}{2}, \frac{3}{2}\right)\right\}$

17. $\{(-2, 4), (1, 1)\}$

19. $\{(-3, 8), (1, 4)\}$

21. $\{(2, -1), (6, 7)\}$

23. $\{(-3, 6), (3, 2)\}$

25. $\left\{\left(-1, \frac{5}{2}\right), (2, 1)\right\}$

27. $\{(2, 1)\}$

29. $\{(-2, 3)\}$

31. $\{(3, -1)\}$

33. $\{(x, y) \mid x - y = 3\}$

35. $\{(x, y) \mid x - y = 1\}$

37. $\{(4 + i\sqrt{13}, 2 - i\sqrt{13}), (4 - i\sqrt{13}, 2 + i\sqrt{13})\}$

39. $\left\{\left(\frac{19 + 12i\sqrt{3}}{13}, \frac{-7 + 12i\sqrt{3}}{13}\right), \left(\frac{19 - 12i\sqrt{3}}{13}, \frac{-7 - 12i\sqrt{3}}{13}\right)\right\}$

Exercise 14.6B, page 459

1. $\{(2, -1), (-1, 2)\}$

3. $\left\{\left(\frac{1}{5}, \frac{7}{5}\right), (1, 1)\right\}$

5. $\left\{(1, -2), \left(\frac{9}{5}, -\frac{2}{5}\right)\right.$

7. $\left\{(1, 1), \left(\frac{11}{10}, \frac{7}{10}\right)\right\}$

9. $\left\{\left(-\frac{7}{10}, -\frac{41}{10}\right), (1, 1)\right\}$

11. $\{(-1, -3), (2, 3)\}$

13. $\{(1, 3), (2, 1)\}$

15. $\left\{\left(-\frac{7}{2}, \frac{5}{2}\right), (1, 1)\right\}$

17. $\left\{\left(-\frac{9}{2}, \frac{13}{2}\right), (3, -1)\right\}$

19. $\left\{\left(-\frac{3}{10}, \frac{49}{10}\right), (2, -2)\right\}$

21. $\left\{(-7, -3), \left(\frac{33}{5}, \frac{19}{5}\right)\right\}$

23. $\left\{\left(\frac{20}{13}, \frac{4}{13}\right), (2, 1)\right\}$

25. $\{(-1, 1)\}$

27. $\{(-1, 3)\}$

29. $\left\{\left(\frac{i\sqrt{2}}{2}, \frac{i\sqrt{2}}{2}\right), \left(-\frac{i\sqrt{2}}{2}, -\frac{i\sqrt{2}}{2}\right)\right\}$

31. $\varnothing$

Exercise 14.6C, page 461

1. $\{(3, 0), (-3, 0)\}$

3. $\{(0, 2), (0, -2)\}$

5. $\{(1, 3), (1, -3), (-1, 3), (-1, -3)\}$

7. $\{(6, 2), (6, -2), (-6, 2), (-6, -2)\}$

9. $\{(5, 2), (5, -2), (-5, 2), (-5, -2)\}$

11. $\left\{\left(\frac{3}{2}, 3\right), \left(\frac{3}{2}, -3\right), \left(-\frac{3}{2}, 3\right), \left(-\frac{3}{2}, -3\right)\right\}$

13. $\{(1, 2\sqrt{3}), (1, -2\sqrt{3}), (-1, 2\sqrt{3}), (-1, -2\sqrt{3})\}$

15. $\{(3\sqrt{2}, \sqrt{5}), (3\sqrt{2}, -\sqrt{5}), (-3\sqrt{2}, \sqrt{5}), (-3\sqrt{2}, -\sqrt{5})\}$

17. $\varnothing$

19. $\{(3i\sqrt{2}, 2\sqrt{3}), (3i\sqrt{2}, -2\sqrt{3}), (-3i\sqrt{2}, 2\sqrt{3}), (-3i\sqrt{2}, -2\sqrt{3})\}$

Exercise 14.6D, page 462

1. $\left\{(-1, 1), (1, -1), \left(\frac{2\sqrt{7}}{7}, \frac{\sqrt{7}}{7}\right), \left(-\frac{2\sqrt{7}}{7}, -\frac{\sqrt{7}}{7}\right)\right\}$

3. $\left\{(-3, -1), (3, 1), \left(-\sqrt{14}, \frac{\sqrt{14}}{2}\right), \left(\sqrt{14}, -\frac{\sqrt{14}}{2}\right)\right\}$

5. $\left\{\left(\frac{\sqrt{2}}{2}, \frac{\sqrt{2}}{4}\right), \left(-\frac{\sqrt{2}}{2}, -\frac{\sqrt{2}}{4}\right), \left(\frac{3}{4}, \frac{1}{4}\right), \left(-\frac{3}{4}, -\frac{1}{4}\right)\right\}$

7. $\left\{\left(\frac{\sqrt{3}}{2}, \frac{\sqrt{3}}{2}\right), \left(-\frac{\sqrt{3}}{2}, -\frac{\sqrt{3}}{2}\right), (2\sqrt{3}, -\sqrt{3}), (-2\sqrt{3}, \sqrt{3})\right\}$

9. $\{(2\sqrt{2}, -\sqrt{2}), (-2\sqrt{2}, \sqrt{2}), (3, -1), (-3, 1)\}$

11. $\left\{\left(\frac{\sqrt{5}}{5}, \frac{\sqrt{5}}{5}\right), \left(-\frac{\sqrt{5}}{5}, -\frac{\sqrt{5}}{5}\right), \left(\sqrt{2}, -\frac{\sqrt{2}}{2}\right), \left(-\sqrt{2}, \frac{\sqrt{2}}{2}\right)\right\}$

13. $\left\{\left(\frac{\sqrt{2}}{2}, \frac{\sqrt{2}}{2}\right), \left(-\frac{\sqrt{2}}{2}, -\frac{\sqrt{2}}{2}\right), \left(\sqrt{3}, -\frac{2\sqrt{3}}{3}\right), \left(-\sqrt{3}, \frac{2\sqrt{3}}{3}\right)\right\}$

15. $\left\{(\sqrt{3}, 2\sqrt{3}), (-\sqrt{3}, -2\sqrt{3}), \left(\frac{\sqrt{33}}{11}, -\frac{3\sqrt{33}}{11}\right), \left(-\frac{\sqrt{33}}{11}, \frac{3\sqrt{33}}{11}\right)\right\}$

17. $\left\{\left(\frac{8\sqrt{33}}{33}, -\frac{2\sqrt{33}}{33}\right), \left(-\frac{8\sqrt{33}}{33}, \frac{2\sqrt{33}}{33}\right), \left(\frac{2\sqrt{35}}{35}, \frac{8\sqrt{35}}{35}\right),\right.$
$\left.\left(-\frac{2\sqrt{35}}{35}, -\frac{8\sqrt{35}}{35}\right)\right\}$

19. $\left\{\left(\frac{\sqrt{3}}{2}, -\sqrt{3}\right), \left(-\frac{\sqrt{3}}{2}, \sqrt{3}\right), \left(\frac{3i\sqrt{102}}{34}, \frac{i\sqrt{102}}{34}\right),\right.$
$\left.\left(-\frac{3i\sqrt{102}}{34}, -\frac{i\sqrt{102}}{34}\right)\right\}$

Chapter 14 Review, page 463

1. $(-3, 4), 9$

3. $\frac{(x-1)^2}{16} + \frac{(y-1)^2}{9} = 1, C(1, 1), 8, 6, F(1 \pm \sqrt{7}, 1)$

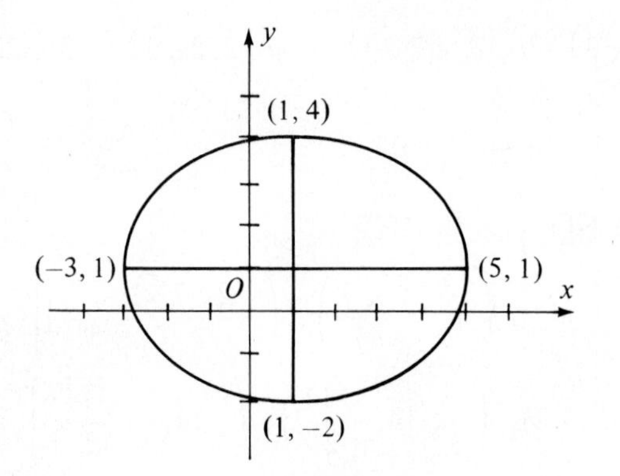

5. $\dfrac{(x+2)^2}{16}+\dfrac{(y+2)^2}{24}=1, C(-2,-2), 4\sqrt{6}, 8, F(-2,-2\pm 2\sqrt{2})$

7. $(-1,-1), (0,-1), x=-2$

9. $(2,4), \left(\dfrac{1}{2}, 4\right), x=\dfrac{7}{2}$

11. $(-1,0), \left(-1, \dfrac{3}{4}\right), y=-\dfrac{3}{4}$

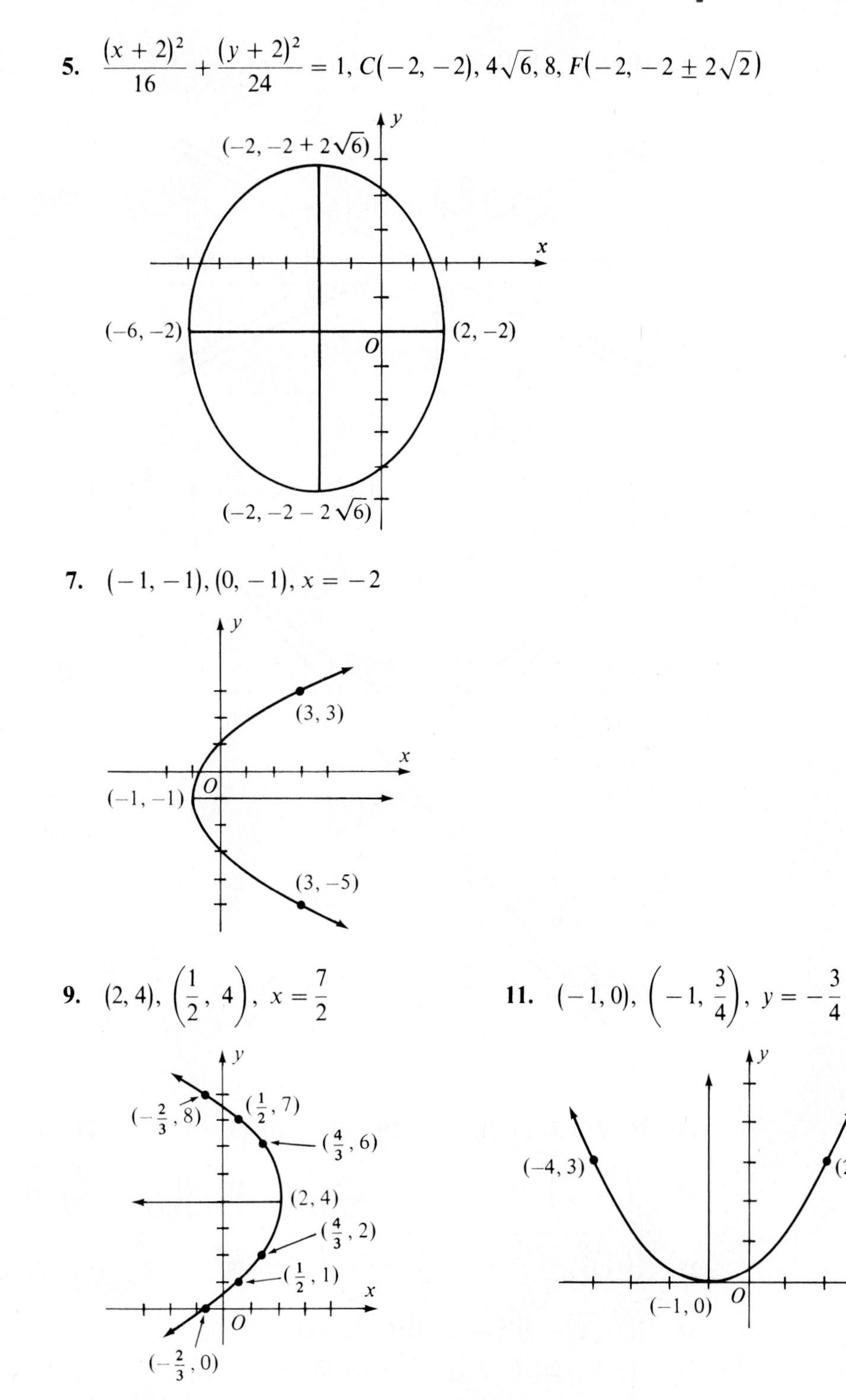

13. $(1, 1), (1 \pm 4, 1), (1 \pm 2\sqrt{5}, 1), y = 1 \pm \frac{1}{2}(x - 1)$

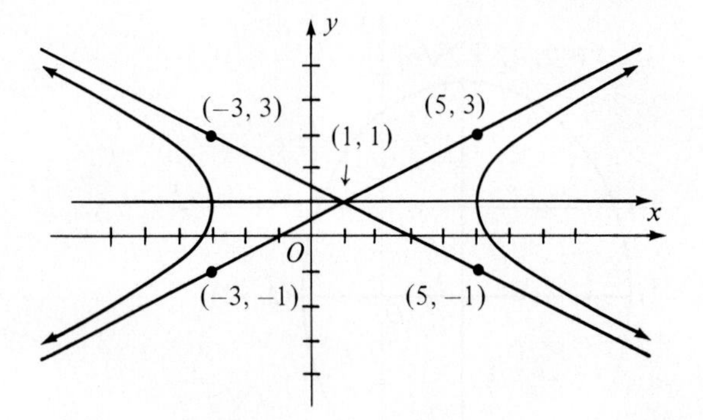

15. $(2, -4), (2, -4 \pm 4), (2, -4 \pm 2\sqrt{7}), y = -4 \pm \frac{2\sqrt{3}}{3}(x - 2)$

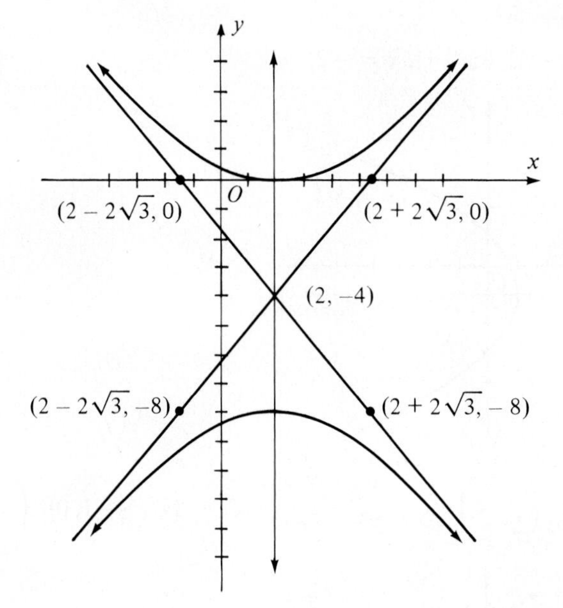

17. $\{(-2, -3), (3, 2)\}$

19. $\{(-2, 6), (3, -4)\}$

21. $\{(-1, -2), (3, 1)\}$

23. $\left\{\left(\frac{1}{5}, \frac{18}{5}\right), (1, 2)\right\}$

25. $\left\{\left(\frac{7}{17}, -\frac{23}{17}\right), (1, 1)\right\}$

27. $\left\{\left(-\frac{3}{2}, \frac{11}{2}\right), (5, -1)\right\}$

29. $\{(-2, -1), (-2, 1), (2, -1), (2, 1)\}$

31. $\{(-\sqrt{2}, -2), (-\sqrt{2}, 2), (\sqrt{2}, -2), (\sqrt{2}, 2)\}$

33. $\{(-3, -2i), (-3, 2i), (3, -2i), (3, 2i)\}$

35. $\left\{(1, 1), (-1, -1), \left(\frac{2\sqrt{5}}{3}, \frac{\sqrt{5}}{3}\right), \left(-\frac{2\sqrt{5}}{3}, -\frac{\sqrt{5}}{3}\right)\right\}$

37. $\left\{(2, -1), (-2, 1), \left(\frac{\sqrt{10}}{5}, -\frac{2\sqrt{10}}{5}\right), \left(-\frac{\sqrt{10}}{5}, \frac{2\sqrt{10}}{5}\right)\right\}$

39. $\left\{\left(\frac{2i}{5}, \frac{6i}{5}\right), \left(-\frac{2i}{5}, -\frac{6i}{5}\right), \left(\frac{i\sqrt{10}}{5}, -\frac{2i\sqrt{10}}{5}\right), \left(-\frac{i\sqrt{10}}{5}, \frac{2i\sqrt{10}}{5}\right)\right\}$

41. 6, 8 **43.** $\frac{3}{2}$, 6 **45.** 16, 28 **47.** 9, 12

49. 11 or 74 **51.** 36 inches, 27 inches. **53.** 40 feet, 28 feet
55. 28 inches, 16 inches. **57.** $2.75

Cumulative Review, page 466

Chapter 12

1. $2i$ **3.** $3i\sqrt{3}$ **5.** $5i\sqrt{2}$ **7.** $5i\sqrt{3}$
9. 1 **11.** $-i$ **13.** -1 **15.** 1
17. $9i$ **19.** $3i\sqrt{2}$ **21.** $-6i\sqrt{6}$
23. $2i\sqrt{2} + 2i\sqrt{3}$ **25.** $16i$ **27.** $2i\sqrt{30}$
29. -6 **31.** -20 **33.** $\frac{i\sqrt{6}}{3}$ **35.** $\frac{3i}{2}$
37. $-\frac{i\sqrt{21}}{3}$ **39.** $-\frac{i\sqrt{30}}{5}$ **41.** $\frac{\sqrt{6}}{3}$ **43.** $\frac{3\sqrt{6}}{8}$
45. $2 + 4i$ **47.** $5 - 3i$ **49.** $25 + 0i$ **51.** $-5 - 15i$
53. $3 - 4i$ **55.** $1 - 2i\sqrt{2}$ **57.** $2 - i$ **59.** $\frac{4}{17} + \frac{1}{17}i$
61. $\frac{3}{5} - \frac{1}{5}i$ **63.** $\frac{1}{13} + \frac{8}{13}i$ **65.** $-\frac{1}{13} + \frac{8}{13}i$
67. $-\frac{2}{5} + \frac{9}{5}i$ **69.** $-\frac{1}{3} + \frac{2\sqrt{2}}{3}i$ **71.** $0 + \frac{1}{2}i$
73. $x = 2; y = 3$ **75.** $x = -1; y = 2$ **77.** $x = 2; y = 0$
79. $x = 3; y = -1$ **81.** $x = -2; y = 3$ **83.** $x = 2; y = \frac{1}{2}$

Chapter 13

85. $\{-5, -2\}$ **87.** $\{-2, 6\}$ **89.** $\{2, 4\}$

91. $\left\{-\frac{3}{2}, -\frac{4}{3}\right\}$ **93.** $\left\{-4, \frac{5}{6}\right\}$ **95.** $\left\{\frac{1}{8}, \frac{3}{2}\right\}$

97. $\{-1-a, -1+a\}$ **99.** $\left\{\frac{1-a}{2}, \frac{1+a}{2}\right\}$ **101.** $\{-a-2, a+2\}$

103. $x^2 - 4x + 3 = 0$ **105.** $x^2 - x - 2 = 0$ **107.** $x^2 + 4x + 4 = 0$

109. $2x^2 - 5x + 2 = 0$ **111.** $x^2 - 2x + 5 = 0$

113. $x^2 - 2\sqrt{2}x + 3 = 0$ **115.** $\{1 - i\sqrt{2}, 1 + i\sqrt{2}\}$

117. $\{-2-2i, -2+2i\}$ **119.** $\left\{-\frac{3}{2} - \frac{\sqrt{17}}{2}, -\frac{3}{2} + \frac{\sqrt{17}}{2}\right\}$

121. $\left\{1 - \frac{\sqrt{10}}{5}, 1 + \frac{\sqrt{10}}{5}\right\}$ **123.** $\left\{\frac{5}{6} - \frac{\sqrt{37}}{6}, \frac{5}{6} - \frac{\sqrt{37}}{6}\right\}$

125. $\left\{\frac{3}{4} - \frac{\sqrt{33}}{12}, \frac{3}{4} + \frac{\sqrt{33}}{12}\right\}$ **127.** $\left\{-\frac{3}{2} + \frac{\sqrt{35}}{2}i, -\frac{3}{2} - \frac{\sqrt{35}}{2}i\right\}$

129. $\left\{\frac{5}{6} + \frac{\sqrt{11}}{6}i, \frac{5}{6} - \frac{\sqrt{11}}{6}i\right\}$ **131.** $\left\{-\frac{3a}{4} - \frac{\sqrt{17}\,a}{4}, -\frac{3a}{4} + \frac{\sqrt{17}\,a}{4}\right\}$

133. $\left\{\frac{2a}{3} + \frac{\sqrt{2}\,a}{3}i, \frac{2a}{3} - \frac{\sqrt{2}\,a}{3}i\right\}$ **135.** $\{2 - 2\sqrt{2}, 2 + 2\sqrt{2}\}$

137. $\left\{\frac{-5-\sqrt{7}}{3}, \frac{-5+\sqrt{7}}{3}\right\}$ **139.** $\left\{\frac{1+i\sqrt{17}}{3}, \frac{1-i\sqrt{17}}{3}\right\}$

141. $\left\{\frac{-\sqrt{2}-\sqrt{22}}{2}, \frac{-\sqrt{2}+\sqrt{22}}{2}\right\}$ **143.** $\left\{\frac{\sqrt{5}-\sqrt{13}}{2}, \frac{\sqrt{5}+\sqrt{13}}{2}\right\}$

145. $\left\{-2\sqrt{3}, \frac{\sqrt{3}}{3}\right\}$ **147.** $k < \frac{9}{4}$ **149.** $k \in R$

151. $k < \frac{25}{12}$ **153.** $k = 4$ **155.** $\varnothing$ **157.** $k = \frac{1}{4}$

159. $k > -\frac{9}{8}$ **161.** $k \in R$ **163.** $k > -\frac{1}{3}$ **165.** 7

167. 2 **169.** 10 **171.** 6 **173.** 11

175. 5 **177.** 3 **179.** $x^2 - 4x - 21 = 0$

181. $4x^2 - 3x - 1 = 0$ **183.** $3x^2 + 8x - 3 = 0$

185. $6x^2 + 5x - 6 = 0$; $6x^2 - 5x - 6 = 0$

187. $3x^2 + 5x - 12 = 0$

189. $6x^2 - 5x - 4 = 0$; $6x^2 + 5x - 4 = 0$

191. $\{2, -1 + i\sqrt{3}, -1 - i\sqrt{3}\}$

193. $\{-4, 2 + 2i\sqrt{3}, 2 - 2i\sqrt{3}\}$

195. $\{3\sqrt{2}, -3\sqrt{2}, 3i\sqrt{2}, -3i\sqrt{2}\}$

197. $\{-3, 3, -2i, 2i\}$

199. $\{-2 - \sqrt{2}, -2 + \sqrt{2}, 2 - \sqrt{2}, 2 + \sqrt{2}\}$

201. $\left\{\frac{-1 + i\sqrt{11}}{6}, \frac{-1 - i\sqrt{11}}{6}, \frac{1 + i\sqrt{11}}{6}, \frac{1 - i\sqrt{11}}{6}\right\}$

203. $\left\{\frac{1}{9}, 4\right\}$

205. $\left\{\frac{4}{9}\right\}$

207. $\{-64, 729\}$

209. $\left\{-\frac{1}{64}, -8\right\}$

211. $\{-4, -3\}$

213. $\{-2, 1\}$

215. $\left\{-\frac{3}{2}, \frac{1}{4}\right\}$

217. $\{1\}$

219. $\{3\}$

221. $\{4\}$

223. $\{8\}$

225. $\{5\}$

227. 6; 10

229. 15 inches

231. 80 feet

233. $V(0, 3)$; $x = 0$

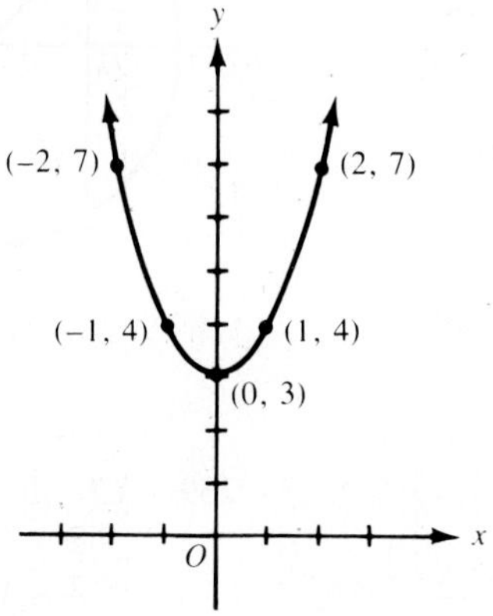

235. $V(-2, 0)$; $x = -2$

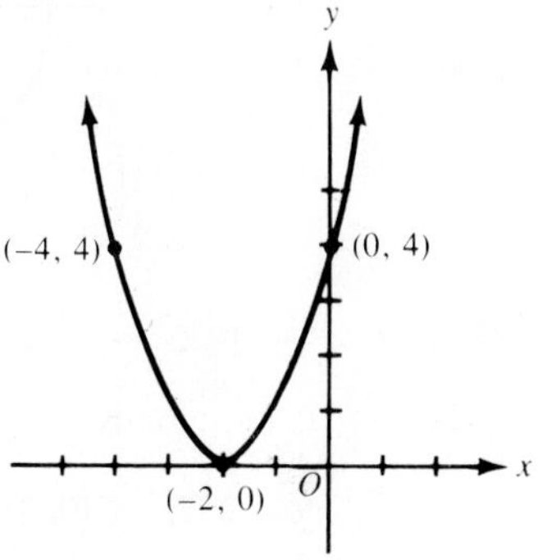

237. $V\left(\frac{1}{2}, -\frac{25}{4}\right)$; $x = \frac{1}{2}$

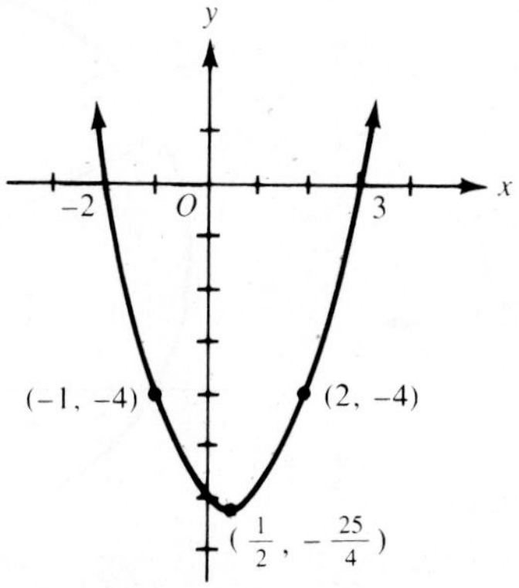

239. $V\left(\frac{3}{2}, \frac{25}{4}\right)$; $x = \frac{3}{2}$

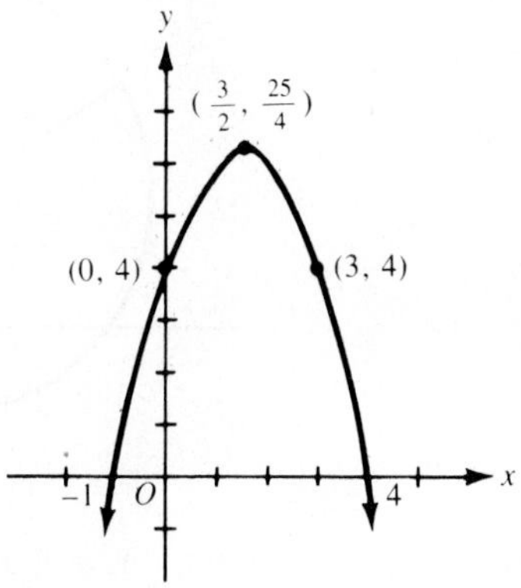

241. $\{x \mid -5 < x < 5\}$

243. $\{x \mid -3 \le x \le -1\}$

245. $\{x \mid -4 < x < 6\}$

247. $\{x \mid x < -4 \text{ or } x > 4\}$

249. $\{x \mid x \le -4 \text{ or } x \ge 1\}$

251. $\{x \mid x < 3 \text{ or } x > 8\}$

Chapter 14

253. $C(0, 0);\ 3\sqrt{3}$
$-3\sqrt{3} \le x \le 3\sqrt{3}$
$-3\sqrt{3} \le y \le 3\sqrt{3}$

255. $C(2, 0);\ 4$
$-2 \le x \le 6$
$-4 \le y \le 4$

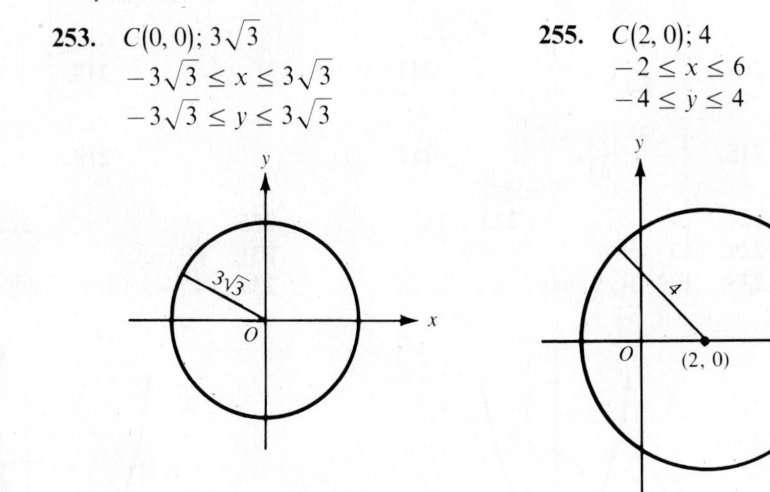

257. $C(0, 3);\ 5$
$-5 \le x \le 5$
$-2 \le y \le 8$

259. $C(-1, 3);\ 6$
$-7 \le x \le 5$
$-3 \le y \le 9$

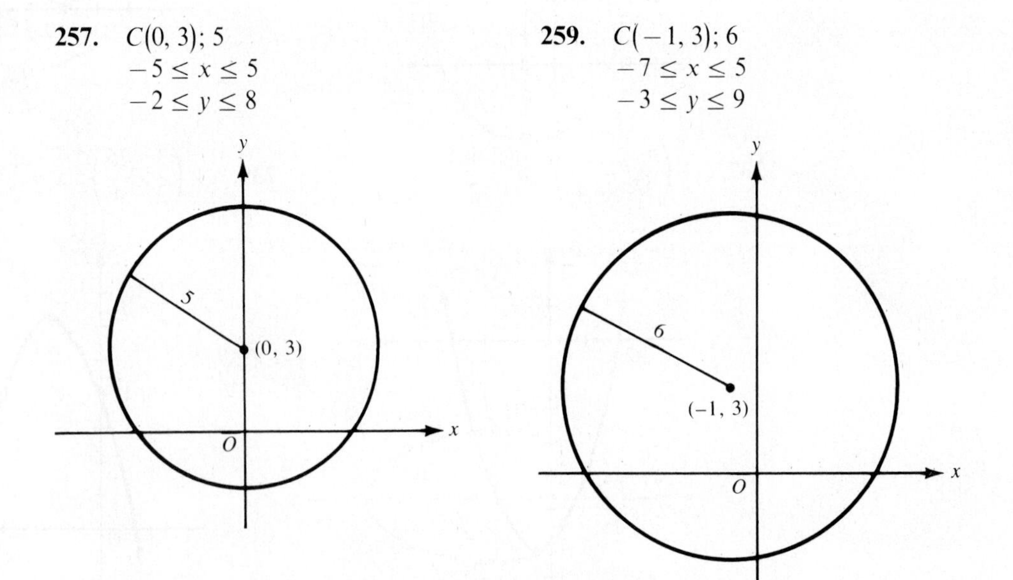

261. $C(3, 1); 3\sqrt{2}$
$3 - 3\sqrt{2} \le x \le 3 + 3\sqrt{2}$
$1 - 3\sqrt{2} \le y \le 1 + 3\sqrt{2}$

263. $C(0, 0); V(\pm 2\sqrt{2}, 0);$
$F(\pm\sqrt{6}, 0)$

265. $C(0, 0), V(0, \pm 3\sqrt{2})$
$F(0, \pm 3)$

267. $C(0, 0); V(\pm 2\sqrt{3}, 0)$
$F(\pm\sqrt{3}, 0)$

269. $C(2, 0); V(2 \pm 4, 0)$
$F(2 \pm 2\sqrt{3}, 0)$

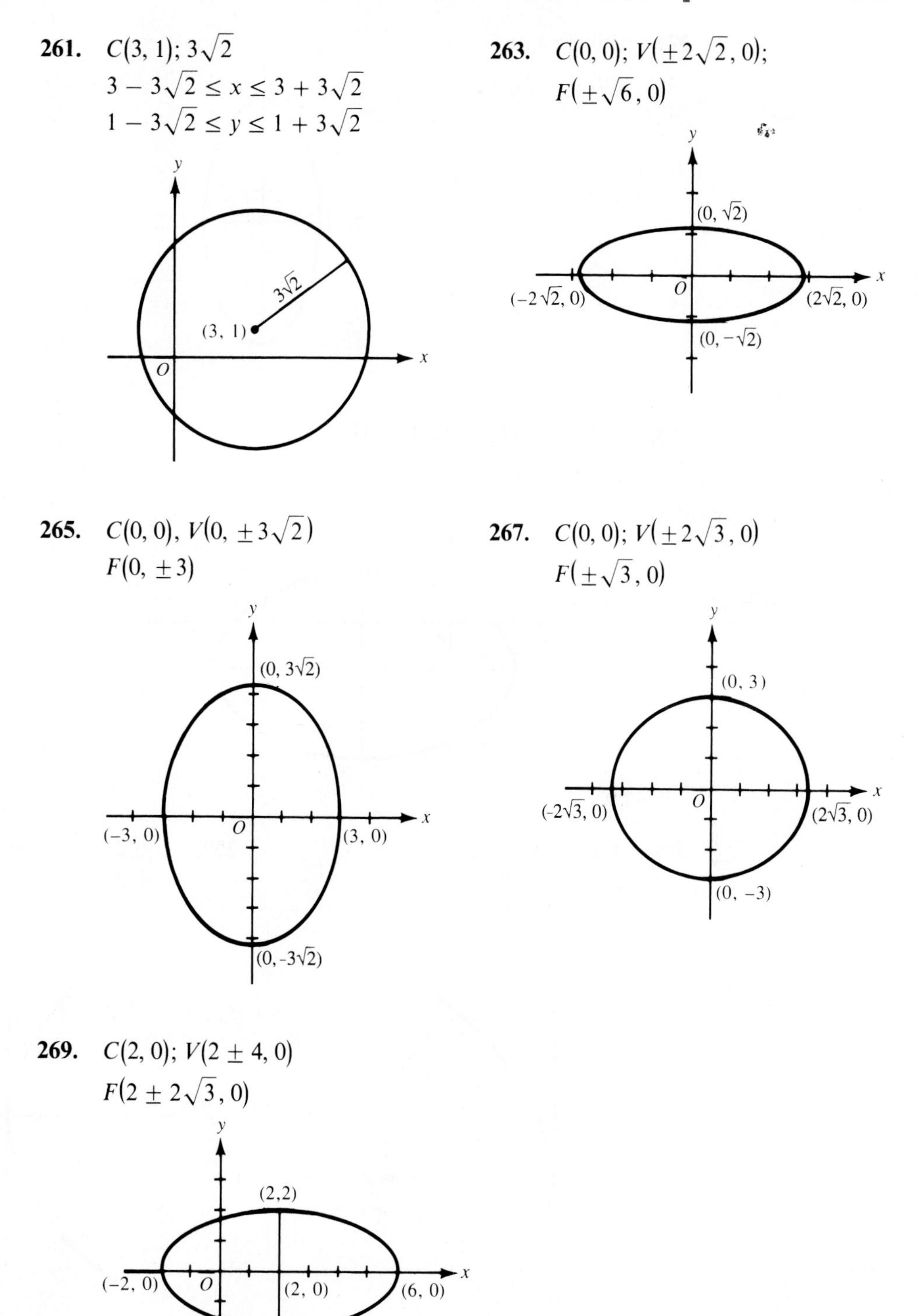

271. $C(0, 2)$; $V(0, 2 \pm 4)$
$F(0, 2 \pm 2\sqrt{3})$

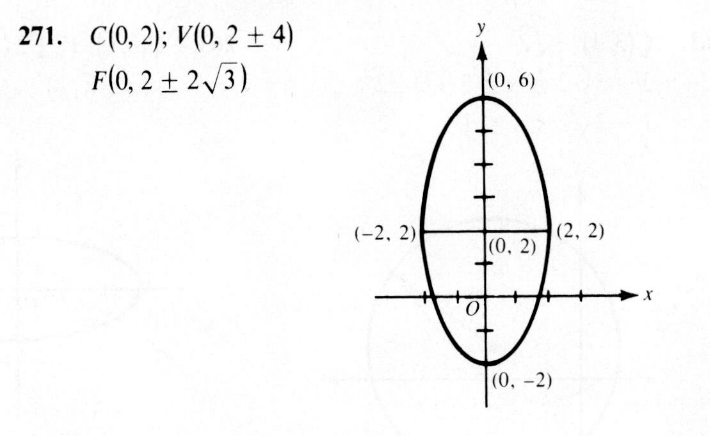

273. $C(2, -1)$; $V(2 \pm 2\sqrt{3}, -1)$
$F(2 \pm 2\sqrt{2}; -1)$

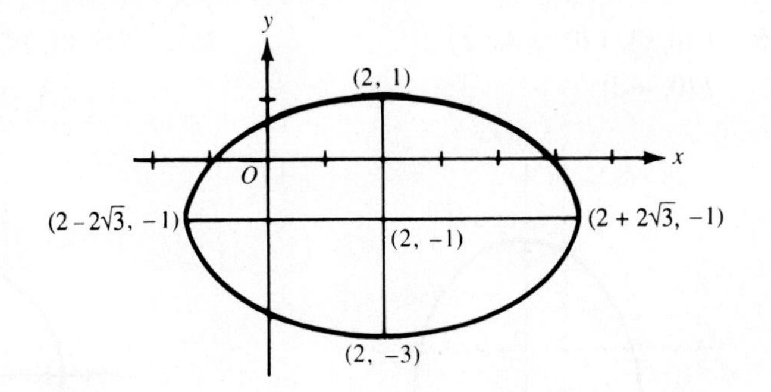

275. $V(0, 0)$; $F\left(\frac{5}{2}, 0\right)$

$x = -\frac{5}{2}$

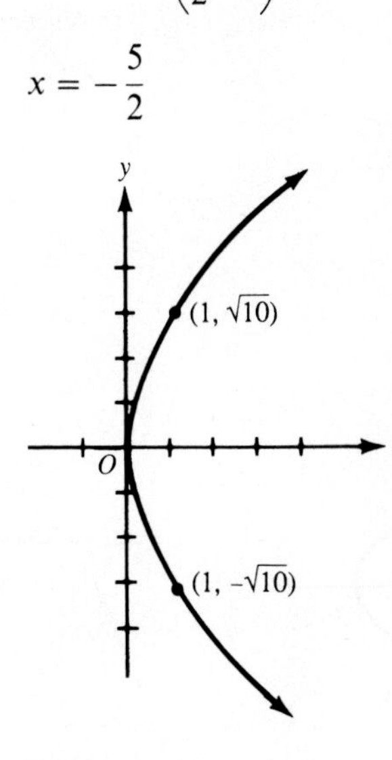

277. $V(0, 0)$; $F(-2, 0)$
$x = 2$

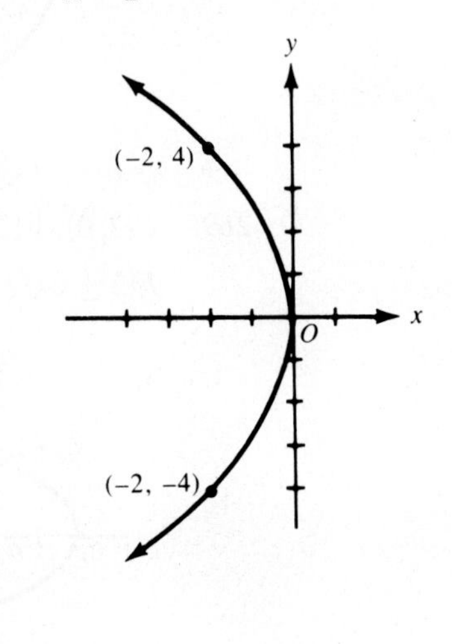

279. $V(0, 0);\ F\left(0, \frac{3}{2}\right)$

$y = -\frac{3}{2}$

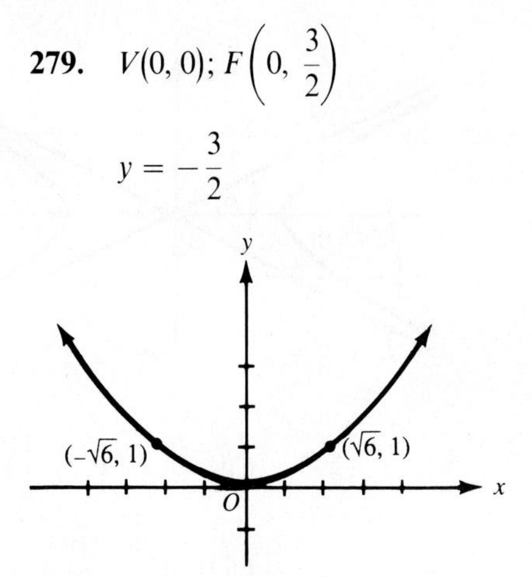

281. $V(0, 0);\ F\left(0, -\frac{5}{2}\right)$

$y = \frac{5}{2}$

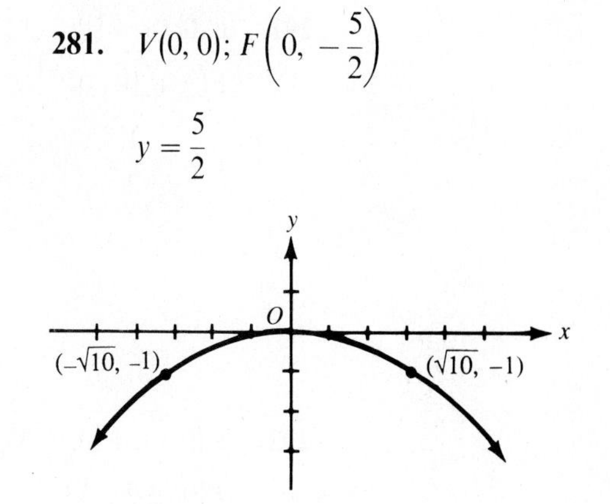

283. $V(-2, 0);\ F(0, 0)$
$x = -4$

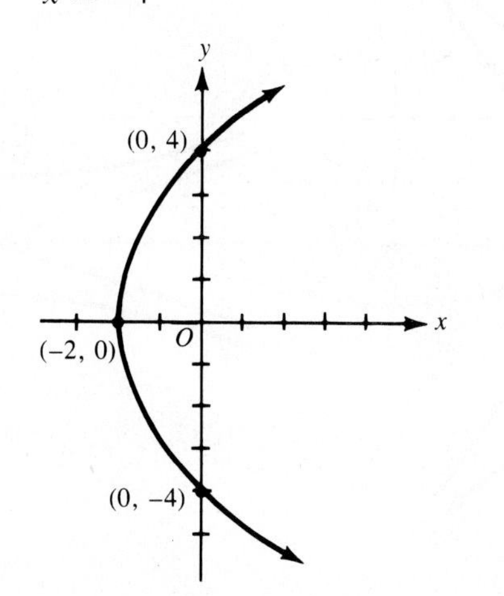

285. $V(1, 3);\ F\left(-\frac{1}{2}, 3\right)$

$x = \frac{5}{2}$

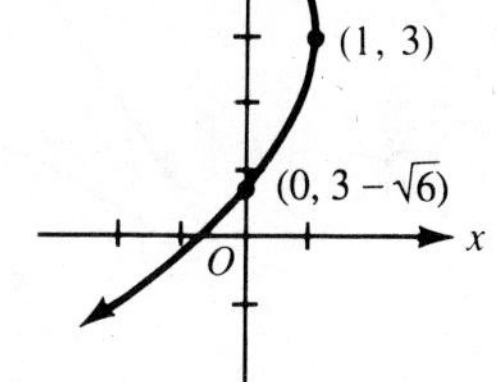

287. $V(0, 2);\ F(0, 4)$
$y = 0$

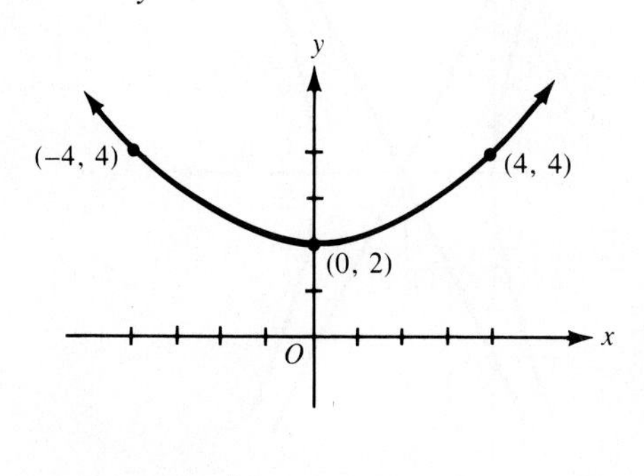

289. $V(-3, 2);\ F(-3, 1)$
$y = 3$

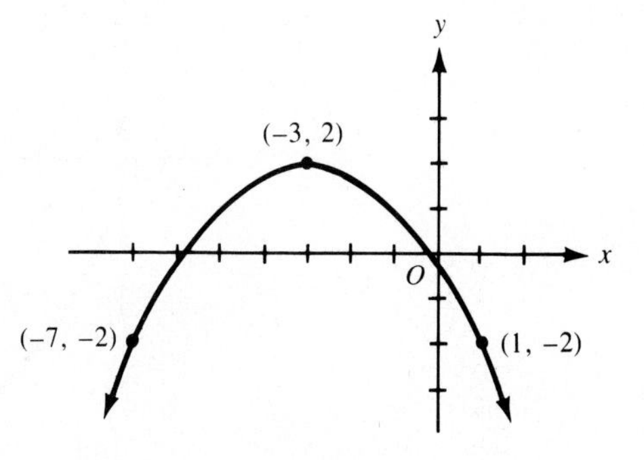

291. $C(0, 0)$; $V(\pm 2\sqrt{2}, 0)$
$F(\pm\sqrt{10}, 0)$
$y = \pm\frac{1}{2}x$

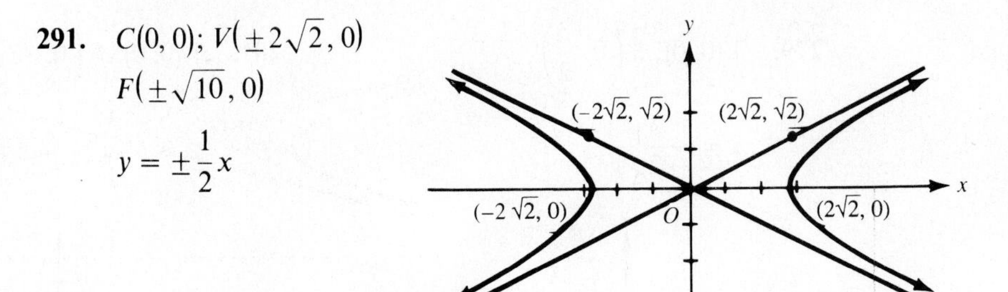

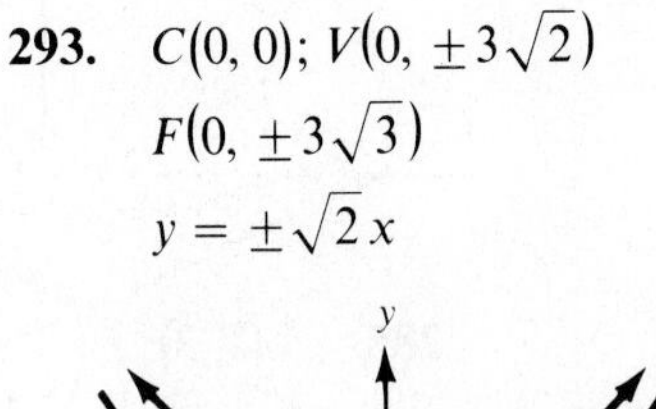

293. $C(0, 0)$; $V(0, \pm 3\sqrt{2})$
$F(0, \pm 3\sqrt{3})$
$y = \pm\sqrt{2}x$

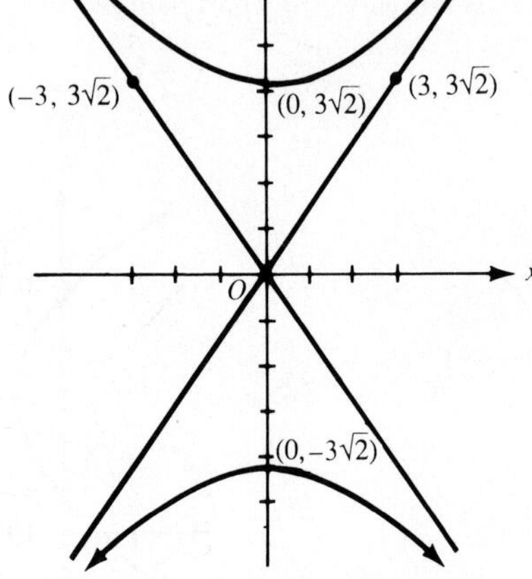

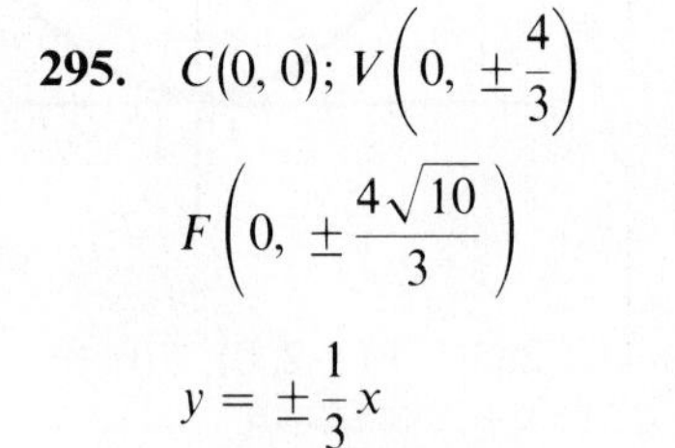

295. $C(0, 0)$; $V\left(0, \pm\frac{4}{3}\right)$
$F\left(0, \pm\frac{4\sqrt{10}}{3}\right)$
$y = \pm\frac{1}{3}x$

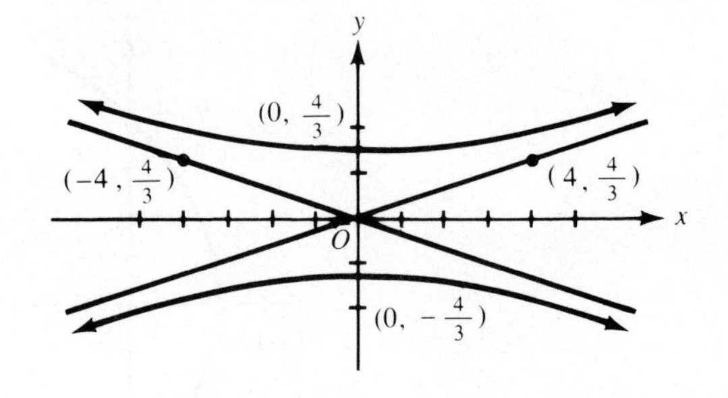

297. $C(2, 0)$; $V(2 \pm 2, 0)$
$F(2 \pm 2\sqrt{5}, 0)$
$y = \pm 2(x - 2)$

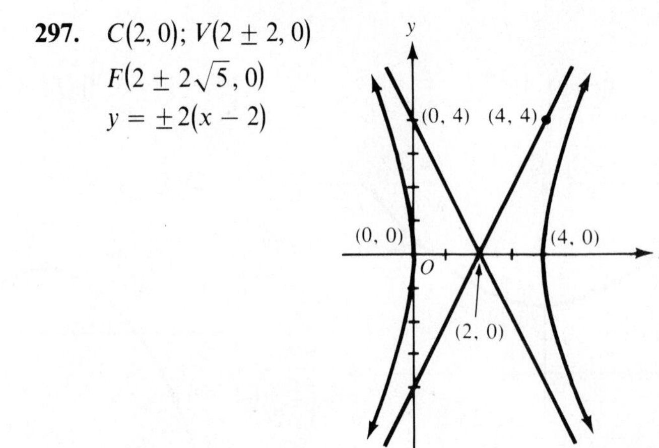

299. $C(-1, 3)$; $V(-1 \pm 4, 3)$
$F(-1 \pm \sqrt{17}, 3)$
$y = 3 \pm \frac{1}{4}(x + 1)$

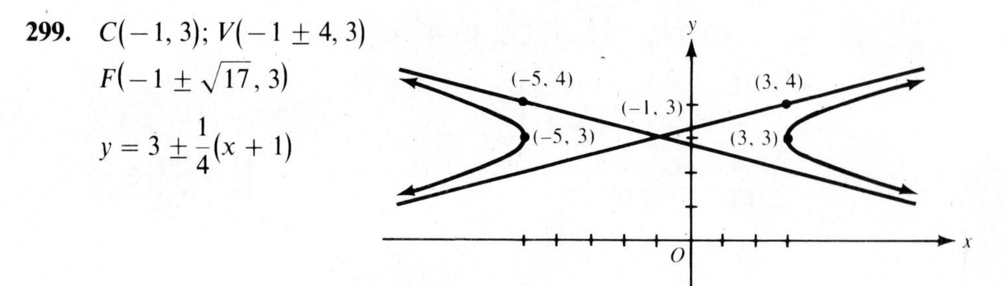

301. $C(-2, -2)$; $V(-2, -2 \pm \sqrt{2})$
$F(-2, -2 \pm \sqrt{11})$
$y = -2 \pm \frac{\sqrt{2}}{3}(x + 2)$

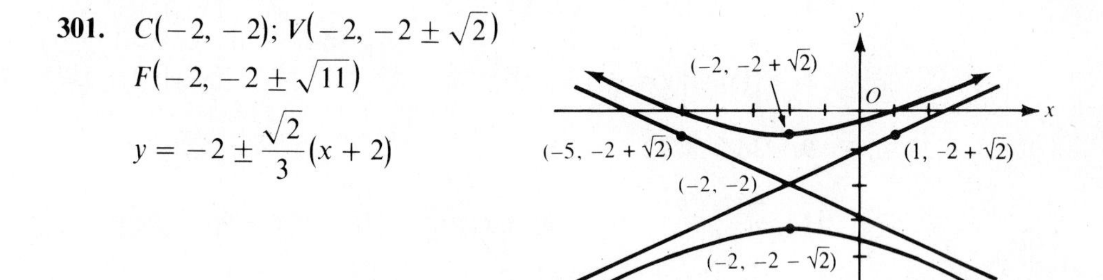

303. $\{(-3, -3), (1, -1)\}$

305. $\left\{(2, 2), \left(3, \frac{3}{2}\right)\right\}$

307. $\left\{\left(1, -\frac{1}{2}\right), \left(4, \frac{1}{4}\right)\right\}$

309. $\left\{(1, 2), \left(\frac{16}{5}, -\frac{12}{5}\right)\right\}$

311. $\left\{\left(\frac{16}{17}, \frac{30}{17}\right), (1, 2)\right\}$

313. $\left\{\left(-\frac{22}{13}, -\frac{45}{13}\right), (2, -1)\right\}$

315. $\left\{\left(-\frac{1}{2}, -\frac{\sqrt{5}}{2}\right), \left(-\frac{1}{2}, \frac{\sqrt{5}}{2}\right), \left(\frac{1}{2}, -\frac{\sqrt{5}}{2}\right), \left(\frac{1}{2}, \frac{\sqrt{5}}{2}\right)\right\}$

317. $\left\{\left(-\frac{\sqrt{30}}{6}, -\frac{\sqrt{6}}{3}\right), \left(-\frac{\sqrt{30}}{6}, \frac{\sqrt{6}}{3}\right), \left(\frac{\sqrt{30}}{6}, -\frac{\sqrt{6}}{3}\right), \left(\frac{\sqrt{30}}{6}, \frac{\sqrt{6}}{3}\right)\right\}$

319. $\left\{\left(-\frac{4\sqrt{2}}{3}, \frac{2\sqrt{2}}{3}\right), \left(\frac{4\sqrt{2}}{3}, -\frac{2\sqrt{2}}{3}\right), (-2, 2), (2, -2)\right\}$

321. $\left\{\left(-\frac{\sqrt{33}}{11}, \frac{2\sqrt{33}}{11}\right), \left(\frac{\sqrt{33}}{11}, -\frac{2\sqrt{33}}{11}\right), \left(-\frac{3\sqrt{66}}{22}, -\frac{\sqrt{66}}{22}\right),\right.$
$\left.\left(\frac{3\sqrt{66}}{22}, \frac{\sqrt{66}}{22}\right)\right\}$

323. $\left\{\left(-\frac{\sqrt{26}}{26}, -\frac{3\sqrt{26}}{26}\right), \left(\frac{\sqrt{26}}{26}, \frac{3\sqrt{26}}{26}\right), \left(-\frac{\sqrt{6}}{6}, \frac{\sqrt{6}}{6}\right), \left(\frac{\sqrt{6}}{6}, -\frac{\sqrt{6}}{6}\right)\right\}$

Exercise 15.1, page 474

1. $\{-3, -2, 0, 3, 5\}, \{0, 2, 4, 6, 7\}$
3. No
5. $\{-2, -1, 0, 1, 2\}, \{0, 2, 4\}$
7. $\{0, 1, 2, 3, \ldots\}, \{1, 3, 5, 7, \ldots\}$
9. $\{1, 2, 3, 4, \ldots\}, \{1, 4, 9, 16, \ldots\}$
11. 2; 3; 4
13. $-10; 0; -7$
15. 0; 1

Exercises 15.2–15.3, page 478

1. $\{(x, y) \mid y = 2x, x \in N\}$
3. $\{(x, y) \mid y = x - 1, x \in N\}$
5. $\left\{(x, y) \,\middle|\, y = \frac{2}{3}x, x \in W\right\}$
7. $\{(x, y) \mid y = 5x, x \in I\}$
9. 36, -8, 22
11. $\frac{4ac - b^2}{4a}$, $ay^2 + by + c$, $ax^2 + (4a + b)x + 4a + 2b + c$
13. $12a^2 - 4a$, 0, $3a^2 + 2a$, 65
15. Not defined, $\frac{3}{11}$, $\frac{2x^2}{4x^2 + 1}$
17. Not true
19. $x^2 + 4x - 3; 4x^3 - x^2 - 8x + 2; 4x^2 - 9; 16x^2 - 8x - 1$

Exercise 15.4, page 481

1. $\{x \mid x \in R\}, \{y \mid y \in R\}$
3. $\{x \mid x \in R\}, \{y \mid y \geq 4\}$
5. $\{x \mid x \in R\}, \{y \mid y \leq 9\}$
7. $\{x \mid x \geq -15\}, \{y \mid y \geq 0\}$
9. $\left\{x \,\middle|\, x \leq \frac{7}{2}\right\}, \{y \mid y \geq 0\}$
11. $\{x \mid x \in R\}, \{y \mid y \geq 2\}$
13. $\{x \mid x \leq -1 \text{ or } x \geq 1\}, \{y \mid y \geq 0\}$
15. $\{x \mid x \leq -4 \text{ or } x \geq 4\}, \{y \mid y \geq 0\}$
17. $\{x \mid -1 \leq x \leq 1\}, \{y \mid 0 \leq y \leq 1\}$
19. $\{x \mid -3 \leq x \leq 3\}, \{y \mid 0 \leq y \leq 3\}$
21. $\{x \mid -2\sqrt{3} \leq x \leq 2\sqrt{3}\}, \{y \mid 0 \leq y \leq 2\sqrt{3}\}$
23. $\{x \mid x \neq 0\}, \{y \mid y \neq 3\}$
25. $\{x \mid x \neq -3\}, \{y \mid y \neq -1\}$
33. $\{x \mid x \in R\}, \{y \mid y \in R\}$

Exercise 15.5, page 484

1. $\{4\}$
3. $\{4\}$
5. $\left\{\frac{2}{3}\right\}$
7. $\left\{\frac{1}{3}\right\}$
9. $\left\{-\frac{2}{5}\right\}$
11. $\left\{-\frac{3}{2}\right\}$

13. $\{-2\}$ **15.** $\{8\}$ **17.** $\left\{\frac{1}{3}\right\}$

19. $\left\{\frac{3}{2}\right\}$ **21.** 1.4, 1.3, 2.6, 2.4 **23.** 3.2, 2.6, 4.6, 4.2

Exercise 15.6, page 488

1. $\{(0, 0), (1, 1), (4, 2), (9, 3), (16, 4)\}$
3. $\{(0, -3), (2, -1), (6, 3), (10, 7), (15, 12)\}$
5. $\{(2, 1), (5, 3), (8, 5), (11, 7), (17, 11)\}$
7. $\{(0, -1), (2, 0), (16, 1), (54, 2), (128, 3)\}$

9. $y = \frac{7 - x}{3}$ **11.** $y = -\frac{x + 7}{6}$ **13.** $y = \frac{3x + 1}{2}$

15. $y = \frac{5 - 3x}{2}$ **17.** $y = -\frac{3x + 4}{4}$ **19.** $y = -\frac{7x + 15}{5}$

21. $y = \frac{1}{2}\sqrt[3]{4x}$; $\{x \mid x \in R\}$; $\{y \mid y \in R\}$

23. $y = \sqrt[3]{x - 3}$; $\{x \mid x \in R\}$; $\{y \mid y \in R\}$
25. $y = \sqrt{x - 2} - 1$; $\{x \mid x \geq 2\}$; $\{y \mid y \geq -1\}$
27. $y = \sqrt{x + 4} - 2$; $\{x \mid x \geq -4\}$; $\{y \mid y \geq -2\}$
29. $y = 3 - \sqrt{x + 1}$; $\{x \mid x \geq -1\}$; $\{y \mid y \leq 3\}$

Chapter 15 Review, page 489

1. -39; 7; 16

3. $c - \frac{b^2}{4}$; $y^2 + by + c$; $x^2 + x(2 - b) + c - b + 1$

5. 0; $4a^2 + 8a$; $a^2 - 4a$; $x^2 - 4x$; $(x + a)^2 - 4(x + a)$
7. Not true
9. $3x - 8$; $2x^2 - 9x + 7$; $2x - 9$; $2x - 8$
11. $\{x \mid x \in R\}, \{y \mid y \in R\}$ **13.** $\{x \mid x \in R\}, \{y \mid y \geq -32\}$
15. $\{x \mid x \geq -9\}, \{y \mid y \geq 0\}$ **17.** $\{x \mid x \leq 4\}, \{y \mid y \geq 0\}$
19. $\{x \mid x \in R\}, \{y \mid y \geq 6\}$ **21.** $\{x \mid x \leq -5 \text{ or } x \geq 5\}, \{y \mid y \geq 0\}$

23. $\left\{x \,\middle|\, -\frac{7}{2} \leq x \leq \frac{7}{2}\right\}$, $\{y \mid -7 \leq y \leq 7\}$

25. $\{x \mid x \neq 0\}, \{y \mid y \neq 5\}$ **29.** $\{-2\}$

31. $\{10\}$ **33.** .7; .8 **35.** $\frac{2x - 1}{3}$ **37.** $\frac{16 - 3x}{10}$

Exercise 16.1, page 494

1. $\log_2 8 = 3$ **3.** $\log_{81} 3 = \frac{1}{4}$ **5.** $\log_6 \frac{1}{36} = -2$

7. $\log_{10} .0001 = -4$ **9.** $\log_{10} 5 = .699$ **11.** $\log_a b = -6$

13. $9 = 3^2$ **15.** $5 = 125^{\frac{1}{3}}$ **17.** $\frac{1}{27} = 3^{-3}$

19. $.01 = 10^{-2}$ **21.** $6 = 10^{.778}$ **23.** $x = y^3$ **25.** 3

27. $\frac{3}{2}$ **29.** $\frac{1}{3}$ **31.** $\frac{1}{2}$ **33.** $\frac{3}{8}$

35. -2 **37.** $-\frac{6}{7}$ **39.** $-\frac{1}{3}$ **41.** -1

43. -2 **45.** -3 **47.** 7 **49.** 8

51. 25 **53.** 9 **55.** 16 **57.** 2

59. 81 **61.** 10 **63.** $\frac{1}{4}$ **65.** $\frac{1}{27}$

67. 64 **69.** 625 **71.** 64 **73.** 128

75. 81 **77.** 3 **79.** 2 **81.** $\frac{1}{10}$

83. $\frac{1}{27}$ **85.** 8 **87.** 25 **89.** $\frac{1}{4}$

Exercises 16.2–16.3, page 500

1. .778 **3.** .176 **5.** $-.176$ **7.** $-.398$

9. .903 **11.** 1.398 **13.** .238 **15.** 1.380

17. 1.602 **19.** 2.574 **21.** .201 **23.** .466

25. $-.602$ **27.** $-.704$ **29.** 3.322 **31.** 1.431

33. 2.322 **35.** 1.465 **37.** .682 **39.** -4.192

41. 1.262 **43.** .774 **45.** 1.721 **47.** 1.013

49. -4.420 **51.** 7.841 **53.** 3.18 **55.** 2.529

Exercise 16.4, page 503

1. .4318461 **3.** 1.7679717 **5.** 2.002598 **7.** 3.5845574

9. $-.3096267$ **11.** -1.7867479 **13.** -2.3754085 **15.** -3.6637404

17. 96.961651 **19.** 5204.0325 **21.** 392.35573 **23.** 2.4646769

25. .2379578 **27.** .0041639 **29.** .0002535

Exercise 16.5, page 505

1. 6.7629607 **3.** 6.128222 **5.** 3.6821317 **7.** 2.0404668
9. .5580145 **11.** $-.8969748$ **13.** -4.1135274 **15.** -4.8056964
17. 112.05614 **19.** 6.5227757 **21.** 5.8299009 **23.** 1.0754079
25. .6914254 **27.** .2426828 **29.** .1293802

Exercise 16.6, page 507

The answers were arrived at from tables of logarithms:

1. 4.298 **3.** 8.33 **5.** 16.36 **7.** 1.902
9. 22.74 **11.** 134.9 **13.** 2.119 **15.** 62.4
17. 1.444 **19.** .004859 **21.** .3509 **23.** 5.736
25. 1.521 **27.** 1.73 **29.** 1.127 **31.** .84
33. -1.376 **35.** $\{2.585\}$ **37.** $\{3.011\}$ **39.** $\{.7094\}$
41. $\{1.718\}$ **43.** $\{3.8\}$ **45.** $\{1.344\}$ **47.** $\{49.3\}$
49. $\{16.4\}$ **51.** $\{12.28\}$ **53.** $\{14.56\}$

Chapter 16 Review, page 508

1. $f^{-1}(x) = \log_2 x$ **3.** $f^{-1}(x) = \frac{1}{3}\log_2 x$ **5.** $f^{-1}(x) = -\frac{1}{2}\log_3 x$

7. $\frac{1}{100}$ **9.** 16 **11.** $\frac{1}{32}$

13. 128 **15.** $\frac{1}{343}$ **17.** 3

19. $\frac{1}{3}$ **21.** $-\frac{3}{2}$ **23.** -2 **25.** $-\frac{2}{3}$

37. 4.344 **39.** 17.87 **41.** 1.37 **43.** 1.034
45. 17.2 **47.** .01094 **49.** 2.369 **51.** 1.535
53. 1.635 **55.** $\{1.674\}$ **57.** $\{11.82\}$ **59.** $\{1.547\}$
61. $\{42.6\}$ **63.** $\{.5933\}$ **65.** 18.8 **67.** 2.48 grams

Exercises 17.1–17.2, page 514

1. 3, 6, 9, 12, 15 **3.** 5, 9, 13, 17, 21
5. $-5, -11, -17, -23, -29$ **7.** 2, 8, 18, 32, 50

9. 2, 11, 26, 47, 74 **11.** $2, 1, \frac{2}{3}, \frac{1}{2}, \frac{2}{5}$ **13.** $\frac{1}{2}, \frac{2}{3}, \frac{3}{4}, \frac{4}{5}, \frac{5}{6}$

15. 1, 3, 9, 27, 81 **17.** $\frac{1}{2}, \frac{1}{4}, \frac{1}{8}, \frac{1}{16}, \frac{1}{32}$

19. $-2, 1, -\frac{2}{3}, \frac{1}{2}, -\frac{2}{5}$ **21.** $-\frac{1}{3}, \frac{1}{7}, -\frac{1}{11}, \frac{1}{15}, -\frac{1}{19}$

23. $1, \frac{1}{8}, \frac{1}{27}, \frac{1}{64}, \frac{1}{125}$ **25.** 3, 7, 12, 18

27. 5, 12, 21, 32 **29.** $-1, -5, -12, -22$ **31.** 2, 7, 17, 34

33. 2, 6, 14, 30 **35.** $\frac{1}{2}, \frac{3}{2}, 3, 5$ **37.** $-1, 1, -2, 2$

39. $-1, -5, -4, -10$ **41.** $1 + 2 + 3 + 4 + 5 + 6$

43. $1 + 3 + 5 + 7 + 9$ **45.** $4 + 5 + 6 + \cdots + (n + 3)$

47. $-\frac{1}{8} + 0 + \frac{1}{8} + \frac{1}{4} + \frac{3}{8}$ **49.** $5 + \frac{19}{3} + \frac{23}{3} + 9 + \frac{31}{3}$

51. $3 \cdot 2 + 6 \cdot 3 + 9 \cdot 4 + 12 \cdot 5$ **53.** $1 \cdot 2 \cdot 3 + 2 \cdot 3 \cdot 4 + 3 \cdot 4 \cdot 5$

55. $1^3 + 2^3 + 3^3 + \cdots + n^3$ **57.** $\frac{1}{1 \cdot 2} + \frac{1}{2 \cdot 3} + \frac{1}{3 \cdot 4} + \frac{1}{4 \cdot 5}$

59. $\frac{1}{2} + \frac{1}{2^2} + \frac{1}{2^3} + \cdots + \frac{1}{2^n}$ **61.** $1 - \frac{1}{2} + \frac{1}{3} - \frac{1}{4} + \frac{1}{5}$

Exercise 17.3, page 520

1. 4 **5.** $2a - b$ **7.** 20, 32 **9.** $19, 56\frac{1}{2}$

11. 250 **13.** -472 **15.** $k(k + 1)$ **17.** $\frac{3}{2}k(k + 1)$

19. 52; 520 **21.** -17; 230 **23.** 13; 247 **25.** 8; 292

27. 2; 66 **29.** -12; -65 **31.** 1; 99 **33.** -40; 100

35. 27; -2 **37.** 30; 1 **39.** -11; 4 **41.** -24; $\frac{5}{2}$

43. 239 **45.** 185 **47.** -274 **49.** 6; 6

51. 5; 15 **53.** 15; 12 **55.** -20, 8

57. 19, 25, 31, 37, 43

59. $8 + 4\sqrt{2}, 6 + 5\sqrt{2}, 4 + 6\sqrt{2}, 2 + 7\sqrt{2}$

61. 35; 2520 **63.** 24; 3444 **65.** 4, 9, 14

67. 6, 10, 14, 18 **69.** 4, 8, 12, 16 **71.** 2, 5, 8, 11, 14

Exercise 17.4, page 527

1. 2 **3.** 3 **5.** $\frac{1}{2}$

9. $-\frac{3}{2}$ **11.** 648; 17,496 **13.** $-\frac{1}{16}, \frac{1}{512}$

15. $192, -768\sqrt{2}$ **17.** $\frac{243}{4}, \frac{665}{4}$ **19.** $\frac{4}{27}, \frac{463}{432}$

21. 12; 192 **23.** 512; -4 **25.** 6; 258 **27.** $-\frac{16}{27}, \frac{133}{54}$

29. $5, \frac{341}{128}$ **31.** $5, \frac{211}{576}$ **33.** 384 **35.** -24

37. 5; 7 **39.** 6; 640 **41.** $\frac{1}{3}, 5$ **43.** $\frac{4}{3}, 6$

45. 5; 9 **47.** 7; 192

49. $5, \frac{1}{2}$ **51.** $18, \frac{1}{3}$ or $288, -\frac{4}{3}$

53. $-1, \frac{2}{3}, -\frac{4}{9}$ or $1, \frac{2}{3}, \frac{4}{9}$ **55.** $-24, 12, -6, 3$

57. 4, 14, 24 or 31, 14, -3 **59.** 20, 31, 42 or 50, 31, 12

Exercises 17.5–17.6, page 533

The answers are rounded to the nearest dollar and nearest year:

1. \$71,632 **3.** \$2314 **5.** \$5671 **7.** 12 years
9. 5% **11.** \$11,638 **13.** \$19,380 **15.** \$6217
17. \$54,044 **19.** \$387

Chapter 17 Review, page 534

1. 3, 7, 11, 15, 19 **3.** $1, -\frac{1}{3}, \frac{1}{5}, -\frac{1}{7}, \frac{1}{9}$ **5.** $-4, -1, 21, 80$

7. $-2, 2, -6, 10$ **9.** $5 + 9 + 13 + 17 + 21$

11. $\frac{1}{2} + \frac{1}{2} + \frac{3}{8} + \frac{1}{4} + \frac{5}{32}$ **13.** 130; 642

15. -3; -18 **17.** $46, -\frac{7}{5}$ **19.** -70

21. 77; 20 **23.** 21; 2268 **25.** $\frac{3}{32}, \frac{381}{32}$

27. $\frac{81}{4}, \frac{665}{12}$ **29.** $\frac{1}{16}$ **31.** $6, -\frac{3}{4}$

33. $14, \frac{1}{7}$ or $896, -\frac{8}{7}$ **35.** 11, 17, 23 or 47, 17, −13

Cumulative Review, page 536

Chapter 15

1. 15; 5; 33 **3.** $0; x^2 - x - 2; 4x^2 - 10x + 4$

5. Not true

7. $x^2 + 2x; 2x^3 + x^2 - 2x - 1; 2x^2 - 1; 4x^2 + 2x$

9. $\{x \mid x \in R\}, \{y \mid y \in R\}$ **11.** $\{x \mid x \in R\}, \{y \mid y \geq 8\}$

13. $\{x \mid x \in R\}, \{y \mid y \geq -1\}$ **15.** $\{x \mid x \in R\}, \{y \mid y \leq 8\}$

17. $\{x \mid x \geq -4\}, \{y \mid y \geq 0\}$ **19.** $\{x \mid x \leq 9\}, \{y \mid y \geq 0\}$

21. $\{x \mid x \in R\}, \{y \mid y \geq 1\}$

23. $\{x \mid x \leq -2\sqrt{2} \text{ or } x \geq 2\sqrt{2}\}, \{y \mid y \geq 0\}$

25. $\left\{x \,\middle|\, x \leq -\frac{1}{2} \text{ or } x \geq \frac{1}{2}\right\}, \{y \mid y \geq 0\}$

27. $\{x \mid -6 \leq x \leq 6\}, \{y \mid 0 \leq y \leq 6\}$

29. $\left\{x \,\middle|\, -\frac{4}{3} \leq x \leq \frac{4}{3}\right\}, \{y \mid 0 \leq y \leq 4\}$

31. $\left\{x \,\middle|\, x \neq -\frac{1}{3}\right\}, \left\{y \,\middle|\, y \neq \frac{1}{3}\right\}$

33. $\{4\}$ **35.** $\{-2\}$ **37.** $\left\{\frac{5}{2}\right\}$ **39.** $\left\{\frac{1}{4}\right\}$

41. $\left\{\frac{1}{2}\right\}$ **43.** $\left\{\frac{3}{2}\right\}$ **45.** $y = \frac{x + 1}{4}$ **47.** $y = \frac{5 - x}{2}$

49. $y = \frac{3x + 5}{2}$ **51.** $y = 4x + 3$ **53.** $y = \frac{4 - 2x}{3}$ **55.** $y = 6 - 5x$

57. $y = \sqrt[3]{x - 1}$ **59.** $y = \sqrt[3]{x + 8}$ **61.** $y = \sqrt[3]{x} + 2$

Chapter 16

63. $\frac{1}{3}$ **65.** $\frac{3}{2}$ **67.** $\frac{5}{2}$ **69.** $-\frac{6}{5}$

71. 64 **73.** 625 **75.** $\frac{1}{5}$ **77.** $\frac{1}{128}$

79. 343 **81.** 25 **83.** $\frac{1}{16}$ **85.** 16

87. $\frac{3a+b}{2a}$ **89.** $\frac{4a+b}{2b}$ **91.** $\frac{3a+b}{a+b}$ **93.** $\frac{5a+b}{2a+b}$

95. $\frac{b-2a}{a+2b}$ **97.** $\frac{3b-4a}{3a+b}$ **107.** .6354838

109. 1.3747483 **111.** 2.3643634 **113.** 3.5819497
115. −.1007268 **117.** −1.2757241 **119.** 5.8348107
121. 4.4391156 **123.** 2.2364453 **125.** −.0618754
127. −2.4191189 **129.** −5.3391394 **131.** 21.877616
133. 426.57952 **135.** 2.9376497 **137.** .0489779
139. .1380384 **141.** 196.36988 **143.** 1.6323162
145. 1.0639623 **147.** .0073725 **149.** 7.8259273
151. 11.613599 **153.** 1.3304865 **155.** 168.23777
157. 3.0796896 **159.** 1.8034976 **161.** {3.4615224}
163. {3.7829486} **165.** {4.9986244} **167.** {3.0028112}
169. {−3.2003979} **171.** {4} **173.** {9}

Chapter 17

175. $5 + 9 + 13 + 17 + 21$ **177.** $2(4) + 4(5) + 6(6) + 8(7)$

179. $\frac{1}{1(3)} + \frac{1}{2(5)} + \frac{1}{3(7)} + \frac{1}{4(9)}$ **181.** 32; 460

183. 24; 204 **185.** −2; −36 **187.** −12; 126 **189.** 18; 4
191. 10; 4 **193.** 214 **195.** 31; 10 or 40; −8

197. 68; 8 or −60; 24 **199.** $\frac{32}{9}$; $\frac{665}{9}$ **201.** 4; 256

203. 16; $\frac{63}{2}$ **205.** 5; $\frac{55}{27}$ **207.** 64 **209.** 6; 8

211. −3; 6 **213.** 7; 384 **215.** $\frac{25}{128}$; −5 or $\frac{1}{8}$; 4

Exercise Appendix B, page A6

1. $x^3 + 3x^2 + 3x + 1$ **3.** $x^4 - 4x^3y + 6x^2y^2 - 4xy^3 + y^4$
5. $x^5 + 5x^4y + 10x^3y^2 + 10x^2y^3 + 5xy^4 + y^5$
7. $x^6 + 6x^5 + 15x^4 + 20x^3 + 15x^2 + 6x + 1$

9. $x^4 - 8x^3 + 24x^2 - 32x + 16$
11. $x^4 + 12x^3 + 54x^2 + 108x + 81$
13. $16x^4 - 32x^3 + 24x^2 - 8x + 1$
15. $16x^4 + 96x^3 + 216x^2 + 216x + 81$
17. $x^4 - 8x^3y + 24x^2y^2 - 32xy^3 + 16y^4$

19. $81x^4 + \frac{108x^3}{y} + \frac{54x^2}{y^2} + \frac{12x}{y^3} + \frac{1}{y^4}$
21. $\frac{1}{x^4} - \frac{4}{x^3y^2} + \frac{6}{x^2y^4} - \frac{4}{xy^6} + \frac{1}{y^8}$

23. $\frac{x^{10}}{y^5} + \frac{5x^7}{y^2} + 10x^4y + 10xy^4 + \frac{5y^7}{x^2} + \frac{y^{10}}{x^5}$

25. $\frac{1}{x^{10}} + \frac{5}{x^8y} + \frac{90}{x^6y^2} + \frac{270}{x^4y^3} + \frac{405}{x^2y^4} + \frac{243}{y^5}$

27. $128x^7 - 448x^5y + 672x^3y^2 - 560xy^3 + \frac{280y^4}{x} - \frac{84y^5}{x^3} + \frac{14y^6}{x^5} - \frac{y^7}{x^7}$

29. $x^8 - 16x^7 + 112x^6 - 448x^5 + 1120x^4 - 1792x^3 + 1792x^2 - 1024x + 256$

31. $x^4 + 2^{\frac{5}{2}}x^3 + 12x^2 + 2^{\frac{7}{2}}x + 4$
33. $x^2 - 8x^{\frac{3}{2}} + 24x - 32x^{\frac{1}{2}} + 16$
35. $x^{10} + 20x^9 + 180x^8 + 960x^7 + 3360x^6$
37. $x^{10} - 10x^8y + 45x^6y^2 - 120x^4y^3 + 210x^2y^4$
39. $x^{12} + 12x^{11} + 66x^{10} + 220x^9 + 495x^8$
41. 1.105
43. 1.194
45. .851

Exercise Appendix C, page A11

1. $6x^3 - 4x^2 + x - 8; -8$
3. $3x^3 - 6x^2 + x + 5; 0$
5. $x^4 + 3x^3 + x^2 - 7x - 3; -3$
7. -4
9. 3
11. 6
13. False
15. True; $(x - 1)(x + \sqrt{3})(x - \sqrt{3})$
17. True; $(x - 4)(x + 1 - i)(x + 1 + i)$
19. -7
21. 0
23. -52

71. 64 **73.** 625 **75.** $\frac{1}{5}$ **77.** $\frac{1}{128}$

79. 343 **81.** 25 **83.** $\frac{1}{16}$ **85.** 16

87. $\frac{3a+b}{2a}$ **89.** $\frac{4a+b}{2b}$ **91.** $\frac{3a+b}{a+b}$ **93.** $\frac{5a+b}{2a+b}$

95. $\frac{b-2a}{a+2b}$ **97.** $\frac{3b-4a}{3a+b}$ **107.** .6354838

109. 1.3747483 **111.** 2.3643634 **113.** 3.5819497
115. $-.1007268$ **117.** -1.2757241 **119.** 5.8348107
121. 4.4391156 **123.** 2.2364453 **125.** $-.0618754$
127. -2.4191189 **129.** -5.3391394 **131.** 21.877616
133. 426.57952 **135.** 2.9376497 **137.** .0489779
139. .1380384 **141.** 196.36988 **143.** 1.6323162
145. 1.0639623 **147.** .0073725 **149.** 7.8259273
151. 11.613599 **153.** 1.3304865 **155.** 168.23777
157. 3.0796896 **159.** 1.8034976 **161.** $\{3.4615224\}$
163. $\{3.7829486\}$ **165.** $\{4.9986244\}$ **167.** $\{3.0028112\}$
169. $\{-3.2003979\}$ **171.** $\{4\}$ **173.** $\{9\}$

Chapter 17

175. $5+9+13+17+21$ **177.** $2(4)+4(5)+6(6)+8(7)$

179. $\frac{1}{1(3)}+\frac{1}{2(5)}+\frac{1}{3(7)}+\frac{1}{4(9)}$ **181.** 32; 460

183. 24; 204 **185.** -2; -36 **187.** -12; 126 **189.** 18; 4
191. 10; 4 **193.** 214 **195.** 31; 10 or 40; -8

197. 68; 8 or -60; 24 **199.** $\frac{32}{9}$; $\frac{665}{9}$ **201.** 4; 256

203. 16; $\frac{63}{2}$ **205.** 5; $\frac{55}{27}$ **207.** 64 **209.** 6; 8

211. -3; 6 **213.** 7; 384 **215.** $\frac{25}{128}$; -5 or $\frac{1}{8}$; 4

Exercise Appendix B, page A6

1. x^3+3x^2+3x+1 **3.** $x^4-4x^3y+6x^2y^2-4xy^3+y^4$
5. $x^5+5x^4y+10x^3y^2+10x^2y^3+5xy^4+y^5$
7. $x^6+6x^5+15x^4+20x^3+15x^2+6x+1$

9. $x^4 - 8x^3 + 24x^2 - 32x + 16$ **11.** $x^4 + 12x^3 + 54x^2 + 108x + 81$
13. $16x^4 - 32x^3 + 24x^2 - 8x + 1$
15. $16x^4 + 96x^3 + 216x^2 + 216x + 81$
17. $x^4 - 8x^3y + 24x^2y^2 - 32xy^3 + 16y^4$

19. $81x^4 + \frac{108x^3}{y} + \frac{54x^2}{y^2} + \frac{12x}{y^3} + \frac{1}{y^4}$ **21.** $\frac{1}{x^4} - \frac{4}{x^3y^2} + \frac{6}{x^2y^4} - \frac{4}{xy^6} + \frac{1}{y^8}$

23. $\frac{x^{10}}{y^5} + \frac{5x^7}{y^2} + 10x^4y + 10xy^4 + \frac{5y^7}{x^2} + \frac{y^{10}}{x^5}$

25. $\frac{1}{x^{10}} + \frac{5}{x^8y} + \frac{90}{x^6y^2} + \frac{270}{x^4y^3} + \frac{405}{x^2y^4} + \frac{243}{y^5}$

27. $128x^7 - 448x^5y + 672x^3y^2 - 560xy^3 + \frac{280y^4}{x} - \frac{84y^5}{x^3} + \frac{14y^6}{x^5} - \frac{y^7}{x^7}$

29. $x^8 - 16x^7 + 112x^6 - 448x^5 + 1120x^4 - 1792x^3 + 1792x^2 - 1024x + 256$

31. $x^4 + 2^{\frac{5}{2}}x^3 + 12x^2 + 2^{\frac{7}{2}}x + 4$ **33.** $x^2 - 8x^{\frac{3}{2}} + 24x - 32x^{\frac{1}{2}} + 16$
35. $x^{10} + 20x^9 + 180x^8 + 960x^7 + 3360x^6$
37. $x^{10} - 10x^8y + 45x^6y^2 - 120x^4y^3 + 210x^2y^4$
39. $x^{12} + 12x^{11} + 66x^{10} + 220x^9 + 495x^8$
41. 1.105 **43.** 1.194 **45.** .851

Exercise Appendix C, page A11

1. $6x^3 - 4x^2 + x - 8; -8$ **3.** $3x^3 - 6x^2 + x + 5; 0$
5. $x^4 + 3x^3 + x^2 - 7x - 3; -3$ **7.** -4
9. 3 **11.** 6
13. False **15.** True; $(x - 1)(x + \sqrt{3})(x - \sqrt{3})$
17. True; $(x - 4)(x + 1 - i)(x + 1 + i)$
19. -7 **21.** 0 **23.** -52

Index

Abcissa, 277
Absolute statements, 108
Absolute value, 116
  equation with, 117
  inequalities with, 124
  properties, 116
Additive identity, 9
Additive inverse, 13, 35
Algebraic fractions, 167
  addition of, 172, 178
  division of, 186
  equations involving, 199
  multiplication of, 183
  reducing, 167
  simplication of, 167
Annuity, 531
  accumulated amount, 531
  payment interval, 531
  periodic rent, 531
  present value, 531
  term, 531
Area
  of a rectangle, 98
  of a square, 98
  of a triangle, 98
Arithmetic means, 519
Arithmetic progression, 515
  common difference, 515
  sum of, 516
Arm of a weight, 96
Associative law
  of addition, 9
  of multiplication, 11
Asymptotes, 453, 483

Base, 39
Binary operations, 8
Binomial, 33
Binomial expansion, A4
Binomial theorem, A4

Cartesian coordinate system, 275
Cartesian coordinates, 276
Celsius scale, 95
Character of the roots, 402
Characteristic, A13
Circle, 440
  center, 440
  equation, 440
  graphing, 440
  radius, 440
  standard form, 440
Closure, 9, 10, 12
Collinear points, 283

Combined operations, 23, 190
Common multiple, 20
Commutative law
  of addition, 9
  of multiplication, 11
Complementary angles, 98
Complex fractions, 192
Complex number(s), 379
  addition, 381
  additive inverse, 381
  Cartesian form, 380
  conjugate, 384
  division, 384
  graphs, 386
  imaginary part, 379
  multiplication, 383
  plane, 386
  products, 383
  real part, 379
  simplified form, 380
  standard form, 380
  subtraction, 381
Composite number, 16
Compound interest, 529
  accumulated amount, 529
  period, 529
  present value, 529
Computations with logarithms, 505
Conditional inequalities, 108
Conic sections, 439
  circle, 440
  ellipse, 443
  hyperbola, 451
  parabola, 447
Consistent, 300
Constant polynomial, 390
Coordinate axes, 276
Coordinates of a point, 3, 276
Counting numbers, 2
Cramer's rule, 355, 361
Cube root, 143, 238

Degree of a polynomial, 56
Denominator, 14
Depressed equation, A9
Depressed polynomial, A9
Determinants, 353
  cofactor of an element, 357
  columns of, 353
  elements, 353
  minor of an element, 357
  *n*th order, 353
  properties, 353
  rows of, 353
  second order, 353
  value of, 357
Difference of two cubes, 144
Difference of two squares, 141
Dimension(s)
  one, 336
  two, 336
  zero, 336
Discriminant, 403
Distance between two points, 280
Distributive law of multiplication, 11, 35, 45
Dividend, 14, 26, 57
Division, 14, 50
  by zero, 15
Divisor, 14, 26, 57
Domain, 275
  of a function, 275
  of a relation, 275
Double root, 391

Ellipse, 443
  center, 443
  equation, 443
  foci, 443
  graphing, 445
  major axis, 444
  minor axis, 444
  standard form, 443
  vertices, 444
Equation(s), 65
  equivalent, 66
  fractional, 310
  involving absolute values, 117
  involving fractions, 70, 74, 199, 310
  involving grouping symbols, 73, 310
  involving radical expressions, 259
  lead to a quadratic, 413
  linear, 65, 285
  literal, 196
  quadratic, 390
    pure, 392
  roots, 65
  solution set, 65
Equation of a line
  intercepts, 296
  one point and slope, 296
  two points, 295
Equivalent fractions, 18
Exponents, 39, 314
  positive fractional, 215
  negative, 224
  zero, 224
Expressions, 33
  evaluation of, 33
Extent, 439
  domain, 439
  range, 439
Extremes, 519, 527

Factors, 10, 33, 137
Factored completely, 137
Factoring, 15, 137
Factoring a binomial, 140
  difference of two cubes, 144
  difference of two squares, 141
  sum of two cubes, 143
Factoring a trinomial, 146, 150
Factoring by completing the square, 156
Factoring by grouping, 159
Factoring four-term polynomials, 159
  grouping as three and one, 160
  grouping in pairs, 161
Factors common to all terms, 137
Fahrenheit scale, 95
False statements, 65
Foci, 443, 447, 451
Fraction(s), 14
  addition of, 172, 178
  algebraic, 176
  complex, 192
  denominator of, 167
  division of, 186
  equivalent, 18, 167
  higher terms, 18
  improper, 26
  lowest terms, 18, 167
  multiplication of, 183
  numerator of, 167
  proper, 26
  reduced, 18, 167
  simplification, 167
Fulcrum, 96
Function(s), 275, 474
  algebra of, 476
  composite, 477
  constant, 476, 480
  defined by equations, 478
  domain, 275, 474
  exponential, 481
  identity, 476
  inverse, 485
  linear, 480
  logarithmic, 492
  notation, 475
  of a function, 477
  product, 477
  quadratic, 480
  range, 275, 474
  sum, 476
Fundamental theorem of arithmetic, 16

Geometric means, 527
Geometric progressions, 522
  common ratio, 522
  sum of, 523
Graphical interpretations, 300, 337, 341
Graphs of
  circles, 440
  ellipses, 445
  exponential functions, 482
  hyperbolas, 454
  linear equations, 286
  linear inequalities, 325
  logarithmic functions, 496
  numbers, 3, 12
  parabolas, 449
  quadratic equations, 423
Greater than, 3, 105
  or equal to, 105
Greatest common divisor, 18
Greatest common factor, 18, 137, 168
Grouping symbols, 37

Half plane, 326
Hindu-Arabic system, 2
Hyperbola, 451
  asymptotes, 453
  axes, 452
  center, 452
  conjugate, 453
  equation, 451, 453
  foci, 451
  graphing, 454
  standard form of, 452
  transverse axis, 452
  vertices, 452

Identity, 9, 65
Identity function, 476
Imaginary numbers, 375
Imaginary unit, 375
Inconsistent, 300
Index, 140, 238
Inequalities, 108
  conditional, 108
  equivalent, 109
  graphs, 325
  involving absolute values, 124
  linear, 104, 325
  quadratic, 429
  solution of, 109, 325
  systems of, 113, 328
Infinity, A2
Integers, 12
  addition of, 13
  division of, 14
  multiplication of, 13
  subtraction of, 13
Intercepts, 289
Interest, 86, 529

Interval, A2
bounded, A2
half closed, A2
half open, A2
infinite, A2
open, A2
Irrational numbers, 29, 141, 239

Kelvin scale, 95

Laws of multiplication, 11
Laws of addition, 9
Least common denominator, 20, 178
Least common multiple, 20, 176
Less than, 3, 105
or equal to, 105
Line of symmetry, 424
equation, 424
Linear and quadratic equation, 456
Linear combination, 305
Linear equation(s)
consistent and dependent, 300
consistent and independent, 300
graphs, 286
inconsistent, 300
in one variable, 65
in three variables, 336, 342
in two variables, 274, 285
Linear equations, solution of
algebraic, 304, 342
determinants, 355, 361
elimination method, 304
graphical method, 302
substitution method, 308
Linear inequalities, 104, 325
solution of, 108
Lines
coincident, 300
equation, 295
intersect, 300
parallel, 297, 300
slope of, 290
Literal equations, 196
Literal numbers, 8, 33
Logarithm, 492
base ten, 502
common, 502
computation with, 505
graphs of, 496
natural, 504
properties, 496
table, use of, A12
Lowest terms, 18, 167

Mantissa, A13
Maximum point, 424
Means, 519, 527
Minimum point, 423
Mixed numbers, 26
Monomial, 33, 50
Multiplication by zero, 10
Multiplication of
complex numbers, 383
integers, 13
monomials, 41
radical expressions, 246
rational numbers, 22
Multiplicative identity, 11
Multiplicative inverse, 22, 186
Multiplicity of a root, 391

Natural numbers, 2
Negative direction, 13
Negative exponents, 224
Negative integers, 12
*n*th root, 238
Null set, 4
Number line, 2
Numerator, 14
Numerical coefficient, 33

One dimension, 336
Order of a radical, 238
Order relation, 105
properties of, 106
Ordered pair, 275
first component of, 275
first coordinate of, 275
second component of, 275
second coordinate of, 275
Ordinate of a point, 277
Origin, 2, 276

Pairs of numbers, components, 275
Parabola, 423, 447
axis, 424, 447
directrix, 447
equation, 448, 449
focus, 447
graphing, 449
standard form, 448
vertex, 423, 448
Percent, 85
Perimeter of a rectangle, 98
Perimeter of a square, 98
Perfect root, 239
Perfect square, 140, 395
Period, 529
Place value system, 2
Planes
coincide, 337

intersect, 337
parallel, 337
Polynomial equation, 390
degree of, 390
depressed, A9
Polynomial(s), 33, 390
addition of, 35
constant, 390
degree, 56, 390
depressed, A9
division of, 50, 54, 56
equation, 390
factoring, 136
linear, 390
multiplication of, 39, 45, 47
quadratic, 390
subtraction of, 35
Positive direction, 13
Power, 39
Present value, 529, 531
Prime factors, 16
Prime numbers, 15
Principal, 86
Principal root, 238
Progressions, 515, 522
Pure imaginary numbers, 375
addition, 375
division, 377
products, 376
subtraction, 375

Quadrants, 276
Quadratic equation in one variable, 390
character of roots, 402
discriminant, 403
graphs, 423
properties of roots, 406
pure quadratic, 392
roots of, 390
solution by completing the square, 395
solution by factoring, 390
solution by graphs, 427
solution by the quadratic formula, 399
standard form of, 390
Quadratic equation in two variables, 439
solution, 456, 458, 460, 461
Quadratic formula, 399
Quadratic inequalities, 429, 432
Quadrants, 276
Quotient, 14, 26, 57

Radical(s), 140, 238
index of, 140, 238
order of, 238
sign, 140
simplifying, 240
standard form of, 239
Radical expressions
combination of, 243
division of, 250
multiplication of, 246
similar, 243
Radicand, 140, 238
Rate of interest, 86, 529
Rational numbers, 17
addition of, 19
decimal form of, 24
division of, 22
multiplication of, 22
subtraction of, 21
Rationalizing factor, 254
Rationalizing the denominator, 254
Real numbers, 29
Reciprocals, 22, 186
Rectangular coordinate system, 275
Reducing fractions, 18, 167
Relation, 275, 474
domain of, 275, 474
range of, 275, 474
Remainder, 26, 57, A8
Remainder theorem, A9
Root, principal, 238
Rounding off numbers, A5

Scale, 2
Scientific notation of a number, 230, A13
Scissors, factoring, 150
Sequence, 512
partial sums, 512
terms of, 512
Series, 512
Set(s), 3
disjoint, 6
element of, 3
equal, 5
integers, 12
intersection of, 6
member of, 3
natural numbers, 3
null, 4
rational numbers, 17
real numbers, 29
union of, 5
whole numbers, 39
Slope of a line, 290
Solving equations, 67
Specifics, 33
Square root, 140, 238
Subsets, 5

Subtraction, 12, 21, 35
Sum of two cubes, 143
Summation
  index, 513
  notation, 513
  range, 513
  symbol, 513
Supplementary angles, 98
Symmetry, 424
Synthetic division, A7
Systems of equations, 300
Systems of inequalities, 113

Terms, 33
  like, 35
  numerical coefficient of, 33
  of a fraction, 14
  of a sequence, 512
  of a sum, 9
Trinomial, 33, 395
Two dimension, 336

Variable, 4

Whole numbers, 2
Word problems, 78, 203, 315, 346, 419
  geometry problems, 98
  lever problems, 96
  motion problems, 93
  number problems, 81
  percentage problems, 85
  temperature problems, 95
  value problems, 90
  work problems, 206

x-axis, 275
x-coordinate, 277
x-intercept, 289

y-axis, 275
y-coordinate, 277
y-intercept, 289

Zero
  dimension, 336
  division by, 15
  exponents, 224

The Prindle, Weber & Schmidt Series in Mathematics

Althoen and Bumcrot, *Introduction to Discrete Mathematics*
Brown and Sherbert, *Introductory Linear Algebra with Applications*
Buchthal and Cameron, *Modern Abstract Algebra*
Burden and Faires, *Numerical Analysis*, Third Edition
Cullen, *Linear Algebra and Differential Equations*
Cullen, *Mathematics for the Biosciences*
Dobyns, Steinbach, and Lunsford, *The Electronic Study Guide for Precalculus Algebra*
Eves, *In Mathematical Circles*
Eves, *Mathematical Circles Adieu*
Eves, *Mathematical Circles Revisited*
Eves, *Mathematical Circles Squared*
Eves, *Return to Mathematical Circles*
Fletcher and Patty, *Foundations of Higher Mathematics*
Geltner and Peterson, *Geometry for College Students*
Gilbert and Gilbert, *Elements of Modern Algebra*, Second Edition
Gobran, *Beginning Algebra*, Fourth Edition
Gobran, *College Algebra*
Gobran, *Intermediate Algebra*, Fourth Edition
Gordon, *Calculus and the Computer*
Hall, *Algebra for College Students*
Hall and Bennett, *College Algebra with Applications*
Hartfiel and Hobbs, *Elementary Linear Algebra*
Hunkins and Mugridge, *Applied Finite Mathematics*, Second Edition
Kaufmann, *Algebra for College Students*, Second Edition
Kaufmann, *Algebra with Trigonometry for College Students*
Kaufmann, *College Algebra*
Kaufmann, *College Algebra and Trigonometry*
Kaufmann, *Elementary Algebra for College Students*, Second Edition
Kaufmann, *Intermediate Algebra for College Students*, Second Edition
Kaufmann, *Precalculus*
Kaufmann, *Trigonometry*
Keisler, *Elementary Calculus: An Infinitesimal Approach*, Second Edition
Konvisser, *Elementary Linear Algebra with Applications*
Laufer, *Discrete Mathematics and Applied Modern Algebra*
Nicholson, *Linear Algebra with Applications*
Pasahow, *Mathematics for Electronics*
Powers, *Elementary Differential Equations*

Powers, *Elementary Differential Equations with Boundary Value Problems*
Powers, *Elementary Differential Equations with Linear Algebra*
Proga, *Arithmetic and Algebra*
Proga, *Basic Mathematics*, Second Edition
Radford, Vavra, and Rychlicki, *Introduction to Technical Mathematics*
Radford, Vavra, and Rychlicki, *Technical Mathematics with Calculus*
Rice and Strange, *Calculus and Analytic Geometry for Engineering Technology*
Rice and Strange, *College Algebra*, Third Edition
Rice and Strange, *Plane Trigonometry*, Fourth Edition
Rice and Strange, *Technical Mathematics*
Rice and Strange, *Technical Mathematics and Calculus*
Schelin and Bange, *Mathematical Analysis for Business and Economics*, Second Edition
Steinbach and Lunsford, *The Electronic Study Guide for Trigonometry*
Strnad, *Introductory Algebra*
Swokowski, *Algebra and Trigonometry with Analytic Geometry*, Sixth Edition
Swokowski, *Calculus with Analytic Geometry*, Second Alternate Edition
Swokowski, *Calculus with Analytic Geometry*, Fourth Edition
Swokowski, *Fundamentals of Algebra and Trigonometry*, Sixth Edition
Swokowski, *Fundamentals of College Algebra*, Sixth Edition
Swokowski, *Fundamentals of Trigonometry*, Sixth Edition
Swokowski, *Precalculus: Functions and Graphs*, Fifth Edition
Tan, *Applied Calculus*
Tan, *Applied Finite Mathematics*, Second Edition
Tan, *Calculus for the Managerial, Life, and Social Sciences*
Tan, *College Mathematics*, Second Edition
Venit and Bishop, *Elementary Linear Algebra*, Second Edition
Venit and Bishop, *Elementary Linear Algebra*, Alternate Second Edition
Willard, *Calculus and Its Applications*, Second Edition
Willerding, *A First Course in College Mathematics*, Fourth Edition
Wood and Capell, *Arithmetic*
Wood, Capell, and Hall, *Developmental Mathematics*, Third Edition
Wood, Capell, and Hall, *Intermediate Algebra*
Wood, Capell, and Hall, *Introductory Algebra*
Zill, *A First Course in Differential Equations with Applications*, Third Edition
Zill, *Calculus with Analytic Geometry*, Second Edition
Zill, *Differential Equations with Boundary-Value Problems*

Table of Measures

METRIC WEIGHTS AND MEASURES

Length

10 millimeters (mm)	= 1 centimeter (cm)	10 meters	= 1 decameter (dam)
10 centimeters	= 1 decimeter (dm)	10 decameters	= 1 hectometer (hm)
10 decimeters	= 1 meter (m)	10 hectometers	= 1 kilometer (km)
			= 1000 meters

Volume

10 milliliters (ml)	= 1 centiliter (cl)	10 liters	= 1 decaliter (dal)
10 centiliters	= 1 deciliter (dl)	10 decaliters	= 1 hectoliter (hl)
10 deciliters	= 1 liter (l)	10 hectoliters	= 1 kiloliter (kl)
			= 1000 liters

Weight

10 milligrams (mg)	= 1 centigram (cg)	10 grams	= 1 decagram (dag)
10 centigrams	= 1 decigram (dg)	10 decagrams	= 1 hectogram (hg)
10 decigrams	= 1 gram (g)	10 hectograms	= 1 kilogram (kg)
			= 1000 grams

1000 kilograms = 1 metric ton (t)

MEASUREMENT CONVERSIONS

Length

1 in.	= 2.5400 cm	1 cm	= 0.3937 in.
1 ft	= 0.3048 m	1 m	= 3.2809 ft
1 yd	= 0.9144 m		= 1.0936 yd
1 mile	= 1.6093 km	1 km	= 0.6214 mile

Volume

1 cubic in.	= 0.0164 liter	1 liter	= 61.0250 cu in.
1 cubic ft	= 28.3162 liters	1 liter	= 0.0353 cu ft
1 gal	= 3.7853 liters	1 liter	= 0.2642 gal
		1 liter	= 1.0567 qt (liquid)

Weight—Mass

1 oz	is the weight of 28.3495 g	1 g weighs 0.0353 oz
1 lb	is the weight of 0.4536 kg	1 kg weighs 2.2046 lb
1 ton (short)	is the weight of 907.1848 kg	1 kg weighs 0.0011 ton (short)